Between-Subjects, One-Way ANOVA (Chapter 10)

Used to compare the means of two or more independent samples.

Source of Variability	Sum of Squares	Degrees of Freedom	Mean Square	*F* ratio
Between groups	$SS_{Between} = \Sigma\left(\frac{(\Sigma X_{Group})^2}{n_{Group}}\right) - \frac{(\Sigma X)^2}{N}$	$k - 1$	$\frac{SS_{Between}}{df_{Between}}$	$\frac{MS_{Between}}{MS_{Within}}$
Within groups	$SS_{Within} = \Sigma\left(\Sigma X^2_{Group} - \frac{(\Sigma X_{Group})^2}{n_{Group}}\right)$	$N - k$	$\frac{SS_{Within}}{df_{Within}}$	
Total	$SS_{Total} = \Sigma X^2 - \frac{(\Sigma X)^2}{N}$	$N - 1$		

Repeated-Measures ANOVA (Within-Subjects, One-Way ANOVA) (Chapter 11)

Used to compare the means of two or more dependent samples.

Source of Variability	Sum of Squares	Degrees of Freedom	Mean Square	*F* ratio
Subjects	See Chapter Appendix.	$n - 1$		
Treatment	See Chapter Appendix.	$k - 1$	$\frac{SS_{Treatment}}{df_{Treatment}}$	$\frac{MS_{Treatment}}{MS_{Residual}}$
Residual	See Chapter Appendix.	$(n - 1)(k - 1)$	$\frac{SS_{Residual}}{df_{Residual}}$	
Total	See Chapter Appendix.	$N - 1$		

Between-Subjects, Two-Way ANOVA (Chapter 12)

Used to compare the means of four or more independent samples when there are two explanatory variables.

Source of Variability	Sum of Squares	Degrees of Freedom	Mean Square	*F* ratio
Between groups	See Chapter Appendix.	$df_{Rows} + df_{Columns} + df_{Interaction}$		
Rows	See Chapter Appendix.	$R - 1$	$\frac{SS_{Rows}}{df_{Rows}}$	$\frac{MS_{Rows}}{MS_{Within}}$
Columns	See Chapter Appendix.	$C - 1$	$\frac{SS_{Columns}}{df_{Columns}}$	$\frac{MS_{Columns}}{MS_{Within}}$
Interaction	See Chapter Appendix.	$df_{Rows} \times df_{Colu}$ [illegible]	$SS_{Interaction}$ [illegible]	$MS_{\ldots raction}$ / [illegible]thin
Within groups	See Chapter Appendix.	$N - (R \times C$ [illegible]	[illegible]	
Total	See Chapter Appendix.	$N - 1$		

Pearson r (Chapter 13)

Used to examine the linear relationship between two interval/ratio variables.

$$r = \frac{\Sigma[(X - M_X)(Y - M_Y)]}{\sqrt{SS_X SS_Y}}$$

Simple Regression (Chapter 14)

Used to predict an outcome variable (Y') from a predictor variable (X).

First, calculate: $b = r\left(\frac{s_Y}{s_X}\right)$

where b = the slope of the regression line
r = the observed correlation between X and Y
s_Y = the standard deviation of the Y scores
s_X = the standard deviation of the X scores

Then, calculate: $a = M_Y - bM_X$

where a = the Y-intercept for the regression line
M_Y = the mean of the Y scores
b = the slope of the regression line
M_X = the mean of the X scores

Finally, calculate the regression line equation:

$$Y' = bX + a$$

where Y' = the predicted value of Y
b = the slope of the regression line
X = the value of X for which one wants to find Y'
a = the Y-intercept of the regression line

Chi-Square Goodness-of-Fit Test (Chapter 15)

Used to compare the distribution of a nominal or ordinal categorical variable in a sample to the variable's expected distribution.

First, calculate expected frequencies:

$$f_{Expected} = \frac{\%_{Expected} \times N}{100}$$

Then, calculate the chi-square value:

$$\chi^2 = \Sigma\frac{(f_{Observed} - f_{Expected})^2}{f_{Expected}}$$

Chi-Square Test of Independence (Chapter 15)

Used to determine whether two or more samples differ on a nominal or ordinal categorical dependent variable.

First, calculate the expected frequencies for a contingency table:

$$f_{Expected} = \frac{N_{Row} \times N_{Column}}{N}$$

Then, calculate the chi-square value:

$$\chi^2 = \Sigma\frac{(f_{Observed} - f_{Expected})^2}{f_{Expected}}$$

Writing a Four-Point Interpretation

1. Recap the study. What was done? Why?
2. Present the main results factually. For example, what were the mean scores for the control and experimental groups? Present the results of the hypothesis test in APA format.
3. Explain what the results mean.
4. Make suggestions for future research. What were the strengths and/or weaknesses of this study? What should be done in the next study?

Using and Interpreting Statistics

THIRD EDITION

Using and Interpreting Statistics

A Practical Text for the Behavioral, Social, and Health Sciences

ERIC W. CORTY
The Pennsylvania State University

Publisher, Psychology and Sociology:	Rachel Losh
Executive Acquisitions Editor:	Daniel McDonough
Development Editor:	Andrew Sylvester
Assistant Editor:	Kimberly Morgan
Executive Marketing Manager:	Katherine Nurre
Marketing Assistant:	Allison Greco
Director, Content Standards:	Rachel Comerford
Associate Media Editor:	Anthony Casciano
Director, Content Management Enhancement:	Tracey Kuehn
Managing Editor, Sciences and Social Sciences:	Lisa Kinne
Senior Project Editor:	Jane O'Neill
Media Producer:	Elizabeth Dougherty
Senior Production Supervisor:	Paul Rohloff
Photo Editors:	Sheena Goldstein, Robin Fadool
Director of Design, Content Management:	Diana Blume
Art Manager:	Matt McAdams
Interior Designer:	Cambraia Fernandez
Cover Designer:	Vicki Tomaselli
Composition:	codeMantra
Printing and Binding:	RR Donnelley
Cover Art:	Leslie Wayne

Library of Congress Control Number: 2015957307

ISBN-13: 978-1-4641-0779-5
ISBN-10: 1-4641-0779-3

Printed in the United States of America

First Printing

Worth Publishers
One New York Plaza
Suite 4500
New York, NY 10004-1562

www.macmillanlearning.com/

For Sara, David, and Paul

ABOUT THE AUTHOR

David Corty

ERIC W. CORTY has a bachelor's degree in psychology from Vassar College, a doctorate in clinical psychology from Indiana University, and two postdoctoral fellowships, one in neuropsychopharmacology (University of Pennsylvania) and one in human sexuality (Case Western Reserve University). Since 1993, Corty has been a member of the psychology faculty at Penn State Erie, the Behrend College. There, he teaches principles of measurements, abnormal psychology, human sexuality, introductory psychology, and, of course, statistics. The quality of his teaching was recognized in 1997 when he received the Council of Fellows Excellence in Teaching Award and in 2001 when he became a Penn State Teaching Fellow. At present, Corty is the Interim Director of the School of Humanities & Social Sciences at Penn State Behrend.

Corty's work has appeared in more than three dozen peer-reviewed publications. His research on ejaculatory latencies received worldwide attention, including being made fun of on the *Late Show with David Letterman.* His statistics textbook was recognized as Book of the Year by the *American Journal of Nursing* in 1997. Corty serves as a member of the editorial board for *The Journal of Sex & Marital Therapy* and previously was on the editorial board for *The Journal of Consulting and Clinical Psychology.*

Corty was born in Wilmington, Delaware, and still celebrates Delaware Day every December 7th. He now lives in Beachwood, Ohio. He likes to eat and to cook, loves to ride his bicycles, and is sad to report that he has not made much progress on his pool game since the second edition.

BRIEF CONTENTS

CONTENTS

PREFACE

TO THE STUDENT

If you are like many students, this is not a course you have been looking forward to taking. And you probably don't feel like reading a long message about how important this course is. So, I'll be brief and say just five things to introduce the book to you:

1. When the semester is over, you'll find that statistics wasn't as hard as you feared. Learning statistics is like learning a foreign language—the concepts build on each other and require regular practice. The best way to practice is just like the best way to eat: Take small bites, chew thoroughly, and swallow thoughtfully. Work through the "Practice Problems" as they pop up. Make sure you can do them before moving on to new material.
2. The "Review Your Knowledge" exercises at the end of the chapter are just that, a review of the chapter. Do them when you've finished the chapter to make sure that you're comfortable with all of the material in the chapter. Then, do them again before a test as a refresher.
3. The questions in the "Apply Your Knowledge" exercises at the end of the chapter have at least two questions on each topic. Each odd-numbered question is followed by an even-numbered question that asks the same thing. Answers for the odd-numbered questions appear in the back of the book. So if you struggle with an odd-numbered question, you can turn there for help. And then, with that guidance, you should be able to work through the even-numbered question on your own.
4. The book is divided into four parts. At the end of each, you'll find a test that covers the techniques from all chapters in that part. Working these problems can be a great way to determine whether you have truly mastered the material.
5. One last thing: In each chapter, you'll find a boxed feature called "DIY." I've crafted these projects as a way for you to gain experience gathering data and conducting experiments. I encourage you to work the projects on your own, even if your teacher doesn't assign them.

I hope you enjoy this book half as much as I enjoyed writing it. If you have any comments about it that you'd like to share with me, please write to me at ewc2@psu.edu.

TO THE INSTRUCTOR

Welcome to the third edition of *Using and Interpreting Statistics.* I wrote this book because I couldn't find a text that presented the right amount of material in a straightforward manner that engaged students. My approach is applied—I want students to walk away from a first course in statistics with an ability to do what I call the "human side" of statistics, the things that computers can't do. Yes, I teach the math of statistics—how to calculate t, F, r, χ^2, confidence intervals, and a variety of effect sizes—but my overall focus is on leading students to an understanding of the logic (and the beauty) of statistics. At the end of the course, I want students to be able to select the appropriate statistical test for a research question. For the statistical tests, I want them to be able to write, in simple language, a complete interpretative statement and to explain what the results mean. In line with the recommendations from the American Statistical Association's *Guidelines for Assessment and Instruction in Statistics Education: College Report* (2010), I aim for a conceptual understanding, not just procedural knowledge.

There are a number of techniques that I use to achieve these goals. The first technique is my clear and approachable writing style, which makes it easier for students to engage with, and actually read, the book. Next, my organization within chapters breaks complex concepts into component parts, so they can be learned in much the same manner as a behavior is shaped. To aid learning, chapters are sprinkled with mnemonic and organizational devices. For example, there are "How to Choose" flowcharts that help students pick the correct statistical procedure, and each statistical test has a series of questions that lead students through the main concepts that need to be covered in an interpretation.

Features of the Book

Stat Sheets Each chapter has a Stat Sheet (collected at the end of the text) that can be pulled out of the book. These tear sheets contain all the formulas, flowcharts, and steps that are necessary to complete a statistical test. The Stat Sheets provide essential guidance when solving problem sets or studying on the go.

Picking the Right Statistical Test Knowing what statistical test to use is an important statistical skill, yet many introductory textbooks devote little time to it. Not true for me, as I give it a whole chapter. I thought long and hard about where to place this chapter content, early or late, and I finally decided late, making this chapter the last in the book. In this position, "Selecting the Right Statistical Test" serves as a coda that brings together all the elements of the course, and it presents a unifying view of statistics.

Six Steps of Hypothesis Testing Continuity from chapter to chapter reinforces understanding and makes procedures second nature. For example, when I introduce hypothesis testing, I teach a six-step procedure for completing hypothesis tests.

- **Step 1**: Pick a test.
- **Step 2**: Check the assumptions.
- **Step 3**: List the hypotheses.
- **Step 4**: Set the decision rule.
- **Step 5**: Calculate the test statistic.
- **Step 6**: Interpret the results.

For every subsequent test taught, I follow the same six steps. This is a repetitive, cookbook approach, but it is purposeful: When you are learning to cook, it pays to follow a recipe. At first the steps are rote, then they become a routine, and finally the steps become internalized.

Interpreting Results Knowing how to calculate the value of a test statistic is vitally important in statistics and each chapter teaches students these skills. But, the learning outcomes for research methods, as spelled out in the *APA Guidelines for the Undergraduate Psychology Major* (Version 2.0, 2013), stress evaluating the appropriateness of conclusions derived from psychological research. To this end, the coverage of each statistical tests ends with a substantial section on interpreting results. This format also aligns with the APA's emphasis on communication and professional development by exposing students early and often to the presentation of results that they will see in professional articles and that will be expected when they present their own research.

For each statistical test, students learn to address a series of questions to gather information for interpreting the results. For the independent-samples *t* test, for example, there are three questions:

1. Was the null hypothesis rejected, and what does this reveal about the direction of the difference between the two populations?
2. How big is the size of the effect?
3. How big or small might the effect be in the population?

Integrating confidence intervals and effect sizes into the interpretation of results for tests, rather than isolating them in a separate chapter, teaches students to use these techniques. Students learn to write an interpretation that addresses four points: (a) why was the study done, (b) what were its main results, (c) what do the results mean, and (d) what are suggestions for future research. This four-point interpretation is used for every test, making a thorough interpretation a natural part of completing a statistical test.

Practice, Practice, Practice Learning statistics is like learning a foreign language—concepts build on each other and are best learned with regular and graduated practice. Because of this, a statistics book lives or dies by its exercises. This book has been constructed with that in mind. All the exercises were written by me, so they are consistent in content and tone with the rest of the text. Importantly, three tiers of exercises—reviewing knowledge, applying knowledge, and expanding knowledge—allow professors to assign questions at different levels and give students the opportunity to push themselves to a deeper understanding.

Opportunities for practice are also presented throughout each chapter:

- **Worked Examples.** There are Worked Examples spaced throughout every chapter that students can work along with the text. These lead students through problems step-by-step. The Worked Examples allow students to make sure they know the correct steps for a statistical test and help to develop their statistical thinking.

- **In-Chapter Practice Problems.** Each major section of a chapter concludes with Practice Problems and solutions so that students can practice the material and assess how well they've learned it.
- **End-of-Chapter Exercises.** The end-of-chapter exercises have three tiers, one for students to review their knowledge, another with applied questions, and a third with more challenging problems to stretch understanding. Students have ample opportunities to practice—the *t* test chapters, for example, have close to 100 questions each, more than other texts on the market. The applied questions are written so that they build on each other, moving to the final goal of calculating and interpreting a statistical test. These applied questions isolate and test each intermediate step in the process—for example, making sure a student knows how to compute a standard error of the difference before computing a *t* value. Finally, the applied questions are paired, with at least two questions for each concept. An answer to the first question is given at the back of the book, so the student can check and correct his or her work.

Part Tests In addition to this in-chapter practice, the third edition has been divided into four parts, with a brief introduction and capstone Part Test. These tests offer challenging problems that require students to pick from the various methods and techniques they've learned from the whole text up to that point. By working through these problems, students will gain a deeper understanding of the material, and will be better prepared for course exams.

More Tools for Students

In addition to the various opportunities for practice, each chapter features an expanded set of learning tools that help students by previewing, reviewing, and rehearsing the chapter's lessons.

Learning Objectives and Summary The Learning Objectives at the start of the chapter set up the key concepts for the chapter. They show up again at the end of the chapter to organize the summary, reinforcing the chapter's framework.

User-Friendly Equations Whenever an equation is introduced, all the symbols in the formula are defined. This makes it easy for students to plug correct values into formulas.

DIYs New to this edition, this feature presents the framework for a do-it-yourself project built around the chapter topic. Appropriate for either group or individual work, the DIYs allow students to create their own data sets and draw their own conclusions.

End-of-Chapter Application Demonstrations Each chapter culminates in an Application Demonstration problem that usually uses real-life data and employs the techniques learned in the chapter to answer questions about the world in which we live.

End-of-Chapter Summary The end-of-chapter summaries are brief journeys back through the main sections of the chapter. Along with the key terms, they provide students with a quick review of the main concepts.

Noteworthy Changes in the Third Edition

For those of you who are familiar with the second edition of this text, you'll notice some significant changes here. I've mentioned a couple of new features above—the part introductions and tests, and the DIY feature—that will provide additional opportunity for your students to engage with the material. Based on reviewer suggestions and my own experience working with this text, I've also made several other organizational and content changes.

- **Chapter 2** now includes a section on stem-and-leaf plots, including coverage in the SPSS guide.
- The discussion of the importance of *z* scores in **Chapter 4** has been expanded. In addition, the review of probability has moved to this chapter, where it follows the discussion of the normal curve.
- **Chapter 5** includes a more in-depth discussion of sampling and a revised introduction to confidence intervals as a practical application of the central limit theorem.
- As a consequence of this revised organization, the discussion of confidence intervals in **Chapter 7** has been simplified, allowing the student to focus on the procedure for the single-sample *t* test. In addition, the discussion of effect size now presents r^2 alongside Cohen's *d*, making for an easier comparison and contrast of these two measures.
- In **Chapter 8**, the formula for pooled variance is now introduced prior to the presentation of the standard error equation, so students can clearly see how pooled variance is used to calculate the estimated standard error.
- **Chapter 9** now addresses the pitfalls of using r^2, as well as Cohen's *d*, with paired-samples *t* tests.
- The coverage of sum of squares in **Chapter 10** now includes the computational formula in addition to the definitional formulas.
- The coverage of one-way, repeated-measures ANOVA and between-subjects, two-way ANOVA in **Chapters 11 and 12**, respectively, has been augmented with the inclusion of in-chapter appendixes that present the formulas for calculating sums of squares for these tests.
- The introduction of one-way, repeated measures ANOVA has been greatly expanded in **Chapter 11** to present the uses of this test.
- The coverage of the Pearson Correlation Coefficient in **Chapter 13** now includes more details about the definitional formula. In addition, a major section has been added to introduce the student to partial correlation.
- The presentation of the formula for calculating cell expected frequencies in **Chapter 15** has been simplified.

Examples and exercise sets throughout the text have been revised and the end-of-chapter SPSS guides thoroughly updated.

MEDIA AND SUPPLEMENTS

LaunchPad with LearningCurve Quizzing

A comprehensive Web resource for teaching and learning statistics

LaunchPad combines Worth Publishers' award-winning media with an innovative platform for easy navigation. For students, it is the ultimate online study guide, with rich interactive tutorials, videos, an e-Book, and the LearningCurve adaptive quizzing system. For instructors, LaunchPad is a full course space where class documents can be posted, quizzes can be easily assigned and graded, and students' progress can be assessed and recorded. Whether you are looking for the most effective study tools or a robust platform for an online course, LaunchPad is a powerful way to enhance your class.

LaunchPad to Accompany *Using and Interpreting Statistics,* Third Edition, can be previewed and purchased at launchpadworks.com.

Using and Interpreting Statistics, Third Edition, and LaunchPad can be ordered together with ISBN-13: 978-1-319-06187-6/ISBN-10: 1-319-06187-7. Individual components of LaunchPad may also be available for separate, standalone purchase.

LaunchPad for *Using and Interpreting Statistics,* Third Edition, includes all the following resources:

- The **LearningCurve** quizzing system was designed based on the latest findings from learning and memory research. It combines adaptive question selection, immediate and valuable feedback, and a game-like interface to engage students in a learning experience that is unique. Each LearningCurve quiz is fully integrated with other resources in LaunchPad through the Personalized Study Plan, so students will be able to review using Worth's extensive library of videos and activities. And state-of-the-art question analysis reports allow instructors to track the progress of their entire class.
- An **interactive e-Book** allows students to highlight, bookmark, and make their own notes, just as they would with a printed textbook. Google-style searching and in-text glossary definitions make the text ready for the digital age.
- **Statistical Video Series** consisting of StatClips, StatClips Examples, and Statistically Speaking "Snapshots." View animated lecture videos, whiteboard lessons, and documentary-style footage that illustrate key statistical concepts and help students visualize statistics in real-world scenarios.
    - **StatClips lecture videos,** created and presented by Alan Dabney, PhD, Texas A&M University, are innovative visual tutorials that illustrate key statistical concepts. In 3 to 5 minutes, each StatClips video combines dynamic animation, data sets, and interesting scenarios to help students understand the concepts in an introductory statistics course.
    - In **StatClips Examples,** Alan Dabney walks students through step-by-step examples related to the StatClips lecture videos to reinforce the concepts through problem solving.
    - **Snapshots** videos are abbreviated, student-friendly versions of the **Statistically Speaking** video series, and they bring the world of statistics into the classroom. In the same vein as the successful PBS series Against All Odds Statistics, Statistically Speaking uses new and updated documentary

footage and interviews that show real people using data analysis to make important decisions in their careers and in their daily lives. From business to medicine, from the environment to understanding the Census, Snapshots focus on why statistics is important for students' careers, and how statistics can be a powerful tool to understand their world.

- **Statistical Applets** allow students to master statistical concepts by manipulating data. They also can be used to solve problems.
- **EESEE Case Studies** are taken from the *Electronic Encyclopedia of Statistical Exercises and Examples* developed by The Ohio State University. EESEE Case Studies offer students additional applied exercises and examples.
- The **Assignment Center** lets instructors easily construct and administer tests and quizzes from the book's Test Bank and course materials. The Test Bank includes a subset of questions from the end-of-chapter exercises with algorithmically generated values, so each student can be assigned a unique version of the question. Assignments can be automatically graded, and the results are recorded in a customizable Gradebook.

Additional Student Supplements

- **SPSS®: A User-Friendly Approach** by Jeffery Aspelmeier and Thomas Pierce of Radford University is a comprehensive introduction to SPSS that is easy to understand and vividly illustrated with cartoon-based scenarios. In the newest edition of the text for SPSS Version 22, the authors go beyond providing instructions on the mechanics of conducting data analysis and develop students' conceptual and applied understanding of quantitative techniques.
- The **iClicker** Classroom Response System is a versatile polling system developed by educators for educators that makes class time more efficient and interactive. iClicker allows you to ask questions and instantly record your students' responses, take attendance, and gauge students' understanding and opinions. iClicker is available at a 10% discount when packaged with *Using and Interpreting Statistics,* Third Edition.

Instructor Supplements

One book alone cannot meet the education needs and teaching expectations of the modern classroom. Therefore, Worth has engaged some skilled teachers and statisticians to create a comprehensive supplements package that brings statistics to life for students and provides instructors with the resources necessary to supplement their successful strategies in the classroom.

- **Instructor's Resources.** This guide offers an Instructor's Resource Manual containing classroom activities, handouts, additional reading suggestions, and online resources. The Instructor's Resources also include lecture slides and all of the book's images in either JPEG or Slideshow format and can be downloaded from the book's catalog page at http://www.macmillanhighered.com/Catalog/product/usingandinterpretingstatistics-thirdedition-corty/instructorresources#tab.

- **Downloadable Test Bank.** Powered by Diploma, the downloadable Test Bank includes hundreds of multiple-choice questions to use in generating quizzes and tests for each chapter of the text. The Diploma software allows instructors to add an unlimited number of new questions; edit questions; format a test; scramble questions; and include figures, graphs, and pictures. The computerized Test Bank also allows instructors to export into a variety of formats compatible with many Internet-based testing products.

ACKNOWLEDGMENTS

Though writing is a solo endeavor, writing a book is not. Many people provided aid and support.

- Let me start with a nod to my parents, Claude and Sue Corty, whose love of numbers (thanks, Dad) and love of words (that's you, Mom) can be found on every page.
- My wife, Sara Douglas, and my sons, David and Paul, keep me grounded. Both of my sons have recently taken statistics courses and both made it a point of honor never to ask their father for any help. Thanks guys, keep keeping me humble.
- Chuck Linsmeier and Dan DeBonis acquired my book for Worth and shepherded it through with grace and ease. When Dan was promoted to Senior Acquisitions Editor, another Dan, Dan McDonough, took over. New Dan was great and I thank him for asking Andrew Sylvester to join the project.
- Andrew Sylvester was my DE. That's publisher talk for development editor, the person in charge of making sure I deliver a manuscript that is up to snuff and on time, the person who is the buffer between me and the rest of the Worth publishing apparatus. Andrew set a bruising pace for the revision and kept track of all the balls that I had in the air, while juggling a fair number himself. He was appreciative when I handed things in on time and gently inquisitive when I didn't. Whenever I had questions or concerns, he had answers. Andrew, if you're reading this, and of course you are—you've read and commented on every word I've written for this book—thank you very much.
- There has been a whole host of statistics instructors who reviewed individual chapters over the course of the past three editions. To make sure they pulled no punches, their reviews were anonymous. And pull no punches they did. As they say, if it doesn't kill you, it makes you stronger and the many issues they raised have made the book better. Now that the book is going to press, I have finally learned their names and I want to thank each of them individually:

Christopher Aberson, *Humboldt State University*
Christina Anguiano, *Arizona State University Polytechnic*
William Ajayi, *Kent State University*
David Alfano, *Community College of Rhode Island*
Jayne Allen, *University of New Hampshire*
Janet Andrews, *Vassar College*
Pamela Ansburg, *Metropolitan State University of Denver*
Stephen Armeli, *Fairleigh Dickinson University—Teaneck*
Matthew R. Atkins, *University of North Texas*
Alison Aylward, *University of Miami*
Marie Balaban, *Eastern Oregon University*
Nicole Ballardini, *Truckee Meadows Community College*
Jonathan Banks, *Nova Southeastern University*
Lucy Barnard-Brak, *Baylor University*
Linda Bastone, *State University of New York—Purchase*

Dennis Berg, *California State University, Fullerton*
Kristin Bernard, *Stony Brook University*
Robert Bertram, *Bradley University*
Joan Bihun, *University of Colorado Denver*
Paul Billings, *Community College of Southern Nevada*
Victor Bissonnette, *Berry College*
Eliane Boucher, *University of Texas of the Permian Basin*
Jacqueline Braun, *Ramapo College of New Jersey*
Hallie Bregman, *University of Miami*
Kelly Brennan-Jones, *State University of New York at Brockport*
James Briggs, *Susquehanna University*
Amanda Brouwer, *Winona State University*
Eric Bruns, *Campbellsville University*
Carrie Bulger, *Quinnipiac University*
Danielle Burchett, *Kent State University*
Christopher Burke, *Lehigh University*
Jessica Cail, *Pepperdine University*
Leslie Cake, *Memorial University of Newfoundland, Grenfell Campus*
Robert Carini, *University of Louisville*
Michael Carlin, *Rider University*
Mary Jo Carnot, *Chadron State College*
Linn E. Carothers, *California State University—Northridge*
H. Chen, *Rosemont College*
Jeremy Cohen, *Xavier University of Louisiana*
Scott Cohn, *Western State Colorado University*
Richard Conti, *Kean University*
Amy Cota-McKinley, *Worcester State University*
Christy Cowan, *Lincoln Memorial College*
Robert Crutcher, *University of Dayton*
Maria Cuddy-Casey, *Immaculata University*
Christopher Daddis, *The Ohio State University*
Marilyn Dantico, *Arizona State University*
Kelly de Moll, *University of Tennessee*
Katherine Demitrakis, *Central New Mexico Community College*
Daniel Denis, *University of Montana*
Jessica Dennis, *California State University, Los Angeles*
Justin DeSimone, *Georgia Institute of Technology*
Darryl Dietrich, *The College of St. Scholastica*
Beth Dietz-Uhler, *Miami University*
Kristen Diliberto-Macaluso, *Berry College*
Stephanie Ding, *Del Mar College*
Nancy Dorr, *College of St. Rose*
Michael Dudley, *Southern Illinois University Edwardsville*
Vera Dunwoody, *Chaffey College*
Charles Earley, *Houston Community College*
Vanessa Edkins, *Florida Institute of Technology*
Jeanne Edman, *Cosumnes River College*
Vicky Elias, *Texas A&M University—San Antonio*
Holger Elischberger, *Albion College*
Domenica Favero, *Lynchburg College*
David Feigley, *Rutgers, The State University of New Jersey*
Kathleen Flannery, *Saint Anselm College*
Jonathon Forbey, *Ball State University*
Michelle Foust, *Baldwin Wallace University*
Mike Frank, *Stockton State College*
Scott Frasard, *University of Georgia*
Andrea Friedrich, *University of Kentucky*
Jacqueline Fulvio, *University of Wisconsin—Madison*
John Galla, *Widener University*
Renee Gallier, *Utah State University*
Brian Garavaglia, *Macomb Community College*
Richard Gardner, *Alliant International University*
Ray Garza, *Texas A&M International University*
Mark Gebert, *University of Kentucky*
Edwin Gomez, *Old Dominion University*
Michael Green, *Lone Start College—Montgomery*
Anthony Greene, *University of Wisconsin—Milwaukee*
Alexis Grosofsky, *Beloit College*
Carrie Hall, *Miami University*
Elizabeth Hannah, *Boise State University*
Christine Hansvick, *Pacific Lutheran University*
Evan Harrington, *The Chicago School of Professional Psychology*
Wayne Harrison, *University of Nebraska at Omaha*
Helen Harton, *University of Northern Iowa*
Christopher Hayashi, *Southwestern College*
Jeremy Heider, *Stephen F. Austin State University*
Linda Henkel, *Fairfield University*
Roberto Heredia, *Texas A&M University*
Heather Hill, *St. Mary's University*
Charles Hinderliter, *University of Pittsburgh at Johnstown*
Brian Hock, *Austin Peay State University*
Jeanne Horst, *James Madison University*
Michael Horvath, *Cleveland State University*
Jay Irwin, *University of Nebraska at Omaha*
Annette Iskra, *Xavier University of Louisiana*
Daniel Ispas, *Illinois State University*
Lora Jacobi, *Stephen F. Austin State University*
Dharma Jairam, *Pennsylvania State University—Erie, The Behrend College*

Rafa Kasim, *Kent State University*
Donald Keller, *George Washington University*
Karl Kelley, *North Central College*
Stephen Kilianski, *Rutgers University*
C. Ryan Kinlaw, *Marist College*
Elizabeth Kudadjie-Gyamfi, *Long Island University*
John Kulas, *St. Cloud State University*
Jonna Kwiatkowski, *Mars Hill College*
Karla Lassonde, *Minnesota State University*
Jeffrey Leitzel, *Bloomsburg University of Pennsylvania*
Ryan Leonard, *Gannon University*
David Lester, *Stockton State College*
Thomson Ling, *Caldwell College*
Elissa Litwin, *Touro College*
William London, *California State University, Los Angeles*
Javier-Jose Lopez-Jimenez, *Minnesota State University—Mankato*
Mark Ludorf, *Stephen S. Austin State University*
Molly Lynch, *Northern Virginia Community College*
Michael Mangan, *University of New Hampshire*
Michael Maniaci, *Florida Atlantic University*
Kelly Marin, *Manhattan College*
Harvey Marmurek, *University of Guelph*
Chandra Mason, *Mary Baldwin College*
Susan Mason, *Niagara University*
Jonathan Mattanah, *Towson University*
Amanda McCleery, *UCLA David Geffen School of Medicine*
Robert McCoy, *Skyline College*
Roselie McDevitt, *Mount Olive College*
Daniel McElwreath, *William Paterson University*
Connor McLennan, *Cleveland State University—Ohio*
Ron Mehiel, *Shippensburg University*
Jackie Miller, *The Ohio State University*
Joe Morrissey, *Binghamton University*
Brendan Morse, *Bridgewater State College*
Daniel Mossler, *Hampden—Sydney College*
Siamak Movahedi, *University of Massachusetts Boston*
Anne Moyer, *Stony Brook University*
Elise Murowchick, *Seattle University*
David Nalbone, *Purdue University, Calumet*
Jeffrey Neuschatz, *University of Alabama in Huntsville*
Ian Newby-Clark, *University of Guelph*
Erik Nilsen, *Lewis and Clark College*
Helga Noice, *Elmhurst College*
Ken Oliver, *Quincy University*
Steve O'Rourke, *College of New Rochelle*
Geoffrey O'Shea, *SUNY—Oneonta*
Melanie Otis, *University of Kentucky*
Jennifer Pacyon, *Temple University*
Jennifer Peszka, *Hendrix College*
John Petrocelli, *Wake Forest University*
John Pfister, *Dartmouth College*
Catherine Phillips, *University of Calgary*
John Pierce, *Philadelphia University*
Angela Pirlott, *University of Wisconsin—Eau Claire*
Gary Popoli, *Harford Community College*
William Price, *North Country Community College*
Jianjian Qin, *California State University, Sacramento*
Laura Rabin, *Brooklyn College*
Krista Ranby, *University of Colorado—Denver*
Jean Raniseski, *Alvin Community College*
Robert Reeves, *Augusta State University*
Heather Rice, *Washington University in St. Louis*
Kim Roberts, *California State University, Sacramento*
Shannon Robertson, *Jacksonville State University*
Dennis Rodriguez, *Indiana University—South Bend*
Craig Rogers, *Campbellsville University*
Bryan Rooney, *Concordia University College of Alberta*
Patrick Rosopa, *Clemson University*
John Ruscio, *The College of New Jersey*
Samantha Russell, *Grand Canyon University*
Ron Salazar, *San Juan College*
Nick Salter, *Ramapo College of New Jersey*
Amy Salvaggio, *University of New Haven*
Erika Sanborne, *University of Massachusetts*
Lowell Gordon Sarty, *University of Saskatchewan*
Michele Schlehofer, *Salisbury University*
Ingo Schlupp, *University of Oklahoma*
Brian Schrader, *Emporia State University*
Carl Scott, *University of Saint Thomas*
Andrea Sell, *University of Kentucky*
Marc Setterlund, *Alma College*
Sandra Sgoutas-Emch, *University of San Diego*
Keith Shafritz, *Hofstra University*
Stephen Shapiro, *Old Dominion University*
Greg Shelley, *Kutztown University*
Mike Sherrick, *Memorial University of Newfoundland*
Matthew Sigal, *York University*
Angela Sikorski, *Texas A&M University—Texarkana*
Ned Silver, *University of Nevada—Las Vegas*
Royce Simpson, *Spring Hill College*
Michael Sliter, *Indiana University—Purdue University Indianapolis*
Lara Sloboda, *Tufts University*

Albert Smith, *Cleveland State University*
Brian William Smith, *St. Edward's University*
Dale Smith, *Olivet Nazarene University*
Linda Solomon, *Marymount Manhattan College*
Hilda Speicher, *Albertus Magnus College*
Mark Stambush, *Muskingum University*
Francis Staskon, *Saint Xavier University*
Ross Steinman, *Widener University*
Mark Stellmack, *University of Minnesota*
Sharon Stevens, *Western Illinois University*
Garrett Strosser, *Southern Utah University*
Colleen Sullivan, *Worcester State University*
Kyle Susa, *University of Texas at El Paso*
Cheryl Terrance, *University of North Dakota*
Heather Terrell, *University of North Dakota*
William Thornton, *University of* Southern Maine
Brian Tilley, *National University*
Patricia Tomich, *Kent State University*
Loren Toussaint, *Luther College*
Sharmin Tunguz, *DePauw University*
Kristine Turko, *University of Mount Union*
Chantal Tusher, *Georgia State University*
Lynne Unikel, *LaSalle University*
Mary Utley, *Drury University*
Katherine Van Giffen, *California State University, Long Beach*
Peter Vernig, *Suffolk University*
Yvonne Vissing, *Salem State University*
William Wagner, *California State University, Channel Islands*
Elizabeth Weiss, *The Ohio State University, Newark*
Gary Welton, *Grove City College*
George Whitehead, *Salisbury University*
Wayne Williams, *Johnson C. Smith University*
Elizabeth Williford, *Belhaven University*
Karen Wilson, *St. Francis College*
Darren Woodlief, *University of South Carolina*
Jean Wynn, *Manchester Community College*
Yi Yang, *James Madison University*
Tammy Zacchilli, *Saint Leo University*
Jennifer Zimmerman, *DePaul University*

- In addition to the reviewers who read single chapters, there have been a handful of individuals who have read every word and checked every number in every chapter over the last two editions. Melanie Maggard and Sherry Serdikoff provided thorough and timely guidance through the second edition. For this edition, Carl Schwarz, James Lapp, and Catherine Matos read through the text and exercises for clarity and accuracy. Andrew can tell you that their attention to detail at times drove me to distraction, but I always appreciated the safety net they provided me. Of course, as I was the last person to read the manuscript as it went to press, any remaining errors are my fault.
- Those reviewers read for content, but Patti Brecht has read the manuscript for grammar and style over the past two editions. She has corrected my quirks and imposed consistency on my irregularities. Because of her work on the second edition, this one went more smoothly. Patti, I can't wait to work with you on the fourth!
- Jane O'Neill and Paul Rohloff worked behind the scenes, overseeing all the details of the production of the book. For the fact that you are holding a copy of such a good-looking book in your hands, you have them to thank.

PART I Basic Concepts

This section, roughly the first third of the book, introduces the basic concepts that form the foundation of statistics. By the time you have finished these five chapters, you should be familiar with statistical terminology and be able to describe a set of data using tables, graphs, and numbers that summarize the average score and the amount of heterogeneity in the scores. The last two chapters in this section cover concepts—the normal distribution, probability, sampling, and confidence intervals—that pave the way for the transition from this section on descriptive statistics to inferential statistics, the topic of the final two thirds of the book.

Introduction to Statistics

LEARNING OBJECTIVES

- Determine if a study is correlational, experimental, or quasi-experimental.
- Classify variables and determine level of measurement.
- Learn the language and rules of statistics.

CHAPTER OVERVIEW

This first chapter is an introduction to statistics. The topics covered—types of experiments, types of variables, levels of measurement, statistical notation, order of operations, and rounding—will play a role in every chapter of this book.

1.1 The Purpose of Statistics

Statistics are techniques used to summarize data in order to answer questions. Statistical techniques were developed because humans are limited information processors. Give a human a lot of numbers at once and that person will focus on just a handful of them—likely the most distinctive numbers, not the most typical ones. If Sara were applying to graduate school, would she want the admissions committee to see a list of grades (including the one D that will stick out like a sore thumb) or her GPA? In this case, GPA, which is a summary score—a statistic—is better than a hodgepodge of individual course grades.

Statistics bring order to chaos (Dodge, 2003). On the following three pages are some data from the Statistical Abstract of the United States, showing the percentage of the 18- to 25-year-old population, for each state, that has engaged in binge drinking at least once in the past 30 days. (Binge drinking is defined as five or more drinks within a couple of hours.) Table 1.1 is unorganized. It is hard to find a specific state, to figure out which state has the most/least binge drinking, or to figure out what the average is.

Such a chaotic arrangement of data is not very useful. Just by alphabetizing the states, as in Table 1.2, order is brought to the data, making it easier to find a state.

In Table 1.3, the binge drinking data are arranged from low to high, bringing a different order to the information. As a result, it is easy to see the range of binge drinking rates. This arrangement may bring up some questions: Care to speculate why Utah has the lowest rate? With the data sorted from low to high, it is even possible to get some sense of what the average score is.

TABLE 1.1 Percentage of 18- to 25-Year-Olds Engaging in Binge Drinking During the Past 30 Days per State, Arranged in Random Order

Oregon	39%		Missouri	41%		South Dakota	46%		New Hampshire	49%
Nevada	37%		Connecticut	44%		New York	41%		Kansas	39%
Oklahoma	38%		Minnesota	45%		Alabama	32%		Vermont	45%
Mississippi	33%		Wisconsin	47%		Arizona	38%		Idaho	32%
Alaska	36%		Delaware	41%		Maryland	39%		Hawaii	39%
California	36%		Ohio	41%		South Carolina	39%		Virginia	41%
Kentucky	40%		Pennsylvania	44%		Florida	35%		North Dakota	54%
Colorado	44%		Georgia	31%		Louisiana	36%		Michigan	42%
Wyoming	45%		New Jersey	41%		Tennessee	30%		Nebraska	42%
Illinois	44%		Indiana	41%		Maine	43%		New Mexico	38%
West Virginia	37%		North Carolina	33%		Montana	45%		Texas	35%
Rhode Island	49%		Washington	39%		Iowa	48%			
Massachusetts	46%		Utah	26%		Arkansas	36%			

Note: When a table has no order–not alphabetical, not ranked from high to low–it is difficult to use.

Source: Substance Abuse and Mental Health Services Administration, *2012–2013 National Survey of Drug Use and Health.*

TABLE 1.2 Percentage of 18- to 25-Year-Olds Engaging in Binge Drinking During the Past 30 Days per State, Arranged in Alphabetical Order

Alabama	32%	Louisiana	36%	Ohio	41%
Alaska	36%	Maine	43%	Oklahoma	38%
Arizona	38%	Maryland	39%	Oregon	39%
Arkansas	36%	Massachusetts	46%	Pennsylvania	44%
California	36%	Michigan	42%	Rhode Island	49%
Colorado	44%	Minnesota	45%	South Carolina	39%
Connecticut	44%	Mississippi	33%	South Dakota	46%
Delaware	41%	Missouri	41%	Tennessee	30%
Florida	35%	Montana	45%	Texas	35%
Georgia	31%	Nebraska	42%	Utah	26%
Hawaii	39%	Nevada	37%	Vermont	45%
Idaho	32%	New Hampshire	49%	Virginia	41%
Illinois	44%	New Jersey	41%	Washington	39%
Indiana	41%	New Mexico	38%	West Virginia	37%
Iowa	48%	New York	41%	Wisconsin	47%
Kansas	39%	North Carolina	33%	Wyoming	45%
Kentucky	40%	North Dakota	54%		

Note: When some order is imposed on a table, it is easier to use. Compared to Table 1.1, it is much easier to locate a state in this alphabetized table.

Source: Substance Abuse and Mental Health Services Administration, *2012–2013 National Survey of Drug Use and Health.*

TABLE 1.3 Percentage of 18- to 25-Year-Olds Engaging in Binge Drinking During the Past 30 Days, Arranged from Low to High by State

Utah	26%		Missouri	41%
Tennessee	30%		Delaware	41%
Georgia	31%		Indiana	41%
Alabama	32%		New Jersey	41%
Idaho	32%		Virginia	41%
Mississippi	33%		New York	41%
North Carolina	33%		Ohio	41%
Florida	35%		Michigan	42%
Texas	35%		Nebraska	42%
California	36%		Maine	43%
Louisiana	36%		Pennsylvania	44%
Arkansas	36%		Colorado	44%
Alaska	36%		Illinois	44%
West Virginia	37%		Connecticut	44%
Nevada	37%		Montana	45%
Arizona	38%		Vermont	45%
Oklahoma	38%		Wyoming	45%
New Mexico	38%		Minnesota	45%
Washington	39%		South Dakota	46%
South Carolina	39%		Massachusetts	46%
Oregon	39%		Wisconsin	47%
Kansas	39%		Iowa	48%
Hawaii	39%		Rhode Island	49%
Maryland	39%		New Hampshire	49%
Kentucky	40%		North Dakota	54%

Note: This table is arranged in order of binge drinking rate. This makes it easy to find the range of rates and possible to get a sense of the average binge drinking rate.

Source: Substance Abuse and Mental Health Services Administration, *2012–2013 National Survey of Drug Use and Health.*

For another way to summarize the data, see Table 1.4. Here, the states are grouped into the four U.S. census regions (and within region from low to high), giving us some idea if geographic differences in binge drinking rates exist across the United States. All three tables (1.2, 1.3, and 1.4) arrange the data differently and answer different questions. Statistics involves organizing and summarizing information to answer a question.

Practice Problems 1.1

Review Your Knowledge

1.01 What is the purpose of statistics?

Apply Your Knowledge

1.02 A researcher has collected IQ scores from a sample of sixth graders. What are some questions she might want to ask of the data? How would the data be arranged to answer the questions?

TABLE 1.4 Percentage of 18- to 25-Year-Olds Engaging in Binge Drinking During the Past 30 Days, by U.S. Census Region

Northeast		Midwest		South		West	
New Jersey	41%	Kansas	39%	Tennessee	30%	Utah	26%
New York	41%	Missouri	41%	Georgia	31%	Idaho	32%
Maine	43%	Indiana	41%	Alabama	32%	California	36%
Pennsylvania	44%	Ohio	41%	Mississippi	33%	Alaska	36%
Connecticut	44%	Michigan	42%	North Carolina	33%	Nevada	37%
Vermont	45%	Nebraska	42%	Florida	35%	Arizona	38%
Massachusetts	46%	Illinois	44%	Texas	35%	New Mexico	38%
Rhode Island	49%	Minnesota	45%	Louisiana	36%	Washington	39%
New Hampshire	49%	South Dakota	46%	Arkansas	36%	Oregon	39%
		Wisconsin	47%	West Virginia	37%	Hawaii	39%
		Iowa	48%	Oklahoma	38%	Colorado	44%
		North Dakota	54%	South Carolina	39%	Montana	45%
				Maryland	39%	Wyoming	45%
				Kentucky	40%		
				Delaware	41%		
				Virginia	41%		

Note: The states in this table are grouped into their geographical census regions and then arranged within a region in order of increasing rates of binge drinking. This makes it possible to examine the table for geographic differences in the binge drinking rate.

Source: Substance Abuse and Mental Health Services Administration, *2012–2013 National Survey of Drug Use and Health.*

1.2 Experiments and Variables

The characteristics measured by researchers are called **variables.** They are called variables for a simple reason—they vary. Some characteristics commonly measured by psychologists are height, weight, intelligence, aggression, and neurotransmitter levels. In any group of people, there are differences on all these variables—people differ in their heights and weights, some are smarter than others, some exhibit more aggressive behavior, and not everyone has the same levels of neurotransmitters. No matter the variable, different people have different amounts of it or different types of it. Variables vary.

The subjects of a study, the objects on whom these variables are being measured, are called **cases.** In many psychological studies, cases are people who participate in the study. But psychologists also study other animals like rats, pigeons, monkeys, and even cockroaches—all of which would be called cases. Sometimes the object of study, the case, is smaller than a whole organism, such as nerve cells. And sometimes the cases being studied are larger, like cities or nations. For example, in the binge drinking tables shown earlier, the cases were states.

The most common type of study done by psychologists asks a question about the relationship between two variables. Here are some examples:

- Is there a relationship between how much violent television children watch and how aggressive they are?

- Is there a relationship between how one studies and how much one learns?
- Is there a relationship between children's attachment to their parents and their future mental health?
- Is there a relationship between the intensity of a stimulus and whether a neuron fires?
- Is there a relationship between sugar consumption and hyperactivity?
- Is there a relationship between the population density of cities and their suicide rates?

To answer a relationship question, a researcher obtains a sample of cases and measures each case's level on each of the two variables. If a social psychologist were investigating a possible relationship between the amount of violent television watched and aggression, she would get a group of children and find some way to measure *both* how much violent TV they watched *and* how aggressive they were. She might, for example, find out what shows the children watched during a week, find out how many violent acts were in each show, and then total the number of violent acts seen during the week. To measure the children's aggression, she might have their teachers complete an aggression inventory on them. In doing so, she would have measured two variables—the number of violent acts seen on television each week, and aggression level as rated by teachers—and could now look for a relationship between them.

Correlational Designs

There is a formal name for the type of study that asks a relationship question—a *correlational design*. In a **correlational design,** the two variables are simply measured, or observed, as they naturally occur. The two variables are not controlled or manipulated by the experimenter. In the TV and aggression study outlined above, for example, the researcher didn't change or limit the type of TV the kids watched and she didn't try to control their aggression. She simply took the cases as they were on the two variables.

In a correlational study, the researcher measures the pairs of variables for the cases in the sample and determines if they vary together in a systematic way. If cases have a lot of one variable, do they also tend to have a lot of the other? In the TV and aggression study, imagine that the researcher found that kids who watched more violent TV were rated as more aggressive by their teachers. This relationship is shown in Figure 1.1. If this researcher thinks that watching violent TV has an influence on aggressive behavior, she would call the amount of violent TV watched the **predictor variable** and the rating of aggression the **criterion variable**. The predictor variable is the one that comes first chronologically or that is believed to drive the relationship. The criterion variable is the outcome variable; it is where the influence of the predictor variable is observed.

A correlational design allows researchers to study relationships between variables as they exist in real life—and that's a big advantage.

A correlational design allows researchers to study relationships between variables as they exist in real life—and that's a big advantage. But, there is a substantial disadvantage to correlational designs as well: correlational designs don't allow a researcher to draw a conclusion about cause and effect, about whether one variable causes the other.

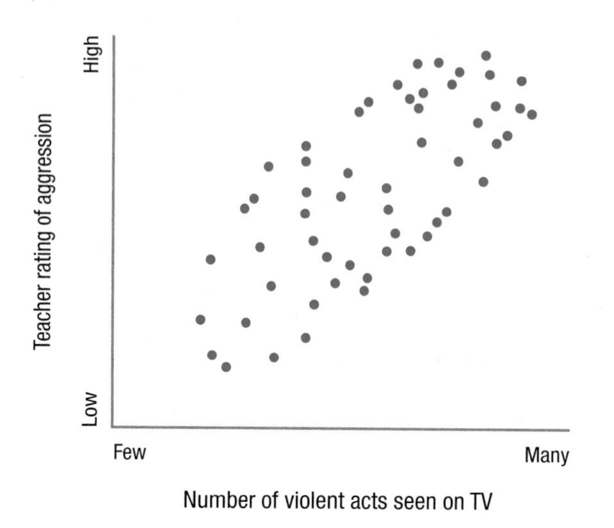

Figure 1.1 Example of Correlational Design Results A correlational design examines the relationship between two variables without any attempt to manipulate them. The hypothetical results of this correlational study show that children who see more violent acts on TV are also rated as more aggressive by their teachers.

Statisticians and college professors love to go around saying, "Correlation is not causation." Here's why that's true. Think about the relationship shown between watching violent TV and being aggressive in Figure 1.1. It makes sense that seeing violence modeled on TV leads one to imitate it and behave aggressively. So, it is tempting to conclude this study means that TV violence *causes* aggression. However, one can't reach such a conclusion with certainty. If two variables (let's call them *X* and *Y*) are correlated, three possible explanations exist. However, the correlation doesn't reveal which of the explanations is the right one.

1. Does *X* cause *Y*?
2. Or, does *Y* cause *X*?
3. Or, does a third variable cause both *X* and *Y*?

Using the labels *X* and *Y*, one can label the amount of violent TV watched as *X* and the amount of aggressive behavior as *Y*. The *X* causes *Y* explanation for the relationship would be that watching violent TV causes children to behave aggressively. As discussed above, this is a plausible explanation. But, isn't the *Y* causes *X* explanation just as plausible? Doesn't it make sense that kids who are more aggressive, who like to hit and hurt other kids, would be drawn to watch TV shows that mimic their behavior? Looking at Figure 1.1, couldn't one conclude that being aggressive leads to choosing to watch more violent TV? A correlation only tells that two variables are related. A correlation doesn't tell anything about the direction of the relationship.

There's one more possible explanation for a relationship between *X* and *Y*, that a third variable causes *both* *X* and *Y*. This type of third variable has a formal name: *confounding variable*. In a correlational design, a **confounding variable** has an impact on both variables (on *X* and on *Y*) and is the reason why the two variables covary. If there is a confounding variable, *X* doesn't cause *Y* and *Y* doesn't cause *X*; the confounding variable causes both.

This can be confusing, so here's an example to help clarify. There are parents who control their children—they limit what type of TV their kids watch, they restrict consumption of junk food, make sure homework is done before playtime, make sure their children are well-mannered, and so on. Wouldn't the children of these parents watch very little violent TV? Wouldn't these children also be more likely to be rated as less aggressive by teachers? If it were found that watching violent TV covaried with aggressiveness, couldn't this be explained by the confounding variable of parenting style? That is, parenting style causes *both* the amount of violent TV watched *and* aggressive behavior.

There can be more than one confounding variable in a correlational study. Sex, whether one is a male or a female, could be another confounding variable for the TV and aggression study. Who will, in general, be rated as more aggressive by teachers, boys or girls? Probably boys. What if boys also tend to watch more violent television? Then, the observed relationship between watching violent TV and being aggressive could be explained by the confounding variable of sex.

A Common Question

Q Can a confounding variable affect just one of the variables?

A No, a researcher has to make a case that the confounding variable has an impact on both *X* and *Y*.

Experimental Designs

Not being able to draw cause-and-effect conclusions is a significant limitation of correlational designs. Luckily, there's another type of study, called an **experimental design,** that allows researchers to draw cause-and-effect conclusions. In an experimental design, one of the variables, called the **independent variable,** is manipulated or controlled by the experimenter. The effect of this manipulation is measured on the other variable, the **dependent variable.** This sounds more complex than it is, so an example should help clarify.

Let's make the TV and aggression study into an experimental design. Suppose a different researcher obtained a group of kids, brought them into his laboratory, and separated them into two groups by flipping a coin for each kid. He called one group the "control" group and the other the "experimental" group. If the coin turned up heads, he assigned the kid to the control group and showed a nonviolent cartoon. If the coin turned up tails, he assigned the kid to the experimental group and showed a violent cartoon. Note that the researcher is controlling what type of cartoon, violent or nonviolent, each kid watches. This, the independent variable, is the only difference between the control group and the experimental group. Thus, the control group provides a comparison group for an examination of the effect of a violent cartoon.

After watching the cartoon, the researcher has each kid play a video game that involves making choices between a cooperative response or an aggressive one. He keeps track of the number of aggressive responses each child makes as the dependent variable.

The researcher completes the study by comparing the average number of aggressive responses in the group of kids who watched the violent cartoon to the average number in the group of kids who watched the nonviolent cartoon. If he found that watching the violent cartoon increased the number of aggressive responses, then the results might look like they do in Figure 1.2.

Note that this experiment differs from the correlational study because the cases are in groups and the groups are being treated differently. What makes this a "true"

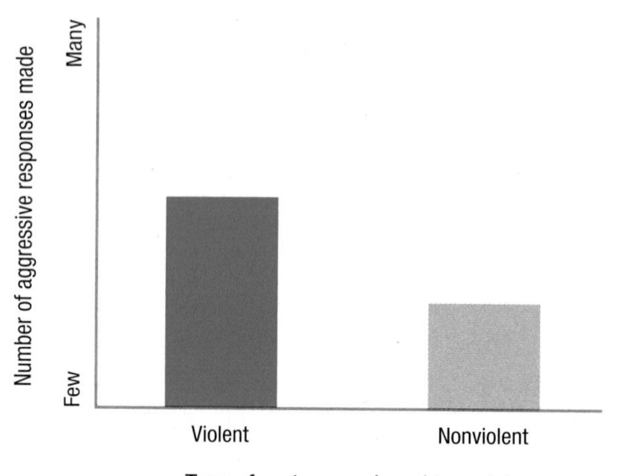

Figure 1.2 Example of Experimental Design Results The hypothetical results of this experimental design study show that children who were randomly assigned to watch a violent cartoon made more aggressive responses in the video game than did children who had been assigned to watch the nonviolent cartoon.

experimental design is how the cases are assigned to groups, by *random assignment.* In **random assignment**, each case has an equal chance of being assigned to either group. When the researcher flipped a coin to determine which type of cartoon each kid would watch, that was random assignment. Random assignment is the hallmark of an experiment.

Why is random assignment so important? Random assignment means that the only difference between the two groups should be the independent variable that is controlled by the researcher.

Why is random assignment so important? Random assignment means that the two groups *should* be similar on all dimensions before the experiment begins. Thus, after the experiment is over, the only difference between the two groups should be the independent variable that was controlled by the researcher. And, if the only difference between the two groups is the independent variable *and* the two groups differ on the dependent variable, then the only cause for the observed difference is the independent variable. Confounding variables, which caused a problem with correlational designs, are taken out of contention by random assignment. Because of this, with experimental designs, one can draw a conclusion about cause and effect.

To clarify how this works, think of the TV and aggression experiment in which random assignment was used to assign kids to groups. Remember the two confounding variables mentioned for the correlational design, parenting style and sex? Thanks to random assignment, these are no longer an issue in an experimental design. If a child has controlling parents, he or she has an equal chance of being assigned to either group. As a result, both groups should have roughly equal numbers of children who have controlling parents, which means parenting style can no longer be used as an explanation for the difference found in the number of aggressive responses. Similarly, the two groups should be roughly equal in the percentage of boys. In fact, no matter what the potential confounding variable is, the two groups should be close to equivalent on it.

Here's the logic of why the independent variable has to be the cause of any observed differences on the dependent variable when we use random assignment:

a. *If* the two groups are alike in all ways except the independent variable (e.g., type of cartoon watched),

b. *and* the two groups are found to differ on the dependent variable (e.g., number of aggressive responses),

c. *then* the only plausible explanation for the difference on the dependent variable is the independent variable.

Quasi-Experimental Designs

There's one more type of study to know about, a mash-up of a correlational design with an experimental design. It looks like an experimental study, but is like a correlational one and is called a *quasi-experimental design.* In a **quasi-experimental design,** the cases are *classified* as being in different groups on the basis of some characteristic they already possess. They are not *assigned* to groups based on a variable that the researcher controls. In a quasi-experiment, the groups are naturally occurring groups. The naturally occurring groups are then compared on an outcome variable.

Again, this sounds more complex than it is. To clarify, imagine a third researcher studying the television and aggression example, one using a quasi-experimental design. This researcher starts the same way the researcher did for the experimental design, by bringing kids to her laboratory. But she treats them differently than he did once they get to the lab. She tells them that they can pick a cartoon to watch from two options: (1) a violent cartoon with lots of hitting and fighting, or (2) a nonviolent cartoon with lots of cooperative acts. After each kid has watched his or her cartoon choice, she has him or her play the video game and records how many aggressive responses were made. This looks like the experimental design because she is comparing two groups. However, she is not controlling any of the variables, so it is actually similar to a correlational design.

The results might turn out as shown in Figure 1.3, indicating that those who watched the violent cartoon engaged in more aggressive behavior. These results might tempt the researcher to draw a cause-and-effect conclusion—that watching violent TV leads to aggressive behavior—but a researcher can't draw a cause-and-effect conclusion for a quasi-experimental design. Because a quasi-experimental design doesn't utilize random assignment, confounding variables could exist. For

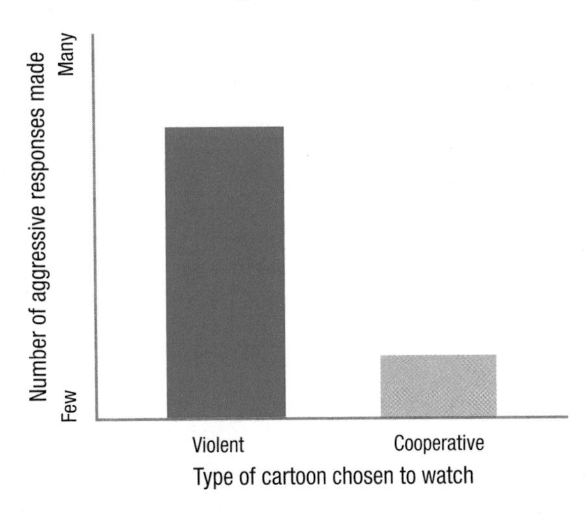

Figure 1.3 Example of Quasi-Experimental Design Results The hypothetical results of the quasi-experimental design show that children who chose to watch the violent cartoon made more aggressive responses in the video game than did children who picked the nonviolent cartoon to watch.

instance, maybe the cooperative cartoon group has more children of controlling parents or maybe it has more girls. These would be confounding variables that could be the real cause of the results.

In quasi-experimental designs, the variable that defines the group in which a participant belongs is called the **grouping variable**. The variable where the influence of the grouping variable is measured is called the **dependent variable**, exactly the same name used in experimental designs. In the case above, cartoon choice is the grouping variable and number of aggressive responses is the dependent variable.

Why would a researcher ever use a correlational design or a quasi-experimental design if they don't allow one to draw a conclusion about cause and effect? Correlational designs and quasi-experimental designs allow researchers to study things that, for ethical or practical reasons, can't be manipulated. If one wanted to know the impact of prenatal maternal cigarette smoking on children's intellectual development, it wouldn't be possible to take pregnant women and randomly assign some to smoke cigarettes and others not to smoke. One could, however, do a quasi-experiment in which women were classified as having smoked or not while they were pregnant and the intelligence of their kids was assessed.

Explanatory and Outcome Variables

The term *independent variable* has a specific meaning: the independent variable is the variable in an experimental study that is controlled by the experimenter and has an effect on the dependent variable. To help remember the difference between independent variable and dependent variable, here's a mnemonic: *iced* (see Figure 1.4). Iced stands for "independent = cause; effect = dependent."

A variable is only an independent variable if it is controlled by the experimenter. However, researchers sometimes use terms casually, generically referring to variables in correlational and quasi-experimental studies as independent and dependent. That can get confusing. In this book, the generic term for the variable that causes, influences, or precedes the other is **explanatory variable. Outcome variable** is the generic term used here for the variable that is caused, influenced, or comes after the

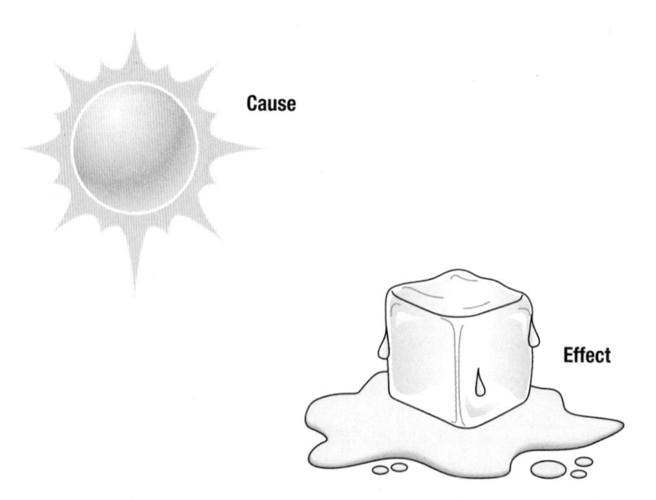

Figure 1.4 ICED: Mnemonic for Types of Variables The mnemonic *iced* (independent = cause, effect = dependent) can be used to remember that the independent variable is the cause and the dependent variable the effect that is measured. Here, the sun is the cause of the melting of the ice cube.

explanatory variable. The explanatory variables for correlational, experimental, and quasi-experimental designs, respectively, are predictor variable, independent variable, and grouping variable. The three outcome variables, respectively, are criterion variable, dependent variable, and dependent variable.

Table 1.5 summarizes the three different types of studies. It is important to know what type of study—correlational, experimental, or quasi-experimental—is being conducted in order to choose the correct statistical test.

To determine the type of study, one needs to know which variable is the explanatory variable and which variable is the outcome variable. Here are some guidelines to help:

- Formulate a cause-and-effect statement (e.g., watching violent TV causes aggression). The "cause" is the explanatory variable and the "effect" is the outcome variable.
- Think about chronological order. Does one variable come before the other? The explanatory variable is the one that comes first (e.g., first one watches a lot of violent TV, then one behaves aggressively).
- Sometimes it is easier to figure out the outcome variable first, the variable where the "effect" is being measured (e.g., what is being measured is how much aggression kids exhibit). The variable "left over" is the explanatory variable.

Once it is known which variable is the explanatory variable, use the flowchart in Figure 1.5 to figure out the type of research design. Remember, if the study is correlational or quasi-experimental, consider the possibility of a confounding variable.

TABLE 1.5 Comparing Correlational, Experimental, and Quasi-Experimental Studies

	Correlational	Experimental	Quasi-Experimental
Explanatory variable is called	Predictor variable	Independent variable	Grouping variable
Outcome variable is called	Criterion variable	Dependent variable	Dependent variable
Cases are . . .	Measured for naturally occurring variability on both variables.	Assigned to groups by experimenter using random assignment.	Classified in groups on basis of naturally occurring status.
Is explanatory variable manipulated/controlled by the experimenter?	No	Yes	No
Can one draw a firm conclusion about cause and effect?	No	Yes	No
Is there a need to worry about confounding variables?	Yes	No	Yes
What is the question being asked by the study?	Is there a relationship between the two variables?	Do the different groups possess different amounts of the dependent variable?	Do the different groups possess different amounts of the dependent variable?
What is an advantage of this type of study?	Researchers can study conditions that can't be manipulated.	Researchers can draw a conclusion about cause and effect.	Researchers can study conditions that can't be manipulated.

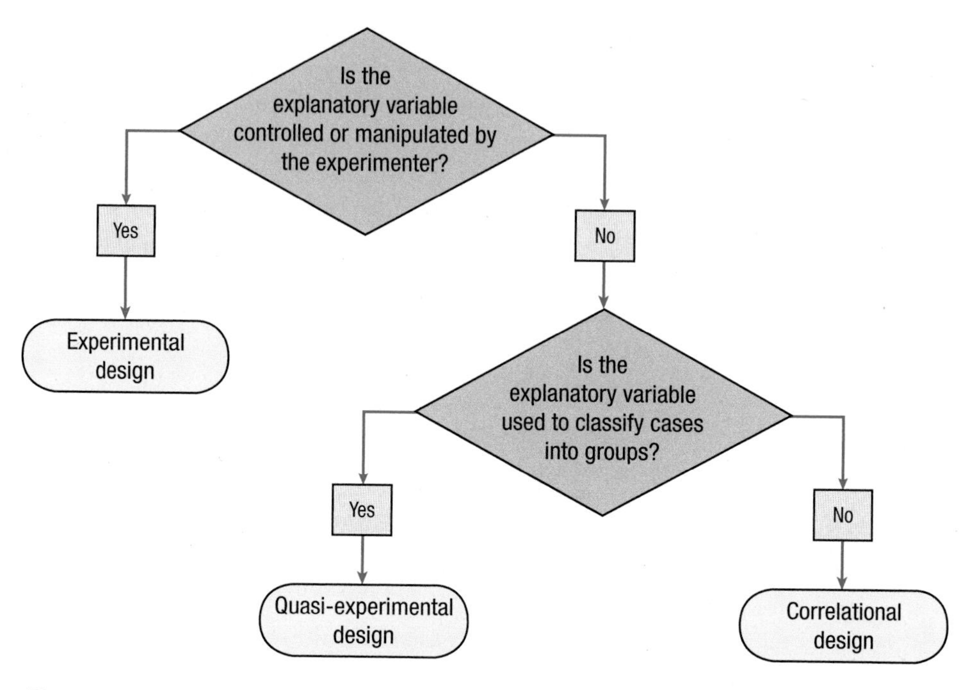

Figure 1.5 How to Choose: Type of Study Once someone has determined which variable is the explanatory variable, this flowchart can be used to determine whether the study is correlational, experimental, or quasi-experimental. It depends on whether the explanatory variable is controlled by the experimenter and, if not, whether cases are classified into groups.

A Common Question

Q What do the different names for explanatory variables in correlational and quasi-experimental designs signify?

A In a quasi-experimental design, the explanatory variable is called a grouping variable because cases are grouped together. Cases aren't put into groups for the predictor variable in a correlational design. Rather, typically, cases have a wide range of individual values on both the predictor variable and the criterion variable.

Worked Example 1.1

Let's practice figuring out (a) which variable is the explanatory variable and which is the outcome variable, and (b) what type of study is being conducted. A health psychologist was interested in studying the effect of life stress on physical health. She administered a life-stress inventory to a group of college students and used their responses to classify them as being high or low on life stress. She also determined how many days in the past year each student had been sick. As shown in Figure 1.6, she found a relationship between the two variables: people in the high life-stress group had more days of illness than those in the low life-stress group.

The two variables are life stress (high vs. low) and number of days of illness. The researcher believes that life stress causes/leads to/affects physical health, so

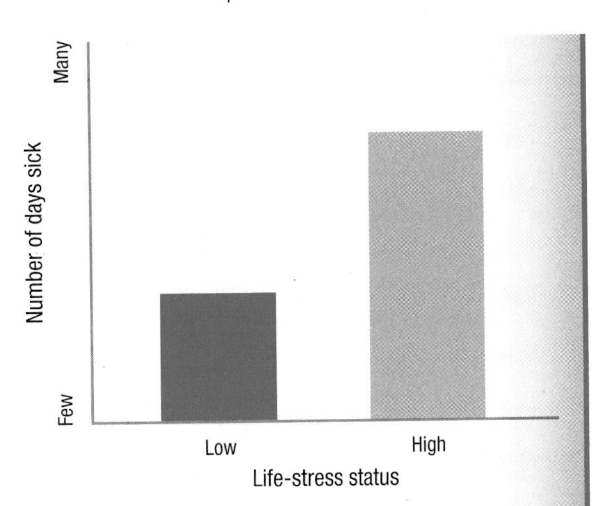

Figure 1.6 Life Stress and Days of Illness Results Hypothetical results showing that those with more life-stress experience more days of physical illness.

the amount of stress is the explanatory variable and the number of sick days is the outcome variable.

Using the flowchart in Figure 1.5, one concludes that the study is quasi-experimental because the participants are placed in groups on the basis of how much life stress they naturally have. In quasi-experimental studies, the explanatory variable is called a grouping variable and the outcome variable a dependent variable.

This is a quasi-experimental study, so the possibility of confounding variables exists. What other variables could affect both the grouping variable and the dependent variable in this study? Perhaps people with poor memories recall less stress and are placed in the low-stress group. They might also "forget" how much they were sick last year. These results would seem to suggest that people with more stress experience more illness, but the data really only show that people with better memories recall more stress and recall more illness. Another possible confounding variable is socioeconomic status—people with more money may have less stress and be healthier, as they can afford better medical care. The bottom line: the results show that the amount of stress and the number of sick days covary, but one can't draw a conclusion about cause and effect from a correlation.

Practice Problems 1.2

Review Your Knowledge

1.03 Name the three different types of studies.

1.04 In which studies are cases assigned to groups?

1.05 In which type of study can one draw a conclusion about cause and effect?

1.06 In which types of studies does one need to worry about confounding variables?

Apply Your Knowledge

For Practice Problems 1.07–1.09:

a. Write a sentence that states the researcher's question.

b. Name the explanatory variable and the outcome variable. Label them as predictor variable and criterion variable, or independent variable and dependent variable, or grouping variable and dependent variable.

Practice Problems 1.2 (Cont.)

c. ***Determine the type of study: correlational, experimental, or quasi-experimental.***

d. ***If the study is correlational or quasi-experimental, name a plausible confounding variable and explain how it affects both of the other variables.***

1.07 A psychologist gets a group of college students to come to his lab and randomly divides them into two groups. He has each group study a list of nonsense syllables for 15 minutes. One group, the massed group, studies for 15 minutes in a row. The other group, the spaced group, studies for three 5-minute periods, each separated by 10 minutes of a distractor task. Immediately after each group completes the 15 minutes of studying, he measures recall of the nonsense syllables.

1.08 An educational researcher is interested in the effects of sleep on school performance. She classifies students as (a) getting an adequate amount of sleep or (b) not getting an adequate amount of sleep. Then she measures school performance by finding out their GPAs for a semester.

1.09 A nutritionist is examining the role of fiber in gastrointestinal (GI) health. He believes that consuming fiber is good for GI health. He gathers a group of college students and asks each one the number of grams of fiber consumed per day. He also asks each one how many episodes of GI distress (upset stomach, nausea, vomiting, diarrhea, etc.) he or she has experienced in the past year.

1.3 Levels of Measurement

Statistical techniques are performed on data: numbers. All numbers may look alike, but there are different types of numbers that vary in how much information they contain. The type of number determines the statistics that can be used, so it is important to know about the different types of numbers, called levels of measurement.

Statisticians divide numbers into four levels of measurement: nominal, ordinal, interval, and ratio. As the numbers move up the hierarchy, from nominal to ratio, they become more complex and contain more information.

Statisticians divide numbers into four levels of measurement: nominal, ordinal, interval, and ratio. As the numbers move up the hierarchy, from nominal to ratio, they become more complex and contain more information. Nominal numbers contain only one piece of information, while ratio numbers contain four pieces of information.

Before learning about the four levels of measurement, Figure 1.7 shows a mnemonic to help remember them in the correct order: *noir.* Noir is French for "black" and allows one to remember nominal, ordinal, interval, and ratio in order from least complex to most complex.

Nominal Numbers

Nominal-level numbers, the simplest level of measurement, sort cases into categories. The numbers used for each category are arbitrary and provide no quantitative information. The numbers simply indicate whether two things are the same or different. Cases are assigned different numbers if they possess different qualities, and they are assigned the same number if they possess the same quality. For example, the race of participants in a study could be measured by assigning a 1 to whites, 2 to blacks, 3 to Native Americans, 4 to Asians, and 5 to Pacific Islanders. If Jack is a 1 and Jill is a 1, then both are white. If Simone is a 1 and Susanne is a 2, then they are of different races.

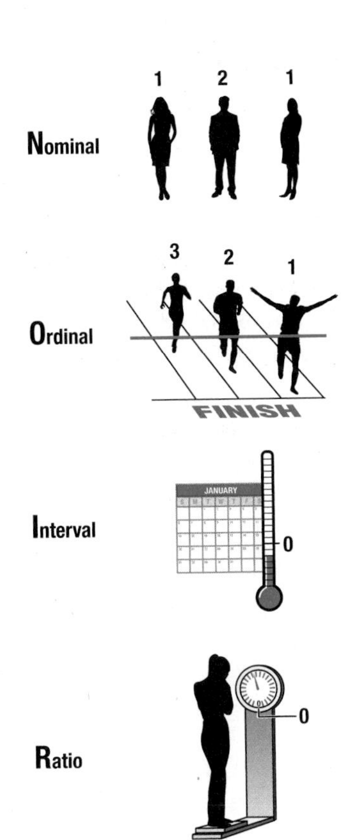

Figure 1.7 Four Levels of Measurement The French word for "black," *noir,* can be used as a mnemonic for the four levels of measurement—nominal, ordinal, interval, ratio.

Race is a variable that social scientists regularly measure. Other variables psychologists use that are measured at the nominal level are factors like sex (male or female), handedness, and psychiatric diagnosis.

There is not much math that can be done with nominal numbers because the numbers are arbitrary. The researcher studying race could just as easily assign the number 7 to whites, 22 to blacks, 0.78 to Native Americans, and so on. The numbers are arbitrary, so it doesn't make sense to find the "average" race of a group of people. However, the researcher could count how many are 1s (white), 2s (black), 3s (Native American), and so on.

If two cases receive different numbers, then those numbers reflect a *difference* in the attribute being measured. If the numbers also tell the *direction* of the difference, which case has more of the attribute and which has less, then one has moved to the next level of measurement, referred to as ordinal.

Ordinal Numbers

Ordinal-level numbers reveal whether cases possess more or less of some characteristic. In ordinal measurement, one might order the objects being measured from lowest to highest (or vice versa) in terms of how much of the characteristic each possesses. Then, ranks are assigned. For example, it is common to say that the person with the fastest time in a race came in first place, the next fastest person was in second place, and so on. Thus, ordinal numbers give information about the *direction* of differences, about which case possesses more or less of the attribute being measured.

But, ordinal numbers don't give any information about how much distance separates cases with different ranks. For example, the most populous state in the United States is California, with about 39 million people. It is followed by Texas (27 million) and Florida (20 million). Note that the distance between rank 1 and 2 is 12 million people and the distance from rank 2 to 3 is 7 million people, yet both are one rank apart. Sometimes the distance between two adjacent scores is large, and sometimes it is small. Ordinal numbers give no information about how far apart cases are.

Though ordinal numbers are still limited in the math one can perform, there are special statistics developed to use with ordinal measures. Here are some ordinal variables psychologists use:

- Class rank and birth order, two variables psychologists use, are ordinal. In fact, most variables with the term *rank* or *order* in them are ordinal.
- Individual items on the Apgar scale, a way of measuring the health of newborns, are ordinal. For example, on the Apgar scale, a score of 0 means the pulse is less than 60, 1 is used if the pulse is from 60 to 100, and 2 means the pulse is greater than 100.

- Many rating scales are ordinal. For example, a question on a survey asking whether one drinks alcohol never (0), rarely (1), every month (2), every week (3), or every day (4) would be ordinal.

Interval Numbers

Interval-level numbers tell *how much* of the characteristic being measured a case possesses. They contain meaningful information about *distance* between cases in addition to information about same/different and more/less. That is, they reveal how much of a difference exists between two cases.

Interval-level numbers can tell how far apart two cases are because there is equality of units. Equality of units means that the interval used on the scale is a consistent width along the whole length of the scale. For example, on a Fahrenheit temperature scale, the distance from 31° to 32° is the same as the distance from 211° to 212°.

If an attribute is measured on an interval scale, one can add and subtract the numbers, as well as calculate averages. However, one can't divide a specific interval-level measure by another interval-level measure. For example, it is not legitimate to say that an 80° summer day is twice as warm as a 40° winter day. Proportions can't be calculated because interval-level scales have arbitrary zero points. On an interval scale, a value of zero does not mean the absence of the attribute being measured. If a person takes a paper-and-pencil IQ test and answers all the questions incorrectly, one wouldn't say that he or she has no intelligence. Similarly, if the outside temperature on a winter day is 0°F, shown in Figure 1.8, it would be silly to state that there is no temperature that day.

A lot of the variables that psychologists measure, factors like intelligence and personality, are interval-level numbers. Statisticians prefer interval-level numbers over ordinal-level numbers and nominal-level numbers because interval-level numbers contain more information and more math can be done on them. For example, as already mentioned, one can find an average for a variable measured on an interval scale. More statistical techniques have been developed for interval-level data than for nominal and ordinal data.

Ratio Numbers

Ratio-level numbers have everything interval-level numbers have, but they also have an absolute zero point, a real zero point. Absolute zero indicates that a score of 0 means none of the attribute being measured is present. A simple example is speed. On a speed scale, such as miles per hour (mph), a value of zero means that the object is not moving. It has zero speed. An absolute zero point means one can divide two ratio-level numbers, to turn them into a proportion. If José is driving a car at 10 mph and Celeste is driving at 20 mph, one can legitimately say that her car is going twice as fast.

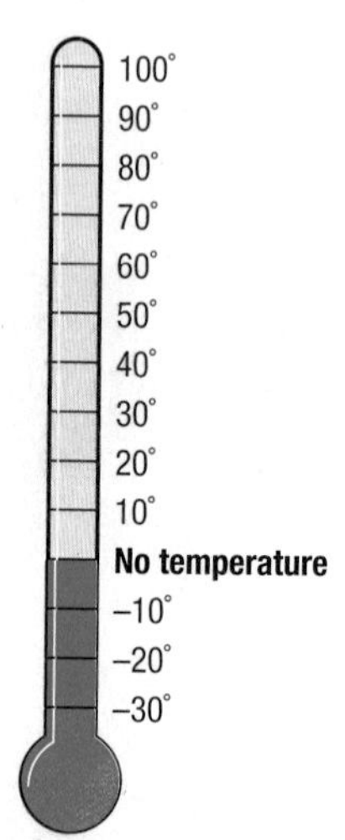

Figure 1.8 Thermometer with Nonsensical Zero Point This thermometer shows that temperature in degrees Fahrenheit has an arbitrary zero point and is an interval measure. It doesn't make sense to consider a temperature of 0 as "no temperature."

Figure 1.9 Ratio Level of Measurement and Negative Numbers Even though they have absolute zero points, some ratio-level measures can have negative values.

Most of the time, absolute zero points exist only for physical variables, not psychological ones. For example, height, weight, speed, and population all have absolute zero points. Also, things like age, cholesterol level, and pollution level have absolute zero points. There are some psychologically relevant variables—heart rate, crime rate, and length of marriage—that are ratio level. But, many other psychological variables—for instance, intelligence, masculinity, and conscientiousness—are not ratio level.

If a zero on a ratio scale means that none of the attribute is present, it seems like it would be impossible to have a negative ratio-level number. But, it is possible for ratio numbers to be negative. Anyone who's ever overdrawn a checking account (see Figure 1.9) and had a negative balance has had firsthand experience with this phenomenon.

An absolute zero point is necessary for a ratio-level measure, but a ratio-level measure must also have equality of units. Table 1.6 shows that there is a hierarchy to the levels of measurement where each more complex level has all the qualities of the preceding one.

Knowing the level of measurement of a variable is important. Figure 1.10 is a flowchart that shows four successive questions that can be used to determine a variable's level of measurement. As long as the answer to a question is yes, go on to ask the next question. When one hits a "no," stop. The last "yes" gives the level of measurement.

A Common Question

Q How can variables like height and weight have absolute zero points?

A No person will ever be zero inches tall or weigh zero pounds, so that value would never be assigned to a case, but that isn't what absolute zero means. An absolute zero point means that the zero *on the scale* represents an absence of the characteristic.

TABLE 1.6 Level of Measurement: Information Contained in Numbers

	Same/Different	Direction of Difference (more/less)	Amount of Distance (equality of units)	Proportion (absolute zero point)
Nominal	✓			
Ordinal	✓	✓		
Interval	✓	✓	✓	
Ratio	✓	✓	✓	✓

Note: As levels of measurement become more complex, numbers contain more information.

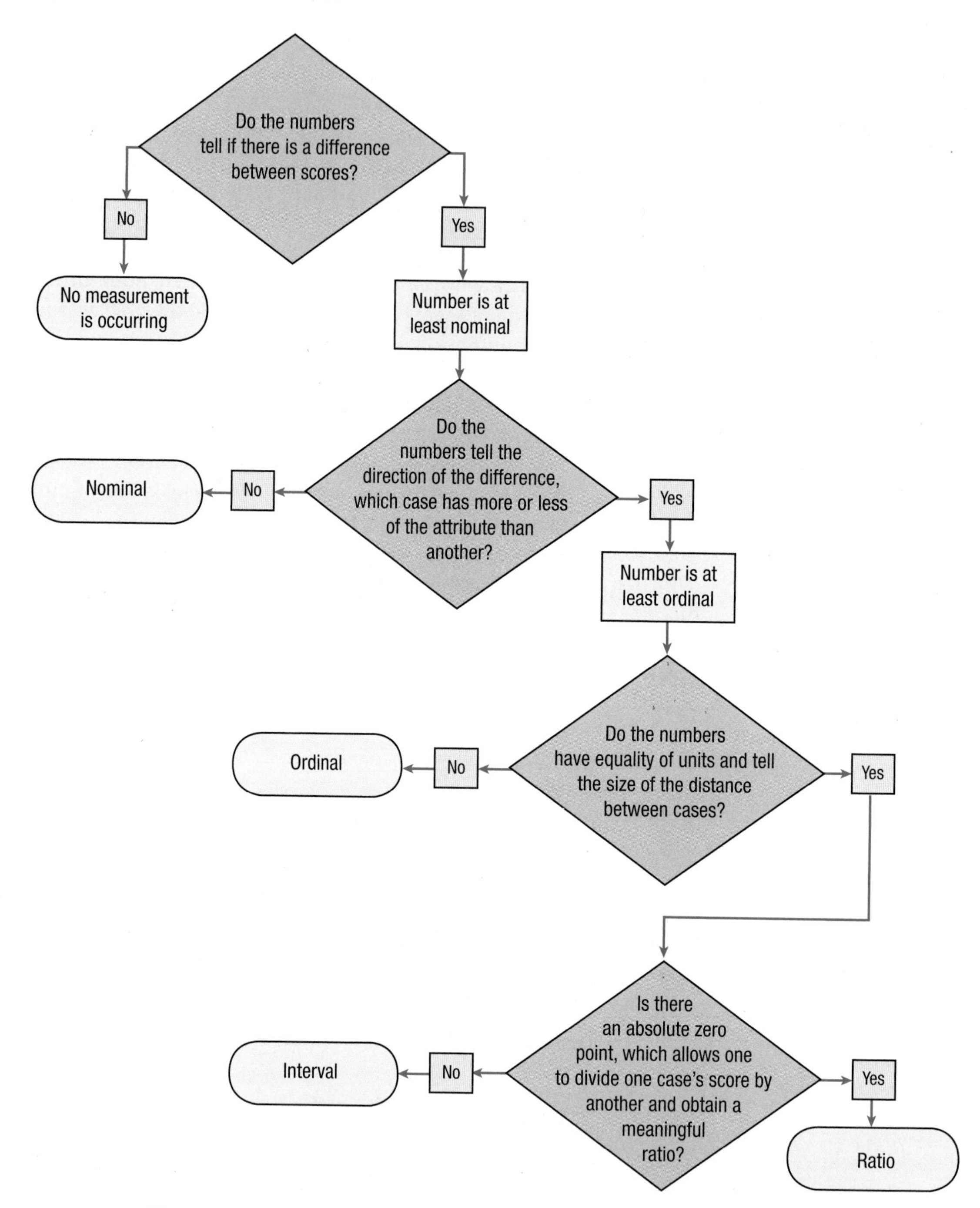

Figure 1.10 How to Choose: Level of Measurement Asking the four questions in this flowchart will lead to a decision regarding the level of measurement of a variable.

Worked Example 1.2

For practice figuring out level of measurement, imagine an end-of-semester test in an introductory psychology class that is meant to measure knowledge of psychology. It is 50 items long and each item is worth 2 points. Is this test measuring at the nominal, ordinal, interval, or ratio level?

To figure out the problem, think of two students, with two different scores on the test, one with an 80 and one with a 90, as shown in **Figure 1.11**. The test scores

Do the scores tell if there is a difference?	**Yes.** So at least nominal.
Do the scores tell the direction of the difference?	**Yes.** So at least ordinal.
Do the scores tell the size of the difference?	**Yes.** So at least interval.
Do the scores have an absolute zero point?	**No.** So interval level.

Figure 1.11 Choosing Level of Measurement To choose the level of measure, think of two cases with different scores. Then ask these four questions: (1) Do the scores tell whether the two cases are the same or different? (2) Do the scores reveal the direction of the difference between the two cases? (3) Do the scores tell the size of the difference between the two cases? (4) Is there an absolute zero point on the scale?

tell same vs. different, 80 is different from 90, so this test is measuring knowledge of psychology at least at the nominal level. One can also tell the direction of the difference, that one student did better than the other (90 is more than 80), so this test is measuring at least at the ordinal level. There was equality of units (each question was worth 2 points), so one can speak meaningfully of how much better one student did than the other. The test is measuring at least at the interval level.

The final question is whether there is an absolute zero point. It was stated earlier that this test was meant to measure knowledge of psychology. If someone got a zero on the test, would that mean this person has zero knowledge of psychology? No! It would just mean that the person didn't answer those 50 questions correctly. There might be 50 other facts that the person knows. That means this test has an arbitrary zero point, not an absolute one. As the last "yes" answer was at the interval level, one should stop there and conclude that this test measures knowledge of psychology at the interval level.

However, be aware that there is room for disagreement. Someone else may use this test to measure how many of the 50 questions were answered correctly. In that case, the zero point is absolute and the test is measuring at the ratio-level variable. Students don't like this, but whether a zero point is real or arbitrary can depend on how a test is being used or interpreted.

Practice Problems 1.3

Review Your Knowledge

1.10 List the four levels of measurement, in order from simplest to most complex.

1.11 When a scale moves from ordinal to interval by adding equality of units, what additional information does that scale now provide?

1.12 If a scale has an absolute zero point, what does a zero on the scale represent?

Apply Your Knowledge

For Practice Problems 1.13–1.16, determine what level of measurement the variable is (nominal, ordinal, interval, or ratio).

1.13 Ivan Pavlov gathers a group of dogs, exposes each to its favorite food for 5 minutes, and measures how many milliliters of saliva each dog produces during the 5-minute period. Salivation is being measured at what level?

1.14 At the end of the semester, a professor surveys his students using the following scale with regard to how much of the assigned reading was done: 0 = none, 1 = a bit, 2 = about half, 3 = most, 4 = all. Amount of assigned reading completed is measured at what level?

Practice Problems 1.3 (Cont.)

1.15 Infants who are being reared by heterosexual couples are observed and classified as (a) being attached to the mother, (b) being attached to the father, (c) being attached to both, or (d) being attached to neither. Attachment is being measured at what level?

1.16 A personality psychologist developed a femininity scale. From the list of hundreds of behaviors that are considered feminine, he selected 10. For each of these 10 behaviors, a person completing the scale indicates if he or she engages in the behavior. Each "yes" answer is worth 1 point. Scores on the scale can range from 0 to 10. Femininity is being measured at what level?

1.4 The Language of Statistics

Like any discipline, statistics defines words in special ways to make communication easier. The names for different types of studies and levels of measurement have already been covered. Now it is time to learn some more terms that will be used throughout the book.

The subjects in a study are drawn from a **population,** which is the larger group of cases a researcher is interested in studying. If a psychologist were interested in studying depression, the population could consist of all people in the world with depression. Such a large population may be too unwieldy to study, so the researcher might limit it by adding other criteria such as age (e.g., depression in adults) or a specific diagnosis (e.g., major depression).

Even if the population is carefully defined (e.g., adult males in industrialized nations who are experiencing a major depression and do not have a substance abuse diagnosis), there is no way that the researcher could find and contact all the people who meet these criteria. As a result, researchers almost always conduct their research on a subset of the population, and this subset is called a *sample.* A **sample** is a group of cases that is selected from the population. A sample is always smaller in size than the population, as shown in Figure 1.12. If a sample is an accurate reflection of the larger population, it is called a representative sample.

The data from a sample or population are often reduced to a single number, like an average, in order to summarize the group. This number has a different name depending on whether it is used to characterize a sample or a population. If the number characterizes a sample, it is called a **statistic.** If the number characterizes a population, it is called a **parameter.**

The difference between a statistic and a parameter is important, so different abbreviations indicate whether a value refers to a sample or a population. Statisticians use Latin letters to symbolize sample statistics and Greek letters to symbolize population parameters. When a researcher calculates the mean for a sample, it is called a statistic and symbolized by the letter M. For the population, a mean is a parameter and symbolized by the Greek letter mu (pronounced as "mew" and written like a script u, μ).

Statisticians also distinguish between descriptive statistics and inferential statistics. A **descriptive statistic** is a summary statement about a set of cases. It involves reducing a set of data to some meaningful value in order to describe the characteristics of that group of observations. If someone reported that 37% of his or her class was male, that would be a descriptive statistic.

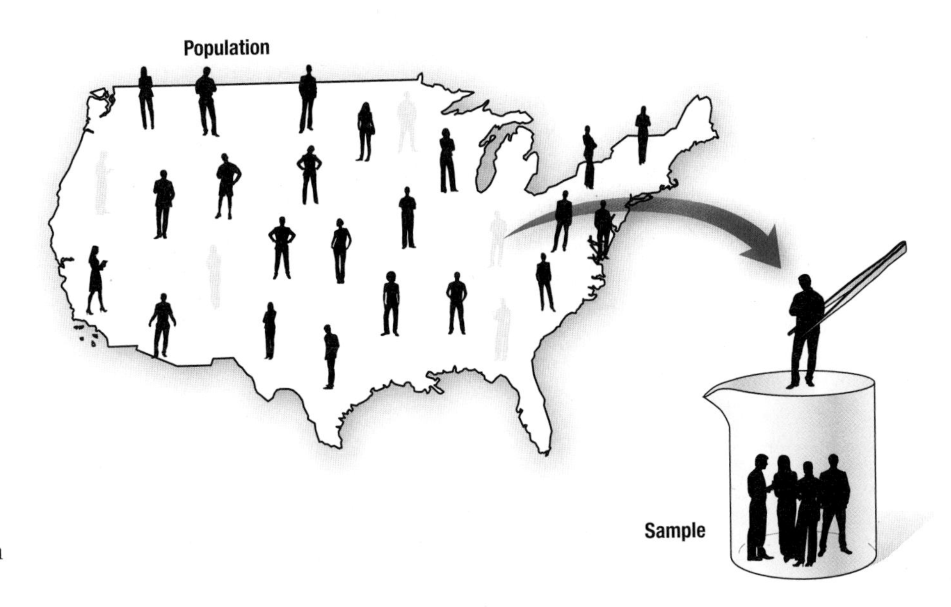

Figure 1.12 Sampling from a Population A sample is a subset drawn from a population.

An **inferential statistic** uses a sample of cases to draw a conclusion about the larger population. Inferential statistics also involve reducing a set of data to a single meaningful value, but it is being done to make inferences about a population (rather than to describe a sample). In other words, an inferential statistic allows one to generalize from a sample to a population.

For example, one might obtain a sample of students at a college, measure their depression level, and calculate the average depression level. If one stops there and makes a statement like, "The average depression level in the sample of students at the college was 12.75," that would be a descriptive statistic. But, if one went on and generalized to college students with a statement such as, "College students have, on average, a depression level of 12.75," that would be an inferential statistic. An inferential statistic is used to draw a conclusion about a larger set of cases.

Practice Problems 1.4

Review Your Knowledge

1.17 What are Greek letters used as abbreviations for? Latin letters?

1.18 What is the name for a statistic that is used to draw a conclusion about a population from a sample?

Apply Your Knowledge

1.19 The athletic director at a college is interested in how well female athletes perform academically. He interviews the women's softball team, determines what their average GPA is, and reports that the women's softball team has a GPA of 3.37. (a) Is this a statistic or a parameter? (b) Is this a descriptive or an inferential statistic?

1.20 A demographer is interested in the annual incomes of people who live in apartments. From the Census Bureau, she obtains a representative sample of 2,500 people living in apartments all across the United States and learns their annual incomes. She then reports that the average annual income of apartment dwellers in the United States is $29,983. (a) Does this number come from a sample or a population? (b) Is it a descriptive or an inferential statistic?

1.5 Statistical Notation and Rounding

The data that statisticians work with are almost always in the form of numbers. When statisticians refer to the outcome variable that the numbers represent, they abbreviate it with the letter X. If they were measuring age in a group of children, they would say $X =$ age. Another common abbreviation is N to indicate the number of cases. If statisticians measured the ages of five children, they would say $N = 5$.

When adding together sets of scores, an uppercase Greek sigma (Σ) is used as a summation sign. If the five kids had ages of 5, 7, 10, 11, and 15, then ΣX means that one should add together all the X scores:

$$\begin{aligned}\Sigma X &= 5 + 7 + 10 + 11 + 15 \\ &= 48\end{aligned}$$

The order of operations tells the order in which math is done. Following the order of operations is important for getting the right answer. Many students remember the acronym PEMDAS from elementary school. PEMDAS stands for parentheses, exponents, multiplication, division, addition, and subtraction. It is the order to be followed in equations. "Please excuse my dear Aunt Sally" is a mnemonic commonly used to remember PEMDAS.

The PEMDAS order of operations means the math within parentheses and brackets is done first, then the exponents (numbers raised to a power like 3^2 and radicals like $\sqrt{\ }$) are computed. Next, the multiplication and division are calculated in order from left to right. Finally, the addition and subtraction are done, again in order from left to right. A problem like $(7 + 3) \times 3^2 \div 2 + 3 - 2 \times 3 \times \sqrt{9}$ would be completed as follows:

- First the parentheses: $(7 + 3) \times 3^2 \div 2 + 3 - 2 \times 3 \times \sqrt{9} =$
  $10 \times 3^2 \div 2 + 3 - 2 \times 3 \times \sqrt{9}$
- Then the exponents and radicals: $10 \times 3^2 \div 2 + 3 - 2 \times 3 \times \sqrt{9} =$
  $10 \times 9 \div 2 + 3 - 2 \times 3 \times 3$
- Then the multiplication and division: $10 \times 9 \div 2 + 3 - 2 \times 3 \times 3 =$
  $45 + 3 - 18$
- Finally, the addition and subtraction: $45 + 3 - 18 = 30$

For another example of following the order of operations, see **Figure 1.13**. Each chalkboard in the figure shows another step in the order of operations.

Keep alert for summation signs in the order of operations. Using the order of operations correctly for the five ages, ΣX^2 means $5^2 + 7^2 + 10^2 + 11^2 + 15^2$, not 48^2. Also, complete summations *before* doing other addition and subtraction. For example, with the age data, $\Sigma X + 1$ means $48 + 1$. It doesn't mean add 1 to each age and then add them all up. That would be $\Sigma(X + 1)$.

Rules of Rounding

It is important to round correctly in order to obtain the right answer. Rounding involves making a number easier to work with by removing or simplifying digits on the right. A rounded number should accurately reflect the unrounded number. If someone reports her salary of $35,400 as $35,000, that's rounding. Note that she rounded her salary to $35,000, not $36,000. That's because $35,000 is a more accurate

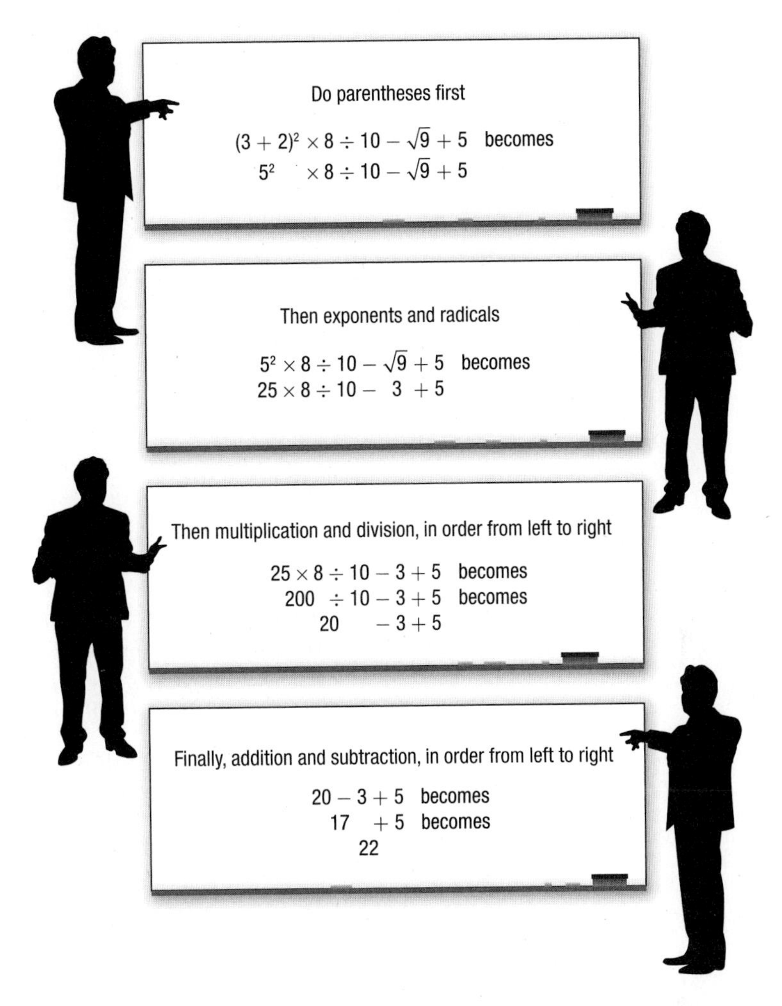

Figure 1.13 Order of Operations Math in equations should be done following these four steps for order of operations: first the math within parentheses, then exponents and radicals, then multiplication and division, finishing with addition and subtraction.

reflection of the unrounded number than is $36,000. It is a more accurate answer because it is closer to the original value. This is shown on a number line in Figure 1.14.

The American Psychological Association recommends reporting results to two decimal places (APA, 2010). So, Rounding Rule 1 is as follows: final answers should be rounded to two decimal places.

Rounding Rule 2 is that numbers shouldn't be rounded until the very end. Carry as many decimal places as possible through every calculation. Sometimes, though, it is unrealistic to carry all decimal places. If that's the case, round intermediate steps to four decimal places, which is two more decimal places than the final answer will have. Problems worked in the book will carry four decimal places.

Here's an example of how not carrying enough decimal places can get one into trouble. Look at this equation and see if, without using a calculator, you can figure out the answer:

$$\frac{123}{789} \times 789 = ?$$

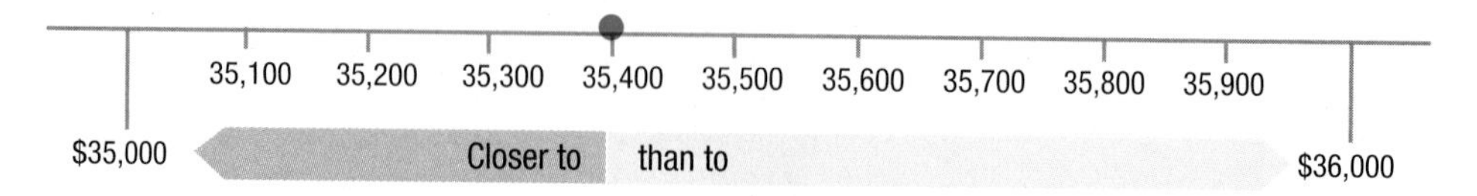

Figure 1.14 Number Line Showing How to Round $35,400 to the Nearest Thousand Dollars The unrounded number ($35,400) is closer to $35,000 than to $36,000, so it is rounded to the closer number.

The divisor (789) and the multiplier (789) will cancel each other out and the result should be 123. If one were to violate the second rule of rounding and round the result of the initial division to two decimal places, the final result would be off:

$$\frac{123}{789} \times 789$$
$$= 0.16 \times 789$$
$$= 126.24$$

In contrast, if one had rounded the first step to four decimal places, the final answer would have been much closer to what it should have been:

$$\frac{123}{789} \times 789$$
$$= 0.1559 \times 789$$
$$= 123.0051$$
$$= 123.01$$

The bottom line: don't round to two decimal places until the end.

Now that the first two rules of rounding have been stated, here's how to round a number to two decimal places. For an example, use 34.568:

- Start by cutting off the number at two decimal places: 34.56. That's one option for the rounded number.
- Get a second rounding option by adding one unit to the final digit, by going one higher in the hundredths column, to 34.57.
- Now it is time to decide which number, 34.56 or 34.57, is closer to the unrounded number. Look at the number line in Figure 1.15. The number 34.56 is 0.008 away from the unrounded number, while 34.57 is only 0.002 away. Because 34.57 is closer to the unrounded number, that is the answer.

If an unrounded number (e.g., 76.125) is exactly centered between the two options (76.12 and 76.13), then round up. That's Rounding Rule 3: when the two options for the rounded number are the same distance from the unrounded number,

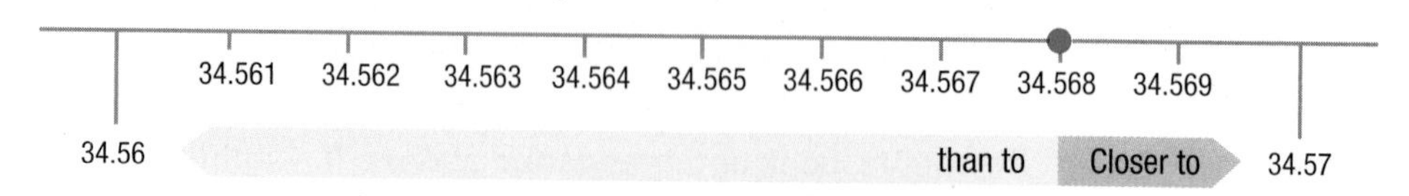

Figure 1.15 Number Line Showing How to Round 34.568 to Two Decimal Places The unrounded number (34.568) is closer to 34.57 than to 34.56, so it is rounded to the closer option.

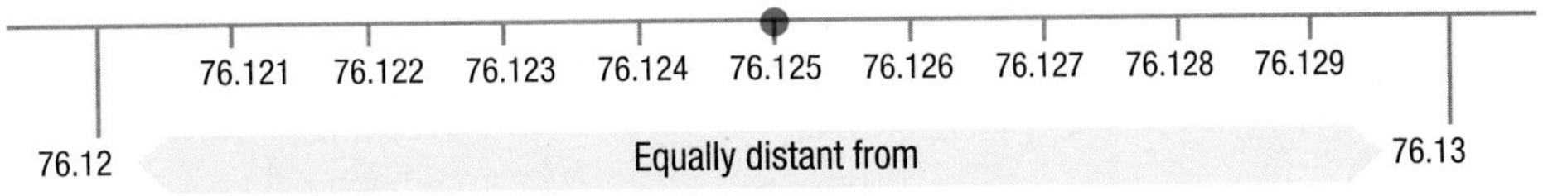

Figure 1.16 Number Line Showing an Unrounded Number Equally Close to Both Rounding Options The unrounded number (76.125) is equally distant from the two rounding options, 76.12 and 76.13, so round up to 76.13.

round up. Thus, 76.125 would be rounded to 76.13, as explained in Figure 1.16. The rules of rounding are summarized in Table 1.7.

TABLE 1.7 The Rules of Rounding
1. Round your final answers to two decimal places.
2. Don't round until the very end. (If you do, round as you go, carry four decimal places.)
3. If the unrounded number is centered exactly between the two rounding options, round up.

A Common Question

Q Why aren't the rules of rounding followed when reporting *N*, sample size?

A The number of cases, *N*, is always a whole number. One can't have 57.8 cases in a study. So, *N* is always reported without any decimal places.

Worked Example 1.3 As practice for order of operations and rounding, here's a tougher problem than we've encountered so far. By following the order of operations and breaking down the equation into pieces, it is possible to get the right answer. Note that after performing math on a number, follow the rules of rounding and carry four decimal places.

$$\sqrt{\left[\frac{(6-1)\times 2.13+(r-1)\times 3.35}{6+5-2}\right]\left[\frac{6+5}{6\times 5}\right]}$$

First come parentheses and brackets. Working from the inside out, doing the interior parentheses first and then the bracket on the right, one gets

$$\sqrt{\left[\frac{(5.0000)\times 2.13+(4.0000)\times 3.35}{6+5-2}\right]\left[\frac{11.0000}{30.0000}\right]}$$

The bracket on the left can't be attacked until the math inside it is done. Inside a parentheses or bracket, one still has to follow the order of operations, so do the multiplication and division first. Doing the multiplication gives

$$\sqrt{\left[\frac{10.6500+13.4000}{6+5-2}\right]\left[\frac{11.0000}{30.0000}\right]}$$

The next step is the addition and subtraction inside the brackets:

$$\sqrt{\left[\frac{24.0500}{9.0000}\right]\left[\frac{11.0000}{30.0000}\right]}$$

Now do the division within the brackets:

$$\sqrt{[2.6722][0.3667]}$$

When brackets or parentheses are next to each other, that means multiplication: (3)(2) is the same thing as 3×2. So, as the next to last step, do the multiplication:

$$\sqrt{0.9799}$$

Exponents were supposed to be completed at the second step in the order of operations. But, the math within the square root sign had to be completed first. So, finally, take the square root of 0.9799 to get 0.9899, which is rounded to 0.99 for the final answer.

DIY

In this do-it-yourself section, you will explore the power of rounding. Get a receipt from a trip to the grocery store, one where a lot of items were purchased. Find the total spent before taxes. Round each item to the nearest dollar. If something cost less than 50 cents, it would round to zero. Now, add up all the rounded numbers. Is the sum of the rounded numbers remarkably close to the sum of the unrounded numbers? Why?

Does this approach still work if you round all items to the nearest \$2? (If something costs less than a dollar, round it to zero. From \$1.00 to \$2.99 rounds to \$2; from \$3.00 to \$4.99 rounds to 4; etc.) How about if you round to the nearest \$5? (\$0.00 to \$2.49 = \$0; \$2.50 to \$7.49 = \$5; etc.) What happens as you round more and more?

Save the receipt because you will need it again in Chapter 3.

Practice Problems 1.5

Review Your Knowledge

1.21 What abbreviation do statisticians use for a summation sign?

1.22 Following the order of operations, what does one do first in an equation?

1.23 According to the American Psychological Association, how many decimal places should a result have?

Apply Your Knowledge

1.24 Given this data set (12, 8, 4, 6), calculate what is requested. Don't forget the rules of rounding.

a. N

b. ΣX

c. ΣX^2

d. $(\Sigma X)^2$

e. $\Sigma X + 1$

f. $\Sigma (X + 1)$

1.25 Round the following appropriately:

a. 17.7854

b. 9.7432

c. 12.9845

d. 8.3450

e. 7.1205

f. $\frac{5}{3}$

g. $\frac{4}{2}$

Application Demonstration

Here's a study about memory that reviews everything from Chapter 1. Two researchers investigated "use it or lose it" in relation to mental ability. They studied whether people lose mental ability if they don't keep using their minds as they age (Rohwedder & Willis, 2010).

The researchers used data from men and women who were 60 to 64 years old and from 13 countries (the United States and 12 European nations). The men and women had taken a recall test in which they were read a list of 10 words and then asked to recall as many as they could. Five minutes later, the participants were again asked to recall the words. The test was scored as the total number of words recalled in both conditions (the immediate and the delayed). So, memory scores could range from 0 to 20, with higher scores indicating better recall. It is important to know that the researchers consider this memory score to be a measure of cognitive functioning, not just a count of the number of words recalled.

The cases in this study are countries, so there were 13 scores, each one being the average score for a sample of men and women in that country. Each country also provided a second score. The men and women in the countries were also asked if they were working for pay. The percentage of the 60- to 64-year-olds not working for pay was calculated for each country. It ranged from about 40% in Sweden to more than 90% in Austria.

The researchers figured that people who were not working for pay were retired and less likely to use their minds in the way that people do at work. They examined the relationship between the percentage of respondents not working for pay and the average recall score and found a relationship—the *higher* the percentage of people not working for pay, the *lower* the cognitive functioning score. This suggests that if people don't use it, they do lose it. That's why the researchers called their paper "Mental Retirement."

Levels of Measurement

First, what are the levels of measurement for the variables in the study? The two variables were cognitive function, as measured by the recall task, and the percentage of people not working. On the recall task, scores can range from 0 to 20. To figure out level of measurement, make up two scores—country A has an average score of 10 and country B a 12. The scores tell same/different, as one can tell that the average cognitive functioning is not the same in the two countries, so the score is at least nominal. One can also tell the direction of the difference, that country B has better average cognitive function than country A, so the score is at least ordinal. There is equality of units—each word recalled correctly is worth the same number of points—so one can tell the size of the difference—citizens of country B recall an average of two more words than citizens of country A—so the measure is at least interval. Now for the absolute zero point question: Does a zero on the scale mean an absence of cognitive functioning? A zero means that the participant recalled no words, which is different from an absence of cognitive functioning. So, there is not an absolute zero point. The answer to the absolute zero point question is a no, so the cognitive functioning score is measured at the interval level.

For the second variable, the percentage of people not working, make up two scores again—country C has a 40 and country D an 80. These two countries have different percentages of 60- to 64-year-olds not working, so the variable is at least nominal. The scores tell the direction of the difference, which country has a higher percentage not working, so the variable is at least ordinal. As percentages have equality of units, one can meaningfully speak of the 40-point distance between the two countries. As one can tell the size of the difference, the variable is at least interval. Finally, percentages have an absolute zero point, so one can form a ratio and say that country D has twice the percentage of people not working as country C; thus, the variable is measured at the ratio level.

Explanatory Variable and Outcome Variable

Which of the two variables is the explanatory variable and which is the outcome variable? The researchers seem to believe that not working, being retired, causes a loss of cognitive functioning, so the percentage of people not working is the explanatory variable and cognitive functioning is the outcome variable. Another way to think about this is to look at chronological order: the explanatory variable, retirement, happens before the outcome variable, cognitive function, is measured. People were working or not working before their recall was measured.

Type of Study

What type of design have the researchers used for their study? Using the flowchart in Figure 1.5, the decision moves down the path of the explanatory variable *not* having been controlled by the researchers. After all, the researchers didn't assign some countries to retire workers at an earlier age and others to retire them at a later age. Next, decide whether the explanatory variable was used to classify cases into groups. It wasn't, so the study used a correlational design. Correlational studies have specific names for the explanatory and outcome variables. Percentage of people not working is the predictor variable and cognitive function is the criterion variable.

Because the study is a correlational design, one can't conclude that retirement is the cause of lower cognitive function. (Remember, correlation is not causation.) Retirement might cause lower cognitive functioning, but the cause also might run the other way. Perhaps people decide to retire because they have declined cognitively. Also, one has to consider the possibility of confounding variables, third variables that affect both retirement and cognitive function. The economy might be a plausible confounding variable—the overall unemployment rate could affect both the percentage of 60- to 64-year-olds who are working and people's mental state. Similarly, the quality of health care in the countries could be a confounding variable—poorer health care could lead to earlier retirement and poorer cognitive function.

The Language of Statistics

Did the researchers use a sample or a population in their study? The countries studied were the United States and 12 European nations, but there are more nations in North America than just the United States and there are more than 12 nations in the European Union. The study didn't consist of every North American and European nation, so it used a sample.

Are the results meant to be descriptive or inferential? Do the researchers' findings describe the relationship between percentage retired and cognitive function for those 13 nations only? Or, do the researchers want to draw a general conclusion about the relationship between these two variables? It seems reasonable that they want the readers of their study to draw a conclusion that goes beyond these 13 nations. The researchers probably mean to suggest, for humans of all nations, that a relationship exists between these two variables. As they indicate the results to be generalized beyond the few cases in the sample, the researchers mean the results to be inferential, not descriptive.

SUMMARY

Determine if a study is correlational, experimental, or quasi-experimental.

- Statistics summarize data collected in studies designed to answer research questions about relationships between variables. There are three types of research designs: correlations, experiments, and quasi-experiments.
- In correlational studies, the relationship is examined without manipulating any of the variables. Correlational studies address real-life questions, but can't draw conclusions about cause and effect.
- Cause-and-effect conclusions can be drawn from experiments because they use random assignment to assign cases to groups. The independent variable, the cause, is controlled by the experimenter; its effect is measured in the dependent variable.
- In quasi-experiments, cases are categorized on the basis of groups they naturally belong to and then compared on the dependent variable. Quasi-experiments look like experiments but have confounding variables like correlational studies.

Classify variables and determine levels of measurement.

- Predictor variables (correlations), independent variables (experiments), and grouping variables (quasi-experiments) are explanatory variables, while criterion variables (correlations) and dependent variables (experiments and quasi-experiments) are outcome variables.
- Variables are measured at nominal, ordinal, interval, or ratio levels. As the level of measurement moves up, information contained in the number increases from qualitative (nominal), to basic quantitative (rank order at the ordinal level), to more advanced quantitative (distance information for interval level), and finally proportionality at the ratio level.

Learn the language and rules of statistics.

- A population is the larger group of cases that a researcher wishes to study. Researchers almost always study samples, which are subsets of populations.
- A statistic is a value calculated for a sample and a parameter is a value calculated for a population. Latin letters are used as abbreviations for statistics; Greek letters as abbreviations for populations.
- Descriptive statistics are numbers used to describe a group of cases; inferential statistics are used to draw conclusions about a population from a sample.
- Following the order of operations for mathematical operations and applying the rules of rounding are necessary to get the right answer.

KEY TERMS

cases – the participants in or subjects of a study.

confounding variable – a third variable in correlational and quasi-experimental designs that is not controlled for and that has an impact on *both* of the other variables.

correlational design – a scientific study in which the relationship between two variables is examined without any attempt to manipulate or control them.

criterion variable – the outcome variable in a correlational design.

dependent variable – the variable where the effect is measured in an experimental or quasi-experimental study.

descriptive statistic – a summary statement about a set of cases.

experimental design – a scientific study in which an explanatory variable is manipulated or controlled by the experimenter and the effect is measured in a dependent variable.

explanatory variable – the variable that causes, predicts, or explains the outcome variable.

grouping variable –the variable that is the explanatory variable in a quasi-experimental design.

independent variable – the variable that is controlled by the experimenter in an experimental design.

inferential statistic – using observations from a sample to draw a conclusion about a population.

interval-level numbers – numbers that provide information about how much of an attribute is possessed, as well as information about same/different and more/less; interval-level numbers have equality of units and an arbitrary zero point.

nominal-level numbers – numbers used to place cases in categories; numbers are arbitrary and only provide information about same/different.

ordinal-level numbers – numbers used to indicate if more or less of an attribute is possessed; numbers provide information about same/different and more/less.

outcome variable – the variable that is caused, predicted, or influenced by the explanatory variable.

parameter – a value that summarizes a population.

population – the larger group of cases a researcher is interested in studying.

predictor variable – the explanatory variable in a correlational design.

quasi-experimental design – a scientific study in which cases are classified into naturally occurring groups and then compared on a dependent variable.

random assignment – every case has an equal chance of being assigned to either group in an experiment; random assignment is the hallmark of an experiment.

ratio-level numbers – numbers that have all the attributes of interval-level numbers, plus a real zero point; numbers that provide information about same/different, more/less, how much of an attribute is possessed, and that can be used to calculate a proportion.

sample – a group of cases selected from a population.

statistic – a value that summarizes data from a sample.

statistics – techniques used to summarize data in order to answer questions.

variables – characteristics measured by researchers.

CHAPTER EXERCISES

Answers to the odd-numbered exercises appear at the back of the book.

Review Your Knowledge

1.01 Statistics ____ data in order to answer questions.

1.02 The characteristics measured by a researcher are called ____.

1.03 The objects being studied by a researcher are called ____.

1.04 In a correlational design, the two variables are simply measured; they are not ____ by the experimenter.

1.05 If two variables, X and Y, are found to be related in a correlational design, the three possible explanations for the relationship are: (a) ____, (b) ____, and (c) ____.

1.06 A third variable that could be the real cause of an apparent relationship between X and Y in correlational research is called ____.

1.07 ____ is the hallmark of an experimental design.

1.08 In an experimental design, the ____ variable is manipulated or controlled by the experimenter.

1.09 The variable where the effect is measured in an experimental design is called the ____.

1.10 Experimental designs allow one to draw a conclusion about ____.

1.11 In a quasi-experimental design, cases are classified into ____ based on characteristics they already possess.

1.12 Though a quasi-experimental design looks like a (an) ____ design, it is really a (an) ____ design.

1.13 The mnemonic *iced* stands for ____.

1.14 The ____ variable in ____ is called the dependent variable.

1.15 The explanatory variable in quasi-experimental designs is called the ____ variable.

1.16 A ratio-level number contains ____ information than a nominal-level number.

1.17 The mnemonic to help remember, in order, the four levels of measurement is ____.

1.18 Nominal-level numbers contain information about ____.

1.19 ____-level numbers contain information about same/different and direction.

1.20 Because interval-level numbers have ____, we can meaningfully speak of the distance between two scores.

1.21 Interval-level numbers have a (an) ____ zero point, and ratio-level numbers have a (an) ____ zero point.

1.22 Proportions can be found for ____-level numbers.

1.23 A (an) ____ is the larger group of cases that a researcher is interested in studying.

1.24 A sample is a (an) ____ of a population.

1.25 A number characterizing a sample is called a (an) ____; a number characterizing a population is called a (an) ____.

1.26 We use ____ letters as abbreviations for sample values and ____ letters as abbreviations for population values.

1.27 If a summary statement is used to describe a group of cases, it is a (an) ____ statistic; if it is used to draw a conclusion about the larger population, it is a (an) ____ statistic.

1.28 The letter we use as an abbreviation for an outcome variable is ____. The abbreviation for the number of cases in a group is ____. The symbol for adding up a group of scores is ____, the uppercase Greek letter sigma.

1.29 According to the American Psychological Association, final results should be rounded to ____ decimal places.

1.30 If you do round as you go, carry at least ____ decimal places.

Apply Your Knowledge

Figuring out types of studies and types of variables

For Exercises 1.31–1.38:

- ***a.*** ***Generate a sentence that states the question the researcher is trying to answer.***
- ***b.*** ***List the variables and label them as explanatory and outcome.***
- ***c.*** ***Determine what type of study is being done: correlational, experimental, or quasi-experimental.***
- ***d.*** ***If the study is correlational or quasi-experimental, come up with a plausible confounding variable and explain how it affects both variables.***

1.31 The local police department has come to a criminologist to help it evaluate a new type of disposable, plastic handcuffs. They are just as effective as metal handcuffs in terms of immobilizing someone who has been arrested and they are cheaper than metal handcuffs, but the police are concerned that the plastic ones might cause more abrasion to the skin of the wrist. The criminologist finds 20 volunteers, randomly divides them into two groups, cuffs one group with metal and the other with plastic, and then rides them around in a squad car for 20 minutes. After this, the criminologist measures the degree of abrasion on their wrists as the percentage of skin that is roughed up.

1.32 Some football players put streaks of black paint under their eyes because they believe that it helps them see better in sunny conditions and react more quickly. A sensory psychologist wants to see if this is really true. He gathers a group of volunteers and randomly divides them in two. Half get black paint applied under their eyes and half get flesh-color. The players are not allowed to look in the mirror, so they don't know which color has been applied below their eyes. The psychologist then gives them a reaction time task to measure, in milliseconds, how quickly they can respond to a change in a stimulus while bright lights are being shined at them.

1.33 Ever notice that some college students buy all the books for class, complete all the readings, do all the homework, and so on? These students usually end up with better grades as well. An education professor decided to investigate if these more conscientious students received better grades because they worked harder or because they were innately smarter. The professor assembled (a) a group of conscientious students from a number of different colleges and (b) a group of nonconscientious students from the same colleges, and compared the two groups in terms of a standardized IQ test.

1.34 A personality psychologist has kindergarten teachers use an empathy scale that ranges from 0 (not at all empathetic) to 100 (extremely high levels of empathy) to rate their students. Thirteen years later, when the students are ready to graduate from high school, he tracks them down and rates their level of mental health on a scale from 0 (very, very poor) to 100 (very, very good).

1.35 An economist believed that as nations become wealthier, they produce more greenhouse gases. He took a country and found both its gross domestic product (GDP) and the total tons of CO_2 emissions for each year over the past 50 years.

1.36 A consumer behavior researcher is curious as to whether, in terms of the monetary value of Christmas presents received, it makes a difference if a child is naughty or nice. She has parents classify their children as naughty or nice, then calculates how much the parents spent on Christmas presents for the two different groups of children.

1.37 A political scientist is curious as to what influences voting behavior on taxes for school districts. She obtains a sample of voters and divides them, randomly, into three groups. One group serves as the control group, nothing is done to them. To one experimental group, she gives information about the school taxes that focuses on the positive—how the levy will improve student performance, make the community more attractive to young families, and so on. To the other experimental group, she gives negative information about the school taxes—how much overall taxes will increase, how school taxes will take away funding from other projects, how wasteful the school district has been, and so on. She then measures, for each group, the percentage voting in favor of the school taxes.

1.38 A neuroscientist believes that proteins form plaques in the brain that cause Alzheimer's disease. He gets a sample of older adults, measures the nanograms per liter of protein in spinal fluid, and measures short-term memory as the percentage of words in a list that are recalled.

Determining level of measurement

1.39 A meteorologist classifies cities in the United States in terms of winter weather: "dreary" (0) or "not dreary" (1). Type of winter weather (0 vs. 1) is measured at what level of measurement?

1.40 A social worker obtains the suicide rates for students at colleges in the United States. If the college has a suicide rate that is below average, he classifies it as −1. If the suicide rate is average, the college gets a 0, and if the suicide rate is above average, it gets a +1. The suicide rate (−1, 0, +1) is being measured at what level of measurement?

1.41 The owner of an automobile shipping company classifies cars in terms of size. If a car is a subcompact, she assigns it the value of 1. A compact car gets a 2, a mid-size car a 3, and a full-size car a 4. At what level is she measuring car size?

1.42 The admissions committee at a college does not distinguish between different types of high school extracurricular activities. As far as it is concerned, being a member of the tiddlywinks club is equivalent to being student council president. On the admission form to the college, applicants are asked to report the number of extracurricular activities in which they were involved in high school. The college is measuring extracurricular activities at what level?

1.43 The same college asks students to submit their SAT scores, but only on the math subtest. Subtest scores on the SAT range from a low of 200 to a high of 800, with 500 representing an average score. The SAT measures math skills at what level?

1.44 A nurse researcher measures how many minutes patients must wait before being seen by the triage nurse after they enter an emergency room. Wait time is measured at what level?

1.45 A housing developer advertises her houses as being fully carpeted (2), partially carpeted (1), or not carpeted (0). Amount of carpeting is measured at what level?

1.46 A person's knowledge of English grammar is measured by a 50-item multiple-choice test. Each correct answer is worth 2 points, so scores can range from 0 to 100. English grammar knowledge is being measured at what level?

1.47 If a person reads books for pleasure, he or she is classified as a "1"; if a person doesn't read books for pleasure, he or she is classified as "0." Whether or not a person reads books for pleasure is being measured at what level?

1.48 A person's depression level is measured on a 20-item inventory where each item is a true/false item and is meant to measure depression. Each item endorsed in the "depressed" direction adds 1 point to the person's score, so scores can range from 0 to 20. Depression level is being measured at what level?

Using statistical terminology

1.49 A college dean wanted to find out which students were smarter: those seeking liberal arts degrees (like English or psychology) or those seeking professional degrees (like nursing, business, or engineering). From all the colleges in the United States, she picked 1,200 liberal arts majors and 1,000 professional degree majors. Each student took an IQ test and she calculated the average IQ for each group.

a. Is the group of 1,200 liberal arts majors a sample or a population?

b. If the college dean uses the averages in statements like, "The average intelligence of liberal arts students in the United States is 115.67," is she treating the averages as statistics or as parameters?

c. If she uses the two averages to answer her question, is this an example of inferential statistics or descriptive statistics?

1.50 A political pollster calls 2,000 registered American voters and finds out whether they plan to vote for the Democratic or Republican candidate in an upcoming election. From this she predicts the outcome of the election. Is she using the information about the sample as a descriptive or as an inferential statistic?

1.51 Every 10 years, the U.S. Census Bureau attempts to collect information from all Americans. Assuming that they are successful, would it be a statistic or a parameter if the Census Bureau reported that 12.2% of Americans identify themselves as of African descent?

1.52 A college president wants to know what the average quantitative SAT is for the first-year class at her college. She calls the registrar and the registrar accesses the database for the entire first-year class to calculate the average. Does the average the registrar calculated correspond to a sample or a population?

Order of operations and rounding

For Exercises 1.53–1.56, use this data set: 8, 9, 5, 4, 7, and 8 to find the following:

1.53 N

1.54 ΣX

1.55 $\Sigma X^2 =$

1.56 $\Sigma X - 1$

For Exercises 1.57–1.60, use this data set: 13, 18, and 11 to find the following:

1.57 ΣX

1.58 ΣX^2

1.59 $\frac{\Sigma X}{N} =$

1.60 $\Sigma(X - 14) =$

For Exercises 1.61–1.68, use the rounding rules to round the following:

1.61 12.6845 =

1.62 189.9895 =

1.63 121.0056 =

1.64 674.064005 =

1.65 22.467 =

1.66 37.97700001 =

1.67 2.53200005 =

1.68 99.995 =

Expand Your Knowledge

1.69 A researcher gives a different amount of X to different subjects and then measures how much Y each subject produces. She finds that subjects who got more X produce more Y, and subjects who got less X produce less Y. She also finds no other variable that can account for the different amounts of Y that the subjects produce. What conclusion should she draw?

a. There is a relationship between X and Y.

b. There is a relationship between Y and X.

c. X causes Y.

d. Y causes X.

e. A confounding variable, Z, causes both X and Y.

f. There is not enough information to draw any conclusion.

1.70 Explanatory variable is to outcome variable as:

a. controlled is to manipulated.

b. relationship is to cause and effect.

c. statistic is to parameter.

d. parameter is to statistic.

e. none of the above

1.71 A sociologist believes that physical distress in cities leads to social distress. She measures physical distress by seeing how much graffiti there is. Based on this, she classifies cities as being high, moderate, or low in terms of physical distress. She randomly samples cities in North America until she has 10 cities in each category. She then measures social distress by obtaining the teenage pregnancy rate for each of these cities.

a. What question is the researcher trying to answer?

b. List the variables and label them as explanatory and outcome variables.

c. Determine what type of study is being done: correlational, experimental, or quasi-experimental.

d. If the study is correlational or quasi-experimental, name a plausible confounding variable and explain how it affects both variables.

1.72 A clinical psychologist administers a list of fears to measure how phobic people are. It has 10 items on it (such as spiders, snakes, height, darkness). For each item, a person answers "yes" or "no" as to whether he or she is afraid of it. Each yes equals 1 point, so scores can range from 0 to 10. The score on the fear survey is measured at what level of measurement?

Frequency Distributions

LEARNING OBJECTIVES

- Make a frequency distribution for a set of data.
- Decide if a number is discrete or continuous.
- Choose and make the appropriate graph for a frequency distribution.
- Describe modality, skewness, and kurtosis for a frequency distribution.
- Make a stem and leaf display.

CHAPTER OVERVIEW

The purpose of statistics is data reduction, taking a mass of numbers and reducing them in some way to bring order to them. Chapter 1 showed how organizing binge drinking rates for 18- to 25-year-olds in the United States—alphabetically or from low to high—made them easier to understand. That just involved reorganizing the data, not reducing it. Chapter 2 covers some data reduction techniques, making tables and graphs. It also introduces another way to think about numbers, whether the number is continuous or discrete, and talks about the "shapes" data can take.

2.1 Frequency Distributions

Frequency distributions can be made for nominal-, ordinal-, interval-, or ratio-level data. A frequency distribution is an intuitive way to organize and reduce data. A **frequency distribution** is simply a count of how often the values of a variable occur in a set of data. For example, to tell someone how many boys and girls are in a class is to make a frequency distribution.

Ungrouped vs. Grouped Frequency Distributions

There are two different types of frequency distribution tables, ungrouped and grouped. An **ungrouped frequency distribution** table is a count of how often each value of a variable occurs in a data set. In a **grouped frequency distribution** table, the frequency counts are for adjacent groupings of values, or intervals, of the variable.

Ungrouped frequency distributions are used when the values a variable can take are limited. For example, if students were surveyed about how many children were in their families, a limited number of responses would exist. Students would most commonly report that there were one, two, or three kids in their families. Almost no one would report ten or more kids. An ungrouped frequency distribution for data like these would be compact, taking up maybe a half-dozen lines on one page, and could be viewed easily.

Now if the same students were surveyed about the size of their high school graduating classes, it would be a very different frequency distribution. The survey would yield answers ranging from 1 (homeschooled) to 500 or more students. If one listed each value separately, it would take many lines and could run to multiple pages. This would be hard to view and not a good summary of the data. In this case, it would make sense to group answers together in intervals (fewer than 100 in the class, 100 to 199 in the class, etc.) to make a more compact presentation.

Grouped frequency distributions should be used when the variable has a large number of values and it is acceptable to lose information by collapsing the values into intervals. If the variable has a large number of values but it is important to retain information about all the unique values, then one should use an ungrouped frequency distribution.

Table 2.1 shows both an ungrouped frequency distribution and a grouped frequency distribution for the binge drinking data from Chapter 1. Note the size of the ungrouped frequency distribution and the many gaps in it. There is too much detail to get a clear picture of the data. Compare that to the compactness of the grouped frequency distribution, which reduces the data to a greater degree and does a better job summarizing and describing it.

Ungrouped Frequency Distributions

Here are some data regarding how many children are in the families of 31 students. Nine of the 31 reported being in one-child families, 14 families had two children, 5 had three, 2 had four, and 1 had six. An ungrouped frequency distribution table for these data can be seen in Table 2.2 on page 42.

There are a number of things to note about Table 2.2:

- This is a basic, no-frills frequency distribution. It just provides the values that the variable takes and how often each value occurs. That's it.
- There is a title! All tables need to have titles that clearly describe the information the table contains.
- The columns are labeled.
- The abbreviation for frequency is *f*.
- The table is "upside down," with the largest value of the variable (six children per family) at the top and the smallest value (one child per family) at the bottom. Why the table is arranged this way will become clear in a moment when cumulative frequencies are introduced.
- There is no value in the table that indicates how many students were in the class, so that information was placed in the title. The sample size, *N*, is commonly reported in tables and graphs.
- There was one value (five children per family) that did not exist in the class. But, this is included anyway with a frequency of zero. Include zero frequency values because that shows breaks in the data set.

TABLE 2.1 Comparison of Ungrouped and Grouped Frequency Distributions for Data Showing Percentage of 18- to 25-Year-Olds per State Who Engaged in Binge Drinking During the Past 30 Days

Ungrouped Frequency Distribution

Percentage of 18- to 25-Year-Olds Per State Who Engaged in Binge Drinking During Past Month	Number of States That Had That Percentage of Binge Drinking 18- to 25-Year-Olds
54	1
53	
52	
51	
50	
49	2
48	1
47	1
46	2
45	4
44	4
43	1
42	2
41	7
40	1
39	6
38	3
37	2
36	4
35	2
34	
33	2
32	2
31	1
30	1
29	
28	
27	
26	1

Grouped Frequency Distribution

Range for Percentage of 18- to 25-Year-Olds Per State Who Engaged in Binge Drinking During Past Month	Number of States That Had a Percentage of Binge Drinking 18- to 25-Year-Olds in That Range
50–54	1
45–49	10
40–44	15
35–39	17
30–34	6
25–29	1

Both of these frequency distributions reduce the data shown in Table 1.3. Which one gives a clearer picture of the binge drinking rates, by state, of 18- to 25-year-olds? The ungrouped version shows the whole range of scores but has a lot of empty categories. The much more compact grouped version shows the big picture but sacrifices detail.

TABLE 2.2 A Basic Ungrouped Frequency Distribution for Number of Children in Families of 31 Students in a Class

Number of Children in Family	Frequency (f)
6	1
5	0
4	2
3	5
2	14
1	9

Note: This is the simplest version of a frequency distribution. The only information it contains is the frequency with which each value occurs.

Table 2.2 is a basic ungrouped frequency distribution. Table 2.3 is one with more bells and whistles. This more complex ungrouped frequency distribution has three new columns. The first, **cumulative frequency,** tells how many cases in a data set have a given value *or* a lower value. For example, there are 23 people who have two *or fewer* children in their family and 28 who have three *or fewer.*

TABLE 2.3 Ungrouped Frequency Distribution for Number of Children in Families of 31 Students in a Class

Number of Children in Family	Frequency (f)	Cumulative Frequency (f_c)	Percentage (%)	Cumulative Percentage ($\%_c$)
6	1	31	3.23	100.00
5	0	30	0.00	96.77
4	2	30	6.45	96.77
3	5	28	16.13	90.32
2	14	23	45.16	74.19
1	9	9	29.03	29.03

Note: This ungrouped frequency distribution contains more information than the basic version in Table 2.2.

A cumulative frequency, abbreviated f_c, is calculated by adding up all the frequencies at or below a given row. It is easier to visualize this than to make it into a formula. Look at Figure 2.1 and note how the cumulative frequencies stair-step up. For the first row, the frequency and the cumulative frequency are the same, 9. Moving up one step, add the frequency at this level, 14, to the cumulative frequency from the level below, 9, to get the cumulative frequency of 23 for this

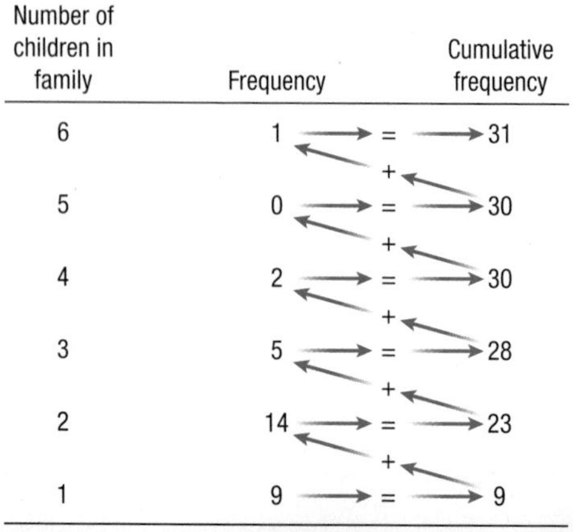

Number of children in family	Frequency		Cumulative frequency
6	1	=	31
		+	
5	0	=	30
		+	
4	2	=	30
		+	
3	5	=	28
		+	
2	14	=	23
		+	
1	9	=	9

Figure 2.1 Calculating Cumulative Frequencies for a Frequency Distribution One can calculate cumulative frequencies by "stair-stepping" up. The cumulative frequency in the bottom row is the same as the frequency for that row. Then add the frequency from the row above to find its cumulative frequency. Continue the process until one reaches the final row. Note that the cumulative frequency in the top row is the same as the number of cases in the data set.

level. Then repeat the process—add the frequency at the new level to the cumulative frequency from one step down—to get the next cumulative frequency.

There are a few other things one should know about cumulative frequencies:

- Frequency distributions are organized upside down, with the biggest value the variable can take on the top row. This is because of cumulative frequency, the number of cases at or *below* a given value. In this table, the row for six children in the family is the top row and the row for one child in the family is the bottom row.
- The cumulative frequency for the top row should be the same as the total number of cases, *N*, in the data set. If they don't match, something is wrong.
- Cumulative frequencies can only be calculated for data that have an order, data where the numbers tell direction. This means that cumulative frequencies can be calculated for ordinal-, interval-, or ratio-level data, but not for nominal data.
- If one has nominal data, one needs to organize the data in some logical fashion to fit one's intended purpose. For example, one could make a frequency distribution for the nominal variable of choice of college major. To draw attention to the relative popularity of majors, it could be organized by ascending or descending frequency.

Cumulative frequencies can only be calculated for data that have an order, data where the numbers tell direction.

The other new columns in Table 2.3 take information that is already there, frequency and cumulative frequency, and present it as percentages. Percentages are a way of transforming scores to put them in context. To see how this works, imagine that a psychologist said that two of the clients he treated in the last year experienced major depression. Does he specialize in the treatment of depression? The answer depends on how many total clients he treated in the last year. If he offered therapy to four patients, then half of them, or 50%, were depressed. If he treated 100 cases, then only 2% were depressed. Whether the psychologist is a depression specialist depends on what percentage of his practice is devoted to treating that illness. Percentages put scores in context.

The percentage column, abbreviated %, takes the information in the frequency column and turns it into percentages. Equation 2.1 shows that this is done by dividing a frequency by the total number of cases in the data set and then multiplying the quotient by 100.

Equation 2.1 Formula for Calculating Frequency Percentage (%) for a Frequency Distribution

$$\% = \frac{f}{N} \times 100$$

where % = frequency percentage
f = frequency
N = total number of cases

For example, in the third row where the frequency is 5, one would calculate

$$\% = \frac{5}{31} \times 100$$

$$= 0.1613 \times 100$$

$$= 16.13\%$$

This means, in plain language, that 16.13% of the students come from families with three children.

The final column, abbreviated $\%_c$, gives the **cumulative percentage,** the percentage of cases with scores at or below a given level. The cumulative percentage is a restatement of the information in the cumulative frequency column in percentage format. For example, in the second row from the bottom of Table 2.3, it is readily apparent that just over 74% of the students come from one-child or two-child families.

Equation 2.2 Formula for Calculating Cumulative Percentage ($\%_c$) for a Frequency Distribution

$$\%_c = \frac{f_c}{N} \times 100$$

where $\%_c$ = cumulative percentage
f_c = cumulative frequency
N = total number of cases

The formula for cumulative percentage is given in Equation 2.2. To find $\%_c$ for a given row in a frequency distribution table, one divides the row's cumulative frequency by the total number of cases and then multiplies the quotient by 100. In the second to last row of Table 2.3, where the cumulative frequency is 23, one would calculate

$$\begin{aligned}\%_c &= \frac{23}{31} \times 100 \\ &= 0.7419 \times 100 \\ &= 74.19\end{aligned}$$

Note that the cumulative percentage for the top row should equal 100%.

Grouped Frequency Distributions

Ungrouped frequency distributions work well in two conditions: (1) when the variable takes a limited set of values, or (2) when one wants to document all the values a variable can take. If one were classifying semester status of students in college, there is a limited number of options (1st, 2nd, 3rd, etc.) and an ungrouped frequency distribution would work well. There are many more options for the number of credit hours a student has completed, and an ungrouped frequency distribution for this variable would only make sense if it were important to keep track of how many students had 15 hours vs. 16 hours vs. 17 hours, and so forth.

When dealing with a variable that has a large range, like number of credit hours, a grouped frequency distribution makes more sense. In a grouped frequency distribution, one finds the frequency with which a variable occurs over a range of values. In the credit hours example, one might count how many students have completed from 0 to 14 hours, or from 15 to 29 hours, and so forth. These ranges are called intervals or bins and are abbreviated with a lowercase *i*.

Grouped frequency distributions work best when there is an order to the values a variable can take, that is, when the variable is measured at the ordinal, interval, or ratio level. Nominal data can be grouped if there is some logical categorization. For example, imagine that a psychologist collected detailed information about the diagnoses of her patients—whether they had major depression, dysthymia, bipolar disorder, obsessive-compulsive disorder, phobias, generalized anxiety disorder, alcoholism, heroin addiction. These responses could be grouped into categories of mood disorders, anxiety disorders, and substance abuse disorders.

TABLE 2.4 Property Crime Rates for the 50 States

State	Rate	State	Rate	State	Rate
South Dakota	17	Wisconsin	28	Kansas	37
North Dakota	19	Wyoming	29	New Mexico	37
New Hampshire	19	Illinois	29	Missouri	37
New York	20	Colorado	30	Nevada	38
New Jersey	22	California	30	Georgia	39
Idaho	22	Minnesota	30	Arkansas	40
Vermont	23	Michigan	31	Alabama	40
Pennsylvania	24	Nebraska	32	Washington	40
Massachusetts	24	Mississippi	32	Louisiana	41
Connecticut	24	Delaware	34	North Carolina	41
Maine	24	Alaska	34	Tennessee	41
Virginia	25	Indiana	34	Florida	41
Kentucky	25	Maryland	34	Texas	41
West Virginia	25	Ohio	35	Hawaii	42
Iowa	26	Utah	35	South Carolina	43
Rhode Island	26	Oregon	35	Arizona	44
Montana	28	Oklahoma	35		

Note: Property crime rates are reported as the number of crimes that occur per every 1,000 people who live in the state. Here, the data are organized in order from low to high.

Source: Statistical Abstract of the United States.

For variables that are measured at the ordinal level or higher, the first step is to decide how many intervals to include in a grouped frequency distribution. There needs to be a balance between the amount of detail presented and the number of intervals. There shouldn't be so few intervals that one can't see important details in the data set, and there shouldn't be so many intervals that the big picture is lost in the details. A rule of thumb is to use psychology's magic number, 7 ± 2, and have from five to nine intervals. But, that's just a rule of thumb—if fewer than five intervals or more than nine intervals do a better job of communication, that's fine. The other thing to keep in mind is convention. It is common to make intervals that are 5, 10, 20, 25, or 100 units wide.

Table 2.4 displays data we'll use to make a grouped frequency distribution. These numbers represent the number of property crimes that occur in a state for every thousand people who live in that state. They are organized from low (17 crimes per 1,000) to high (44 crimes per 1,000) to make it easy to construct a grouped frequency distribution.

The first task is to decide how many intervals to include. One needs to have intervals that will capture all data and that don't overlap. One option is to have 10-point-wide intervals—one for the values in the 10s, one for the 20s, the 30s, and the 40s. But that means only four intervals and so few intervals would lose too much detail. Instead, it makes sense here to opt for 5-point-wide intervals. Note that the first interval starts at 15, not 17, because starting an interval at a multiple of 5 is a convention commonly followed. The first interval is 15–19. (If you don't believe it is five points wide, count it off on your fingers.)

There's one other thing to mention—all intervals should be the same width so that the frequency in one interval can be compared to the frequency in another. So, there will be six intervals, ranging from 15–19 on the bottom row to 40–44 on the top.

TABLE 2.5 Grouped Frequency Distribution for State Property Crime Rates per 1,000 Population (Interval Width = 5)

Crime Rate Interval	Midpoint (m)	Frequency (f)	Cumulative Frequency (f_c)	Percentage (%)	Cumulative Percentage ($\%_c$)
40–44	42.00	11	50	22.00	100.00
35–39	37.00	9	39	18.00	78.00
30–34	32.00	10	30	20.00	60.00
25–29	27.00	9	20	18.00	40.00
20–24	22.00	8	11	16.00	22.00
15–19	17.00	3	3	6.00	6.00

Note: This table organizes the data from all 50 states in Table 2.4 into six 5-point intervals, giving an overview of the data.

Table 2.5 shows what a completed grouped frequency distribution looks like for these state crime data.

As with an ungrouped frequency distribution, a grouped frequency distribution has a title, labeled columns, and is upside down. A grouped frequency distribution also presents the same information—frequency, cumulative frequency, percentage, and cumulative percentage. This grouped frequency distribution does have one new column, m, which gives information about the *midpoint* of each interval. The **midpoint** is just what it sounds like, the middle point of the interval. One can verify that the midpoint of the first interval is 17.00 by counting through the interval, 15–19, on one's fingers. Or, find the average of 15 and 19, the endpoints of the interval

$$\frac{15 + 19}{2} = 17.00$$

There are two reasons for having a midpoint. First, the midpoint is, in a sense, a summary of the interval. Imagine a compass drawn as a circle with only four headings—north, east, south, and west (see Figure 2.2). North is at the top, labeled 360°, east is to the right, labeled 90°, and so on. If Sharon were heading off to the right at exactly 90°, one would say that she was heading east. But when else would one say she was heading east? If the only options were to say a person was heading north, east, south, or west, one would say "east" when that person was heading anywhere from 45° to 135°. East is the midpoint of that interval $\left(\frac{45 + 135}{2} = 90.00\right)$, a summary of the direction for all the points in that interval.

There's another reason for having a midpoint. Suppose a grouped frequency distribution, like that in Table 2.5, was the only information one had. There was no access to the original data. One would know that there are three cases in the 15–19 interval, but their specific values wouldn't be known. Were they all 15s? All 19s? Spread throughout the interval? Statisticians have developed the convention that when the exact value for a case is unknown, it is assigned the value of the midpoint of the interval within which it falls.

Figure 2.2 Four Cardinal Compass Headings This figure shows the degrees associated with the four main compass headings—north, east, south, and west—as well as the degrees halfway between them. Though a heading of east is exactly 90°, we would say a person is heading east anywhere from 45° to 135°. "East" is the midpoint of that interval.

A lot of material was covered in this section, so here are some summaries. Figure 2.3 is a flowchart that leads one through the process of deciding whether to make an ungrouped or a grouped frequency distribution. Table 2.6 summarizes what material to include when making a table.

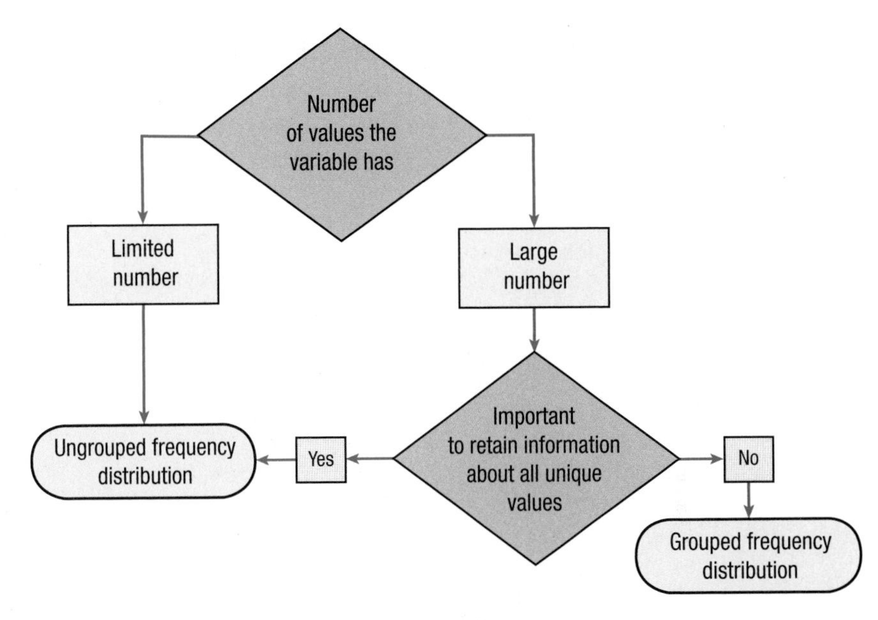

Figure 2.3 How to Choose: Deciding Whether to Make an Ungrouped Frequency Distribution or a Grouped Frequency Distribution for an Ordinal-, Interval-, or Ratio-Level Variable This flowchart leads one through the decision process of whether to make an ungrouped frequency distribution or a grouped frequency distribution for a variable that is measured at a level higher than nominal.

TABLE 2.6 How to Choose: Making a Frequency Distribution Table

Step 1	Decide whether to make a grouped or ungrouped frequency distribution: • Grouped, if the variable takes a wide range of values. • Ungrouped, if important to maintain information about all values.
Step 2	If grouped, decide the width and number of intervals: • Aim to have 5–9 intervals. • Make sure the intervals capture all values, have the same width, and don't overlap.
Step 3	Organize the data: • For ordinal, interval, or ratio variables, organize the table "upside down," with the largest value/interval on the top row and the rest of the values following in descending order. • For nominal variables, apply some order to the values (e.g., alphabetical, like categories together, ascending or descending frequencies).
Step 4	Decide what information to include in the table: • At a minimum, the values of the variable and their respective frequencies should be included. • Optional: Cumulative frequencies (not for nominal variables), percentage, cumulative percentage, interval midpoints.
Step 5	Communicate clearly: • Have a descriptive title. • Label all columns. • If abbreviations are used, include a note of explanation.

Worked Example 2.1 For practice making a frequency distribution table, imagine a school district that wanted to get a better picture of the intelligence of its students. It hired a psychologist to administer IQ tests to a random sample of sixth-grade students. Now the district needs a summary of the results.

Table 2.7 shows the raw data for the students tested. The first thing to note is that there is a lot of data—four columns of 15 and one of 8, so $N = 68$. The second thing to note is that a wide range of IQ scores exists, from the 60s to the 150s. Given the wide range of scores and given that it doesn't seem necessary to maintain information about all the unique scores, a grouped frequency distribution is a more sensible option than an ungrouped frequency distribution to summarize the data.

TABLE 2.7 Sample Data: IQ Test Results for 68 Sixth Graders

74	109	100	76	101	134	97	67
126	105	134	95	90	152	142	101
111	111	147	119	108	89	122	90
121	105	109	119	129	128	109	102
73	98	106	82	75	68	98	115
95	102	97	85	84	80	122	
111	72	85	148	111	112	136	
92	94	105	93	116	88	90	
80	107	118	79	103	94	128	

Note: These IQ test results are in no particular order.

The first step in making a grouped frequency distribution is organizing the data. Table 2.8 shows the data arranged in ascending order.

TABLE 2.8 Sixth-Grade IQ Data Arranged in Order

67	80	90	98	105	111	121	136
68	82	92	98	105	111	122	142
72	84	93	100	106	111	122	147
73	85	94	101	107	112	126	148
74	85	94	101	108	115	128	152
75	88	95	102	109	116	128	
76	89	95	102	109	118	129	
79	90	97	103	109	119	134	
80	90	97	105	111	119	134	

Note: These are the same data as in Table 2.7, just arranged in ascending order of IQ score.

The next decision is how many intervals to have and/or how wide they should be. The range of scores is from 67 to 152. Grouping scores by tens into 60s, 70s, 80s, and so on seems like a reasonable approach. That would mean having 10 intervals, more than the five-to-nine-intervals rule of thumb suggests. But, it is not many more and having an interval width (10) that is familiar is important.

Next, one needs to decide what information to put into the frequency distribution. The bare minimum is the interval and the frequency, but tables are more useful if they also include cumulative frequency, percentage, and cumulative percentage. Table 2.9 shows a table, empty except for title, column labels, and interval information. Note that the intervals appear in the reverse order they did in Table 2.8. And, note that there are enough non-overlapping intervals so that each value falls in one, and only one, interval.

TABLE 2.9 Template for Grouped Frequency Distribution for IQ Scores of a Random Sample of Sixth Graders (Interval Width = 10)

IQ Score Interval	Midpoint (m)	Frequency (f)	Cumulative Frequency (f_c)	Percentage (%)	Cumulative Percentage ($\%_c$)
150–159					
140–149					
130–139					
120–129					
110–119					
100–109					
90–99					
80–89					
70–79					
60–69					

Note: This empty table is ready to be filled with data. Note that there is already a clear title, all columns are labeled, the interval with the lowest scores is on the bottom row, the intervals are all the same width, and the intervals are non-overlapping.

The completed table, Table 2.10, will provide the school district with a good descriptive summary of the intelligence of its sixth graders. It shows the range of scores, from the 60s to the 150s, and the most common scores, from 90 to 119.

TABLE 2.10 Grouped Frequency Distribution for IQ Scores of a Random Sample of Sixth Graders (Interval Width = 10)

IQ Score Interval	Midpoint (m)	Frequency (f)	Cumulative Frequency (f_c)	Percentage (%)	Cumulative Percentage ($\%_c$)
150–159	154.5	1	68	1.47	100.00
140–149	144.5	3	67	4.41	98.53
130–139	134.5	3	64	4.41	94.12
120–129	124.5	7	61	10.29	89.71
110–119	114.5	10	54	14.71	79.41
100–109	104.5	15	44	22.06	64.71
90–99	94.5	13	29	19.12	42.65
80–89	84.5	8	16	11.76	23.53
70–79	74.5	6	8	8.82	11.76
60–69	64.5	2	2	2.94	2.94

Note: This grouped frequency distribution makes it easier to see the range of scores and what the most common scores are.

Practice Problems 2.1

Review Your Knowledge

2.01 Under what conditions should an ungrouped frequency distribution be made? A grouped frequency distribution?

2.02 What information does a cumulative frequency provide?

2.03 How should a frequency distribution be organized for a nominal variable like type of religion?

2.04 How many intervals should a grouped frequency distribution have?

2.05 What is the formula to calculate a midpoint for an interval in a grouped frequency distribution?

Apply Your Knowledge

2.06 An exercise physiologist has first-year college students exercise and then measures how many minutes it takes for their heart rates to return to normal. Make an ungrouped frequency distribution that shows frequency, cumulative frequency, percentage, and cumulative percentage for the following data:

1, 3, 7, 2, 8, 12, 11, 3, 5, 6, 7, 4, 14, 8, 2, 3, 5, 8, 11, 10, 9, 8, 4, 3, 2, 3, 4, 2, 6, and 7

2.07 Below are final grades, in order, from an upper-level psychology class. Make a grouped frequency distribution to show the distribution of grades. Use an interval width of 10 and start the lowest interval at 50. Be sure to include midpoint, frequency, cumulative frequency, percentage, and cumulative percentage for the data:

53, 54, 56, 59, 60, 60, 60, 62, 63, 63, 66, 67, 67, 70, 70, 70, 71, 72, 73, 74, 74, 75, 75, 77, 77, 77, 78, 79, 79, 80, 80, 81, 81, 81, 81, 81, 83, 83, 84, 84, 85, 85, 85, 85, 86, 86, 87, 87, 88, 91, 91, 92, 93, 94, 96, 98, 98, 99

2.2 Discrete Numbers and Continuous Numbers

Chapter 1 covered levels of measurement, classifying numbers as nominal, ordinal, interval, or ratio. Now let's learn another way to classify numbers, as discrete or continuous. Knowing whether a number is discrete or continuous will be important in the next section of this chapter, which addresses how to graph a frequency distribution.

Discrete numbers answer the question "How many?" For example, the answers to questions like how many siblings one has, how many neurons are in a spinal cord, or how many jeans are in a closet are all discrete numbers. Discrete numbers take whole number values only and have no "in-between," or fractional, values. No matter how raggedy a person's favorite pair of jeans is, it's still one pair of jeans. One would never say that a pair of jeans with a lot of holes is 0.79 of a pair of jeans. Nominal- and ordinal-level numbers are always discrete. Sometimes, interval- and ratio-level numbers are discrete.

Because fractional values for continuous numbers represent distance, continuous numbers are always interval- or ratio-level numbers. But not all interval- or ratio-level numbers are continuous.

Continuous numbers answer the question "How much?" For example, the answers to questions like how much aggression a person has, how much intelligence one has, or how much a person weighs would be continuous numbers. Continuous numbers can take on values between whole numbers, that is, they may have fractional values. The number of decimal places reported for continuous numbers, how specific they are, depends on the measuring instrument used. More precise measuring instruments allow attributes to be measured more exactly.

Because fractional values for continuous numbers represent distance, continuous numbers are always interval- or ratio-level numbers. Keep in mind, though,

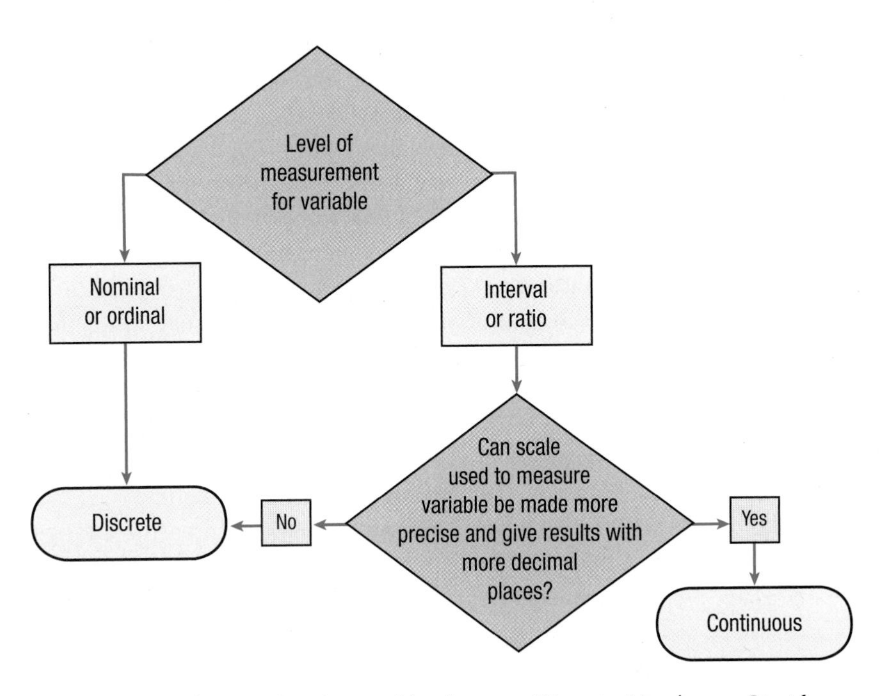

Figure 2.4 How to Choose: Continuous Numbers vs. Discrete Numbers Start by using Figure 1.10 on page 20 to determine the variable's level of measurement. Then, depending on whether the variable is (a) nominal or ordinal or (b) interval or ratio, follow the flowchart to determine if the variable is continuous or discrete.

that not all interval- or ratio-level numbers are continuous. Figure 2.4 is a flowchart that leads one through the process of determining whether a number is discrete or continuous.

A Common Question

Q If a number, like how many children in a family, is discrete and only takes whole number values, why is the average number of children in families reported as a decimal value such as 2.37?

A Individual cases can only take on whole number values if a variable is discrete, but math can be done on these whole numbers to obtain meaningful fractional values.

An example should make it clearer how continuous numbers can take on in-between values. Weight is a continuous number. Suppose Tyrone stepped on a digital scale, one that weighs to the nearest pound, and reported, "I weigh 175 pounds." Does Tyrone weigh *exactly* 175 pounds? If there were a more precise scale, one that weighed to the nearest tenth of a pound, his weight might turn out to be 175.3 pounds. And, with an even more precise scale, one that measured to the nearest hundredth of a pound, his weight might now be found to be 175.34 pounds. How much Tyrone weighs depends on the precision of the scale used to measure him. Theoretically, continuous numbers can always be made more exact by using a better—more precise—measuring instrument.

Tyrone could weigh exactly 175 pounds, but he probably doesn't. Statisticians think of his weight as falling within an *interval* that has a midpoint of 175. The question is this: How wide is the interval? How much or how little does Tyrone really weigh?

The answer is that Tyrone weighs somewhere from 174.5 to 175.5 pounds. The scale reports weight to the nearest pound, so the unit of measurement is 1 pound. Tyrone's real weight falls somewhere in the interval from half a unit of measurement below his reported weight to half a unit of measurement above his reported weight. That's from $175 - 0.5 = 174.5$ to $175 + 0.5 = 175.5$.

If Tyrone weighed less than 174.5, say, 174.2, then the scale would have reported his weight as 174, not 175 pounds. And if he weighed more than 175.5, say, 175.9, it would have reported his weight as 176 pounds. It is not clear where in the range from 174.5 to 175.5 his weight really falls, but his weight does fall in that range. His weight—a continuous number—is reported as the midpoint (175) of a range of values, any of which could be his real weight.

A single continuous number represents a range of values. To help understand this, think back to high school chemistry and measuring how acidic or basic a liquid was. This was done by dipping pH paper into the liquid and comparing the color of the pH paper to a key. An example of a key for pH is shown in Figure 2.5.

Suppose one dipped a piece of the pH paper from Figure 2.5 into a solution and it turned the shade associated with a pH of 6. Is the pH of the solution exactly 6? The pH is closer to 6 than to 4, or else the paper would have turned the lighter shade for 4. It is also closer to 6 than to 8, because the paper was not the darker shade for 8. The pH of the solution is closer to 6 than to 4 or 8, but it probably is not exactly 6.

In what range does the pH of the solution fall? This pH paper measures to the nearest 2 points so that is the unit of measurement. Half of the unit of measurement is 1, so subtracting that from 6 and adding it to 6 give the range within which the pH of the solution really falls: between 5 and 7. When this pH paper says 6, it really means somewhere between 5 and 7.

Those pH values, 5 and 7, are called the real limits of the interval. **Real limits** are the upper and lower bounds of a single continuous number or of an interval in a grouped frequency distribution for a continuous variable.

Table 2.10, the grouped frequency distribution of IQ for the 68 sixth graders, can be used to examine the real limits for an interval in a grouped frequency distribution. Here, IQ is measured to the nearest whole number. However, it would be possible

Figure 2.5 Key for pH Paper pH is a continuous measure. If this pH paper turns the middle shade when dipped into a solution, the pH is called 6. But the pH of the solution may not be exactly 6. It is just closer to 6 than it is to 4 or 8.

to make a test that measures IQ to the nearest tenth or hundredth of a point, so IQ is a continuous variable. In Table 2.10, the bottom IQ interval ranges from 60 to 69, what are called the apparent limits for the interval. These are called **apparent limits** because they represent how wide the interval appears to be. But an IQ score of 60, at the bottom of the interval, really ranges from 59.5 to 60.5 and a score of 69, at the top of the interval, really ranges from 68.5 to 69.5. So, the scores in the 60 to 69 interval really fall somewhere in the range from 59.5, the real lower limit of the interval, to 69.5, the real upper limit. For continuous measures, real limits make researchers aware of how (im)precise their measures are.

Worked Example 2.2

The next section of this chapter turns to making graphs. Differentiating between discrete numbers and continuous numbers is important when figuring out what graph to make, so here's more practice.

I. A professor surveys a group of college students about how much time, in minutes, they spend on schoolwork per week. She makes a grouped frequency distribution that has intervals of 0–59 minutes, 60–119 minutes, 120–179 minutes, and so on. Is this a discrete measure or a continuous measure? What are the real limits for the 60–119 interval? For that interval, what are the interval width and the midpoint?

To figure this out, follow the flowchart in Figure 2.4. Minutes is a ratio-level variable, so it can be either discrete or continuous. To decide, one needs to think if the scale could be made more precise and to measure minutes to fractional values. The answer is yes. So, the amount of time spent studying is a continuous variable. The real lower limit for the 60–119 interval is 59.5, which is half a unit of measurement below the apparent lower limit of the interval. Similarly, the real upper limit for the interval is 119.5, half a unit of measurement above the apparent upper limit. The interval width is 60, the distance between the real limits of an interval, calculated by 119.5 – 59.5. And the midpoint of the interval is 89.5, halfway between 60 and 119, calculated by $\frac{60 + 119}{2}$.

II. A developmental psychologist finds out, from a national sample of teenagers, how many texts they send per week. He ends up making a grouped frequency distribution that has intervals of 1–25, 26–50, 51–75, and so forth. Is this a continuous or a discrete measure? For the second interval, 26–50, what are the real limits? For that interval, what are the interval width and the midpoint?

Again, use the flowchart in Figure 2.4 to figure out whether this is a continuous or a discrete measure. The number of texts sent is a ratio-level number, so it could be either continuous or discrete. A text is a text is a text, whether it is one character long or 160. One can't measure the number of texts more precisely than with whole numbers, which means that this variable, number of texts sent, is a discrete number. If a person sends 26 texts, he or she has sent 26 texts, not somewhere from 25.5 to 26.5 texts. This means that the real limits of the interval are the same as the apparent limits: 26–50. The interval width is 25 and the midpoint is 38, halfway between 26 and 50, calculated by way of $\frac{26 + 50}{2}$.

Practice Problems 2.2

Review Your Knowledge

2.08 How does a continuous number differ from a discrete number?

2.09 If there were five cases in the interval 45–49 of a grouped frequency distribution for a continuous variable and their original raw scores were unknown, what value(s) should be assigned to them?

Apply Your Knowledge

2.10 For each variable, determine if it is continuous or discrete.

a. The amount of snowfall, in inches, recorded during winter in a city

b. The number of days, in the past 30, on which a person has consumed any alcohol

c. A person's level of depression, as measured by the percentage of questions on a scale that was endorsed in the depressed direction

2.11 Below are sets of consecutive intervals from frequency distributions for continuous variables. For the second interval in each set, the one in bold, tell the real lower limit; the real upper limit; the interval width, *i*; and the interval midpoint, *m*.

a. 15–19; **20–24;** 25–29

b. 205–245; **255–295;** 305–345

c. 1.1–1.2; **1.3–1.4;** 1.5–1.6

d. 1,000–2,000; **3,000–4,000;** 5,000–6,000

2.3 Graphing Frequency Distributions

"A picture is worth a thousand words" is a common saying. It's time to put that saying into practice and explore visual displays of frequency distributions. Graphs, the subject of this section, can be used to display frequency information. With graphs, the information leaps out, whereas one has to work harder to find the information in a frequency distribution table.

Choosing which graph to use depends on whether the numbers are discrete or continuous. If the numbers are discrete, use a bar graph. If the data are continuous, use a histogram or a frequency polygon.

This chapter covers three different graphs for showing frequency: (1) bar graphs, (2) histograms, and (3) frequency polygons. Choosing which graph to use depends on whether the numbers are discrete or continuous. If the numbers are discrete, use a bar graph. If the data are continuous, use a histogram or a frequency polygon.

Decisions about which graph to use are often based on the level of measurement for the data—nominal, ordinal, interval, or ratio. Nominal- and ordinal-level data are discrete and should be graphed with a bar graph. Interval- and ratio-level data may be discrete or continuous. If the variable is discrete, use a bar graph. If the variable is continuous, use a histogram or frequency polygon. In actual practice, histograms and frequency polygons are often made for interval variables and ratio variables, whether the variable is continuous or not. In these instances, histograms or frequency polygons are used even if the data are discrete. Figure 2.6 is a flowchart summarizing the decision rules about how to choose the correct graph for a frequency distribution.

Bar Graphs

Bar graphs are used to demonstrate the frequency with which the different values of discrete variables occur. Sex, whether one is male or female, is a discrete variable. Table 2.11 shows a frequency distribution for the sex of students in a psychology class.

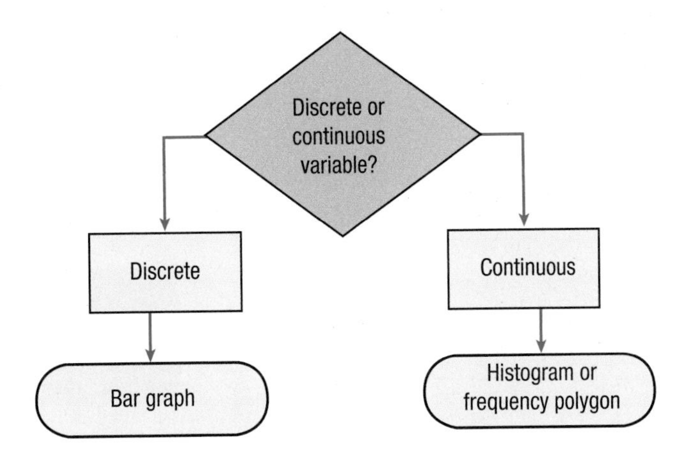

Figure 2.6 How to Choose: What Graph Should One Make for a Frequency Distribution? Start by using Figure 2.4 to determine whether the variable is continuous or discrete. The type of graph depends on this decision. No matter which graph one makes, be sure to give the graph a title and label all axes. If the variable being graphed is measured at the ordinal, interval, or ratio level, one can comment on the shape of the graph.

TABLE 2.11 Frequency Distribution of the Sex of 65 Students in an Upper-Level Psychology Class

Sex	Frequency (f)	Percentage (%)
Male	19	29.23
Female	46	70.77

Note: A frequency distribution for a nominal variable only provides frequency information and the percentage in each category. Because order is arbitrary for nominal variables, do not calculate cumulative frequencies or cumulative percentages.

Here's how to turn this table into a bar graph. Graphs are usually wider than tall, so start by making the X-axis longer than the Y-axis. **Figure 2.7** shows the template for the graph. The different categories of the variable, in this case male and female, go on the X-axis. The X-axis is labeled "Sex" and is marked "Male" and "Female."

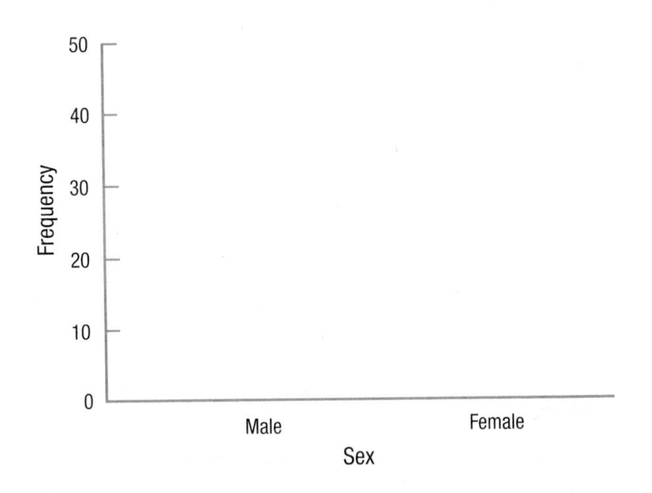

Figure 2.7 Template for Bar Graph Showing the Sex of Students ($N = 65$) in Upper-Level Psychology Class This "empty" graph shows how to set up a bar graph. Note that the graph is wider than tall, all axes are labeled, frequency goes on the Y-axis, and the categories being counted go on the X-axis.

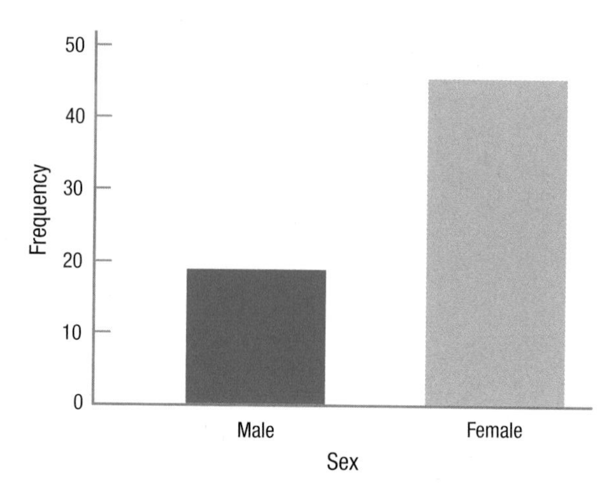

Figure 2.8 Bar Graph Showing the Sex of Students ($N = 65$) in Upper-Level Psychology Class Note that the bars don't touch each other in bar graphs because the variable on the *X*-axis is discrete.

Frequency goes on the *Y*-axis, so it is labeled "Frequency" and marked off in equal intervals, by 10s, from 0 to 50. Why go up to 50? Because the largest frequency was 46 and the axis has to accommodate that value.

Note that, just as with tables, the graph has a detailed title, *N* is mentioned, and everything is labeled clearly. Obviously, it is a good idea to use graph paper and a ruler.

To complete the bar graph, all one needs to do is draw a bar above each category on the *X*-axis. Note that the bar goes up as high as the frequency of the category in Figure 2.8, the completed figure. This is called a bar graph because of the bars; note that all of the bars are the same width and they don't touch each other. They don't touch each other because the variable is discrete.

The advantage of a picture over words, of the graph (Figure 2.8) over the table (Table 2.11), should be obvious. In this bar graph, it jumps out that there were a lot more women than men in this class.

Histograms

A **histogram** is a graphic display of a frequency distribution for continuous data. It is like a bar graph in that bars rise above the *X*-axis, with the height of the bars representing the frequencies in the intervals. However, a histogram differs from a bar graph in that the bars touch each other, representing the fact that the variable is continuous, not discrete. In Figure 2.9, a template for a histogram for the sixth-grade IQ data from earlier in the chapter is presented.

There are several things to note in this histogram:

- Even though the histogram is "empty" right now, it already has a detailed title.
- Both axes are clearly labeled.
- Frequency goes on the *Y*-axis and the values of the variable (in this case, IQ) go on the *X*-axis.
- The *Y*-axis starts at zero and is marked off by 5s to a height of 20. This accommodates the largest frequency in the data set, 15.

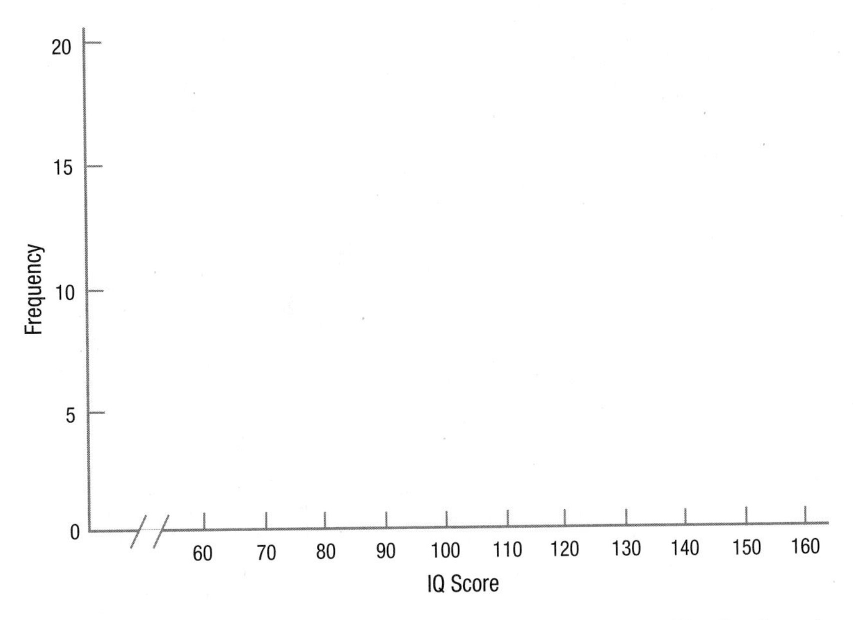

Figure 2.9 Template for Histogram Showing Grouped Frequency Distribution of IQ Scores for 68 Sixth Graders (Interval Width = 10) This "empty" graph shows how to set up a histogram. Note that frequency goes on the *Y*-axis, values of the variable being graphed on the *X*-axis, and that both axes are clearly labeled.

- The smallest value for IQ is 67, which is far away from a value of zero. If the *X*-axis started at zero, there would be a lot of blank space before the first interval. Instead, the *X*-axis starts at 60, the apparent lower limit of the first interval.
- Note the discontinuity mark on the *X*-axis, which alerts people viewing the graph that the axis doesn't start at zero.
- The highest IQ value is 152, so the *X*-axis ends at 160, what would be the start of the next higher interval.

Figure 2.10 shows the histogram once it has been completed. Note:

- The bars go up as high as the frequency in the interval, 15 for the highest frequency.
- Unlike a bar graph, the bars in a histogram touch each other because the variable being represented is continuous.
- There is one tricky thing in a histogram—the bars go up and come down at the real limits of the interval, not the apparent limits. The first interval, 60 to 69, for example, really covers scores from 59.5 to 69.5 and the bar reflects that. Look at the enlarged section of Figure 2.10 to get a clear picture of this.

Frequency Polygon

A **frequency polygon,** sometimes called a line graph, displays the frequency distribution of a continuous variable as a line, not with bars. A frequency polygon differs from a histogram in another way: the frequencies "go to zero" at the beginning (the far left) and the end (the far right) of the graph. Whether one uses a histogram or a frequency

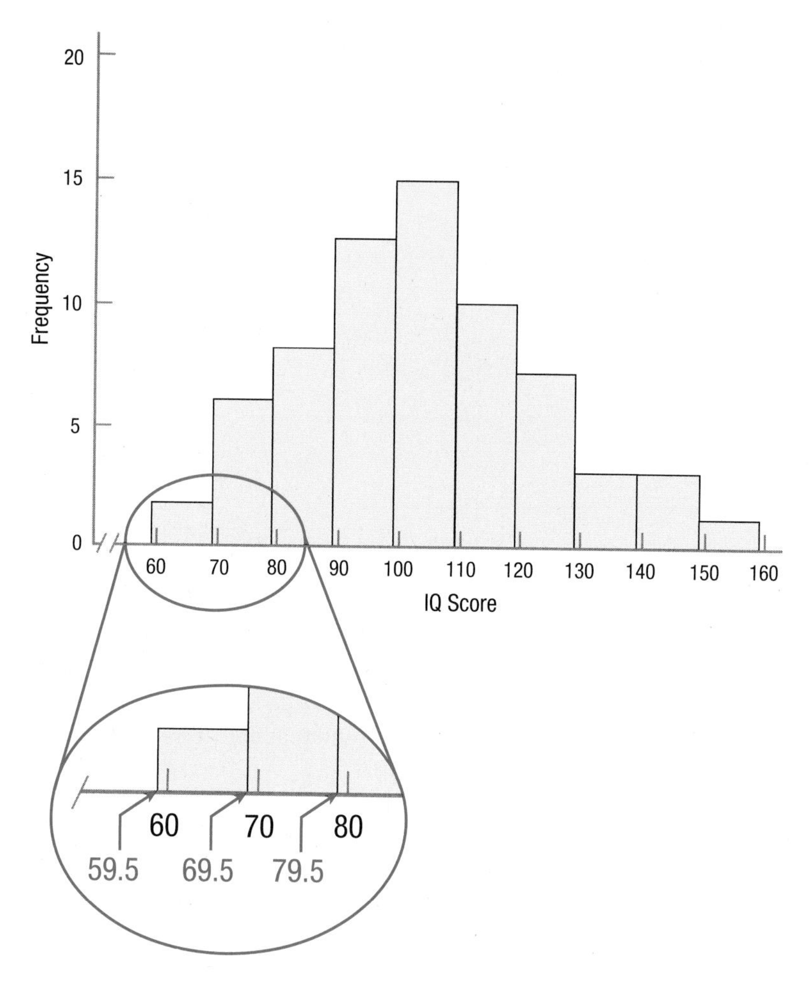

Figure 2.10 Histogram Showing Grouped Frequency Distribution of IQ Scores for 68 Sixth Graders (Interval Width = 10) Histograms graph the frequency of continuous variables, so the bars touch each other. Note that the bars are as wide as the real limits of each interval, not the apparent limits. This is highlighted in the enlarged section of the graph.

polygon to represent a frequency distribution is a matter of personal preference—they both are legitimate options for graphing frequencies for a continuous variable.

Figure 2.11 shows a template for a frequency polygon for the sixth-grade IQ data:

- Of course, there is a title and both axes are labeled.
- Frequency, on the *Y*-axis, starts at zero and goes up, in equal intervals, to the first interval above the largest possible frequency.
- The *X*-axis shows the midpoints of the intervals for the variable being graphed.
- The lowest IQ score in the data set is in the interval with a midpoint of 64.5 and the highest in an interval with a midpoint of 154.5. But, two "extra" midpoints are displayed on the *X*-axis, one (54.5) below the lowest interval and one (164.5) above the highest interval.

Figure 2.12 shows the completed frequency polygon. It was completed by placing a dot above each midpoint at the level of its frequency and then connecting the dots.

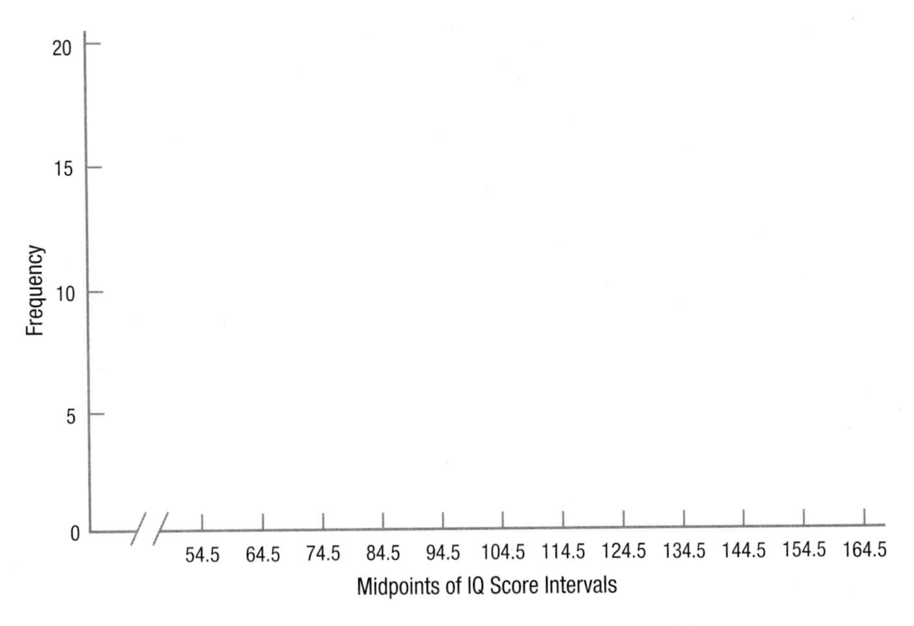

Figure 2.11 Template for Frequency Polygon Showing Grouped Frequency Distribution of IQ Scores for 68 Sixth Graders (Interval Width = 10) This "empty" graph shows how a frequency polygon should be set up. Frequency is marked on the *Y*-axis and midpoints of the IQ intervals on the *X*-axis. Both axes are clearly labeled and the graph is wider than tall.

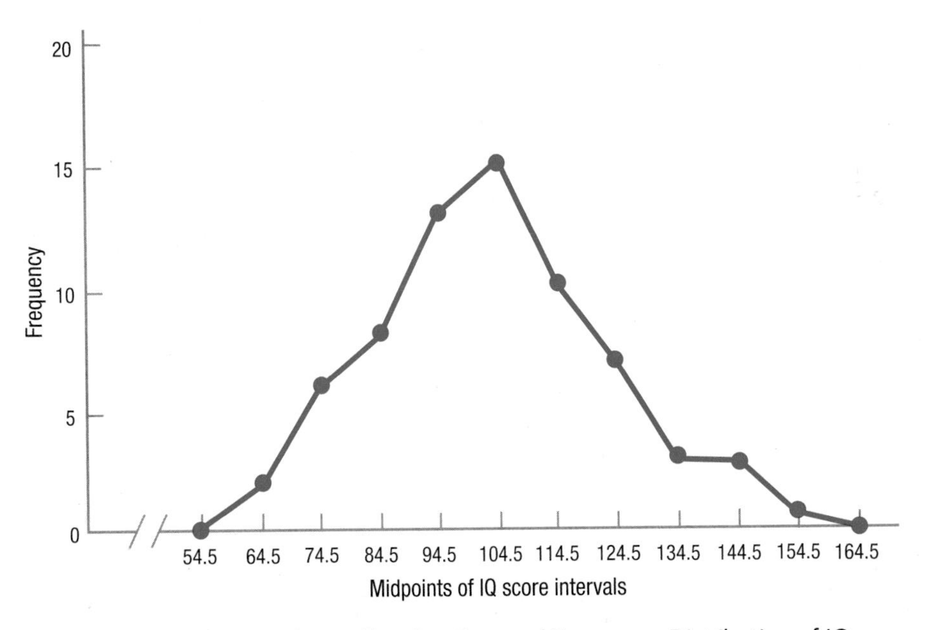

Figure 2.12 Frequency Polygon Showing Grouped Frequency Distribution of IQ Scores for 68 Sixth Graders (Interval Width = 10) Frequency polygons are made for continuous variables. The frequencies are marked by dots at the appropriate height at the midpoints of the intervals. Then the dots are connected by lines. Note that the graph starts and ends at intervals with a frequency of zero.

Note that the frequency for the very bottom and very top interval is zero. This is what was meant by saying that frequency polygons "go to zero" at the beginning and end of the graph.

There are a number of differences among the three graphs. Table 2.12 summarizes the differences among bar graphs, histograms, and frequency polygons.

TABLE 2.12 Key Differences Among Graphs

	Bar Graphs	Histograms	Frequency Polygons
Type of data	Discrete	Continuous	Continuous
Physical characteristics of the graph	✓ Bars don't touch each other.	✓ Bars touch each other. ✓ Bars go up and down at the real limits of the interval.	✓ Frequencies are marked with dots at the midpoints of the intervals. ✓ Dots are connected by lines. ✓ Frequencies "go to zero" at the far left and far right of the graph.

Worked Example 2.3 What type of graph could one make for the property crime rate frequency distribution in Table 2.5. The data are rates, so they are continuous and one could make either a histogram or a frequency polygon. This author is partial to histograms, but there would be nothing wrong with choosing to make a frequency polygon instead. The completed histogram—wider than tall, with crime rate on the *X*-axis and frequency on the *Y*, all axes labeled and with a clear title—is shown in Figure 2.13.

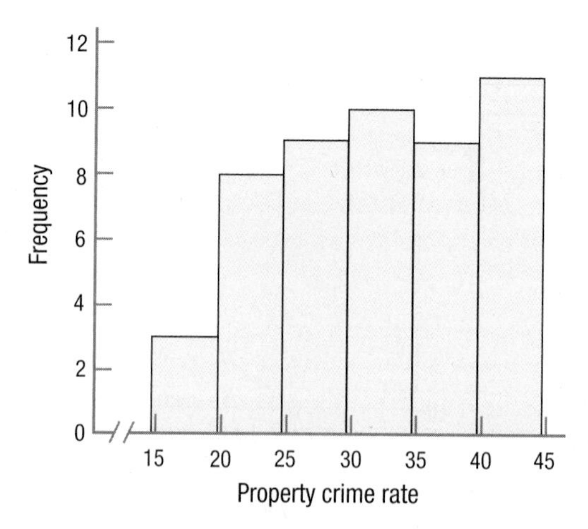

Figure 2.13 Histogram Showing Property Crime Rates per 1,000 Population for the 50 States This frequency distribution, which is for a continuous variable, could also be graphed as a frequency polygon.

Practice Problems 2.3

Review Your Knowledge

2.12 Match these two types of data, continuous and discrete, with the three types of graphs (bar graphs, histograms, and frequency polygons).

2.13 Where do the bars go up and come down for each interval on a histogram?

Apply Your Knowledge

2.14 A religion professor at a large university took a sample of 53 students and asked them what their religious faith was. Nine reported they were Muslim, 4 Buddhist, 2 Hindu, 16 Christian, 10 Jewish, 6 atheist, and 6 reported being other religions. Make a graph for these data.

2.15 A psychologist administered a continuous measure of depression to a representative sample of 772 residents of a midwestern state. Scores on the depression scale can range from 0 to 60 and higher scores indicate more depression. There were 423 with scores in the 0–9 range, 210 in the 10–19 range, 72 in the 20–29 range, 37 in the 30–39 range, 18 in the 40–49 range, and 12 in the 50–59 range. Make a graph for the data.

2.4 Shapes of Frequency Distributions

There are three aspects of the shape of frequency distributions to focus on—modality, skewness, and kurtosis.

Now that frequency distributions are being graphed, it is time to look at the shapes that data sets can take. There are three aspects of the shape of frequency distributions to focus on—modality, skewness, and kurtosis—and they are summarized in Table 2.13.

It is important to know the shape of a distribution of data. The shape will determine whether certain statistics can be used. In the next chapter, for instance, it will be shown that it's inappropriate to calculate a mean if a data set has certain irregular shapes.

TABLE 2.13 Three Aspects of Shape of Frequency Distribution Curves to Describe

Modality	Skewness	Kurtosis
How many high points there are in a data set	Whether the data set is symmetric	Whether a data set is peaked or flat

The shape of a data set only matters if the variable is measured at the ordinal, interval, or ratio level. With nominal variables, factors like race or religion, the order in which the categories are arranged is arbitrary. For example, Practice Problem 2.14 involved making a bar graph for the frequencies of different religions. The shape of this graph will vary, depending on whether one organizes the religions alphabetically, by ascending frequency, or by descending frequency. In contrast, one can't legitimately alter the order of the depression variable, graphed with either a histogram or frequency polygon, in Practice Problem 2.15.

One example of a common shape is Figure 2.14. Many call this a bell-shaped curve, but statisticians call it a normal curve or normal distribution. This curve has one highest point, in the middle, and the frequencies decrease symmetrically as the values move away from the midpoint. (Being symmetrical means that the left side of the curve is a mirror image of the right.) The frequency polygon of the sixth-grade IQ data in Figure 2.12 has a "normalish" shape.

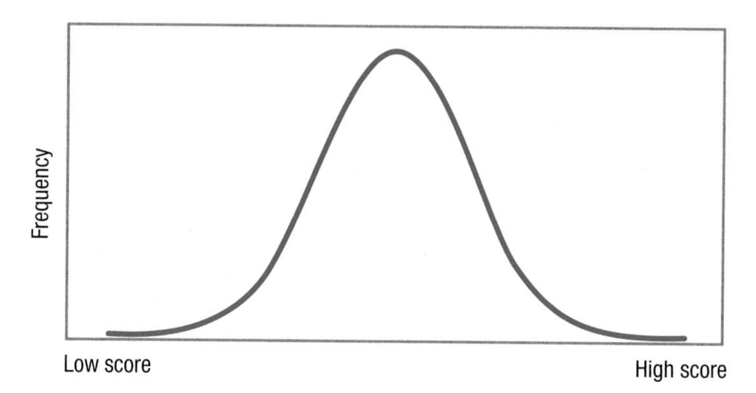

Figure 2.14 The Normal Distribution The normal distribution, sometimes called the bell-shaped curve, is highest at its midpoint and then shows symmetrical decreases in frequency as it moves to the tails.

Modality, the first of the three aspects used to describe shape, refers to how many peaks exist in the curve of the frequency distribution. A peak is a high point, also called a mode, and it represents the score or interval with the largest frequency. The normal curve has one peak, in the center of the distribution, and is called unimodal. A distribution can have two peaks, in which case it is called bimodal. If a distribution has three or more peaks, it is called multimodal. Figure 2.15 shows what distributions with different modalities look like.

Another characteristic of the shape of frequency distributions is **skewness,** which is a measure of how symmetric they are. A distribution is considered skewed if it is not symmetric. The normal curve is perfectly symmetric, as the left side and

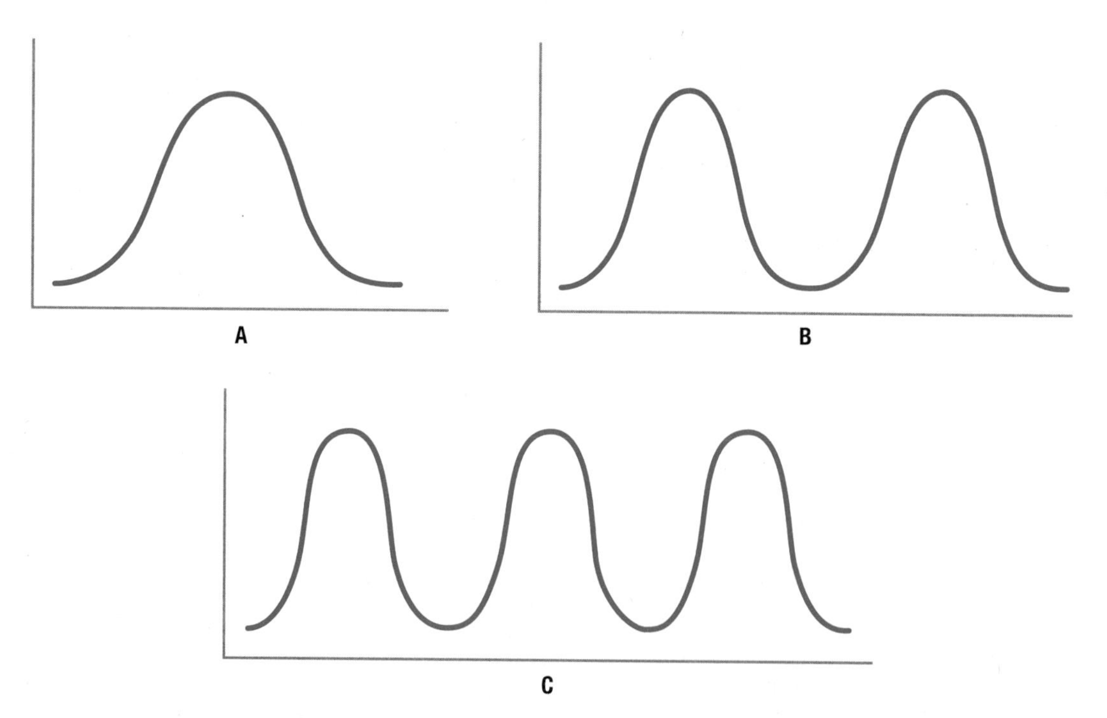

Figure 2.15 Examples of Modality Panel A shows a unimodal data set, panel B a bimodal data set, and panel C a multimodal data set.

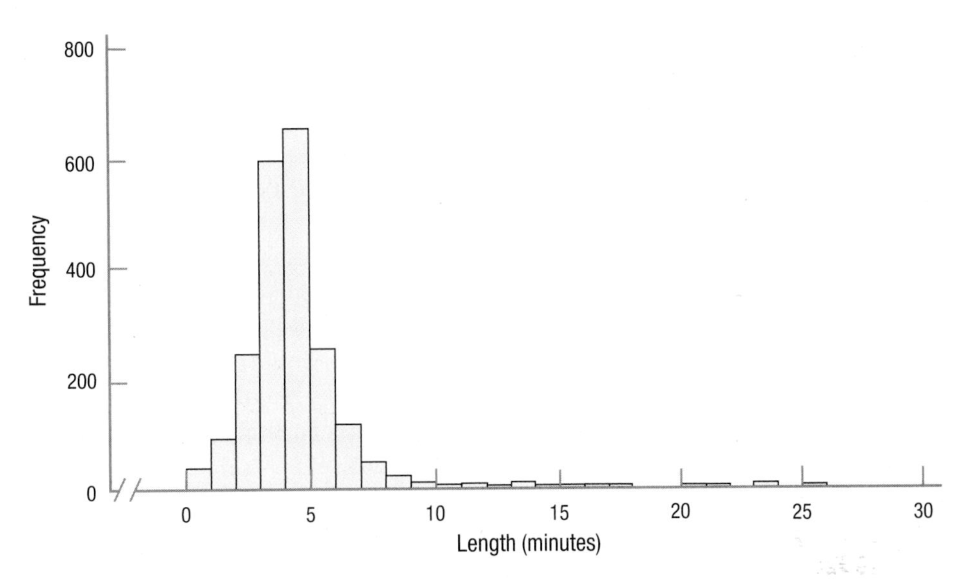

Figure 2.16 Length of Songs in an iTunes Library This histogram is a good example of positive skew, the data set not being symmetric but trailing off to the right. The "spike" around 4 minutes makes this a good example of a data set that is quite peaked. (Thanks to Samantha DeDionisio for collecting these data.)

right side are mirror images of each other. If there is asymmetry and the data tail off to the right, it is called **positive skew.** If the data tail off to the left, it is called **negative skew.** Positive and negative don't have good and bad connotations here. Statisticians simply use positive and negative to describe which side of the *X*-axis the tail is on, the right side for positive numbers and the left side for negative numbers.

For a good example of skewness, look at Figure 2.16. It shows the length of all 2,132 songs in a student's iTunes library. Most songs are around 3 or 4 minutes in length, but there are a few approximately 25 minutes long, giving this frequency distribution positive skew.

The third aspect of a frequency distribution is **kurtosis,** which is just a fancy term for how peaked or flat the distribution is. In Figure 2.14, the normal curve is neither too peaked nor too flat, so it has a normal level of kurtosis. Some sets of data have higher peaks, like the iTunes data in Figure 2.16. Other distributions are more flat, like the one shown in Figure 2.17 of the results from a single die rolled multiple times.

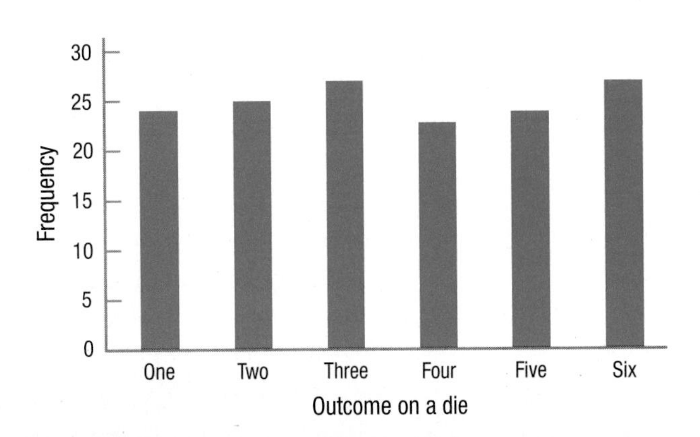

Figure 2.17 Frequency Distribution of Outcomes on a Die Rolled 150 Times Kurtosis refers to how peaked or flat a data set is. This data set is flat.

Stem-and-Leaf Displays

The final topic in this chapter, stem-and-leaf displays, is a wonderful way to summarize the whole chapter. A **stem-and-leaf display** is a combination of a table and a graph. It contains all the original data like an ungrouped frequency distribution table, summarizes them in intervals like a grouped frequency distribution, and "pictures" the data like a graph.

A stem-and-leaf display divides numbers into "stems" and "leaves." The leaves are the last digit on the right of a number. The stems are all the preceding digits. For the data in Table 2.4, the stem for South Dakota, North Dakota, and New Hampshire would be the tens digit 1, and the leaves would be, respectively, the ones digits 7, 9, and 9. With New York, the stem changes to the tens digit 2 and the leaf would be 0. All the numbers in Table 2.4 are two-digit numbers, ranging from numbers in the teens to numbers in the 40s, so the leaves are 1, 2, 3, and 4. Table 2.14 shows the first step in making a stem-and-leaf display, listing all the possible stems, followed by a vertical line.

TABLE 2.14 First Step in Constructing a Stem-and-Leaf Display

Stem	Leaves
1	
2	
3	
4	

The first step in constructing a stem-and-leaf display is placing the stems.

The next step is to start tallying the leaves for each stem. When the leaves for all 50 states have been added, the results will look like Table 2.15. Note that the leaves for each stem are in order from low to high. Stem-and-leaf displays are a great way to organize data in preparation for making a grouped frequency distribution.

TABLE 2.15 Stem-and-Leaf Display Showing Property Crime Rates per 1,000 Population for the 50 United States

Stem	Leaves
1	799
2	02234444555668899
3	0001224444555577789
4	00011111234

The three values in the top row are 17, 19, and 19.

Worked Example 2.4

For practice using a graph to figure out the shape of a data set, here are some data collected by a student, Neil Rufenacht. He took a ruler to a fast-food restaurant, bought several orders of french fries, and measured how long they were to the nearest tenth of a centimeter (cm). His results are shown as a stem-and-leaf display in Table 2.16 and as a histogram in Figure 2.18.

It is apparent in the histogram that the smallest french fries fell in the interval with a midpoint of 2 cm and the largest in the 14-cm interval. (That's from about three quarters of an inch to five and a half inches for those who prefer the English system of measurement.)

TABLE 2.16 Stem-and-Leaf Display for French Fry Data

Stem	Leaf
1	79
2	5
3	345
4	000146667889
5	0012234445668
6	033333555668999
7	0001444567778889999
8	1234
9	00000235678
10	133444555
11	23466
12	0569
13	
14	2

This displays the same data as seen in the histogram in Figure 2.18. It has the same "normalish" shape. But, because the intervals are narrower, it shows more detail.

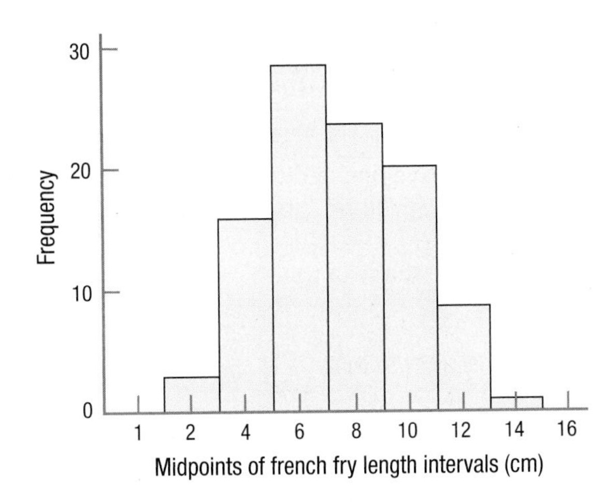

Figure 2.18 Histogram for Length of French Fries With one mode near the center and somewhat symmetrical tailing off of frequencies from the mode, this data set has a "normalish" shape.

The histogram has one mode, near the center, that doesn't seem to go up unusually high. This means that the distribution is unimodal and neither too peaked nor flat. In addition, the frequencies tail off, in a somewhat symmetrical fashion, as the lengths move away from the mode. The shape is not a perfect normal distribution, but it is "normalish." It seems reasonable to conclude that lengths of french fries, at least at this restaurant, resemble a normal distribution.

The stem-and-leaf display shows the same general shape, but offers more details. One can tell, for example, that the smallest french fry is 1.7 cm long.

DIY

Find something you can measure 100 times. Here are some ideas:

- Light a match and time how long it takes to burn down. Get another match, light it, and time the burning. Do this 98 more times.
- Find the prices of 100 stocks.
- Find the length, in seconds, of 100 songs from your iTunes library.
- Stick your tongue in a bowl of Cheerios and count how many stick to your tongue.
- Dip a tablespoon into a bowl of change and tally the value of the coins you picked up.
- Weigh, in grams, 100 eggs.

After you have collected your 100 data points, graph them and examine the shape. Given what you measured, does it make sense? For example, if you recorded the length of 100 songs from your music library and the distribution was positively skewed, does that, on reflection, seem reasonable?

Practice Problems 2.4

Review Your Knowledge

2.16 For what levels of measurement can one describe the shape of a frequency distribution?

2.17 Describe the normal curve in terms of symmetry, modality, and kurtosis.

2.18 What does it mean if a frequency distribution is positively skewed?

Apply Your Knowledge

2.19 Here is a graph for the religious affiliation of the 53 students surveyed in Practice Problem 2.14. Determine the shape of the data set.

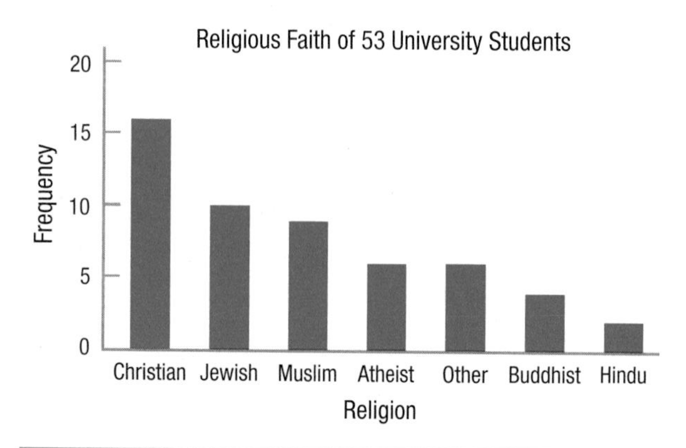

2.20 This graph shows the depression scores of the 772 residents surveyed in Practice Problem 2.15. Determine the shape of the data set.

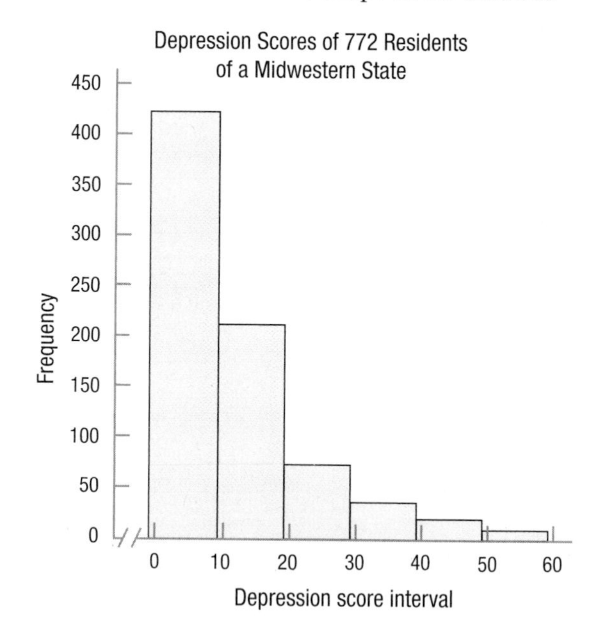

2.21 Make a stem-and-leaf display and comment on the shape of the distribution for these data: 183, 192, 203, 203, 211, 219, 220, 227, 228, 228, 229, 230, 233, 234, 234, 248, 249, 254, and 266.

Application Demonstration

Data showing the suicide rates for all 50 U.S. states were obtained from the *Statistical Abstract of the United States.* The rates are expressed as the number of deaths per 100,000 population. The rates range from a low of 6.5 deaths per 100,000 population in New Jersey to a high of 21.7 per 100,000 population in Wyoming. To put that in perspective, 0.0065% of the population of New Jersey committed suicide in 2006 vs. a rate more than three times higher in Wyoming, 0.0217%.

The data are arranged in ascending order in Table 2.17 in preparation for making a frequency distribution. Note that the rates are reported to the nearest tenth, so that is the unit of measurement. There are a large number of unique values for the 50 states, but all values don't need to be maintained. Instead, those close to each other can be grouped together into intervals without losing vital information. Following the flowchart in Figure 2.3, a grouped frequency distribution makes sense.

TABLE 2.17 2006 U.S. Suicide Rate Data Arranged in Order

6.5	10.3	11.2	13.0	15.2
6.6	10.4	11.4	13.3	15.6
6.7	10.6	11.6	13.5	15.8
7.8	10.8	11.9	13.6	16.0
8.0	11.0	11.9	13.8	16.0
8.1	11.0	12.0	14.1	18.0
8.6	11.1	12.2	14.2	19.5
9.2	11.1	12.3	14.6	19.7
9.2	11.1	12.4	15.0	20.0
10.0	11.2	12.6	15.2	21.7

Note: Though the range of scores is not large, there are many unique values for these scores. A grouped frequency distribution makes more sense for these data than an ungrouped frequency distribution.

The next question is how many intervals to include. With a range of values from 6.5 to 21.7, the high score and low score are 15.2 points apart. An interval width of 2 points would mean having eight intervals. This fits in with the rule of thumb of including from five to nine intervals suggested in Table 2.5.

Intervals can't overlap. If the bottom interval starts at 6 and is 2 points wide, then the next interval starts at 8. Suicide rates are reported to the nearest tenth, so the apparent limits of the first interval are from 6.0 to 7.9. The bottom interval starts at 6. The real limits of the interval run from half a unit of measurement below the apparent bottom of the interval to half a unit of measurement above the apparent top of the interval. That is, from 5.95 to 7.95. *Note:* The distance between the real limits is the interval width.

The midpoints of the intervals are the points halfway between the interval's limits. The midpoint for the first interval, which ranges from 6.0 to 7.9, is calculated:

$$\frac{6.0 + 7.9}{2} = 6.95$$

The next midpoint is one interval width higher:

$$6.95 + 2.00 = 8.95$$

After calculating the other midpoints, one finds the frequencies by counting the number of cases in each interval. These frequencies are placed in a grouped frequency distribution table, as shown in Table 2.18.

TABLE 2.18 Partially Completed Grouped Frequency Distribution for Suicide Rate Data for the United States (Interval Width = 2.00)

Suicide Rate (Deaths per 100,000)	Interval Midpoint (*m*)	Frequency (*f*)	Cumulative Frequency (f_c)	Percentage (%)	Cumulative Percentage ($\%_c$)
20.0–21.9	20.95	2			
18.0–19.9	18.95	3			
16.0–17.9	16.95	2			
14.0–15.9	14.95	8			
12.0–13.9	12.95	10			
10.0–11.9	10.95	16			
8.0–9.9	8.95	5			
6.0–7.9	6.95	4			

Note: This partially completed frequency distribution contains the basic information.

Table 2.18 contains only the intervals, the midpoints, and the frequencies. Using Figure 2.1 as a guide, stair-step up to find the cumulative frequencies. Then use Equation 2.1 to transform frequencies into percentages and Equation 2.2 to transform cumulative frequencies into cumulative percentages in order to complete the grouped frequency distribution (see **Table 2.19**).

TABLE 2.19 Grouped Frequency Distribution for Suicide Rate Data for the United States (Interval Width = 2.00)

Suicide Rate (Deaths per 100,000)	Interval Midpoint (*m*)	Frequency (*f*)	Cumulative Frequency (f_c)	Percentage (%)	Cumulative Percentage ($\%_c$)
20.0–21.9	20.95	2	50	4.00	100.000
18.0–19.9	18.95	3	48	6.00	96.00
16.0–17.9	16.95	2	45	4.00	90.00
14.0–15.9	14.95	8	43	16.00	86.00
12.0–13.9	12.95	10	35	20.00	70.00
10.0–11.9	10.95	16	25	32.00	50.00
8.0–9.9	8.95	5	9	10.00	18.00
6.0–7.9	6.95	4	4	8.00	8.0

Note: This table summarizes the suicide rate, at a state level, in the United States.

The next step is to make a graph of the data. Use Figure 2.4 to determine whether the data are continuous and Figure 2.6 to determine whether a histogram or frequency polygon should be made. **Figure 2.19** displays a histogram.

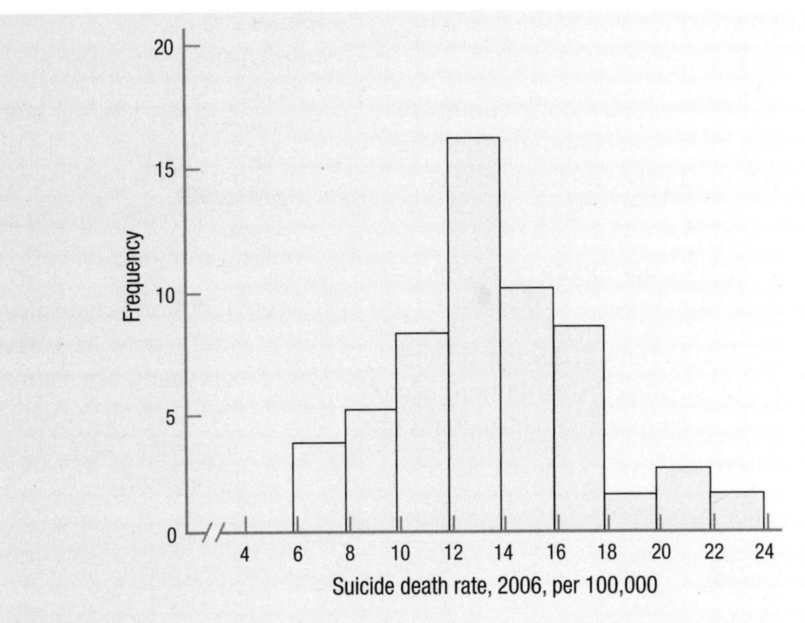

Figure 2.19 Histogram for U.S. Suicide Rates This histogram, with an interval width of 2.00, is unimodal and seems a little more peaked than normal. It might have a little positive skew.

This is a histogram, so the bars go up and come down at the real limits of the intervals, not the apparent limits. It may be hard to see, but the first bar goes up at 5.95, not 6.00, and comes down at 7.95, not 8.00.

DIY

Every year the U.S. Census Bureau used to publish a compendium of odd facts about America and Americans. Want to know the dollar value of all the crops and livestock produced in a state? The *Statistical Abstract of the United States* will tell you that it ranges from $32 million in Alaska to $35 billion in California. California also leads in the number of federal and state prisoners (171,000), while Hawaii has the longest life expectancy, 77 years.

Though the last *Statistical Abstract* was published in 2012, it is still available on the Web and in libraries. Make a reference librarian happy and ask to see a copy. Or Google the term and explore it online. Whatever approach you take, get hold of a copy, find a table that has data of interest to you, and reduce that data into a frequency distribution and graph. By doing so, what did you learn about your variable?

SUMMARY

Make a frequency distribution for a set of data.

- Frequency distributions summarize a set of data by tallying how often the values, or ranges of values, occur. For all but nominal-level data, they may also display information about cumulative frequency, percentage, and cumulative percentage.

Decide if a number is discrete or continuous.

- Discrete numbers are whole numbers that answer the question "How many?" Continuous numbers can be fractional and answer the question "How much?" The number of decimal places reported for a continuous variable depends on the precision of the measuring instrument.

- The range within which a single continuous value, or an interval of continuous numbers, falls is bounded by the real limits of the interval.

Choose and make the appropriate graph for a frequency distribution.

- Frequency distributions of discrete data are graphed with bar graphs; distributions of continuous data are graphed with histograms or frequency polygons.

Describe modality, skewness, and kurtosis for a frequency distribution.

- Modality refers to how many high points there are in a data set. If a distribution is asymmetric, it is skewed, while kurtosis refers to how peaked or flat a distribution is.

Make a stem-and-leaf display.

- Stem-and-leaf displays are a great way to summarize a set of data. They present the data compactly, like a grouped frequency distribution; keep all the details like an ungrouped frequency distribution; and show the shape, like a graph.

KEY TERMS

apparent limits – what seem to be the upper and lower bounds of an interval in a grouped frequency distribution.

bar graph – a graph of a frequency distribution for discrete data that uses the heights of bars to indicate frequency; the bars do not touch.

continuous numbers – answer the question "how much" and can have "in-between" values; the specificity of the number, the number of decimal places reported, depends on the precision of the measuring instrument.

cumulative frequency – a count of how often a given value, or a lower value, occurs in a set of data.

cumulative percentage – cumulative frequency expressed as a percentage of the number of cases in the data set.

discrete numbers – answer the question "how many," take whole number values, and have no "in-between" values.

frequency distribution – a tally of how often different values of a variable occur in a set of data.

frequency polygon – a frequency distribution for continuous data, displayed in graphical format, using a line connecting dots, above interval midpoints, that indicate frequency.

grouped frequency distribution – a count of how often the values of a variable, grouped into intervals, occur in a set of data.

histogram – a frequency distribution for continuous data, displayed in graph form, using the heights of bars to indicate frequency; the bars touch each other.

kurtosis – how peaked or flat a frequency distribution is.

midpoint – the middle of an interval in a grouped frequency distribution.

modality – the number of peaks that exist in a frequency distribution.

negative skew – an asymmetrical frequency distribution in which the tail extends to the left, in the direction of lower scores.

positive skew – an asymmetrical frequency distribution in which the tail extends to the right, in the direction of higher scores.

real limits – what are really the upper and lower bounds of a single continuous number or of an interval in a grouped frequency distribution.

stem-and-leaf display – a data summary technique that combines features of a table and a graph.

skewness – deviation from symmetry in a frequency distribution, which means the left and right sides of the distribution are not mirror images of each other.

ungrouped frequency distribution – a count of how often each individual value of a variable occurs in a set of data.

CHAPTER EXERCISES

Answers to the odd-numbered exercises appear at the back of the book.

Review Your Knowledge

2.01 A frequency distribution is a ____ of how often the values of a variable occur in a data set.

2.02 There are two types of frequency distribution tables: ____ and ____.

2.03 A basic frequency distribution just contains information about what ____ occur in a data set and what their ____ are.

2.04 ____ tells how many cases in a data set have a given value or a lower one.

2.05 The abbreviation for cumulative frequency is____.

2.06 The cumulative frequency for the top row in a frequency distribution is equal to ____.

2.07 If one has ____ -level data in a frequency distribution, it should be organized in some logical fashion.

2.08 The ____ column in a frequency distribution is a restatement of the information in the cumulative frequency column.

2.09 When dealing with a variable that has a large range, a ____ frequency distribution usually makes more sense than an ungrouped frequency distribution.

2.10 As a general rule of thumb, grouped frequency distributions have from ____ to ____ intervals.

2.11 Having a small number of intervals may show the big picture, but the danger exists of losing sight of the ____ in the data set.

2.12 All intervals in a grouped frequency distribution should be the same ____, so the frequencies in different intervals can be compared meaningfully.

2.13 A ____ is the middle point of an interval.

2.14 If the value for a case in an interval is unknown, assign it the value associated with the ____.

2.15 Discrete numbers answer the question ____.

2.16 Discrete numbers take on ____ number values only.

2.17 ____- and ____-level numbers are discrete numbers.

2.18 ____ numbers answer the question "How much?"

2.19 Continuous numbers can have ____ values.

2.20 The specificity of a ____ depends on the precision of the instrument used to measure it.

2.21 Continuous numbers are always ____- or ____-level numbers.

2.22 Interval- and ratio-level numbers can be ____ or ____.

2.23 A single value of a continuous number really represents a ____ of values.

2.24 The ____ are the upper and lower bounds of a single continuous number or of an interval in a grouped frequency distribution for continuous numbers.

2.25 The distance between the real limits of an interval is ____.

2.26 Which graph is used to display a frequency distribution depends on whether the numbers in the frequency distribution are ____ or ____.

2.27 The graph used for a frequency distribution of discrete data is a ____.

2.28 To make a graph of a frequency distribution of continuous data, one can use a ____ or a ____.

2.29 In terms of size, graphs are usually ____ than ____.

2.30 Graphs should have a descriptive title and the ____ should be clearly labeled.

2.31 Frequency is marked on the ____-axis of the graph of a frequency distribution.

2.32 The intervals for the values of the variable being graphed are marked on the ____ -axis of a histogram.

2.33 In a histogram, the data are continuous, so the bars ____ each other.

2.34 In a frequency polygon, frequency is marked with a dot placed at the appropriate height above the ____ of an interval.

2.35 One shouldn't look at the shape of a graph for ____-level data.

2.36 The ____ is symmetric, has the highest point in the middle, and includes frequencies that decrease as one moves away from the midpoint.

2.37 The three aspects of the shape of a frequency distribution that are described are ____, ____, and ____.

2.38 ____ describes how many peaks exist in a frequency distribution.

2.39 If a frequency distribution has two peaks, it is called ____.

2.40 A frequency distribution that tails off on one side is said to be ____.

2.41 If a frequency distribution tails off on the left-hand side, it has ____ skewness.

2.42 Kurtosis refers to how ____ or ____ a frequency distribution is.

2.43 A stem-and-leaf display summarizes a data set like a __________ frequency distribution.

2.44 In a stem-and-leaf display for 21, 21, 24, 25, 28, 30, 31, 34, and 39, the numbers 2 and 3 would be the _____s and 0, 1, 4, 5, 8, and 9 would be the _______s.

Apply Your Knowledge

Making frequency distribution tables

2.45 A psychologist completes a survey in which residents in a psychiatric hospital are given diagnostic interviews in order to determine their psychiatric diagnoses. After each interview, the researcher counts how many diagnoses each resident has. Given the following data, make an ungrouped frequency distribution showing frequency, cumulative frequency, percentage, and cumulative percentage: 2, 3, 2, 1, 0, 1, 1, 4, 2, 3, 1, 0, 1, 2, 4, 3, 2, 2, 2, 2, 1, 2, 3, 2, 5, 2, 3, 2, 1, 1, 2, 3, 3, 3, 2, 2, 1, 1, 2, 2, 3, 1, 1, 2, and 2.

2.46 A cognitive psychologist wanted to investigate the stage of moral development reached by college students. She believed that people progressed, in order, through six stages of moral reasoning, from Stage I to Stage VI. She obtained a representative sample of college students in the United States and administered an inventory to classify stage of moral development. Here are the numbers of people classified at the different stages: I = 17, II = 34, III = 78, IV = 187, V = 112, and VI = 88. Make a frequency distribution table for these data, showing frequency, cumulative frequency, percentage, and cumulative percentage.

2.47 A college registrar completes a survey of classrooms on campus in order to find out how many usable seats there are in each one. Make a grouped frequency distribution for her data, using an interval width of 20 and an apparent lower limit of 10 for the bottom interval. Report midpoint, frequency, cumulative frequency, percentage, and cumulative percentage. Here are the data the registrar collected: 12, 26, 18, 17, 102, 20, 35, 46, 50, 28, 29, 53, 75, 30, 37, 45, 58, 43, 42, 36, 50, 60, 55, 45, 40, 23, 28, 38, 39, 40, 50, 60, 45, 36, 28, 40, 54, 62, 38, 58, and 24.

2.48 A professor of education was curious about students' expectations of academic success. She obtained a sample of college students at the start of their college careers and asked them what they thought their GPAs would be for the first semester. Below are the data she obtained. Make a grouped frequency distribution for the data, using an interval width of 0.5 and an apparent lower limit of 2.1 for the bottom interval. Report midpoint, frequency, cumulative frequency, percentage, and cumulative percentage.

3.9, 4.0, 3.0, 3.5, 3.0, 3.5, 3.5, 3.9, 3.5, 2.5, 3.5, 3.8, 3.9, 3.0, 3.9, 3.5, 3.9, 3.0, 3.5, 2.8, 2.4, 2.7, 2.6, 3.4, 3.6, 2.5, 3.8, 2.9, 2.6, 2.4

Discrete vs. continuous numbers

2.49 For each variable, decide if it is discrete or continuous:

a. The number of cases of flu diagnosed at a college in a given semester
b. The weight, in grams, of a rat's hypothalamus
c. The number of teachers, including substitutes, in a school district who are certified to teach high school math
d. The number of times that a rat pushes a lever in a Skinner box during a 15-minute period
e. How long, in seconds, it takes applause to die down at the end of a school assembly

2.50 For each variable, decide if it is continuous or discrete:
a. The depth, in inches, a person can drive a 3-inch nail with one hammer blow
b. The number of students in a classroom who are absent for at least one day during the month of February
c. The total number of days that students in a classroom are absent during the month of February
d. How many friends one has on Facebook
e. The amount of time, measured in seconds, that a person in a driving simulator keeps his or her eyes on the road while engaging in a cell phone conversation

Real vs. apparent limits

2.51 Below are sets of consecutive numbers, or intervals, from frequency distributions for continuous variables. For the second number (or interval) in each set, the one in bold, identify the real lower limit, real upper limit, interval width, and midpoint.
a. 27–30; **31–34;** 35–38
b. 10–20; **30–40;** 50–60
c. 2.00–2.49; **2.50–2.99;** 3.00–3.49
d. 10; **11;** 12

2.52 For each scenario, tell what the real limits of the number/interval are:
a. If a person's temperature is reported as 100.3 (and not 100.2 or 100.4), what is the interval within which the person's temperature really falls?
b. If a family has 7 children, not 6 or 8, what is the interval within which the number of children they have really falls?
c. If a person's IQ is reported as 115 (and not 114 or 116), what is the interval within which the person's IQ really falls?
d. If nations' populations are reported to the nearest million and the United States has a population of 321 million, not 320 or 322, how many people really live in the United States?
e. If a man has 3 televisions in his house, not 4 or 5, what is the interval within which the number of televisions he has really falls?

Graphs and shapes of distributions

2.53 A developmental psychologist observes how children interact with their parents and classifies the children's attachment as secure, avoidant, resistant, or disorganized. He classifies 33 children as securely attached, 17 as avoidant, 13 as resistant, and 21 as disorganized. Graph this frequency distribution and comment on its shape.

2.54 A music teacher held an assembly at an elementary school during which he explained the different types of instruments to the children. Afterwards, he surveyed them as to what kind of instrument each would like to play. Here are the results: percussion = 12, brass = 39, wind = 42, and string = 28. Graph the results and comment on the shape of the graph.

2.55 A psychologist administered an aggression scale to a sample of high school girls. Scores on the scale can range from 20 to 80. Fifty is considered an average score; higher scores mean more aggression. Twelve girls had scores in the 20–29 range; 33 were 30–39; 57 were 40–49; 62 were 50–59; 19 were 60–69; and 8 had scores in the 70–79 range. Graph the results and comment on the shape of the graph.

2.56 A researcher from a search engine company runs Internet searches and times how long each one takes. She found that 17 searches took 0.01 to 0.05 seconds; 57 took 0.06 to 0.10 seconds; 134 took 0.11 to 0.15 seconds; 146 took 0.16 to 0.20 seconds; 398 took 0.21 to 0.25 seconds; 82 took 0.26 to 0.30 seconds;

and 56 took 0.31 to 0.35 seconds. Graph the frequency distribution and comment on its shape.

Making stem-and-leaf displays

2.57 Make a stem-and-leaf display for these data and comment on the shape of the distribution: 11, 12, 21, 23, 27, 27, 29, 30, 30, 33, 34, 35, 39, 43, 45, 47, 53, 53, 67, 75, 84, and 96.

2.58 Make a stem-and-leaf display for these data and comment on the shape of the distribution: 8, 9, 13, 13, 16, 21, 22, 24, 25, 25, 29, 33, 33, 45, 48, 55, 58, 61, 63, 64, 66, 66, 69, 71, 83, 92, 93, and 95.

Expand Your Knowledge

2.59 If the real limits and apparent limits for an interval are the same, then the variable is:

a. continuous.
b. discrete.
c. interval.
d. ratio.
e. Real and apparent limits are never the same.

2.60 For a row in the middle of a frequency distribution, either ungrouped or grouped, which statement is probably true?

a. $f > f_c$
b. $f_c > f$
c. $f = f_c$
d. None of these statements is ever true.

2.61 If a researcher has a grouped frequency distribution for a continuous variable and there are five people in the interval ranging from 50 to 54, what *X* values should he assign them if he wants to assign values?

a. He should randomly select values from 50 to 54.
b. He should randomly select values from 49.5 to 54.5.
c. As there are five people and there are five integers from 50 through 54, he should assign them values of 50, 51, 52, 53, and 54.
d. 50
e. 52
f. 52.5
g. 54

2.62 The distribution of prices of all the new cars sold in a year, including Ferraris, Lamborghinis, and other high-end cars, is probably:

a. normally distributed.
b. positively skewed.
c. negatively skewed.
d. not skewed.

2.63 Diane wears a digital watch that reports the time to the nearest minute. At 10:16 A.M., by her watch, she left her office. She returned at 10:21 A.M. To the nearest minute, what is the shortest amount of time that she was out of her office? What is the longest?

2.64 Make a stem-and-leaf display for these data. Use an interval width of 5.
120, 122, 123, 123, 124, 125, 126, 125, 126, 130, 134, 135, 135, 138, 147

SPSS

SPSS can be used to generate frequency distribution tables and graphs like those created in this chapter. The "Frequencies" command is used to generate an ungrouped frequency distribution. Figure 2.20 shows how to access the Frequencies command by clicking on "Analyze" and then "Descriptive Statistics."

Once the menu shown in Figure 2.21 has been opened, use the arrow key to shift the variable one wants to use from the box on the left to the "Variable(s)" box. Here, the variable name is "Children."

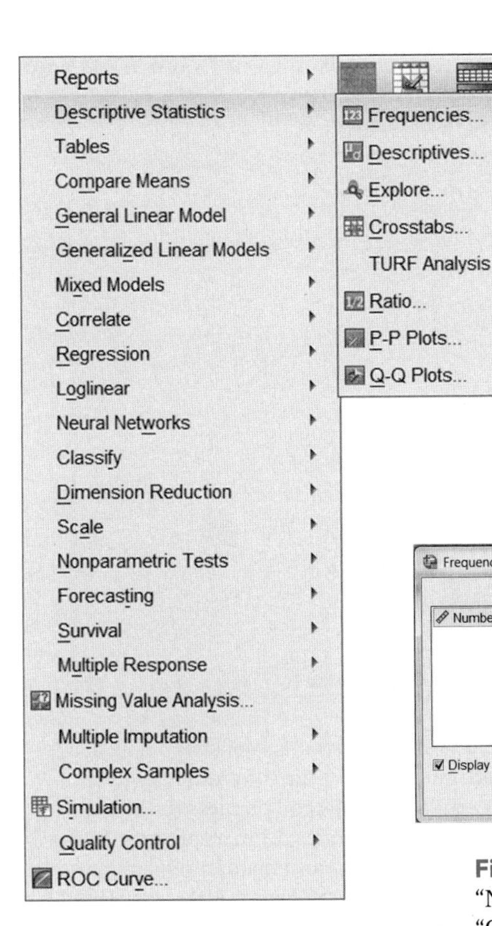

Figure 2.20 Starting the Frequencies Command The "Frequencies" command is accessed under "Analyze" and "Descriptive Statistics." (Source: SPSS)

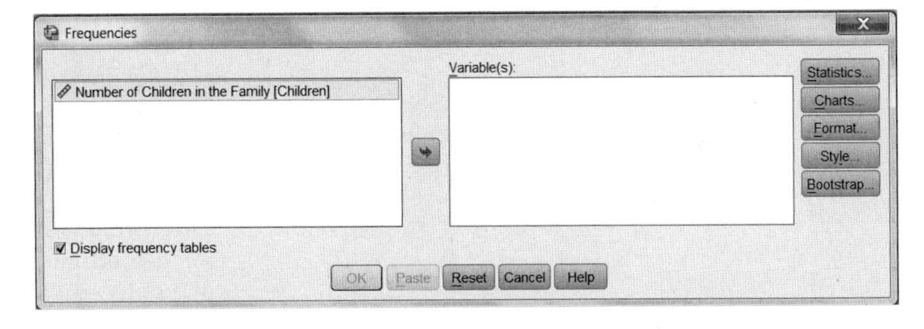

Figure 2.21 The Frequencies Command Menu Once a variable like "Number of children" has been moved into the "Variable(s)" box, the "OK" button on the lower right becomes active. (Source: SPSS)

Once there's a variable in the Variable(s) box, click on "OK" to generate an ungrouped frequency distribution. The Frequencies output can be seen in Figure 2.22.

Making a grouped frequency distribution in SPSS is more involved. One has to use "Transform" commands like "Recode" or "Visual binning" to place the data into intervals before using Frequencies.

Graphing is easy with SPSS. To start making a graph, click on "Graphs" as shown in Figure 2.23.

Number of Children in the Family

		Frequency	Percent	Valid Percent	Cumulative Percent
Valid	1.00	9	29.0	29.0	29.0
	2.00	14	45.2	45.2	74.2
	3.00	5	16.1	16.1	90.3
	4.00	2	6.5	6.5	96.8
	6.00	1	3.2	3.2	100.0
	Total	31	100.0	100.0	

Figure 2.22 Example of Frequencies Output from SPSS Compare this SPSS output to Table 2.3, for the same data. Note that Table 2.3 was arranged upside down, compared to SPSS. Table 2.3 also contained additional information—cumulative frequency. Finally, Table 2.3 included values, such as five children in the family, for which the frequency was zero. (Source: SPSS)

Clicking on "Chart Builder. . ." opens up the screen shown in Figure 2.24. Note that the list of graphs SPSS can make may be seen on the bottom left.

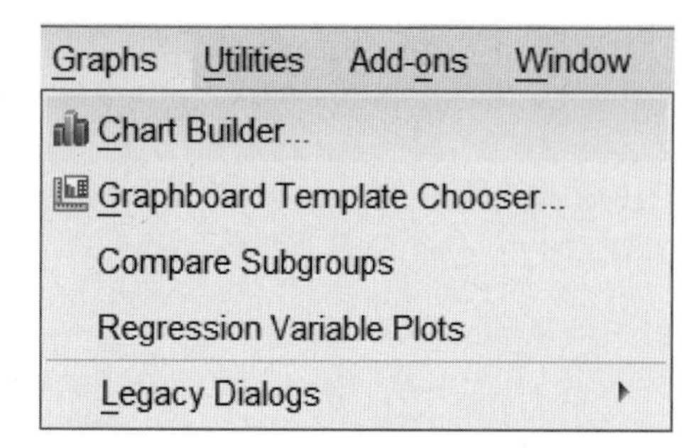

Figure 2.23 Starting the Graph Command in SPSS To start a graph, click on "Graphs" and then on "Chart Builder. . . ." (Source: SPSS)

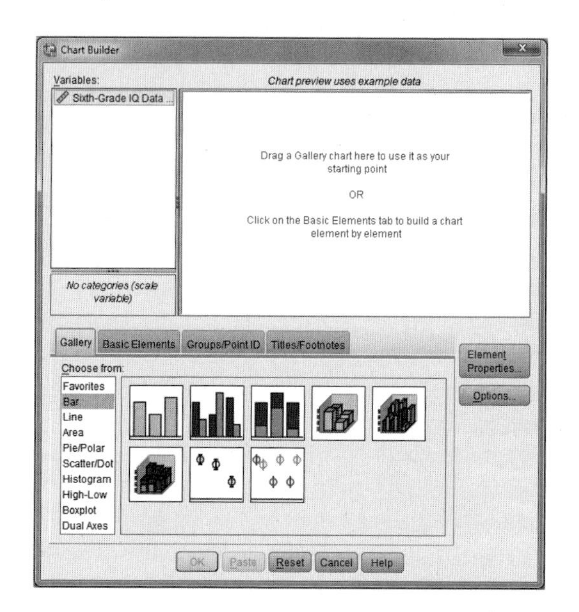

Figure 2.24 Chart Builder Screenshot The list on the bottom left shows the different categories of graphs available, while the pictures to the right show the different options within a category. In the list, there are bar graphs and histograms; look under "Line" to find frequency polygons. (Source: SPSS)

Figure 2.25 shows how to make a histogram. Histogram was highlighted in the list of graphs and the mouse was used to drag the basic histogram from the gallery of histograms to the preview pane.

To finish the histogram commands, see Figure 2.26. Note that the variable for which the frequencies are being graphed, IQ, has been highlighted and dragged to the *X*-axis.

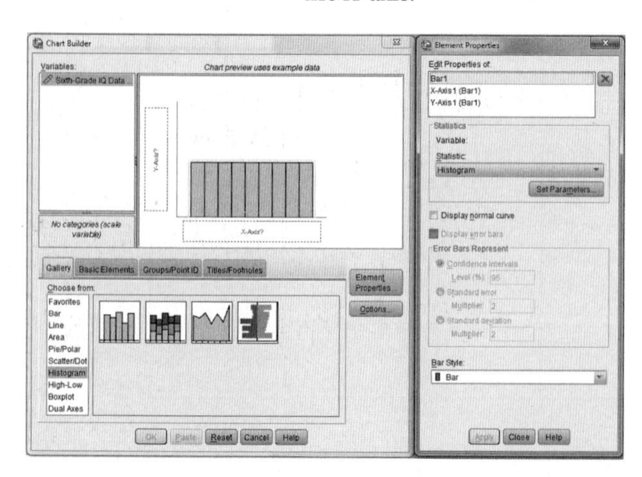

Figure 2.25 SPSS Commands for Making a Histogram Histogram was highlighted in the gallery. Then the basic histogram was clicked and dragged into the preview pane. (Source: SPSS)

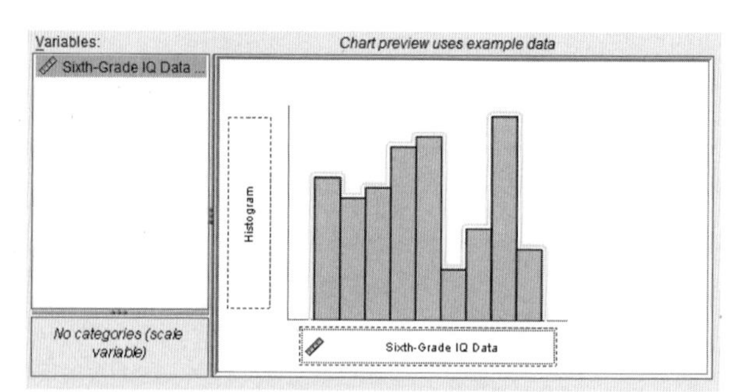

Figure 2.26 Finishing the Histogram Commands As the final step, highlight the variable for which the histogram is being completed (here, the IQ score) and drag it to the *X*-axis. (Source: SPSS)

Once the "OK" button has been clicked, SPSS generates a histogram as seen in Figure 2.27. This is called the "default" version because SPSS decides how wide the intervals should be.

A major advantage of using SPSS to make a graph over making it by hand is that it is easy to edit a graph in SPSS. Double-click on the graph to open up the editor. With just a few clicks of the mouse, one can change the interval width, the scale on the *Y*-axis, the labels on the axes, or the size of the graph. Figure 2.28 shows an edited version of a histogram for the IQ data.

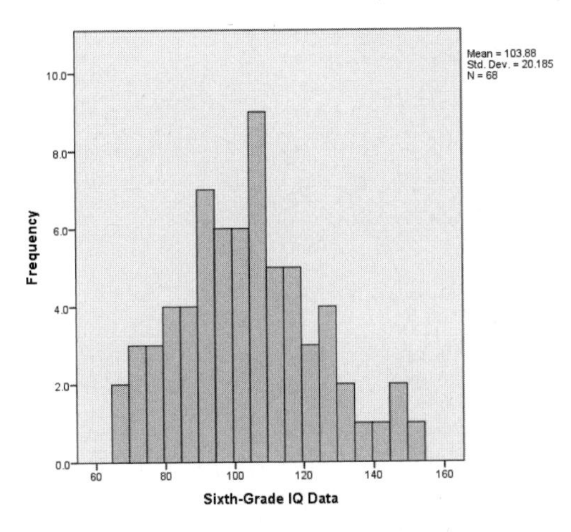

Figure 2.27 Histogram Generated by SPSS This is the histogram generated by SPSS when the "OK" button was pressed. (Source: SPSS)

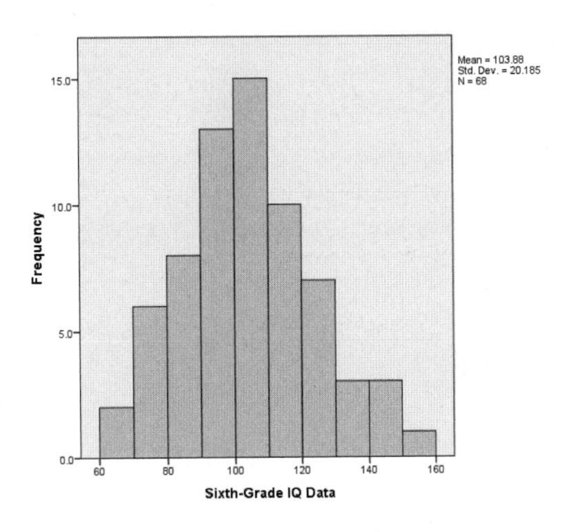

Figure 2.28 Edited Version of Histogram Editing graphs is easy in SPSS. Compare this version to the default graph shown in Figure 2.27. (Source: SPSS)

Stem-and-leaf plots can also be produced easily using SPSS. From the Analyze menu, choose "Descriptive Statistics" and then the "Explore" command as shown in Figure 2.29.

As shown in Figure 2.30, move the variable you want to analyze into the "Dependent List" by highlighting the variable and the clicking the arrow. Next,

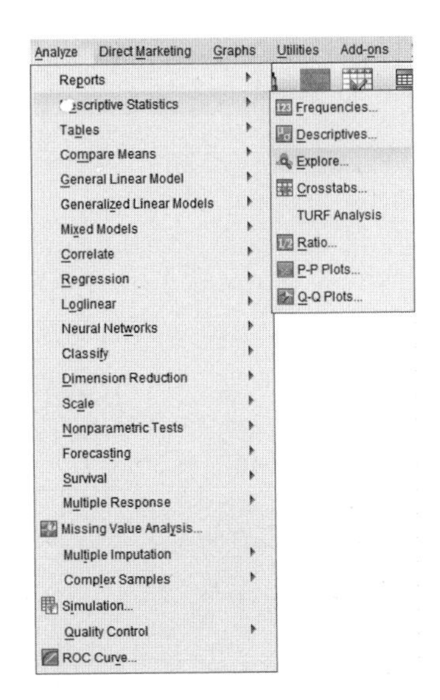

Figure 2.29 Starting Stem-and-Leaf Plot Commands Stem-and-leaf plots are accessed through the "Explore" command. (Source: SPSS)

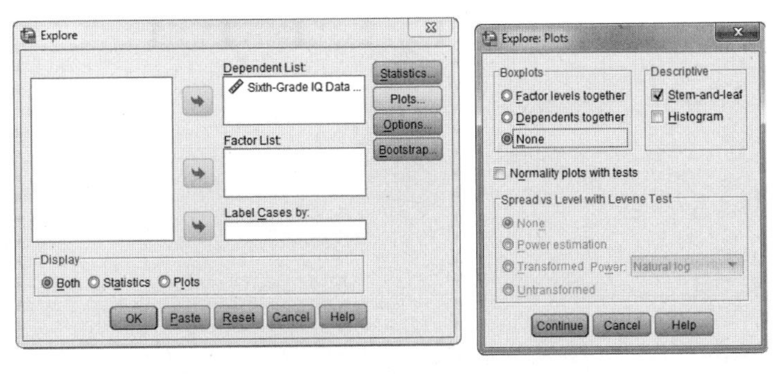

Figure 2.30 Commands for Making a Stem-and-Leaf Plot The variable to be plotted has been moved to "Dependent List" and "Stem-and-leaf" has been chosen. (Source: SPSS)

```
Sixth-Grade IQ Data Stem-and-Leaf Plot

 Frequency    Stem &  Leaf

     2.00        6 .  78
     6.00        7 .  234569
     8.00        8 .  00245589
    13.00        9 .  0002344557788
    15.00       10 .  011223555678999
    10.00       11 .  1111256899
     7.00       12 .  1226889
     3.00       13 .  446
     3.00       14 .  278
     1.00       15 .  2

 Stem width:        10
 Each leaf:        1 case(s)
```

Figure 2.31 SPSS Output of a Stem-and-Leaf Display SPSS provides an output that includes the stem and leaf values as well as the frequencies. (Source: SPSS)

click "Plots" and make sure that under the Descriptive option "Stem-and-leaf" is checked. This is the default. You can change the default option under "Boxplots" to "None" because we are not running a boxplot.

After clicking "Continue" and "OK," the output in Figure 2.31 will be produced.

Measures of Central Tendency and Variability

LEARNING OBJECTIVES

- Define and know when to calculate three measures of central tendency.
- Define variability and know how to calculate four different measures of it.

CHAPTER OVERVIEW

The last chapter explored how to summarize a set of data using a frequency distribution and/or a graph. The current chapter uses descriptive statistics to summarize a whole set of data with just one or two numbers.

The numbers summarize different aspects of a set of scores. One of the numbers describes what statisticians call **central tendency**, a single value used to represent the typical score in a set of scores. Another number used to summarize a set of data, a measure of **variability,** summarizes how much variety exists in a set of scores.

Both measurements—central tendency and variability—are important. Imagine two students with GPAs of 3.00, as seen in Table 3.1. A measure of central tendency, like GPA, gives an overall summary of how well the students are doing academically. Here, both have the same GPA, so the two students look alike—Darren and Marcie are doing equally well in school. Variability, however, provides a different perspective on the students. Marcie achieved her 3.00 average by getting B's in every course. Darren achieved his 3.00 GPA by getting A's in half of his classes and C's in the other half. Both students have the same grade point average, a 3.00, but they differ in variability. Here, variability provides important information. Variability reveals that Marcie is more consistent in her academic work than Darren.

3.1 Central Tendency

Central tendency tells the typical or average score in a set. Statisticians use central tendency as a summary value for a set of scores, much the way the midpoint was used for intervals in a grouped frequency distribution. Usually central tendency is the value at the center of a set of scores. Mean, median, and mode are the three most common measures of central tendency.

TABLE 3.1 Two Students with the Same Central Tendency and Different Variability

	Marcie	Darren
English	3.00	4.00
Math	3.00	2.00
Spanish	3.00	4.00
Art	3.00	2.00
Speech	3.00	4.00
Political Science	3.00	2.00
History	3.00	4.00
Psychology	3.00	2.00
GPA	**3.00**	**3.00**

Two students with the same GPA achieved it with very different patterns of scores. Darren shows more variability in performance than Marcie.

Mean

The **mean** is what most people think of when they contemplate the average. It is the sum of all the values in a set of data divided by the number of cases. The formula for *M*, the sample mean, is shown in Equation 3.1.

Equation 3.1 Formula for Sample Mean (*M*)

$$M = \frac{\Sigma X}{N}$$

where M = the mean of a sample

Σ = summation sign

X = the values of X for the cases in the sample

N = the number of cases in the sample

For practice in using Equation 3.1, imagine that a demographer has selected five adults from the United States and measured their heights in inches. Here are the data she collected: 62, 65, 66, 69, and 73. Applying Equation 3.1, she would calculate the sample mean:

$$\begin{aligned} M &= \frac{\Sigma X}{N} \\ &= \frac{62 + 65 + 66 + 69 + 73}{5} \\ &= \frac{335.0000}{5} \\ &= 67.0000 \\ &= 67.00 \end{aligned}$$

The demographer would report, "The mean height for the sample of five Americans is 67.00″ or 5′ 7″."

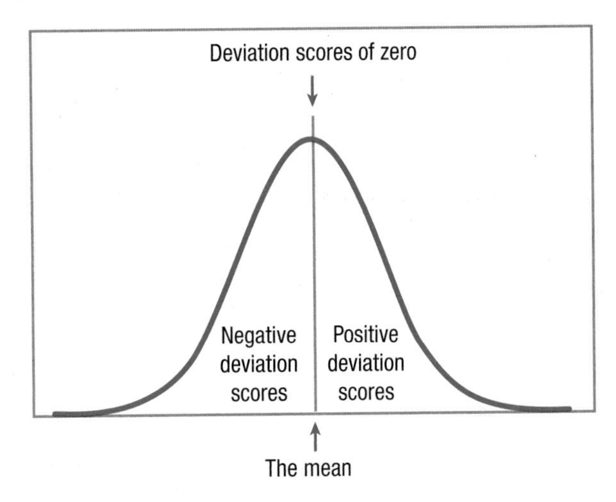

Figure 3.1 Deviation Scores Deviation scores are calculated by subtracting a raw score from the mean. Raw scores above the mean have positive deviation scores and those below the mean have negative deviation scores. Scores exactly at the mean have deviation scores of zero. The farther the raw score is from the mean, the bigger the deviation score.

An additional way that a mean can be used is to find out how much a single score, *X*, deviates from it. This is called a **deviation score**. A deviation score is calculated by subtracting the mean from a score:

$$\text{Deviation score} = X - M$$

Deviation scores in a normal distribution are shown in Figure 3.1. No matter the shape of a distribution of scores, a positive deviation score means that the score is above the mean, a negative deviation score means that the score is below the mean, and a deviation score of zero means that the score is right at the mean. Table 3.2 shows the deviation scores for the heights of the five Americans.

Note, in Table 3.2, that the sum of the deviation scores equals zero. This is always true: the sum of a set of deviation scores is zero. The mean is the central score in a set of scores because it balances the negative deviation scores on one side of it with the positive deviation scores on the other side of it. This is shown in Figure 3.2, where the mean is the fulcrum of a seesaw balancing the five height deviation scores. Deviation

TABLE 3.2 Deviation Scores for Heights of Five Americans

Height in Inches	Deviation Score ($X - M$)
62	−5.00
65	−2.00
66	−1.00
69	2.00
73	6.00
	$\Sigma = 0.00$

These are the deviation scores, calculated as the raw score minus the mean of 67.00 for the height data. For example, the deviation score for the person 62″ tall is 62 − 67.00 = −5.00. When all the deviation scores for a set of scores are added together, they sum to zero.

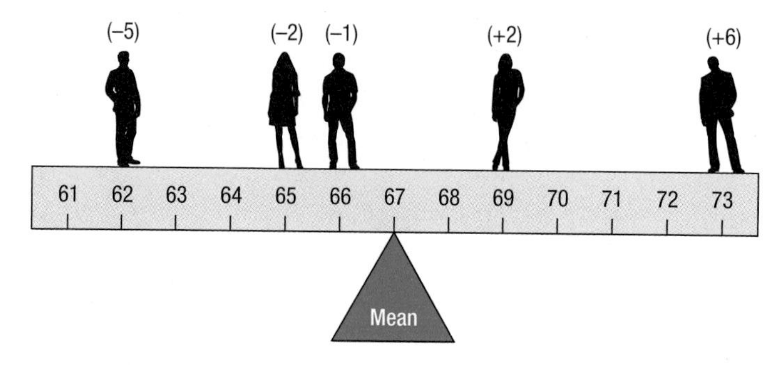

Figure 3.2 Mean as Balancing Point The numbers in parentheses are deviation scores. The mean is the balancing point for the deviation scores, balancing a sum of deviation scores of −8.00 on one side with a total of +8.00 on the other side. Because the mean takes distance between cases into account, it can only be used with interval- or ratio-level numbers.

scores rely on information about distance between scores, so the mean can only be used with interval- or ratio-level data.

One problem with the mean is that it can be influenced by an **outlier**, an extreme score that falls far away from the rest of the scores in a data set. Robert Wadlow, the world's tallest man, is an example of an outlier. Wadlow was born in 1918 and had an overactive pituitary gland that caused him to grow to 8′ 11″. In Figure 3.3, you can see Robert Wadlow standing next to two average height women. In terms of height, Robert Wadlow is an outlier.

Imagine Robert Wadlow, at 107″, being added to the demographer's sample of five Americans, making it a sample of six. The mean, which had been 67.00 when there were only five cases, would now be

$$\begin{aligned} M &= \frac{\Sigma X}{N} \\ &= \frac{62 + 65 + 66 + 69 + 73 + 107}{6} \\ &= \frac{442.0000}{6} \\ &= 73.6667 \\ &= 73.67 \end{aligned}$$

Adding one outlier, Robert Wadlow, causes the average height to jump from 5′ 7″ to about 6′ 2″, a jump of almost 7″. That's a big impact for a single case to have on the mean. Being influenced by outliers is a problem for the mean. The next measure of central tendency, the median, is not as influenced by outliers.

Figure 3.3 Robert Wadlow, an Outlier Robert Wadlow at 8′ 11″ towers over two women of average height. Because of his height, Wadlow is an outlier, an extreme score that falls far away from the rest of the scores in a data set. (NYPL/Science Source/Getty Images)

Median

The **median** is the middle score, the score associated with the case that separates the top half of the scores from the bottom half. The median focuses on direction (more/less) and ignores information about the distance between scores. Because of this, the median can be used with ordinal data (unlike the mean). The abbreviation for median is *Mdn*.

The easiest way to calculate the median is to use the counting method, shown in Equation 3.2. By this equation, the median is the score associated with case number $\frac{N+1}{2}$.

Equation 3.2 Formula for the Median (*Mdn*)

Step 1 Put the scores in order from low to high and number them (1, 2, 3, etc.).

Step 2 Find the X value associated with the score number $\frac{N+1}{2}$

where X = raw score

N = number of cases in the data set

Here's how to calculate the median for the original height data set, the one with only five cases:

- In Table 3.3, the five heights are listed in order and the numbers 1–5 are assigned to them.
- According to Equation 3.2, the median is the raw score associated with the score number $\frac{N+1}{2}$.
- Calculate $\frac{5+1}{2} = \frac{6}{2} = 3$.
- The median is the score associated with the third case.
- Looking in Table 3.3, one can see that the third case has an X value of 66.
- So, the median is 66.00. One could say, "The median height for these five Americans is 5′ 6″."

TABLE 3.3 Calculating the Median for the Height Data, N = 5

Score Number	Height in Inches
1	62
2	65
3	**66**
4	69
5	73

The median, 66, is bolded. It is calculated by finding the value associated with the score number $\frac{N+1}{2}$. In this case, where $N = 5$, that is score number 3, which appears in bold.

Figure 3.4 shows how the median is also the central score in a set of scores, but in a different way than the mean is. Just as many cases fall below the median as fall above it. Notice how the distance between cases is ignored with the median.

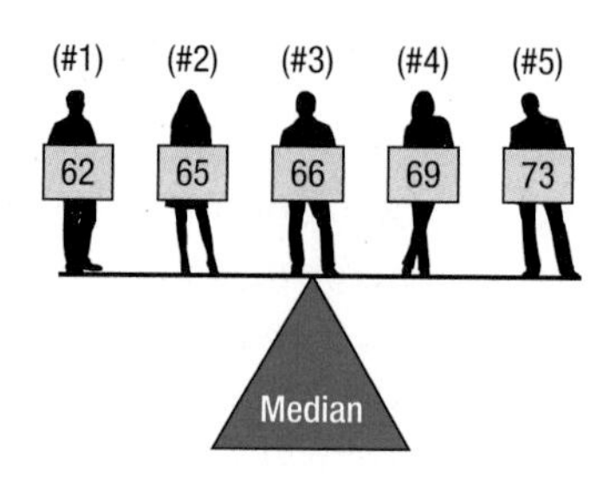

Figure 3.4 Median as Balancing Point The numbers in parentheses above the scores are their ranks. The median is the balancing point, the center, for the *number* of scores. The mean (Figure 3.2) is the balancing point for the *distance* between scores. The distance between cases is irrelevant for the median.

One advantage of the median over the mean is that it can be used with ordinal-level data. Another is that the median is less influenced by outliers. Let's see what happens to 66.00, the median for the five scores, when the outlier, 107″ Robert Wadlow, is added. According to Equation 3.2, now that there are six scores, the median case is score number 3.5 and the median is the value associated with this case:

$$\frac{N+1}{2} = \frac{6+1}{2} = 3.5$$

Table 3.4 shows the six cases and there is no case numbered 3.5. There's a case number 3, with a height of 66″, and a case number 4, with a height of 69″. Statisticians are happy to have fractional cases, so case number 3.5 is halfway between case 3 and case 4. Similarly, the X value associated with case 3.5 is halfway between the raw scores associated with those two cases, 66 and 69. Another way of saying this is that the median in such a case is the mean of the two values:

$$\frac{66+69}{2} = 67.50$$

The median for the six cases is 67.50.

Before the outlier was added, the median was 66.00. Adding the outlier changed the median, but doing so only moved it 1.5″ higher. Compare this to the almost 7″ that the same outlier added to the mean. Medians are less affected by outliers than are means because they don't take distance information into account. This is an advantage for the median.

TABLE 3.4 Calculating the Median for the Height Data with Outlier Added

Score Number	Height in Inches
1	62
2	65
3	66
•	***X*** (67.50)
4	69
5	73
6	107

The dot in the first column indicates where case number 3.5 falls. The X in the second column indicates the raw score, 67.50, that is associated with this case.

Mode

The third measure of central tendency is the **mode**, the score that occurs with the greatest frequency. For the height data, there is no mode. Each value occurs with the same frequency, once, and there is no score that is the most common. Looking back at Table 2.15, the data for the sex of students in a psychology class, the most common value is female, so that is the mode. This points out an advantage of the mode: the mode can be used for nominal data (unlike the mean or the median).

Choosing a Measure of Central Tendency

It is important to know how to select the correct measure of central tendency to represent a given set of data. Table 3.5 shows which measure of central tendency can be used with which level of measurement:

- If one has nominal data, there is only one option for a measure of central tendency, the mode.
- With ordinal data, there are two options for a measure of central tendency, the mode or the median.
- If one has interval- or ratio-level data, there are three options: mode, median, or mean.

When there are multiple options for a measure of central tendency, choose the measure of central tendency that takes into account the most information:

- The mode only takes same/different information into account.
- The median takes same/different information into account, as well as direction information.
- The mean takes same/different, direction, and distance information into account.

Remember: Choose the measure of central tendency that conveys the most information. So, with interval- or ratio-level data, the "go to" option is the mean. However, there are conditions where it is better to fall back to a different measure of central tendency for interval- or ratio-level data.

Choose the measure of central tendency that conveys the most information.

To decide which measure of central tendency to choose for interval or ratio data, one should make a frequency distribution and think about the shape of the data. If

TABLE 3.5 How to Choose: Which Measure of Central Tendency for Which Level of Measurement

	Measure of Central Tendency		
Level of Measurement	Mode	Median	Mean
Nominal	✓		
Ordinal	✓	✓	
Interval or ratio	✓	✓	✓

Not all measures of central tendency can be used with all levels of measurement. When more than one measure of central tendency may be used, choose the measure that uses more of the information available in the numbers. If planning to calculate a mean, be sure to check the shape of the data set for skewness and modality.

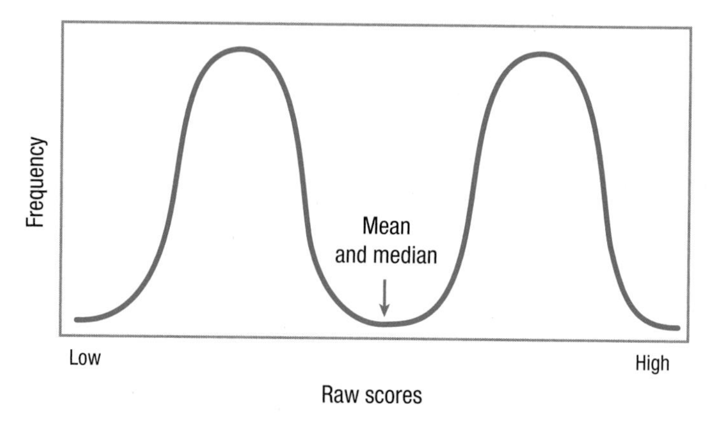

Figure 3.5 **Central Tendency for a Bimodal Distribution** It is not appropriate to calculate a mean or a median for bimodal data. In this case, central tendency should be reported as two modes.

there are outliers or if the data set is skewed, the mean will be pulled in the direction of the outliers or the skew and won't be an accurate reflection of central tendency. In these situations, the median is a better option. If the data set is bimodal or multimodal, a mean or a median may not be appropriate. In the bimodal data in Figure 3.5, the mean and median fall in a no-man's land where there are few cases. Does a score in this region typify the data set? No. In this situation, reporting the multiple modes makes more sense.

A Common Question

Q What does it mean if the mean and the median for a set of data are very different from each other?

A It could mean that the data set is skewed. When the mean is bigger than the median, that suggests positive skew. When the mean is smaller than the median, that suggests negative skew.

Worked Example 3.1

Here's a small data set for practice calculating central tendency. Suppose a psychologist decided to base some research on the famous "Bobo doll" experiment (Bandura, Ross, & Ross, 1961). In that study, children saw an adult interact with a Bobo doll, an inflated doll with a weighted base that bounces back up when it is punched. Some of the kids saw the adult behave aggressively toward Bobo and other kids didn't. The kids who saw aggression modeled later behaved more aggressively toward Bobo.

Our researcher, Dr. Gorham, put together a sample of 10 third graders and had each of them watch an adult *physically* attack Bobo. Each child was then left alone in a room with Bobo and other toys for 5 minutes. Dr. Gorham observed and counted how many times each child *verbally* insulted Bobo. Each time a child used a negative word to address Bobo, for instance, called him "stupid" or "ugly," was counted. Here are the totals for number of aggressive comments: 2, 9, 3, 4, 6, 7, 5, 5, 4, and 4. What value should Dr. Gorham use to represent the average number of aggressive comments for this sample?

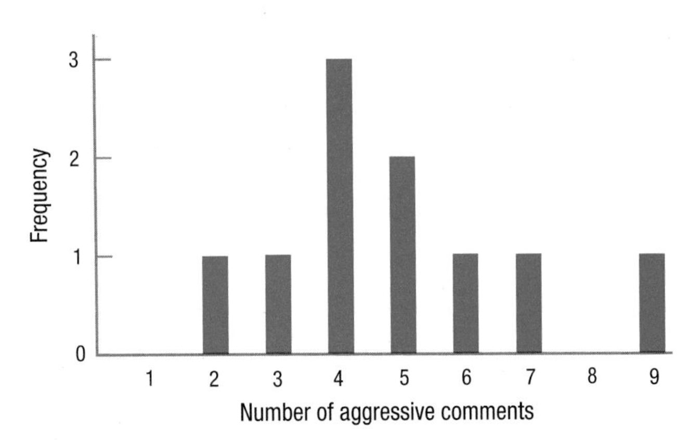

Figure 3.6 Number of Aggressive Comments Made Against the Bobo Doll by 10 Third Graders When sample size is small, as it is here, it is difficult to draw conclusions about the shape of a data set. This data set isn't multimodal and doesn't look skewed, so calculating the mean seems OK.

The first thing to do is figure out which measure of central tendency is most appropriate. The data, number of aggressive comments, are measured at the ratio level, so Dr. Gorham could use mean, median, or mode (see Table 3.5). His initial plan should be to use the mean, which utilizes all the information in the numbers. To make sure the mean will work, he needs to check the shape of data and confirm that it is not skewed or multimodal.

The distribution, shown in Figure 3.6, doesn't look unusual in terms of modality or skewness, so the mean still seems appropriate. However, to compare all three, Dr. Gorham calculates all three measures of central tendency.

He calculates the mode first, because it is the easiest. Looking at Figure 3.6, he sees one peak, at 4, so he can report that the modal number of aggressive comments is 4.00 in his sample.

Next, he calculates the mean using Equation 3.1:

$$
\begin{aligned}
M &= \frac{2 + 9 + 3 + 4 + 6 + 7 + 5 + 5 + 4 + 4}{10} \\
&= \frac{49.0000}{10} \\
&= 4.9000 \\
&= 4.90
\end{aligned}
$$

The mean number of aggressive comments is 4.90.

To calculate the median, following Equation 3.2, he arranges the scores in order and numbers them as shown in Table 3.6. $N = 10$, so the median is the value associated with score number $\frac{10 + 1}{2}$, the 5.5th score.

The 5.5th score is the score between the 5th score, which has a value of 4, and the 6th score, which has a value of 5. The mean of those two values, the halfway point between them, is calculated as $\frac{4 + 5}{2} = 4.50$. Dr. Gorham would report $Mdn = 4.50$.

TABLE 3.6 Calculating the Median Number of Aggressive Comments Made Against the Bobo Doll

Score Number	Number of Aggressive Comments
1	2
2	3
3	4
4	4
5	4
6	5
7	5
8	6
9	7
10	9

With 10 scores, the median is the score for case number 5.5. This falls halfway between case 5 and case 6.

He calculated the mean (4.90), the median (4.50), and the mode (4.00). The mean is a little above the median, which suggests there is some positive skew in the data set. Now, primed to see it, Figure 3.6 does look like it has a slightly positive skew. Because of this, Dr. Gorham could decide that the median is the best measure of central tendency to report. However, the discrepancy between the mean and the median is not very large, so it still makes sense to use the mean. The mean uses more of the information in the numbers, so this is the measure of central tendency that Dr. Gorham decides to report. Here's what he said, "The children in the sample made a mean of 4.90 aggressive comments toward a Bobo doll after having seen an adult commit physical aggression against it."

Practice Problems 3.1

Review Your Knowledge

3.01 List the three measures of central tendency.

3.02 Which measure of central tendency can be used (a) for nominal-level data? (b) For ordinal-level data? (c) For interval- or ratio-level data?

3.03 What impact does a skewed data set have on the mean?

Apply Your Knowledge

For Practice Problems 3.04–3.06, calculate the appropriate measure of central tendency for each sample.

3.04 A school psychologist measured the IQs of seven children as 94, 100, 110, 112, 100, 98, and 100.

3.05 A neuroscientist measured how often a sample of neurons fired during a specified period of time. The 10 values were 1, 11, 10, 10, 2, 2, 3, 10, 9, and 2.

3.06 A sensory psychologist measured whether taste buds were primarily sensitive to (1) bitter, (2) salty, (3) savory, (4) sour, or (5) sweet. Here are the data for her sample of taste buds: 1, 2, 3, 4, 5, 5, 5, 3, 2, 3, 4, 5, 1, 5, 4, and 5.

3.2 Variability

Statisticians define variability as how much spread there is in a set of scores. Spread refers to how much the scores cluster together or how much they stretch out over a wider range. For an example of this, see Figure 3.7. The curve on the left shows a more tightly clustered set of scores, and the curve on the right shows a data set with more variability in scores.

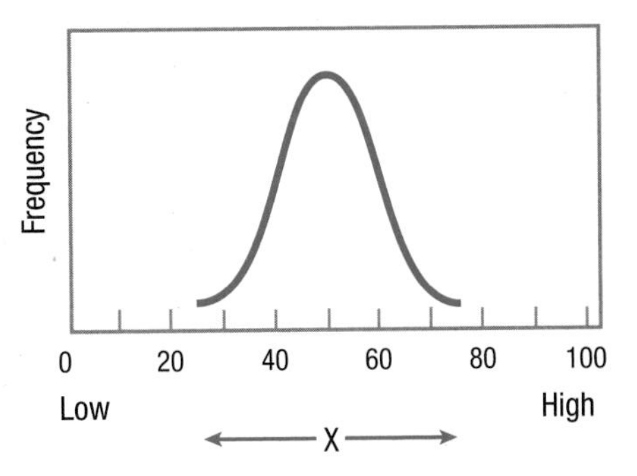

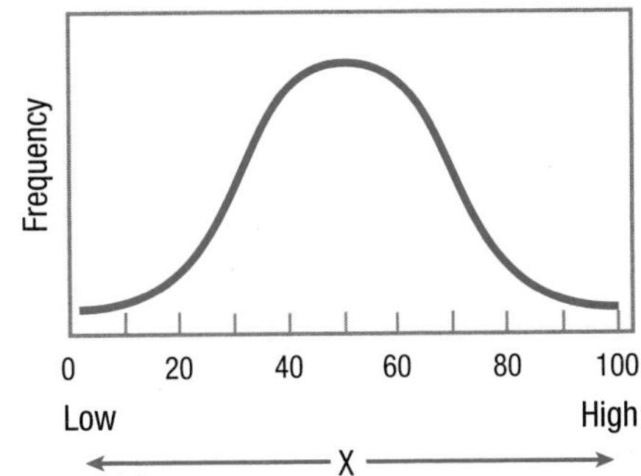

Figure 3.7 Differing Amounts of Variability The range of scores in the set of scores on the right is greater than for the set on the left. The set on the right has more variability.

Both measurements—central tendency and variability—are important in describing a set of scores. The need for both measurements was made clear at the start of the chapter with Darren and Marcie, the two students with the same GPA. There are two facts about them—(1) their GPAs were 3.00, and (2) they achieved this either by getting all B's or by a mixture of A's and C's. The first statement gives information about central tendency, and the second gives information about variability. Both statements offer useful information.

Statisticians have developed a number of ways to summarize variability. The remainder of this chapter will be used to explore four measures of variability: range, interquartile range, variance, and standard deviation. These four measures of variability all make use of the distance between scores, so they are appropriate for use with interval- or ratio-level data, but not with nominal or ordinal data.

Range and Interquartile Range

The simplest measure of variability is the **range**, the distance from the lowest score to the highest score. In Figure 3.7, notice how the range of scores, marked by a double-headed arrow below the *X*-axis, is greater for the set of scores on the right. The formula for the range, which says to subtract the smallest score from the largest score, appears in Equation 3.3.

Equation 3.3 Formula for Range

$$\text{Range} = X_{\text{High}} - X_{\text{Low}}$$

where Range = distance from lowest score to highest score

X_{Low} = the smallest value of X in the data set

X_{High} = the largest value of X in the data set

For the height data, here is the calculation for range:

$$\text{Range} = 73 - 62 = 11.00$$

The range is a single number, so the range for the height data would be reported as 11.00. This makes it easy to compare amounts of variability from one set of data to another. If the value for the range is greater for data set A than it is for data set B, then data set A has more variability.

There is a problem with using the range as a measure of variability. The range depends only on two scores in the data set, the top score and the bottom score, so most of the data are ignored. This means that the range is influenced by extreme scores even more than the mean. If all 107″ of Robert Wadlow joined the sample, the range would jump from 11.00″ to 45.00″. That's a 34-inch change in the range as the result of adding just one case.

The influence of outliers can be reduced by removing extreme scores before calculating the range. The only question is how many scores to trim. The most common solution is to trim the bottom 25% of scores and the top 25% of scores. Then, the range is calculated for the middle 50% of the scores. This is called the **interquartile range** (abbreviated *IQR*).

The interquartile range is called the inter*quartile* range because data sets can be divided into four equal chunks, called quartiles. Each quartile contains 25% of the cases. The interquartile range represents the distance covered by the middle two quartiles, as shown in Figure 3.8.

As with the range, the interquartile range is a single number. However, range is often reported as two scores—the lowest value and the highest value in the data set. Following a similar format—reporting the interquartile range as two scores—makes it a statistic that does double duty. The two scores provide information about *both* variability *and* central tendency. Knowing how wide the interval is indicates how much variability is present—the wider the interval, the larger the amount of variability. Knowing the values of the interval tells us where the average scores fall. To illustrate this point, let's look at an example that will be familiar to most students.

In the search for colleges, students often encounter the interquartile range without realizing it. Many colleges use the interquartile range in reporting SAT or ACT scores for their students, often calling it something like "25th–75th percentile" or "middle 50%."

When the interquartile is reported as a single number, it gives useful comparative information about variability. In a list of the best research universities, the interquartile ranges for the combined SAT scores of students at the schools were

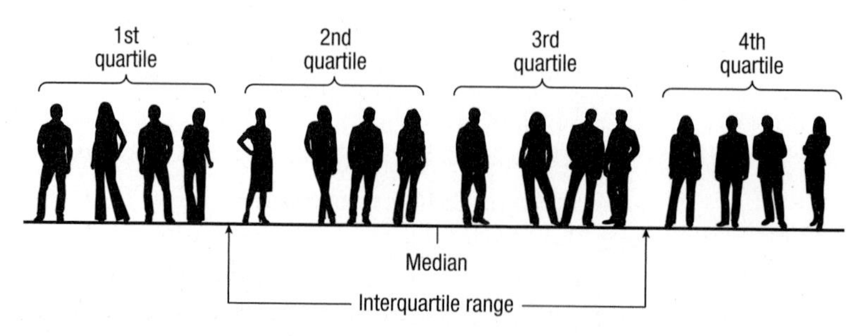

Figure 3.8 Four Quartiles of a Data Set Data sets can be divided into quartiles, four equal sections, each with 25% of the cases. The interquartile range is the distance covered by the two middle quartiles.

reported (www.satscores.us). Schools at the top of the list had interquartile ranges about 280 SAT points wide and those at the bottom of the list had interquartile ranges about 360 SAT points wide.* This indicates that more variability in SAT performance exists among the students at the lower-ranked schools than among students at the upper-ranked schools. Professors at the lower-ranked schools can expect a wider range of skills in their classrooms than would be found at an upper-ranked school.

When the interquartile range is reported as two scores, it functions as a measure of central tendency because it reveals the interval that captures the average students—the middle 50%. For example, the upper-ranked schools have an interquartile range for combined SAT scores that ranges from 2,100 to 2,380, while the lower-ranked schools' interquartile range goes from 1,170 to 1,530. One can see that the average students at the top schools did almost 1,000 points better in terms of combined SAT scores than the average students at the lower-ranked schools. The interquartile range is a helpful descriptive statistic because it provides information about variability *and* central tendency.

Variability in Populations

So far, variability has only been calculated for samples of cases. Now, let's consider variability in populations, the larger groups from which samples are drawn. For example, the demographer's sample of five adult Americans is a subset of the more than 200 million adult Americans. Just as there is variability for height in the sample of five adults, there is variability for height in the population. In fact, more variability should occur in the population than in the sample. The shortest person in the sample is 5′ 2″ and the tallest is 6′ 1″. Aren't there American adults who are shorter than 5′ 2″ and who are taller than 6′ 1″? Of course. Populations almost always have more variability than samples do.

Variability exists in populations, but there is a practical problem in measuring it. When the population is large, one can't measure everyone. If it took just a minute to measure a person's height, it would take more than 300 years—working 24/7—to measure the heights of all 200 million American adults. That would be impossible. Except in rare situations, variability can't be calculated for a population. Nonetheless, as we'll see in the next section, the *idea* that there is a population value for variability is an important one.

Population Variance

Remember deviation scores from earlier in this chapter? Deviation scores are used to calculate variance and standard deviation. The **variance** is the mean squared deviation score and a **standard deviation** is the square root of the variance. In essence, a standard deviation tells the average distance by which raw scores deviate from the mean.

The variance and the standard deviation use information from all the cases in the data set. As a result, they are better measures of variability in the whole data set.

The variance and standard deviation have a big advantage over the range and interquartile range. The range and interquartile range use limited information from a data set. For example, the range only uses the case with the highest score and the case with the lowest score, ignoring variability information from all the other cases. In contrast, the variance and the standard deviation use information from all the cases in the data set. As a result, they are better measures of variability in the whole data set.

* Note: These figures are based on the three-section SAT in use prior to 2016. Beginning March 2016, the SAT includes only two sections, Reading/Writing and Math.

Deviation scores represent the amount of distance a score falls from the mean. With data from a whole population, deviation scores would be calculated by subtracting the population mean, μ, from the raw scores:

$$\text{Population deviation score} = X - \mu$$

To understand how deviation scores measure variability, examine the two distributions in Figure 3.7 on page 89. The distribution on the right has more variability, a greater range of scores, than is found in the distribution on the left. Said another way, the scores are less tightly clustered around the mean in the distribution on the right than they are in the distribution on the left. The distribution on the right has some cases that fall farther away from the mean and these will have bigger deviation scores. Deviation scores serve as a measure of variability, and bigger deviation scores mean more variability.

If deviation scores are measures of variability and there is a deviation score for each case in the population, then how is the size of the deviation scores summarized? The obvious approach is to find the average deviation score by adding together all the deviation scores and dividing this sum by the number of scores. Unfortunately, this doesn't work. Remember, the mean balances the deviation scores, so the sum of a set of deviation scores is always zero. As a result, the average deviation score will always be zero no matter how much variability there was in the data set.

To get around this problem, statisticians square the deviation scores—making them all positive—and then find the average of the squared deviation scores. The result is the variance, the mean squared deviation score. For a population, variance is abbreviated as σ^2 (σ is the lowercase version of the Greek letter sigma; σ^2 is pronounced "sigma squared"). The formula for calculating population variance is shown in Equation 3.4.

Equation 3.4 Formula for Population Variance (σ^2)

$$\sigma^2 = \frac{\Sigma(X - \mu)^2}{N}$$

where σ^2 = population variance
X = raw score
μ = the population mean
N = the number of cases in the population

This formula for calculating population variance requires four steps: (1) first, create deviation scores for each case in the population by subtracting the population mean from each raw score; (2) then, square each of these deviation scores; (3) next, add up all the squared deviation scores; and (4) finally, divide this sum by the number of cases in the population. The result is σ^2, the population variance.

Population Standard Deviation

Interpreting variance can be confusing because it is based on squared scores. The solution is simple: find the square root of the variance. The square root of the variance is called the standard deviation, and it transforms the variance back into the original unit of measurement. The standard deviation is the most commonly reported measure of variability. For a population, the standard deviation is abbreviated as σ. The formula for calculating a population standard deviation is shown in Equation 3.5.

Equation 3.5 Formula for Population Standard Deviation (σ)

$$\sigma = \sqrt{\sigma^2}$$

where σ = population standard deviation
σ^2 = population variance (Equation 3.4)

A standard deviation gives information about the average distance that scores fall from the mean.

A standard deviation tells how much spread, or variability, there is in a set of scores. If the standard deviation is small, then the scores fall relatively close to the mean. As standard deviations grow larger, the scores are scattered farther away from the mean. A standard deviation gives information about the average distance that scores fall from the mean.

Calculating Variance and Standard Deviation for a Sample

Population variance and population standard deviation are almost never known because it is rare that a researcher has access to all the cases in a population. Researchers study samples, but they want to draw conclusions about populations. In order to do so, they need to use population values in their equations. To make this possible, statisticians have developed a way to calculate variance and standard deviation for a sample and to "correct" them so that they are better approximations of the population values. The correction makes the sample variance and sample standard deviation a little bit larger.

Why does the correction increase the size of the sample variance and sample standard deviation? Because there is more variability in a population than in a sample. To visualize this, imagine a big jar of jellybeans of different colors and flavors. The jar is the population. Fernando comes along, dips his hand in, and pulls out a handful. The handful is the sample. Now, Fernando is asked how many different colors he has in his hand. The number of colors is a measure of variability. Fernando's handful probably includes a lot of different colors. But, given the size of the jar, it is almost certain that there are some colors in the jar that are missing from his sample. More variability exists in the population than was found in the sample, so a sample measure of variability has to be corrected to reflect this.

The formula for calculating sample variance is shown in Equation 3.6. Sample variance is abbreviated s^2, pronounced "s squared" and sample standard deviation is abbreviated *s*. (*s* is used because sigma, σ, is the Greek letter "s.")

Equation 3.6 Formula for Sample Variance (s^2)

$$s^2 = \frac{\Sigma(X - M)^2}{N - 1}$$

where s^2 = sample variance
X = raw score
M = the sample mean
N = the number of cases in the sample

To calculate sample variance, make a table with all the data in it. As seen in Table 3.7, the data go in the first column, with each raw score on a separate row. It doesn't matter whether the data are in order or not. The table should have three columns. Once the table is ready, follow this four-step procedure to calculate s^2, the sample variance. Here's an example finding the variance for the heights in the demographer's sample of five adult Americans:

Step 1 Subtract the mean from each score in order to calculate deviation scores. This is shown in the second column in Table 3.7. For the height data, the mean is 67.00. Here is the calculation of the deviation score for the case in the first row with a raw score of 62:

$$62 - 67.00 = -5.00$$

Step 2 Take each deviation score and square it. This is shown in the third column in Table 3.7. In the first row, the deviation score of −5.00 becomes 25.00 when squared.

Step 3 Add up all the squared deviation scores. This is called a **sum of squares**, abbreviated *SS*, because that is just what it is—a sum of squared scores. At the bottom of the third column is Σ, a summation sign. Adding together all the squared deviation scores in Table 3.7 totals to 70.00.

TABLE 3.7 Deviation Scores and Squared Deviation Scores for Height Data with a Mean of 67.00

Height (*X*)	Deviation Score(*X* − *M*)	Squared Deviation Score $(X - M)^2$
62	−5.00	25.00
65	−2.00	4.00
66	−1.00	1.00
69	2.00	4.00
73	6.00	36.00
		$\Sigma = 70.00$

Whether a deviation score is positive or negative, the squared deviation score is always positive.

Step 4 The final step involves taking the sum of the squared deviation scores (70.00) and dividing it by the number of cases minus 1 to find the sample variance.

$$\begin{aligned} s^2 &= \frac{70.00}{5-1} \\ &= \frac{70.00}{4} \\ &= 17.5000 \\ &= 17.50 \end{aligned}$$

And, that's the answer. The demographer would report the sample variance for the five Americans as $s^2 = 17.50$. With variances, bigger numbers mean more variability. Without another variance for comparison, it is hard to say whether this sample variance of 17.50 means there is a lot of variability or a little variability in this sample.

Wondering where the "correction" is that makes this sample variance larger and a better estimate of the population value? It is in the denominator, where 1 is subtracted from *N*, the number of cases. This subtraction makes the denominator smaller, which makes the quotient, s^2, larger, making the sample variance a better estimate of σ^2.

Once the sample variance, s^2, has been calculated, it is straightforward to calculate s, the sample standard deviation. The formula for the sample standard deviation is shown in Equation 3.7.

Equation 3.7 Formula for Sample Standard Deviation (s)

$$s = \sqrt{s^2}$$

where s = sample standard deviation
s^2 = sample variance (Equation 3.6)

As $s^2 = 17.50$ for the five heights, here is the calculation for the standard deviation for the sample:

$$\begin{aligned} s &= \sqrt{17.50} \\ &= 4.1833 \\ &= 4.18 \end{aligned}$$

The demographer would report the standard deviation of the sample as $s = 4.18$. Often, means and standard deviations are used together to describe central tendency and variability for data sets. For the sample of five Americans, the demographer would describe them as having a mean height of 67.00 inches with a standard deviation of 4.18 inches, or she might write $M = 67.00''$ and $s = 4.18''$.

Means are easy to interpret—readers of the demographer's report will have some sense of what an average height of 5′ 7″ looks like—but most people don't have an intuitive sense of how to interpret standard deviations. One can think of a standard deviation as revealing the distance a score, on average, falls from the mean, but does a standard deviation of 4.18″ mean there is a lot of variability in the sample or a little? For now, just remember that bigger standard deviations mean more variability. If someone else had a different sample and reported that $s = 5.32$, then his sample would have more variability in height than the original demographer's sample, where $s = 4.18$. The scores in the second sample would tend to fall, on average, a little farther away from the mean.

Worked Example 3.2

Remember Dr. Gorham, the psychologist who collected data about the number of aggressive comments that third graders made to a Bobo doll after seeing an adult physically attack it? The raw scores were 2, 3, 4, 4, 4, 5, 5, 6, 7, 9 and the mean was 4.90. Let's use these data to practice calculating the sample variance and sample standard deviation.

Once the data are in a table, calculating the sample variance follows four steps. Here, the data in Table 3.8 are listed in order from low to high, but the procedure still works if they are not in order.

TABLE 3.8 Raw Scores, Deviation Scores, and Squared Deviation Scores for Number of Aggressive Comments Made Against the Bobo Doll ($M = 4.90$)

Number of Aggressive Comments	Deviation Score $(X - M)$	Squared Deviation Score $(X - M)^2$
2	−2.90	8.41
3	−1.90	3.61
4	−0.90	0.81
4	−0.90	0.81
4	−0.90	0.81
5	0.10	0.01
5	0.10	0.01
6	1.10	1.21
7	2.10	4.41
9	4.10	16.81
		$\Sigma = 36.90$

This table contains the information necessary to calculate the variance for the number of aggressive comments made against the Bobo doll by the 10 children.

1. The first step involves calculating deviation scores (e.g., $2 - 4.90 = -2.90$).
2. The next step is squaring the deviation scores. For the first row, this is $-2.90^2 = 8.41$.
3. For the third step, sum the squared deviation scores (e.g., $8.41 + 3.61 + 0.81$, etc.) and find the total ($\Sigma = 36.90$, shown at the bottom of column 3).
4. Finally, for Step 4, divide the sum of the squared deviations by the number of cases minus 1 to find the sample variance:

$$s^2 = \frac{36.90}{10 - 1}$$

$$= \frac{36.90}{9}$$

$$= 4.1000$$

$$= 4.10$$

Once the variance has been calculated, it is easy to find the standard deviation by taking its square root:

$$s = \sqrt{4.10}$$
$$= 2.0248$$
$$= 2.02$$

When asked to use central tendency and variability to describe this sample, Dr. Gorham would say that these 10 third graders made a mean of 4.90 aggressive comments to Bobo, with a standard deviation of 2.02 comments. He could report this as $M = 4.90$ and $s = 2.02$.

Practice Problems 3.2

Review Your Knowledge

3.07 What does it mean when one set of scores has more variability than another set?

3.08 What is a disadvantage of the range as a measure of variability?

3.09 What is an advantage of the interquartile range as a measure of variability?

3.10 What is an advantage of the variance and standard deviation over the range and interquartile range as measures of variability?

Apply Your Knowledge

3.11 A clinical psychologist took 12 people who were diagnosed with depression and gave them an interval-level depression scale. On this scale, scores above 70 are considered to be indicative of clinical depression. Here are the 12 scores: 65, 81, 66, 83, 70, 70, 72, 60, 78, 78, 79, and 84. What is the range?

3.12 The American Association of Organic Apple Growers took a sample of five family farms and determined how large, in acres, the orchards were. The results were 23, 8, 22, 10, and 32 with a mean of 19.00. Calculate (a) sample variance and (b) sample standard deviation for the size, in acres, of the five orchards.

Application Demonstration

A psychologist took a sample of eight rats and trained them to run a maze in order to find food in a goal box. After they were well trained, she timed how long it took them, in seconds, to run the maze. Here are the data she collected: 4.73, 6.54, 5.88, 4.68, 5.82, 6.01, 5.54, and 6.24. She wants to report a measure of central tendency and variability for the sample.

Central Tendency

The data, time in seconds, are measured at the ratio level, so she can use a mean as her measure of central tendency. But, she should consider the shape of the data set before committing to the mean. Figure 3.9 shows a histogram for the maze running times.

Figure 3.9 doesn't illustrate anything unusual in terms of shape. The sample size is small, but there is nothing obviously odd in terms of modality or skew, so

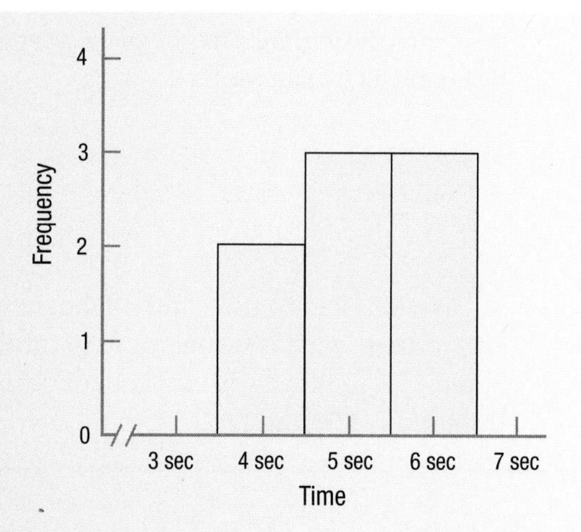

Figure 3.9 Histogram of Maze Running Times for Eight Rats When sample size is small, it is difficult to judge the shape of a distribution. This histogram does not show obvious skewness, so calculating the mean seems reasonable.

she can proceed with calculating the mean. Using Equation 3.1, she should sum all the scores and divide by the number of cases to find the mean:

$$\begin{aligned} M &= \frac{\Sigma X}{N} \\ &= \frac{4.73 + 6.54 + 5.88 + 4.68 + 5.82 + 6.01 + 5.54 + 6.24}{8} \\ &= \frac{45.44}{8} \\ &= 5.6800 \\ &= 5.68 \end{aligned}$$

The mean time to run the maze is 5.68 seconds.

Variability

The most often reported measure of variability is the standard deviation, s. To calculate s, she completes Table 3.9 to lead her through the first four steps involved in calculating the sample variance (Equation 3.4).

Step 1 involves listing the scores and that is easy to do. Using the mean, 5.68, she can complete the second step and calculate deviation scores (see the second column in Table 3.9). Column 3 shows the squared deviation scores, whereas the bottom of column 3 gives the sum of the squared deviation scores, 3.1438.

For Step 5, she uses the sum of the squared deviation scores to calculate the sample variance:

$$\begin{aligned} s^2 &= \frac{3.1438}{8 - 1} \\ &= \frac{3.1438}{7} \\ &= 0.4491 \\ &= 0.45 \end{aligned}$$

TABLE 3.9 Squared Deviation Scores to Be Used for Calculating Variance for Maze Running Times

Raw Score	Deviation Score $(X - M)$	Squared Deviation Score $(X - M)^2$
4.73	−0.95	0.9025
6.54	0.86	0.7396
5.88	0.20	0.0400
4.68	−1.00	1.0000
5.82	0.14	0.0196
6.01	0.33	0.1089
5.54	−0.14	0.0196
6.24	0.56	0.3136
		$\Sigma = 3.1438$

This table contains the information necessary to calculate the variance for the number of seconds it takes to run the maze.

To find the sample standard deviation, she applies Equation 3.7:

$$\begin{aligned} s &= \sqrt{s^2} \\ &= \sqrt{0.45} \\ &= 0.6708 \\ &= 0.67 \end{aligned}$$

She would report central tendency as a mean maze running time of 5.68 seconds, with a sample standard deviation of 0.67. Following APA format, she could write $M = 5.68''$ and $s = 0.67''$.

DIY

Did you save the grocery receipt you used in the rounding DIY for Chapter 1? Now we are going to use it to calculate how much your average grocery item costs. Price is a ratio-level variable, so you could calculate mean, median, or mode as an average. Which should you do?

We would like to calculate the mean because it uses interval and ratio data that contain more information, but does the shape of the sample allow that? What should you do if most of your sample falls in the \$2–\$5 range and one or two items are in the \$10–\$15 range?

Calculate all three measures and compare them. Could you calculate a mode? Which measure should you report and why?

And, what should you do for a measure of variability?

SUMMARY

Define and know when to calculate three measures of central tendency.

- The measure of central tendency that is calculated for a data set is determined by the level of measurement and the shape of the data set. A mean, *M*, can be calculated for interval- or ratio-level data that are not skewed or multimodal. The median, *Mdn*, is the score associated with the case that separates the top half of scores from the bottom half of scores. Medians can be calculated for ordinal-, interval-, or ratio-level data. The mode, the score that occurs most frequently, is the only measure of central tendency that can be calculated for nominal-level data. Modes can also be calculated for ordinal-, interval-, or ratio-level data. When there are multiple options for which measure of central tendency to choose, select the one that conveys the most information and takes the shape of the data set into consideration.

Define variability and know how to calculate four different measures of it.

- Variability refers to how much spread there is in a set of scores. All four measures of variability are for interval- or ratio-level data. The range tells the distance from the smallest score to the largest score and is heavily influenced by outliers. The interquartile range, *IQR*, tells the distance covered by the middle 50% of scores. The *IQR* provides information about both central tendency and variability. Variance uses the mean squared deviation score as a measure of variability; standard deviation is the square root of variance and gives the average distance that scores fall from the mean. The larger the standard deviation, the less clustered scores are around the mean.
- There is usually more variability in the population than in a sample. Population variance and standard deviation are abbreviated as σ^2 and σ; sample variance and standard deviation as s^2 and s. The sample variance and standard deviation are "corrected" to approximate population values more accurately.

KEY TERMS

central tendency – a value used to summarize a set of scores; also known as the average.

deviation score – a measure of how far a score falls from the mean, calculated by subtracting the mean from the score.

interquartile range – a measure of variability for interval- or ratio-level data; the distance covered by the middle 50% of scores; abbreviated *IQR*.

mean – an average calculated for interval- or ratio-level data by summing all the values in a data set and dividing by the number of cases; abbreviated *M*.

median – an average calculated by finding the score associated with the middle case, the case that separates the top half of scores from the bottom half; abbreviated *Mdn*; can be calculated for ordinal-, interval-, or ratio-level data.

mode – the score that occurs with the greatest frequency.

outlier – an extreme (unusual) score that falls far away from the rest of the scores in a set of data.

range – a measure of variability for interval- or ratio-level data; the distance from the lowest score to the highest score.

standard deviation – a measure of variability for interval- or ratio-level data; the square root of the variance; a measure of the average distance that scores fall from the mean.

sum of squares – squaring a set of scores and then adding together the squared scores; abbreviated *SS*.

variability – how much variety (spread or dispersion) there is in a set of scores.

variance – a measure of variability for interval- or ratio-level data; the mean of the squared deviation scores.

CHAPTER EXERCISES

Answers to the odd-numbered exercises appear at the back of the book.

Review Your Knowledge

3.01 The two dimensions used in this chapter to summarize a set of data are measures of ____ and ____.

3.02 The three measures of central tendency covered in this chapter are ____, ____, and ____.

3.03 Two sets of scores may be alike in central tendency but can still differ in ____.

3.04 The formula for a ____ is $X - M$.

3.05 The sum of deviation scores equals ____.

3.06 A mean may be influenced by ____, which are extreme scores.

3.07 The median is ____ influenced by outliers than is the mean.

3.08 The median separates the top ____% of scores from the bottom ____% of scores.

3.09 If one has interval or ratio data, the first choice for a measure of central tendency to calculate should be a ____.

3.10 If one has interval or ratio data and the data set is skewed, one should use a ____ as the measure of central tendency.

3.11 If one has interval or ratio data and the data set is multimodal, one should use ____ as the measure of central tendency.

3.12 If one has nominal data, one should use a ____ as the measure of central tendency.

3.13 The range, interquartile range, variance, and standard deviation can only be used with variables measured at the ____ or ____ level.

3.14 The range tells the distance from the ____ score to the ____ score.

3.15 The range *is/is not* influenced by outliers.

3.16 If the top 25% of scores and the bottom 25% of scores are trimmed off the range, it is called the ____.

3.17 If the interquartile range for student A's grades is wider than it is for student B's grades, then student A has ____ variability in his or her grades.

3.18 The interquartile range can be used as a measure of ____ as well as variability.

3.19 There is more variability in a ____ than in a ____.

3.20 Because populations are so ____, it is practically impossible to measure variability in them.

3.21 The ____ is the mean squared deviation score.

3.22 The ____ is the square root of the ____ and tells the average distance by which the raw scores deviate from the mean.

3.23 The variance and standard deviation, unlike the range and the interquartile range, use information from all of the ____.

3.24 The information used from each case, in calculating variance and standard deviation, is its ____ score.

3.25 If scores are ____ clustered around the mean, then there is less variability in that set of scores.

3.26 The average deviation score is not useful as a measure of variability because the ____ of the deviation scores is always ____.

3.27 To use deviation scores to measure variability, statisticians ____ them before summing them.

3.28 ____ is the abbreviation for population variance.

3.29 ____ is the abbreviation for population standard deviation.

3.30 If the standard deviation is small, then the scores fall relatively close to the ____.

3.31 Sample values of variance and standard deviation are ____, so they are better approximations of the ____ values.

3.32 The abbreviation for sample variance is ____; for sample standard deviation, it is ____.

3.33 The first step in calculating a sample variance is to calculate a ____ for each case.

3.34 If s for one set of data is 12.98 and it is 18.54 for a second set of data, then the first set of data has ____ variability than the second.

Apply Your Knowledge

Selecting and calculating the appropriate measure of central tendency

3.35 An industrial/organizational psychologist measured the number of errors seven research participants made in a driving simulator while talking on their cell phones. Here are her data: 12, 6, 5, 4, 4, 3, and 1.

3.36 A cognitive psychologist at a state university noted whether her research participants had been raised in a non–English-speaking household (0), an English-only–speaking household (1), or a multilingual, including English, household (2). Here are her data: 0, 0, 1, 1, 1, 1, 1, 1, 1, 1, 1, 1, 1, 2, 2, 2, and 2.

3.37 A cardiologist wanted to measure the increase in heart rate in response to pain. She directed healthy volunteers to immerse their dominant hands into a bucket of ice water for 1 minute. Here are the increases in heart rate for her participants: 20, 23, 25, 19, 28, and 24.

3.38 An educational psychologist was teaching a graduate seminar with seven students. He was curious to know if the amount of sleep had an impact on performance. On the first exam, he asked how many hours the students had slept the night before. Here are the data he obtained: 8, 9, 7, 6, 8, 4, and 7.

Calculating the mean

3.39 Given the following ratio-level numbers, calculate the mean: 80, 88, 76, 65, 59, and 77.

3.40 Given the following interval-level numbers, calculate the mean: 0.13, 0.28, 0.42, 0.36, and 0.26.

Calculating median and mode

3.41 Find the median and the mode for the following data: 65, 66, 66, 70, 71, 72, 72, 72, 78, 83, 85, 86, 87, 87, 88, 88, 92, 93, 95, 95, 99, 100, 102, 102, 102, 102, 102, 103, 104, 108, 111, 113, 118, 119, 119, 119, 121, 121, 122, 128, 130, 131, 134, 136, 138, 139, and 145.

3.42 Find the median and the mode for the following data: 30, 30, 32, 33, 35, 35, 38, 41, 42, 43, 45, 45, 48, 49, 50, 50, 50, 51, 52, 53, 53, 53, 53, 53, 53, 55, 55, 58, 59, 61, 62, 62, 64, 65, 66, 66, 67, 68, 70, 71, 71, 72, 72, 73, 73, 75, 76, 78, 79, and 80.

Range and interquartile range

3.43 Here is a listing, to the nearest million dollars, of the total payrolls for the 30 Major League baseball teams in 2009.

a. Calculate the range.

b. Which is a better measure of variability to report, the range or the interquartile range? Why?

201	100	78	63
149	99	75	62
135	97	74	60
122	96	74	49
115	83	71	44
114	82	68	37
113	81	67	
103	80	65	

3.44 **a.** Find the range for the data in Exercise 3.41.

b. Which is a better measure of variability to report, the range or the interquartile range? Why?

Calculating variability

3.45 A researcher obtained a sample from a population and measured an interval-level variable on each case. Here are the values he obtained: 47, 53, 67, 45, and 38. He found $M = 50.00$. Find the following: (a) N, (b) $\Sigma(X - M)$, (c) $\Sigma(X - M)^2$.

3.46 Given this data set (12, 9, 15, 13, and 11) with a mean of 12.00, find the following: (a) N, (b) $\Sigma(X - M)$, (c) $\Sigma(X - M)^2$.

3.47 A clinical psychologist took a sample of four people from the population of people with anxiety disorders and gave them an interval-level anxiety scale. Their scores were 57, 60, 67, and 68 with a mean of 63.00. Calculate s^2.

3.48 A nurse practitioner took a sample of five people from the population of people with high blood pressure and measured their systolic blood pressures as 140, 144, 148, 156, and 162 with a mean of 150.00. Calculate s^2.

3.49 An interval-level measure was obtained on a sample of four cases. Their raw scores were 21, 18, 12, and 9. Calculate s^2.

3.50 An interval-level measure was obtained on five cases. Their raw scores were 108.5, 95.5, 98.0, 112.0, and 112.5. Calculate s^2.

3.51 If $s^2 = 6.55$, what is s?

3.52 If $s^2 = 256.38$, what is s?

3.53 Calculate s for these data, which have a mean of 37.20: 24, 36, 42, 50, and 34.

3.54 Calculate *s* for these data, which have a mean of 3.30: 1.4, 2.2, 4.5, 4.1, and 4.3.

Expand Your Knowledge

3.55 A librarian measured the number of pages in five books. No two had the same number of pages. Here are the results from four of the books: 150, 100, 210, and 330. Using the counting method, he calculated the median for all five books as 150. In the list below, what is the *largest* value possible for the number of pages in the fifth book?

- **a.** 99
- **b.** 101
- **c.** 125
- **d.** 149
- **e.** 151
- **f.** 185
- **g.** 209
- **h.** 211
- **i.** 270
- **j.** 329
- **k.** 331
- **l.** It would be possible to determine this, but the value is not listed above.
- **m.** It is not possible to determine this from the information in the question.

3.56 A teacher has a set of 200 numbers, where the mean is dramatically greater than the median. What does this suggest about the shape of the distribution?

- **a.** The distribution is probably normal.
- **b.** The distribution is probably bimodal.
- **c.** The distribution is probably flat.
- **d.** The distribution is probably unusually peaked.
- **e.** The distribution is probably negatively skewed.
- **f.** The distribution is probably positively skewed.
- **g.** Comparing the mean to the median provides no information about a distribution's shape.

3.57 A researcher has a set of positive numbers. No numbers are duplicates. The mean is greater than the median. Which of the following statements is true?

- **a.** It is likely that more of the numbers are greater than the mean than are less than the mean.
- **b.** It is likely that more of the numbers are less than the mean than are greater than the mean.
- **c.** It is likely that more of the numbers are greater than the median than are less than the median.
- **d.** It is likely that more of the numbers are less than the median than are greater than the median.
- **e.** None of these statements is likely true.

3.58 Select the appropriate measure of central tendency for each scenario:

- **a.** A researcher obtains a sample of men who have been married for 7 years and asks them to rate their levels of marital happiness on an ordinal scale that ranges from −7 (extremely unhappy) to +7 (extremely happy).
- **b.** An anatomist obtains a sample of athletes that consists of 75 jockeys (who are extremely short) and 75 basketball players (who are extremely tall). She measures the height of all 150 and wants to report the overall average.
- **c.** A social studies teacher, with a classroom of students who have roughly the same level of ability, administered a final exam of multiple-choice questions on the social studies facts they had learned.
- **d.** An astronomer classifies celestial objects as 1 (stars), 2 (planets), 3 (dwarf planets), 4 (asteroids), 5 (moons), 6 (comets), 7 (meteoroids), and 8 (other). He wants to report the average celestial object in a sample of the universe.
- **e.** A college psychology program has 12 faculty members. One of them is a full professor, three are associate professors, and eight are assistant professors. (Full professors have substantially higher salaries than associate professors, who have substantially higher salaries than assistant professors.) The chair of the program wants to report the average salary of the 12 faculty members.

3.59 A researcher takes two samples, one with $N = 10$ and the other with $N = 100$, from the same population.

a. Which sample, $N = 10$ or $N = 100$, will have to correct s^2 less in order to approximate the population variance?

b. Why is it sensible that Equation 3.6 works this way?

3.60 A meteorologist measured the temperature every 2 hours for a day in January and for a day in June. Here are the temperatures, in degrees Fahrenheit, for the January day: 23, 22, 21, 21, 23, 27, 28, 31, 32, 32, 30, and 26. The June temperatures were: 56, 56, 59, 62, 64, 67, 73, 75, 77, 76, 75, and 74. On which day were the temperatures more varied? Support the conclusion by calculating a measure of variability.

SPSS

SPSS has multiple ways to calculate central tendency and variability. The "Frequencies," "Descriptives," and "Explore" commands all provide some measures of central tendency and variability, but they differ in what they provide. Table 3.10 shows which SPSS command provides which measures of central tendency and variability.

No matter which command one chooses—Frequencies, Descriptives, or Explore—they all start the same way, under the "Analyze/Descriptive Statistics" drop-down menu. This is illustrated in Figure 3.10.

Figure 3.11 shows the initial screen for the Descriptives command. The arrow key is used to move a variable, here "Maze Running Time," from the box on the left to the "Variable(s)" box.

Clicking on the "Options" button in Figure 3.11 opens up the menu seen in Figure 3.12. Here, select the statistics for SPSS to calculate.

TABLE 3.10 List of SPSS Commands and the Descriptive Statistics Provided by Each

	Frequencies	Descriptives	Explore
Mean	✓	✓	✓
Median	✓		✓
Mode	✓		
Range	✓	✓	✓
Interquartile range		✓	✓
Standard deviation (s)	✓	✓	✓
Variance (s^2)	✓	✓	✓

Though all three of these SPSS commands can calculate means, ranges, variances, and standard deviations, the other measures of central tendency and variability can only be calculated by some commands.

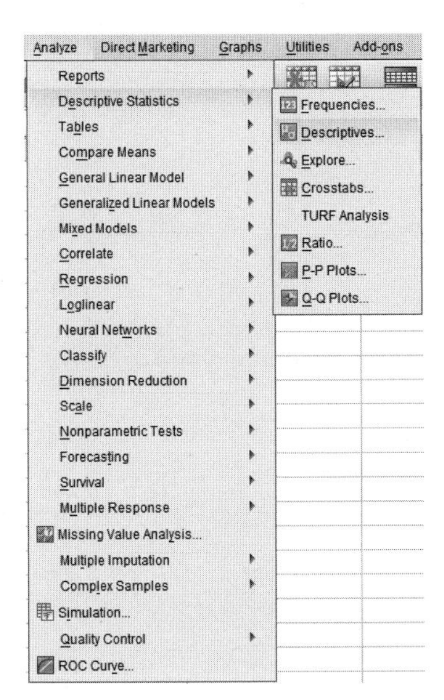

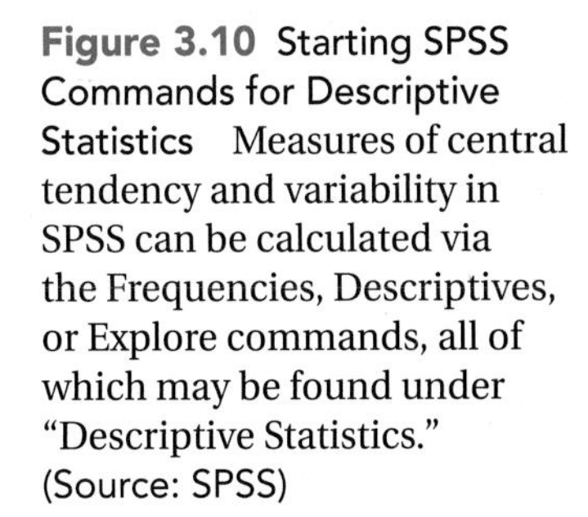

Figure 3.10 Starting SPSS Commands for Descriptive Statistics Measures of central tendency and variability in SPSS can be calculated via the Frequencies, Descriptives, or Explore commands, all of which may be found under "Descriptive Statistics." (Source: SPSS)

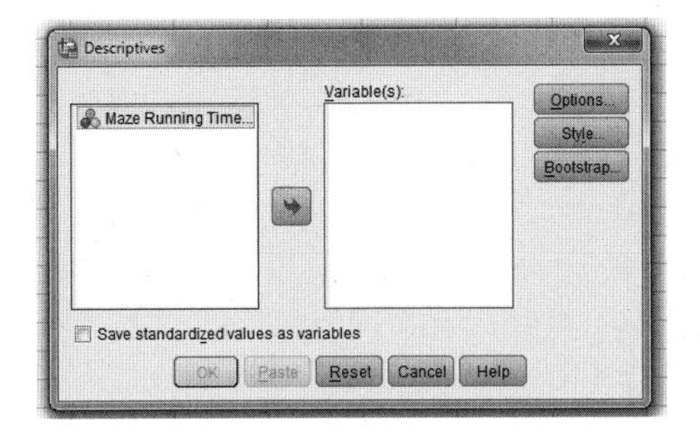

Figure 3.11 The Descriptives Command in SPSS Once a variable has been moved from the box on the left to the box labeled "Variable(s)," the "OK" button on the bottom left becomes active. (Source: SPSS)

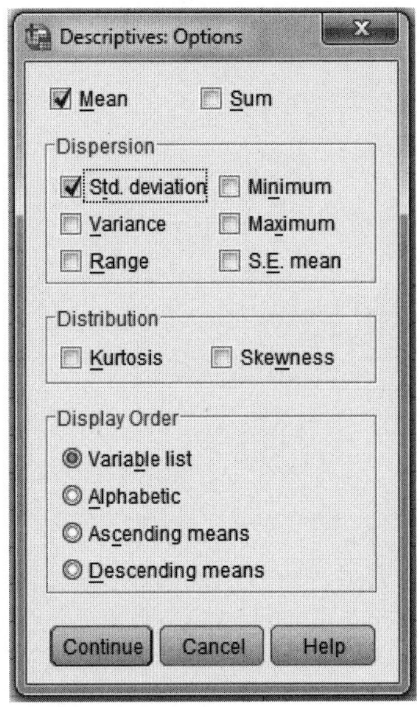

Figure 3.12 Selecting Descriptive Statistics Clicking on the "Options" button in Figure 3.12 opens up this menu, which allows one to select the desired descriptive statistics. (Source: SPSS)

For an example of the output produced by SPSS, see Figure 3.13. Only the mean and the standard deviation were requested and this is all that was calculated.

Descriptive Statistics

	N	Mean	Std. Deviation
Maze Running Time (in seconds)	8	5.6800	.67016
Valid N (listwise)	8		

Figure 3.13 SPSS Output for Descriptive Statistics An example of SPSS output. Only the mean and standard deviation were requested, and this is all that was calculated. (Source: SPSS)

Standard Scores, the Normal Distribution, and Probability

LEARNING OBJECTIVES

- Transform raw scores into standard scores (z scores) and vice versa.
- Describe the normal curve.
- Transform raw scores and standard scores into percentile ranks and vice versa.
- Calculate the probability of an outcome falling in a specified area under the normal curve.

CHAPTER OVERVIEW

This chapter covers standard scores, normal curves, and probability. Most people know a little about bell-shaped curves, what statisticians call normal curves. This chapter explains why normal curves are called normal, what their characteristics are, and how they are used in statistics.

Standard scores, also called *z* scores, are new territory for most students. Standard scores are useful in statistics because raw scores can be transformed into standard scores (and vice versa). Standard scores allow different kinds of measurements to be compared on a common scale—such as the juiciness of an orange vs. the crispiness of an apple. Standard scores are also useful for measuring distance on the normal curve, which shows how big the pieces of the bell curve are in terms of what percentage of cases fall in different sections of it. This allows scores to be expressed as percentile ranks, which tell what percentage of cases fall at or below a given score. Breaking the normal curve into chunks allows researchers to talk about how common or rare different scores should be, how probable they are. This excursion into probability opens the way to understanding how statistical tests work in upcoming chapters.

4.1 Standard Scores (*z* Scores)

People often say that you cannot compare apples and oranges. In this chapter, we'll see this can be done thanks to something called a standard score. We'll also see that standard scores allow one to quantify how common or how rare a score is.

Imagine that two fruit growers, Anne who grows apples and Oliver who grows oranges, get into a good-natured argument over whose fruit is better. Anne claims

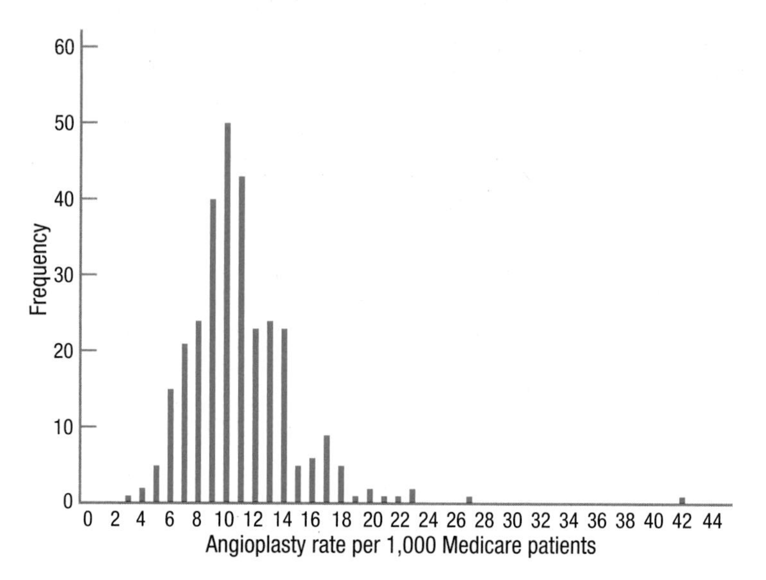

Figure 4.1 Angioplasty Rates for Medicare Patients by City There is one city, far out on the right-hand side that is very different from the others. A standard score, a z score, will quantify how unusual this score is. (Data from *The New York Times*, August 18, 2006.)

that the crispiness of apples is what makes them so good. Oliver cites the juiciness of oranges as their best feature. How can one objectively compare the crispiness of an apple to the juiciness of an orange to decide which fruit has more of its desired quality? Standard scores.

Look at the graph in Figure 4.1. It shows the rate with which Medicare patients receive angioplasty in different geographic hospital regions of the United States. It should be apparent that most hospitals cluster around rates of 6 per thousand patients to 14 per thousand patients. And then, there is one city, way off by itself on the right-hand side, treating an unusually large percentage of Medicare patients with angioplasty. How can we describe how unusual that city is? Standard scores.

A **standard score** is a raw score expressed in terms of how many standard deviations it is away from the mean. Standard scores are commonly called ***z* scores**. The formula for calculating a z score is shown in Equations 4.1 and 4.2.

Equation 4.1 Formula for Calculating Standard Scores (z Scores) for a Population

$$z = \frac{X - \mu}{\sigma}$$

where z = the standard score
X = the raw score
μ = the population mean
σ = the population standard deviation

Equation 4.1 is used when the population mean, μ, and population standard deviation, σ, are known. As it is relatively rare that they are known, Equation 4.2 is more commonly used.

Equation 4.2 Formula for Calculating Standard Scores (z Scores) for a Sample

$$z = \frac{X - M}{s}$$

where $z =$ the standard score
$X =$ the raw score
$M =$ the population mean
$s =$ the sample standard deviation

Each of these two formulas has two steps:

Step 1 Subtract the mean from the raw score, just as was done in Chapter 3 when calculating deviation scores. Be sure to keep track of the sign, whether the deviation score is positive, negative, or zero.

Step 2 Divide the deviation score by the standard deviation and round to two decimal places. Again, be sure to keep track of the sign, whether the z score is positive, negative, or zero.

Chapter 3 introduced a sample of five Americans where the mean height, in inches, was 67.00 and the standard deviation was 4.18. One of the members of this sample was 62″ tall. For this person, one would calculate the standard score like this:

$$\begin{aligned} z &= \frac{X - M}{s} \\ &= \frac{62 - 67.00}{4.18} \\ &= \frac{-5.0000}{4.18} \\ &= -1.1962 \\ &= -1.20 \end{aligned}$$

This person's z score was -1.20, meaning that his or her score fell 1.20 standard deviations *below* the mean. The scores for all five members of the sample, as z scores, are shown in Table 4.1.

TABLE 4.1 Heights of Five Americans, Expressed as Standard Scores

Height (X)	Deviation Score ($X - M$)	z Score $\left(\frac{X - M}{s}\right)$
62	−5.00	−1.20
65	−2.00	−0.48
66	−1.00	−0.24
69	2.00	0.48
73	6.00	1.44
	$\Sigma = 0.00$	$\Sigma = 0.00$

In this data set, $M = 67.00$ and $s = 4.18$. Standard scores, or z scores, tell the distance from the mean in a standard unit of measurement, the standard deviation. Since z scores are transformed deviation scores, they add up to zero for a data set.

There are two things to note in this table:

1. The z scores quickly show where a person's score falls in relation to the mean. As shown in Figure 4.2, positive z scores indicate a score above the mean, negative z scores a score below the mean, and a z score of 0 is a score right at the mean.
2. If the z scores for a data set are added together, they sum to zero.

z scores standardize scores. They transform scores to a common unit of measurement, the standard deviation. This allows researchers to compare variables that are measuring different things on different scales.

The same information found in z scores is contained in the regular deviation scores that are shown in the middle column of Table 4.1. So, what is the advantage of z scores? z scores *standardize* scores. They transform scores to a common unit of measurement, the standard deviation. This allows researchers to compare variables that are measuring different things on different scales.

For example, consider Carlos, who is very smart and very extroverted. Which does Carlos have more of, intelligence or extroversion? To answer this question, Carlos takes two tests. One is an intelligence test and the other is an extroversion scale. Carlos gets a score of 130 on the IQ test and a 75 on the extroversion scale. Do these two scores answer the question of whether Carlos is more unusual in terms of his intelligence or his extroversion?

These scores are out of context, so the question can't be answered. Knowing the means will help put the scores in context. The mean for the intelligence test is 100, and the mean for the extroversion scale is 50. With the means, one can tell that Carlos is above average in terms of intelligence and in terms of extroversion. The next step is to calculate the deviation scores to see how far above average he is. Carlos's deviation score for intelligence is $130 - 100 = 30.00$. For extroversion, his deviation score is $75 - 50 = 25.00$. Based on the deviation scores, it seems as if Carlos is more unusual in terms of intelligence than he is in terms of extroversion. After all, he scored 30 points above the mean on intelligence, but only 25 points above the mean on extroversion. But that would be a premature conclusion. A true comparison can't be made until standard deviation information is considered and z scores are calculated.

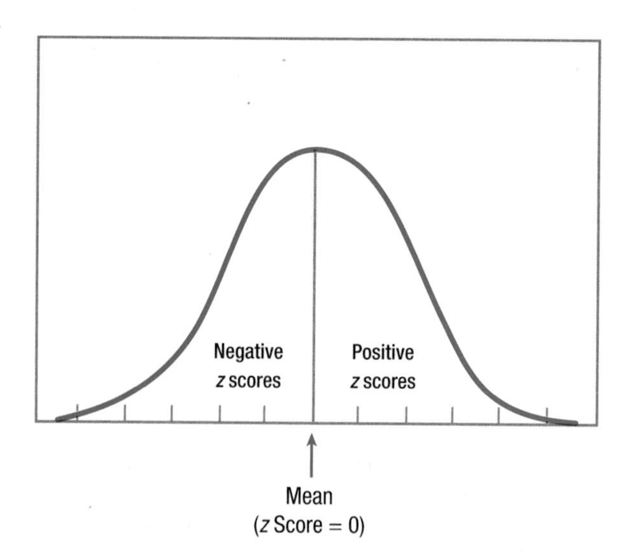

Figure 4.2 Positive, Negative, and Zero z Scores Another term for standard score is *z score.* A positive value for a z score means that the score falls above the mean, a negative value means that the score falls below the mean, and a value of 0 means that the score falls right at the mean.

Here are the standard deviations: 15 for the IQ test and 10 for the extroversion scale. Now, Equation 4.2 can be used to calculate *z* scores:

$$z_{\text{IQ}} = \frac{130 - 100}{15}$$
$$= \frac{30.0000}{15}$$
$$= 2.0000$$
$$= 2.00$$

$$Z_{\text{Extroversion}} = \frac{75 - 50}{10}$$
$$= \frac{25.0000}{10}$$
$$= 2.5000$$
$$= 2.50$$

Now the question can be answered using a standard unit, the amount of variability in a set of scores in terms of standard deviation units. Carlos's intelligence test score falls 2 standard deviations above the mean, but his extroversion score falls 2.5 standard deviations above the mean. As shown in Figure 4.3, Carlos deviates more from the mean in terms of extroversion than he does in terms of intelligence. His level of extroversion is more unusual than his level of intelligence.

With the *z* scores, two different variables measured on two different scales can be compared in a statement like this, "Carlos is more above average in his level of extroversion than he is in his level of intelligence." The comparison can be made because the scores have been standardized.

In the same way, the apple and orange example from the start of the chapter would work. If there were a juiciness scale, Oliver's orange could be measured. And, with a crispiness scale, Anne's apple could be measured. Then, each measurement would be converted to a standard score and the one with the higher score would be the winner.

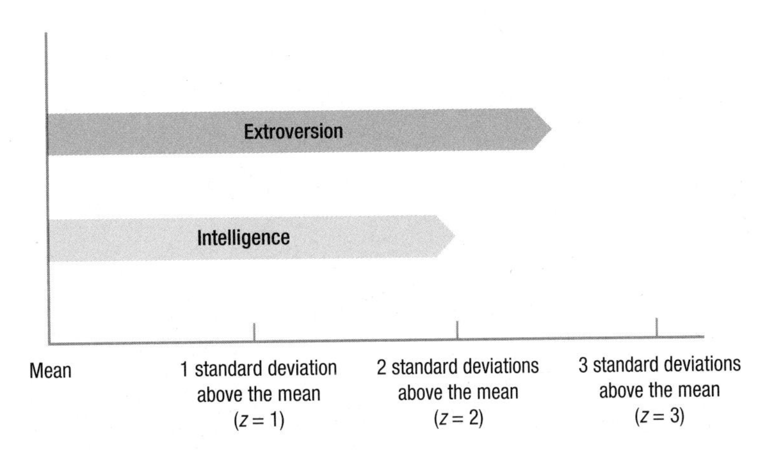

Figure 4.3 Carlos's Scores on an Extroversion Scale and an Intelligence Test
When the extroversion and intelligence scores are transformed into a standard unit of measurement, *z* scores, it becomes clear that Carlos's score on the extroversion scale is higher than his score on the intelligence test.

Just as raw scores can be turned into standard scores, it is possible to reverse direction and turn z scores into raw scores. Suppose Mei-Li was also extroverted and that her z score was 1.00 on the extroversion scale. What was her raw score?

Here's what to do. A z score of 1.00 is a positive score, which indicates that her score was above the mean. The mean on the extroversion scale is 50.00, so her score will be greater than this. A z score of 1.00 says that Mei-Li's score was 1 standard deviation above the mean. A standard deviation on the extroversion scale is 10.00, so her score was 10 points above the mean. Adding 10.00 (1 standard deviation) to 50.00 (the mean) leads to the conclusion that Mei-Li's score on the test was 60.00. Equation 4.3 formalizes how to turn z scores into raw scores.

Equation 4.3 Formula for Calculating a Raw Score (X) from a Standard Score (z)

$$X = M + (z \times s)$$

where X = the raw score
M = the mean
z = the standard score for which the raw score is being calculated
s = the standard deviation

Following this formula, here is how to calculate Mei-Li's score:

$$\begin{aligned} X &= M + (z \times s) \\ &= 50.00 + (1.00 \times 10.00) \\ &= 50.00 + 10.0000 \\ &= 60.0000 \\ &= 60.00 \end{aligned}$$

Here's one more example of converting a z score into a raw score, this one with a negative z score. Suppose that Tabitha's z score on the IQ test was −0.75. What was her raw score? Using Equation 4.3, here are the calculations:

$$\begin{aligned} X &= 100.00 + (-0.75 \times 15.00) \\ &= 100.00 + (-11.2500) \\ &= 100.00 - 11.2500 \\ &= 88.7500 \\ &= 88.75 \end{aligned}$$

On the IQ test, Tabitha's score was 88.75. Note that the raw score is reported to two decimal places and does not have to be a whole number.

Worked Example 4.1

SAT subtests are handy for practice in converting from raw scores to z scores and vice versa. As of March 2016, there are two parts on the SAT: a subtest that measures math, and a subtest that measures reading and writing. Both subtests are scored on the same scale, where 500 is the average (mean) score and the standard deviation is 100. Suppose a high school senior takes the SAT and gets a 460 on the reading and writing subtest. What is her score as a standard score?

Applying Equation 4.2, here are the calculations:

$$z = \frac{460 - 500}{100}$$

$$= \frac{-40.0000}{100}$$

$$= -0.4000$$

$$= -0.40$$

She scored 0.40 standard deviations below the mean, so her z score would be reported as -0.40.

Suppose she had a friend who took the SAT and did very well on the math section. Her friend's score on the math section, expressed as a z score, was 1.80. What was her friend's score?

Applying Equation 4.3, here are the calculations:

$$X = 500 + (1.80 \times 100)$$

$$= 500 + 180.0000$$

$$= 680.0000$$

$$= 680.00$$

This student, who had scored 1.80 standard deviations above the mean, obtained a score of 680.00 on the math subtest of the SAT.

Practice Problems 4.1

Review Your Knowledge

4.01 In what units do z scores express raw scores?

4.02 What is the sum of all z scores in a data set?

4.03 If a z score is positive, what does that mean?

Apply Your Knowledge

4.04 The average GPA in a class is 2.75 with a standard deviation of 0.40. One student has a GPA of 3.20. Express this GPA as a z score?

4.05 Another student's GPA, expressed as a z score, is -2.30. What is his GPA?

4.2 The Normal Distribution

Now that calculating and interpreting z scores have been covered, it is time to explore the normal distribution. The **normal distribution** (often called the bell curve) is a specific symmetrical distribution whose highest point occurs in the middle and whose frequencies decrease as the values on the X-axis move away from the midpoint. An example of a normal distribution is shown in Figure 4.4.

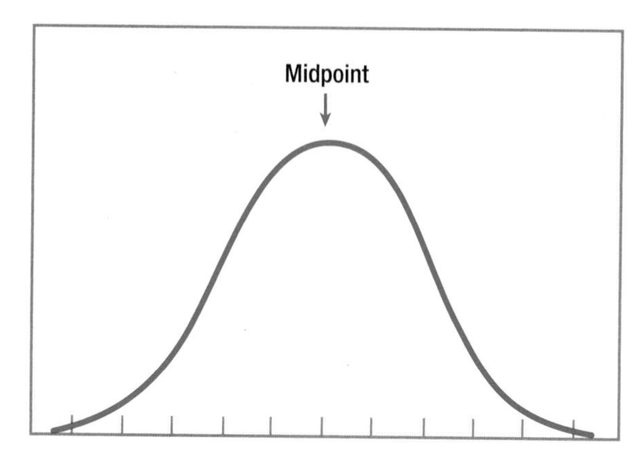

Figure 4.4 The Normal Distribution In a normal distribution, the midpoint is the mean, median, and mode. As scores move away from the midpoint, the frequency of their occurrence decreases symmetrically.

There are several things to note about the shape of a normal distribution:

- The frequencies decrease as the curve moves away from the midpoint, so the midpoint is the mode.
- The distribution is symmetric, so the midpoint is the median—half the cases fall above the midpoint and half fall below it.
- The fact that the distribution is symmetrical also means that the midpoint is the mean. It is the spot that balances the deviation scores.

Figure 4.5 shows a variety of bell-shaped curves that meet these criteria, but not all of them are normal distributions. There's another criterion that must be met for a bell-shaped curve to be considered a normal distribution: a specific percentage of cases has to fall in each region of the curve.

In Figure 4.6, the regions under the normal curve are marked off in *z* score units. In the normal distribution, 34.13% of the cases fall from the mean to 1 standard deviation above the mean, 13.59% in the next standard deviation, and decreasing percentages in subsequent standard deviations. Because the normal distribution is symmetrical, the same percentages fall in the equivalent regions below the mean. Figure 4.7 shows the percentages of cases that fall in each standard deviation unit of the normal distribution as the curve moves away from the mean.

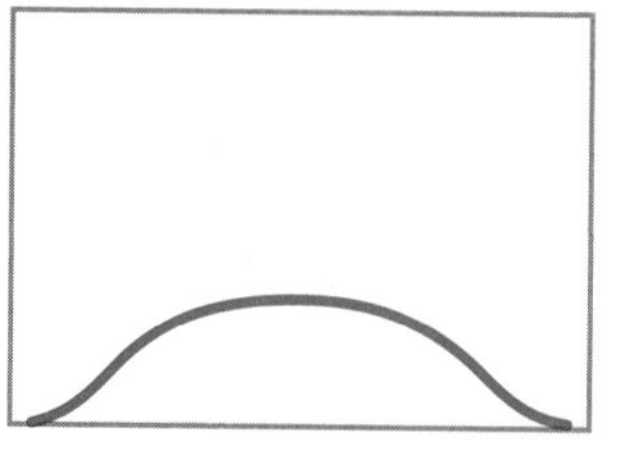
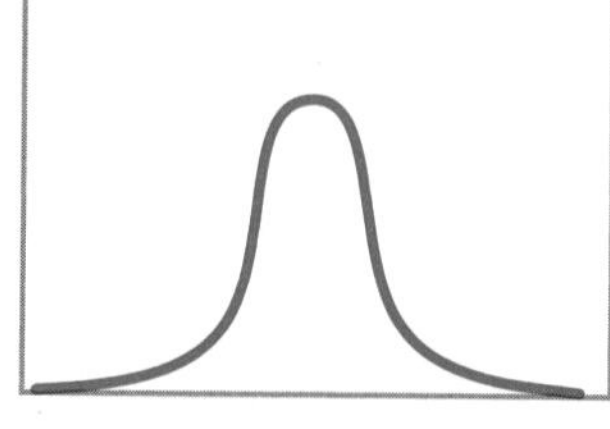
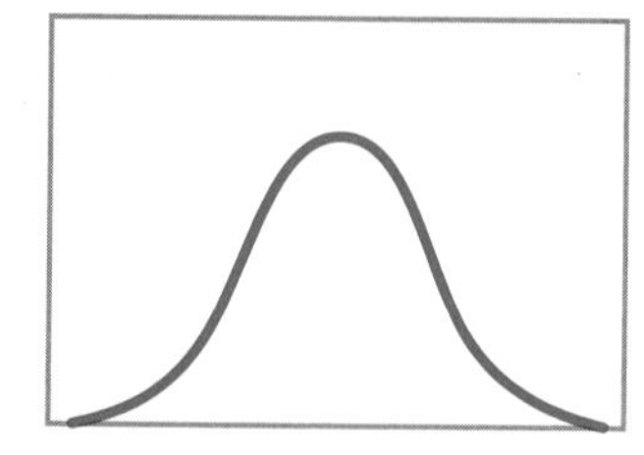

Figure 4.5 Not All Bell-Shaped Curves Are Normal Distributions In bell-shaped curves, the midpoint is the mean, median, and mode, and frequencies decrease symmetrically as values move away from the midpoint. Though all the curves shown here are bell-shaped, not all bell-shaped curves are normal distributions. The normal distribution is a specific bell-shaped curve, defined by the percentage of cases that fall within specified regions. Only the last curve is a normal distribution.

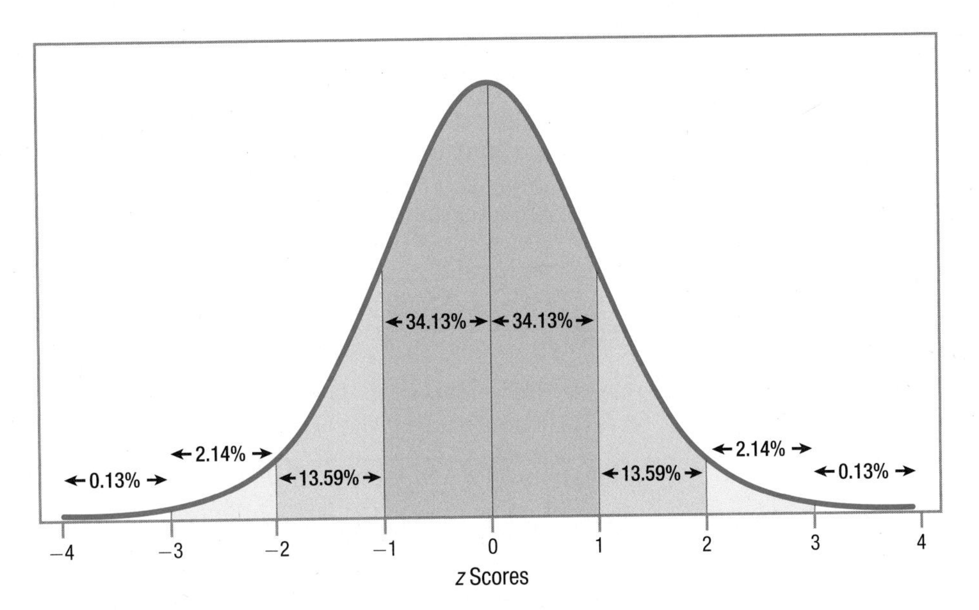

Figure 4.6 Percentage of Cases Falling in Specified Regions of the Normal Distribution The normal distribution is defined by the percentage of cases that fall within specified regions. This figure shows the percentage of cases that fall in each standard deviation as one moves away from the mean. Note the symmetrical nature and how the percentage of cases drops off dramatically as one moves away from the mean. Theoretically, the curve extends out to infinity, with fewer and fewer cases in each standard deviation.

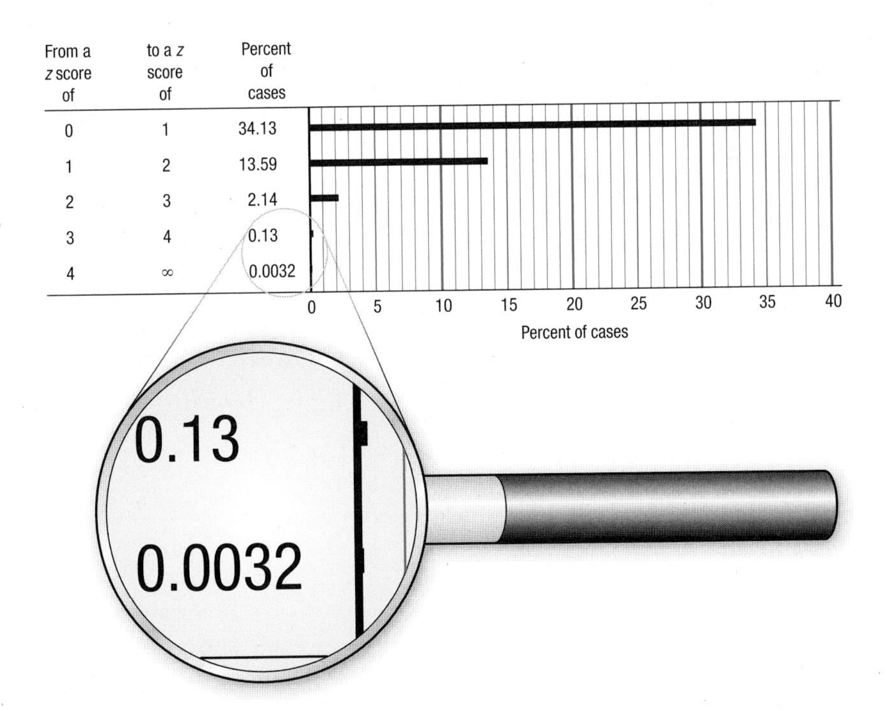

Figure 4.7 Percentage of Cases Falling in Each Standard Deviation of the Normal Distribution as It Moves Away from the Mean Note how quickly the percentage of cases falling in each standard deviation decreases as one moves away from the mean. Only a very small percentage of cases falls more than 3 standard deviations away from the mean.

One important thing to note in Figure 4.7 is how quickly the percentage of cases in a standard deviation drops off as the curve moves away from the mean. The first standard deviation contains about 34% of the cases, the next about 14%, then about 2% in the third, and only about 0.1% in the fourth standard deviation. This distribution has an important implication for where the majority of cases fall:

- About two-thirds of the cases fall within 1 standard deviation of the mean. That's from 1 standard deviation below the mean to 1 standard deviation above it, from $z = -1.00$ to $z = 1.00$. (The exact percentage of cases that falls in this region is 68.26%.)
- About 95% of the cases fall within 2 standard deviations of the mean, from $z = -2.00$ to $z = 2.00$. (The exact percentage is 95.44%.)
- More than 99% of the cases fall within 3 standard deviations, from $z = -3.00$ to $z = 3.00$. In fact, close to 100% of the cases—almost all—fall within 3 standard deviations of the mean. (The exact percentage is 99.73%.)

It is rare, in a normal distribution, for a case to have a score that is more than 3 standard deviations away from the mean. In practical terms, these scores don't happen, so when graphing a normal distribution, setting up the z scores on the X-axis so they range from -3.00 to $z = 3.00$ is usually sufficient.

Here's a simplified version of the normal distribution. Look at Figure 4.8 and note the three numbers: 34, 14, and 2. These are the approximate percentages of cases that fall in each standard deviation as the curve moves away from the mean:

- $\approx$34% of cases fall from the mean to 1 standard deviation above it. The same is true for the area from the mean to 1 standard deviation below it. (This symbol, $\approx$, means approximately.)

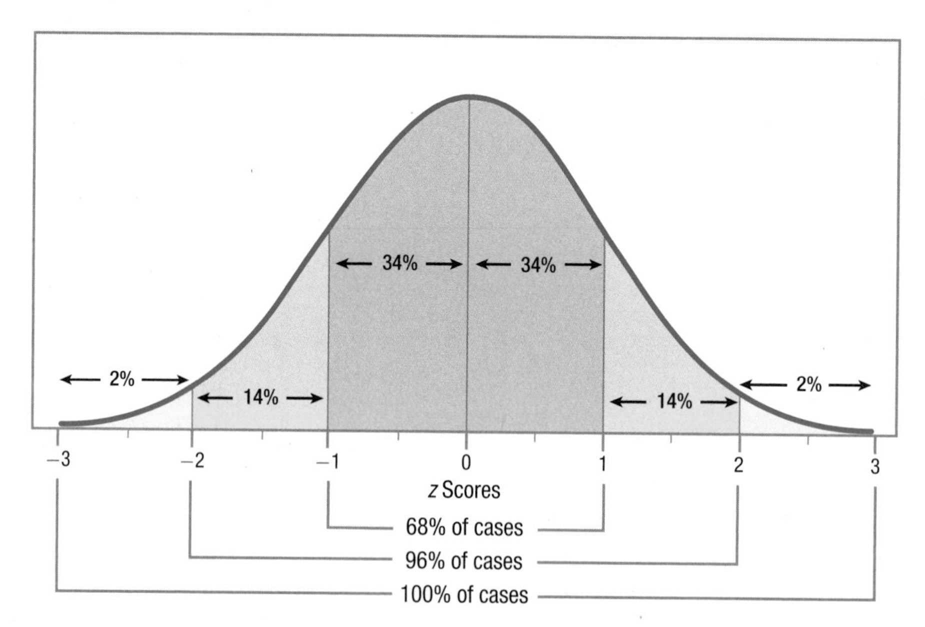

Figure 4.8 Simplified Version of the Normal Distribution This simplified version of the normal distribution takes advantage of the fact that almost all cases fall within 3 standard deviations of the mean and limits the normal distribution to z scores ranging from -3 to $+3$. It uses rounded percentages (34%, 14%, and 2%) for the percentage of cases falling in each standard deviation unit.

- ≈14% of cases fall from 1 standard deviation above the mean to 2 standard deviations above it. Ditto for the area from 1 standard deviation below to 2 standard deviations below.
- ≈2% of cases fall from 2 standard deviations above the mean to 3 standard deviations above it. Again, ditto for the same area below the mean.

Here's a real example of how scores only range from 3 standard deviations below the mean to 3 standard deviations above it. Think of the SAT where subtests have a mean of 500 and a standard deviation of 100. A *z* score of −3.00 is a raw score of 200 on a subtest. This can be calculated using Equation 4.3:

$$\begin{aligned} X &= 500 + (-3.00 \times 100) \\ &= 500 + (-300.0000) \\ &= 500 - 300.0000 \\ &= 200.0000 \\ &= 200.00 \end{aligned}$$

A *z* score of 3.00 is a raw score of 800 on a subtest. This can be calculated using Equation 4.3:

$$\begin{aligned} X &= 500 + (3.00 \times 100) \\ &= 500 + 300.0000 \\ &= 800.0000 \\ &= 800.00 \end{aligned}$$

This means that almost all scores on SAT subtests should range from 200 to 800. And that is how the SAT is scored. No one receives a score lower than a 200 or higher than an 800 on an SAT subtest.

The normal distribution is called "normal" because it is a naturally occurring distribution that occurs as a result of random processes. For example, consider taking 10 coins, tossing them, and counting how many turn up heads. It could be anywhere from 0 to 10. The most likely outcome is 5 heads and the least likely outcome is either 10 heads or no heads. If one tossed the 10 coins thousands of times and graphed the frequency distribution, it would approximate a normal distribution (see Figure 4.9).

Psychologists don't care that much about coins, but they do care about variables like intelligence, conscientiousness, openness to experience, neuroticism,

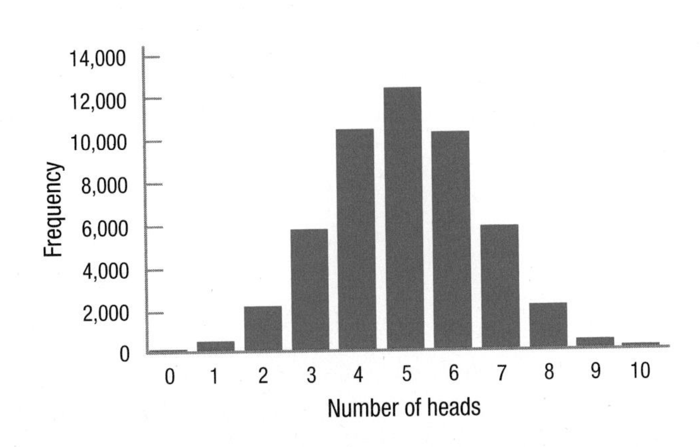

Figure 4.9 Approximation of Normal Distribution for Number of Heads for 10 Coins Tossed Together 50,000 Times If one tossed 10 coins 50,000 times and counted the number of heads, the distribution might look like this. Though this is a bar graph for discrete data, it has a normal shape like that seen for a continuous variable in Figure 4.5.

It is often assumed that psychological variables are normally distributed.

agreeableness, and extroversion. It is often assumed that these psychological variables (and most other psychological variables) are normally distributed. Here's an example that shows how helpful this assumption is:

1. If 34.13% of the cases fall from the mean to a standard deviation above the mean, and
2. if intelligence is normally distributed, and
3. if intelligence has a mean of 100 and a standard deviation of 15,
4. then 34.13% of people have IQs that fall in the range from 100 to 115, from the mean to 1 standard deviation above it.

If a person were picked at random, there's a 34.13% chance that he or she would have an IQ that falls in the range from 100 to 115. That's true as long as intelligence is normally distributed.

Finding percentages using the normal distribution wouldn't be useful if the percentages were only known for each whole standard deviation. Luckily, the normal curve has been sliced into small segments and the area in each segment has been calculated. Appendix Table 1, called a z score table, contains this information. A small piece of it is shown in Table 4.2.

There are a number of things to note about this z score table:

- It is organized by z scores to two decimal places. Each row is for a z score that differs by 0.01 from the row above or below.
- There are no negative z scores listed. Because the normal distribution is symmetrical, the same row can be used whether looking up the area under the curve for $z = 0.50$ or -0.50.
- The columns report the percentage of area under the curve that falls in different sections. This is the same as the percentage of cases that have scores in that region.

TABLE 4.2 From Appendix Table 1: Area Under the Normal Distribution

	A	B	C
z Score	Below +z or Above −z	From Mean to z	Above +z or Below −z
0.50	69.15%	19.15%	30.85%
0.51	69.50%	19.50%	30.50%
0.52	69.85%	19.85%	30.15%
0.53	70.19%	20.19%	29.81%
0.54	70.54%	20.54%	29.46%
0.55	70.88%	20.88%	29.12%
0.56	71.23%	21.23%	28.77%
0.57	71.57%	21.57%	28.43%
0.58	71.90%	21.90%	28.10%
0.59	72.24%	22.24%	27.76%

This is a section of Appendix Table 1. Note that there are three columns of information for each z score.

- The percentages reported are always greater than 0%. In any region of the normal distribution, there is always some percentage of cases that fall in it. Sometimes the percentage is very small, but it is always greater than zero.
- The percentages reported can't be greater than 100%. In any region, there can't be more than 100% of the cases.
- There are three columns, A, B, and C, each containing information about the percentage of cases that fall in a different section of the normal distribution. For each row in the table, there are three different views for each z score.

Column A

For a *positive* z score, column A tells the percentage of cases that have scores *at* or *below* that value. A positive z score falls *above* the midpoint, so the area *below* a positive z score also includes the 50% of cases that fall below the midpoint. This means the area below a positive z score will always be greater than 50% as it includes both the 50% of cases that fall below the mean, and an additional percentage that falls above the mean. Look at **Figure 4.10**, where A marks the area from the bottom of the curve to a z score of 0.50. This area is shaded /// and it should be apparent that it covers more than 50% of the area under the curve. According to column A in Appendix Table 1, the exact percentage of cases that fall in the region below a z score of 0.50 is 69.15%.

Because the normal distribution is symmetric, column A also tells the percentage of cases that fall *at* or *above* a *negative* z score. **Figure 4.11** marks off area A for $z = -0.50$. Now compare Figure 4.10 to Figure 4.11, which are mirror images of each other: the same areas are marked off, just on different sides of the normal distribution. So, 69.15% of cases fall at or above a z score of -0.50.

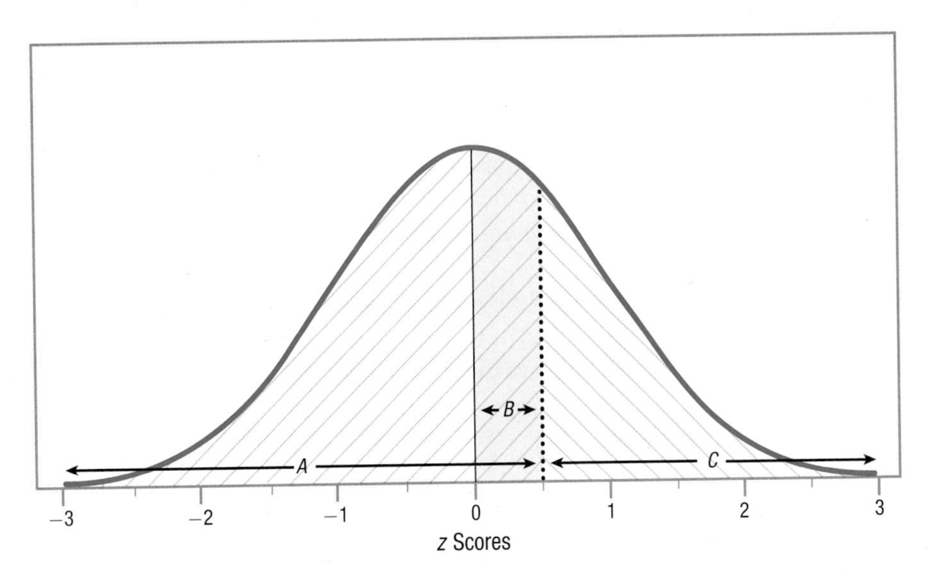

Figure 4.10 Finding Areas Under the Curve for a z Score of 0.50 Area A, marked ///, is the portion of the normal curve that falls below $z = 0.50$. Area B, the shaded area, is the section from the mean to $z = 0.50$. Area C, marked \\\, is the section that falls above $z = 0.50$. These areas, A, B, and C, correspond to the columns A, B, and C in Appendix Table 1.

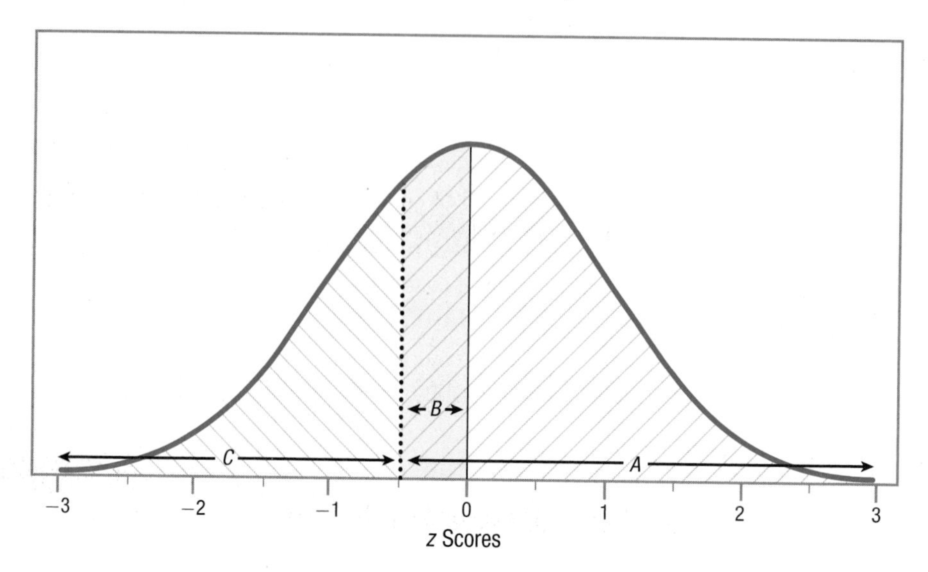

Figure 4.11 Finding Areas Under the Curve for a z Score of −0.50 Compare this figure to Figure 4.10. Note that the area *A* is the same in Figure 4.10 as here in Figure 4.11, but the side of the normal distribution it is on has changed as the sign of the *z* score changes. This shows the symmetry between positive and negative *z* scores. The same is true for area *B* in the two figures and for area *C*.

Column B

In Appendix Table 1, column B tells the percentage of cases that fall from the mean to a *z* score, whether positive or negative. This value can never be greater than 50% because it can't include more than half of the normal distribution. Figures 4.10 and 4.11 show the shaded area *B* for a *z* score of ±0.50. Looking in Table 4.2, one can see that this area captures 19.15% of the normal distribution.

Column C

Column C tells the percentage of cases that fall *at* or *above* a *positive z* score. Because of the symmetrical nature of the normal distribution, column C also tells the percentage of cases that fall *at* or *below* a *negative z* score. Look at Figures 4.10 and 4.11, which show area *C*, marked \\\, for a *z* score of ±0.50. It should be apparent that area *C* values will always be less than 50%. The *z* score table in Appendix Table 1 shows that the exact value is 30.85% for the percentage of cases that fall above a *z* score of 0.50 or below a *z* score of −0.50.

There are three questions that the *z* score table in Appendix Table 1 can be used directly to answer about a normal distribution:

1. What percentage of cases fall above a *z* score?
2. What percentage of cases fall from the mean to a *z* score?
3. What percentage of cases fall below a *z* score?

Figure 4.12 is a flowchart that walks through using the *z* score table in Appendix Table 1 to answer those questions.

Here's an example of the first question: "What is the percentage of cases in a normal distribution that have scores above a *z* score of −2.30?" Before turning to the

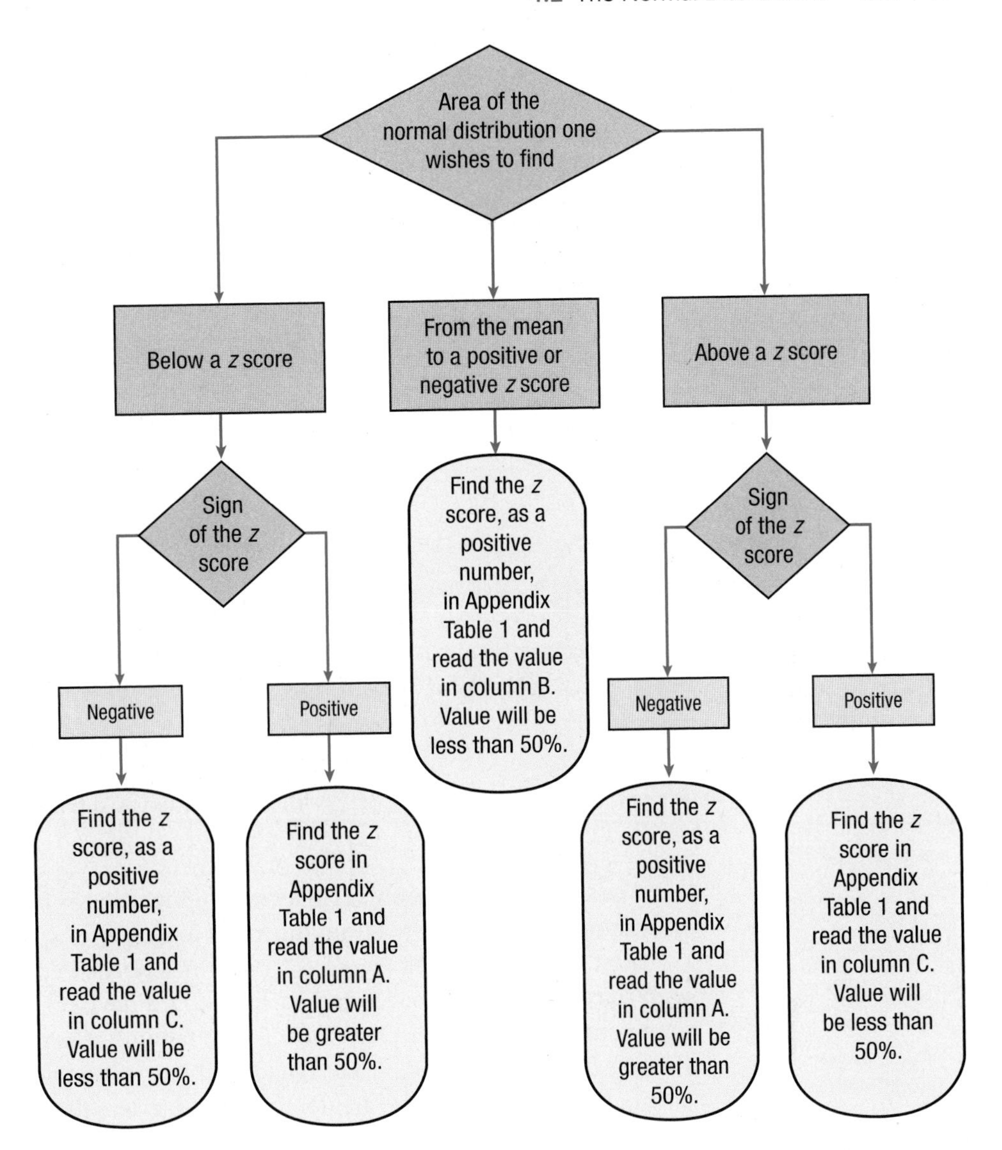

Figure 4.12 How to Choose: Flowchart for Finding Areas Under the Normal Curve It is always a good idea to make a sketch of the normal distribution and shade in the area one wishes to find in order to make a rough estimate. Then, use this flowchart to determine which column (A, B, or C) of Appendix Table 1 will lead to the exact answer.

flowchart in Figure 4.12, it is helpful to make a quick sketch of the area one is trying to find. **Figure 4.13** is a normal distribution, where the *X*-axis is marked from $z = -3$ to $z = 3$. A vertical line is drawn at $z = -2.30$, and the area to the right of the line all the way to the far right side of the distribution is shaded in. That's the area above a *z* score of -2.30. The percentage of cases that fall in this shaded in area is greater than 50% and is probably fairly close to 100%.

Now, turn to the flowchart in Figure 4.12. It directs one to use column A of the *z* score table to find the exact answer. Turn to Appendix Table 1 and go down the column of *z* scores to 2.30. Then look in column A, which tells the area above a negative *z* score, to find the value of 98.93%. And that's the answer: 98.93% of cases in a normal distribution fall above a *z* score of -2.30.

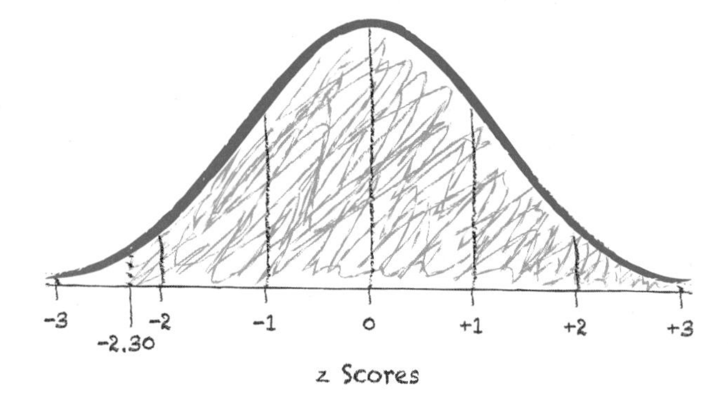

Figure 4.13 Sketch for Finding Area Above a z Score of −2.30 This figure marks the area in a normal distribution that falls above a z score of −2.30. The area includes the section to the right of the midpoint, so the total has to be greater than 50%. Inspection of the figure suggests that the area will be close to 100%.

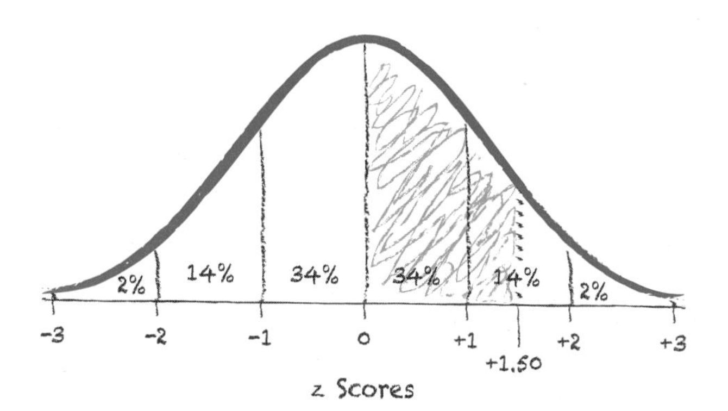

Figure 4.14 Sketch for Finding Area from the Mean to a z Score of 1.50 This figure marks the area in a normal distribution falling from the mean to a z score of 1.50. The numbers (34%, 14%, and 2%) that represent the approximate percentages in each standard deviation help estimate how much area the shaded area represents.

The second type of problem involves finding the area from the mean to a z score. For example, what's the area from the mean to a z score of 1.50?

Start with a sketch. In Figure 4.14, there's a normal curve. The X-axis is labeled with z scores from −3 to +3, and each standard deviation unit is marked with the approximate percentages (34, 14, or 2) that fall in it. There's a line at $z = 1.50$, and the area from $z = 0.00$ to $z = 1.50$ is shaded in. Inspection of this figure shows that the area being looked for will be greater than 34% but less than 48%.

To find the exact answer, use the flowchart in Figure 4.12: the answer can be found in column B of the z score table. The flowchart also indicates that the answer will be less than 50% (which is consistent with what was found in Figure 4.14). Finally, turn to the intersection of the row for $z = 1.50$ and column B in Appendix Table 1 for the answer: 43.32%. In a normal distribution, 43.32% of the cases fall from the mean to 1.50 standard deviations above it.

The third type of question asks one to find the area below a z score. For example, what is the area below a z score of −0.75?

As always, start with a sketch. Figure 4.15 shows a normal distribution, filled in with the approximate areas in each standard deviation. There's a vertical line at −0.75, and the area to the left of this line (the area below this z score) is shaded in. That's the area to be found. This area will be larger than 16% because it includes the bottom 2%, the next 14%, and then some more.

To find the exact answer, turn to the flowchart in Figure 4.12. The flowchart says to use column C to find the z score (a positive number) in Appendix Table 1. Looking at the intersection of the row where $z = 0.75$ and column C, we see that the answer is 22.66%. In a normal distribution, 22.66% of the scores are lower than a z score of −0.75.

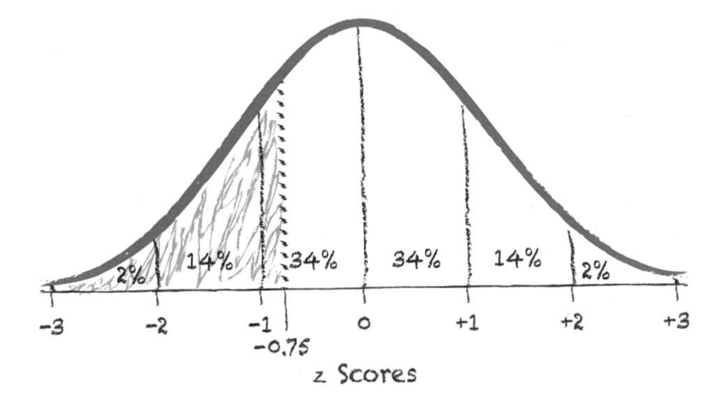

Figure 4.15 Sketch for Finding Area Less Than a z Score of −0.75 This figure marks the area in a normal distribution falling below a *z* score of −0.75. The numbers (34%, 14%, and 2%) that represent the approximate percentages in each standard deviation help estimate how much area the shaded area represents. The shaded area will be a little bit more than 16%.

After some practice finding the area under the normal distribution curve, we will move on to two new ways to use *z* scores and the normal curve. One, percentile ranks, is a new way to describe a person's performance. The other, probability, underpins how statistical tests work.

Worked Example 4.2

Here's some practice finding the area under the normal distribution using a real-life example. The expected length of a pregnancy from the day of ovulation to childbirth is 266 days, with a standard deviation of 16 days. Assuming that the length of pregnancy is normally distributed, what percentage of women give birth at least two weeks later than average? Two weeks is equivalent to 14 days, so the question is asking for the percentage of women who deliver their babies on day 280 or later.

The first step is to transform 280 into a *z* score:

$$z = \frac{280 - 266}{16}$$

$$= \frac{14.0000}{16}$$

$$= 0.8750$$

$$= 0.88$$

Next, make a sketch to isolate the area we're looking for. In Figure 4.16, the area above a *z* score of 0.88 is shaded in. This area will be less than 50%. The area includes the distance from 2 standard deviations above the mean to 3 standard deviations above it (2%), the distance from 1 standard deviation above the mean to 2 standard deviations above it (14%), and then a little bit more. The area will be a bit more than 16%.

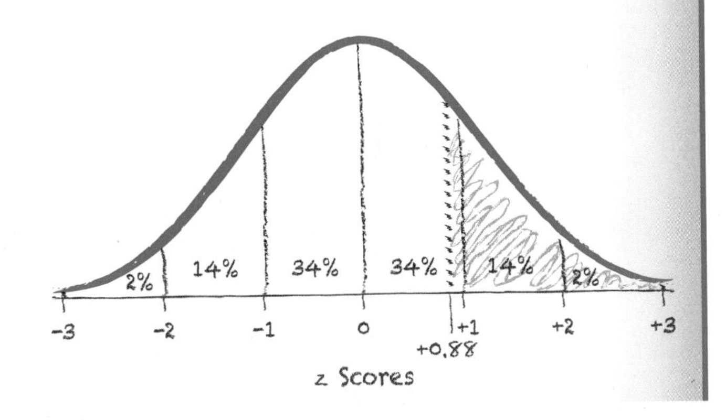

Figure 4.16 Sketch for Finding Probability of Having a Pregnancy That Lasts Two or More Weeks Longer Than Average The dotted line at $z = 0.88$ marks the area where deliveries two weeks later than average start. The numbers (34%, 14%, and 2%) that represent the approximate percentages in each standard deviation help estimate how much area the shaded area represents. The area will be a little bit more than 16%.

The area being looked for is the area above a z score. Consulting Figure 4.12, it is clear that the answer may be found in column C of Appendix Table 1. There, a z score of 0.88 has the corresponding value of 18.94%. And that's the conclusion—almost 19% of pregnant women deliver two weeks or more later than the average length of pregnancy.

Practice Problems 4.2

Review Your Knowledge

4.06 Describe the shape of a normal distribution.

4.07 What measure of central tendency is the midpoint of a normal distribution?

4.08 What do numbers 34, 14, and 2 represent in a normal curve?

Apply Your Knowledge

4.09 What percentage of area in a normal distribution falls at or above a z score of -1.33?

4.10 What percentage of area in a normal distribution falls at or above a z score of 2.67?

4.11 What percentage of area in a normal distribution falls from the mean to a z score of -0.85?

4.3 Percentile Ranks

So far, the z score table in Appendix Table 1 has been used as it was designed. First, the z score is found in the table and then the area in the normal distribution is found above, below, or to that z score. Now, the table will be used to calculate percentile ranks.

A **percentile rank** tells the percentage of cases whose scores are at or below a given level in a frequency distribution. Percentile ranks provide a new way of thinking about a person's score. Imagine a student, Leah, who obtained a score of 600 on an SAT subtest. (Remember, SAT subtests have a mean of 500 and a standard deviation of 100.) There have been two ways thus far of expressing Leah's score: (1) as a raw score ($X = 600$), or (2) as a standard score ($z = 1.00$).

Here's a third way, as a percentile rank. Using the z score of 1.00 and looking in column A of Appendix Table 1, 84.13% of the area under the normal curve falls at or below Leah's score. Leah's score as a percentile rank, abbreviated *PR*, is 84.13. Leah scored higher on this SAT subtest than about 84% of students. When her score is expressed in percentile rank form, it's clear that Leah did well on this test.

Now let's say that Joshua took the same test and his percentile rank was 2. Joshua didn't do very well—he only performed better than 2% of other students. Comparing Leah to Joshua shows a major advantage of percentile ranks—they put scores into an easily interpretable context.

Figure 4.17 shows a normal distribution that is marked off (1) with the approximate percentages in each standard deviation (34%, 14%, and 2%), (2) with z scores (from -3 to $+3$) on the X-axis, and (3) with approximate percentile ranks, also on the X-axis. There are several things to note in this figure:

- In this simplified version of the normal distribution, there are no cases that fall more than 3 standard deviations below the mean and none that fall more than 3 standard deviations above it.
- This means that the percentile rank associated with a z score of -3 is 0, and the percentile rank associated with a z score of $+3$ is 100.

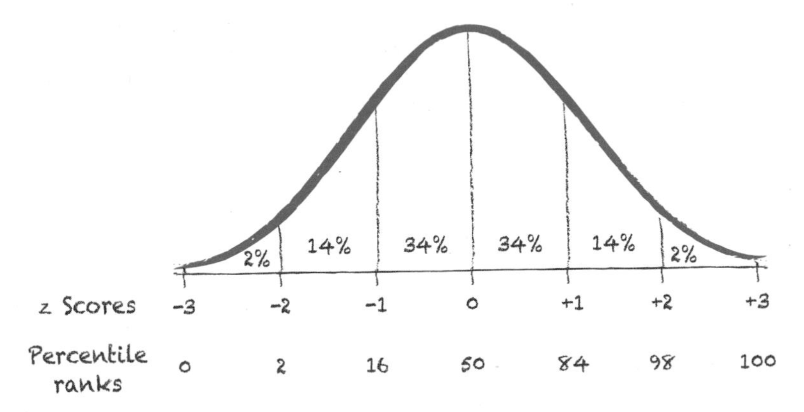

Figure 4.17 Relationship Among the Area Under the Normal Curve, z Scores, and Percentile Ranks Percentile ranks, which can be computed from *z* scores, express the same information but in a different format.

- The midpoint is the median. This means that a score at the midpoint has a percentile rank of 50.
- ≈2% of cases fall from the bottom of the normal curve to $z = -2$, so the percentile rank for $z = -2$ is 2.
- Moving up the normal distribution, it's clear that percentile ranks grow as one moves to the right along the *X*-axis.

A Common Question

Q What is the level of measurement for percentile ranks?

A Percentile ranks don't have equality of units, so they are ordinal. Look at Figure 4.16 and note that the distance (in *z* score units) from a *PR* of 2 to a *PR* of 16 is the same as the distance from a *PR* of 16 to a *PR* of 50. In one case, a single *z* score covers 14 *PR* units and in the other it covers 34 *PR* units.

A percentile rank can be calculated from a raw score or a *z* score. Alternatively, a raw score or a *z* score can be calculated from a percentile rank by using Appendix Table 1. The flowchart in Figure 4.18 shows how to do this.

Imagine another student, Kelli, who took the same SAT subtest and whose score had a percentile rank of 40. What is her *z* score? A sketch always helps. Figure 4.19 shows a normal distribution. The *X*-axis is marked with *z* scores, percentile ranks, and SAT subtest scores, which shows how these three scores are equivalent. For example, a *z* score of 1 is equivalent to a percentile rank score of 84 and an SAT subtest score of 600. There is also an X on the *X*-axis about where a percentile rank of 40 would be. The X shows that this percentile rank will have an associated *z* score that falls somewhere between −1 and 0 and an associated SAT score that falls somewhere between 400 and 500.

Now, it is time to calculate Kelli's *z* score more exactly. According to the flowchart in Figure 4.18, to find the *z* score associated with a percentile rank less than 50, go to column C in Appendix Table 1, which shows that the *z* score will be negative. Move down column C. The row closest to a value of 40.00 shows a *z* score of 0.25. This score needs to be treated as a negative number, so the answer is −0.25. Kelli's grade on the test, as a *z* score, is −0.25.

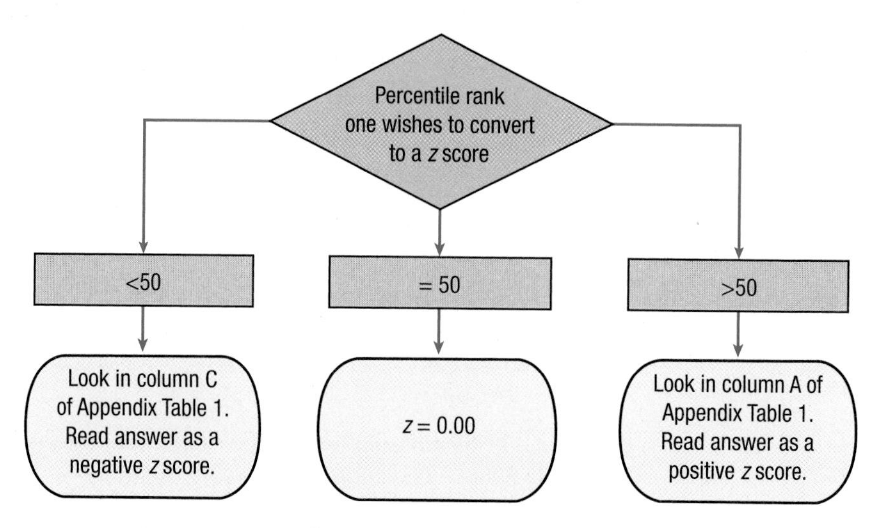

Figure 4.18 How to Choose: Flowchart for Calculating *z* Scores from Percentile Ranks Before calculating a *z* score for a percentile rank, it is a good idea to sketch out a normal distribution with both *z* scores and percentile ranks on the *X*-axis in order to estimate the answer. Once the answer is in *z* score format, it can be transformed to a raw score using Equation 4.3.

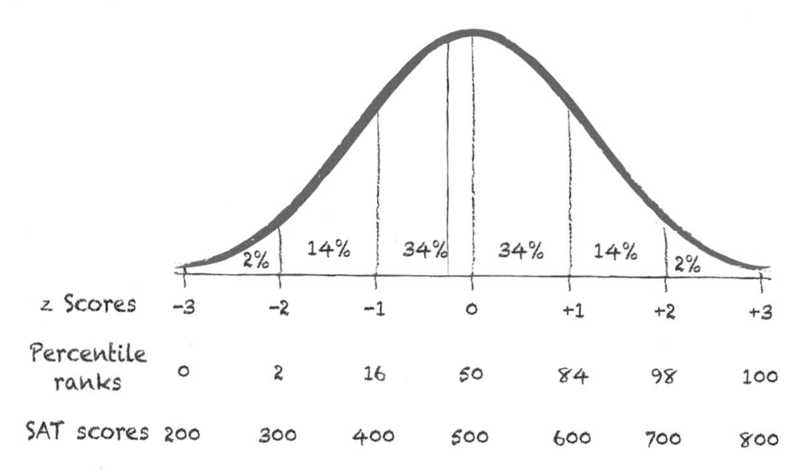

Figure 4.19 Sketch for Figuring Out the SAT Score for a Student with a Percentile Rank of 40 By labeling the *X*-axis with percentile rank scores as well as with SAT scores, it is possible to make a rough estimate as to what SAT score corresponds to a given percentile rank score.

Kelli's grade is now known in two formats: as a percentile rank (40) and as a *z* score (−0.25). What is her score as a raw score on the SAT subtest? To answer that question, use Equation 4.3:

$$
\begin{aligned}
X &= 500 + (-0.25 \times 100) \\
&= 500 + (-25.0000) \\
&= 500 - 25.0000 \\
&= 475.0000 \\
&= 475.00
\end{aligned}
$$

Kelli obtained a 475.00 on this SAT subtest.

Worked Example 4.3 Imagine that Clayton takes an IQ test, with a mean of 100 and a standard deviation of 15, and does better than 95% of the population. If he did better than 95%, then his score, as a percentile rank, is 95. What was his IQ score if his score was a 95 as a percentile rank?

First, make a sketch. In **Figure 4.20**, the *X*-axis is marked with *z* scores, percentile ranks, and IQ scores and a dotted line is drawn about where *PR* 95 should be. The line is between *z* scores of 1 and 2, closer to 2 than to 1. In IQ units, it is between 115 and 130, closer to 130.

To calculate the raw score, use the flowchart in Figure 4.18. Clayton's score has a percentile rank above 50, so turn to column A of Appendix Table 1. In that column, look for the value closest to 95%. In this case, there are two values, 1.64 and 1.65, that are equally close to 95.00. Logic can be used to decide which of these to select:

- A *z* score of 1.64 is higher than 94.95% of scores. This doesn't quite capture the 95% value that is Clayton's score.
- A *z* score of 1.65 is higher than 95.05% of scores. This captures the 95% value that is Clayton's score.

To select the score that is as good or better than 95% of the cases in a normal distribution, go with 1.65. Clayton's score, as a *z* score, is 1.65.

To convert that *z* score to an IQ score, use Equation 4.3:

$$\begin{aligned} X_{IQ} &= 100 + (1.65 \times 15) \\ &= 100 + 24.7500 \\ &= 124.7500 \\ &= 124.75 \end{aligned}$$

Clayton's score on the IQ test is 124.75, a score that was better than that of 95% of the population.

Remember the angioplasty graph from the beginning of the chapter? It showed that most cities had "normal" rates of angioplasty and one city was unusually high. Let's see how unusual that city is.

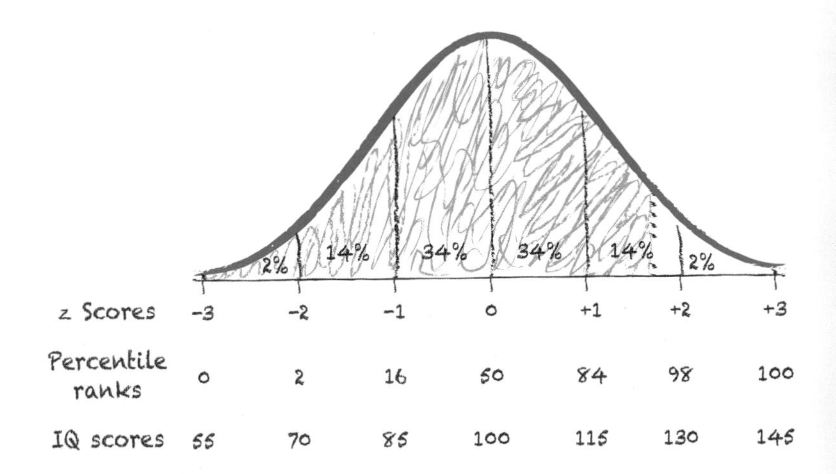

Figure 4.20 Sketch for Finding the IQ Score Associated with a Percentile Rank of 95 One can use the percentile ranks marked on the *x*-axis to estimate where to draw the line for the sketch for the IQ score associated with a percentile rank of 95.

To do so, we need to know the mean angioplasty rate for the sample of cities (11) and the standard deviation (4). This allows us to calculate a z score for the outlier city that has an angioplasty rate of 42:

$$z = \frac{X - M}{s} = \frac{42 - 11}{4} = 7.75$$

This city has an angioplasty rate that is 7.75 standard deviations above the national average. This is a very extreme score. The z score table in the back of the book only goes up to a z score of 5.00; and "only" 0.00003% of cases fall above this. A score of 7.75 is farther out in the tail than this, so it is an incredibly rare event. To a statistician's mind, such extreme events don't just happen by chance. There's something fishy going on in that city, something that has inflated the rate at which angioplasties are performed.

Practice Problems 4.3

Review Your Knowledge

4.12 What is a percentile rank?

4.13 What is the percentile rank for a score at the midpoint of a set of normally distributed scores?

Apply Your Knowledge

4.14 If a person's score, as a percentile rank, is 80, what is her z score?

4.15 What is the percentile rank associated with a z score of -0.45?

4.4 Probability

The focus of this chapter now shifts to probability. To understand why probability is important, consider a study in which a sample of people with depression is divided into two groups, one of which is given psychotherapy and the other isn't. After treatment, the two groups are assessed on a depression scale and the psychotherapy participants are found to be less depressed than the no-psychotherapy participants. Which conclusion, A or B, would a statistician draw?

A. Psychotherapy is effective in the treatment of depression.

B. Psychotherapy is probably effective in the treatment of depression.

A statistician would pick statement B, that psychotherapy is *probably* effective in the treatment of depression. Why? Because she would know that psychotherapy was effective for the *sample* tested, but she couldn't be sure it would be effective for all the depressed people not in the sample, for the larger *population* of depressed people.

Statisticians' conclusions involve *probability,* just as weather forecasters say it will "probably" rain tomorrow. The conditions may be right to predict that it will rain, but rain is not guaranteed. **Probability** concerns how likely it is that an event or outcome will occur. Probability is defined as the number of ways a specific outcome (or set of outcomes) can occur, divided by the total number of possible outcomes. The formula, with probability abbreviated as p and the specific outcome abbreviated as A, is given in Equation 4.4.

Equation 4.4 Formula for Calculating Probability

$$p(A) = \frac{\text{Number of ways outcome } A \text{ can occur}}{\text{Total number of possible outcomes}}$$

where $p =$ probability
$A =$ specific outcome

Coins are often used to teach probability. A coin has two possible outcomes, heads or tails, and they are *mutually exclusive.* This means the coin can only land heads or tails—not both—on a given toss. (*Fairness,* that each outcome is equally likely, is going to be assumed for all examples.) Applying Equation 4.4, if the coin has two possible outcomes, heads or tails, and there is only one way it can land heads, the probability of it landing heads is $p(\text{heads}) = \frac{1}{2}$. This could be reported a number of ways:

- Probability can be reported as a fraction. In this case, one would say, "There is a 1 out of 2 chance that a coin will turn up heads."
- Probability can be reported as a proportion by doing the math in the fraction, by dividing the numerator by the denominator: $\frac{1}{2} = .50$. One could report, "The probability of a coin turning up heads is .50." Or, one could say, "$p(\text{heads}) = .50$."
- Finally, by multiplying the proportion by 100, the probability can be turned into a percentage: $.50 \times 100 = 50$. One could say, "There is a 50% chance that a coin will land heads."

To get a little more complex than a single coin, let's toss two coins so that two outcomes occur at once. We'll assume that the two outcomes are *independent.* **Independence** means that the occurrence of one outcome has no impact on the other outcome. This means that how the first coin turns out, heads or tails, has no impact on how the second coin turns out.

As shown in Figure 4.21, there are four possible outcomes for the two coins: (1) the first is heads and the second is heads (HH), (2) the first is heads and the second is tails (HT), (3) the first is tails and the second is heads (TH), and (4) the first is tails and the second is tails (TT).

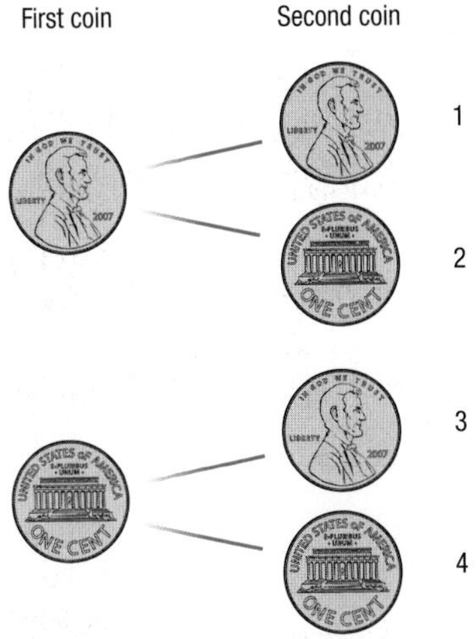

First coin
Second coin
1
2
3
4

Figure 4.21 Four Possible Outcomes for Two Coins Being Tossed The first coin can be either heads or tails and so can the second coin, leading to four unique outcomes. Each of the four outcomes has one chance in four of occurring, so the probability of each outcome is $\frac{1}{4}$.

TABLE 4.3 Probabilities, Expressed as Proportions, for the Three Outcomes for Two Coins Being Tossed

Outcome	Probability
Two heads (HH)	.25
One heads and one tails (HT or TH)	.50
Two tails (TT)	.25
	$\Sigma = 1.00$

There are three possible outcomes if two coins are tossed. The sum of the probabilities for all three outcomes is 1.00.

What is the probability of obtaining two heads? There's one way such a result can occur out of four possible outcomes, so $p(\text{two heads}) = \frac{1}{4} = .25$. What is the probability of getting a heads and a tails? There are two ways this can occur (HT or TH) out of four possible outcomes, so $p(\text{one heads and one tails}) = \frac{2}{4} = .50$. Finally, there's one other outcome possible for the two coins: $p(\text{two tails}) = \frac{1}{4} = .25$. Table 4.3 summarizes the probabilities of the three outcomes for the two coins.

The probability for any given outcome can't be less than zero or greater than 1.

If a probability for an outcome is 1.00, or expressed as a percentage 100%, that means this outcome is a sure thing. In fact, any given outcome can't have a probability higher than 1.00, and the sum of the probabilities of all the possible outcomes is 1.00. If two coins are tossed, it is guaranteed that one of these outcomes—two heads, *or* a heads and a tails, *or* two tails—will turn out. In addition, the probability of an outcome occurring can't be less than .00, or 0%. A probability of zero means that an outcome can't happen. Putting together the two facts in this paragraph, the probability for any given outcome can't be less than zero or greater than 1. Stated more formally,

$$.00 \leq p(A) \leq 1.00$$

A Common Question

Q When probabilities are reported as a proportion, they are written without a leading zero, for example, as .25 not as 0.25. Why?

A APA format says to put a zero before the decimal point for numbers that are less than 1 only when the numbers can be greater than 1. Probabilities can't exceed 1, so they don't get a leading zero.

The most common way that statisticians calculate and use probability involves the normal distribution. They often ask, if a variable is normally distributed and a case is selected at random from a population, how likely is it—how probable is it—that this case would have a score that falls in a certain range?

The material covered in Section 4.2 of this chapter can be construed in just this way. For example, if 50% of cases fall at or above the mean, then the probability of a case selected at random having a score at or above the mean is .50. If approximately 84% of the cases fall in the region at or below a z score of 1, then the probability of picking a case at random and it having a z score less than or equal to 1 is .84. (To move from a percentage to a probability, move the decimal two places to the left; to move from a probability to a percentage, move the decimal place two spots to the right.)

Statisticians use probabilities to determine if a result is common or rare. By *common,* they mean the result occurs frequently, that it falls somewhere near the middle of a normal distribution. Typically, statisticians consider something that has a 95% chance of occurring as common. A *rare* result is one that is unlikely to occur, one that falls in the ends, the tails, of the distribution. As will be seen in Chapter 6, how rare something must be to be considered rare is a decision that should be made on an experiment-by-experiment basis. But, usually, statisticians consider something rare if it has a 5% or less chance of occurring. Phrased in terms of probability, if $p < .05$, it is usually considered a rare outcome to a statistician.

Let's find the z scores that mark off the middle 95% of scores in a normal distribution. This will lead us to one of the most important numbers in statistics, 1.96.

Follow the procedure in Table 4.4 to find the z scores that mark off the middle 95% of cases. The first step is to take the percentage, here 95%, and split it in two:

$$\frac{95}{2} = 47.5000 = 47.50$$

TABLE 4.4 How to Choose: Finding z Scores Associated with Middle and Extreme Percentages of the Normal Distribution

	Finding z Scores Associated with a Middle Percentage of the Normal Distribution
Step 1	Take the middle percentage and cut it in half.
Step 2	Use column B of Appendix Table 1 to find the percentage closest to the ½ value calculated in Step 1.
Step 3	Report the z score for the percentage in ± format.
	Finding z Scores Associated with an Extreme Percentage of the Normal Distribution
Step 1	Take the extreme percentage and cut it in half.
Step 2	Use column C of Appendix Table 1 to find the percentage closest to the ½ value calculated in Step 1.
Step 3	Report the z score for the percentage in ± format.

These guidelines help one calculate the percentage of cases, in a normal distribution, that fall in the middle or in the two tails.

Why split it in two? Because that tells us how much of the area is above the mean and how much is below it. We now know that the 95% consists of 47.5% (half) above the mean and 47.5% (the other half) below the mean.

The second step is to look in column B of Appendix Table 1 to find the value closest to this percentage. Column B contains the exact value of 47.50, and the z score associated with it is 1.96. The final step in Table 4.4 is to report the results as a range from a negative z score to a positive z score: in a normal distribution, the middle 95% of cases fall from a z score of -1.96 to a z score of 1.96 (see Figure 4.22). Or, thinking in terms of probability, if one took a case at random from the population, the probability is .95 that it came from the interval ranging from $z = -1.96$ to $z = 1.96$.

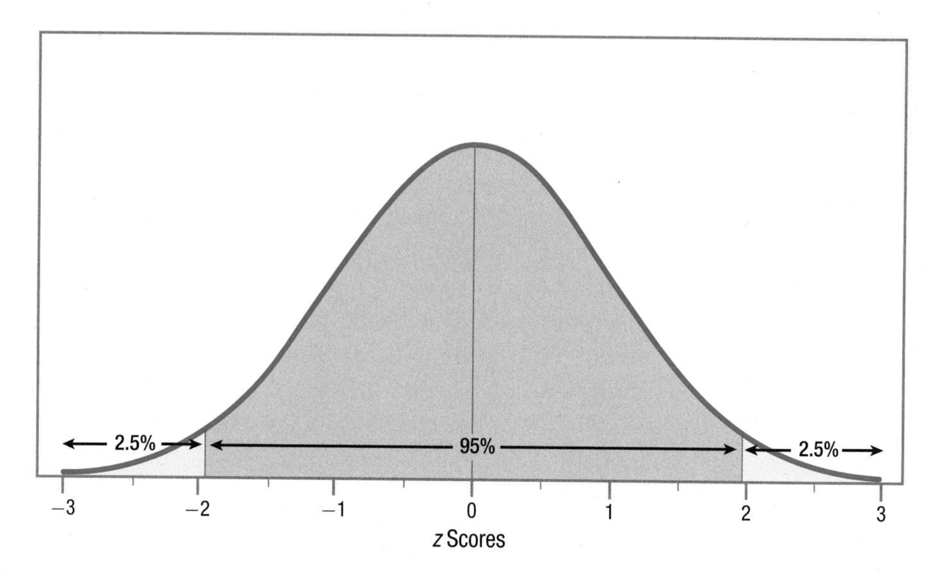

Figure 4.22 Middle and Extreme Sections of the Normal Curve Associated with z scores of ± 1.96 The same z scores, ± 1.96, that mark off the middle 95% of scores also mark off the extreme 5% of scores.

Look at Figure 4.22 that shows the middle 95%. If 95% of the cases fall from a z score of -1.96 to a z score of 1.96, isn't that the same as saying 5% of the cases don't fall in that region? The complement to a **common zone**, the middle section, is the **rare zone**, the section in the ends, the two tails, of the distribution. This section in the extremes of the distribution is evenly divided between the two tails. The extreme 5%, for example, consists of the 2.5% at the very bottom of the normal distribution (the far left side in Figure 4.22) *and* the 2.5% at the very top of it (the far right side in Figure 4.22). Of course, this can also be expressed as a probability: $p = .05$ that a case selected at random from the population falls below -1.96 or above 1.96 in a normal distribution. (Table 4.4, which shows how to calculate a **middle percentage**, also contains guidelines on how to calculate an **extreme percentage**.)

The z scores, ±1.96, that mark off the middle 95% of scores in a normal distribution are also the z scores that mark off the extreme 5% of scores in a normal distribution.

The z scores, ± 1.96, that mark off the middle 95% of scores in a normal distribution are also the z scores that mark off the extreme 5% of scores in a normal distribution. Keep your eyes open, these values will appear a lot more in later chapters. You'll see 1.96 used in formulas; you'll see 95% used in confidence intervals; and you'll see two probabilities, $p < .05$ and $p > .05$, to indicate, respectively, whether a result is rare or common.

Worked Example 4.4

Mount Pleasant High School instituted a new math curriculum. The principal was not worried about how the average students would fare, but was concerned about the poorer and the better students. He wanted to assess the impact on the extreme 50% of students as measured by IQ. He needed to find the IQ scores that mark off the extreme 50% of cases, the 25% at the top and the 25% at the bottom.

Figure 4.23, where the X-axis is marked off with z scores and IQ scores, is the sketch he made. The approximate areas falling in each standard deviation are noted, and dotted lines are drawn about where the cut-off points will be. Based on the sketch, one cut-off point is expected to occur between z scores of

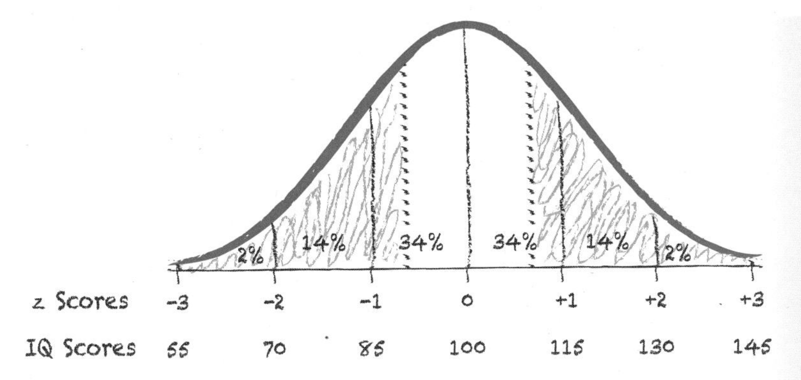

Figure 4.23 Sketch for Finding the Extreme 50% of IQ Scores The extreme 50% of scores consist of the bottom 25% of scores, the shaded area on the left of the curve, and the top 25%, the shaded area on its right.

−1 and 0. The other cut-off point will occur, symmetrically, on the positive side of the curve.

How did he know where to sketch the cut-off points? The objective is to have half of the 50% of extreme scores, or 25%, in each tail. He knew that below, or to the left of, a z score of −1, about 16% of the scores fall. That isn't 25%. So, he needed to move partway into the next standard deviation, enough to add 9% to the existing 16%. This puts the z score somewhere between $z = -1$ and $z = 0$.

Now, with a rough idea of what the answer was, he followed the guidelines in Table 4.4 to find the exact answer. First, he cut the percentage in half. Then, he used column C of Appendix Table 1 to find the z score associated with the half percentage value. Looking in column C of Appendix Table 1, he found that the value closest to 25% was 25.14 and the z score associated with it was 0.67. In z score format, he had his answer: below a z score of −0.67 *and* above a z score of 0.67, the extreme 50% of cases fall.

But, to answer the original question, he needed to turn the z scores into IQ scores using Equation 4.3:

$$\begin{aligned} X &= 100 + (-0.67 \times 15) \\ &= 100 + (-10.0500) \\ &= 100 - 10.0500 \\ &= 89.9500 \\ &= 89.95 \end{aligned}$$

$$\begin{aligned} X &= 100 + (0.67 \times 15) \\ &= 100 + 10.0500 \\ &= 110.0500 \\ &= 110.05 \end{aligned}$$

He now had his answer: The IQ scores that mark off the extreme 50% of scores are 89.95 and 110.05. He would need to follow the progress of students with IQs lower than 90 or higher than 110 to see how the new math curriculum affected the extreme 50% of students.

Practice Problems 4.4

Review Your Knowledge

4.16 What is the definition of probability?

4.17 What's the smallest possible probability for an outcome? The largest?

4.18 If the z scores that mark off the middle 60% of scores are $\pm.84$, what are the z scores that mark off the extreme 40% of scores?

Apply Your Knowledge

For Practice Problems 4.19–4.20, assume that in any given year, there are 52 weeks in a year, 365 days in a year, and equal numbers of every day of the week.

4.19 If a researcher picked one date at random out of a year, what is the probability of it being a Monday?

4.20 If a researcher picked one date at random out of a year, what is the probability of it not being a Monday?

4.21 What are the z scores that mark the middle 34% of scores?

4.22 If someone is picked at random, what is the probability that her IQ will be 123 or higher? (Use a mean of 100 and a standard deviation of 15.)

Application Demonstration

Let's use some historical data about IQ to see material from this chapter in action. In the past, intelligence was used as the sole criterion of whether a person was classified as having an intellectual disability or being gifted. These days, multiple criteria are used. Let's see what the implications of the former criteria are, based on an IQ test with a mean of 100 and a standard deviation of 15.

Let's start with intellectual disability. At one point, people were considered intellectually disabled if their IQs were less than 70. Now, adaptive behavior—the ability to function in everyday life—is also taken into account. But, using the old definition, if a person were picked at random, what is the probability that he or she would be classified as intellectually disabled?

An IQ score below 70 means an IQ score less than or equal to 69, so our question becomes, "What is the probability of having an IQ score $\leq$ 69?" It always helps to visualize what one is doing, so sketch out a normal distribution. Figure 4.24 shows a normal distribution with the X-axis marked in both IQ scores and z scores. For the IQ scores, the midpoint is 100 and then the standard deviation is added or subtracted in both directions 3 times. Thus, the IQ scores range from 55 to 145. There is a dotted vertical line marking an IQ score of 69 and the area below the line, to the left of it, is shaded in. This is the area to be found. It should be apparent that this is not a large percentage of cases. In fact, it should be close to 2%.

Now, it is time to calculate more exactly. Using Equation 4.2, convert the IQ score to a z score:

$$\begin{aligned} z_{\text{IQ}} &= \frac{X - M}{s} \\ &= \frac{69 - 100}{15} \\ &= \frac{-31.0000}{15} \\ &= -2.0667 \\ &= -2.07 \end{aligned}$$

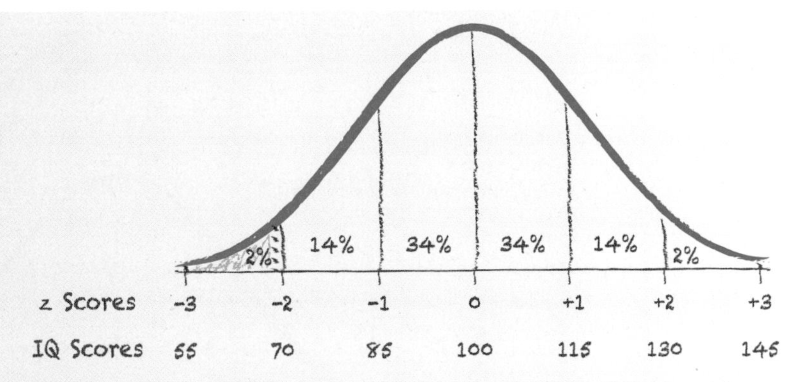

Figure 4.24 Sketch for Finding Percent of Cases with IQ Scores at or Below 69
Shading in the area at or below an IQ score of 69 allows one to estimate that less than 2% of the population falls in that area.

An IQ score of 69 is equivalent to a *z* score of −2.07. (Remember, the fact that the *z* score is negative means that it falls below the mean.) Now, take that *z* score and find the percentage of cases at or below it. The flowchart in Figure 4.12 directs one to column C of Appendix Table 1. Turn to that table and find the row with a *z* score of 2.07. (Remember, the normal distribution is symmetrical, so one can consider 2.07 the same as −2.07.) The answer as a percentage, 1.92%, is found in column C. We can turn 1.92% into a probability of .0192 by moving the decimal place two places to the left. And that's the answer—if a person were picked at random out of the world, the probability is .0192 that he or she would be classified as intellectually disabled if that classification were based entirely on IQ.

Here's one more way to use this probability. If the population of the United States is 320 million, how many have IQs of 69 or lower? Answer this by taking the probability, .0192, and multiplying it by the population:

$$.0192 \times 320{,}000{,}000 = 6{,}144{,}000$$

That's a lot of people! Using the old standard, more than 6 million people in the United States should have IQs of 69 or lower and would be classified as intellectually disabled.

Let's look at the other side of the intelligence continuum—genius. How common are geniuses? One cut-off for genius, on a standard IQ test (mean = 100 and standard deviation = 15), is a score of 145 or higher. This is sketched in Figure 4.25. It should be apparent that this is the upper limit of the simplified version of the normal distribution. There are very few people with IQ scores of 145 or higher.

To find out what percentage of people could be classified as geniuses, use Equation 4.2 to turn the IQ score into a *z* score:

$$\begin{aligned} Z_{IQ} &= \frac{145 - 100}{15} \\ &= \frac{45.0000}{15} \\ &= 3.0000 \\ &= 3.00 \end{aligned}$$

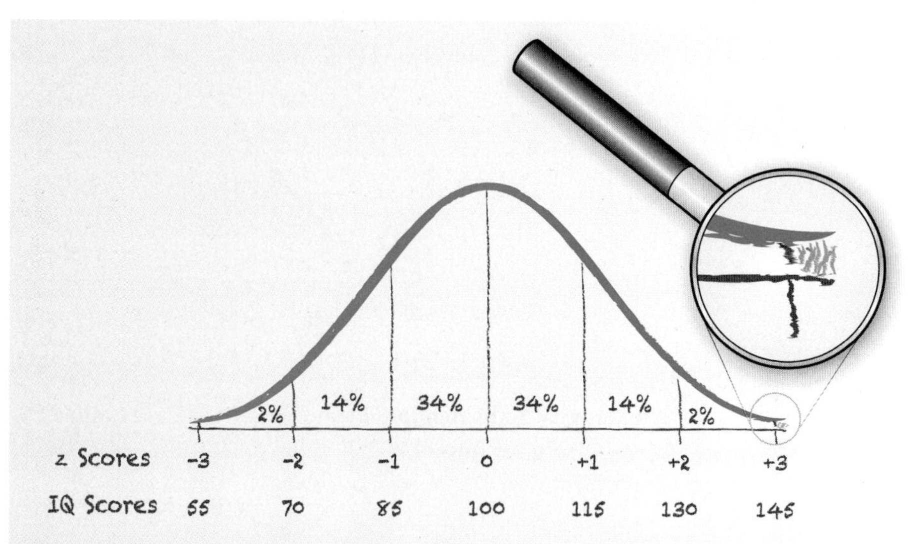

Figure 4.25 Sketch for Finding Percentage of Population with IQs of 145 or Higher Trying to shade in the area to the right of an IQ score of 145 makes one aware of how few scores fall in that area.

The flowchart in Figure 4.11 indicates that the exact percentage may be found in column C of Appendix Table 1. Reading across the row in the z score table for a z score of 3.00, one finds that the value in column C is 0.135%. And that's the answer: assuming intelligence is normally distributed and considering anyone with an IQ of 145 or higher a genius, then 0.135% of the population will be geniuses.

What does that mean for the 320 million Americans? Turn 0.135% into a proportion (.00135) and multiply that by the population:

$$.00135 \times 320{,}000{,}000 = 432{,}000$$

Fewer than half a million Americans, only 432,000 to be exact, should have IQs high enough to be classified as geniuses.

One last thing, let's suppose someone takes an IQ test and gets a 145. What's his or her score as a percentile rank? The flowchart in Figure 4.18 says to consult column A of Appendix Table 1. Reading column A for a z score of 3.00, we see that the person's score is 99.865, as a percentile rank. Not a bad percentile rank to have.

DIY

Dig out the standardized test scores, either the SAT or the ACT, that you used on your college application. How did you do? Can you turn your score into a z score and a percentile rank? For the combined SAT score, which is the sum of the math, writing, and critical reading tests, use a mean of 1,500 and a standard deviation of 300 (If you have taken the new SAT, which has only two subtests, use a mean of 1000 and a standard deviation of 200). For the ACT composite score, use a mean of 20.5 and a standard deviation of 5.5.

SUMMARY

Transform raw scores into standard scores (z scores) and vice versa.

- A standard score, called a *z* score, transforms a raw score so that it is expressed in terms of how many standard deviations it falls away from the mean. A positive *z* score means the score falls above the mean, a negative *z* score means it falls below the mean, and a *z* score of zero means the score falls right at the mean. *z* scores standardize scores, allowing different variables to be expressed in a common unit of measurement.

Describe the normal curve and calculate the likelihood of an outcome falling in specified areas of it.

- The normal distribution is important because it is believed that many psychological variables are normally distributed. The normal curve is a specific bell-shaped curve, defined by the percentage of cases that fall in specified areas. About 34% of cases fall from the mean to 1 standard deviation above the mean, ≈14% fall from 1 to 2 standard deviations above the mean, and ≈2% fall from 2 standard deviations above to 3 above the mean. Because the normal distribution is symmetric, the same percentages fall below the mean. It is rare that a case has a score more than 3 standard deviations from the mean.
- Appendix Table 1 lists the percentage of cases that fall in specified segments of the normal distribution. A flowchart, Figure 4.12, can be used as a guide to find the area under a normal curve that is above a *z* score, below a *z* score, or from the mean to a *z* score.

Transform raw scores and standard scores into percentile ranks and vice versa.

- Percentile ranks tell the percentage of cases in a distribution that have scores at or below a given level. They provide easy-to-interpret information about how well a person performed on a test.

Calculate the probability of an outcome falling in a specified area under the normal curve.

- Conclusions that statisticians draw are probabilistic.
- Probability is defined as the number of ways a specific outcome or event can occur, divided by the total number of possible outcomes.
- The normal curve can be divided into a middle section and extreme sections. A middle percentage of scores is symmetric around the midpoint, while an extreme percentage is evenly divided between the two tails of the distribution. The same *z* score, for example, defines the middle 60% and the extreme 40%. The middle 60% consists of the first 30% above the midpoint *and* the first 30% below the midpoint, while the extreme 40% consists of the 20% of scores above the middle 60% *and* the 20% below it.
- *z* scores can be used to find the probability that a score, selected at random, falls in a certain section of the normal curve. For example, $p = .50$ that a score selected at random has a score at or below the mean.
- The rare zone of a distribution is the part in the tails where scores rarely fall. Commonly, statisticians consider something rare if it happens less than 5% of the time. If rare is written as $p < .05$, then a common outcome is written as $p > .05$, meaning it happens more than 5% of the time. The common zone of a distribution is the central part where scores commonly fall.

KEY TERMS

common zone – the section of a distribution where scores usually fall; commonly set to be the middle 95%.

extreme percentage – percentage of the normal distribution that is found in the two tails and is evenly divided between them.

independence – in probability, when the occurrence of one outcome does not have any impact on the occurrence of a second outcome.

middle percentage – percentage of the normal distribution found around the midpoint, evenly divided into two parts, one just above the mean and one just below it.

normal distribution – also called the normal curve; a specific bell-shaped curve defined by the percentage of cases that fall in specific areas under the curve.

percentile rank – percentage of cases with scores at or below a given level in a frequency distribution.

probability – how likely an outcome is; the number of ways a specific outcome can occur, divided by the total number of possible outcomes.

rare zone – section of a distribution where scores do not usually fall; in most instances, set to be the extreme 5%.

standard score – raw score expressed in terms of how many standard deviations it falls away from the mean; also known as a *z* score.

***z* score** – raw score expressed in terms of how many standard deviations it falls away from the mean; also known as a standard score.

CHAPTER EXERCISES

Answers to the odd-numbered exercises appear at the back of the book.

Review Your Knowledge

4.01 A standard score is a ____ score expressed in terms of how many standard deviations it is away from the mean.

4.02 ____ is another term for standard score.

4.03 To calculate a *z* score, one needs to know the raw score, the ____, and the ____.

4.04 A positive *z* score means the score is ____ the mean; a negative *z* score means the score is ____ the mean.

4.05 If a raw score is right at the mean, it has a *z* score of ____.

4.06 The sum of *z* scores for a data set is ____.

4.07 Given a *z* score, it is possible to turn it into a raw score as long as one knows the sample ____ and ____.

4.08 The normal curve is symmetrical, has the highest point in the ____, and has frequencies that ____ as one moves away from the midpoint.

4.09 Because the normal curve is symmetric, the midpoint is also the ____ and the ____.

4.10 In a normal curve, 34.13% of the cases fall from the mean to ____ above the mean.

4.11 As one moves away from the mean in a normal distribution, the percentage of cases that fall in each standard deviation ____.

4.12 Within 2 standard deviations of the mean in a normal distribution, about ____% of the cases fall.

4.13 In a normal distribution, it is ____ to have a *z* score that is greater than 3.

4.14 Three numbers that are good to remember for estimating areas in the first 3 standard deviations of a normal distribution are ____, ____, and ____.

4.15 The normal distribution is a naturally occurring distribution that is the result of ____ processes.

4.16 Most psychological variables are considered to be ____.

4.17 Appendix Table 1 slices the normal distribution into segments that are ____ z score units wide.

4.18 The area that falls above a positive z score is the same as the area that falls ____ a negative z score.

4.19 The area that falls from the mean to a positive z score is ____ as the area from the mean to the same value as a negative z score.

4.20 The area that falls from the mean to a negative z score can't exceed ____%.

4.21 The area that falls below a negative z score will always be ____ 50%.

4.22 One can use the z score table in reverse, to find a ____ when given a percentile rank.

4.23 A percentile rank tells the percentage of cases whose scores are ____ a given level.

4.24 In a normal distribution, the percentile rank at the ____ is 50.

4.25 Statistical conclusions involve ____.

4.26 ____ concerns how likely an outcome is to occur.

4.27 Probability is defined as the number of ways a specific outcome can occur divided by ____.

4.28 If two outcomes are ____, then only one outcome can occur at a time.

4.29 Probability can be reported as a ____, a ____, or a ____.

4.30 If two outcomes are ____, then what the first outcome is has no impact on what the second outcome is.

4.31 The highest possible probability for an outcome is ____ and the lowest possible probability is ____.

4.32 If 45% of the scores fall in a certain section of the normal distribution, then $p =$ ____ that a case selected at random from the population would have a score that falls in that section.

4.33 Statisticians usually think of something that has a probability of ____ as rare and something with a probability of .95 as ____.

4.34 The middle 16% of scores in a normal distribution consist of ____% just above the mean and ____% just below the mean.

4.35 The extreme 6% of scores in a normal distribution consist of ____% in each tail of the distribution.

4.36 ____ is a very important number in statistics.

Apply Your Knowledge

Calculating standard scores

4.37 If $M = 7$, $s = 2$, and $X = 9.5$, what is z?

4.38 If $X = 17.34$, $s = 5.45$, and $M = 24.88$, what is z?

Calculating raw scores

4.39 If $z = 1.45$, $s = 3.33$, and $M = 12.75$, what is X?

4.40 If $s = 25$, $M = 150$, and $z = -0.75$, what is X?

Calculating z scores

4.41 If John's score on an IQ test is 73, what is his score as a z score? (For this and subsequent IQ questions, use a mean of 100 and a standard deviation of 15.)

4.42 If Ebony's score on an IQ test is 113, what is her score as a z score?

Calculating raw scores

4.43 Chantelle's score on an IQ test, expressed as a z score, is 0. What was her score?

4.44 Sven's score on an IQ test, expressed as a z score, is $-.38$. What was his score?

Comparing scores

4.45 Hillary's score on the math subtest of the SAT was 620. On a spelling test ($M = 60$, $s = 15$), she got a 72. Is she better at math or spelling? (On subtests of the SAT, the mean is 500 and the standard deviation is 100.)

4.46 Jong-Il was feeling depressed and anxious. His therapist gave him a depression test ($M = 10$, $s = 3$, higher scores mean more depression) and an anxiety inventory ($M = 80$, $s = 25$, higher scores mean more anxiety). His score was a 15 on the depression test and a 110 on the anxiety inventory. Is Jong-Il more depressed than anxious, or more anxious than depressed?

Finding the area above, below, or to a z score

4.47 What percentage of scores in a normal distribution fall at or above a z score of 1.34?

4.48 What percentage of cases in a normal distribution fall at or below a z score of 2.34?

4.49 What percentage of cases in a normal distribution fall at or above a z score of -0.85?

4.50 What percentage of scores in a normal distribution fall at or below a z score of -2.57?

4.51 What percentage of scores in a normal distribution fall from the mean to a z score of -1.96?

4.52 What percentage of cases in a normal distribution fall from the mean to a z score of 2.58?

4.53 The mean diastolic blood pressure in the United States is 77, with a standard deviation of 11. If diastolic blood pressure is normally distributed and a diastolic blood pressure of 90 or higher is considered high blood pressure, what percentage of Americans have high diastolic blood pressure?

4.54 The average American male is 5′ 9″. The standard deviation for height is 3″. If a basketball coach only wants to have American men on his team who are at least 6′ 6″, what percentage of the U.S. male population is eligible for recruitment?

Solving for percentile rank

4.55 If $z = -2.12$, what is the score as a percentile rank?

4.56 What is the percentile rank for a z score of 1.57?

4.57 If Gabrielle's score on an IQ test is 113, what is her score as a percentile rank?

4.58 Werner's score on an SAT subtest was 480. What is his score as a percentile rank?

Moving from percentile rank to raw score

4.59 Justin's score as a percentile rank on an SAT subtest was 88.5. What is his score on the SAT subtest?

4.60 Assume systolic blood pressure is normally distributed, with a mean of 124 and a standard deviation of 16. Sarita's systolic blood pressure, as a percentile rank, is 20. What is her systolic blood pressure?

Calculating probabilities

4.61 If there are 28 students in a classroom and 18 of them are boys, what is the probability—if a student is picked at random—of selecting a boy?

4.62 Paolo, a kindergarten student was given a pack of colored construction paper with 10 pages each of red, orange, yellow, green, blue, indigo, violet, black, and white. His favorite color is green. What is the probability, if he selects a page at random, of picking a green one?

Finding standard scores for middle and extreme sections of the normal curve

4.63 What are the z scores associated with the middle 10% of scores?

4.64 What are the z scores associated with the middle 54% of scores?

4.65 What are the IQ scores associated with the middle 84% of scores?

4.66 What are the IQ scores associated with the middle 61% of scores?

4.67 What are the z scores associated with the extreme 4% of scores?

4.68 What are the z scores associated with the extreme 18% of scores?

4.69 What are the SAT subtest scores associated with the extreme 15% of scores?

4.70 What are the SAT subtest scores associated with the extreme 8% of scores?

Finding probabilities

4.71 What is the probability, for a score picked at random from a normal distribution, that it falls at or above 3 standard deviations above the mean.

4.72 What is the probability, for a score picked at random from a normal distribution, that it falls at or below 2.7 standard deviations below the mean.

4.73 What is the probability, for a score picked at random from a normal distribution, that it falls within .38 standard deviations of the mean.

4.74 What is the probability, for a score picked at random from a normal distribution, that it falls within 2.1 standard deviations of the mean?

4.75 What is the probability, for a score picked at random from a normal distribution, that it does not fall within half a standard deviation of the mean?

4.76 What is the probability, for a score picked at random from a normal distribution, that it falls at least 4 standard deviations away from the mean?

Defining common and rare

4.77 A researcher decides to call something rare if it should happen no more than 10% of the time. What *z* scores should she use as cut-off scores?

4.78 Another researcher decides to call something rare if it should happen no more than 1% of the time. What *z* scores should he use as cut-off scores?

4.79 Dr. Noyes wants his rare zone to cover an area with a probability of .05. He would like this whole area to be on the positive side of the normal curve. What *z* score will be the cut-off for this area?

4.80 Dr. Hicks will call something rare if it happens less than or equal to 3% of the time and is greater than the mean. Expressed as a *z* score, what is the cut-off score?

Expand Your Knowledge

4.81 What is the standard deviation of a set of *z* scores?

4.82 Answer this without consulting Appendix Table 1. In a normal distribution, does a higher percentage of scores fall between *z* scores of 1.1 and 1.2, or between *z* scores of 2.1 and 2.2?

a. 1.1 and 1.2
b. 2.1 and 2.2
c. The percentages are the same.
d. This can't be answered without consulting Appendix Table 1.

4.83 A researcher obtained a set of scores and has made a frequency polygon for the distribution. The distribution is symmetric, has the highest percentage of scores occurring at the midpoint, and the frequency of scores decreases as it moves away from the midpoint. The distribution is which of the following?

a. Normally distributed
b. Not normally distributed
c. May be normally distributed
d. Not enough information to tell

4.84 Marilyn vos Savant claims that her IQ was once measured at 228 and that she is the smartest person in the world. (vos Savant is a real person and writes a weekly column for *Parade* magazine.) Let's assume that she took a standard IQ test. How credible is her claim that her IQ is 228?

4.85 A researcher has a sample of five cases. She has used the sample mean and sample standard deviation to calculate *z* scores. Four of the *z* scores are 0.50, 1.00, 1.50, and 2.00. What is the missing *z* score?

4.86 A researcher wants to study people who are very jealous. She believes that jealousy is normally distributed. She uses a jealousy scale that has a mean of 75 and a standard deviation of 25. She administers the jealousy scale as a screening device to a large group of people. For her study, she will only select people with scores of 125 or higher. How many people will she need to screen in order to end up with 50 people in her sample?

4.87 Assume that there is an animal intelligence test. A psychologist administers it to a representative sample of cats and a representative sample of dogs. Both species turn out to be, on average, equally smart. (That is, $M_{\text{Dogs}} = M_{\text{Cats}}$.) But, more variability occurred in the IQs of dogs than cats. (That is, $s_{\text{Dogs}} > s_{\text{Cats}}$.) If we consider animals with IQs above 110 to be "geniuses," will there be more genius cats or more genius dogs?

4.88 If a person is selected at random, what is the probability that he is in the top 7% of the world in terms of height?

4.89 A scale to measure perseverance has been developed. Perseverance is normally distributed and scores on the measure range from 15 to 45. It is very, very rare that anyone scores lower than 15 or higher than 45. What are the scale's mean and standard deviation?

4.90 What percentage of cases in the normal distribution have *z* scores in the range from 1.25 through 1.75?

SPSS

SPSS doesn't do a lot with *z* scores or percentile ranks. It does, however, offer a way of converting raw scores to *z* scores and saving them in the data file so that they can be used in other analyses. To do so, go to the menu under "Analyze," then "Descriptive Statistics," and finally "Descriptives" as seen in Figure 4.26.

Clicking on Descriptives opens up the menu seen in Figure 4.27. Notice that the variable "height" has already been moved to the "Variable(s)" box and that the box for "Save standardized values as variables" has been checked. Clicking "OK" starts the analysis.

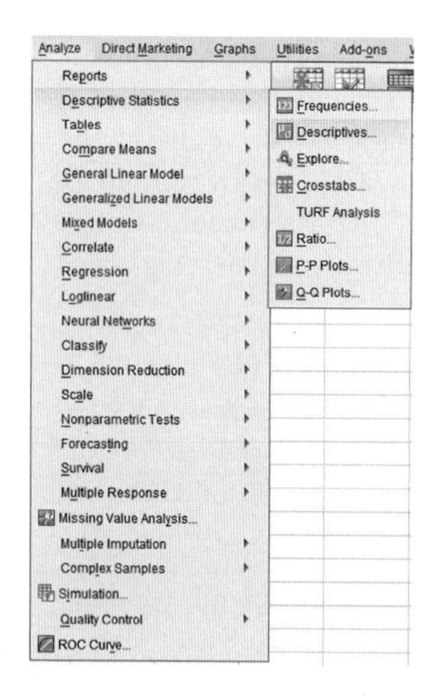

Figure 4.26 Commands for Saving *z* Scores in SPSS The "Descriptives" command can be found under "Analyze" and then "Descriptive Statistics." (Source: SPSS)

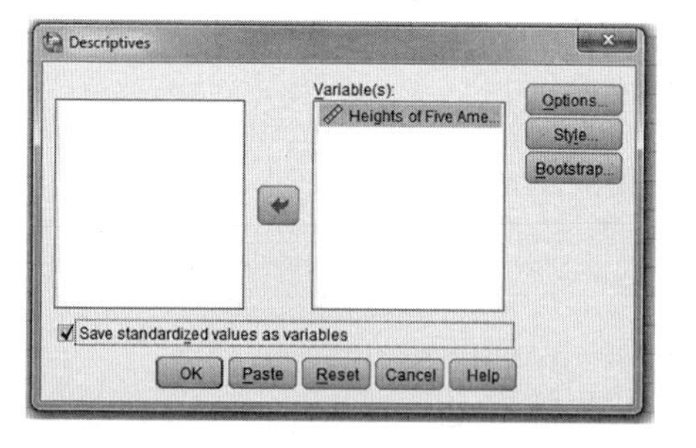

Figure 4.27 Menu for the "Descriptives" Command Clicking on "Descriptives" in Figure 4.26 opens up this menu. Note that the variable for which the *z* scores, or "standardized values" as SPSS calls them here, are to be calculated has been moved to the "Variable(s)" box. And, importantly, the box has been checked to "Save standardized values as variables." (Source: SPSS)

Height	ZHeight
62	-1.19523
65	-.47809
66	-.23905
69	.47809
73	1.43427

Figure 4.28 SPSS Output for *z* scores for Heights of Five Americans Compare the number of variables in this data file to the number seen in the upper left of Figure 4.27. The z scores for height have been given the name Zheight and added to the data set. (Source: SPSS)

Look at Figure 4.28, which shows the data editor for the demographer's sample of five Americans. Notice that there are two variables listed: height and Zheight. Zheight is the name assigned to the *z* score for height by SPSS. The *z* scores that SPSS calculates carry more decimal places throughout the calculations, but they match up well to the ones found for these data earlier in this chapter.

Sampling and Confidence Intervals

LEARNING OBJECTIVES

- Define a "good" sample and how to obtain one.
- List three facts derived from the central limit theorem.
- Calculate the range within which a population mean probably falls.

CHAPTER OVERVIEW

This chapter starts by expanding on concepts introduced in previous chapters and revisiting the differences between populations and samples. The discussion then turns to the different ways that samples are gathered and to the criteria for a "good" sample. Next, sampling distributions and the central limit theorem are introduced, two concepts that are part of the foundation for statistical decision making, which is introduced in the next chapter. To cap off this chapter and to presage the logic of hypothesis testing, a practical application of the central limit theorem, the 95% confidence interval for the population mean is introduced. This confidence interval uses the sample mean to calculate a range, an interval, that we are reasonably certain captures the population mean.

5.1 Sampling and Sampling Error

Let's start by reviewing some terminology from Chapter 1. A population is the larger group of cases a researcher is interested in studying, and a sample is a subset of cases from the population. Samples are used in research because populations are usually large, and it is impractical to study all members of the population.

Types of Samples

The way a sample is selected has an impact on whether the sample is a good one. What is meant by a "good" sample? A good sample is one that is *representative* of the population it came from. **Representative** means that all the attributes in the population are in the sample in the same proportions by which they are present in the population. Imagine obtaining a sample of students from a college and no one in the sample was a psychology major. That wouldn't be a representative sample. If 3% of students at the college are psychology majors, then, for the sample to be

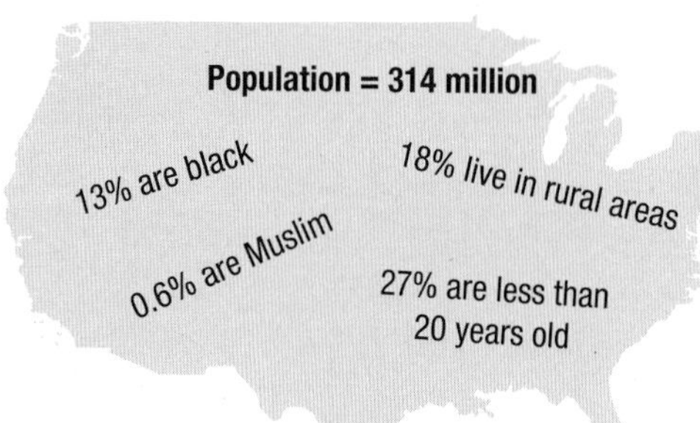

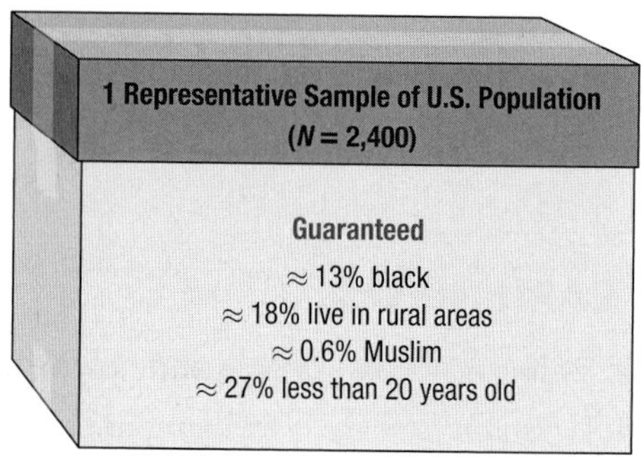

Figure 5.1 A sample is smaller than the population Here, a sample of 2,400 people is used to represent a population of 314,000,000. If the sample is representative, it has all the attributes that are present in the population in roughly the same degree.

representative, ≈3% of the sample should be psychology majors. For a representative sample, this would have to be true for all the other attributes of the students at the college—for race, religion, having an eating disorder, being an athlete, sexual orientation, wearing glasses, having divorced parents, and so on. A representative sample is like a microcosm of the population (see Figure 5.1).

When a sample is representative, the results of the study can be *generalized* from the sample back to the population. This is very handy. If a physician wants to know how much cholesterol is circulating in a patient's bloodstream, she doesn't have to drain all 10 pints of the patient's blood. Instead, she takes a small test tube full of blood, measures the amount of cholesterol in that, and generalizes the results to the entire blood supply. This can be done because the blood in the test tube is a representative sample of the population of blood in the body.

Let's move from medicine to psychology. How are samples obtained for psychology studies? Many use college students as participants. So, let's make up a study in which a psychologist is studying students' sexual attitudes and behaviors.

The psychologist goes to the student union at lunchtime, and as students pass through, he asks them to complete his survey. This is called a **convenience sample,** because it consists of easily gathered cases. Convenience samples are easy to obtain, but they are unlikely to be representative. Students who aren't on campus that day, or who have classes at lunchtime, or who don't walk through the student union can't be in the sample. So, this sampling plan would not provide a representative sample of the student population.

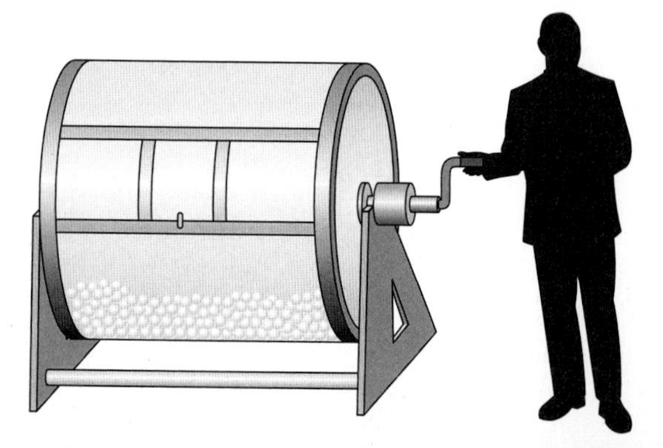

A solution is something called *random sampling.* In a simple **random sample**, all the cases in the population have an equal chance of being selected. To assemble a random sample, the psychologist might go to the registrar, get a list of the names of all registered students, put each name on a slip of paper, place all the slips in a barrel, mix it up, draw a name, mix the barrel again, draw another name, and so on. This way all students, whether they are on campus that day or not, whether they walk through the student union or not, have a chance of being in the researcher's sample.

TABLE 5.1 Random Number Table

	A	B	C	D	E	F
1	8607	1887	5432	2039	5502	3174
2	5574	4576	5273	8582	1424	9439
3	5515	8367	6317	6974	3452	2639
4	0296	8870	3197	4853	4434	1571
5	0149	1919	8684	9082	0335	6276

The random digits in this table can be used to select a random sample from a population.

Of course, it is rare that researchers draw names out of a hat. Typically, researchers use a computer or a random number table to select a random sample. A random number table is a list of digits in random order, where the number 1 is just as likely to be followed by a 2 as it is by a 1 or a 9 or any other digit. Appendix Table 2 is a random number table and a section of it is reproduced in Table 5.1.

Suppose there are 4703 students at the psychologist's college. Here's how the random number table could be used to draw a random sample of 50 from the population of 4703. The first thing the researcher does is assign sequential numbers to the pool of participants. That is, he makes a list of all the students and numbers them from 1 to 4703. Next, he goes to the random number table and moves across it, looking for values of 4703 or lower. Each time such a value is encountered, that person will go in his sample. The first number in Table 5.1 is "8607." There's no case numbered 8607 in the sample, so he'll skip this value. The next value, "1887," is a hit, and the case numbered 1887 will become part of his sample. With the review of the first row complete, case numbers 2039 and 3174 are also part of the sample. The researcher will continue this process until he has randomly selected 50 cases numbered 4703 or lower.

A Common Question

Q What happens if the same number is selected again?

A If the same number occurs again and is included in the sample a second time, that is called sampling with replacement. Though in terms of the mathematics of probability, this is fair to do, no researcher ever does it when drawing a sample for a study. Sampling without replacement is almost always done.

Random selection is great, but it doesn't guarantee a representative sample. Sample size, N, is also important. The law of large numbers states that other things being equal, larger samples are more likely to represent the population. Imagine that a researcher drew two random samples from a college, found the number of sexual partners for each person, and then wanted to use the mean number of sexual partners in the sample to represent the average number of sexual partners for students at the school. Which sample would provide a more accurate picture? One where N is 5, or one where N is 50? The larger sample, the one with 50 cases, is the better choice because it has a greater chance of capturing the characteristics of the population. A larger sample is more likely to provide a sample value close to the population value because it is more likely to contain the range of values that are in the population.

Problems with Sampling

With random samples, there are two problems to be concerned about. One is called *self-selection bias*, and the other is called *sampling error.* **Self-selection bias** occurs when not everyone who is asked to participate in a study agrees to do so. For example, in the study of students' sexual attitudes and behavior, the researcher needs to describe the study to the randomly selected potential subjects and then let them decide if they wish to participate. That's called informed consent. This study is about sex, a sensitive and private topic, so not everyone will wish to participate and the psychologist will end up with a self-selected sample. If the people who choose to participate differ in some way from those who choose not to participate (perhaps their attitudes and behaviors are more permissive), then the sample is no longer representative of the population. If too much self-selection occurs, the researcher can no longer generalize results from the sample to the population.

One can often tell when self-selection bias has occurred by looking at **consent rate**, the percentage of targeted subjects who agree to participate. If 100% of potential subjects agree to participate, then no problem with self-selection exists. If only 5% agree, then there is something unusual about those who agree and they shouldn't be used to represent the population. But where does one draw the line between these two extreme situations? One rule of thumb is that self-selection bias isn't a problem as long as the consent rate is above 70% (Babbie, 1973). Problems with sampling and self-selection bias are why researchers need to pay careful attention to how samples are obtained and whether they are representative.

Problems with sampling and self-selection bias are why researchers need to pay careful attention to how samples are obtained and whether they are representative.

Even when a sample is large, randomly obtained, and has a high consent rate, it is almost certainly not going to be an exact replica of the population. Let's continue with the example of taking a random sample of college students and collecting information about sexual attitudes and behavior. Further, pretend that, somehow, each student at the college—the entire population of students—has reported how many sexual partners he or she has had. From this, it is possible to calculate a population mean (μ). The sample mean (M) should be close to that value, but it would be surprising if it were exactly the same. And, if the researcher took another random sample, it would be surprising if M_2 were exactly the same as M_1. These discrepancies between sample values and the population value are called **sampling error**.

Sampling error is the result of random factors, which means there is nothing systematic about it. Sampling error is thought of as normally distributed. This means that about half the time sampling error will result in a sample value, for example, a mean, that is larger than the population mean, and half the time sampling error will result in a sample mean smaller than the population mean. Because in a normal distribution most of the scores are bunched around the midpoint, more often than not, sampling error will be small. This means that usually the sample mean will be relatively close to the population mean. But, occasionally, sampling error will be large, resulting in a sample mean that is dramatically different from the population mean.

Worked Example 5.1

A familiar example, M&Ms, should help clarify sampling error. Every day millions of M&Ms are made in different colors. At the factory, the colors are mechanically mixed together and packaged for sale. Each package of M&Ms is like a random sample from the population of M&Ms manufactured that day.

Charles Taylor/Shutterstock

Several years ago, I bought 500 of the single-serving bags of M&Ms, those that are sold at checkout counters. At the time, the Mars Company reported that 20% of the M&Ms it produced every day were red. This means that each bag, each random sample from the population of M&Ms, should contain 20% red M&Ms. However, due to sampling error, one would expect discrepancies between the population value, 20% red, and the percentage of red M&Ms found in the samples, the purchased bags. Figure 5.2 is a histogram showing the distribution of the percentage of red M&Ms for these 500 random samples. There are several things to note in this figure:

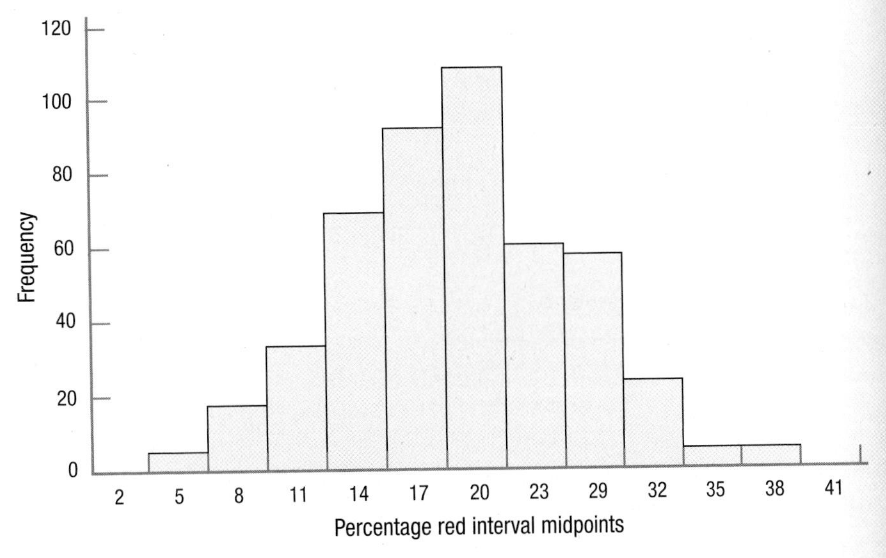

Figure 5.2 **Percentage of Red M&Ms in 500 Single-Serving Bags** Sampling error is random, uncontrolled error that causes samples to look different from populations. It explains why there is variability in the percentage of red M&Ms in each of these bags and why each one doesn't contain exactly 20% red M&Ms.

- Though the most frequently occurring sample value is around 20% (the population value), the majority of samples don't have 20% red M&Ms. That is, the majority of samples show evidence of sampling error.
- A large percentage of the other values fall near the population value of 20%. So, sampling error is usually not large.
- Occasionally, sampling error is large and leads to a sample value that is very different from the population value. Instead of being near 20%, there were bags of M&Ms that had only 5% red M&Ms and others that had up to 38% red M&Ms.
- The distribution is symmetric. About half the cases have "positive" sampling error, resulting in a sample value that is greater than the population value, and about half have "negative" sampling error.
- In fact, the distribution looks very much like a normal distribution, centered around the population value of 20% red.

Perhaps the most important thing to remember is that sampling error just happens by accident. There isn't anything malicious going on. Random factors—for example, the way the M&Ms are mixed together—cause sampling error. It just happens. That's why samples, even if they are random and large, are rarely exact replicas of populations.

When a researcher takes a single sample from a population and uses it to represent the population, he or she needs to bear in mind that sampling error is almost certainly present. The sampling error may be a positive amount or a negative amount. There probably is not too much sampling error, but it is also possible that—due just to random events and bad luck—there is a lot of sampling error. If the latter is the case, then the researcher shouldn't generalize from the sample to the population. The trouble is, the researcher doesn't know if there is a little or a lot of sampling error. If a Martian came to Earth, bought a bag of M&Ms, counted the colors, and found that 5% of the M&Ms were red, then it would probably conclude that red was an uncommon M&M color. And, as red actually makes up 20% of M&Ms, its conclusion would be wrong. But, the Martian wouldn't know.

Practice Problems 5.1

Review Your Knowledge

5.01 What does it mean if a sample is representative of a population?

5.02 What is the consent rate for a sample?

5.03 What causes sampling error?

Apply Your Knowledge

5.04 There are 50 students in a class and they count off from 1 to 50. Describe how to draw a random sample of 10 students.

5.05 How can one minimize sampling error?

5.2 Sampling Distributions and the Central Limit Theorem

Sampling Distributions

The histogram for the percentage of red M&Ms in 500 bags of M&Ms, Figure 5.2, is an example of a *sampling distribution.* A **sampling distribution** is generated by (1) taking repeated, random samples of a specified size from a population; (2) calculating some statistic (like a mean or the percentage red M&Ms) for each sample; and (3) making a frequency distribution of those values.

To understand sampling distributions, let's use an example with a very small population. This example involves a small town in Texas that has a population of only five people (Diekhoff, 1996). Each person is given an IQ test and their five scores can be seen in Figure 5.3. The frequency distribution for these data forms a flat line and does not look like a normal distribution.

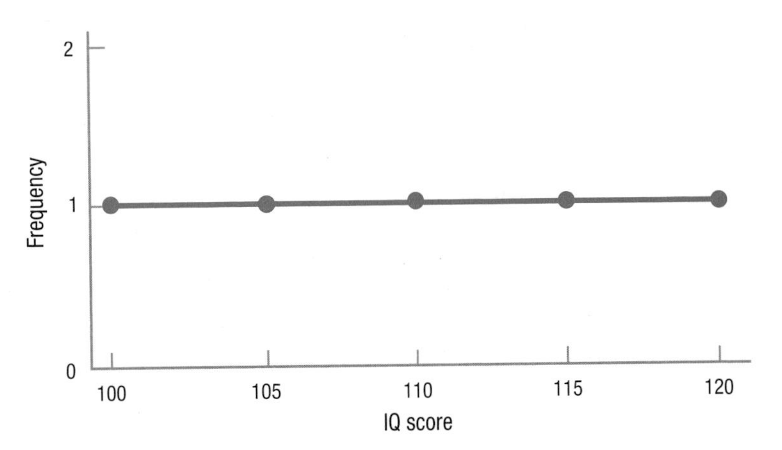

Figure 5.3 Distribution of IQ Scores in Small Texas Town Only five people live in this town and each one has a different IQ score. The distribution of IQ scores in this population is flat.

These five people make up the entire population of the town. As there is access to all the cases in the population, this is one of those rare instances where one can calculate a population mean:

$$
\begin{aligned}
\mu &= \frac{\Sigma X}{N} \\
&= \frac{100 + 105 + 110 + 115 + 120}{5} \\
&= \frac{550.0000}{5} \\
&= 110.0000 \\
&= 110.00
\end{aligned}
$$

To make a sampling distribution of the mean for this population: (1) take repeated, random samples from the population; (2) for each sample, calculate a mean; (3) make a frequency distribution of the sample means. That's a sampling distribution of the mean.

There are two things to be aware of with regard to sampling distributions:

- First, sampling occurs with replacement. This means after a person is selected at random and his or her IQ is recorded, the person is put back into the population, giving him or her a chance to be in the sample again.
- Second, the order in which cases are drawn doesn't matter. A sample with person A drawn first and person B second is the same as person B drawn first and A second.

Our sampling distribution of the mean will have samples of size $N = 2$. There are 15 possible unique samples of size $N = 2$ for this Texas town. They are shown in the first panel in Table 5.2.

Table 5.2 also shows the pairs of IQ scores for each of the samples (panel 2), as well as the mean IQ score for each sample (panel 3). Note that not all of the sample

TABLE 5.2 Samples ($N = 2$) From Population of Five Cases

All Possible Unique Samples				
A, A	A, B	A, C	A, D	A, E
B, B	B, C	B, D	B, E	
C, C	C, D	C, E		
D, D	D, E			
E, E				
IQs for Samples				
100, 100	100, 105	100, 110	100, 115	100, 120
105, 105	105, 110	105, 115	105, 120	
110, 110	110, 115	110, 120		
115, 115	115, 120			
120, 120				
Mean IQs for Samples				
100.00	102.50	105.00	107.50	110.00
105.00	107.50	110.00	112.50	
110.00	112.50	115.00		
115.00	117.50			
120.00				

The letters in the first panel represent all possible samples, of size $N = 2$ and sampling with replacement, from a population of 5. The second panel replaces the letters with the IQs of the participants. The third panel reports the mean IQ for each pair of participants.

means are the same. As there is variability in the means, it is possible to calculate a measure of variability (like a standard deviation) for the sampling distribution. **Standard error of the mean** (abbreviated σ_M) is the term used for the standard deviation of a sampling distribution of the mean. The standard error of the mean tells how much variability there is from sample mean to sample mean.

Figure 5.4 shows the sampling distribution for the 15 means from the bottom panel of Table 5.2. The first thing to note is the shape. The population (see Figure 5.3) was flat, but the sampling distribution is starting to assume a normal shape. This normal shape is important because statisticians know how to use *z* scores to calculate the likelihood of a score falling in a specified segment of the normal distribution.

The mean of this sampling distribution is abbreviated as μ_M because it is a *population* mean of sample means. Calculating it leads to an interesting observation—the mean of the sampling distribution is the same as the population mean:

$$\mu_M = \frac{100 + 102.5 + 105 + 107.5 + 110 + 105 + 107.5 + 110 + 112.5 + 110 + 112.5 + 115 + 115 + 117.5 + 120}{15}$$

$$= \frac{1{,}650.0000}{15}$$

$$= 110.0000$$

$$= 110.00$$

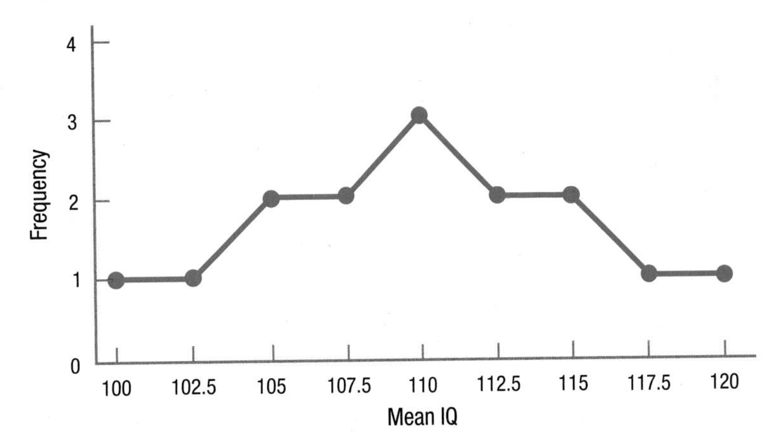

Figure 5.4 Sampling Distribution of Means for Repeated, Random Samples of Size $N = 2$ from Small Texas Town The central limit theorem states that the sampling distribution of the mean will be normally distributed, no matter what the shape of the parent population is, as long as the sample size is large. Large means $N \geq 30$. Here, even though the N for each sample is small ($N = 2$), the sampling distribution is starting to assume the shape of a normal distribution.

The Central Limit Theorem

Having made a number of observations from the sampling distribution for IQ in the small Texas town, it is time to introduce the *central limit theorem.* The **central limit theorem** is a description of the shape of a sampling distribution of the mean when the size of the samples is large and every possible sample is obtained.

Imagine someone had put together a sample of 100 Americans and found the mean IQ score for it. What would the sampling distribution of the mean look like for repeated, random samples with $N = 100$? With more than 300 million people in the United States, it would be impossible to obtain every possible sample with $N = 100$. That's where the central limit theorem steps in—it provides a mathematical description of what the sampling distribution would look like if a researcher obtained every possible sample of size N from a population.

Keep in mind that the central limit theorem works when the size of the sample is large. So which number is the one that needs to be large?

A. Is it the size of the population, which is 5 for the Texas town example?

B. Is it the size of the repeated, random samples that are drawn from the population? These have $N = 2$ for the town in Texas.

C. Is it the number of repeated, random samples that are drawn from the population? This was 15 for the Texas town.

The answer is B: the large number needs to be the number of cases in the sample.

How large is large? An N of 2 is certainly not large. Usually, an N of 30 is considered to be large enough. So, the central limit theorem applies when the size of the samples that make up the sampling distribution is 30 or larger. This means that the researcher with a sample of 100 Americans can use the central limit theorem.

The central limit theorem is important because it says three things:

1. If N is large, then the sampling distribution of the mean will be normally distributed, no matter what the shape of the population is. (In the small town IQ example, the population was flat, but the sampling distribution was starting to look normal.)
2. If N is large, then the mean of the sampling distribution is the same as the mean of the population from which the samples were selected. (This was true for our small town example.)
3. If N is large, then a statistician can compute the standard error of the mean (the standard deviation of the sampling distribution) using Equation 5.1.

Equation 5.1 Formula for Calculating the Standard Error of the Mean

$$\sigma_M = \frac{\sigma}{\sqrt{N}}$$

where σ_M = the standard error of the mean
σ = the standard deviation of the population
N = the number of cases in the sample

Given the sample of 100 Americans who were administered an IQ test that had a standard deviation of 15, the standard error of the mean would be calculated as follows:

$$\sigma_M = \frac{\sigma}{\sqrt{N}}$$

$$= \frac{15}{\sqrt{100}}$$

$$= \frac{15}{10{,}000}$$

$$= 1.50$$

When σ Is Not Known

Access to the entire population is rare and it is rare that σ, the population standard deviation, is known. So, how can the standard error of the mean be calculated without σ? In such a situation, one uses the sample standard deviation s, an estimate of the population standard deviation, to calculate an *estimated* standard error of the mean. The formula for the estimated standard error of the mean, abbreviated s_M, is shown in Equation 5.2.

Equation 5.2 Formula for Estimated Standard Error of the Mean

$$s_M = \frac{s}{\sqrt{N}}$$

where s_M = the estimated standard error of the mean
s = the sample standard deviation (Equation 3.7)
N = the number of cases in the sample

Suppose a nurse practitioner has taken a random sample of 83 American adults, measured their diastolic blood pressure, and calculated s as 11. Using Equation 5.2, he would estimate the standard error of the mean as

$$\begin{aligned} s_M &= \frac{s}{\sqrt{N}} \\ &= \frac{11}{\sqrt{83}} \\ &= \frac{11}{9.1104} \\ &= 1.2074 \\ &= 1.21 \end{aligned}$$

A reasonable question to ask right about now is: What is the big deal about the central limit theorem? How is it useful? Thanks to the central limit theorem, a researcher doesn't need to worry about the shape of the population from which a sample is drawn. As long as the sample size is large enough, the sampling distribution of the mean will be normally distributed even if the population isn't. This is handy, because the percentages of cases that fall in different parts of the normal distribution is known.

Look at the shape of the population displayed in Figure 5.5. It is far from normal. Yet, if one were to take repeated random samples from this population, calculate a mean for each sample, and make a sampling distribution of the means, then that sampling distribution would look normal as long as the sample sizes were large. This is advantageous because when hypothesis testing is introduced in the next chapter, the hypotheses being tested turn out to be about sampling distributions. If the shape of a sampling distribution is normal, then it is possible to make predictions about how often a particular value will occur.

Another benefit of the central limit theorem is that it allows us to calculate the standard error of the mean from a single sample. What's important about the standard error of the mean? A smaller standard error of the mean indicates that

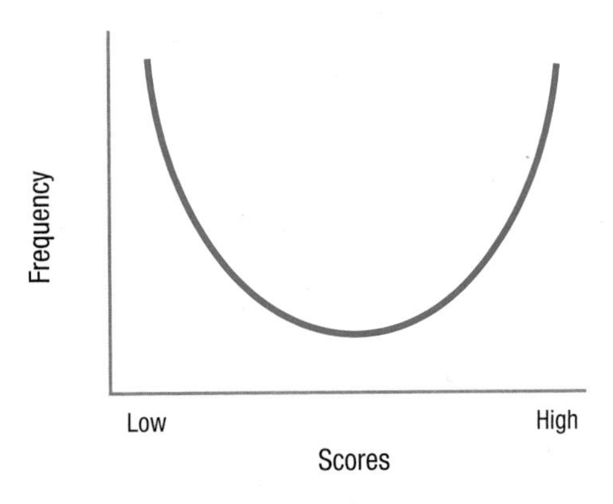

Figure 5.5 A Population That Is Not Normally Distributed This graph shows a population with a non-normal shape. According to the central limit theorem, as long as the size of the samples drawn from this population is large enough, a sampling distribution of the mean for this non-normally shaped population will have a normal distribution.

the means in a sampling distribution are packed more closely together. This tells us that there is less sampling error, that the sample means tend to be closer to the population mean. If the standard error of the mean is small, then a sample mean is probably a more accurate reflection of the population mean because it likely falls close to the population mean.

In a sense, a sampling distribution is a representation of sampling error. Figure 5.6 shows this graphically. Note how the distribution with a larger sample size has less variability and is packed more tightly around the population value. It has less sampling error.

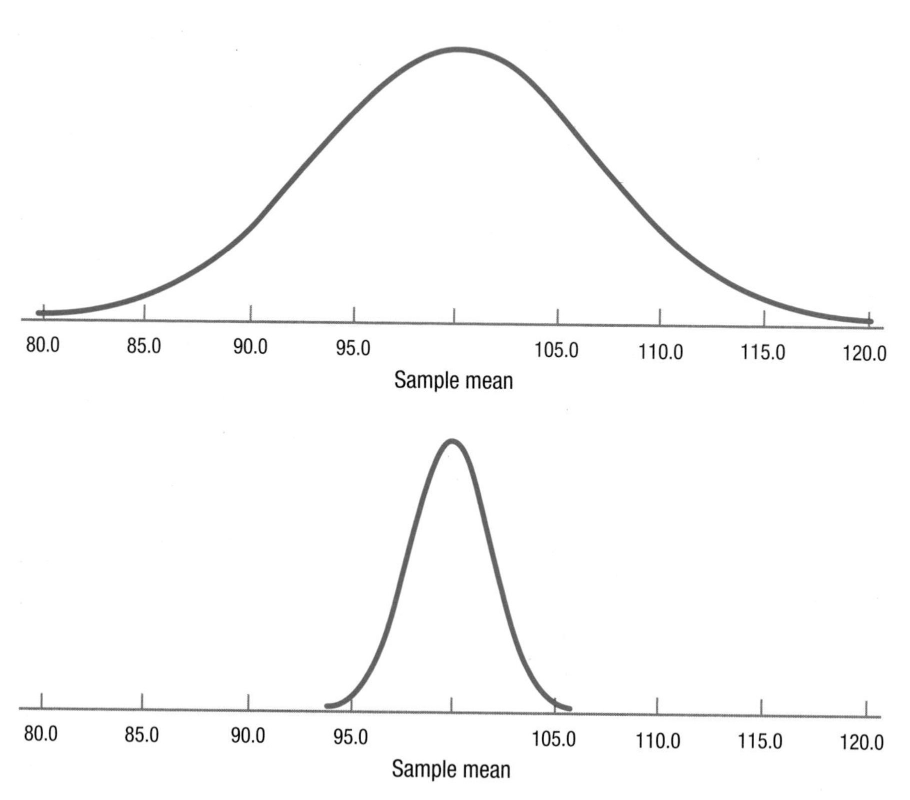

Figure 5.6 Effect of Size of Standard Error of the Mean on Sampling Distributions Both sampling distributions are from populations where $\mu = 100$ and $\sigma = 15$. In the top panel, the sample size is smaller ($N = 9$), so the standard error of the mean is 5.00. In the bottom panel, the sample size is larger ($N = 100$), so the standard error is 1.50. Where the standard error of the mean is smaller, notice how the sampling distribution is clustered more tightly around the population mean of 100. Less sampling error occurs in the bottom panel.

A Common Question

Q Are sampling distributions only for means?

A No, a sampling distribution can be constructed for any statistic. There could be a sampling distribution of standard deviations or of medians. In future chapters, sampling distributions of statistics called *t*, *F*, and *r* will be encountered.

Worked Example 5.2 Sampling distribution generators (available online) draw thousands of samples at a time and form a sampling distribution in a matter of seconds right on the computer screen. A researcher can change parameters—like the shape of the parent population, the size of the samples, or the number of samples—and see the impact on the shape of the sampling distribution. Nothing beats playing with a sampling distribution generator for gaining a deeper understanding of the central limit theorem. Google "Rice Virtual Lab in Statistics," click on "Simulations/ Demonstrations," and play with the "sampling distribution simulation." For those who prefer a guided tour to a self-guided one, read on.

Figure 5.7 uses the Rice simulation to show a population constructed *not* to be normal. In this population, scores range from 0 to 32, $\mu = 15.00$, $\sigma = 11.40$, and the midpoint does not have the greatest frequency.

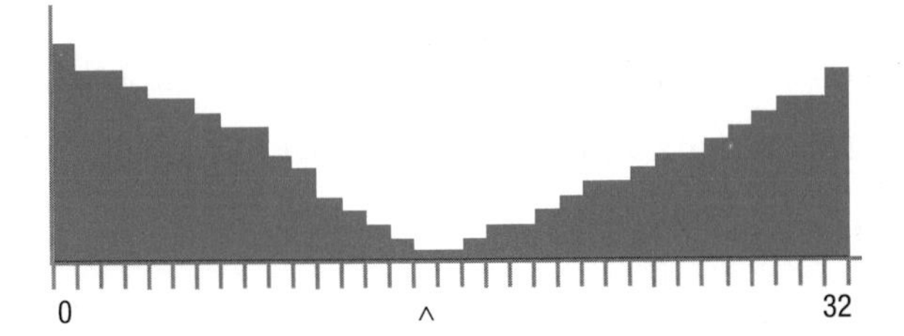

Figure 5.7 Histogram for Values in Entire Population The range of values in this population is from 0 to 32, and the shape is decidedly not normal. The caret marks the population mean, $\mu = 15.00$. (The population is generated through Rice Virtual Lab in Statistics.)

The Rice simulator allows one to control how large the samples are and how many samples one wishes to take from the population. Figure 5.8 illustrates one random sample of size $N = 5$ from the population. The left panel in Figure 5.8 shows the five cases that were randomly selected and the right panel the mean of these five cases, the first sample. Note that the five cases are scattered about and that the sample mean ($M = 16.00$) is in the ballpark of the population mean ($\mu = 15.00$), but is not an exact match.

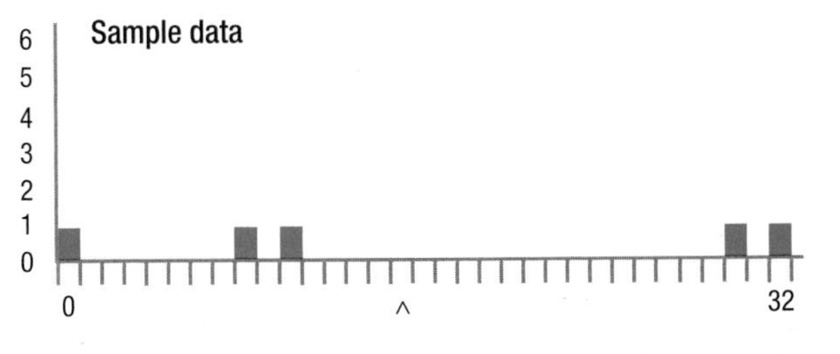

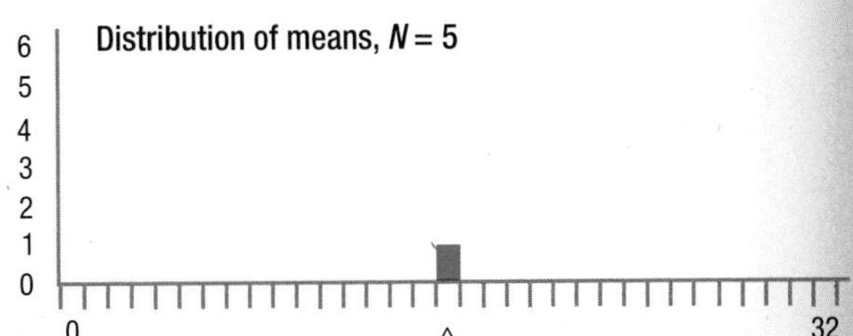

Figure 5.8 Random Sample of Five Cases The left panel shows a random sample of five cases from the population illustrated in Figure 5.7. The caret marks the population mean. The right panel shows the mean of these five cases. (Screenshot from Rice Virtual Lab in Statistics.)

Figure 5.9 shows the sampling distribution after 1,000 random samples were drawn, after 10,000 random samples were drawn, and after 100,000 random samples were drawn. There are several things to note:

- All three of the sampling distributions have shapes that are similar to those of a normal distribution, despite the fact that the parent population is distinctly non-normal. This is predicted by the central limit theorem.
- As the number of samples increases, the distribution becomes smoother and more symmetrical. But, even with 100,000 samples, the sampling distribution is not perfectly normal.

- The central limit theorem states that the mean of a sampling distribution is the mean of the population. Here, the population mean is 15.00, and the mean of the sampling distribution gets closer to the population mean as the number of samples in the distribution grows larger. With 1,000 samples $M = 15.15$, with 10,000 $M = 15.02$, and with 100,000 it is 15.00.
- Notice the wide range of the means in the sampling distribution—some are at each end of the distribution. With a small sample size, like $N = 5$, one will occasionally draw a sample that is not representative of the population and that has a mean far away from the population mean. This is due entirely to the random nature of sampling error.
- The central limit theorem states that the standard error of the mean, which is the standard deviation of the sampling distribution, can be calculated from the population standard deviation and the size of the samples: $\sigma_M = \frac{\sigma}{\sqrt{N}} = \frac{11.40}{\sqrt{5}} = 5.10$. This is quite accurate—the standard deviations of the three sampling distributions are 5.06, 5.19, and 5.12.

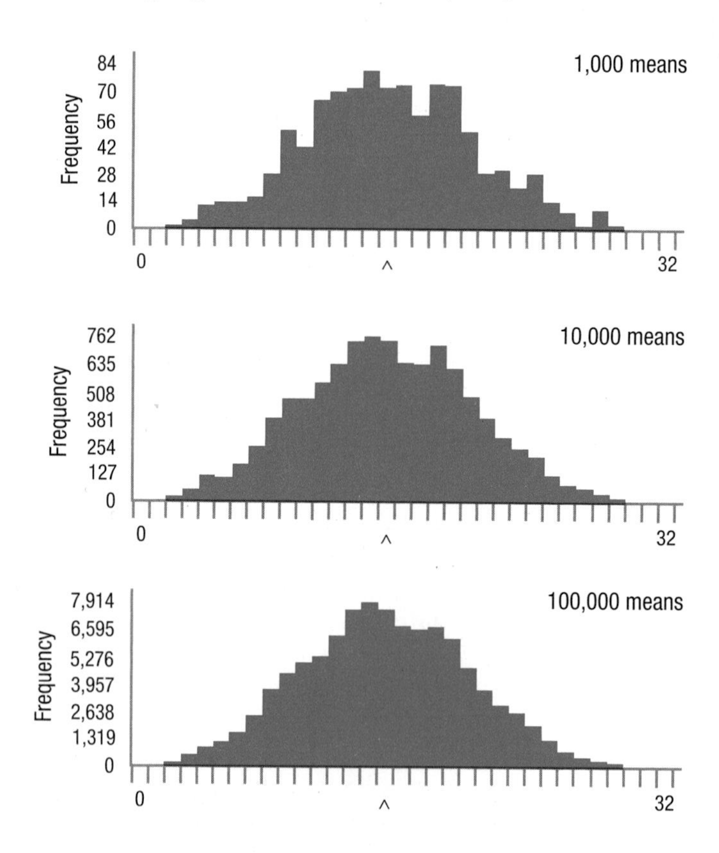

Figure 5.9 Sampling Distributions for Increasing Number of Samples of Five Cases Repeated, random samples of size $N = 5$ were taken from the population shown in Figure 5.7. The caret marks the population mean. A mean was calculated for each sample. The top panel shows the sampling distribution for 1,000 means, the middle panel for 10,000 means, and the bottom panel for 100,000 means. Note that the shape becomes more normal and more regular as the number of means increases. (Sampling distributions generated online at Rice Virtual Lab in Statistics.)

When the size of the random samples increases from $N = 5$ to $N = 25$, some things change about the shape of the sampling distribution, as can be seen in Figure 5.10. It, like Figure 5.9, shows sampling distributions with 1,000, 10,000, and 100,000 samples. But, unlike Figure 5.9, the changes in the sampling distribution as the numbers of samples increase are more subtle. All three sampling distributions have a normal shape, though as the number of samples increases from 1,000 to 10,000, there is a noticeable increase in symmetry.

The most noticeable difference between the sampling distributions based on smaller samples, seen in Figure 5.9, and those based on larger samples, seen in Figure 5.10, lies in the range of means. In Figure 5.9, the means ranged from 2 to 29, while in Figure 5.10, they range from 9 to 22. There is less variability in the sampling distribution based on the larger samples.

This decreased variability is mirrored in the standard deviations. (Remember, the standard deviation of the sampling distribution is the standard error of measurement. When 100,000 samples of size $N = 5$ are taken, the standard deviation of the sampling distribution was 5.12. When the same number of samples is taken, but the size of each sample is 25, not 5, the standard deviation of the sampling

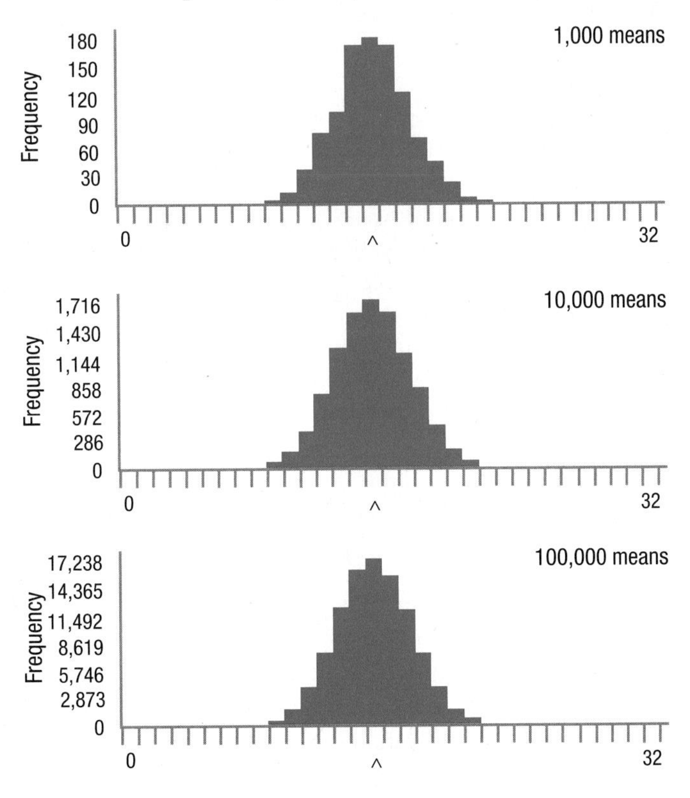

Figure 5.10 Sampling Distributions for Increasing Numbers of Samples of 25 Cases Repeated, random samples of size $N = 25$ were taken from the population shown in Figure 5.7. The caret marks the population mean. A mean was calculated for each sample. The top panel shows the sampling distribution for 1,000 means, the middle panel for 10,000 means, and the bottom panel for 100,000 means. Compare these sampling distributions to those in Figure 5.9, and it is apparent that a larger number of cases in each sample yields a more regularly shaped sampling distribution and a narrower range of values. (Sampling distributions generated online at Rice Virtual Lab in Statistics.)

distribution falls to 2.28. This is exactly what is predicted for the standard error of the mean by the central limit theorem:

$$\sigma_M = \frac{\sigma}{\sqrt{N}} = \frac{11.40}{\sqrt{25}} = 2.28$$

As was mentioned above, when the standard error of the mean is smaller, the means are packed more tightly together and less sampling error exists. A larger sample size means that there is less sampling error and that the sample is more likely to provide a better representation of the population.

Practice Problems 5.2

Review Your Knowledge

5.06 What is a sampling distribution?

5.07 What are three facts derived from the central limit theorem?

Apply Your Knowledge

5.08 There's a small town in Ohio that has a population of 6 and each person has his or her blood pressure measured. The people are labeled as A, B, C, D, E, and F. If one were to draw repeated, random samples of size $N = 2$ to make a sampling distribution of the mean, how many unique samples are there?

5.09 Researcher X takes repeated, random samples of size $N = 10$ from a population, calculates a mean for each sample, and constructs a sampling distribution of the mean. Researcher Y takes repeated, random samples of size $N = 100$ from the same population, calculates a mean for each sample, and constructs a sampling distribution of the mean. What can one conclude about the shapes of the two sampling distributions?

5.10 If $\sigma = 12$ and $N = 78$, what is σ_M?

5.3 The 95% Confidence Interval for a Population Mean

All the pieces are now in place for the culmination of this chapter—calculating a confidence interval for the mean. A **confidence interval** is a range, based on a sample value, within which a researcher estimates a population value falls. To use the language of statistics from the first chapter, confidence intervals use statistics to estimate parameters. Confidence intervals are useful because population values are rarely known, but researchers like to draw conclusions about populations.

Here's how it works. A researcher takes a sample from a population and calculates a mean, *M*. Then, based on that sample mean, the researcher constructs a range around it, an interval, that is likely to contain the population mean, μ. This interval is called a confidence interval.

For example, suppose a researcher wanted to know how many fears the average American had. Clearly, it would be impossible to survey more than 300 million Americans to find the population mean, μ. But, the researcher could obtain a representative sample of, say, 1,000 Americans and ask them this question. Suppose he did so and found that the mean number of fears in the sample, *M*, was 2.78. If someone asked the researcher how many fears the average American had, the researcher shouldn't say, "2.78." Because of sampling error, it is unlikely that M is exactly the same as μ. M should be close to μ, but it is unlikely to be the same down to the last decimal point.

Sampling error, which is a random factor, could inflate the number of fears reported, in which case 2.78 is an overestimate, or it could lead to underreporting, in which case the average American has more than 2.78 fears. The wisest thing for the researcher to do is to add and subtract some amount to his sample mean, say, 2.78 $\pm$1.25, so that he has a range within which it is likely that M falls. Said another way, there will be an interval (based on the sample value) that the researcher is fairly confident will capture the population value. This is a confidence interval.

Though confidence intervals may sound exotic, most people are already comfortable with them, albeit under a different name. Polls regularly report a margin of error, a confidence interval by another name. For example, a NYT/CBS poll reported that 66% of a sample of 1,002 adult Americans, interviewed by phone, believe that the distribution of money and wealth should be more even in the United States (Scheiber and Sussman, 2015). Because the pollsters went to some effort to gather a representative sample, they conclude that 66% of adult Americans feel the current income distribution is unfair. But, this poll has a margin of error of $\pm$3%, so it is probably not exactly 66% who feel that income distribution is unfair. If asked what percentage of Americans feel it is unfair, the reporters should respond with a confidence interval: "The percentage of adult Americans who believe the distribution of wealth should be more even is somewhere from 63% to 69%."

A confidence interval is an **interval estimate** for a population value, not a **point estimate.** A point estimate is just a single value as an estimate of a population value. An example of a point estimate is s, which is the estimated population value for a standard deviation. Interval estimates are ranges and are better than point estimates, because they are more likely to be true. Suppose a compulsive gambler was trying to guess the average GPA at your school. With which guess do you think she would have a better chance of being right?

A. The average GPA is 3.19, or

B. The average GPA is somewhere between 3.00 and 3.40.

She'd have to be awfully lucky if the average GPA at your school were exactly 3.19, a point estimate, so she is more likely to capture the actual population value with the interval estimate. A researcher is more likely to be correct with an interval estimate, but an interval estimate is less specific than a point estimate.

The math for calculating a confidence interval is not difficult, but the logic of why a confidence interval works is a little tricky. But, don't worry. Confidence intervals will crop up regularly from here on out. I'm confident that at some point in the next 11 chapters, you will have an "aha" experience.

Let's start by imagining that a researcher has taken all possible unique, random samples of size N from some population. For each sample, she has calculated a mean and she has made a sampling distribution of the means as shown in Figure 5.11.

There are several things to note about this figure:

- Thanks to the central limit theorem, it is normally distributed.
- Thanks to the central limit theorem, the mean of all the sample means, which is the midpoint of the distribution, is also the mean of the population. It's marked as μ in Figure 5.11.

- Thanks to the central limit theorem, it is possible to calculate the standard error of the mean. The *X*-axis has been marked off in units of the standard error of the mean, ranging from −3 to 3. This indicates how many standard errors of the mean each sample mean falls away from μ.

To understand how confidence intervals work, look at Figure 5.12. There, the mean for the sample falls 1.5 standard errors of the mean above μ.

Let's build an interval around the mean, adding 1.96 standard errors of the mean to it and subtracting 1.96 standard errors of the mean from it. Mathematically, that is

$$M \pm 1.96\sigma_M$$

Look at the brackets which extend $1.96\sigma_M$ above and below *M*. Does the interval within the brackets capture μ? The answer is yes.

In Figure 5.13 is an example that doesn't capture μ. Figure 5.13 shows another sample with brackets extended $1.96\sigma_M$ above and below *M*. The mean of the sample

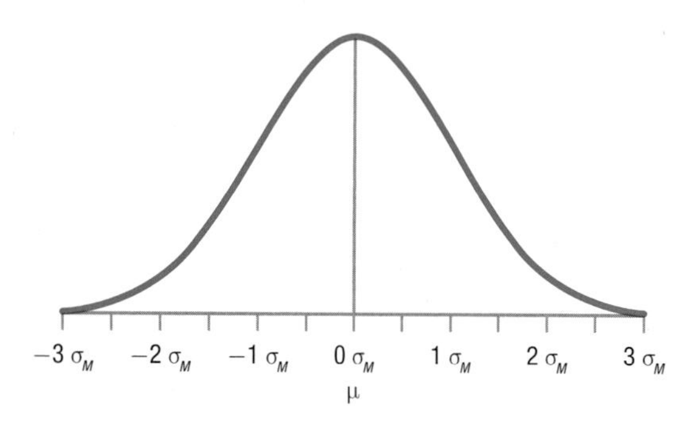

Figure 5.11 A Sampling Distribution of the Mean By the central limit theorem, (a) the sampling distribution of the mean is normally distributed, (b) the midpoint of a sampling distribution of the mean is the mean of the population from which the samples are drawn, and (c) the standard deviation of the sampling distribution of the mean is the standard error of the mean (σ_M).

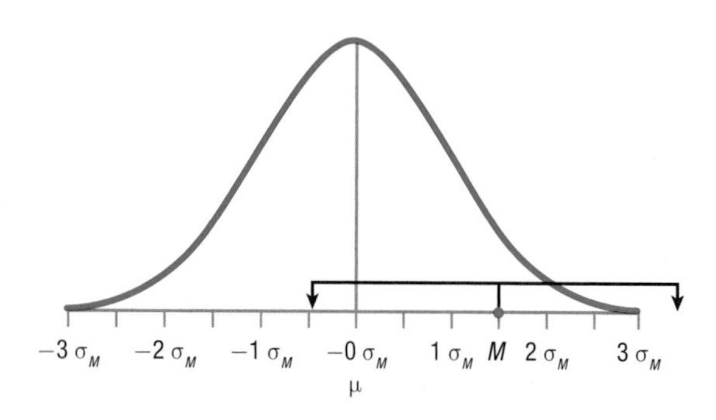

Figure 5.12 $\pm 1.96\ \sigma_M$ Brackets Built Around a Sample Mean The 95% confidence interval extends from $1.96\ \sigma_M$ below *M* to $1.96\ \sigma_M$ above it. Notice that this interval captures μ, the population mean.

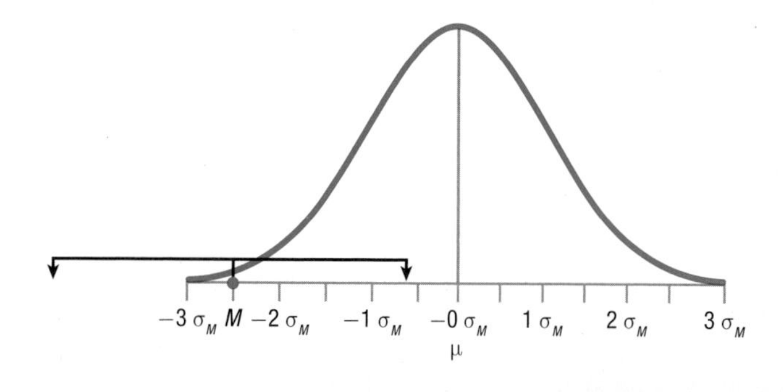

Figure 5.13 $\pm 1.96\ \sigma_M$ Brackets Built Around a Sample Mean The 95% confidence interval extends from $1.96\ \sigma_M$ below *M* to $1.96\ \sigma_M$ above it. Notice that this interval fails to capture μ, the population mean.

in Figure 5.13 is 2.5 standard errors of the mean below μ, so μ doesn't fall within the brackets around the sample mean.

That value of 1.96 was not arbitrarily chosen for these examples. 1.96 was developed in Chapter 4 as the cut-off score, in a normal distribution, for the middle 95% of cases.

Thanks to the central limit theorem, it is safe to assume that the sampling distribution of the mean is normally shaped. Also, the standard error of the mean is the standard deviation of the sampling distribution, so 95% of the sample means in a sampling distribution will fall from 1.96 standard errors of the mean below the midpoint to 1.96 standard errors of the mean above the midpoint. That is an important point, so here it is again and more succinctly: in a sampling distribution of the mean, 95% of the means fall within 1.96 σ_M of the midpoint. This is shown in Figure 5.14.

Now imagine any sample mean in the shaded region in Figure 5.14. Will the *population* mean (μ) fall within brackets extended 1.96 σ_M to the right and to the left of the *sample* mean? The answer is yes. This also means that any sample mean more than 1.96 σ_M away from the midpoint will fail to capture μ if the brackets are extended 1.96 σ_M around it.

Let's put all the pieces together and define the 95% confidence interval for the population mean. Here's what we know:

For any mean picked at random from the sampling distribution, there's a 95% chance that 1.96 σ_M brackets extended symmetrically around it will capture μ.

- If a sample mean is picked at random from a sampling distribution of the mean, there's a 95% chance that the sample mean falls in the shaded region shown in Figure 5.14.
- Brackets that extend $\pm 1.96\ \sigma_M$ capture μ for every sample mean in the shaded region.
- For any mean picked at random from the sampling distribution, there's a 95% chance that 1.96 σ_M brackets extended symmetrically around it will capture μ.

We've just developed the 95% confidence interval for the population mean: take a sample mean, create an interval around it that is $\pm 1.96\ \sigma_M$, and there is a 95% chance that this interval captures μ. The formula is shown in Equation 5.3.

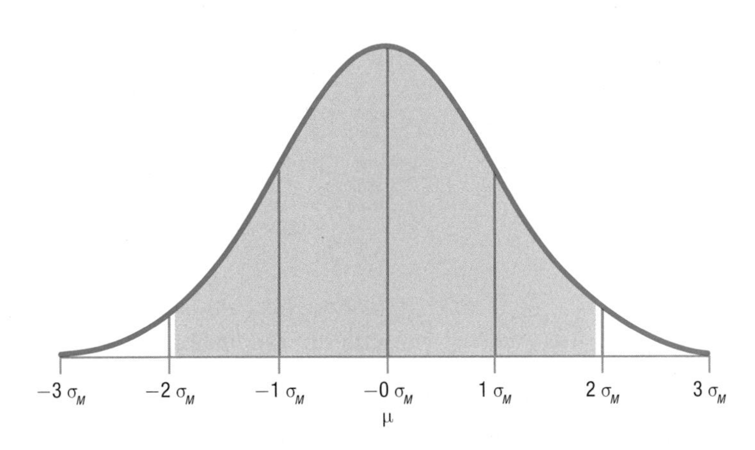

Figure 5.14 Area Within Which 95% of the Sample Means Fall in a Sampling Distribution of the Mean 95% of sample means in a sampling distribution of the mean fall in the area from −1.96 σ_M below the midpoint to −1.96 σ_M above the midpoint. 95% confidence intervals that are built around means in the shaded area will successfully capture the population mean, μ.

Equation 5.3 Formula for 95% Confidence Interval for the Population Mean

$$95\%\text{CI}_{\mu} = M \pm (1.96 \times \sigma_M)$$

where $95\%\text{CI}_{\mu}$ = the 95% confidence interval for the population mean being calculated
M = sample mean
σ_M = standard error of the mean (Equation 5.1). [If σ_M is unknown, substitute s_M (Equation 5.2).]

Let's put this equation into practice and use the results to think about what a 95% confidence interval means. Dr. Saskia obtained a random sample of 37 students at an elite college, gave them an IQ test ($\sigma = 15$), and found $M = 122$. What can she conclude from this sample mean about the average IQ of the entire student body at that college. That is, what can she conclude about μ? This calls for a confidence interval.

All the pieces are present that are needed to calculate a confidence interval: M, σ, and N. σ and N are needed to calculate σ_M using Equation 5.1:

$$\begin{aligned}
\sigma_M &= \frac{\sigma}{\sqrt{N}} \\
&= \frac{15}{\sqrt{37}} \\
&= \frac{15}{6.0828} \\
&= 2.4660 \\
&= 2.47
\end{aligned}$$

Now that the standard error of the mean is known, Equation 5.3 can be used to calculate the 95% confidence interval:

$$\begin{aligned}
95\%\text{CI}_{\mu} &= M \pm (1.96 \times \sigma_M) \\
&= 122 \pm (1.96 \times 2.4660) \\
&= 122 \pm 4.8837 \\
&= \text{ranges from } 117.1166 \text{ to } 126.8834 \\
&= \text{from } 117.12 \text{ to } 126.88
\end{aligned}$$

Note that the subtraction was done first as confidence intervals are reported from the lower number to the higher number. The confidence interval ranges from 117.12 to 126.88. Note that the whole confidence interval is 9.76 points wide and that each side of it is almost 5 points wide.

The American Psychological Association (2010) has a format for reporting confidence intervals. APA format asks that we do three things: (a) use CI as an abbreviation for confidence interval, (b) report what type of confidence interval (e.g., 95%), and (c) put in brackets first the lower limit and then the upper limit of the confidence

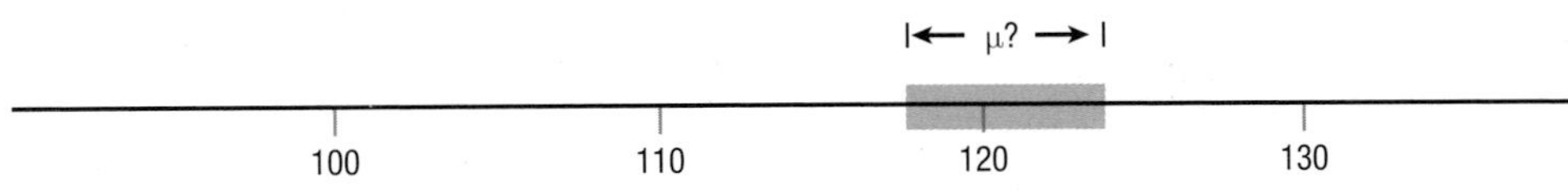

Figure 5.15 Using a Confidence Interval to Indicate the Range Within Which a Population Value Likely Falls The shaded area indicates the range within which the mean IQ of all students at the elite college probably falls.

interval. To report the mean and confidence interval for the IQ data, one would write $M = 118.00$, 95%CI [117.12, 126.88].

What does this confidence interval reveal to us? *Officially*, a 95% confidence interval only tells us that if we repeated the process of getting a sample and calculating a confidence interval, 95 out of 100 times, the confidence interval would capture the population mean, μ. *Officially*, a confidence interval doesn't say whether any particular calculation of the confidence interval is one of those 95 times, and, *officially*, it doesn't mean that there's a 95% chance μ falls in this confidence interval (Cumming & Finch, 2005).

That's officially. In reality, most people interpret a confidence interval as indicating that they are fairly certain the population mean falls in such a range. Figure 5.15 shows that with our IQ example we would conclude there's a 95% chance that the mean (μ) IQ score of the students at this college falls somewhere in the range from 117.12 to 126.88. As new statisticians, the interpretation that a confidence interval gives a range in which the population parameter falls is fine. Just remember, there's also a chance that the population parameter doesn't fall in the confidence interval.

Worked Example 5.3

Earlier in this chapter, mention was made of a sample of 83 American adults in whom diastolic blood pressure was measured and where the standard error of the mean, s_M, was 1.21. If the mean blood pressure in the sample was 81, what can one conclude about the mean diastolic blood pressure of American adults?

Using Equation 5.3 to calculate the 95% confidence interval for the population mean:

$$
\begin{aligned}
95\%\text{CI}_\mu &= M \pm (1.96 \times s_M) \\
&= 81 \pm (1.96 \times 1.21) \\
&= 81 \pm 2.3716 \\
&= 81 \pm 2.37 \\
&= \text{from } 78.63 \text{ to } 83.37
\end{aligned}
$$

The confidence interval of 78.63 to 83.37, which is a total of 4.34 points wide, has a 95% probability of capturing the mean diastolic blood pressure of adult Americans. One can conclude that the mean diastolic blood pressure of American adults is probably somewhere from 78.63 to 83.37. In APA format, one would write $M = 81.00$, 95%CI [78.63, 83.37].

Practice Problems 5.3

Review Your Knowledge

5.11 What is the difference between a point estimate and an interval estimate for a population value?

5.12 How often will a 95% confidence interval for μ capture the population mean?

5.13 How often will a 95% confidence interval for μ fail to capture the population mean?

Apply Your Knowledge

5.14 Given $M = 17$, $\sigma = 8$, and $N = 55$, calculate the 95% confidence interval for μ and report it in APA format.

5.15 Given $M = 250$, $s = 60$, and $N = 180$, calculate the 95% confidence interval for μ and report it in APA format.

Application Demonstration

A hypothetical survey about sexual attitudes and behaviors was used as an example in this chapter. To understand sampling better, let's examine two real studies about human sexuality, one with such a poor sample that the results are meaningless and one that shows how much effort goes into obtaining a representative sample.

Shere Hite is famous for her "Hite Reports" on male and female sexuality. In 1987 she published *Women and Love,* known as *The Hite Report on Love, Passion, and Emotional Violence.* It was a survey of about 4,500 women from across the United States and it generated a number of thought-provoking findings. One result that gathered a lot of media attention was the rate of adultery among married women—Hite found that 70% of women who had been married five years reported extramarital affairs. Seventy percent!

This number is quite high and the fact that the sample size is so large, 4,500 women, gives the finding credibility. Further, the procedure Hite used to put together her sample was interesting. Hite was a feminist and didn't want to be accused of only including feminists in her sample. So, she sent her survey to a broad cross section of women's groups in 43 states—church groups, garden clubs, and so on. If 70% of these mainstream American women were having affairs, that's surprising.

And, it turns out that that 70% figure can be safely ignored because of one important factor—the consent rate. Hite sent out more than 100,000 surveys and only 4,500 usable surveys were returned. That's a consent rate of 4.5%, well below the necessary rate of 70%.

If a woman received a Shere Hite survey, what was the normal response? For 95.5% of women, the normal response was not returning the survey. The odd response came from the 4.5% of women who completed and returned the survey. Maybe the odd 4.5% of women who returned the survey are also odd in terms of marital fidelity? The self-selection bias is so great in Hite's survey that the results can't be taken seriously.

To gather a representative sample for a survey is difficult, but it can be done. Here's an example of how to do it correctly. In 1994 the sociologist Edward Laumann and three colleagues published a book that reported on sexual behaviors and attitudes in the United States. It was based on a sample of about 3,400 Americans and also offered some interesting findings. Take the frequency with which people have sex, for example. They found that Americans fell into three groups with regard to how much sex they had in the past year:

- About a third had sex infrequently, either having no sex in the past year or just a couple of times.
- About a third had sex a few times a month.
- And, about a third had sex frequently, two or more times a week.

Laumann and colleagues went to a lot of trouble to get a representative sample. Using a computer, they generated addresses in randomly selected neighborhoods in randomly selected cities, towns, and rural areas, from randomly selected geographic regions in the United States. This yielded almost 4,400 valid addresses. The people living at the address who spoke English and were between 18–59 years old were eligible for participation. From the eligible people at the address, the researchers randomly selected one person, approached him or her, explained the study, got permission to do the interview, and completed it. About 4 out of every 5 targeted participants agreed to be interviewed. That's a consent rate of 80%, well above the minimum rate of 70%.

Such a large and randomly selected sample is likely to be representative of the U.S. population, but the researchers still checked their results. One action they took was to compare the characteristics of the sample to census data about the United States. Figure 5.16 compares data from the Census Bureau to results from their survey. The numbers don't match exactly, but they are close and suggest that Laumann and colleagues achieved their goal of obtaining a sample that was representative of the United States. The sample mirrors America, so the study provides one of the best pictures we have ever had of sexual behavior in this nation.

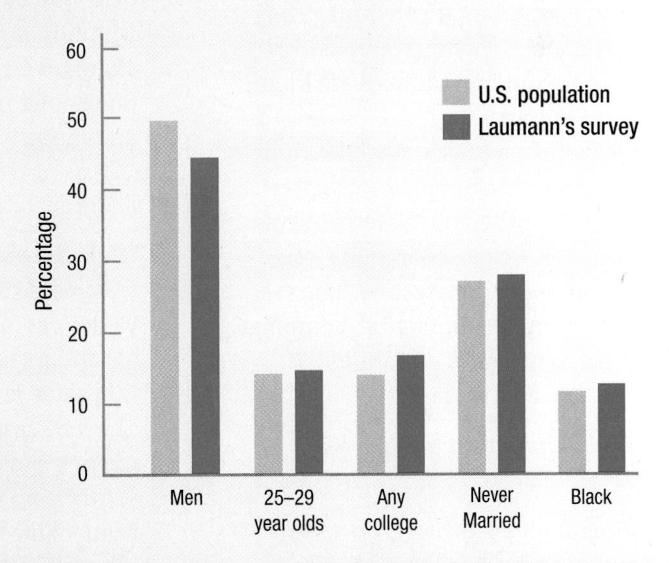

Figure 5.16 Comparison of Demographic Characteristics of Survey Sample to U.S. Population This bar chart compares the percentages of respondents in Laumann's survey of sexual behaviors who meet selected demographic characteristics to the percentages with which the same characteristics are found in the U.S. population. The matches are very close, supporting the ability to generalize from the survey results to the larger U.S. population (Michael, Gagnon, Laumann, & Kolata, 1994).

DIY

Get your music player and pick one of your biggest playlists, say, one with hundreds of songs. That will be your population. Find out how many songs are on the playlist and how many total minutes of music it contains. Divide the total number of minutes of music by the number of songs to find the average length of a song. That is, μ. Next, use the shuffle function to select $\approx$10 songs from the playlist. This is your random sample from the population. Note the time that each song lasts and calculate the mean, *M*, and the standard deviation, *s*. Use these to create a 95% confidence interval for the population mean. Does it capture μ? If so, why? If not, why?

SUMMARY

Define a "good" sample and how to obtain one.

- A good sample is representative of the larger population, so results can be generalized from the sample to the population. A sample is representative if it contains all the attributes in the population, in the same proportions that they are present in the population.
- Random sampling, where all the cases in the population have an equal chance of being selected, makes a representative sample likely but does not guarantee it. Sample size is also important as larger random samples are more likely to be representative.
- Two factors that affect representativeness are self-selection bias and sampling error. Self-selection bias occurs when not all targeted participants agree to participate. Sampling error, which is the result of random factors, causes a sample to differ from the population. Sampling error, thought of as normally distributed, is more likely to be small than large.

List three facts derived from the central limit theorem.

- The central limit theorem is based on a sampling distribution of the mean. A sampling distribution of the mean is obtained by (a) taking repeated, random samples of size *N* from a population; (b) calculating a mean for each sample; and then (c) making a frequency distribution of the mean. It demonstrates how much sampling error exists in the samples.
- The central limit theorem makes three predictions about a sampling distribution of the mean when the number of cases in each sample is large: (1) The sampling distribution is normally distributed; (2) The mean of the sampling distribution is the mean of the population; (3) The standard error of the mean can be calculated if one knows the population standard deviation and the size of the sample.

Calculate the 95% confidence interval for μ.

- Though less precise than a point estimate, which is a single value estimate of a population value, a confidence interval is a range, built around a sample value, within which a population value is thought to be likely to fall.
- The 95% confidence interval for the population mean is built by taking the sample mean and subtracting 1.96 standard errors of the mean from it and adding 1.96 standard errors of the mean to it. An interval constructed this way will capture μ 95% of the time.

KEY TERMS

central limit theorem – a statement about the shape that a sampling distribution of the mean takes if the size of the samples is large and every possible sample were obtained.

confidence interval – a range within which it is estimated that a population value falls.

consent rate – the percentage of targeted subjects who agree to participate in a study.

convenience sample – a sampling strategy in which cases are selected for study based on the ease with which they can be obtained.

interval estimate – an estimate of a population value that says the population value falls somewhere within a range of values.

point estimate – an estimate of a population value that is a single value.

random sample – a sampling strategy in which each case in the population has an equal chance of being selected.

representative – all the attributes of the population are present in the sample in approximately the same proportion as in the population.

sampling distribution – a frequency distribution generated by taking repeated, random samples from a population and generating some value, like a mean, for each sample.

sampling error – discrepancies, due to random factors, between sample statistic and a population parameter.

self-selection bias – a nonrepresentative sample that may occur when the subjects who agree to participate in a research study differ from those who choose not to participate.

standard error of the mean – the standard deviation of a sampling distribution of the mean.

CHAPTER EXERCISES

Answers to the odd-numbered exercises appear at the back of the book.

Review Your Knowledge

5.01 A ____ is the larger group of cases a researcher is interested in studying, and a sample is a ____ of these cases.

5.02 A good sample is ____ of the population.

5.03 A representative sample contains all the ____ found in the population in the same ____ as in the population.

5.04 When a sample is representative, we can ____ the results from the sample to the ____.

5.05 A ____ sample is composed of easily obtained cases.

5.06 In a ____, all the cases in the population have an equal chance of being selected.

5.07 A ____ can be used to draw a random sample.

5.08 A representative sample ____ guaranteed if one has a random sample.

5.09 Larger samples provide sample values closer to ____ values.

5.10 If not everyone targeted for inclusion in the sample agrees to participate, ____ may occur.

5.11 Self-selection bias occurs if the people who agree to participate in a study ____ in some way from those who opt not to participate.

5.12 The ____ is the percentage of targeted subjects who agree to participate.

5.13 As long as the consent rate is greater than or equal to ____%, concern with self-selection bias isn't too great.

5.14 Discrepancies between randomly drawn samples and a population can be explained by ____.

5.15 Sampling error is caused by ____ factors.

5.16 Sampling error is thought to be ____ distributed.

5.17 Usually, the amount of error caused by sampling error is ____, but occasionally, it is ____.

5.18 If one takes repeated random samples from a population, calculates some statistic for each sample, and then makes a frequency distribution for that statistic, it is called a ____.

5.19 The standard deviation of the sampling distribution of the mean is called the ____.

5.20 The ____ describes the shape of the sampling distribution of the mean when the size of the samples is ____ and every possible sample is obtained.

5.21 A large sample, as is required for the central limit theorem, has at least ____ cases.

5.22 According to the central limit theorem, the sampling distribution of the mean will be normally distributed, no matter what the ____ of the parent population is.

5.23 According to the central limit theorem, the mean of a sampling distribution of the mean is the same as the ____ of the parent population.

5.24 According to the central limit theorem, we can calculate the standard error of the mean if we know the ____ of the population and the ____ of the sample.

5.25 If we don't know σ, we can use ____ to estimate the standard error of the mean.

5.26 If the standard error of the mean is small, that's an indication there is ____ sampling error.

5.27 A confidence interval is a ____, based on a ____ value, within which it is estimated that the ____ value falls.

5.28 A confidence interval is an interval ____, not a ____ estimate.

5.29 Within ____ standard errors of the mean of the midpoint of a sampling distribution of the mean, 95% of the means fall.

5.30 A mean picked at random from a sampling distribution of the mean has a ____% chance of capturing the population mean if 1.96 σ_M brackets are extended from it.

5.31 If a 95% confidence interval ranges from 7.32 to 13.68, it would be written in APA format as ____.

5.32 If a 95% confidence interval is calculated, there is a ____% chance that it does not capture the population mean.

Apply Your Knowledge

Using the random number table

5.33 There are 50 people in a class. The teacher wants a random sample of 10 students. She numbers the students consecutively (1, 2, 3, 4, . . . 50) and uses the random number table to sample 10 cases without replacement. She starts on the first line of the table and uses it like a book, moving from left to right on each line. She divides each four-digit random number into two two-digit numbers. If the first number were 1273, she would read it as 12 and 73. List the numbers of the 10 cases in her sample.

5.34 A dean wants a random sample, without replacement, of 10 students from the first-year class. There are 423 students in the first-year class. Assign numbers consecutively to the students, use the random number table starting at row 6, and use the last three digits of a number (e.g., 1273 would be read as 273). List the numbers of the 10 cases in her sample.

Evaluating samples

5.35 At an alcohol treatment center, about 10 people with alcohol problems complete treatment each month. The director wants to know their status a year after treatment. By 2014, 112 people had completed treatment. In 2015 she hires a researcher to track down all 112 and assess their status one year after treatment. He manages to locate 86 of the 112, or 77%. Of those 86, 66 agree to be interviewed. 66 is 77% of 86 and 59% of 112. Should the director pay attention to the results of the survey? Why or why not?

5.36 The mayor of a large city wants to know how optimistic new mothers are about their children's future in the city. There were 1,189 live births in the city in 2015. The mayor commissions a researcher who randomly samples 120 of the new mothers. That's a target sample of a little more than 10% of the population. The researcher approaches the target sample, explains the study to them, and tries to obtain their consent to participate. Ninety-four (78%) of those approached give consent and participate. Should the mayor pay attention to the results of the survey? Why or why not?

Calculating the standard error of the mean

5.37 Given $\sigma = 4$ and $N = 88$, calculate σ_M.

5.38 Given $\sigma = 12$ and $N = 88$, calculate σ_M.

5.39 Given $s = 7.50$ and $N = 72$, calculate s_M.

5.40 Given $s = 7.50$ and $N = 225$, calculate s_M.

Calculating confidence intervals (Report answers in APA format.)

5.41 If $\sigma_M = 4$ and $M = 17$, what is the 95% confidence interval for the mean?

5.42 If $s_M = 8$, and $M = -12$, what is the 95% confidence interval for the mean?

5.43 If $s = 17$, $N = 72$, and $M = -12$, what is the 95% confidence interval for the mean?

5.44 If $\sigma = 11$, $N = 33$, and $M = 52$, what is the 95% confidence interval for the mean?

Expand Your Knowledge

5.45 According to the central limit theorem, a sampling distribution of the mean will approach a normal distribution as long as which of the following is true?

a. At least 30 random, repeated samples are drawn.
b. The population has at least 30 cases.
c. Each repeated, random sample has at least 30 cases.
d. The size of the population standard deviation is at least 30.
e. The size of the population standard deviation is less than 30.
f. σ, not s, is used to calculate the standard error of the mean.

5.46 Charlotte obtains a random sample of 120 students from a university, finds out each person's age, and calculates M and s. Why is M not exactly equal to μ?

a. The sample is random.
b. The sample size is >50.
c. The sample is not large enough.
d. This is due to sampling error.
e. The population value for μ was wrong.
f. The central limit theorem does not apply when μ is known.
g. This could not occur.

5.47 If sample size is held constant, how does the size of the population standard deviation affect the size of the standard error of the mean?

5.48 If standard deviation is held constant, how does the size of the sample affect the size of the standard error of the mean?

5.49 Given $N = 81$, $\sigma = 12$, and $M = 100$, calculate a 90% confidence interval for the population mean.

5.50 Which would be narrower, a 90% confidence interval or a 99% confidence interval?

5.51 What can be done to make a 95% confidence interval narrower?

SPSS

There's not much that SPSS does with regard to sampling distributions. All the descriptive statistics procedures covered in Chapter 3—frequencies, descriptive, and explore—do calculate the standard error of the mean. To develop a sense of how much sampling error exists in a sample, use descriptive statistics to find the standard error of the mean.

SPSS can calculate a confidence interval for the mean with whatever percentage of confidence is desired. To do so, go to "Analyze," then "Descriptive Statistics," then "Explore." See Figure 5.17. This opens up a new menu box, as shown in Figure 5.18. Once the variable for which the confidence interval is desired has been moved into the "Dependent List" box, click on "Statistics. . . ." Another box opens up, with the default value of a 95% confidence interval already selected. This can be changed to any value, though the most common other ones are 90% and 99%, as seen in Figure 5.19.

Figure 5.17 First Step in SPSS to Calculate a Confidence Interval. (Source: SPSS)

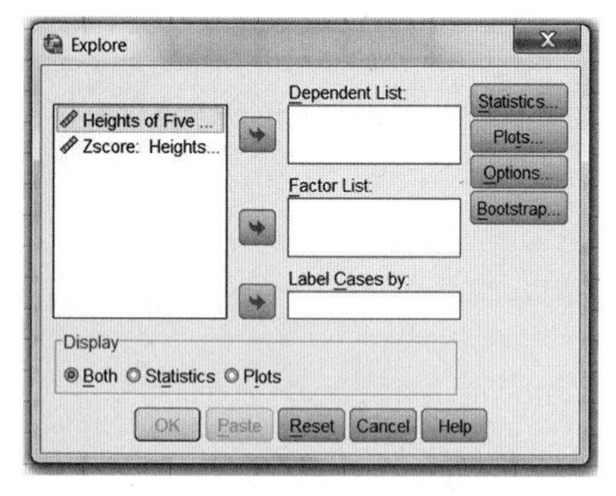

Figure 5.18 Explore Menu in SPSS. (Source: SPSS)

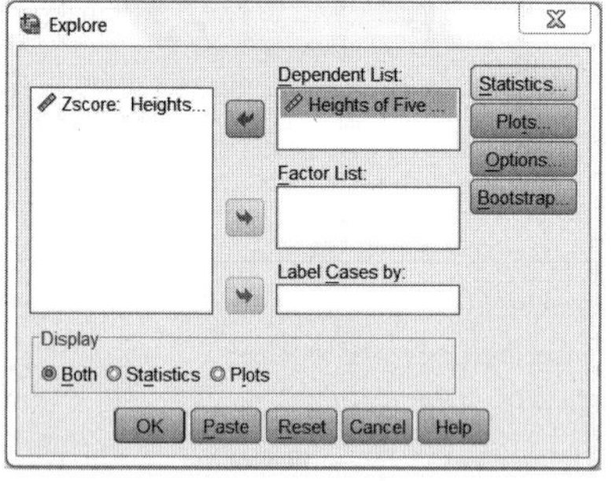

Figure 5.19 Setting the Confidence Interval Value in SPSS. (Source: SPSS)

PART I Test Your Knowledge

These questions are meant to probe your understanding of the material covered in Chapters 1–5. The questions are not in the order of the chapters. Some of them are phrased differently or approach the material from a different direction. A few of them ask you to use the material in ways above and beyond what was covered in the book. This "test" is challenging. But, if you do well on it or puzzle out the answers using the key in the back of the book, you should feel comfortable that you are grasping the material.

1. Answer each part. All questions refer to a normal distribution. For IQ, $\mu = 100$ and $\sigma = 15$.

a. What percentage of scores fall from the mean to 2.33 standard deviations below the mean?

b. What percentage of scores fall at or above $z = 0.93$?

c. What percentage of scores fall at or above $z = -1.37$?

d. What percentage of scores fall from $z = -0.23$ to $z = 1.84$?

e. What percentage of the world's population would be classified as geniuses if to be a genius one must have an IQ ≥ 150?

f. What was Buffy's score on the IQ test if her score, as a PR, were 33.

g. On each SAT subtest, Skip's score was 670. He is planning to take the ACT. If he performs as well on the ACT as he did on the SAT, then what would his ACT score be? (Assume the ACT has a mean of 21.50 and a standard deviation of 5.)

h. What percentage of scores fall from $z = 1.23$ to $z = 2.23$?

2. Answer each part.

a. Make a stem-and-leaf plot for these data: 11, 31, 61, 62, 44, 54, 33, 32, 36, 45, 47, 55, 52, 19, 22, 33, 24, 28, 48, 26, 58, 44, 33, 36

Based on the frequency distribution for the Irrational Beliefs Scale shown in the following table, answer the following questions:

b. $i =$ ____________

c. What are the apparent limits for interval D?

d. What is the real lower limit for interval G?

e. What is the real upper limit for interval H?

f. What is the midpoint for interval E?

g. $N =$ ____________

h. What is the unit of measurement for the Irrational Beliefs Scale?

Frequency Distribution: Irrational Beliefs Scale

	Interval	Frequency	Cumulative Frequency
A	37–40	4	150
B	33–36	10	146
C	29–32	17	136
D	25–28	27	119
E	21–24	30	92
F	17–20	25	62
G	13–16	16	37
H	9–12	11	21
I	5–8	7	10
J	1–4	3	3

3. Answer all parts.

Given 7, 9, 4, 11, and 14,

a. $\Sigma X =$ ____________

b. $M =$ ____________

c. $\Sigma(X - M) =$ ____________

d. $\Sigma(X - M)^2 =$ ____________

e. $s =$ ____________

f. Mdn = ____________

4. a. For a sample of 50 cases from a population, put these in order from smallest to largest: σ, σ^2, s, s^2: ____________

b. Assume that conscientiousness is normally distributed. A researcher develops a new way to measure it. If almost 100% of the scores fall from 15 to 45, then $s =$ __________.

5. Answer both parts.

a. A researcher obtained a random sample of 75 teenagers from the United States. She had each teenager keep track of how many times he or she checked his or her phone for texts or tweets during one 24-hour period. Given $M = 157$ and $s = 32$, calculate the 95% confidence interval for the population mean for the number of times a teenager checks.

b. Explain what this confidence interval means.

6. For each scenario, indicate whether the design is correlational (C), experimental (E), or quasi-experimental (QE). For the variable in **bold,** indicate whether it is a predictor variable (PV), criterion variable (CV), independent variable (IV), dependent variable (DV), or grouping variable (GV).

a. A psychologist is curious if **sex** has an impact on anxiety. She observes male and female nurses as they give shots to patients. Just before the injection, she asks the patients to rate how anxious they are.

b. A researcher is curious how **color** affects children's activity levels. He takes a noncolorful breakfast cereal and dyes it in bright colors. When a child comes to his lab, he randomly assigns him or her to receive the colorful or noncolorful version. He then lets the child play in a ball pit and times, to the nearest second, how long he or she plays.

c. A researcher thinks it is **eyes** that control how attractive a person is. She gets a sample of college students and takes high-resolution images of their faces. She prepares two sets of photos, one of their full faces and one of just their eyes. She goes to another college and assembles two panels of raters. One panel rates how attractive the eyes are and the other panel rates the faces. Each panel rates on a scale of 0 to 100.

d. Can humans echolocate? A researcher advertises for subjects who don't mind getting bruises and 20 volunteers show up. He randomly assigns them to two groups. He teaches one group to click their tongues and listen for the sound to bounce back. He teaches nothing to the other group. He then puts each person in a totally dark room that is filled with obstacles like chairs and tables for ten minutes. The volunteer is instructed to walk around and form a mental map of the room. While each person is walking around the room, the researcher counts **how many times he or she bumps into an object.**

e. Does fussiness as a baby predict fussiness as an adult? A psychologist has mothers classify their babies as fussy or not. Thirty years later, she tracks down the babies and has their spouses rate them on a scale that measures **how difficult they are to please.**

7. Given an interval-level variable, make a grouped frequency distribution. Use $i = 10$, start the lowest interval at 20, and report the midpoint, f, f_c, %, and $\%_c$. The data are given below—note that they are already organized from low to high.

26	40	47	51	55	61
30	40	47	51	56	61
33	41	47	52	57	62
33	42	48	53	57	64
33	43	48	53	57	65
36	45	48	54	59	65
38	45	48	54	59	67
39	45	50	54	59	72
39	45	50	54	59	73
40	46	50	54	61	75

8. For each scenario below, determine the level of measurement of the variable in **bold.**

a. Adopted children are compared to non-adopted children on the 25-item Smith Attachment Checklist to see whether **adoption status** affects attachment level.

b. A researcher believed that amount of sleep influenced learning ability. He had participants keep a sleep diary for a week, so he could calculate the average number of hours each slept per night. He then had each subject memorize a list of 25 words, gave him or her a distractor task, and observed **how many of the 25 words each could recall.**

c. A researcher thought that amount of sleep influenced learning ability. She got 20 volunteers and randomly assigned them to two groups. Group A slept for 8 hours a night, for three nights in a row. Group B slept for 4 hours a night, for three nights in a row. After the three nights, each participant read a newspaper article about a chef who was mugged while walking home from work. Then, each person took a surprise recall test about 20 facts from the article. It was scored based on **the percentage of facts correctly recalled**.

d. A researcher is curious how well a student's score on an initial statistics test predicts how much statistics he or she ends up knowing, as measured by **the final point total in the class**. He takes a random sample of 100 students from previous semesters, determines each person's score on the first test, and then finds his or her final point total.

e. A sexual assault researcher believes that alcohol consumption leads to being in situations where assault is more possible. To investigate this assumption, she obtains a representative sample of 3,500 U.S. college students and classifies them as (0) non-drinkers, (1) people who drink 1–3 days per month, (2) 4–7 days per month, (3) 8–13 days per month, (4) 14–24 days per month, (5) 25–29 days per month, and (6) every day. Each participant also reports how many sexual experiences he or she had under the influence that he or she now regrets. The researcher compares the number of regretted sexual experiences across the different **classifications of drinking days**.

9. Round the following to two decimal places.

a. 12.4550

b. 13.4105

c. 14.7812345

d. 15.9937

e. 16.99

f. 2/3

g. 3/2

h. 10/2

i. 4.00/2.00

10. Graph the following data: Intro psych students were surveyed regarding textbook use. Ten bought a new textbook; 32 bought a used version; 21 bought a looseleaf version; 23 were renting their copy; and 7 did not have a textbook.

11. Determine the shape of each distribution.

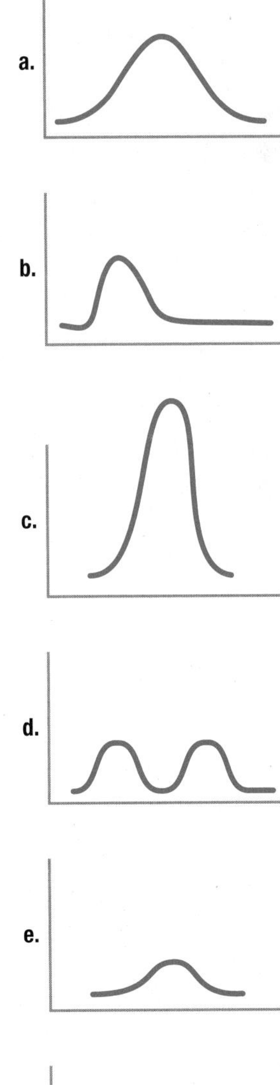

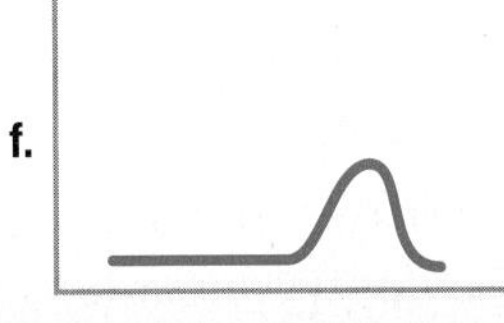

PART II One- and Two-Sample Difference Tests

The four chapters in this section form the core of statistics for psychology. First, hypothesis testing is introduced. Hypothesis testing is the procedure that statisticians use to make decisions that are objective and data-based, not subjective and emotion-based.

Then the most commonly used tool in the psychologist's toolbox, the difference test, is introduced in several varieties. Single-sample difference tests are used to answer questions such as the following: Is the average GPA in this *sample* of students different from the mean GPA for the entire *population* of students? Two-sample difference tests are used to answer questions like those posed by classic experiments that have one experimental group and one control group. In this section, we'll learn about two, the independent-samples *t* test and the paired-samples *t* test. In the next section, we'll cover ANOVA, tests used in more complex situations when experiments include three or more groups.

Introduction to Hypothesis Testing

LEARNING OBJECTIVES

- Explain how hypothesis testing works.
- List the six steps to be followed in completing a hypothesis test.
- Explain and complete a single-sample *z* test.
- Explain the decisions that can be made in hypothesis testing.

CHAPTER OVERVIEW

This chapter makes the transition from descriptive statistics to inferential statistics. In inferential statistics, *specific* information from a sample is used to draw a *general* conclusion about a population. For example, this allows a researcher to study one group of people with depression and to draw a conclusion that applies to people with depression in general. This transition to inferential statistics starts with the logic that statisticians use to reach decisions, a process called hypothesis testing.

After this prelude, the chapter takes a pragmatic turn with coverage of the single-sample *z* test, which we'll use to demonstrate how hypothesis testing works. Psychology researchers rely on statistics to help them make thoughtful, data-driven decisions. The single-sample *z* test allows researchers to determine if a sample mean is statistically significantly different from a population mean or some other specified value.

Completing a hypothesis test is a lot like baking. Before you can put the cake in the oven, you must go to the store to buy the ingredients, turn on the oven, measure out the ingredients, and mix them together in the right order. This chapter introduces a six-step procedure that can be used for all hypothesis tests to make sure that the right steps happen in the right order. We even offer a mnemonic—Tom and Harry despise crabby infants—to keep everything straight.

The chapter ends with an exploration of the different types of correct and incorrect conclusions that can occur. The incorrect conclusions have names: Type I error and Type II error. For a single-sample *z* test, a Type I error occurs when one erroneously concludes that a sample mean differs from a population mean. Type II error is the opposite—it occurs when the researcher indicates there is no evidence that the sample mean differs from the population mean and the sample really does differ.

6.1 The Logic of Hypothesis Testing

A **hypothesis** is a proposed explanation for observed facts. If, for example, a psychologist noted that people living in sunny climates were happy and people living in cloudy climates were sad, this might lead to the hypothesis that a lack of sun leads to depression.

Science involves gathering observations, generating hypotheses to explain them, and then testing the hypotheses to see if the predictions they make hold true. In statistics, **hypothesis testing** is the procedure in which data from a *sample* are used to evaluate a hypothesis about a *population.*

Here's how hypothesis testing works. A cognitive psychologist might hypothesize that the mean IQ in a population is 100. If she gathered a representative sample of people from that population and found that the mean IQ of her sample was 100 or close to 100, there would be little reason to question the hypothesis. However, if the observed mean in the sample were far from the expected mean of 100, then she would have to wonder if the population mean were really 100.

Here's another example that shows how humans already use hypothesis testing in everyday life. Amie has a boyfriend, Allen, and she believes (hypothesizes) that Allen loves her. Based on this hypothesis, she expects that Allen will behave in certain ways. If he loves her, then he should want to spend time with her, do nice things for her, hold her hand, and so on. And if Allen does these things, then there's no reason for Amie to question his love for her. But, if Allen doesn't do these things—if the observed behavior doesn't match the expected behavior—then she'll to start wondering if he truly loves her.

Ten Facts About Hypotheses and Hypothesis Testing

The love example shows how humans use hypothesis testing intuitively. Now, let's see how scientists have formalized the process with 10 facts about hypothesis testing (Table 6.1):

1. A hypothesis is a statement about a *population,* not a sample. This is because a researcher wants to draw a conclusion about the larger population and not about the specific sample being studied.
2. There will be two hypotheses—one called the null hypothesis, H_0, and the other called the alternative hypothesis, H_1.
3. The two hypotheses must be *all-inclusive,* which means they cover all possible outcomes. For example, if one hypothesis said that all cars in the world were white and the other that all cars were black, those two hypotheses wouldn't be all-inclusive because red and green and yellow cars also exist. However, if one hypothesis stated, "All cars are white" and the other hypothesis, "Not all cars are white," then the two hypotheses would be all-inclusive.
4. The two hypotheses must be *mutually exclusive,* which means that only one hypothesis at a time can be true. For example, a coin can land on heads or on tails, but not on both sides at once. Heads and tails are mutually exclusive outcomes.
5. The null hypothesis is a negative statement. The **null hypothesis** says that in the population the explanatory variable does *not* have an impact on the outcome variable. For example, if a researcher were testing a technique to improve

TABLE 6.1 Ten Facts About Hypotheses and Hypothesis Testing

1. A hypothesis is a statement about a *population*, not a sample.
2. In hypothesis testing, there are two hypotheses, the null hypothesis (H_0) and the alternative hypothesis (H_1).
3. The two hypotheses must be all-inclusive.
4. The two hypotheses must be mutually exclusive.
5. The null hypothesis is a negative statement and makes a specific prediction.
6. The alternative hypothesis, which is what the researcher believes is true, isn't a specific prediction.
7. One can't prove a negative statement.
8. But, one can *disprove* a negative statement. And all it takes is one example.
9. If the null hypothesis is disproved (rejected), then the researcher has to accept the mutually exclusive alternative hypothesis.
10. If the researcher fails to disprove the null hypothesis, he or she can't say that the null hypothesis is true. The best that can be said is that one hasn't found enough evidence to reject it.

intelligence, the null hypothesis would state that the technique does *not* improve intelligence. Because the null hypothesis makes a specific prediction, it would say that the technique has zero impact and does not improve intelligence *at all.*

6. The **alternative hypothesis** is what the researcher believes is true; it states that, in the population, the explanatory variable has an impact on the outcome variable. For the researcher testing the intelligence-improving technique, the alternative hypothesis would state that the technique has *some* impact on intelligence. Notice that the alternative hypothesis doesn't specify how much impact. Unlike the null hypothesis, the alternative hypothesis doesn't make a specific prediction.

7. A negative statement, like the null hypothesis, can't be proven true. Suppose an adult wants to prove the negative statement "There are no unicorns" to a child who believes in unicorns. No matter what evidence the adult offers (e.g., "I looked all through Ohio and didn't find one"), the child will counter that the adult wasn't looking in the right place, or that the unicorns heard the adult coming and fled. A negative can't be proven.

8. However, a negative can be disproven. It just takes one example. If the child ever walks up leading a unicorn on a leash, the adult can no longer claim, "There are no unicorns." Because the null hypothesis makes a specific prediction, it is used to predict how a researcher's experiment will turn out. If the experiment doesn't turn out as the null hypothesis predicted, the researcher disproves, or "rejects," the null hypothesis. The null hypothesis can be nullified.

9. If the null hypothesis is rejected, then one has to accept the mutually exclusive alternative hypothesis as true. This happens because the null hypothesis and the alternative hypothesis are all-inclusive and mutually exclusive. If one hypothesis is not true, then the other hypothesis has to be true. As the researcher usually believes that the alternative hypothesis is true, the objective of a study is almost always to reject the null hypothesis.

10. However, if a researcher fails to disprove the null hypothesis, he or she can't say that the null hypothesis is true. Remember, a negative statement, like the null hypothesis, can't be proven true. The best the researcher can say is that he or she hasn't found enough evidence to reject the null hypothesis. This is different from saying the null hypothesis is true, just as a not guilty verdict in a trial is different from saying that the defendant is innocent. Not guilty just means that there wasn't enough evidence to persuade the jury that the defendant was guilty.

It is impossible to prove a negative hypothesis, but one positive example will disprove it.

Practice Problems 6.1

Review Your Knowledge

6.01 How do the null hypothesis and the alternative hypothesis differ?

Apply Your Knowledge

6.02 Explain this statement: "Hypothesis testing involves comparing what is observed to happen in an experiment to what is expected to happen."

6.2 Hypothesis Testing in Action

Let's put the 10 facts about hypothesis testing to work. Let's suppose a psychologist, Dr. Pell, wonders whether children who are adopted differ in intelligence from non-adopted children. To explore this, she obtains a random sample of 72 children from the population of adopted children in the United States and gives each of them an IQ test. She finds that the mean IQ of these 72 children is 104.5. Knowing that the average IQ for children in the United States is 100 with a standard deviation of 15, can she conclude that adopted kids differ in intelligence from the general population?

It is tempting to say that the answer is obvious. 104.5 is a bigger number than 100, so it is true that adopted kids differ from the general population of kids. But, that's not how statisticians think.

Remember sampling error, introduced in Chapter 5? Because of sampling error, researchers don't expect a sample mean to be exactly the same as the population mean. With a population mean for IQ of 100, Dr. Pell can expect the sample mean to be close to 100, but not exactly 100.

So, here's the question for Dr. Pell to ask: Is 104.5 close enough to 100 that sampling error can explain it? If so, then there's no reason to think adopted kids differ from the average IQ. However, if sampling error fails as a reasonable explanation for the difference, then she can conclude that adopted kids differ in intelligence.

The Six Steps of Hypothesis Testing

Learning how to complete a statistical test is like learning how to cook—it's better to follow a recipe. That's why we're going to use a six-step "recipe" for hypothesis testing. Here are the six steps:

Step 1 Test: Pick the right statistical test.

Step 2 Assumptions: Check the assumptions to make sure it is OK to do the test.

Step 3 Hypotheses: List the null and alternative hypotheses.

Step 4 Decision rule: Find the critical value of the statistic that determines when to reject the null hypothesis.

Step 5 Calculation: Calculate the value of the test statistic.

Step 6 Interpretation: State in plain language what the results mean.

Here's a mnemonic to help remember the six steps in order: "Tom and Harry despise crabby infants." The first letters of the six words in the mnemonic stand for "**T**est," "**A**ssumptions," "**H**ypotheses," "**D**ecision rule," "**C**alculation," and "**I**nterpretation." First, let's walk through the steps, then follow Dr. Pell as she applies them to the adoption IQ study.

A mnemonic to remember the six steps of hypothesis testing (**t**est, **a**ssumptions, **h**ypotheses, **d**ecision rule, **c**alculation, **i**nterpretation): "Tom and Harry despise crabby infants."

Picking the right test requires thought.

Step 1 Pick a Test

The first step in hypothesis testing is picking which test to use. There are hundreds of statistical tests, each designed for a specific purpose. Choosing the correct test depends on a variety of factors such as the question being asked, the type of study being done, and the level of measurement of the data. In Chapter 16, there are flowcharts that guide one through choosing the correct hypothesis test. Feel free to peek ahead and take a look at them.

Step 2 Check the Assumptions

All statistical tests have assumptions, conditions that need to be met before a test is completed. If the assumptions aren't met, then researchers can't be sure what the results of the test mean.

Here's a nonstatistics explanation about the role of assumptions. In this day and age, athletes are tested for performance-enhancing substances. The drug test depends on a number of assumptions. For the test to be meaningful, it is assumed that the sample being tested is the athlete's, that the sample was stored at the right temperature after it was taken, that no one has tampered with the sample, and that the machine being used to test it is correctly calibrated.

Imagine that an athlete wins a race, provides a urine sample, and it is tested on an incorrectly calibrated machine. The assumption that the machine is calibrated correctly has been violated. Suppose the results indicate that the urine does not test positive for performance-enhancing substances. Is that true? Maybe it is, maybe it isn't. When an assumption is violated, it is still physically possible to complete the test. But, one should not because it is impossible to interpret the results.

As hypothesis tests are covered, their assumptions will be listed to make sure one knows the conditions that must be met to proceed with the test. There are two types of assumptions: *not robust* and *robust*. A **nonrobust assumption** has to be met for the test to proceed. If a nonrobust assumption is violated, a researcher should stop proceeding with the planned statistical test. A **robust assumption** can be violated, to some degree, and the test can still be completed and interpreted.

Here is a way to think about the difference between the two assumptions. People can be described as being in "robust" health or not. Imagine that Carl is not in robust health—he has a compromised immune system. In fact, his health is so bad that he is in an isolation ward in the hospital. One day he gets a visitor who has the flu and who sneezes while in the room. Carl, whose health has just been violated, is likely to get the flu. In contrast, Bethany is in robust health. One day her roommate, who has the flu, sneezes directly in her face. Bethany, with a strong immune system, is able to fight off this violation and stay healthy. However, there is a limit to what Bethany's immune system can handle. If one roommate with the flu sneezes in her face, by mistake Bethany uses the toothbrush of another roommate with the flu, and it turns out that Bethany's significant other—whom she spends a lot of time kissing—also has the flu, well, that may be too many violations for her immune system to handle. Even a robust assumption will break if it is violated too much.

Step 3 List the Hypotheses

Step 3 involves listing the null hypothesis and the alternative hypothesis. Hypotheses can be for *two-tailed tests* (also called nondirectional tests) or *one-tailed*

Before one can proceed with a test, one needs to make sure that its assumptions have been met.

tests (directional tests). A **two-tailed hypothesis test** doesn't indicate whether the explanatory variable (adoption in our example) has a positive or negative impact on the outcome variable (IQ), just that it has an impact. An advantage of a two-tailed test is that it allows a researcher to test for an effect in either direction, that children who are adopted end up as more intelligent than average or as less intelligent than average. Two-tailed tests are the norm and are used much more frequently than one-tailed tests.

The null and alternative hypotheses need to be specified for every statistical test.

When, in advance of collecting any data, a researcher has an expectation about the direction of the impact of the explanatory variable, a **one-tailed hypothesis test** is called for. One-tailed tests predict that the results will turn out in a certain direction. An advantage to using one-tailed tests is that it is easier to reject the null hypothesis and be forced to accept the alternative hypothesis.

With the current example, it seems reasonable to think a one-tailed test is called for because it is already known that the mean IQ of this sample of adopted children is above average. But the question—do adopted children differ in intelligence from the average—was formulated before any data were collected. The question doesn't specify a direction for the difference, so a two-tailed test is called for. Changing the question after looking at the data is not how statisticians operate.

Step 4 Set the Decision Rule

Setting the decision rule involves finding the *critical value* of the test statistic. The **critical value** is the value that the test statistic must meet or exceed in order to reject the null hypothesis. What the critical value is depends on a number of

Setting the decision rule in advance means that one knows what to do when a choice point is reached.

factors, such as how willing the researcher is to draw the wrong conclusion and how many cases there are.

Just as it is not fair to change from a two-tailed test to a one-tailed test after looking at the results, the decision rule is made in advance. That way, if the results are just short of the point where the null hypothesis is rejected, the researcher won't be tempted to slide the critical value over a bit to get the results he or she desires.

Calculating the value for a test statistic is the fifth step in hypothesis testing.

Step 5 Calculate the Test Statistic

Calculating the test statistic is the most straightforward of the six steps for hypothesis testing. Plug the right numbers into a formula, and push the right buttons in the right order on the calculator: that's how to calculate the test statistic.

Step 6 Interpret the Results

Interpreting the results is the reason statistical tests are done. In Step 6, the researcher explains, in plain language, what the results are and what they mean. Interpretation is a human skill.

Interpretation involves answering questions about the results. For example, in this chapter, we'll ask whether the null hypothesis was rejected. In future chapters, more interpretation questions will be added and the questions will change slightly from hypothesis test to hypothesis test. When there are multiple questions, they build on each other, so a researcher gains a greater understanding of the results. It is possible to stop after any question with enough information to offer an interpretation, but the more questions answered, the better a researcher will understand the results.

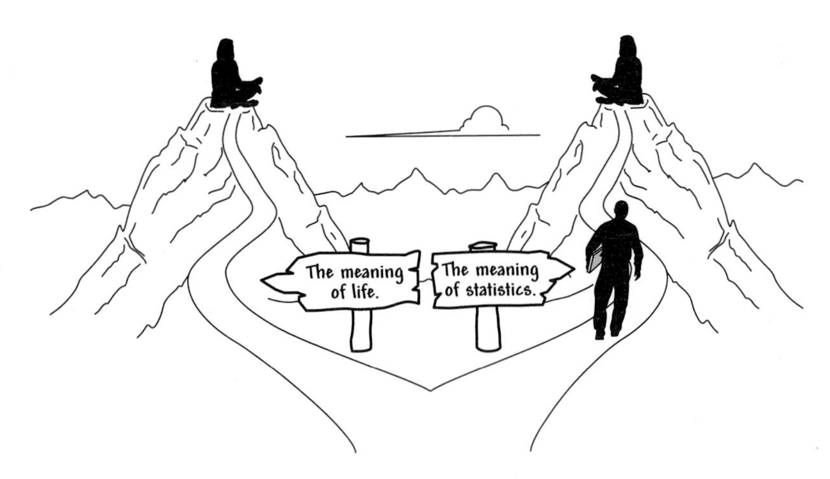

Interpretation, explaining what the results mean, is the goal of statistics.

There are four parts to writing an interpretation:

1. Recap why the study was done.
2. Provide the main factual results, for example, the mean scores for the control group and the experimental group.
3. Explain what the results mean.
4. Make suggestions for future research.

This four-part guide to writing an interpretation is presented in Table 6.2. It may look like interpretations will be long and detailed, but short and clear is better. For many studies, a good interpretation can be accomplished in one paragraph of four or five sentences.

TABLE 6.2 Template for Writing a Four-Point Interpretation Paragraph

1. Recap the study. What was done? Why?
2. Present the main results factually. For example, what were the mean scores for the control and experimental groups? Present the results of the hypothesis test in APA format.
3. Explain what the results mean.
4. Make suggestions for future research. What were the strengths and/or weaknesses of this study? What should be done in the next study?

A thorough interpretation can often be accomplished in one paragraph. In an interpretation, the results are presented both objectively (e.g., group means) and subjectively (the researcher's explanation).

The Single-Sample *z* Test

Now that we've covered the general logic behind hypothesis testing, let's move on to the specific steps of the single-sample z test.

Step 1 Pick a Test. The single-sample z test is the test to use to see whether adopted children differ in intelligence from the general population. Dr. Pell has selected it because the single-sample z test is used to compare a sample mean to a population mean when the standard deviation of the population is known. (It is known that the population standard deviation, σ, for IQ is 15.)

Step 2 Check the Assumptions. The assumptions for the single-sample z test are listed in Table 6.3. The table also notes whether the assumptions are robust or not.

TABLE 6.3 Assumptions for the Single-Sample *z* Test

	Assumption	Robustness
1. Random sample	The sample is a random sample from the population.	Robust
2. Independence of observations	The cases in the sample are independent of each other.	Not robust
3. Normality	The dependent variable is normally distributed in the population.	Robust

If a nonrobust assumption is violated, one needs to find a different statistical test, one with different assumptions, to use.

- **Random sample:** The first assumption, that the sample is a random sample from the population, is not violated. In the IQ example, the sample was a random sample from the population of adopted children. This is a robust assumption, so even if the sample were not a random one, we could still proceed with the test. One just needs to be careful about the population to which one generalizes the results.
- **Independence of observations:** The second assumption is that the observations within the group are independent. This assumption means that the scores of cases in the sample aren't influenced by other cases in the sample. In the IQ example, the participants were randomly sampled, each case was in the sample only once, and each case was tested individually, so this assumption was not violated. This is not a robust assumption, so if the cases were not independent, we would not be able to proceed with the test.
- **Normality:** The third assumption is that the dependent variable is normally distributed in the population. Intelligence is one of the many variables that psychologists assume to be normally distributed, so that assumption is not violated. This is also a robust assumption. So if it were violated, as long as the violation is not too great, we would still be able to proceed with the test.

With no assumptions violated, Dr. Pell can proceed with the planned test.

A Common Question

Q Suppose a teacher has developed a new way of teaching math and then tests the new method with three different groups, each with five students being taught together. When he uses the final exam to evaluate his new method, should he treat the data as 15 scores, one from each student, or three scores, one from each group?

A To avoid violating the independence of observations assumption, this should be treated as three scores, each score the mean of the five students taught together in a group. However, few researchers are this rigorous. Researchers worry more about violating this assumption by including the same participant in a study twice.

Step 3 List the Hypotheses. With a two-tailed test, it is easier to generate the null hypothesis first and the alternative hypothesis second. The null hypothesis is going to be (1) a statement about the population, (2) a negative statement, and (3) a specific statement. All of these conditions are met by

$$H_0: \mu_{\text{AdoptedChildren}} = 100$$

Dr. Pell's null hypothesis says that the population of adopted children has a mean IQ of 100.

- Because it is about μ, the population mean, it is a statement about a population.
- It is a negative statement because it says that the intelligence of adopted children is *not* different from the mean IQ for the population of children in general.
- It is a specific statement because it says the population mean is exactly 100.

Next, Dr. Pell needs to state the alternative hypothesis. The alternative hypothesis has to be mutually exclusive to the null hypothesis, and the two hypotheses together have to be all-inclusive. Further, the alternative hypothesis is not going to make a specific prediction. The alternative hypothesis is just going to say that the null hypothesis is wrong. This means that the alternative hypothesis states the population mean for adopted children is something other than 100:

$$H_1: \mu_{\text{AdoptedChildren}} \neq 100$$

Step 4 Set the Decision Rule. Hypothesis testing works by assuming that the null hypothesis is true. Assume that the null hypothesis, $\mu = 100$, is true and imagine taking hundreds and hundreds of repeated random samples of size 72 from the population of adopted children. For each sample, calculate a mean and then make a sampling distribution for all the means from all the samples. Thanks to the central limit theorem, three things are known about this sampling distribution: (1) the sampling distribution of the mean will be centered at 100, (2) it will have a normal shape, and (3) its standard deviation, called the standard error of the mean, can be calculated using Equation 5.1. For the adoption IQ example, this would be

$$\begin{aligned}\sigma_M &= \frac{\sigma}{\sqrt{N}} \\ &= \frac{15}{\sqrt{72}} \\ &= \frac{15}{8.4853} \\ &= 1.7678 \\ &= 1.77\end{aligned}$$

The sampling distribution of the mean that should occur if the null hypothesis is true is shown in Figure 6.1. Note that it has a normal shape, is centered around the population mean of 100, and that IQ scores on the *X*-axis are marked off by the size of the standard error of the mean, 1.77.

Now, divide the sampling distribution into two parts (Figure 6.2). The middle section is called the **common zone,** because it is the section in which sample means commonly fall. The two extreme sections form the **rare zone** because it is the part of the sampling distribution in which it is rare that a sample mean falls.

Next, place the observed sample mean, 104.5, in the sampling distribution of the mean.

- If the observed sample mean falls in the common zone, there's no reason to question the null hypothesis. A common result, what is expected to happen if the null hypothesis is true, did happen.

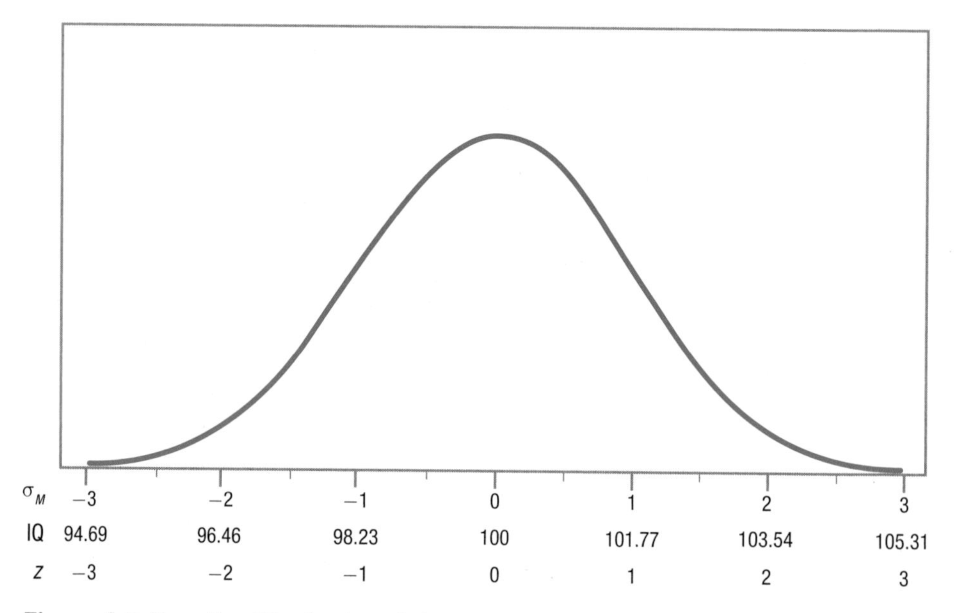

Figure 6.1 Sampling Distribution of the Mean This distribution shows what the sampling distribution of the mean would look like for repeated random samples of size 72 from a population with a mean IQ of 100. Note that it is normally distributed and centered around the population mean. The standard error of the mean, σ_M, is 1.77 and is being used to mark off IQ scores by standard deviation units on the *X*-axis.

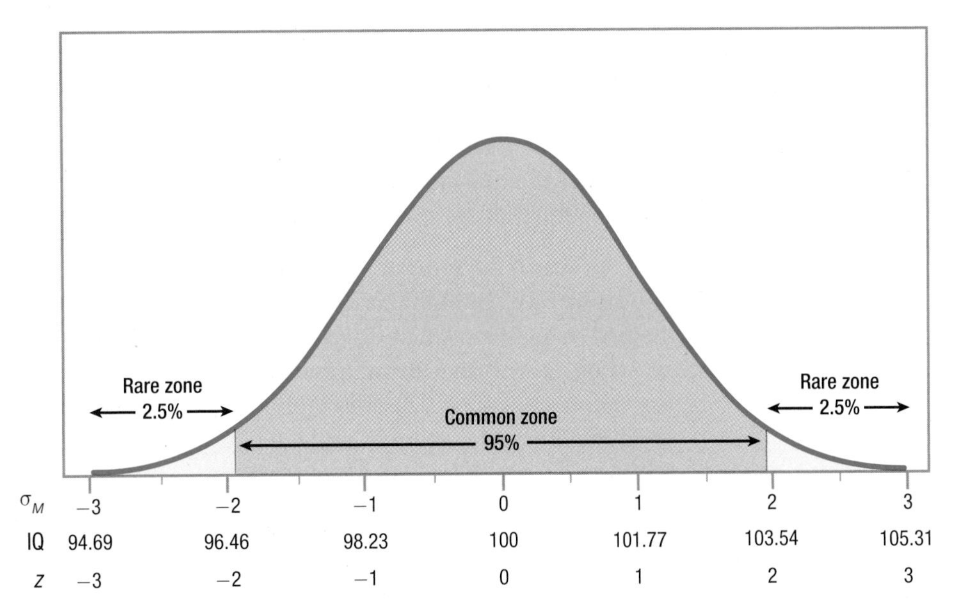

Figure 6.2 Sampling Distribution of the Mean with the Common Zone and Rare Zone Marked for IQ Data When the alpha level is set at .05 and the hypotheses are nondirectional, the common zone captures the middle 95% of the sampling distribution, and the rare zone captures the extreme 5%. The border between the rare zone and common zone occurs at *z* scores of −1.96 and 1.96, respectively, because they are the *z* values that separate the middle 95% of a normal distribution from the extreme 5%.

- If the observed sample mean falls in the rare zone, then an unusual event has happened. In this situation, by the logic of hypothesis testing, the null hypothesis is rejected and the researcher is forced to accept the alternative hypothesis as true. As the alternative hypothesis is what the researcher believes to be true, he or she is happy to be "forced" to accept it.

The line drawn between the common zone and the rare zone is called, as was noted above, the critical value. The critical value of *z*, abbreviated z_{cv}, depends on how small the researcher wants the rare zone to be. The convention most commonly used in statistics is to say that something that happens 95% of the time is common and something that happens 5% of the time or less is rare. This means that the common zone is the middle 95% of the sampling distribution, and the rare zone is the extreme 5% of the sampling distribution. Notice that the rare zone is two-tailed, with 2.5% at the very top (the right-hand side) of the sampling distribution and 2.5% at the very bottom (the left-hand side) of the distribution. This is why the test is called a two-tailed test—the rare zone falls in both sides (tails) of the sampling distribution.

In Chapter 4, it was found that the *z* scores of ±1.96 marked off the middle 95% of a normal distribution from the extreme 5%. Now, the *z* scores of ±1.96 will be used as the critical values, z_{cv}, for the single-sample *z* test. Figure 6.2 shows how the critical values divide the sampling distribution of the expected outcomes into common and rare zones. Here is the decision rule:

- If the *z* score calculated in Step 5 (the next step) is less than or equal to −1.96 *or* if the *z* score is greater than or equal to 1.96, the researcher will reject the null hypothesis.
    - Written mathematically, this is: if $z \leq -1.96$ *or* if $z \geq 1.96$, then reject H_0.

- If the z score calculated in Step 5 is greater than -1.96 *and* less than 1.96, then the researcher will fail to reject the null hypothesis.
    - Written mathematically: if $-1.96 < z < 1.96$, then fail to reject H_0.
- Notice that the critical value itself is part of the rare zone. If $z = -1.96$ or $z = 1.96$, then the researcher will reject the null hypothesis.

The size of the rare zone, expressed as a probability, is the *alpha level* of the test. The **alpha level,** or **alpha,** is the probability that a result will fall in the rare zone and the null hypothesis will be rejected when the null hypothesis is really true. Alpha levels are also called **significance levels.**

When statisticians discuss alpha levels, they use proportions, not percentages. So in this example, alpha is set at .05. Alpha is abbreviated with a lowercase Greek letter, α, so one would write, "$\alpha = .05$." This means that the researcher considers a rare event something that happens no more than 5% of the time.

The critical value that is the dividing line between the rare zone and the common zone depends on (1) whether the researcher is doing a one-tailed or a two-tailed test and (2) what alpha level is selected. The most commonly used options for a single-sample z test are shown in Table 6.4.

TABLE 6.4 Critical Values of z, One-Tailed and Two-Tailed, for Commonly Used Alpha Levels for Single-Sample z Tests

	Critical Values of z			
df	$\alpha = .10$, two-tailed or $\alpha = .05$, one-tailed	$\alpha = .05$, two-tailed or $\alpha = .025$, one-tailed	$\alpha = .02$, two-tailed or $\alpha = .01$, one-tailed	$\alpha = .01$, two-tailed or $\alpha = .005$, one-tailed
	$z_{cv} = 1.65$	$z_{cv} = 1.96$	$z_{cv} = 2.33$	$z_{cv} = 2.58$

Note that as the alpha level gets smaller, the critical value of z moves farther away from zero, making the rare zone smaller.

Step 5 Calculate the Test Statistic. The test statistic to be calculated is a z value. So, Step 5 involves turning the observed sample mean of 104.50 into a z score. The formula for calculating the z value for a single-sample test is shown in Equation 6.1.

Equation 6.1 Formula for the Single-Sample z Test

$$z = \frac{M - \mu}{\sigma_M}$$

where z = the z score
M = the sample mean
μ = the population mean
σ_M = the standard error of the mean (Equation 5.1)

Earlier, using Equation 5.1, the standard error of the mean for the adoption IQ example was calculated as $\sigma_M = 1.77$. With that value, and with $M = 104.5$ and

$\mu = 100$, Dr. Pell can calculate the value of z for the single-sample z test for the adoption IQ study:

$$z = \frac{M - \mu}{\sigma_M}$$

$$= \frac{104.50 - 100}{1.77}$$

$$= \frac{4.5000}{1.77}$$

$$= 2.5424$$

$$= 2.54$$

Having calculated $z = 2.54$, this step is over.

Step 6 Interpret the Results. In interpreting the results, plain language is used to explain what the results mean. In this chapter, we'll start with the most basic interpretative question, "Was the null hypothesis rejected?"

This is addressed by comparing the observed value of the test statistic, 2.54, to the critical value, ±1.96. Which of the following statements generated in Step 4 is true?

- Is either $2.54 \leq -1.96$, or is $2.54 \geq 1.96$?
- Is $-1.96 < 2.54 < 1.96$?

The first statement is true as 2.54 is greater than or equal to 1.96. This means that the results fall in the rare zone and the null hypothesis is rejected. (To help visualize the results falling in the rare zone, see Figure 6.3.) By rejecting the null hypothesis, Dr. Pell must accept the alternative hypothesis and conclude that the population mean is something other than 100.

It is possible to go a step beyond just saying that the population mean of IQ for adopted children is something other than 100. It is possible to comment on the direction of the difference by comparing the sample mean to the population mean. The sample mean of 104.5 was above the general population mean of 100, so Dr. Pell

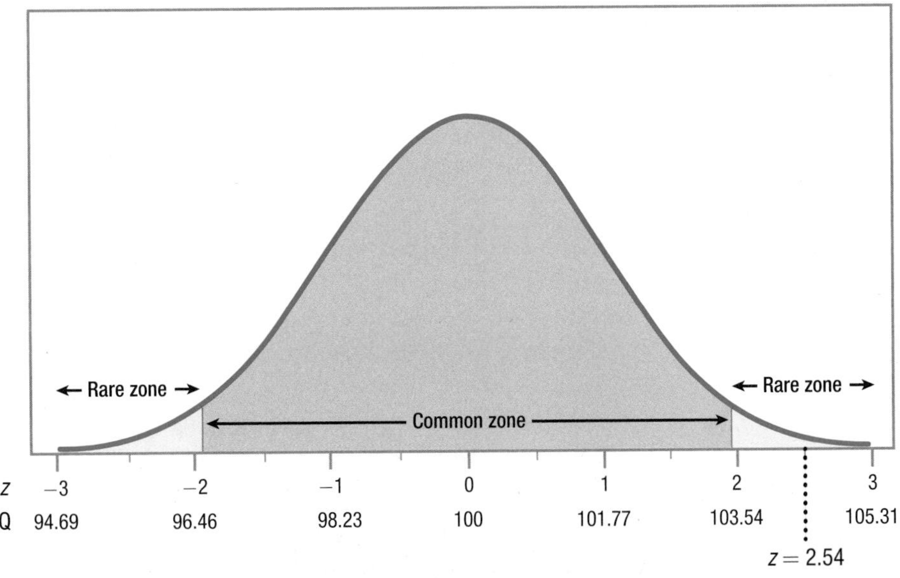

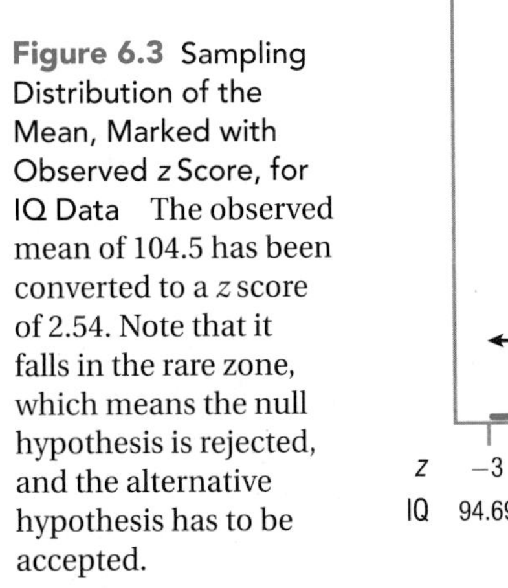
Figure 6.3 Sampling Distribution of the Mean, Marked with Observed z Score, for IQ Data The observed mean of 104.5 has been converted to a z score of 2.54. Note that it falls in the rare zone, which means the null hypothesis is rejected, and the alternative hypothesis has to be accepted.

can conclude that the population of adopted children has an average intelligence that is higher than 100.

The terms "statistically significant" and "not statistically significant" are commonly used in reporting results. **Statistically significant** means that a researcher concludes the observed sample results are more different from the null-hypothesized population value than would be expected by chance. With the adoption IQ example, Dr. Pell could state that the observed mean of 104.5 is "statistically significant" or "statistically different from 100."

It is also common to report the results in APA format. APA format indicates what test was done, how many cases there were, what the value of the test statistic was, what alpha level was selected, and whether the null hypothesis was rejected. In APA format, Dr. Pell would report the results for the adoption IQ study as

$$z\,(N = 72) = 2.54,\ p < .05$$

- z tells what test was done, a z test.
- $N = 72$ gives the sample size.
- 2.54 is the value of the test statistic that was calculated.
- .05 refers to the alpha level. (Because it is associated in APA format with the letter p, for probability, it is often referred to as the ***p* value.**)
- $p < .05$ informs the reader that the null hypothesis was rejected.

A Common Question

Q What does $p < .05$ mean?

A $p < .05$ means that the observed result is a rare result as the probability of it happening is less than .05 if the null hypothesis is true. If a researcher fails to reject the null hypothesis, he or she would write $p > .05$. This means that the result is a common occurrence—it happens more than 5% of the time—when the null hypothesis is true.

Here is what the psychologist, Dr. Pell, wrote for her interpretation. Note how she follows the four-point interpretation template from Table 6.2: (1) indicating what was done, (2) providing some facts (the sample and population means as well as the z test results), (3) telling what the results mean, and (4) making a suggestion for future research:

> In this study, the intelligence of adopted children was compared to the IQ of the general population of children in the United States. The mean IQ of a random sample of 72 adopted children in the United States was 104.5. Using a single-sample z test, their mean of 104.5 was statistically significantly different from the population mean of 100 ($z\,(N = 72) = 2.54$, $p < .05$). This study shows that adopted children have a higher average IQ than children in general. As adoptive parents are carefully screened before they are allowed to adopt, future research may want to explore the role that this plays in the higher intelligence of their children.

Worked Example 6.1

Let's practice with hypothesis testing and the single-sample z test, this time with an example where the null hypothesis is not rejected. Imagine a psychic who claimed he could "read" blood pressure. A public health researcher, Dr. Levine, tested the man's claim by asking him to select people with abnormal blood pressure. The psychic picked out 81 people. Dr. Levine took their blood pressures and the average systolic blood pressure for the sample, M, was 127. From previous research, Dr. Levine knew that the population mean, μ, for systolic blood pressure was 124 with a population standard deviation, σ, of 18.

The sample mean in this study, 127, is three points higher than the population mean. Is this different enough from normal blood pressure to support the psychic's claim that he can read blood pressure? Did he find people with higher than average blood pressure? Or, can the deviation of 127 from 124 be explained by sampling error?

Step 1 Pick a Test. Comparing a sample mean to a population mean when the population standard deviation is known calls for a single-sample z test.

Step 2 Check the Assumptions. Table 6.3 lists the assumptions for the single-sample z test:

- The sample is not a random sample from the population of people the psychic considers to have high blood pressure, so the first assumption is violated. This is a robust assumption, however, so it can be violated and the test still completed. Dr. Levine will need to be careful about the population to which he generalizes the results.
- Each participant takes part in the study only once. There's no evidence that the 81 observations influence the measurement of each other's blood pressure. The second assumption, independence of observations, is not violated.
- Eighty-one cases is a large enough sample to graph its frequency distribution. If the distribution looks normal-ish, it seems reasonable to assume that blood pressure is normally distributed in the larger population. In this scenario, the third assumption, normality, is not violated.

With no nonrobust assumptions violated, Dr. Levine can proceed with the planned test.

Step 3 List the Hypotheses. The null hypothesis states that the psychic is *not* able to read blood pressure. As a result, the mean blood pressure of the people he picks should be no different from the mean blood pressure of those with normal pressure. This is stated mathematically as

$$H_0: \mu_{\text{PsychicSelected}} = 124$$

The null hypothesis and the alternative hypothesis, together, have to be all-inclusive and mutually exclusive. The alternative hypothesis will state that the mean blood pressure in the population of people the psychic selects is different from the mean normal systolic blood pressure of 124:

$$H_1: \mu_{\text{PsychicSelected}} \neq 124$$

Step 4 Set the Decision Rule. Dr. Levine wants to do a two-tailed test and has set alpha at .05. According to Table 6.4, the critical value of z is ± 1.96. If the value of z calculated in the next step is less than or equal to -1.96 *or* greater than or equal to 1.96, the researcher will reject the null hypothesis. If z is greater than -1.96 *and* less than 1.96, he will fail to reject the null hypothesis. Figure 6.4 displays the rare and common zones for this decision.

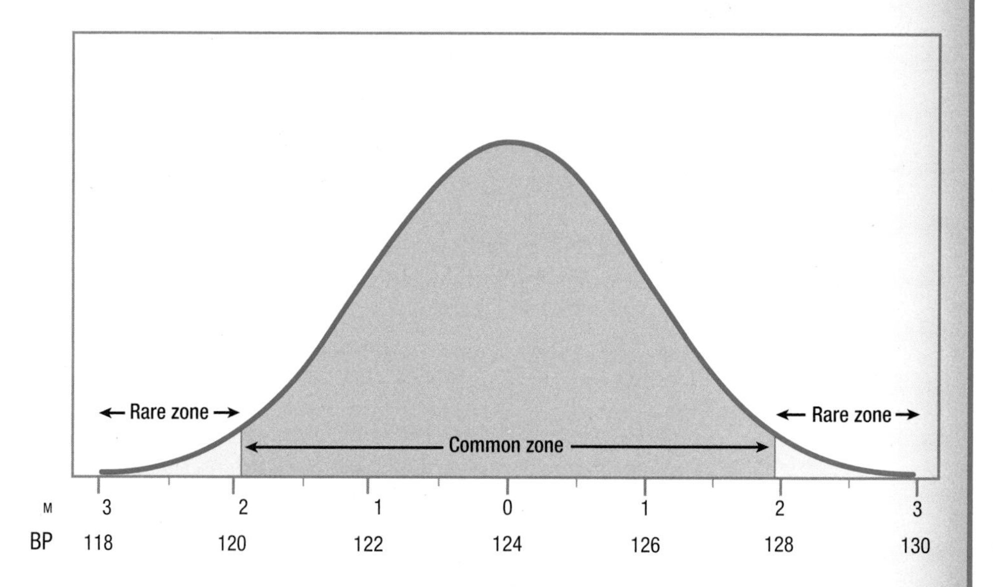

Figure 6.4 Sampling Distribution of the Mean, Marked with the Rare Zone and Common Zone, for Psychic Blood Pressure Data This is the sampling distribution of the mean for the blood pressure data with a population mean of 124 and a standard error of the mean of 2.00.

Step 5 Calculate the Test Statistic. In order to calculate the z value, Dr. Levine will first use Equation 5.1 to calculate the standard error of the mean:

$$\begin{aligned}\sigma_M &= \frac{\sigma}{\sqrt{N}} \\ &= \frac{18}{\sqrt{81}} \\ &= \frac{18}{9.0000} \\ &= 2.0000 \\ &= 2.00\end{aligned}$$

Now that the standard error of the mean is known, Dr. Levine can use Equation 6.1 to calculate the z value:

$$\begin{aligned}z &= \frac{M - \mu}{\sigma_M} \\ &= \frac{127 - 124}{2.00} \\ &= \frac{3.0000}{2.00} \\ &= 1.5000 \\ &= 1.50\end{aligned}$$

Step 6 Interpret the Results. Which of the following two statements is true?

- Is either $1.50 \leq -1.96$, or is $1.50 \geq 1.96$?
- Or, is $-1.96 < 1.50 < 1.96$?

The second statement is true as 1.50 falls between −1.96 and 1.96. Dr. Levine has failed to reject the null hypothesis. Figure 6.5 shows how the observed value of the test statistic, 1.50, falls in the common zone.

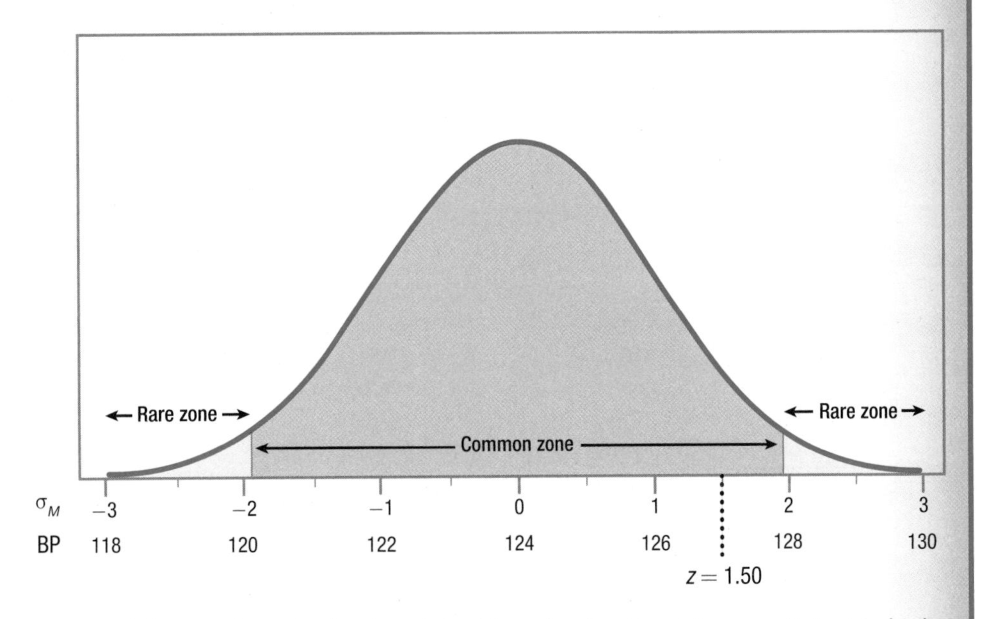

Figure 6.5 Sampling Distribution of the Mean for the Blood Pressure Data, Marked with Observed Value of *z* The sample mean of 127 is equivalent to a *z* value of 1.50. Note that this falls in the common zone, meaning that one fails to reject the null hypothesis.

The results are not statistically significant. Not enough evidence exists to conclude that this sample mean, 127, is statistically different from 124. Because there is not sufficient evidence of a difference, the direction of the difference doesn't matter.

In APA format, a researcher indicates failure to reject the null hypothesis by writing "$p > .05$." (.05 is used because alpha was set at .05.) This indicates that the value of the test statistic fell in the common zone as it is a value that happens more than 5% of the time when the null hypothesis is true. In APA format for these results, Dr. Levine would write

$$z\,(N = 81) = 1.50,\ p > .05$$

Here is what Dr. Levine wrote for an interpretation. Note that he follows the template in Table 6.2: (1) telling what was done, (2) providing some facts,

(3) explaining what the results mean, and (4) making a suggestion for future research. Again, the interpretation takes just a paragraph:

> This study was set up to determine if a psychic could "read" blood pressure as he claimed. The psychic selected 81 people whom he believed had abnormal blood pressure. Their systolic blood pressures were measured ($M = 127$) and compared to the population mean ($\mu = 124$) using a single-sample z test. No statistically significant difference was found ($z(N = 81) = 1.50$, $p > .05$), indicating there was no evidence that this group of people selected for having high blood pressure had above-average blood pressure. Based on these data, there is not enough evidence to suggest that this psychic has any ability to read blood pressure. Though the results from this study seem conclusive, if one wanted to test this psychic's ability again, it would be advisable to use a larger sample size.

Practice Problems 6.2

Review Your Knowledge

6.03 What are the six steps of hypothesis testing?

Apply Your Knowledge

6.04 If $N = 55$ and $\sigma = 12$, what is σ_M?

6.05 If $M = 19.40$, μ is 22.80, and $\sigma_M = 4.60$, what is z?

6.06 A researcher believes the population mean is 20 and the population standard deviation is 4. He takes a random sample of 64 cases from the population and calculates $M = 20.75$. (a) Do the calculations for a single-sample z test and (b) report the results in APA format.

6.3 Type I Error, Type II Error, Beta, and Power

Imagine a study in which a clinical psychologist, Dr. Cipriani, is investigating if a new treatment for depression works. One way that her study would be successful is if the treatment were an effective one and the data she collected showed that. But, her study would also be successful if the treatment did not help depression and her data showed that outcome.

Those are "good" outcomes; there are also two "bad" outcomes. Suppose the treatment really is an effective one, but her sample does not show this. As a result of such an outcome, this effective treatment for depression might never be "discovered." Another bad outcome would occur if the treatment were in reality an ineffective one, but for some odd reason it worked on her subjects. As a result, this ineffective treatment would be offered to people who are depressed and they would not get better.

Table 6.5 shows the four possible outcomes for hypothesis tests. The two columns represent the reality in the population. The column on the left says the null hypothesis is really true. In Dr. Cipriani's terminology, this means treatment has no impact. The column on the right says the treatment does make a difference; the null hypothesis is not true. The two rows represent the conclusions based on the sample. The top row says that the results fell in the common zone, meaning that the researcher will fail to

TABLE 6.5 The Four Possible Outcomes for a Hypothesis Test

	H_0 is really true	H_0 is really not true
	A	B
We fail to reject H_0	Results fall in common zone. Correctly say, "Fail to reject H_0"	Results fall in common zone. Erroneously say, "Fail to reject H_0"
	C	D
We reject H_0	Results fall in rare zone. Erroneously say, "Reject H_0"	Results fall in rare zone. Correctly say, "Reject H_0"

Cells A and D are correct decisions, while B and C are erroneous decisions.

reject the null hypothesis. (Dr. Cipriani will conclude that her new treatment does not work.) The bottom row says that the results fell in the rare zone, so the researcher will reject the null hypothesis. (Dr. Cipriani will conclude that her treatment works.)

The two scenarios in which a study would be considered a success are Outcome A and Outcome D. In Outcome A, in the terminology of hypothesis testing, a researcher correctly fails to reject the null hypothesis. For Dr. Cipriani, this outcome means that treatment is ineffective and the evidence of her sample supports this. In Outcome D, the null hypothesis is correctly rejected. For Dr. Cipriani, the treatment does work and she finds evidence that it does.

The two other outcomes, Outcome B and Outcome C, are bad outcomes for a hypothesis test because the conclusions would be wrong. In Outcome B, the researcher fails to reject the null hypothesis when it should be rejected. If this happened to Dr. Cipriani, she would say that there is insufficient evidence to indicate the treatment works and she would be wrong as the treatment really does work. In Outcome C, a researcher erroneously rejects the null hypothesis. If Dr. Cipriani concluded that treatment works, but it really doesn't, she would have erroneously rejected the null hypothesis.

A problem with hypothesis testing is that because the decision relies on probability, a researcher can't be sure if the conclusion about the null hypothesis is right (Outcomes A or D) or wrong (Outcomes B or C). Luckily, a researcher can calculate the probability that the conclusion is in error and know how *probable* it is that the conclusion is true. With hypothesis testing, one can't be sure the conclusion is true, but one can have a known degree of certainty.

Type I Error

Let's start the exploration of errors in hypothesis testing with Outcome C, a wrong decision. When a researcher rejects the null hypothesis and shouldn't have, this is called a **Type I error.** In the depression treatment example, a Type I error occurs if Dr. Cipriani concludes that the treatment makes a difference when it really doesn't. If that happened, psychologists would end up prescribing a treatment that doesn't help.

Routinely, researchers decide they are willing to make a Type I error 5% of the time.

Routinely, researchers decide they are willing to make a Type I error 5% of the time. Five percent is an arbitrarily chosen value, but over the years it has become the standard value. One doesn't need to use 5%. If the consequences of making a Type I error seem particularly costly, one might be willing only to risk a 1% chance of this mistake. For example, imagine using a test to select technicians for a nuclear

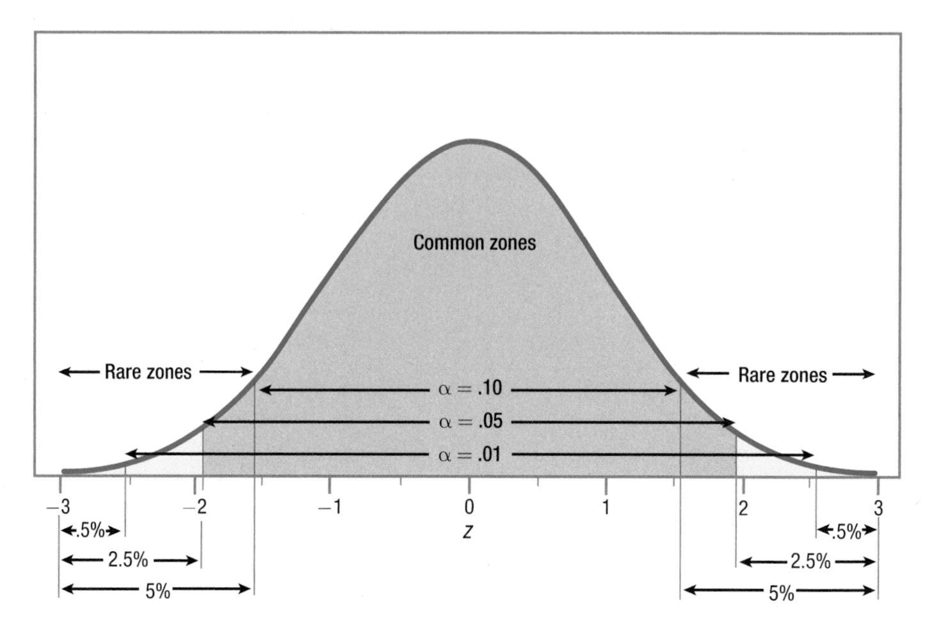

Figure 6.6 The Size of the Rare Zone and the Common Zone for Different Alpha Levels As the alpha level decreases from .10 to .05 to .01, the common zone increases in size and the rare zone shrinks. As alpha gets smaller, it is more difficult to reject the null hypothesis.

power plant. The potential catastrophe if the wrong person is hired is so great that the cut-off score on the test should be very high. On the other hand, if making a Type I error doesn't seem consequential—for example, if one is hiring people for a task that doesn't have serious consequences for failure—then one might be willing to live with a greater chance of this error.

The likelihood of Type I error is determined by setting the alpha level. As alpha gets larger, so does the rare zone. A larger rare zone makes it easier to reject the null hypothesis. Figure 6.6 shows the common and rare zones for a single-sample z test for three common alpha levels ($\alpha = .10$, $\alpha = .05$, and $\alpha = .01$). Note that as alpha decreases, so does the size of the rare zone, making it more difficult to reject the null hypothesis.

Setting alpha at .05 gives a modest chance of Type I error and a reasonable chance of being able to reject the null hypothesis.

Making errors is never a good idea, so why isn't alpha always set at .01 or even lower? The reason is simple. When alpha is set low, it is more difficult to reject the null hypothesis because the rare zone is smaller. Rejecting the null hypothesis is almost always the goal of a research study, so researchers would be working against themselves if they made it too hard to reject the null hypothesis. When setting an alpha level, a researcher tries to balance two competing objectives: (1) avoiding Type I error and (2) being able to reject the null hypothesis. The more likely one is to avoid Type I error, the less likely one is to reject the null hypothesis. The solution is a compromise. Setting alpha at .05 gives a modest chance of Type I error and a reasonable chance of being able to reject the null hypothesis.

Type II Error and Beta

There's another type of error possible, *Type II error*. This is Outcome B in Table 6.5. A **Type II error** occurs when the null hypothesis should be rejected, but it isn't. If Dr. Cipriani's new treatment for depression really was effective, but she didn't find evidence that it was effective, that would be a Type II error.

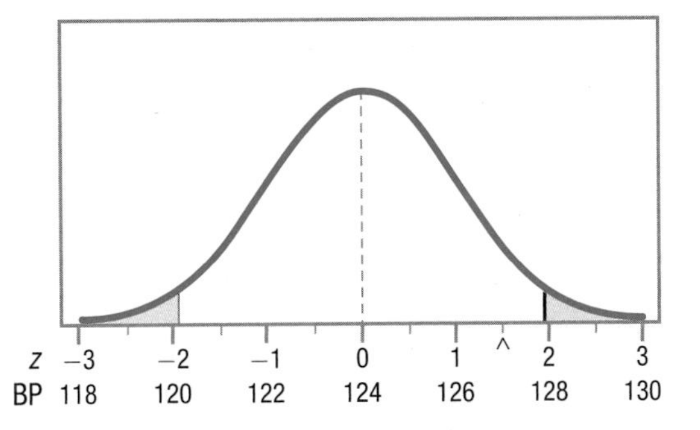

Figure 6.7 Sampling Distribution of the Mean for Blood Pressure if $\mu = 124$ The sampling distribution of the mean is centered at 124 and is normally distributed. The rare zones, where $z \leq -1.96$ and $z \geq 1.96$, are shaded in.

The probability of making a Type II error is known as **beta,** abbreviated with the lowercase Greek letter β. The consequences of making a Type II error can be as serious as the consequences of making a Type I error. But, when statisticians attend to beta, they commonly set it at .20. This gives a 20% chance for this error, not a 5% chance as is usually set for alpha.

A researcher only needs to worry about beta when failing to reject the null hypothesis. Let's use the psychic blood pressure example from Worked Example 6.1—where the null hypothesis wasn't rejected—to examine how Type II error may occur.

In that study, Dr. Levine studied a sample of 81 people whom a psychic believed had abnormal blood pressure. Their mean blood pressure was 127. The population mean was 124, with $\sigma = 18$. Dr. Levine tested the null hypothesis (H_0: $\mu = 124$) vs. the alternative hypothesis (H_1: $\mu \neq 124$) using a two-tailed single-sample z test with $\alpha = .05$. The critical value of z was ± 1.96. After calculating $\sigma_M = 2.00$, he calculated $z = 1.50$. As z fell in the common zone, he failed to reject the null hypothesis. There was insufficient evidence to say that the psychic could discern people with abnormal blood pressure.

The null hypothesis was tested by assuming it was true and building a sampling distribution of the mean centered around $\mu = 124$. In Figure 6.7, note these four things about the sampling distribution:

1. The midpoint—the vertical line in the middle—is marked on the X-axis both with a blood pressure of 124 and a z score of 0.
2. The other blood pressures marked on the X-axis are two points apart because $\sigma_M = 2$. The X-axis is also marked at these spots with z scores from −3 to 3.
3. The rare zone, the area less than or equal to a z value of −1.96 and greater than or equal to a z value of 1.96, is shaded. If a result falls in the rare zone, the null hypothesis is rejected.
4. A caret, ^, marks the spot where the sample mean, 127, falls. This is equivalent to a z score of 1.50 and falls in the common zone. The null hypothesis is not rejected.

Now, imagine that the null hypothesis is not true. Imagine that the population mean is 127 instead of 124. Why pick 127? Because that was the sample mean and is the only objective evidence that we have for what the population mean might be for people the psychic says have high blood pressure.

With this in mind, look at Figure 6.8. The top panel (A) is a repeat of the sampling distribution shown in Figure 6.7, the distribution centered around the mean of 124, the mean hypothesized by the null hypothesis. The bottom panel (B), the dotted line

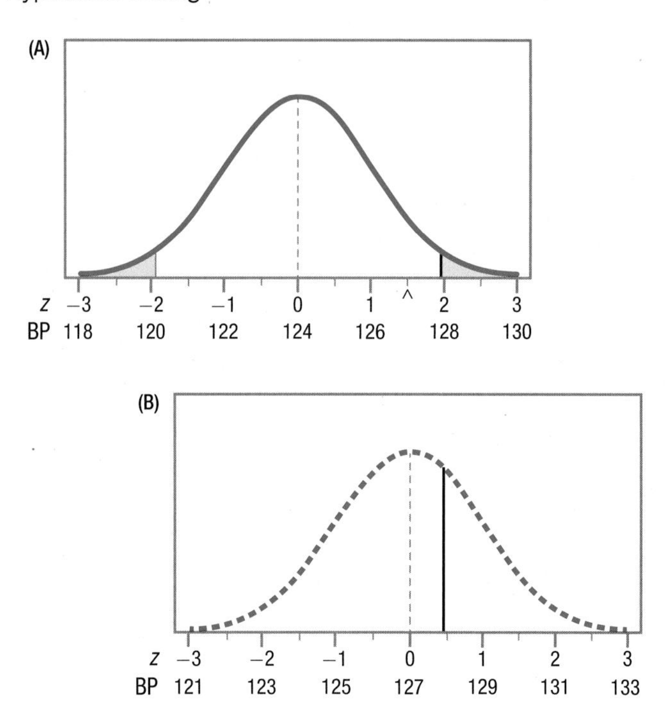

Figure 6.8 Sampling Distribution if $\mu = 127$ The top panel (A), centered at 124, is the same as the sampling distribution in Figure 6.7. The bottom panel (B) is centered at a blood pressure of 127 and shows what the sampling distribution would look like if that were the population mean. The vertical line in the bottom panel is an extension of the critical value, $z = 1.96$, from the top panel.

distribution, is new. This is what the sampling distribution would look like if $\mu = 127$. Note these four characteristics about the sampling distribution in the bottom panel:

1. The dotted line sampling distribution has exactly the same shape as the one in the top panel; it is just shifted to the right so that the midpoint, represented by a dotted vertical line, occurs at a blood pressure of 127.

2. This midpoint, 127, is right under the spot marked by a caret in the top panel. That caret represents where the sample mean, 127, fell.

3. The other points on the *X*-axis are marked off by blood pressure scores ranging from 121 to 133, and *z* scores ranging from −3 to 3.

4. There is a solid vertical line in the lower panel of Figure 6.8. The vertical line is drawn at the same point as the *z* score of 1.96 was in the top panel. This vertical line marks the point that was one of the critical values of *z* in the top panel. In the top panel, scores that fell to the left of this line fell in the common zone and scores that fell on or to the right of the line fell in the rare zone.

Figure 6.9 takes Figure 6.8 and hatches in the area to the left of the vertical line with /// in the bottom panel (B). This area, which is more than 50% of the area under the curve, indicates the likelihood of a Type II error. How so? Well, if μ were really 127, then the null hypothesis that the population mean is 124 should be rejected! However, looking at the top panel (A) to make this decision, a researcher wouldn't reject the null hypothesis after obtaining a mean that falls in the area hatched like ///.

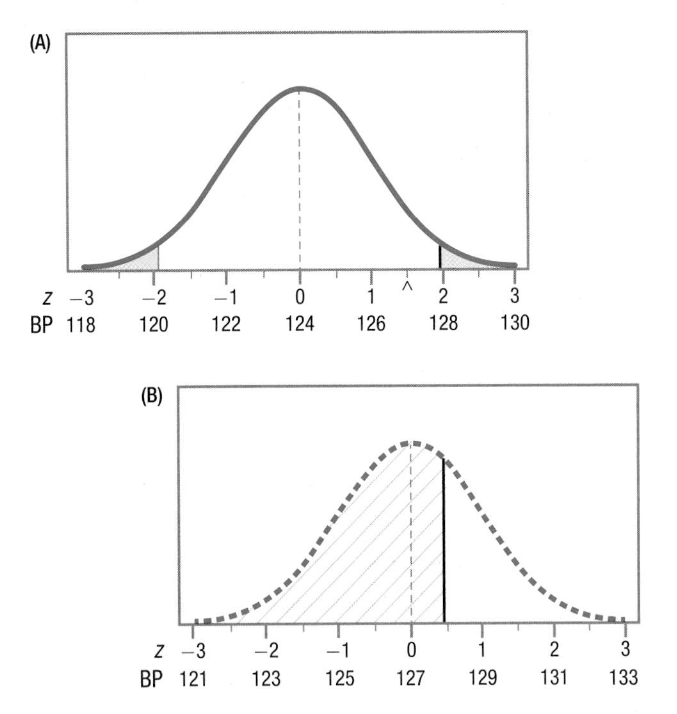

Figure 6.9 The Probability of Type II Error (Beta) if $\mu = 127$ This figure is the same as Figure 6.8, but the area to the left of the solid vertical line in the bottom panel (B) has been hatched in. Note that the solid vertical line is directly below the critical value of *z*, 1.96, in the top panel (A). Any sample mean that falls in the hatched area of the bottom panel would fall in the common zone of the top panel, and the null hypothesis would not be rejected. But, as the bottom distribution has a population mean of 127, not 124, the null hypothesis of $\mu = 124$ should be rejected. Note that more than half of the distribution is shaded in, meaning there is a large probability of Type II error if we hypothesize $\mu = 124$ but the population mean is really 127.

(Remember, the sample means will be spread around the population mean of 127 because of sampling error.) Whenever a sample mean falls in this hatched-in area, a Type II error is committed and a researcher fails to reject a null hypothesis that should be rejected.

Under these circumstances, the researcher would commit a Type II error fairly frequently if the population mean were really 127, but the null hypothesis claimed it was 124. The researcher would commit a Type II error less frequently if the population mean were greater than 127. Right now, the difference between the null hypothesis value (124) and the value that may be the population value (127) is fairly small, only 3 points. This suggests that if the psychic can read blood pressure, he has only a small ability to do so. If the effect were larger, say, the dotted line distribution shifted to a midpoint of 130, the size of the effect, the psychic's ability to read blood pressure, would be larger. The bottom panel (C) of Figure 6.10 shows the sampling distribution made with a dashed line. This is the sampling distribution that would exist if the population mean were 130. Notice how much smaller the hatched-in error area is in this sampling distribution, indicating a much smaller probability of Type II error.

The point is this: the probability of Type II error depends on the size of the effect, the difference between what the null hypothesis says is the population parameter and what the population parameter really is. When the difference between these

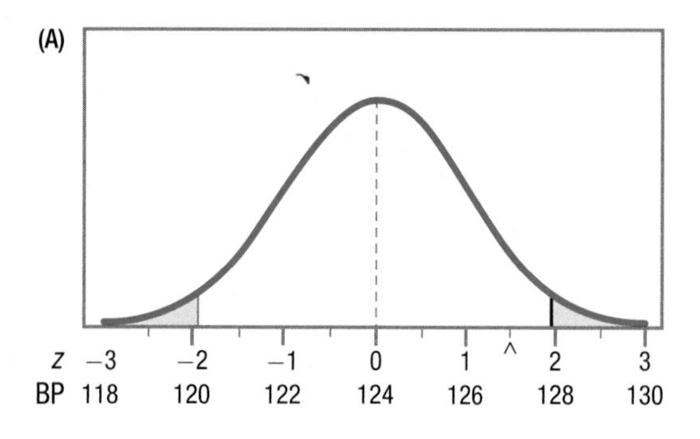

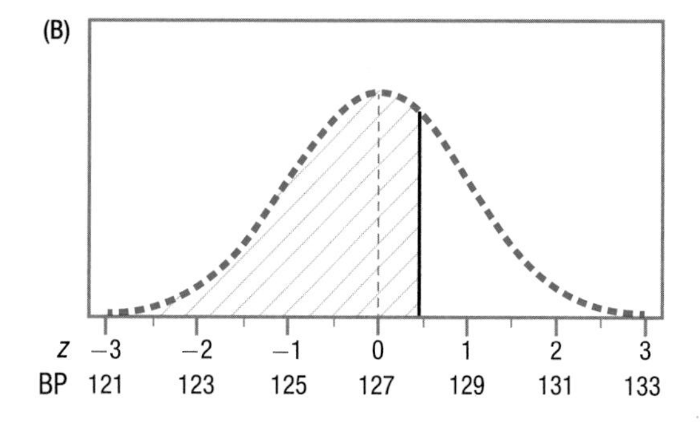

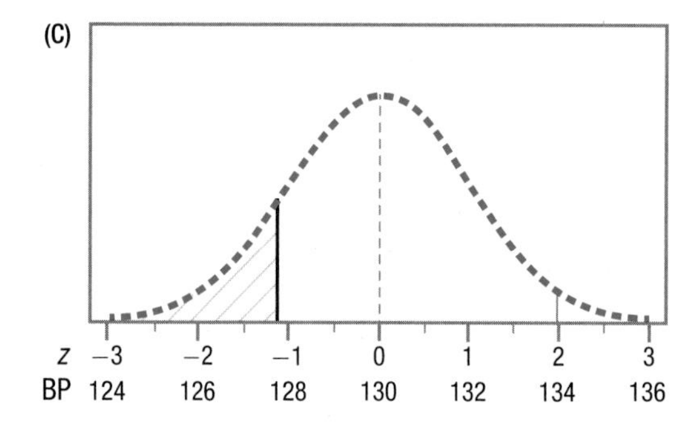

Figure 6.10 The Impact of Effect Size on the Probability of Type II Error The top panel (A) in this figure shows the sampling distribution if $\mu = 124$, the middle panel (B) if $\mu = 127$, and the bottom panel (C) if $\mu = 130$. The shaded portion in the middle panel reflects beta, the probability of Type II error if it is hypothesized that the population mean is 124 but it really is 127. Note that the shaded portion in the bottom distribution, which represents beta if $\mu = 130$, is smaller. When the effect is bigger, the probability of Type II error decreases.

values is small, the probability of a Type II error is large. Type II error can still occur when the effect size is large, but it is less likely.

When one is starting out in statistics, it is hard to remember the differences between Type I error and Type II error. Table 6.6 summarizes the differences between the two.

Power

There's one more concept to introduce before this chapter draws to a close and that is *power.* **Power** refers to the probability of rejecting the null hypothesis when the null hypothesis should be rejected. Because the goal of most studies is to reject the null hypothesis and be forced to accept the alternative hypothesis (which is what the researcher really believes is true), researchers want power to be as high as possible.

The area that is hatched \\\ in the bottom panel (C) in Figure 6.11 demonstrates the power of the single-sample z test in Dr. Levine's psychic blood pressure study. If the population mean is really 127, whenever a sample mean falls in this hatched-in

TABLE 6.6 How to Choose: Type I Error vs. Type II Error

	Type I Error	Type II Error
Definition	One rejects the null hypothesis when one should have failed to reject the null hypothesis.	One fails to reject the null hypothesis when the null hypothesis should have been rejected.
Erroneous conclusion reached	Reject the null hypothesis.	Fail to reject the null hypothesis.
Example of erroneous conclusion	Conclude that a treatment makes a difference in the outcome, but it really doesn't.	Conclude there's no evidence a treatment makes a difference in outcome, but it really does.
Probability of this error	α	β
Need to worry about making this error	If the null hypothesis is rejected.	If one fails to reject the null hypothesis.

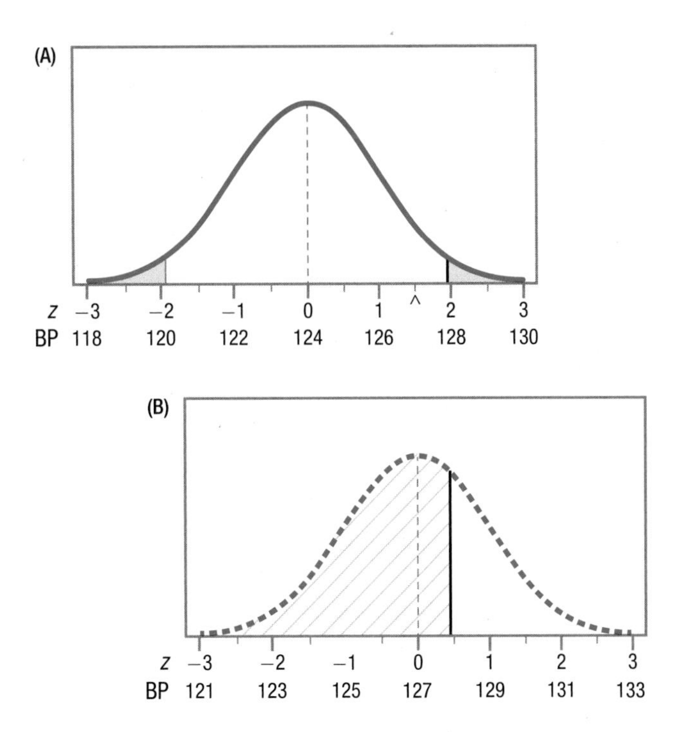

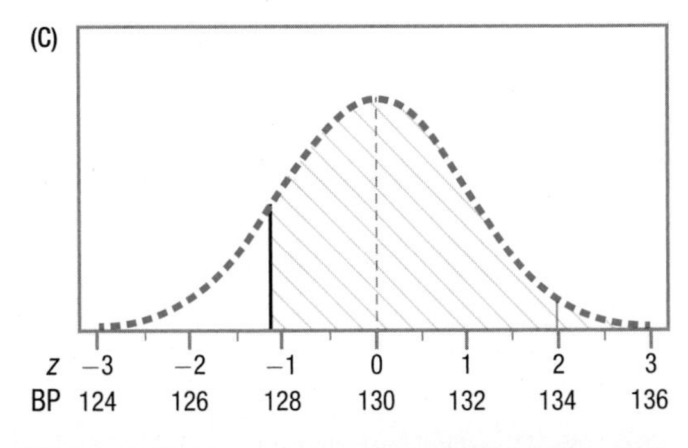

Figure 6.11 The Relationship Between Beta and Power The shaded area on the top panel (A) is the rare zone. If a result falls there, the null hypothesis will be rejected. The area hatched /// on the middle panel (B) represents beta, the probability of Type II error. The area shaded \\\ on the bottom panel (C) represents power, the likelihood if the population mean is really 127, that the null hypothesis, which claims it is 124, will be rejected. Note that the areas shaded /// and \\\ incorporate the whole distribution. (The probability of beta plus the probability of power equals 1.)

area, the researcher will reject the null hypothesis. Why? Because a population mean in this hatched-in area falls in the shaded-in rare zone shown in the top panel (A).

Look at the area hatched in /// in the middle panel (B) of Figure 6.11 and the area hatched in \\\ in the bottom panel (C) of Figure 6.11. The hatched-in area in the middle panel represents beta, the probability of Type II error, while the hatched-in area in the bottom panel represents power. Together, these two hatched-in areas shade in the entire dotted-line curve. This shows that beta and power are flip sides of the same coin. When the null hypothesis should be rejected, beta plus power equals 100% of the area under the dotted-line curve.

Statisticians refer to beta and power using probabilities, not percentages. They would say: $\beta + \text{power} = 1.00$. If either beta or power is known, the other can be figured out. If beta is set at .20, then power is .80. If power is .95, then beta is .05.

Type I error and Type II error are errors of the conclusion, not errors of the facts on which conclusions are based.

There's more to be learned about Type I and Type II errors, beta, and power in later chapters. For now, here's a final point. If a Type I error or a Type II error occurs, this doesn't mean the results of the study are wrong. After all, the sample mean *is* the mean that was found for the sample. What a Type I or Type II error means is that the *conclusion* drawn from the sample mean about the population mean is wrong. Type I error and Type II error are errors of the conclusion, not errors of the facts on which conclusions are based.

Practice Problems 6.3

Review Your Knowledge

6.07 When does Type II error occur?

6.08 What is power?

Apply Your Knowledge

6.09 What is a correct conclusion in hypothesis testing?

6.10 What is an incorrect conclusion in hypothesis testing?

Application Demonstration

To see hypothesis testing in action, let's leave numbers behind and turn to a simplified version of one way the sex of a baby is identified before it is born. Around the fifth month of pregnancy, an ultrasound can be used to identify whether a fetus is a boy or a girl. The test relies on hypothesis testing to decide "boy" or "girl," and it is not 100% accurate. Both Type I errors and Type II errors are made. Here's how it works.

The null hypothesis, which is set up to be rejected, says that the fetus is a girl. Null hypotheses state a negative, and the negative statement here is that the fetus has no penis. Together, the null hypothesis and the alternative hypothesis have to be all-inclusive and mutually exclusive, so the alternative hypothesis is that the fetus has a penis.

The sonographer, then, starts with the hypothesis that the fetus is a girl. If the sonographer sees a penis, the null hypothesis is rejected and the alternative hypothesis is accepted, that the fetus is a boy.

Of course, mistakes are made. Suppose the fetus is a girl and a bit of her umbilical cord is artfully arranged so that it masquerades as a penis. The sonographer would say "Boy" and be wrong. This is a Type I error, where the null hypothesis is wrongly rejected.

A Type II error can be made as well. Suppose the fetus is a shy boy who arranges his legs or hands to block the view. The sonographer doesn't get a clear view of the area, says "Girl," and would be wrong. This is a Type II error, wrongly failing to reject the null hypothesis.

However, in this case, a smart sonographer doesn't say "Girl." Instead, the sonographer, failing to see male genitalia, indicates that there's not enough evidence to conclude the fetus is a boy. That is, there's not enough evidence to reject the hypothesis that it's a girl. Absence of evidence is not evidence of absence.

DIY

Grab a handful of coins, say, 25, and let's investigate Type I error. Type I error occurs when one erroneously rejects the null hypothesis. In this exercise, the null hypothesis is going to be that a coin is a fair coin. This means that each coin has a heads and a tails, and there is a 50-50 chance on each toss that a heads will turn up.

For each coin in your handful, you will determine if it is a fair coin by tossing it, by observing its behavior. Imagine you toss the coin once and it turns out to be heads. Does that give you any information whether it is a fair coin or not? No, it doesn't. A fair coin could certainly turn up heads on the first toss. So, you toss it again and again it turns up heads. Could a fair coin turn up heads 2 times in a row? Yes. There are four outcomes (HH, HT, TH, and TT), one of which is two heads, so two heads in a row should happen 25% of the time. How about three heads in a row? That will happen 12.5% of the time with a fair coin. Four heads? 6.25% of the time. Finally, if there are five heads in a row, we will step over the 5% rule. Five heads in a row, or five tails in a row, will happen only 3.125% of the time with a fair coin. It could happen, but it is a rare event, and when a rare event occurs—one that is unlikely to occur if the null hypothesis is true—we reject the null hypothesis.

So, here's what you do. Test one coin at a time from your handful. Toss each coin 5 times in a row. If the result is a mixture of heads and tails, which is what you'd expect if the test were fair, there is not sufficient evidence to reject the null hypothesis. But, if it turns up five heads in a row or five tails in a row, you have to reject the null hypothesis and conclude the coin is not fair. If that happens to you, then pick up the coin and examine it. Does it have a heads and a tails? Toss it 5 more times. Did it turn up five heads or five tails again? This time, it probably behaved more like a fair coin.

There are two lessons here. First, occasionally, about 3% of the time, a fair coin will turn up heads 5 times in a row. If that happens, by the rules of hypothesis testing, we would declare the coin unfair. That conclusion would be wrong; it would be a Type I error. If you stopped there, if you did not replicate the experiment, you'd never know that you made a mistake. And that is a danger of hypothesis testing—your conclusion can be wrong and you don't know it. So here's the second lesson: replicate. When the same results occur 2 times in a row, you have a lot more faith in them.

SUMMARY

Explain how hypothesis testing works.

- Hypothesis testing uses data from a sample to evaluate a hypothesis about a population. If what is observed in the sample is close to what was expected based on the hypothesis, then there is no reason to question the hypothesis.
- The null hypothesis is paired with a mutually exclusive alternative hypothesis, which is what the researcher believes is true. The researcher's goal is to disprove (reject) the null hypothesis and be "forced" to accept the alternative hypothesis. If the null hypothesis is not rejected, then the researcher doesn't say it was proven, but states that there wasn't enough evidence to reject it. This is like a "not guilty" verdict because there wasn't enough evidence to make a convincing case.

List the six steps to be followed in completing a hypothesis testing.

- To complete a hypothesis test, a researcher (1) picks an appropriate test, (2) checks its assumptions to make sure it can be used, (3) lists the null and alternative hypotheses, (4) sets the decision rule, (5) calculates the value of the test statistic, and (6) interprets the results.

Explain and complete a single-sample *z* test.

- A single-sample *z* test is used to compare a sample mean to a population mean, or a specified value, when the population standard deviation is known. The decision rule says that if the deviation of the sample mean from the specified value can't be explained by sampling error, then the null hypothesis is rejected.

Explain the decisions that can be made in hypothesis testing.

- The conclusion from a hypothesis test may or may not be correct. It's a correct decision (1) if the null hypothesis should be rejected and it is, or (2) if the null hypothesis should not be rejected and it is not. The potential incorrect decisions are (3) the null hypothesis should not be rejected and it is (Type I error), or (4) the null hypothesis should be rejected and it is not (Type II error).
- Error can't always be avoided, but the probability of one occurring can be determined. The probability of a Type I error, alpha, is usually set, so the error occurs no more than 5% of the time. The probability of Type II error is called beta. Power is the probability of making a correct decision "1." That is, power is the probability of rejecting the null hypothesis when it should be rejected. Since the goal of research is usually to reject the null hypothesis, researchers want power to be as high as possible.

KEY TERMS

alpha or alpha level – the probability of making a Type I error; the probability that a result will fall in the rare zone and the null hypothesis will be rejected when the null hypothesis is true; often called significance level; abbreviated α; usually set at .05 or 5%.

alternative hypothesis – abbreviated H_1; a statement that the explanatory variable has an effect on the outcome variable in the population; usually, a statement of what the researcher believes to be true.

beta – the probability of making a Type II error; abbreviated β.

common zone – the section of the sampling distribution of a test statistic in which the observed outcome should fall if the null hypothesis is true; typically, 95% of the sampling distribution.

critical value – the value of the test statistic that forms the boundary between the rare zone and the common zone of sampling distribution of the test statistic.

hypothesis – a proposed explanation for observed facts; a statement or prediction about a population value.

hypothesis testing – a statistical procedure in which data from a sample are used to evaluate a hypothesis about a population.

nonrobust assumption – an assumption for a statistical test that must be met in order to proceed with the test.

null hypothesis – abbreviated H_0; a statement that in the population the explanatory variable has no impact on the outcome variable.

one-tailed hypothesis test – hypothesis that predicts the explanatory variable has an impact on the outcome variable in a specific direction.

***p* value** – the probability of Type I error; the same as alpha level or significance level.

power – the probability of rejecting the null hypothesis when the null hypothesis should be rejected.

rare zone – the section of the sampling distribution of a test statistic in which it is unlikely an observed outcome will fall if the null hypothesis is true; typically, 5% of the sampling distribution.

robust assumption – an assumption for a statistical test that can be violated to some degree and it is still OK to proceed with the test.

significance level – the probability of Type I error; the same as alpha level or *p* value.

statistically significant – when a researcher concludes that the observed sample results are different from the null-hypothesized population value.

two-tailed hypothesis test – hypothesis that predicts the explanatory variable has an impact on the outcome variable but doesn't predict the direction of the impact.

Type I error – the error that occurs when the null hypothesis is true but is rejected; $p(\text{Type I error}) = \alpha$.

Type II error – the error that occurs when we fail to reject the null hypothesis but should have rejected it; $p(\text{Type II error}) = \beta$.

CHAPTER EXERCISES

Answers to the odd-numbered exercises appear at the back of the book.

Review Your Knowledge

6.01 A hypothesis is a proposed ____ for observed ____.

6.02 The procedure by which the observation of a ____ is used to evaluate a hypothesis about a ____ is called ____.

6.03 If what is observed in a sample is close to what is expected if the hypothesis is true, there is little reason to question the ____.

6.04 A hypothesis is a statement about a ____ not a ____.

6.05 The null hypothesis is abbreviated as ____ and the alternative hypothesis as ____.

6.06 The null and alternative hypotheses must be ____ and ____.

6.07 The null hypothesis is a ____ prediction and a ____ statement.

6.08 The null hypothesis says that the explanatory variable *does/does not* have an impact on the outcome variable.

6.09 The hypothesis a researcher believes is really true is the ____ hypothesis.

6.10 One ____ prove that a negative statement is true.

6.11 It takes just one example to ____ a negative statement.

6.12 When the null hypothesis is rejected, the researcher is forced to accept the ____.

6.13 Because the null and alternative hypotheses are mutually exclusive, if one is not true, then the other is ____.

6.14 If one fails to disprove the null hypothesis, one can't say it has been ____.

6.15 One shouldn't expect the sample mean to be exactly the same as the population mean because of ____.

6.16 The mnemonic to remember the six steps of hypothesis testing is ____.

6.17 The six steps of hypothesis testing, in order, are ____.

6.18 In the first step of hypothesis testing, one picks a ____.

6.19 If the ____ of a hypothesis test aren't met, one can't be sure what the results mean.

6.20 If a robust assumption is violated, one ____ proceed with the test.

6.21 A two-tailed test has ____ hypotheses.

6.22 A two-tailed test allows one to test for a positive or a negative effect of the ____ on the ____.

6.23 It is easier to reject the null hypothesis with a ____-tailed test than a ____-tailed test.

6.24 Once the data are collected, it is *OK/not OK* to change from a two-tailed test to a one-tailed test.

6.25 The decision rule involves finding the ____ of the test statistic.

6.26 When the value of the test statistic meets or exceeds the critical value of the test statistic, one ____ the null hypothesis.

6.27 Explaining, in plain language, what the results of a statistical test mean is called ____.

6.28 A single-sample z test is used to compare a ____ mean to a population ____.

6.29 In order to use a single-sample z test, one must know the ____ standard deviation.

6.30 The random sample assumption for a single-sample z test says that the ____ is a random sample from the ____.

6.31 Independence of observations within a group means that the cases don't ____ each other.

6.32 The normality assumption states that the ____ is normally distributed in the ____.

6.33 The null hypothesis for a single-sample z test says that the ____ is a specific value.

6.34 The alternative hypothesis for a single-sample z test says that the population mean is not what the ____ indicated it was.

6.35 The common zone of the sampling distribution of the mean is centered around a z score of ____.

6.36 Sample means will commonly fall in the ____ of the sampling distribution of the mean.

6.37 It is rare that a sample mean will fall in the ____ of the sampling distribution of the mean.

6.38 If the observed mean falls in the common zone, then what was expected to happen if the ____ is true did happen.

6.39 If the sample mean falls in the rare zone, then this is a ____ event if the null hypothesis is true.

6.40 Statisticians say that something that happens more than ____% of the time is common and less than or equal to ____% of the time is rare.

6.41 The z scores that are the critical values for a two-tailed, single-sample z test with alpha set at .05 are ____ and ____.

6.42 If ____ $\leq$ ____, reject the null hypothesis.

6.43 If ____ $\geq$ ____, reject the null hypothesis.

6.44 The alpha level is the probability that an outcome that is ____ to occur if the null hypothesis is true does occur.

6.45 ____ is the abbreviation for alpha.

6.46 If alpha equals ____, then a rare event is something that happens at most only 5% of the time.

6.47 The numerator in calculating a single-sample z test is the difference between the ____ and the ____.

6.48 The denominator in calculating a single-sample z test is ____.

6.49 The first question to be addressed in an interpretation is whether one ____ the null hypothesis.

6.50 If one rejects the null hypothesis, one can decide the direction of the difference by comparing the ____ to the ____.

6.51 If the result of a single-sample z test is statistically significant, that means the sample mean is ____ from the population mean.

6.52 APA format indicates what statistical test was done, how many ____ there were, what the value of the ____ was, what ____ was selected, and whether the null hypothesis was ____.

6.53 ____, in APA format, means the null hypothesis was rejected.

6.54 ____, in APA format, means the null hypothesis was not rejected.

6.55 If one fails to reject the null hypothesis for a single-sample z test, one *does/does not* need to be concerned about the direction of the difference between the sample mean and the population mean.

6.56 If one fails to reject the null hypothesis, one says there is ____ evidence to conclude that the independent variable affects the dependent variable.

6.57 It is a correct conclusion in hypothesis testing if one rejects the null hypothesis and the null hypothesis should ____.

6.58 It is an incorrect conclusion in hypothesis testing if one ____ the null hypothesis and it should have been rejected.

6.59 With hypothesis testing, one *can / can't* be sure that the conclusion about the null hypothesis is correct.

6.60 Type ____ error occurs when one rejects the null hypothesis but shouldn't have.

6.61 If $\alpha = .05$, then the probability of making a Type I error is ____.

6.62 If the cost of making a Type I error is high, one might set alpha at ____.

6.63 Compared to $\alpha = .01$, $\alpha = .10$ has a ____ rare zone, making it ____ to reject the null hypothesis.

6.64 When alpha is set low, say, at .01, the chance of being able to reject the null hypothesis is *larger/smaller.*

6.65 In hypothesis testing, one wants to keep the probability of Type I error ____ and still have a reasonable chance to ____ the null hypothesis.

6.66 The term for the error that occurs when the null hypothesis should be rejected but isn't is ____.

6.67 The probability of Type I error is usually set at ____.

6.68 The probability of Type II error is commonly set at ____.

6.69 If one fails to reject the null hypothesis, one needs to worry about ____ error but not ____ error.

6.70 As the size of the effect increases, the probability of Type II error ____.

6.71 Power is the probability of rejecting the null hypothesis when ____.

6.72 $1.00 =$ ____ + power.

6.73 If one rejects the null hypothesis, one needs to worry about ____ error but not ____ error.

6.74 If one makes a Type I error or a Type II error, then the conclusion about the ____ is wrong.

Apply Your Knowledge

Select the right statistical test.

6.75 A scientific supply company has developed a new breed of lab rat, which it claims weighs the same as the classic white rat. The population mean (and standard deviation) for the classic white rat is 485 grams (50 grams). A researcher obtained a sample of 76 of the new breed of rats, weighed them, and found $M = 515$ grams. What test should he do to see if the company's claim is true?

6.76 The mean vacancy rate for apartment rentals in the United States is 10%, with a standard deviation of 4.6. An urban studies major obtained a sample of 15 rustbelt cities and found that the mean vacancy rate was 13.3%. What statistical test should she use to see if the mean vacancy rate for these cities differs from the U.S. average?

Check the assumptions.

6.77 A researcher has a first-grade readiness test that is administered to kindergarten students and scored at the interval level. The population mean is 60, with a standard deviation of 10. He has administered it, individually, to a random sample of 58 kindergarten students in a city, $M = 66$, and wants to use a single-sample z test, two-tailed with $\alpha = .05$, in order to see whether first-grade readiness in this city differs from the national level. Check the assumptions and decide if it is OK for him to proceed with the single-sample z test.

6.78 A Veterans Administration researcher has developed a test that is meant to predict combat soldiers' vulnerability to developing post-traumatic stress disorder (PTSD). She has developed the test so that $\mu = 45$ and $\sigma = 15$. She is curious if victims of violent crime are as much at risk for PTSD as combat veterans. So, she obtains a sample of 122 recent victims of violent crime and administers the test to each one of them. She's planning to use a single-sample z test, two-tailed and with alpha set at .05, to compare the sample mean (42.8) to the population mean. Check the assumptions and decide if it is OK to proceed with the single-sample z test.

List the hypotheses.

6.79 List the null and alternative hypotheses for Exercise 6.77.

6.80 List the null and alternative hypotheses for Exercise 6.78.

State the decision rule.

6.81 State the decision rule for Exercise 6.77.

6.82 State the decision rule for Exercise 6.78.

Calculate σ_M.

6.83 Calculate σ_M using the data from Exercise 6.77.

6.84 Calculate σ_M using the data from Exercise 6.78.

Calculate z.

6.85 Calculate z for $M = 100$, $\mu = 120$, and $\sigma_M = 17.5$.

6.86 Calculate z for $M = 97$, $\mu = 85$, and $\sigma_M = 4.5$.

Calculate σ_M. Use it to calculate z.

6.87 Use the following information to calculate (a) σ_M and (b) z. $M = 12$, $\mu = 10$, $\sigma = 5$, and $N = 28$.

6.88 Use the following information to calculate (a) σ_M and (b) z. $M = 15$, $\mu = 21$, $\sigma = 1.5$, and $N = 63$.

Determine if the null hypothesis was rejected and use APA format.

6.89 Given $N = 23$ and $z = 2.37$, (a) decide if the null hypothesis was rejected, and (b) report the results in APA format. Use $\alpha = .05$, two-tailed.

6.90 Given $N = 87$ and $z = -1.96$, (a) decide if the null hypothesis was rejected, and (b) report the results in APA format. Use $\alpha = .05$, two-tailed.

Determine if the difference was statistically significant and the direction of the difference.

6.91 $M = 16$, $\mu = 11$, and the results were reported in APA format as $p < .05$. (a) Was there a statistically significant difference between the sample mean and the population mean? (b) What was the direction of the difference?

6.92 $M = 20$, $\mu = 24$, and the results were reported in APA format as $p > .05$. (a) Was there a statistically significant difference between the sample mean and the population mean? (b) What was the direction of the difference?

Given the results, interpret them. Be sure to tell what was done in the study, give some facts, and indicate what the results mean.

6.93 A researcher obtained a sample of 123 American women who said they wanted to lose weight. She weighed each of them and found $M = 178$ pounds. The mean weight for women in the United States is 164 pounds and the population standard deviation is known. The researcher used a single-sample z test and found $z (N = 123) = 3.68$, $p < .05$. Interpret her results to see if women who want to lose weight differ from the general population in terms of weight.

6.94 In the population of children in a school district, the mean number of days tardy per year is 2.8. A sociologist obtained a sample of children from single-parent families and found the mean number of days tardy was 3.2. He used a single-sample z test to analyze the data, finding $z(N = 28) = 1.68$, $p > .05$. Interpret the results to see if coming from a single-parent family is related to tardiness.

Type I vs. Type II error

6.95 A journalist was comparing the horsepower of a sample of contemporary American cars to the population value of horsepower for American cars of the 1970s. He concluded that there was a statistically significant difference, such that contemporary cars had more horsepower. Unfortunately, his conclusion was in error. (a) What type of error did the journalist make? (b) What conclusion should he have reached?

6.96 A Verizon researcher compared the number of text messages sent by a sample of teenage boys to the population mean for all Verizon users. She found no evidence to conclude that there was a difference. Unfortunately, her conclusion was in error. (a) What type of error did she make? (b) What conclusion should she have reached?

Given β or power, calculate the other.

6.97 If $\beta = .75$, power = ____.

6.98 If power = .90, β = ____.

Doing a complete statistical test

6.99 A dietitian wondered if being on a diet was related to sodium intake. She knew that the mean daily sodium intake in the United States was 3,400 mg, with a standard deviation of 270. She obtained a random sample of 172 dieting Americans and found, for sodium consumption, $M = 2{,}900$ mg. Complete all six steps of hypothesis testing.

6.100 At a large state university, the population data show that the average number of times that students meet with their academic advisors is 4.2 with $\sigma = 1.8$. The dean of student activities at this university wondered what the relation was between being involved in student clubs and organizations (such as the band, student government, or the ski club) and being involved academically in the university. She assumed that meeting with an academic advisor indicated students took their academics seriously. From the students who had been involved in student clubs and organizations for all four years of college, she obtained a random sample of 76 students, interviewed them individually, and found $M = 4.8$ for the number of times they met with their academic advisors. Complete all six steps of hypothesis testing.

Expand Your Knowledge

For Exercises 6.101 to 6.104, decide which option has a higher likelihood of being able to reject the null hypothesis.

6.101 (a) $\beta = .60$; (b) power = .60

6.102 (a) $\beta = .30$; (b) power = .50

6.103 (a) $\mu_0 = 10, \mu_1 = 15$; (b) $\mu_0 = 10, \mu_1 = 20$

6.104 (a) $\mu_0 = -17, \mu_1 = -23$; (b) $\mu_0 = -17; \mu_1 = -18$

For Exercises 6.105 to 6.108, label each conclusion as "correct" or "incorrect." If incorrect, label it as a "Type I error" or a "Type II error." Use the following scenario. A biochemist has developed a test to determine if food is contaminated or not. The test operates with the null hypothesis that a food is not contaminated.

6.105 Noncontaminated food is tested and it is called "noncontaminated."

6.106 Noncontaminated food is tested and it is called "contaminated."

6.107 Contaminated food is tested and it is called "noncontaminated."

6.108 Contaminated food is tested and it is called "contaminated."

6.109 A researcher is testing the null hypothesis that $\mu = 35$. She is doing a two-tailed test, has set alpha at .05, and has 121 cases in her sample. What is beta?

6.110 A researcher is testing the null hypothesis that $\mu = 120$, with $\sigma_M = 20$. In reality, the population mean is 140. (a) Draw a figure like Figure 6.12. Designate in the figure the areas representing Type II error and power. (b) Does one need to worry about Type I error?

For 6.111 and 6.112, determine how many tails there are.

6.111 A health science researcher believes that athletes have a lower resting heart rate than the general population. He knows, for the general population, that $\mu = 78$ and $\sigma = 9$. He is planning to do a study in which he obtains a random sample of athletes, measures their resting heart rates, and uses a single-sample z test to see if the mean heart rate for the sample of athletes is lower than the general population mean. (a) Should he do a one-tailed test or a two-tailed test? (b) Write the null and alternative hypotheses.

6.112 A lightbulb manufacturer believes that her compact fluorescent bulbs last longer than incandescent bulbs. She knows that the mean number of hours is 2,350 hours for a 60-watt incandescent bulb, with a standard deviation of 130 hours (these are population values). She gets a random sample of her compact fluorescent bulbs and measures the mean number of hours they last. She is going to use a single-sample z test to compare the sample mean to the population mean. (a) Should she do a one-tailed test or a two-tailed test? (b) Write the null and alternative hypotheses.

6.113 The population of random, two-digit numbers ranges from 00 to 99, has $\mu = 49.50$ and $\sigma = 28.87$. A statistician takes a random sample from this population and wants to see if his sample is representative. He figures that if it is representative, the sample mean will be close to 49.50. He plans to use a single-sample z test to test the sample against the population. Can this test be used?

SPSS

Sorry. SPSS doesn't do single-sample z tests. In Chapter 7, we'll learn about single-sample t tests, and SPSS will return!

The Single-Sample *t* Test

LEARNING OBJECTIVES

- Choose when to use a single-sample *t* test.
- Calculate the test statistic for a single-sample *t* test.
- Interpret the results of a single-sample *t* test.

CHAPTER OVERVIEW

In the last chapter, the logic of hypothesis testing was introduced with the single-sample *z* test. The single-sample *z* test was great as a prototype of hypothesis testing, but it is rarely used because it requires knowing the population standard deviation of the dependent variable. For most psychological variables—for example, reaction time in a driving simulator, number of words recalled from a list of 20, how close one is willing to get to a snake—the population standard deviation isn't known. In this chapter, we'll learn the single-sample *t* test, a test that like the single-sample *z* test allows one to compare a sample mean to a population mean. The single-sample *t* test, unlike the single-sample *z* test, can be used when a researcher doesn't know the population standard deviation. Thus, it is a more commonly used test.

7.1 Calculating the Single-Sample *t* Test

The **single-sample *t* test** is used to compare a sample mean to a specific value like a population mean. As it deals with means, it is used when the dependent variable is measured at the interval or ratio level. It does the same thing as a single-sample *z* test, but it is more commonly used because it doesn't require knowing the population standard deviation, σ. Instead, the single-sample *t* test uses the sample standard deviation, s. As long as there is access to the sample, it is possible to calculate s.

For an example, imagine the following scenario. Dr. Farshad, a clinical psychologist, wondered whether adults with attention deficit hyperactivity disorder (ADHD) had reflexes that differed in speed from those of the general population. She located a test for reaction time that was normed on adults in the United States. The average reaction time was 200 milliseconds (msec). This is a population mean, so the Greek letter mu is used to represent it and one would write $\mu = 200$.

From the various ADHD treatment centers in her home state of Illinois, Dr. Farshad obtained a random sample of 141 adults who had been diagnosed with

ADHD. Each adult was tested individually on the reaction-time test. The mean reaction time for the *sample* of 141 was 220 msec ($M = 220$), with a standard deviation of 27 msec ($s = 27$). (Note that because these are values calculated from a sample, the abbreviations for the mean and standard deviation are Roman, not Greek, letters.)

Dr. Farshad's question is whether adults with ADHD differ in reaction time from the general population. 220 msec is definitely different from 200 msec, so the mean reaction time of this sample of adults with ADHD is different from the population mean. However, Dr. Farshad doesn't know if the difference is a statistically significant one! It's possible that the sample mean is different from the population mean due to sampling error. To answer her question, she's going to need hypothesis testing.

The Six Steps of Hypothesis Testing

Let's follow Dr. Farshad through the six steps of hypothesis testing for the single-sample *t* test: (1) picking a *test;* (2) checking the *assumptions;* (3) listing the *hypotheses;* (4) setting the *decision rule;* (5) *calculating* the test statistic; and (6) *interpreting* the results. Remember the mnemonic: "Tom and Harry despise crabby infants."

The six steps for hypothesis testing are captured in the mnemonic "**T**om **a**nd **H**arry **d**espise **c**rabby **i**nfants."

Step 1 Pick a Test

The first step in hypothesis testing involves selecting the appropriate statistical test. Dr. Farshad is comparing the mean of a sample to the mean of a population, so she could use either a single-sample *z* test or a single-sample *t* test. However, she doesn't know the population standard deviation. She'll have to use the single-sample *t* test to figure out whether the mean reaction time of the sample of adults with ADHD is statistically different from the mean reaction time of the general population.

Step 2 Check the Assumptions

To determine if the sample mean is statistically different from the population mean, Dr. Farshad plans to use a single-sample *t* test. However, a single-sample *t* test can only be used if its assumptions are met. The three assumptions for the single-sample *t* test are the same as they were for the single-sample *z* test and are listed in **Table 7.1**.

TABLE 7.1 Assumptions for the Single-Sample t Test

Assumption	Explanation	Robustness
Random sample	The sample is a random sample from the population.	Robust if violated.
Independence of observations	Cases within the sample don't influence each other.	Not robust to violations.
Normality	The dependent variable is normally distributed in the population.	Robust to violations if the sample size is large.

Note: If a nonrobust assumption is violated, a researcher needs to use a different statistical test, one with different assumptions.

The first assumption involves whether the sample is a random sample from the population. This is a robust assumption. So if it is violated—and it often is—the analysis can still be completed. One just has to be careful about the population to which one generalizes the results. In this example, the sample is a random sample from the population of adults with ADHD in one state. This means that Dr. Farshad should only generalize her results to that state.

The second assumption is that the observations within the sample are independent. This assumption is not robust. If it is violated, one cannot proceed with the single-sample t test. In this example, each participant is in the sample only once and the reaction time of each case is not influenced by any other case. The cases were selected randomly and tested individually, so they are independent and the second assumption is not violated.

The third assumption is that the dependent variable, reaction time, is normally distributed in the population. This assumption is robust and the analysis can be completed if the assumption is violated as long as the sample size is large, say, 30 or more, and the deviation from normality is not too large. How does one test this assumption? One way is simply assuming that it is true—it is generally accepted that psychological characteristics (like personality) and physical characteristics (like height) are normally distributed. So, Dr. Farshad is willing to assume that reaction time is normally distributed. Another way is to make a graph of the data. In Figure 7.1, the histogram that Dr. Farshad made for the frequency distribution of the reaction-time data can be seen. Though not perfectly normal, it has a normal-ish shape. That, combined with a large sample size, leads Dr. Farshad to feel sure that this assumption has not been violated.

A Common Question

Q Why does the normality assumption exist?

A Hypothesis testing works by comparing the calculated value of the statistic to the expected value. The expected value comes from the sampling distribution, so the normality assumption is really about the shape of the sampling distribution. We know from the central limit theorem that the sampling distribution of the mean will be normal if N is large. If N is small, the sampling distribution will be normal if the population from which the samples are drawn is normal. And that is why the normality assumption exists.

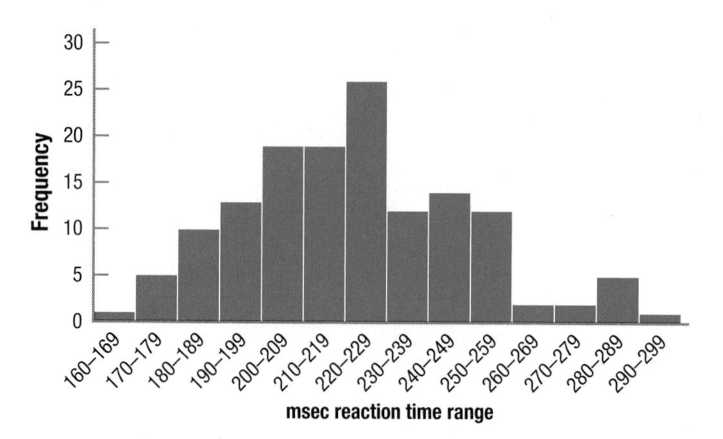

Figure 7.1 Histogram Showing Reaction Time for 141 Adults with Diagnosis of ADHD One way to check the normality assumption, if the sample size is large enough, is to make a histogram and examine it to see if it appears normally distributed. This graph of Dr. Farshad's reaction-time data looks reasonably normal-ish.

The assumptions have been met, so she can proceed with the planned statistical test, the single-sample *t* test.

Step 3 List the Hypotheses

Before writing the hypotheses, the researcher has to know if he or she is doing a one-tailed test or a two-tailed test. Two-tailed tests, called nondirectional tests, are more commonly used and they are a bit more conservative, in the sense that they make it harder to reject the null hypothesis. So, a two-tailed test is the "default" option—if in doubt, choose to do a two-tailed test. But, here's the difference between the two types of tests:

- A two-tailed test is used when the researcher is testing for a difference in either direction. Before collecting her data, Dr. Farshad didn't know whether adults with ADHD had faster or slower reaction times. This calls for a nondirectional test and nondirectional hypotheses.
- A one-tailed test is used when the researcher has a prediction about the direction of the difference *in advance* of collecting the data. As long as the difference is in the expected direction, a one-tailed test makes it easier to obtain statistically significant results.

The hypotheses for the nondirectional (two-tailed) single-sample *t* test for Dr. Farshad's study are

$$H_0: \mu_{\text{ADHD-Adults}} = 200$$

$$H_1: \mu_{\text{ADHD-Adults}} \neq 200$$

The null hypothesis (H_0) says that the mean reaction time for the population of adults with ADHD is *not* different from some specified value. In this case, that value is the mean reaction time, 200 msec, of Americans in general. (The null hypothesis could be phrased as $\mu_{\text{ADHD-Adults}} = \mu_{\text{Americans}} = 200$.)

The alternative hypothesis (H_1) says that the mean reaction time for the population of adults with ADHD is something other than the specified value of 200 msec. It doesn't say whether it is faster or slower, just that the mean reaction time is different. This means that the observed sample mean should be different enough from 200 msec that sampling error does not explain the difference. (The alternative hypothesis could be phrased as $\mu_{\text{ADHD-Adults}} \neq \mu_{\text{Americans}}$.)

If Dr. Farshad had believed that adults with ADHD had, for example, slower reaction times than the general population, then she would have planned a one-tailed test. In such a situation, the hypotheses would have been

$$H_0: \mu_{\text{ADHD-Adults}} \leq 200$$

$$H_1: \mu_{\text{ADHD-Adults}} > 200$$

Never forget that the alternative hypothesis expresses what the researcher believes to be true and that the researcher wants to be forced to reject the null hypothesis, so he or she has to accept the alternative hypothesis. In this one-tailed test example, the researcher's belief is that the reaction time is longer (i.e., slower) for people with ADHD. Once the alternative hypothesis is formed, the null hypothesis is written to make both hypotheses all-inclusive and mutually exclusive. Here, the null hypothesis would be that people with ADHD have reaction times the same as or faster than those of the general population.

Step 4 Set the Decision Rule

Setting the decision rule involves determining when to reject the null hypothesis and when to fail to reject the null hypothesis. To set the decision rule, find the critical value of the test statistic, the boundary between the rare zone and the common zone of the sampling distribution. For the single-sample *t* test, this will be a critical value of *t*.

What is *t*? *t* is the statistic that is calculated in the next step of the six-step hypothesis test procedure. The statistic *t* is a lot like the statistic *z* in the single-sample *z* test:

- If the null hypothesis is true and the sample mean is *exactly* equal to the specified value, *t* will equal zero.
- As the difference between the sample mean and the specified value grows, so does the *t* value.
- When the value of *t* that is calculated (the observed value of *t*) differs enough from zero (the expected value of *t*), then the null hypothesis is rejected.

The Critical Value of t

The point that separates "differs enough from zero" from "doesn't differ enough from zero" is called the **critical value of *t*,** abbreviated t_{cv}. To find t_{cv}, three pieces of information are needed:

1. Is the test one-tailed or two-tailed?
2. How willing is one to make a Type I error?
3. How large is the sample size?

The first question, whether a one-tailed or a two-tailed test is being done, was already answered when writing the hypotheses. Dr. Farshad is examining whether adults with ADHD have faster *or* slower reaction times than the general population. The reaction-time study calls for a two-tailed test because the hypotheses didn't specify a direction for the difference.

The second question in determining t_{cv} involves how willing one is to make a Type I error. A Type I error occurs when the researcher concludes, mistakenly, that the null hypothesis should be rejected. A common convention in statistics is to choose to have no more than a 5% chance of making a Type I error. As alpha (α) is the probability of a Type I error, statisticians phrase this as "setting alpha at .05" or as "$\alpha = .05$." Dr. Farshad has chosen to follow convention and set alpha at .05.

Overall, she wants her chance of making a Type I error to be no more than 5%. She has chosen to do a two-tailed test, so she will need to split that 5% in two and put 2.5% in each tail of the sampling distribution. This means Dr. Farshad will have two critical values of *t*, one positive and one negative. (If the test were one-tailed, all 5% of the rare zone would fall on one side and there would be only one critical value of *t*.)

The third question for determining t_{cv}, how large the sample size is, matters because the shape of the *t* distribution changes as the sample size changes. The tail of a *t* distribution is larger when the sample size is smaller. As a result, the critical value of *t* is farther away from zero when the sample size is small.

This is difficult to visualize without a concrete example, so look at Figure 7.2 in which two sampling distributions of *t* are superimposed—one for $N = 6$ (the dotted line) and one for $N = 60$ (the solid line). Which line is on top—the dotted line or the solid line—depends on whether one is looking at the center of the distribution or the tails.

Look at the tails. There, the dotted line (the one for the *t* distribution when $N = 6$) is above the solid line for the *t* distribution when $N = 60$. What does it mean for one line to be on top of the other? The *y*-axis measures frequency, so when one line is above another, this means it has a higher frequency at that point.

Focus on the positive side, the right-hand side, of the distribution. (Because the *t* distribution is symmetric, both sides are the same.) Note that around $t = 1.5$, the two lines cross. From that point on, the frequencies at each *t* value are higher for the $N = 6$ distribution than for the $N = 60$ distribution. Which distribution has more cases with *t* values above 1.50, $N = 6$ or $N = 60$? The answer is that the $N = 6$ distribution has a higher percentage of cases in the tail.

The implication of this is important: the total frequency of scores in the tail is higher for the distribution with the smaller sample size ($N = 6$). To cut off the extreme 2.5% of the scores for each distribution, the cut-off point will fall farther away from

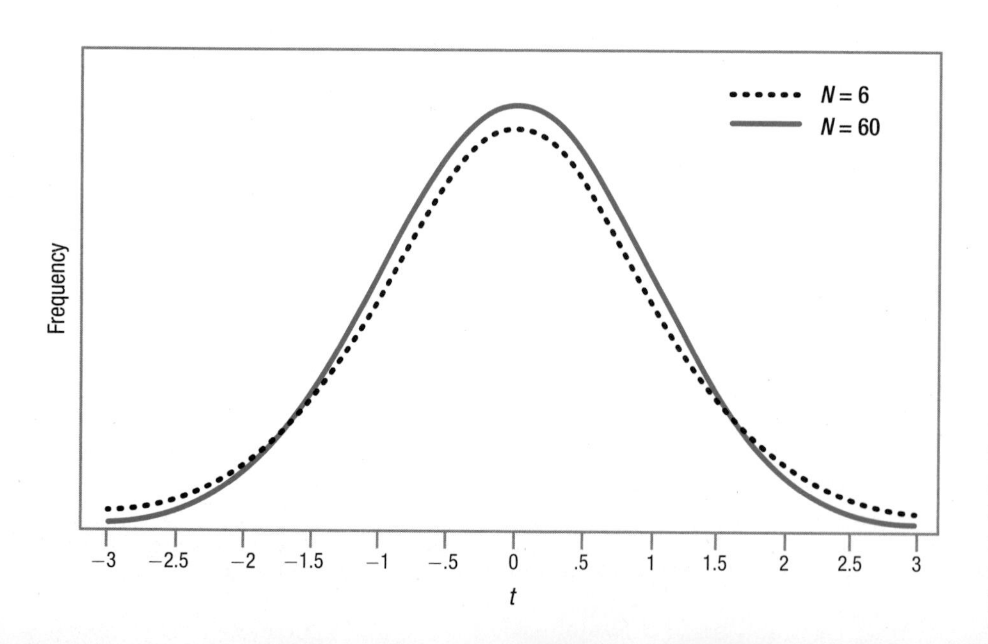

Figure 7.2 Shape of Sampling Distribution of t as a Function of Sample Size This figure shows how the shape of the sampling distribution of *t* changes with the sample size. When the sample size is small, there is more variability and more samples have means that fall farther out in a tail. As a result, the critical value of *t* falls farther away from the midpoint when the sample size is small, making it more difficult to reject the null hypothesis.

zero for the $N = 6$ distribution than for the $N = 60$ distribution. In this example, the cut-off points (the critical values of *t*) are 2.57 for $N = 6$ and 2.00 for $N = 60$. The critical value falls closer to zero when the sample size is larger, and it falls farther away from zero when the sample size is smaller.

Sample size affects the critical value of *t*, and the critical value determines the rare and common zones in the sampling distribution. (Remember, when a result falls in the rare zone, the null hypothesis is rejected; when a result falls in the common zone, the null hypothesis is not rejected.) So, sample size affects the ability to reject the null hypothesis:

- When the sample size is small, the critical value of *t* falls farther away from zero, making the rare zone harder to reach, making it more difficult to reject the null hypothesis.
- Larger sample sizes have the opposite effect. When a sample size is large, the critical value of *t* falls closer to zero. As a result, the rare zone is easier to reach, which makes it easier to reject the null hypothesis.

The goal of research almost always is to reject the null hypothesis, so having a larger sample size is an advantage.

To find the critical value of *t* for a given sample size, use Appendix Table 3, a table of critical values of *t*. A portion of Appendix Table 3 is shown in Table 7.2.

There are several things to note in Table 7.2. First, note that there are different *rows* for the critical values. The rows represent different critical values of *t* based on sample size. The heading for the rows is "*df*," which stands for *degrees of freedom*. **Degrees of freedom** represent the number of values in a sample that are free to vary. For example, if the mean of three cases is 10, then the values for only two cases are

TABLE 7.2 Critical Values of t (Appendix Table 3)

	Critical values of t			
df	$\alpha = .05$, one-tailed or $\alpha = .10$, two-tailed	$\alpha = .025$, one-tailed or $\alpha = .05$, two-tailed	$\alpha = .01$, one-tailed or $\alpha = .02$, two-tailed	$\alpha = .005$, one-tailed or $\alpha = .01$, two-tailed
1	6.314	**12.706**	31.821	63.657
2	2.920	**4.303**	6.965	9.925
3	2.353	**3.182**	4.541	5.841
4	2.132	**2.776**	3.747	4.604
5	2.015	**2.571**	3.365	4.032
6	1.943	**2.447**	3.143	3.707
7	1.895	**2.365**	2.998	3.499
8	1.860	**2.306**	2.896	3.355
9	1.833	**2.262**	2.821	3.250
10	1.812	**2.228**	2.764	3.169

Note: The critical value of *t* is the boundary that separates the rare zone of the sampling distribution of *t* from the common zone. For two-tailed tests, the critical values are both positive and negative values. The α level is the probability of making a Type I error. A one-tailed test is used with directional hypotheses and a two-tailed test with nondirectional hypotheses. *df* stands for degrees of freedom. For a single-sample *t* test, $df = N - 1$. The bold numbers represent the critical values of *t* most commonly used, those for a nondirectional (two-tailed) test with a 5% chance ($\alpha = .05$) of making a Type I error.

free to vary. If one case has a score of 9 and another case has a score of 11, then the third case has to have a score of 10. In that example, there are three values, and 2 degrees of freedom—once two values are known, the third is determined.

Degrees of freedom and sample size are yoked together. As the sample size becomes larger, the degrees of freedom increase. And, other things being equal, larger sample sizes are better because there is a greater likelihood that the sample represents the population. In fact, when the sample size is infinitely large, the *t* distribution is the same as the *z* distribution because the whole population is being sampled.

For a single-sample *t* test, the degrees of freedom are calculated as the sample size minus 1. This is shown in Equation 7.1.

Equation 7.1 Degrees of Freedom (*df*) for a Single-Sample *t* Test

$$df = N - 1$$

where df = degrees of freedom
N = sample size

For the reaction-time study, there are 141 participants in the sample, so degrees of freedom are calculated like this:

$$\begin{aligned} df &= 141 - 1 \\ &= 140 \end{aligned}$$

Look at Table 7.2 again. The second thing to note in the table of critical values of *t* is that there are four columns of critical values. Which column to use depends on two factors: (1) if a one-tailed test or two-tailed test is being done, and (2) where alpha, the willingness to make a Type I error, is set.

The critical values of *t* in the column for a two-tailed test with a 5% chance of Type I error (i.e., $\alpha = .05$) have been bolded to make them easier to find because they are the most commonly used.

With the reaction-time study as a two-tailed test, Dr. Farshad would reject the null hypothesis if adults with ADHD had slower reaction times than the general population *or* if adults with ADHD had faster reaction times than the general population. That is what a two-tailed test means.

Under these conditions ($df = 140$, $\alpha = .05$, two-tailed), Dr. Farshad finds that the critical value of *t* is ± 1.977 (i.e., -1.977 and 1.977). These values are marked in Figure 7.3, along with the rare and common zones.

The critical value of *t* would be different if this were a one-tailed test. If Dr. Farshad had reason, in advance of collecting data, to believe that adults with ADHD had slower reaction times, then she could do a one-tailed test. In such a case, she would only reject the null hypothesis if adults with ADHD had slower reaction times than the general population. For $df = 140$ and with $\alpha = .05$, the one-tailed critical value of *t* would be 1.656. This critical value of 1.656 is closer to zero, the midpoint of the sampling distribution, than was the critical value (1.977) for the two-tailed test. This means the rare zone is more easily reached and makes it easier to reject the null hypothesis for the one-tailed test.

Figure 7.4 shows the larger rare zone on the right for the one-tailed test. The rare zone for the two-tailed test is marked with /// and the rare zone for the one-tailed test is marked with \\\. Though the total *percentage* of the curve that is the rare zone is the same for the two tests, notice how more *area* of the rare zone is

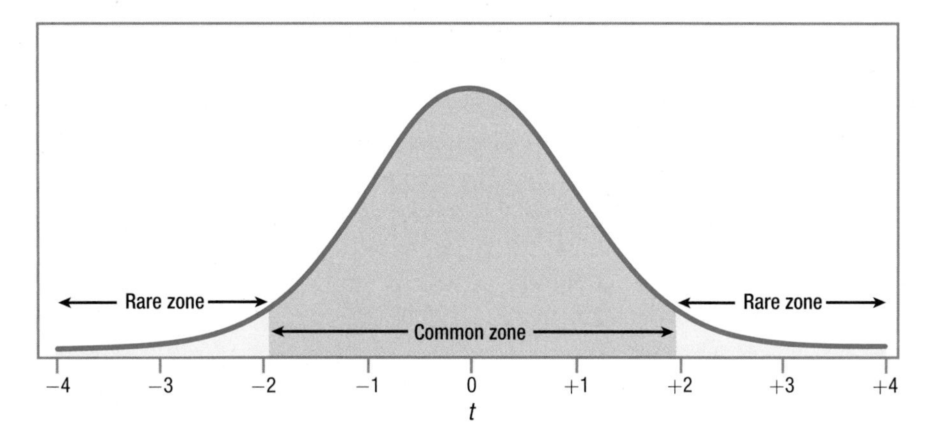

Figure 7.3 Setting the Decision Rule: Two-Tailed, Single-Sample *t* Test In this sampling distribution of *t* values ($df = 140$, $\alpha = .05$, two-tailed), the border between the rare and common zones is ± 1.977. If the observed value of *t* falls in the rare zone, the null hypothesis is rejected; one fails to reject it if the observed value falls in the common zone. Note that the rare zone is split into two parts, half in each tail of the distribution.

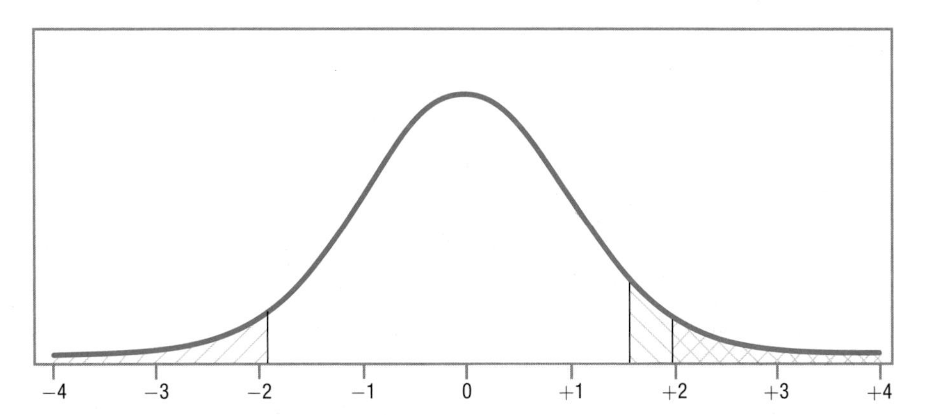

Figure 7.4 Comparing Rare Zones for One-Tailed and Two-Tailed Tests The rare zone of the sampling distribution of *t* with $df = 140$ and $\alpha = .05$ for a one-tailed test is the section marked \\\ to the right of $t = 1.656$. For a two-tailed test, if the difference between the sample and population mean falls in the same direction, it is the area marked /// to the right of $t = 1.977$ (The rare zone for the two-tailed test also includes the /// area on the left). Notice the area of the rare zone is larger for the one-tailed test on one side of the curve, making it that much easier to reject the null hypothesis for a one-tailed test.

marked off on one side by the one-tailed test (\\\) than by the two-tailed test (///). As a result, it is easier to reject the null hypothesis with a one-tailed test than a two-tailed test, as long as the difference is in the expected direction. This is an advantage for one-tailed tests. The advantage disappears if there is a difference, but it is not in the hypothesized direction.

Dr. Farshad was conducting an exploratory study to see if adults with ADHD differed in reaction time—in either direction—from the general population. So, she's doing a two-tailed test. Here's the general version of the decision rule for a two-tailed test:

$$\text{If } t \leq -t_{cv} \text{ or if } t \geq t_{cv}, \text{ reject } H_0.$$
$$\text{If } -t_{cv} < t < t_{cv}, \text{ fail to reject } H_0.$$

When the first statement is true, the observed value of t falls in the rare zone, and the null hypothesis is rejected. When the second statement is true, t falls in the common zone, and the researcher will fail to reject the null hypothesis.

For the reaction-time study, the specific decision rule is:

- Reject the null hypothesis if $t \leq -1.977$ *or* if $t \geq 1.977$.
- Fail to reject the null hypothesis if $-1.977 < t < 1.977$.

A Common Question

Q Appendix Table 3 doesn't contain degrees of freedom for all possible values. What does one do if $df = 54$, for example?

A In the TV game show *The Price Is Right,* the person who comes closest to guessing the price of an object without exceeding the price wins. Follow that rule and use the degrees of freedom that are closest without going over. If $df = 54$ and the table only contains critical values of t for 50 and 55 degrees of freedom, use t_{cv} for $df = 50$.

Step 5 Calculate the Test Statistic

It is time to calculate the t value that compares the sample mean for reaction time (220 msec) to the specified value, the population mean for a reaction time of 200 msec. The formula is shown in Equation 7.2.

Equation 7.2 Formula for Calculating a Single-Sample t Test

$$t = \frac{M - \mu}{s_M}$$

where $t = t$ value

$M =$ sample mean

$\mu =$ population mean (or a specified value)

$s_M =$ estimated standard error of the mean (Equation 5.2)

The numerator in the single-sample t test formula subtracts the population mean (μ) from the sample mean (M). The difference is then divided by the estimated standard error of the mean (s_M). Before using Equation 7.2, one needs to know the estimated standard error of the mean. This is calculated using Equation 5.2:

$$\begin{aligned} s_M &= \frac{s}{\sqrt{N}} \\ &= \frac{27}{\sqrt{141}} \\ &= \frac{27}{11.8743} \\ &= 2.2738 \\ &= 2.27 \end{aligned}$$

Once the estimated standard error of the mean has been calculated, all the values needed to complete Equation 7.2 are available: $M = 220$, $\mu = 200$, and $s_M = 2.27$. Here are the calculations to find the t value:

$$t = \frac{M - \mu}{s_M}$$

$$= \frac{220 - 200}{2.27}$$

$$= \frac{20.0000}{2.27}$$

$$= 8.8106$$

$$= 8.81$$

Step 5 is done and Dr. Farshad knows the t value: $t = 8.81$. The sixth and final step of hypothesis testing is interpretation. We'll turn to that after a little more practice with the first five steps of hypothesis testing in Worked Example 7.1.

A Common Question

Q The equation for t, $t = \frac{M - \mu}{s_M}$ looks a lot like the equation for z, $z = \frac{X - \mu}{s}$. Are they similar?

A Yes they are. Both serve to standardize deviation scores, so we can tell how common they are.

Worked Example 7.1

Adjusting to the first year of college can be hard, especially if something traumatic happens back at home. A veterinarian in an imaginary college town, Dr. Richman, wondered if the GPA went down for students who lost a family pet while they were away at college. The vet found 11 students at the college who indicated that their pets at home had died during the year. He had each student report his or her GPA for the year and found the mean was 2.58, with a standard deviation of 0.50. From the college registrar, Dr. Richman learned that the mean GPA for all students for the year was 2.68. Thus, $M = 2.58$ and $\mu = 2.68$. Does losing a pet have a negative impact on GPA?

The answer appears to be yes, because the sample of students who have lost a pet has a GPA lower than the population's GPA. But, isn't it possible, due to sampling error, that a random sample of 11 people from a population where $\mu = 2.68$ could have a sample mean of 2.58? The vet is going to need hypothesis testing to determine if the difference between the sample and the population is statistically significant.

Step 1 Pick a Test. Either a single-sample z test or a single-sample t test can be used to compare the mean of a sample to a specified value like the mean of a population. However, when the population standard deviation is unknown, a single-sample t test must be used. This situation, with σ unknown, calls for a single-sample t test.

Step 2 Check the Assumptions. The population is college students who have lost their pets. The sample is not a random sample from the population because all participants self-selected and come from only one college. It is possible that the students who lost pets and who chose not to participate differed in some way from those who volunteered. So, the first assumption is violated. Luckily, the first assumption is robust and can be violated, though Dr. Richman will need to be careful about generalizing from the results.

There's no reason to believe that the second assumption is violated. The observations seem to be independent. No participants are siblings who would have lost the same pet. Further, participants did not come from a support group for people who had lost a pet, where people might have influenced each other in coping with their losses. And each participant was in the sample only once.

The third assumption is not violated because Dr. Richman is willing to assume that GPA, like intelligence, is normally distributed. (The sample size, 11, is too small to provide a meaningful histogram.) None of the nonrobust assumptions was violated, so Dr. Richman can proceed with the single-sample *t* test.

Step 3 List the Hypotheses. Dr. Richman believes that losing a pet will harm adjustment, so he is doing a one-tailed test. Here are the null and alternative hypotheses:

$$H_0: \mu_{StudentswithPetLoss} \geq 2.68$$

$$H_1: \mu_{StudentswithPetLoss} < 2.68$$

The null hypothesis (H_0) says that the mean GPA of the population of first-year students who have lost a pet during the year is the same or greater than the mean GPA of students in general for that year. If the null hypothesis is true, when the vet takes a sample of college students who have lost a pet, their mean should be close enough to 2.68 that the difference can be explained by sampling error.

The alternative hypothesis (H_1) says that what the veterinarian believes to be true is true: the population of students with pet loss has a mean GPA lower than the mean GPA of all students. Therefore, a mean for a sample of pet-loss students should be far enough below 2.68 that sampling error is not a plausible explanation for the difference.

Step 4 Set the Decision Rule. Dr. Richman has directional hypotheses and is using the default value of 5% for his willingness to make a Type I error. This means that it is a one-tailed test and that alpha is set at .05. The sample size is 11, and the vet will use Equation 7.1 to determine the degrees of freedom:

$$\begin{aligned} df &= N - 1 \\ &= 11 - 1 \\ &= 10 \end{aligned}$$

Using Appendix Table 3, Dr. Richman finds $t_{cv} = 1.812$. Because he is doing a one-tailed test, he has to decide whether the critical value of t is -1.812 or $+1.812$. If his theory is correct and losing a pet hurts students' GPAs, then the numerator of the single-sample t test formula (Equation 7.2), which is $M - \mu$, will be a negative number because M should be below 2.68. Hence, t_{cv} is a negative value, -1.812. Here, and in Figure 7.5, is the vet's decision rule:

- If $t \leq -1.812$, reject H_0.
- If $t > -1.812$, fail to reject H_0.

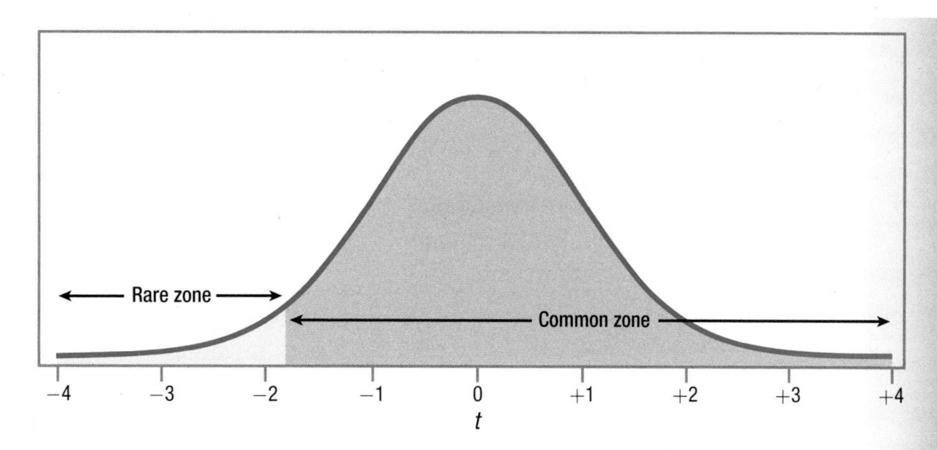

Figure 7.5 One-Tailed Decision Rule for Sampling Distribution of t, $df = 10$
This is a one-tailed test with 10 degrees of freedom. The critical value of t is -1.812. Note that the entire rare zone falls on one side of the sampling distribution.

Step 5 Calculate the Test Statistic. Before calculating t, the vet needs to calculate s_M using Equation 5.2:

$$\begin{aligned} s_M &= \frac{s}{\sqrt{N}} \\ &= \frac{0.50}{\sqrt{11}} \\ &= \frac{0.50}{3.3166} \\ &= 0.1508 \\ &= 0.15 \end{aligned}$$

Next, he'll use s_M in Equation 7.2 to calculate t:

$$\begin{aligned} t &= \frac{M - \mu}{s_M} \\ &= \frac{2.58 - 2.69}{0.15} \\ &= \frac{-0.1000}{0.15} \\ &= -0.6667 \\ &= -0.67 \end{aligned}$$

Step 5 is completed and the vet now knows the value of the test statistic: $t = -0.67$. In the next section, we'll learn how to explain what this means with regard to the impact of the loss of a pet on GPA.

Practice Problems 7.1

Review Your Knowledge

7.01 A researcher draws a sample from a population and finds that the sample mean is different from what he thought the population mean was. One explanation for the discrepancy is that he was wrong about what the population mean is. What's another explanation for the discrepancy?

7.02 What are the six steps to be followed in conducting a hypothesis test?

7.03 When should a single-sample *t* test be used?

7.04 What are the assumptions for a single-sample *t* test?

7.05 Which rare zone is larger if the observed difference is in the expected direction—for a one-tailed test or a two-tailed test?

Apply Your Knowledge

7.06 A researcher has a sample of Nobel Prize winners. She thinks that they may be smarter than average. If the average IQ is 100, what are the null and alternative hypotheses for a single-sample *t* test?

7.07 A researcher has drawn a random sample of 48 cases from a population. He plans to use a single-sample *t* test to compare the sample mean to the population mean. He has set α at .05 and is doing a two-tailed test. Write out the decision rule regarding the null hypothesis.

7.08 If $N = 17$ and $s = 6$, what is s_M?

7.09 If $M = 24$, $\mu = 30$, and $s_M = 8$, what is t?

7.2 Interpreting the Single-Sample *t* Test

Computers are great for statistics. They can crunch numbers and do math faster and more accurately than humans can. But even a NASA supercomputer couldn't do what we are about to do: use common sense to explain the results of a single-sample *t* test in plain English. Interpretation is the most human part of statistics.

Interpretation is the most human part of statistics.

To some degree, interpretation is a subjective process. The researcher takes objective facts—like the means and the test statistic—and explains them in his or her own words. How one researcher interprets the results of a hypothesis test might differ from how another researcher interprets them. Reasonable people can disagree. But, there are guidelines that researchers need to follow. An interpretation needs to be supported by facts. One person may perceive a glass of water as half full and another as half empty, but they should agree that it contains roughly equal volumes of water and air.

Interpretation is subjective, but should be based on facts. Confidence intervals, introduced later in the chapter, provide a little wiggle room. (Photo courtesy of Paul Sahre.)

Interpreting the results of a hypothesis test starts with questions to be answered. The answers will provide the material to be used in a written interpretation. For a single-sample *t* test, our three questions are:

1. Was the null hypothesis rejected?
2. How big is the effect?
3. How wide is the confidence interval?

The questions are sequential. Each one gives new information that is useful in understanding the results from a different perspective. Answering only the first question, as was done in the last chapter, will provide enough information for completing a basic interpretation. However, answering all three questions leads to a more nuanced understanding of the results and a better interpretation.

Let's follow Dr. Farshad as she interprets the results of her study, question by question. But, first, let's review what she did. Dr. Farshad wondered if adults with ADHD differed in reaction time from the general population. To test this, she obtained a random sample, from her state, of 141 adults who had been diagnosed with ADHD. She had each participant take a reaction time test and found the sample mean and sample standard deviation ($M = 220$ msec, $s = 27$ msec). She planned to use a single-sample *t* test to compare the sample mean to the population mean for adult Americans ($\mu = 200$ msec).

Dr. Farshad's hypotheses for the single-sample *t* test were nondirectional as she was testing whether adults with ADHD had faster or slower reaction times than the general population. She was willing to make a Type I error no more than 5% of the time, she was doing a two-tailed test, and she had 140 degrees of freedom, so the critical value of *t* was ± 1.977. The first step in calculating *t* was to find the estimated standard error of the mean ($s_M = 2.21$) and she used that in order to go on and find $t = 8.81$.

Step 6 Interpret the Results

Was the Null Hypothesis Rejected?

In Dr. Farshad's study, the *t* value was calculated as 8.81. Now it is time to decide whether the null hypothesis was rejected or not. To do so, Dr. Farshad puts the *t* value, 8.81, into the decision rule that was generated for the critical value of *t*, ± 1.977, in Step 4. Which of the following statements is true?

- Is $8.81 \leq -1.977$ *or* is $8.81 \geq 1.977$?
- Is $-1.977 < 8.81 < 1.977$?

The second part of the first statement is true: 8.81 is greater than or equal to 1.977. The observed value of *t*, 8.81, falls in the rare zone of the sampling distribution, so the null hypothesis is rejected (see Figure 7.6). She can conclude that there is a statistically significant difference between the sample mean of the adults with ADHD and the general population mean.

Rejecting the null hypothesis means that Dr. Farshad has to accept the alternative hypothesis and conclude that the mean reaction time for adults with ADHD is *different* from the mean reaction time for the general population. Now she needs to determine the *direction* of the difference. By comparing the sample mean (220) to the population mean (200), she can conclude that adults with ADHD take a longer time to react and so have a *slower* reaction time than adults in general.

Dr. Farshad should also report the results in APA format. APA format provides five pieces of information: (1) what test was done, (2) the number of cases, (3) the value

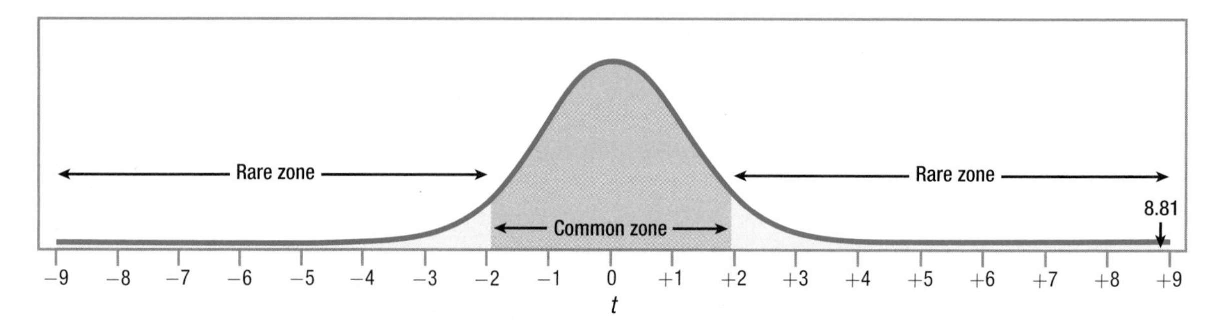

Figure 7.6 Single-Sample *t* Test: Results from ADHD Reaction-Time Study The value of the test statistic (t) is 8.81. This is greater than or equal to the critical value of t of 1.977, so it falls in the rare zone. The null hypothesis is rejected.

of the test statistic, (4) the alpha level used, and (5) whether the null hypothesis was or wasn't rejected.

In APA format, the results would be $t(140) = 8.81, p < .05$:

1. The initial t says that the statistical test was a t test.
2. The sample size, 141, is present in disguised form. The number 140, in parentheses, is the degrees of freedom for the t test. For a single-sample t test, $df = N - 1$. That means $N = df + 1$. So, if $df = 140$, $N = 140 + 1 = 141$.
3. The observed t value, 8.81, is reported. This number is the value of t calculated, *not* the critical value of t found in Appendix Table 3. Note that APA format requires the value to be reported to two decimal places, no more and no fewer.
4. The .05 tells that alpha was set at .05.
5. The final part, $p < .05$, reveals that the null hypothesis was rejected. It means that the observed result (8.81) is a rare result—it has a probability of less than .05 of occurring when the null hypothesis is true.

If Dr. Farshad stopped after answering only the first interpretation question, this is what she could write for an interpretation:

> There is a statistically significant difference between the mean reaction time of adults in Illinois with ADHD and the reaction time found in the general U.S. population, $t(140) = 8.81, p < .05$. Adults with ADHD ($M = 220$ msec) have a slower mean reaction time than does the general public ($\mu = 200$).

Practice Problems 7.2

Apply Your Knowledge

For these problems, use $\alpha = .05$, two-tailed.

7.10 If $N = 19$ and $t = 2.231$, write the results in APA format.

7.11 If $N = 7$ and $t = 2.309$, write the results in APA format.

7.12 If $N = 36$ and $t = 2.030$, write the results in APA format.

7.13 If $N = 340$ and $t = 3.678$, write the results in APA format.

How Big Is the Effect?

All we know so far is that there is a statistically significant difference, such that we can conclude that adults with ADHD in Illinois have slower reaction times than the general American adult population does. We know there's a 20-msec difference, but we are hard-pressed to know how much of a difference that really is.

Thus, the next question to address when interpreting the results is **effect size,** how large the impact of the explanatory variable is on the outcome variable. With the reaction-time study, the explanatory variable is ADHD status (adults with ADHD vs. the general population) and the outcome variable is reaction time. Asking what the effect size is, is asking how much impact ADHD has on a person's reaction time.

In this section, we cover two ways to measure effect size: Cohen's d and r^2. Both can be used to determine if the effect is small, medium, or large. Both standardize the effect size, so outcome variables measured on different metrics can be compared. A 2-point change in GPA would be huge, while a 2-point change in total SAT score would be trivial. Both Cohen's d and r^2 take the unit of measurement into account. And, finally, both will be used as measures of effect size for other statistical tests in other chapters.

Cohen's *d*

The formula for **Cohen's *d*** is shown in Equation 7.3. It takes the difference between the two means and standardizes it by dividing it by the sample standard deviation. Thus, d is like a z score—it is a standard score that allows different effects measured by different variables in different studies to be expressed—and compared—with a common unit of measurement.

Equation 7.3 Formula for Cohen's d for a Single-Sample t Test

$$d = \frac{M - \mu}{s}$$

where d = the effect size
M = sample mean
μ = hypothesized population mean
s = sample standard deviation

To calculate d, Dr. Farshad needs to know the sample mean, the population mean, and the sample standard deviation. For the reaction-time data, $M = 220$, $\mu = 200$, and $s = 27$. Here are her calculations for d:

$$\begin{aligned} d &= \frac{M - \mu}{s} \\ &= \frac{220 - 200}{27} \\ &= \frac{20.0000}{27} \\ &= 0.7407 \\ &= 0.74 \end{aligned}$$

Cohen's $d = 0.74$ for the ADHD reaction-time study. Follow APA format and use two decimal places when reporting d. And, because d values can be greater than 1, values of d from -0.99 to 0.99 get zeros before the decimal point.

In terms of the *size* of the effect, it doesn't matter whether d is positive or negative. A d value of 0.74 indicates that the two means are 0.74 standardized units apart. This is an equally strong effect whether it is 0.74 or -0.74. But the sign associated with d is important for knowing the direction of the difference, so make sure to keep it straight.

Here's how Cohen's d works:

- A value of 0 for Cohen's d means that the explanatory variable has absolutely no effect on the outcome variable.
- As Cohen's d gets farther away from zero, the size of the effect increases.

Cohen (1988), the developer of d, has offered standards for what small, medium, and large effect sizes are in the social and behavioral sciences. Cohen's d values for these are shown in Table 7.3. In Figure 7.7, the different effect sizes are illustrated by comparing the IQ of a control group ($M = 100$) to the IQ of an experimental group that is smarter by the amount of a small effect (3 IQ points), a medium effect (7.5 IQ points), or a large effect (12 IQ points).

TABLE 7.3 Effect Sizes in the Social and Behavioral Sciences

Size of effect	Cohen's d
None	≈ 0.00
Small	≈ 0.20
Medium	≈ 0.50
Large	>0.80

Note: Cohen's d is calculated with Equation 7.3. The sign of Cohen's d doesn't matter. A Cohen's d of -0.50 has the same degree of effect as a Cohen's d of 0.50, just in the opposite direction.

Cohen describes a small effect as a d of around 0.20. This occurs when there is a small difference between means, a difference in performance that would not be readily apparent in casual observation. Look at the top panel in Figure 7.7. That's what a small effect size looks like. Though the two groups differ by 3 points on intelligence ($M = 100$ vs. $M = 103$), it wouldn't be obvious that one group is smarter than the other without using a sensitive measure like an IQ test (Cohen, 1988).

Another way to visualize the size of the effect is to consider how much overlap exists between the two groups. If the curve for one group fits exactly over the curve for the other, then the two groups are exactly the same and there is no effect. As the amount of overlap decreases, the effect size increases. For the small effect in Figure 7.7, about 85% of the total area overlaps.

A medium effect size, a Cohen's d of around 0.50, is large enough to be observable, according to Cohen. Look at the middle panel in Figure 7.7. One group has a mean IQ of 100, and the other group has a mean IQ of 107.5, a 7.5 IQ point difference. Notice the amount of differentiation between groups for a medium effect size. In this example, about two-thirds of the total area overlaps, a decrease from the 85% overlap seen with a small effect size.

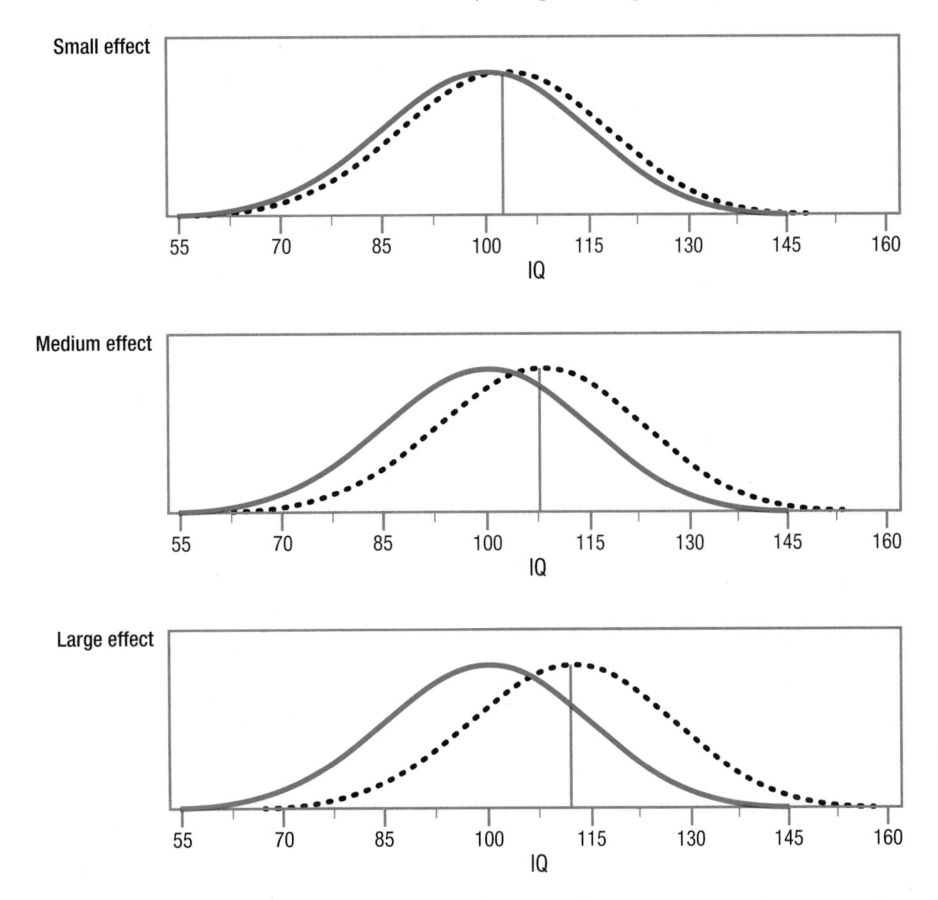

Figure 7.7 Examples of Small, Medium, and Large Effect Sizes This figure uses the distribution of IQ scores for two different groups to show effect sizes. In each panel, the solid-line curve shows the set of scores for the control group and the dotted-line curve shows the set of scores for the experimental group. In each panel, the experimental group has a higher mean IQ. The top panel (Small effect) shows what Cohen (1988) calls a small effect size ($d = 0.20$), the middle panel (Medium effect) a medium effect size ($d = 0.50$), and the bottom panel (Large effect) a large effect size ($d = 0.80$). These are mean differences, respectively, of 3, 7.5, and 12 IQ points. Notice how the differentiation between the two groups in each panel increases as the size of the effect increases, both in terms of increasing distance between the two means and decreasing overlap between the two distributions.

A large effect size is a Cohen's d of 0.80 or larger. This is shown in the bottom panel of Figure 7.7. One group has a mean IQ of 100, and the other group has a mean IQ of 112, a 12 IQ point difference. Here, there is a lot of differentiation between the two groups with only about half of the area overlapping. The increased differentiation can be seen in the large section of the control group that has IQs lower than the experimental group and the large section of the experimental group that has IQs higher than the control group.

The Cohen's d calculated for the reaction-time study by Dr. Farshad was 0.74, which means it is a medium to large effect. If she stopped her interpretation after calculating this effect size, she could add the following sentence to her interpretation: "The effect of ADHD status on reaction time falls in the medium to large range, suggesting that the slower mean reaction time associated with having ADHD may impair performance."

r Squared

The other commonly used measure of effect size is r^2. Its formal name is coefficient of determination, but everyone calls it *r* squared. $\mathbf{r^2}$**,** like *d*, tells how much impact the explanatory variable has on the outcome variable.

Equation 7.4 Formula for r^2, the Percentage of Variability in the Outcome Variable Accounted for by the Explanatory Variable

$$r^2 = \frac{t^2}{t^2 + df} \times 100$$

where r^2 = the percentage of variability in the outcome variable that is accounted for by the explanatory variable

t^2 = the squared value of *t* from Equation 7.2

df = the degrees of freedom for the *t* value

This formula says that r^2 is calculated as the squared *t* value divided by the sum of the squared *t* value plus the degrees of freedom for the *t* value. Then, to turn it into a percentage, the ratio is multiplied by 100. r^2 can range from 0% to 100%. These calculations reveal the percentage of variability in the outcome scores that is accounted for (or predicted) by the explanatory variable. For the ADHD data, these calculations would lead to the conclusion that $r^2 = 36\%$:

$$\begin{aligned} r^2 &= \frac{t^2}{t^2 + df} \times 100 \\ &= \frac{8.81^2}{8.81^2 + 140} \times 100 \\ &= \frac{77.6161}{77.6161 + 140} \times 100 \\ &= \frac{77.6161}{217.6161} \times 100 \\ &= .3567 \times 100 \\ &= 35.67\% \end{aligned}$$

r^2 tells the percentage of variability in the outcome variable that is accounted for (or predicted) by the explanatory variable.

What does r^2 tell us? Imagine that someone took a large sample of people and timed them running a mile. There would be a lot of variability in time, with some runners taking 5 minutes and others 20 minutes or more. What are some factors that influence running speed? Certainly, physical fitness and age play important roles, as do physical health and weight. Which of these four factors has the most influence on running speed? That could be determined by calculating r^2 for each variable. The factor that has the largest r^2, the one that explains the largest percentage of variability in time, has the most influence.

The question addressed by r^2 in the reaction-time study is how much of the variability in reaction time is accounted for by the explanatory variable (ADHD status) and how much is left unaccounted for:

- The closer r^2 is to 100%, the stronger the effect of the explanatory variable is and the less variability in the outcome variable remains to be explained by other variables.

- The closer r^2 is to 0%, the weaker the effect of the explanatory variable and the more variability in the dependent variable exists to be explained by other variables.

Cohen (1988), who provided standards for *d* values, also provides standards for r^2:

- A small effect is an $r^2 \approx 1\%$.
- A medium effect is an $r^2 \approx 9\%$.
- A large effect is an $r^2 \approx 25\%$.

By Cohen's standards, an r^2 of 36%, as is the case in our ADHD study, is a large effect. This means that as far as explanatory variables in the social and behavioral sciences go, this one accounts for a lot of the variability in the outcome variable. Figure 7.8 is a visual demonstration of what explaining 36% of the variability means. Note that although a majority of the variability, 64%, in reaction time remains unaccounted for, this is still considered a large effect. Here is what Dr. Farshad could add to her interpretation now that r^2 is known: "Having ADHD has a large effect on one's reaction time. In the present study, knowing ADHD status explains more than a third of the variability in reaction time."

One important thing to note is that even though Dr. Farshad went to the trouble of calculating both *d* and r^2, it is not appropriate to report both. These two provide overlapping information and it does not make a case stronger to say that the effect was a really strong one, because both *d* and r^2 are large. This would be like testing boiling water and reporting that it was really, really boiling because it registered 212 degrees on a Fahrenheit thermometer and 100 degrees on a Celsius thermometer.

Here's a heads up for future chapters and for reading results sections in psychology articles—there are other measures of effect size that are similar to r^2. Both η^2 (eta squared) and ω^2 (omega squared) provide the same information, how much of the variability in the outcome variable is explained by the explanatory variable.

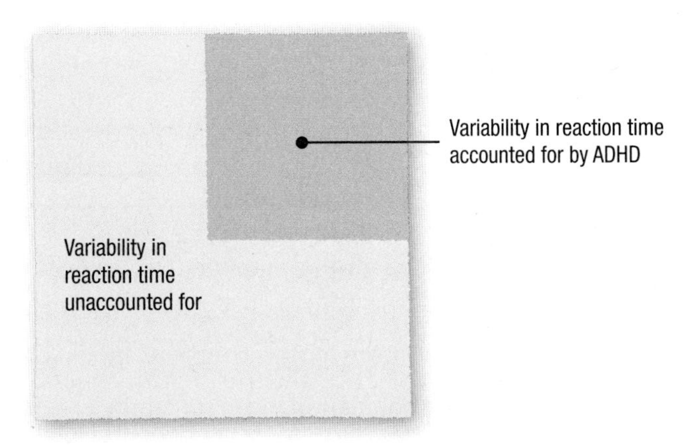

Figure 7.8 Percentage of Variability in Reaction Time Accounted for by ADHD Status The darkly shaded section of this square represents the 36% of the variability in reaction time that is accounted for by ADHD status. The nonshaded part of the square represents the 64% of variability that is still not accounted for.

Practice Problems 7.3

Apply Your Knowledge

7.14 If $M = 66$, $\mu = 50$, and $s = 10$, what is d?

7.15 If $M = 93$, $\mu = 100$, and $s = 30$, what is d?

7.16 If $N = 18$ and $t = 2.37$, what is r^2?

7.17 If $r = .32$, how much of the variability in the outcome variable, Y, is accounted for by the explanatory variable, X?

How Wide Is the Confidence Interval?

In Chapter 5, we encountered confidence intervals for the first time. There, they were used to take a sample mean, M, and estimate a range of values that was likely to capture the population mean, μ. Now, for the single-sample t test, the confidence interval will be used to calculate the range for how big the difference is between two population means. Two populations? Dr. Farshad's confidence interval, in the ADHD reaction-time study, will tell how large (or small) the mean difference in reaction time might be between the *population of adults with ADHD* and the general population of Americans.

The difference between population means can be thought of as an effect size. If the distance between two population means is small, then the size of the effect (i.e., the impact of ADHD on reaction time) is not great. If the two population means are far apart, then the size of the effect is large.

Equation 7.5 is the formula for calculating a confidence interval for the difference between two population means. It calculates the most commonly used confidence interval, the 95% confidence interval.

Equation 7.5 95% Confidence Interval for the Difference Between Population Means

$$95\%\ \mathrm{CI}\mu_{\mathrm{Diff}} = (M - \mu) \pm (t_{cv} \times s_M)$$

where $95\%\ \mathrm{CI}\mu_{\mathrm{Diff}}$ = the 95% confidence interval for the difference between two population means

M = sample mean from one population

μ = mean for other population

t_{cv} = critical value of t, two-tailed, $\alpha = .05$, $df = N - 1$ (Appendix Table 3)

s_M = estimated standard error of the mean (Equation 5.2)

Here are all the numbers Dr. Farshad needs to calculate the 95% confidence interval:

- M, the mean for the sample of adults with ADHD, is 220.
- μ, the population mean for adults in general, is 200.
- $s_M = 2.27$, a value obtained earlier via Equation 5.2 for use in Equation 7.2.
- t_{cv}, the critical value of t, is 1.977.

Applying Equation 7.5, she would calculate

$$\begin{aligned}95\%\ \text{CI}\mu_{\text{Diff}} &= (M - \mu) \pm (t_{cv} \times s_M)\\ &= (220 - 200) \pm (1.977 \times 2.27)\\ &= 20.0000 \pm 4.4878\\ &= \text{from } 15.5122 \text{ to } 24.4878\\ &= \text{from } 15.51 \text{ to } 24.49\end{aligned}$$

The 95% confidence interval for the difference between population means ranges from 15.51 msec to 24.49 msec. In APA format, this confidence interval would be reported as 95% CI [15.51, 24.49].

What information does a confidence interval provide? The technical definition is that if one drew sample after sample and calculated a 95% confidence interval for each one, 95% of the confidence intervals would capture the population value. But, that's not very useful as an interpretative statement. Two confidence interval experts have offered their thoughts on interpreting confidence intervals (Cumming and Finch, 2005). With a 95% confidence interval, one interpretation is that a researcher can be 95% confident that the population value falls within the interval. Another way of saying this is that the confidence interval gives a range of *plausible* values for the population value. That means it is possible, but unlikely, that the population value falls outside of the confidence interval. A final implication is that the ends of the confidence interval are reasonable estimates of how large or how small the population value is. The end of the confidence interval closer to zero can be taken as an estimate of how small the population value might be, while the end of the confidence interval further from zero estimates how large the population value might be.

Dr. Farshad's confidence interval tells her that there's a 95% chance that the difference between the two population means falls somewhere in the interval from 15.51 msec to 24.49 msec. It means that the best prediction of how much slower reaction times are for the population of adults with ADHD than for the general population of Americans ranges from 15.51 msec slower to 24.49 msec slower (Figure 7.9).

Our interpretation of confidence intervals will focus on three aspects: (1) whether the value of 0 falls within the confidence interval; (2) how close the confidence interval comes to zero or how far from zero it goes, and (3) how wide the confidence interval is. These three aspects are explained below and summarized in Table 7.4.

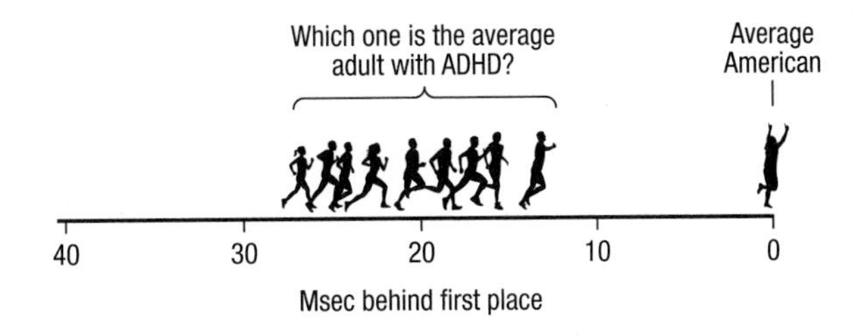

Figure 7.9 95% Confidence Interval for the Difference Between Population Means for the Reaction-Time Study Imagine a race in which a person representing the average American crosses the finish line first, followed by a person representing the average adult with ADHD. How much slower is the average adult with ADHD? The confidence interval says that the average adult with ADHD *probably* trails the average American by anywhere from 15.51 msec to 24.49 msec, but it doesn't specify where in that range the average adult with ADHD is.

TABLE 7.4 How to Choose: Interpreting a Confidence Interval for the Difference Between Population Means

	Confidence Interval Captures Zero	Confidence Interval Is Near Zero	Confidence Interval Is Far from Zero
Confidence Interval Is Narrow	There is not enough evidence to conclude an effect exists. A researcher can't say the two population means are different.	The effect is likely weak. It is plausible that the two population means are different. Cohen's *d* or r^2 represents the size of the effect.	The effect is likely strong. It is plausible that the two population means are different. Cohen's *d* or r^2 represents the size of the effect.
Confidence Interval Is Wide	There is not enough evidence to conclude an effect exists. There is little information about whether the two population means are different. Replicate the study with a larger sample.	The effect is likely weak to moderate. It is plausible that the two population means are different. Calculate Cohen's *d* for both ends of the CI. Replicate the study with a larger sample.	The effect is likely moderate to strong. It is plausible that the two population means are different. Calculate Cohen's *d* for both ends of the CI. Replicate the study with a larger sample.

Note: A narrow confidence interval gives a precise estimate of the size of the difference between population means. A wide confidence interval doesn't provide much guidance on whether the difference between population means is large or small. Whether the confidence interval is "near zero" or falls "far from zero" provides information regarding how large the effect might be in the population.

1. Does the confidence interval capture zero? The 95% confidence interval for the difference between population means gives the range within which it is likely that the actual difference between two population means lies. If the null hypothesis is rejected for a single-sample *t* test, the conclusion reached is that this sample probably did not come from that population; rather, that there are two populations with two different means. In such a situation, the difference between the two population means is not thought to be zero, and the confidence interval shouldn't include zero. When the null hypothesis is rejected, the 95% confidence interval won't capture zero. (This assumes that the researcher was using a two-tailed test with $\alpha = .05$.)

   However, when a researcher fails to reject the null hypothesis, the confidence interval should include zero in its range. All that a zero falling in the range of the confidence interval means is that it is *possible* the difference between the two population means is zero. It doesn't mean that the difference *is* zero, just that it may be. In a similar fashion, when one fails to reject the null hypothesis, the conclusion is that there isn't enough evidence to say the null hypothesis is wrong. It may be right, it may be wrong; the researcher just can't say.

   Having already determined earlier that the null hypothesis was rejected and the results were statistically significant, Dr. Farshad knew that her confidence interval wouldn't capture zero. And, as the interval ranged from 15.51 to 24.49, she was right.

2. How close does the confidence interval come to zero? How far away from zero does it go? If the confidence interval doesn't capture zero, how close one end of the confidence interval comes to zero can be thought of as providing information about how weak the effect may be. If one end of the confidence interval is very close to zero, then the effect size may be small, as there could be little difference between the population means.

   Similarly, how far away the other end of the confidence interval is from zero can be thought of as providing information about how strong the effect may be. The farther away from zero the confidence interval ranges, the larger the effect size may be.

3. How wide is the confidence interval? The width of the confidence interval also provides information. Narrower confidence intervals provide a more precise estimate of the population value and are more useful.

   When the confidence interval is wide, the size of the effect in the population is uncertain. It might be a small effect, a large effect, or anywhere in between. In these instances, replication of the study with a larger sample size will yield a narrower confidence interval.

   When evaluating the width of a confidence interval, one should take into account the variable being measured. A confidence interval that is 1 point wide would be wide if the variable were GPA, but narrow if the variable were SAT. Interpretation relies on our human common sense.

   Dr. Farshad's confidence interval ranges from 15.51 to 24.49. The width can be determined by subtracting the lower limit from the upper limit, as shown in Equation 7.6.

Equation 7.6 Formula for Calculating the Width of a Confidence Interval

$$CI_W = CI_{UL} - CI_{LL}$$

where CI_W = the width of the confidence interval
CI_{UL} = the upper limit of the confidence interval
CI_{LL} = the lower limit of the confidence interval

Applying Equation 7.6 to her data, Dr. Farshad calculates 24.49 – 15.51 = 8.98. The confidence interval is almost 9 msec wide. Is this wide or narrow?

Evaluating the width of a confidence interval often requires expertise. It takes a reaction-time researcher like Dr. Farshad to tell us whether a 9-msec range for a confidence interval is a narrow range and sufficiently precise. Here, given the width of the confidence interval, she feels little need to replicate the study with a larger sample size to narrow the confidence interval.

Putting It All Together

Dr. Farshad has completed her study using a single-sample *t* test to see if adults with ADHD had a reaction time that differed from that for the general population. By addressing the three questions in Table 7.5, Dr. Farshad has gathered all the pieces she needs to write an interpretation. There are four points she addresses in her interpretation:

1. Dr. Farshad starts with a brief explanation of the study.
2. She presents some facts, such as the means of the sample and the population. But, she does not report all the values she calculated just because she calculated them. Rather, she is selective and only reports what she believes is most relevant.
3. She explains what she believes the results mean.
4. Finally, she offers some suggestions for future research. Having been involved in the study from start to finish, she knows its strengths and weaknesses better than anyone else. She is in a perfect position to offer advice to other researchers about ways to redress the limitations of her study. In her suggestions, she uses the word "replicate." To **replicate** a study is to repeat it, usually introducing some change in procedure to make it better.

TABLE 7.5 Three Questions for Interpreting a Single-Sample t Test

1. Was the null hypothesis rejected?
   - Decide by comparing the calculated value of t to the critical value of t, t_{cv}, using the decision rule generated in Step 4.
   - Was H_0 rejected?
     - If yes, (1) call the results statistically significant, and (2) compare M to μ to determine the direction of the difference.
     - If no, (1) say the results are not statistically significant, and (2) conclude there is not enough evidence to say a difference exists.
   - Report the results in APA format:
     - If H_0 is rejected, report the results as "$p < .05$."
     - If H_0 is not rejected, report the results as "$p > .05$."
2. How big is the effect?
   - Calculate Cohen's d (Equation 7.3) or r^2 (Equation 7.4).
     - No effect: $d = 0.00$; $r^2 = 0$.
     - Small effect: $d = 0.20$; $r^2 = 1\%$.
     - Medium effect: $d = 0.50$; $r^2 = 9\%$.
     - Large effect: $d \geq 0.80$; $r^2 \geq 25\%$.
3. How wide is the confidence interval?
   - Calculate the 95% confidence interval for the difference between population means (Equation 7.5).
     - Interpret the confidence interval based on (1) whether it captures zero, (2) how close to/far from zero it comes, and (3) how wide it is. (See Table 7.4.)

Here is Dr. Farshad's interpretation:

A study compared the reaction time of a random sample of Illinois adults who had been diagnosed with ADHD ($M = 220$ msec) to the known reaction time for the American population ($\mu = 200$ msec). The reaction time of adults with ADHD was statistically significantly slower than the reaction time found in the general population [$t(140) = 8.81$, $p < .05$]. The size of the difference in the larger population probably ranges on this task from a 16- to a 24-msec decrement in performance. This is not a small difference—these results suggest that ADHD in adults is associated with a medium to large level of impairment on tasks that require fast reactions. If one were to replicate this study, it would be advisable to obtain a broader sample of adults with ADHD, not just limiting it to one state. This would increase the generalizability of the results.

Worked Example 7.2

For practice in interpreting the results of a single-sample t test when the null hypothesis is not rejected, let's return to Dr. Richman's study.

In his study, Dr. Richman studied the effect of losing a pet during the year on college performance. Dr. Richman located 11 students who had lost a pet. Their mean GPA for the year was 2.58 ($s = 0.50$) compared to a population mean of 2.68. Because he expected pet loss to have a negative effect, Dr. Richman used a one-tailed test. With the alpha set at .05, t_{cv} was -1.812. He calculated $s_M = 0.15$ and found $t = -0.67$. Now it is time for Dr. Richman to interpret the results.

Was the null hypothesis rejected? The vet was doing a one-tailed test and his hypotheses were

$$H_0: \mu_{\text{StudentsWithPetLoss}} \geq 2.68$$
$$H_1: \mu_{\text{StudentsWithPetLoss}} < 2.68$$

Inserting the observed value of *t*, −0.67, into the decision rule he had generated in Step 4, he has to decide which statement is true:

Is $-0.67 \leq -1.812$? If so, reject H_0.
Is $-0.67 > -1.812$? If so, fail to reject H_0.

As the second statement is true, Dr. Richman has failed to reject the null hypothesis. Figure 7.10 shows how the *t* value of −0.67 falls in the common zone of the sampling distribution.

Having failed to reject the null hypothesis, the results are called not statistically significant. Just like finding a defendant not guilty doesn't mean that the defendant is innocent, failing to reject the null hypothesis does not mean that it is true. All the vet can conclude is that there's not enough evidence to conclude that college students who lose a pet do worse in school that year. Because he failed to reject the null hypothesis, there's no reason to believe a difference exists between the two populations, and there's no reason for him to worry about the direction of the difference between them.

In APA format, the results would be written as $t(10) = -0.67, p > .05$ (one-tailed):

- The *t* tells what statistical test was done.
- The 10 in parentheses, which is the degrees of freedom, reveals there were 11 cases as $N = df + 1$ for a single-sample *t* test.
- −0.67 is the observed value of the statistic.
- The .05 indicates that this was the alpha level selected.
- $p > .05$ indicates that the researcher failed to reject the null hypothesis. The observed *t* value is a common one when the null hypothesis is true. "Common" is defined as occurring more than 5% of the time.

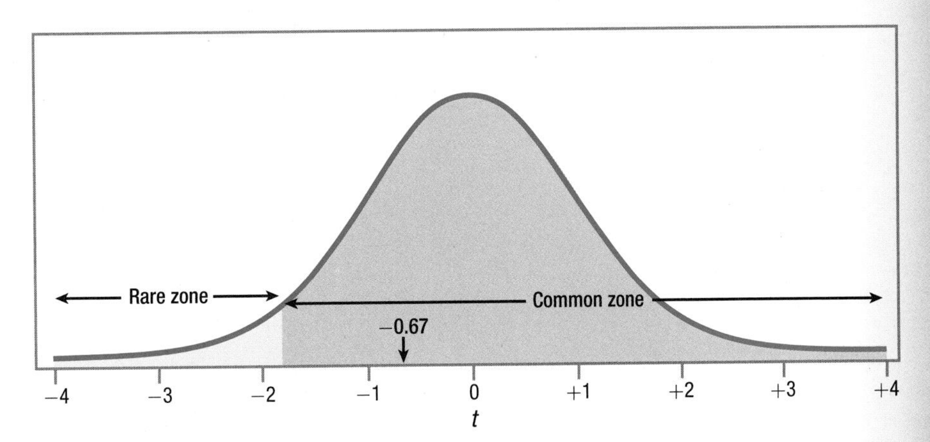

Figure 7.10 Failing to Reject the Null Hypothesis, One-Tailed Test The critical value of *t* (one-tailed, $\alpha = .05$, $df = 10$) is −1.812. This means that the value of the test statistic, −0.67, falls in the common zone, and the null hypothesis is not rejected.

- The final parenthetical expression (one-tailed) is something new. It indicates, obviously, that the test was a one-tailed test. Unless told that the test is a one-tailed test, assume it is two-tailed.

How big is the effect? Calculating an effect size like Cohen's d or r^2 when one has failed to reject the null hypothesis is a controversial topic in statistics. If one fails to reject the null hypothesis, not enough evidence exists to state there's an effect. Some researchers say that there is no need to calculate an effect size if no evidence that an effect exists is found. Other researchers believe that examining the effect size when one has failed to reject the null hypothesis alerts the researcher to the possibility of a Type II error (Cohen, 1994; Wilkinson, 1999). (Remember, a Type II error occurs when the null hypothesis should be rejected, but it isn't.) In this book, we're going to side with calculating an effect size when one fails to reject the null hypothesis as doing so gives additional information useful in understanding the results.

In the pet-loss study, the mean GPA for the 11 students who had lost a pet was 2.58 ($s = 0.50$) and the population mean for all first-year students was 2.68. Applying Equation 7.3, here are the calculations for d:

$$d = \frac{M - \mu}{s}$$

$$= \frac{2.58 - 2.68}{0.50}$$

$$= \frac{-0.1000}{0.50}$$

$$= -0.2000$$

$$= -0.20$$

And, applying Equation 7.4, here are the calculations for r^2:

$$r^2 = \frac{t^2}{t^2 + df} \times 100$$

$$= \frac{-0.67^2}{-0.67^2 + 10} \times 100$$

$$= \frac{0.4489}{0.4489 + 10} \times 100$$

$$= \frac{0.4489}{10.4489} \times 100$$

$$= 0.0430 \times 100$$

$$= 4.30$$

The effect size d, -0.20, is a small effect according to Cohen, while r^2 of 4.30% would be classified as a small to medium effect. Why, when we fail to reject the null hypothesis, does there seem to be some effect? There are two explanations for this:

1. Even if the null hypothesis is true, in which case the size of the effect in the population is zero, it is possible, due to sampling error, that a nonzero effect would be observed in the sample. So, it shouldn't be surprising to find a small effect.

2. It is also possible that the observed effect size represents the size of the effect in the population. If this were true, it would mean a Type II error is being made.

When a researcher has concern about Type II error, he or she can suggest replication with a larger sample size. A larger sample size has a number of benefits. First, it makes it easier to reject the null hypothesis because a larger sample size, with more degrees of freedom, moves the rare zone closer to the midpoint. It is also easier to reject the null hypothesis because a larger sample makes the standard error of the mean smaller, which makes the t value larger. In fact, if the sample size had been 75 in the pet-loss study, not 11, d would still be -0.20, but the vet would have rejected the null hypothesis.

A statistician would call this study underpowered. To be **underpowered** means that a study doesn't include enough cases to have a reasonable chance of detecting an effect that exists. As was mentioned in Chapter 6, if power is low, then beta, the chance of a Type II error, is high. So, raising a concern about one (Type II error) raises a concern about the other (power).

How wide is the confidence interval? The final calculation Dr. Richman should make in order to understand his results is to calculate a 95% confidence interval for the difference between population means. This requires Equation 7.5:

$$\begin{aligned}
95\%\ \text{CI}\mu_{\text{Diff}} &= (M - \mu) \pm (t_{cv} \times s_M) \\
&= (2.58 - 2.68) \pm (2.228 \times 0.15) \\
&= -0.1000 \pm 0.3342 \\
&= \text{from } -0.4342 \text{ to } 0.2342 \\
&= \text{from } -0.43 \text{ to } 0.23
\end{aligned}$$

Before interpreting this confidence interval, remember that the null hypothesis was not rejected and there is no evidence that pet loss has an effect on GPA. The confidence interval should show this, and it does.

This 95% confidence interval ranges from -0.43 to 0.23. First, note that it includes the value of 0. A value of 0 indicates no difference between the two population means and so it is a *possibility*, like the null hypothesis said, that there is no difference between the two populations. The confidence interval for the difference between population means should capture zero whenever the null hypothesis is not rejected. (This is true for a 95% confidence interval as long as the hypothesis test is two-tailed with $\alpha = .05$.)

Next, Dr. Richman looks at the upper and lower ends of the confidence interval, -0.43 and 0.23. This confidence interval tells him that it is possible the average GPA for the population of students who have lost a pet could be as low as 0.43 points worse than the general student average, or as high as 0.23 points better than the general student average. This doesn't provide much information about the effect of pet loss on academic performance—maybe it helps, maybe it hurts.

Finally, he looks at the width of the confidence interval using Equation 7.6:

$$\begin{aligned}
CI_W &= CI_{UL} - CI_{LL} \\
&= 0.23 - (-0.43) \\
&= 0.23 + 0.43 \\
&= 0.6600 \\
&= 0.66
\end{aligned}$$

The confidence interval is 0.66 points wide. This is a fairly wide confidence interval for a variable like GPA, a variable that has a maximum range of 4 points.

A confidence interval is better when it is narrower, because a narrower one gives a more precise estimate of the population value. How does a researcher narrow a confidence interval? By increasing sample size. Dr. Richman should recommend replicating with a larger sample size to get a better sense of what the effect of pet loss is in the larger population. Increasing the sample size also increases the power, making it more likely that an effect will be found statistically significant.

Putting it all together. Here's Dr. Richman's interpretation. Note that (1) he starts with a brief explanation of the study, (2) reports some facts but doesn't report everything he calculated, (3) gives his interpretation of the results, and (4) makes suggestions for improving the study.

This study explored whether losing a pet while at college had a negative impact on academic performance. The GPA of 11 students who lost a pet ($M = 2.58$) was compared to the GPA of the population of students at that college ($\mu = 2.68$). The students who had lost a pet did not have GPAs that were statistically significantly lower [$t(10) = -0.67$, $p > .05$ (one-tailed)].

Though there was not sufficient evidence in this study to show that loss of a pet has a negative impact on college performance, the sample size was small and the study did not have enough power to find a small effect. Therefore, it would be advisable to replicate the study with a larger sample size to have a better chance of determining the effect of pet loss and to get a better estimate of the size of the effect.

Practice Problems 7.4

Apply Your Knowledge

7.18 If $M = 70$, $\mu = 60$, $t_{cv} = \pm 2.086$, and $s_M = 4.36$, what is the 95% confidence interval for the difference between population means?

7.19 If $M = 55$, $\mu = 50$, $t_{cv} = \pm 2.052$, and $s_M = 1.89$, what is the 95% confidence interval for the difference between population means?

7.20 A college president has obtained a sample of 81 students at her school. She plans to survey them regarding some potential changes in academic policies. But, first, she wants to make sure that the sample is representative of the school in terms of academic performance. She knows from the registrar that the mean GPA for the entire school is 3.02. She calculates the mean of her sample as 3.16, with a standard deviation of 0.36. She used a single-sample *t* test to compare the sample mean (3.16) to the population mean (3.02). Using her findings below, determine if the sample is representative of the population in terms of academic performance:

- $s_M = 0.04$
- $t = 3.50$
- $d = 0.39$
- $r^2 = 13.28\%$
- 95% $CI\mu_{Diff}$ = from 0.06 to 0.22

Application Demonstration

Chips Ahoy is a great cookie. In 1997 Nabisco came out with a clever advertising campaign, the Chips Ahoy Challenge. Its cookies had so many chocolate chips that Nabisco guaranteed there were more than a thousand chips in every bag. And the company challenged consumers to count. It's almost two decade later, but do Chips Ahoy cookies still have more than a thousand chocolate chips in every bag? Let's investigate.

It is easier to count the number of chips in a single cookie than to count the number of chips in a whole bag of cookies. If a bag has at least 1,000 chips and there are 36 cookies in a bag, then each cookie should have an average of 27.78 chips. So, here is the challenge rephrased: "Chips Ahoy cookies are so full of chocolate chips that each one contains an average of 27.78 chips. We challenge you to count them."

Step 1 This challenge calls for a statistical test. The population value is known: $\mu = 27.78$. All that is left is to obtain a sample of cookies; find the sample mean, M; and see if there is a statistically significant difference between M and μ. Unfortunately, Nabisco has not reported σ, the population standard deviation, so the single-sample z test can't be used. However, it is possible to calculate the standard deviation (s) from a sample, which means the single-sample t test can be used.

Ten cookies were taken from a bag of Chips Ahoy cookies. Each cookie was soaked in cold water, the cookie part washed away, and the chips counted. The cookies contained from 22 to 29 chips, with a mean of 26.10 and a standard deviation of 2.69.

Step 2 The next step is to check the assumptions. The random sample assumption was violated as the sample was not a random sample from the population of Chips Ahoy cookies being manufactured. Perhaps the cookies being manufactured on the day this bag was produced were odd. Given the focus on quality control that a manufacturer like Nabisco maintains, this seems unlikely. So, though this assumption is violated, it still seems reasonable to generalize the results to the larger population of Chips Ahoy cookies.

We'll operationalize the independence of observations assumption as most researchers do and consider it not violated as no case is in the sample twice and each cookie is assessed individually. Similarly, the normality assumption will be considered not violated under the belief that characteristics controlled by random processes are normally distributed.

Step 3 The hypotheses should be formulated before any data are collected. It seems reasonable to do a nondirectional (two-tailed) test in order to allow for the possibilities of more chips than promised as well as fewer chips than promised. Here are the two hypotheses:

$$H_0: \mu_{\text{Chips}} = 27.28$$

$$H_1: \mu_{\text{Chips}} \neq 27.78$$

Step 4 The consequences of making a Type I error in this study are not catastrophic, so it seems reasonable to use the traditional alpha level of .05. As already decided, a two-tailed test is being used. Thus, the critical value of t, with 9 degrees of freedom, is ± 2.262. Here is the decision rule:

If $t \leq -2.262$ *or* if $t \geq 2.262$, reject H_0.

If $-2.262 < t < 2.262$, fail to reject H_0.

Step 5 The first step in calculating a single-sample t test is to calculate what will be used as the denominator, the estimated standard error of the mean:

$$s_M = \frac{s}{\sqrt{N}} = \frac{2.69}{\sqrt{10}} = 0.85$$

The estimated standard error of the mean is then used to calculate the test statistic, t:

$$t = \frac{M - \mu}{s_M} = \frac{26.10 - 27.78}{0.85} = -1.98$$

Step 6 The first step in interpreting the results is to determine if the null hypothesis should be rejected. The second statement in the decision rule is true: $-2.262 < -1.98 < 2.262$, the null hypothesis is not rejected. The difference between 26.10, the mean number of chips found in the cookies, and 27.78, the number of chips expected per cookie, is not statistically significant. From this study, even though the cookies in the bag at $M = 26.10$ chips per cookie fell short of the expected 27.78 chips per cookie, not enough evidence exists to question Nabisco's assertion that there are 27.78 chips per cookie and a thousand chips in every bag. And, there's not enough evidence to suggest that, since 1997, Nabisco's recipe has changed and/or its quality control has slipped.

Having answered the challenge, it is tempting to stop here. But forging ahead—going on to find Cohen's d, r^2, and the 95% confidence interval—will give additional information and help clarify the results.

Finding Cohen's d, the size of the effect, provides a different perspective on the results:

$$\begin{aligned} d &= \frac{M - \mu}{s} \\ &= \frac{26.10 - 27.78}{2.69} \\ &= \frac{-1,6800}{2.69} \\ &= -0.6245 \\ &= -0.62 \end{aligned}$$

Though the t test says there is not enough evidence to find an effect, Cohen's d says $d = -0.62$, a medium effect. This seems contradictory—is there an effect or isn't there? Perhaps the study doesn't have enough power and a Type II error is being made. This means that the study should be replicated with a larger sample size before concluding that Chips Ahoy hasn't failed the thousand chip challenge. Calculating $r^2 = 30\%$ leads to a similar conclusion.

Does the confidence interval tell a similar story? Here are the calculations for the confidence interval:

$$\begin{aligned} 95\%\ \text{CI}\mu_{\text{Diff}} &= (M - \mu) \pm (t_{cv} \times s_M) \\ &= (26.10 - 27.78) \pm (2.262 \times 0.85) \\ &= -1.6800 \pm 1.9227 \\ &= \text{from } -3.6027 \text{ to } 0.2427 \\ &= \text{from } -3.60 \text{ to } 0.24 \end{aligned}$$

The confidence interval says that the range from −3.60 to 0.24 probably contains the difference between 27.78 (the expected number of chips per cookie) and the number really found in each cookie. As this range captures zero, it is possible that there is no difference, the null hypothesis is true, and bags of cookies really do contain 1,000 chips.

Positive numbers in the confidence interval suggest that the difference lies in the direction of there being more than 27.78 chips per cookie. Negative numbers suggest the difference is in the direction of there being fewer than 27.78 chips per cookie. Most of the confidence interval lies in negative territory. A betting person would wager that the difference is more likely to rest with bags containing fewer than a thousand chips.

The *t* test left open the possibility that Nabisco still ruled the 1,000 chip challenge. But, thanks to going beyond *t*, to calculating *d* and a confidence interval, it no longer seems clear that Nabisco would win the Chips Ahoy challenge. For a complete interpretation of the results of this Chips Ahoy challenge, see **Figure 7.11**.

PENNSTATE

Erie The Behrend College

School of Humanities and Social Sciences
Penn State Erie, The Behrend College
170 Irvin Kochel Center
4951 College Drive
Erie, PA 16563-1501

814-898-6108
Fax: 814-898-6032
behrend.psu.edu

Irene Rosenfeld
Chairman and Chief Executive Officer, Kraft Foods
3 Lakes Drive
Northfield, IL 60093

June 1, 2012

Dear Ms. Rosenfeld:

I am writing to you because Kraft Foods owns Nabisco, the maker of Chips Ahoy cookies. My letter contains both good news and bad news

Back in 1997, Nabisco issued the Chips Ahoy Challenge. They said that each bag of Chips Ahoy cookies contained at least 1,000 chips and they dared consumers to count.

I'm a college professor, working on a revision of a statistics textbook, and I decided to use this challenge as an application of a single-sample *t* test. I was curious if 15 years later there were still 1,000 chips per bag.

I bought a bag of Chips Ahoy cookies, found that it contained 36 cookies, and calculated that if each cookie contained an average of 27.78 chips, the bag would contain 1,000 chips. I then selected ten cookies, dissolved them in water, and counted the number of chips in each. I found a mean of 26.10 chips with a standard deviation of 2.69.

First, the good news. Using a single-sample *t* test, I found that 26.10 was not statistically, significantly, different from 27.78. In plain language, this means that, based on this study, it's plausible that a bag of Chips Ahoy cookies contains 1,000 chips.

Unfortunately, bad news follows the good. By calculating something called an effect size and something called a confidence interval, I became concerned that my conclusion that the difference was not statistically significant was erroneous. My study was what statisticians call "underpowered." It didn't have enough cookies in the sample to have a fair chance of finding a difference if there were a difference. With more cookies in my sample, I probably would have found that it is unlikely that bags contains 1,000 chips.

I plan to replicate the study with a larger sample size.

But, until I do, I am curious – do Kraft and Nabisco still stand behind the Chips Ahoy Challenge?

Sincerely,

Eric W. Corty, Ph.D.
Professor of Psychology

An Equal Opportunity University

Figure 7.11 Interpretation of Results of the Chips Ahoy Challenge

SUMMARY

Choose when to use a single-sample *t* test.

- A single-sample *t* test is used to compare a sample mean to some specified value, like a population mean, when the population standard deviation is not known. Like all statistical tests, it has assumptions that must be met, hypotheses to be listed, and a decision rule to be set in advance of calculating the value of the test statistic.

Calculate the test statistic for a single-sample *t* test.

- The single-sample *t* test calculates a *t* value. *t* is distributed very much like *z*—if the null hypothesis for a two-tailed test is true, the *t* distribution is symmetrical and centered on zero. The *t* value is based on the size of the difference between the sample mean and the specified value. If the difference, when standardized as a *t* score, is too large to be explained by sampling error, the null hypothesis is rejected.

Interpret the results of a single-sample *t* test.

- Interpretation explains the results of a statistical test in plain language. An interpretation is subjective, but it is based on facts. Interpretation proceeds by asking questions of the data (e.g., was the null hypothesis rejected, how big is the effect) and then using the answers to address four points: (1) what was the study about, (2) what were the results, (3) what do the results mean, and (4) what should be done in future research.
- Cohen's *d* and r^2 were introduced in this chapter and confidence intervals made a return appearance. All can be used to measure the size of the effect, the impact of the explanatory variable on the outcome variable. A confidence interval uses a sample value to estimate the range within which a population value falls; narrower confidence intervals give a more precise estimate. Cohen's *d* takes the difference between two means and standardizes it. Cohen has suggested *d* values, for the social and behavioral sciences, representing small, medium, and large effects. r^2 tells how much of the variability in the outcome variable is explained by or accounted for by, the explanatory variable.

DIY

Every year the Centers for Disease Control (CDC) conducts a survey of health practices and risk behaviors in American adults. The survey is called the BRFSS, the Behavioral Risk Factor Surveillance System. In 2013, the most recent year for which data are available, the sample consisted of almost 500,000 Americans, 18 and older, from all 50 states, Washington, D.C., Guam, and Puerto Rico.

I picked three variables in the data set—number of hours of sleep per night, weight in pounds, and number of alcoholic drinks consumed per day during the past month, on days that at least one drink was consumed—and found the mean for each for men, for women, and for both sexes combined.

	Hours of Sleep	Weight in Pounds	Drinks per Drinking Day
Men and women	7.05	176.54	2.21
Men	7.03	196.88	2.66
Women	7.06	161.79	1.80

Pick one of these variables and survey about 10 of your friends. Calculate the mean and the standard deviation. Then complete a single-sample *t* test to see whether your sample differs in its behavior from the U.S. population.

KEY TERMS

Cohen's *d* – a standardized measure of effect used to measure the difference between means.

critical value of *t* – value of *t* used to determine whether a null hypothesis is rejected or not; abbreviated t_{cv}.

degrees of freedom (*df*) – the number of values in a sample that are free to vary.

effect size (*d*) – a measure of the degree of impact of the independent variable on the dependent variable.

replicate – to repeat a study, usually introducing some change in procedure to make it better.

r^2 – an effect size that calculates how much of the variability in the outcome variable is accounted for by the predictor variable.

single-sample *t* test – a statistical test that compares a sample mean to a population mean when the population standard deviation is not known.

underpowered – term for a study with a sample size too small for the study to have a reasonable chance to reject the null hypothesis given the size of the effect.

CHAPTER EXERCISES

Answers to the odd-numbered exercises appear at the back of the book.

Review Your Knowledge

7.01 In order to use a single-sample *t* test, one *does / does not* need to know the population standard deviation.

7.02 The single-sample *t* test compares a sample ____ to a population ____.

7.03 A sample, selected at random from a population, may have a sample mean that differs from the population mean due to ____.

7.04 Tom ____ Harry ____ ____ infants.

7.05 One assumption of the single-sample *t* test is that the sample is a random sample from the population. This *is / is not* robust to violation.

7.06 The population that the sample comes from determines the population to which the results can be ____.

7.07 A second assumption of the single-sample *t* test is that observations within the sample are ____.

7.08 The third assumption of the single-sample *t* test is called the ____ assumption for short.

7.09 If a ____ assumption is violated, a researcher can still proceed with the test as long as the violation is not too great.

7.10 In order to write the null and alternative hypotheses for a single-sample *t* test, the researcher needs to know whether the test has one or two ____.

7.11 The default option in hypothesis testing is a ____-tailed test.

7.12 If a researcher is doing a one-tailed test, he or she should predict the ____ of the results before collecting any data.

7.13 The null hypothesis for a nondirectional, single-sample *t* test says that the sample *does / does not* come from the population.

7.14 The null hypothesis for a two-tailed, single-sample *t* test could be written as ____ $= \mu_2$.

7.15 The alternative hypothesis for a two-tailed, single-sample *t* test could be written as μ_1 ____ μ_2.

7.16 If the null hypothesis for a two-tailed, single-sample *t* test is true, $t =$ ____.

7.17 As the distance between the sample mean and the population mean grows, the value of t ____.

7.18 The abbreviation for the critical value of t is ____.

7.19 Determining the critical value depends on (a) how many ____ the test has, (b) how willing one is to make a Type ____ error, and (c) the ____.

7.20 If the hypotheses are nondirectional, then the researcher is doing a ____-tailed test.

7.21 If the alternative hypothesis is $\mu > 173$, then the test is a ____-tailed test.

7.22 If the researcher wants a 5% chance of Type I error, then alpha is set at ____.

7.23 For a two-tailed test with $\alpha = .05$, the rare zone has ____% of the sampling distribution in each tail.

7.24 When the sample size is large, the rare zone gets ____ and it is ____ to reject the null hypothesis.

7.25 Degrees of freedom, for a single-sample t test, equal ____ minus 1.

7.26 For a two-tailed, single-sample t test, the null hypothesis is rejected if $t \leq$ ____ or if $t \geq$ ____.

7.27 t for a single-sample t test is calculated by dividing the difference between M and μ by ____.

7.28 Interpretation uses human ____ to make meaning out of the results.

7.29 Interpretation is subjective, but needs to be supported by ____.

7.30 The first question asked in interpretation is about whether the ____ hypothesis is ____.

7.31 For the second interpretation question, one calculates ____, and for the third, one calculates a ____.

7.32 A decision is made about rejecting the null hypothesis by comparing the ____ value of t to the value calculated from the sample mean.

7.33 If a researcher rejects the null hypothesis, then he or she is forced to accept the ____.

7.34 If the null hypothesis is rejected, it is concluded that the mean for the population the ____ came from differs from the hypothesized value.

7.35 To determine the direction of a statistically significant difference, compare the ____ to the ____.

7.36 Sample size is reported in APA format for a single-sample t test by reporting ____.

7.37 APA format uses the inequality ____ to indicate that the null hypothesis was rejected when $\alpha = .05$.

7.38 If a result is reported in APA format as $p < .05$, that means the observed value of the test statistic fell in the ____ zone.

7.39 If one fails to reject the null hypothesis, one can say that there is ____ to conclude a difference exists between the population means.

7.40 APA format uses the inequality ____ to indicate that the null hypothesis was not rejected when $\alpha = .05$.

7.41 Effect sizes are used to quantify the impact of the ____ on the ____.

7.42 Cohen's d is an ____.

7.43 A value of 0 for Cohen's d means that the independent variable had ____ impact on the dependent variable.

7.44 Cohen considers a d of ____ a small effect, ____ a medium effect, and ____ or higher a large effect.

7.45 As the effect size d increases, the degree of overlap between the distributions for two populations ____.

7.46 When one fails to reject the null hypothesis, one should ____ calculate Cohen's d.

7.47 Calculating d when one has failed to reject the null hypothesis alerts one to the possibility of a Type ____ error.

7.48 To ____ a study is to repeat it.

7.49 r^2 measures how much variability in the ____ is explained by the ____.

7.50 d and r^2 should lead to similar conclusions about the size of an effect. Though both may have been calculated, it is ____ to report both.

7.51 If a researcher calculates a 95% confidence interval, he or she can be ____% confident that it captures the ____ value.

7.52 The 95% confidence interval for the difference between population means tells how far apart or how close the two ____ means might be.

7.53 The size of the difference between population means can be thought of as another ____.

7.54 If a 95% confidence interval for the difference between population means does not capture zero, and the researcher is doing a two-tailed test with $\alpha = .05$, then the null hypothesis *was / was not* rejected.

7.55 If the 95% confidence interval for the difference between population means falls close to zero, this means the size of the effect may be ____.

7.56 A wide 95% confidence interval for the difference between population means leaves a researcher unsure of ____.

Apply Your Knowledge

Picking the right test

7.57 A researcher has a sample ($N = 38$, $M = 35$, $s = 7$) that he thinks came from a population where $\mu = 42$. What statistical test should he use?

7.58 A researcher has a sample ($N = 52$, $M = 17$, $s = 3$) that she believes came from a population where $\mu = 20$ and $\sigma = 4$. What statistical test should she use?

Checking the assumptions

7.59 A researcher wants to compare the mean weight of a convenience sample of students from a college to the national mean weight of 18- to 22-year-olds. (a) Check the assumptions and decide whether it is OK to proceed with a single-sample *t* test. (b) Can the researcher generalize the results to all the students at the college?

7.60 There is a random sample of students from a large public high school. Each person, individually, takes a paper-and-pencil measure of introversion. (a) Check the assumptions and decide whether it is OK to proceed with a single-sample *t* test to compare the sample mean to the population mean of introversion for U.S. teenagers. (b) To what population can one generalize the results? (The same introversion measure was used for the national sample.)

Writing nondirectional hypotheses

7.61 The population mean on a test of paranoia is 25. A psychologist obtained a random sample of nuns and found $M = 22$. Write the null and alternative hypotheses.

7.62 A researcher wants to compare a random sample of left-handed people in terms of IQ to the population mean of IQ. Given $M = 108$ and assuming $\mu = 100$, write the null and alternative hypotheses.

Writing directional hypotheses

7.63 In America, the average length of time the flu lasts is 6.30 days. An infectious disease physician has developed a treatment that he believes will treat the flu more quickly. Write the null and alternative hypotheses.

7.64 An SAT-tutoring company claims that its students perform above the national average on SAT subtests. If the national average on SAT subtests is 500 and the tutoring company obtains SAT scores from a random sample of 626 of its students, write the null and alternative hypotheses.

Finding t_{cv} (assume the test is two-tailed and alpha is set at .05)

7.65 If $N = 17$, find t_{cv} and use it to draw a *t* distribution with the rare and common zones labeled.

7.66 If $N = 48$, find t_{cv} and use it to draw a *t* distribution with the rare and common zones labeled.

Writing the decision rule (assume the test is two-tailed and alpha is set at .05)

7.67 If $N = 64$, write the decision rules for a single-sample *t* test.

7.68 If $N = 56$, write the decision rules for a single-sample t test.

Calculating s_M

7.69 If $N = 23$ and $s = 12$, calculate s_M.

7.70 If $N = 44$ and $s = 7$, calculate s_M.

Given s_M, calculating t

7.71 If $M = 10$, $\mu = 12$, and $s_M = 1.25$, what is t?

7.72 If $M = 8$, $\mu = 6$, and $s_M = 0.68$, what is t?

Calculating t

7.73 If $N = 18$, $M = 12$, $\mu = 10$, and $s = 1$, what is t?

7.74 If $N = 25$, $M = 18$, $\mu = 13$, and $s = 2$, what is t?

Was the null hypothesis rejected? (Assume the test is two-tailed.)

7.75 If $t_{cv} = \pm 2.012$ and $t = -8.31$, is the null hypothesis rejected?

7.76 If $t_{cv} = \pm 2.030$ and $t = 2.16$, is the null hypothesis rejected?

7.77 If $t_{cv} = \pm 2.776$ and $t = 1.12$, is the null hypothesis rejected?

7.78 If $t_{cv} = \pm 1.984$ and $t = -1.00$, is the null hypothesis rejected?

Writing results in APA format (Assume the test is two-tailed and alpha is set at .05.)

7.79 Given $N = 15$ and $t = 2.145$, write the results in APA format.

7.80 Given $N = 28$ and $t = 2.050$, write the results in APA format.

7.81 Given $N = 69$ and $t = 1.992$, write the results in APA format.

7.82 Given $N = 84$ and $t = 1.998$, write the results in APA format.

Calculating Cohen's d

7.83 Given $M = 90$, $\mu = 100$, and $s = 15$, calculate Cohen's d.

7.84 Given $M = 98$, $\mu = 100$, and $s = 15$, calculate Cohen's d.

Calculating r^2

7.85 Given $N = 17$ and $t = 3.45$, what is r^2?

7.86 If $N = 29$ and $t = 1.64$, what is r^2?

Calculating a 95% confidence interval

7.87 Given $M = 45$, $\mu = 50$, $t_{cv} = \pm 2.093$, and $s_M = 2.24$, calculate the 95% confidence interval for the difference between population means.

7.88 Given $M = 55$, $\mu = 50$, $t_{cv} = \pm 2.093$, and $s_M = 2.24$, calculate the 95% confidence interval for the difference between population means.

Given effect size and confidence interval, interpret the results. Be sure to (1) tell what was done, (2) present some facts, (3) interpret the results, and (4) make a suggestion for future research. (Assume the test is two-tailed and $\alpha = .05$.)

7.89 Given the supplied information, interpret the results. A nurse practitioner has compared the blood pressure of a sample ($N = 24$) of people who are heavy salt users ($M = 138$, $s = 16$) to blood pressure in the general population ($\mu = 120$) to see if high salt consumption were related to raised or lowered blood pressure. She found:

- $t_{cv} = 2.069$
- $t = 5.50$
- $d = 1.13$
- $r^2 = 57\%$
- 95% CI [11.23, 24.77]

7.90 Given the supplied information, interpret the results. A dean compared the GPA of a sample ($N = 20$) of students who spent more than two hours a night on homework ($M = 3.20$, $s = 0.50$) to the average GPA at her college ($\mu = 2.80$). She was curious if homework had any relationship, either positive or negative, with GPA. She found:

- $t_{cv} = 2.093$
- $t = 3.64$
- $d = 0.80$
- $r^2 = 41\%$
- 95% CI [0.17, 0.63]

Completing all six steps of a hypothesis test

7.91 An educational psychologist was interested in time management by students. She had a theory that students who did well in school spent less time involved with online social media. She found out, from the American Social Media

Research Collective, that the average American high school student spends 18.68 hours per week using online social media. She then obtained a sample of 31 students, each of whom had been named the valedictorian of his or her high school. These valedictorians spent an average of 16.24 hours using online social media every week. Their standard deviation was 6.80. Complete the analyses and write a paragraph of interpretation.

7.92 A psychology professor was curious how psychology majors fared economically compared to business majors. Did they do better, or did they do worse five years after graduation? She did some research and learned that the national mean for the salary of business majors five years after graduation was $55,000. She surveyed 22 recent psychology graduates at her school and found $M = \$43,000$, $s = 12,000$. Complete the analyses and write a paragraph of interpretation.

Expand Your Knowledge

7.93 Imagine a researcher has taken two random samples from two populations (A and B). Each sample is the same size ($N = 71$), has the same sample mean ($M = 50$), and comes from a population with the same mean ($\mu = 52$). The two populations differ in how much variability exists. As a result, one sample has a smaller standard deviation ($s = 2$) than the other ($s = 12$). The researcher went on to calculate single-sample t values for each sample. Based on the information provided below, how does the size of the sample standard deviation affect the results of a single-sample t test?

	s_M	t_{cv}	t	d	r^2	CI	Width of CI
A: Less variability ($s = 2$)	0.24	1.994	−8.33	−1.00	50%	−2.48 to −1.52	0.96
B: More variability ($s = 12$)	1.42	1.994	−1.41	−0.17	3%	−4.83 to 0.83	5.66

7.94 Another researcher selected two random samples from one population. This population has a mean of 63 ($\mu = 63$). It turned out that each sample had the same mean ($M = 60$) and standard deviation ($s = 5$). The only way the two samples differed was in terms of size: one, C, was smaller ($N = 10$) and one, D, was larger ($N = 50$). The researcher went on to conduct a single-sample t test for each sample. Based on the information provided below, how does sample size affect the results of a single-sample t test?

	s_M	t_{cv}	t	d	r^2	CI	Width of CI
C: Smaller N ($N = 10$)	1.58	2.262	−1.90	−0.60	29%	−6.57 to 0.57	7.14
D: Larger N ($N = 50$)	0.71	2.010	−4.23	−0.60	27%	−4.43 to −1.57	2.86

7.95 A third researcher obtained two random samples from another population. The mean for this population is 50 ($\mu = 50$). Each sample was the same size ($N = 10$) and each had the same standard deviation ($s = 20$). But one, sample E, had a mean of 80 ($M = 80$) and the other, sample F, a mean of 60 ($M = 60$). Based on the information provided below, how does the distance from the sample mean to the population mean affect the results of a single-sample t test?

	s_M	t_{cv}	t	d	r^2	CI	Width of CI
E: M farther from μ ($M = 80$)	6.32	2.262	4.75	1.50	71%	15.70 to 44.30	28.60
F: M closer to μ ($M = 60$)	6.32	2.262	1.58	0.50	22%	−4.30 to 24.30	28.60

7.96 Based on the answers to Exercises 7.93 to 7.95, what factors have an impact on a researcher's ability to reject the null hypothesis? Which one(s) can he or she control?

7.97 A researcher is conducting a two-tailed, single-sample t test with alpha set at .05. What is the largest value of t that one could have that, no matter how big N is, will guarantee failing to reject the null hypothesis?

7.98 If t_{cv} for a two-tailed test with $\alpha = .05$ is ± 2.228, what would the alpha level be for the critical value of 2.228 as a one-tailed test?

7.99 If $N = 21$ and $s_M = 1$, write as much of the equation as possible for calculating the 90% $CI\mu_{Diff}$ and the 99% $CI\mu_{Diff}$. Use Equation 7.4 as a guide.

7.100 Which confidence interval in Exercise 7.95 is wider: 90% or 99%? Explain why.

SPSS

Let's use SPSS to analyze the Chips Ahoy data from the Application Demonstration at the end of the chapter. The first step is data entry. Figure 7.12 shows the data for the 10 cookies. Note that the variable is labeled "Num_chips" at the top of the column, and that the value for each case appears in a separate row.

	Num_chips
1	27.00
2	23.00
3	28.00
4	29.00
5	27.00
6	28.00
7	22.00
8	22.00
9	28.00
10	27.00

Figure 7.12 Data Entry for Single-Sample *t* Test in SPSS The number of chips in each cookie is entered on its own row. (Source: SPSS)

SPSS calls a single-sample *t* test the "One-Sample T Test." It is found by clicking on "Analyze" on the top line (Figure 7.13). Then click on "Compare Means" and "One-Sample T Test. . . ."

This pulls up the box seen in Figure 7.14. Note that the variable "Num_chips," which we wish to analyze, has already been moved over from the box on the left

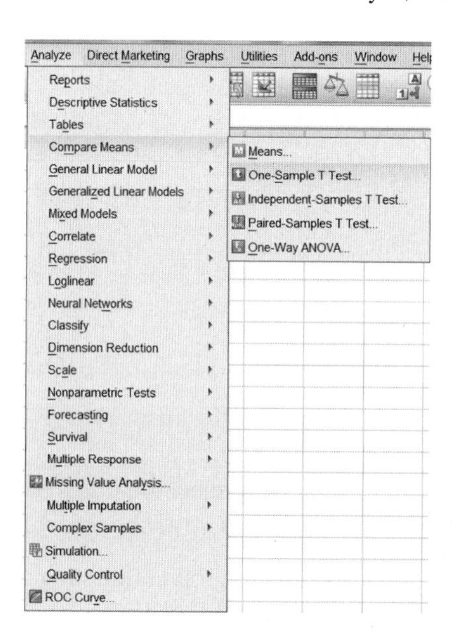

Figure 7.13 The Single-Sample *t* Test in SPSS SPSS calls the single-sample *t* test the "One-Sample T Test." (Source: SPSS)

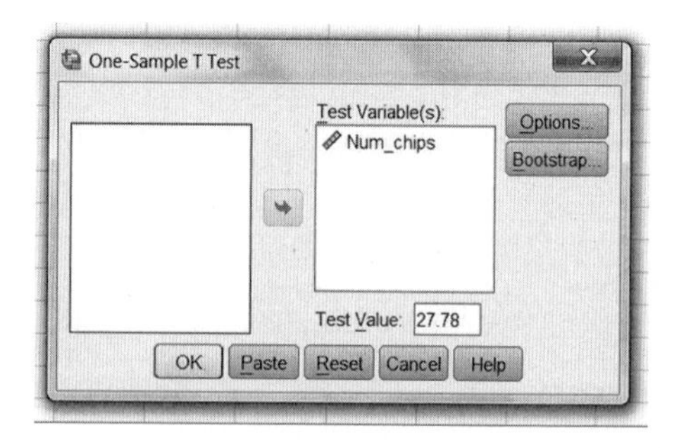

Figure 7.14 Selecting the Dependent Variable and Entering the Population Mean for a Single-Sample *t* Test in SPSS The dependent variable being tested is listed in the box labeled "Test Variable(s)." The population value it is being compared to is entered in the box labeled "Test Value." (Source: SPSS)

(which is now empty) to the box labeled "Test Variable(s)." The population mean, 27.78, has been entered into the box labeled "Test Value." Once this is done, it is time to click the "OK" button on the lower right.

Figure 7.15 shows the output that SPSS produces. There are a number of things to note:

One-Sample Statistics

	N	Mean	Std. Deviation	Std. Error Mean
Num_chips	10	26.1000	2.68535	.84918

One-Sample Test

	Test Value = 27.78					
					95% Confidence Interval of the Difference	
	t	df	Sig. (2-tailed)	Mean Difference	Lower	Upper
Num_chips	-1.978	9	.079	-1.68000	-3.6010	.2410

Figure 7.15 SPSS Output for a Single-Sample *t* Test SPSS provides descriptive statistics, the *t* value, an exact significance level, and the 95% confidence interval for the difference between population means. It does not calculate Cohen's *d* or r^2 but provides enough information that these can be done by hand.
(Source: SPSS)

- The first box provides descriptive statistics, including the standard error of the mean, s_M, the denominator for *t*.
- At the top of the second output box, "Test Value = 27.78" indicates that this is the value against which the sample mean is being compared.
- Next, SPSS reports the *t* value (−1.978) and the degrees of freedom (9).
- SPSS then reports what it calls "Sig. (2-tailed)" as .079.
  - The important thing for us is whether this value is $\leq .05$ or $> .05$.
    - If it is $\leq .05$, the null hypothesis is rejected and the results are reported in APA format as $p < .05$.
    - If it is $> .05$, the null hypothesis is not rejected and the results are reported in APA format as $p > .05$.
  - This value, .079, is the exact significance level for this test with these data. It indicates what the two-tailed probability is of obtaining a *t* value of −1.978 *or larger* if the null hypothesis is true. In the present situation, it says that a *t* value of −1.978 is a common one—it happens 7.9% of the time when 10 cases are sampled from a population where $\mu = 27.78$. APA format calls for the exact significance level to be used. Thus, these results should be reported as $t(9) = -1.98, p = .079$.
  - The next bit of output reports the mean difference, −1.68, between the sample mean (26.10) and the test value (27.78).
  - SPSS reports what it calls the "95% Confidence Interval of the Difference," what is called in the book the 95% confidence interval for the difference between population means, ranging from a "Lower" bound of −3.60 to an "Upper" bound of 0.24.
  - Note that SPSS does not report Cohen's *d*. Similarly, there is not enough information to calculate r^2. However, it provides enough information, the mean difference (−1.68) and the standard deviation (2.685), that it can be done by hand with Equation 7.3.

Independent-Samples *t* Test

LEARNING OBJECTIVES

- Differentiate between independent samples and paired samples.
- Conduct the steps for an independent-samples *t* test.
- Interpret an independent-samples *t* test.

CHAPTER OVERVIEW

Chapters 6 and 7 covered the single-sample *z* test and the single-sample *t* test. Both are difference tests, used to compare the mean of a sample to the mean of a population. They were the first tests covered because they are good for introducing the logic of hypothesis testing and the six-step procedure for completing a hypothesis test. But researchers rarely ask whether this sample mean differs from that population mean.

This chapter introduces a test, the independent-samples *t* test, that researchers do use regularly. The independent-samples *t* test is a two-sample difference test. This is just what it sounds like—a test that is used to see if the average score in one population is better or worse than the average score in a second population. Why is it commonly used? Because classic experiments involve two groups—an experimental group and a control group. Nothing is done to the control group, something is done to the experimental group, and then the outcomes of the two groups are compared to see if the independent variable (the "something") had an effect on the dependent variable.

Tom **a**nd **H**arry **d**espise **c**rabby **i**nfants

8.1 Types of Two-Sample *t* Tests

Two-sample *t* tests are used to compare the mean of one population to the mean of another population. To conduct a two-sample *t* test, a researcher needs two samples of cases and an interval- or ratio-level dependent variable. Many experiments in psychology follow this format. For example, a group of researchers was interested in seeing how stereotypes affect behavior (Bargh, Chen, & Burrows, 1996). They divided undergraduate students into two groups, a control group and an experimental group. Both groups were given words that they had to put into sentences. The control group received neutral words and the experimental group was given words like *Florida*, *wrinkle*, and *forgetful*—words that were meant to prime a stereotype of elderly people. When the participants thought the experiment was over, they exited through a hallway and the experimenters secretly timed how long each person took to walk down the hall. The experimenters believed that being primed with words that are associated with the elderly would lead participants to act like elderly people and walk more slowly.

The results showed that the control group walked down a 30-foot hall, on average, 1 second faster than the experimental group. One second does not sound like much of a difference, but when the experimenters analyzed the data using a two-sample *t* test, they found that the difference was a statistically significant one.

The difference meant that by the time an average person in the control group reached the end of the hall, an average person in the experimental group wasn't yet 90% of the way down the hall (Figure 8.1). The experimental group averaged 2.6 mph compared to 2.9 mph in the control group. Being primed with an elderly stereotype did lead participants to act more like elderly people and a two-sample *t* test helped the researchers reach this conclusion.

There are two different types of two-sample *t* tests, the independent-samples *t* test (the focus of this chapter) and the paired-samples *t* test (see Chapter 9). Obviously, they differ in terms of the samples being analyzed, whether the samples are independent or paired. What's the difference between *independent samples* and *paired samples*?

With **independent samples,** how cases are selected for one sample has no influence on (*is independent of*) the case selection for the other sample. Each sample could be a random sample from its population. With **paired samples,** often called dependent samples, the cases selected for one sample are connected to (*depend on*) the cases in the other sample. The cases in the samples are pairs of cases, yoked together in some way.

To clarify the difference between independent samples and paired samples, here is an example of two different ways to compare the intelligence of men and women to see which sex is smarter:

1. Dr. Smith obtains a random sample of men from the population of men in the world and a random sample of women from the population of women. Each person in the samples takes an IQ test, and Dr. Smith compares the mean of the men to the mean of the women.

2. Dr. Jones gets a random sample of married heterosexual couples. Each man and each woman takes an IQ test, and Dr. Jones compares the mean of the men to the mean of the women.

In the first example, the cases selected for one group have no influence on the cases selected for the other group, so that is an example of independent samples.

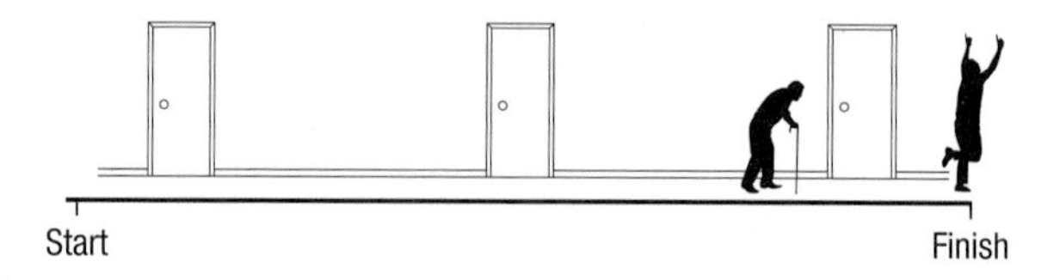

Figure 8.1 Results of a Study with a Control Group and an Experimental Group
By the time the average person in the neutral condition had walked the length of the hallway, the average person in the experimental condition had walked less than 90% of the same distance. Being primed with words that activate a stereotype of elderly people led participants to walk more slowly (Bargh, Chen, & Burrows, 1996).

In the second scenario, who the women are is dependent on the men, so that is an example of paired samples.

It is important to know if the two samples being analyzed are independent or paired so that the researcher can choose the right two-sample test. The example above is one way that subjects can be paired. Table 8.1 contains some guidelines for determining if subjects are paired.

TABLE 8.1 How to Choose: Guidelines for Determining if Samples Are Independent or Paired

Independent Samples	Paired Samples
If each sample is a random sample from its respective population *or* If $n_1 \neq n_2$, that is, each sample has a different number of cases.*	If samples consist of the same cases measured at more than one point in time or in more than one condition *or* If the selection of cases for one sample determines the selection of cases for another sample *or* If the cases in the samples are matched, yoked, or paired together in some way.

Note: These guidelines help decide if samples are independent or paired.
*For paired-samples the two sample sizes must be equal. As they can be equal for independent samples, having equal sample sizes doesn't provide any information about whether the samples are paired or independent.

Practice Problems 8.1

Apply Your Knowledge

8.01 A gym owner is offering a strength training course for women. Eight women have signed up. The gym owner measures how many pounds they can bench press at the first class. He plans to measure this again, 12 weeks later, at the last class, in order to see if their strength has changed. Should he use an independent-samples *t* test or a paired-samples *t* test?

8.02 A public health researcher wants to compare the rate of cigarette smoking for a sample of eastern states vs. a sample of western states. Should she use an independent-samples *t* test or a paired-samples *t* test?

8.03 A nutritionist wants to compare calories for meals at restaurants with tablecloths to restaurants without tablecloths. He gets a sample of each type of restaurant and finds out the calorie count for the most popular meal at each restaurant. Should he use an independent-samples *t* test or a paired-samples *t* test?

8.2 Calculating the Independent-Samples t Test

The two-sample *t* test covered in this chapter is the independent-samples *t* test. The **independent-samples *t* test** is used when seeing if there is a difference between the means of two independent populations. For our example, let's imagine following Dr. Villanova as he replicates part of a classic experiment about factors that influence how well one remembers information (Craik & Tulving, 1975).

Dr. Villanova's participants were 38 introductory psychology students, randomly assigned to two groups. He ended up with 18 in the control group (n_1) and 20 in the experimental group (n_2). The lowercase n is a new abbreviation that will be used to indicate the sample size for a specific group. Subscripts are used to distinguish the two groups. Here, one might say $n_1 = 18$ and $n_2 = 20$ or $n_{\text{Control}} = 18$ and $n_{\text{Experimental}} = 20$. An uppercase N will still be used to indicate the total sample size. For this experiment, $N = n_1 + n_2 = 18 + 20 = 38$.

Each participant was tested individually. All participants were shown 20 words (e.g., *giraffe, DOG, mirror*), one at a time, and asked a question about each word. The control group was asked whether the word appeared in capital letters or not. The experimental group was asked whether the word would make sense in this sentence, "The passenger carried a ______ onto the bus." The first question doesn't require much thought, so the control group was called the "shallow" processing group. Answering the second question required more mental effort, so the experimental group was called the "deep" processing group. After all 20 words had been presented, the participants were asked to write down as many words as they could remember. This recall task was unexpected and the number of words recalled was the dependent variable. The shallow processing group recalled a mean of 3.50 words ($s = 1.54$ words), and the deep processing group a mean of 8.30 words ($s = 2.74$ words). Here is Dr. Villanova's research question: Does deep processing lead to better recall than shallow processing?

Box-and-whisker plots are an excellent way to provide a visual comparison of two or more groups. Box-and-whisker plots are data-rich, providing information about central tendency and variability. In Figure 8.2, the number of words recalled is the dependent variable on the Y-axis and the two groups, shallow and deep, are on the X-axis. Both groups are in the same graph, making it easy to compare them on three things, the median, the interquartile range, and the range:

- The line in the middle of the box is the median, a measure of central tendency. The median number of words recalled is around 8 for the deep processing group and around 4 for the shallow processing group.

- The box that surrounds the line for the median represents the interquartile range, the range within which the middle 50% of scores fall. These middle 50% of scores are often referred to as the average scores for a group, so the interquartile range is a measure of central tendency. Note that in this study, no overlap occurs between these average scores for the two groups.

- The interquartile range is also a measure of variability, measured by the distance from the bottom to the top of the box. Taller boxes mean more variability. The box for the deep processing group looks a little taller than the box for the shallow processing group, indicating more variability for the experimental group than for the control group.

- The whiskers that extend from the box represent the range of scores, another measure of variability. One can see that participants in the shallow processing group recalled from 1 to 6 words and those in the deep processing group recalled from 4 to 12 words. The whiskers show that there is some overlap in the number of words recalled between the two groups and that there is more variability in the deep processing group than in the shallow processing group.

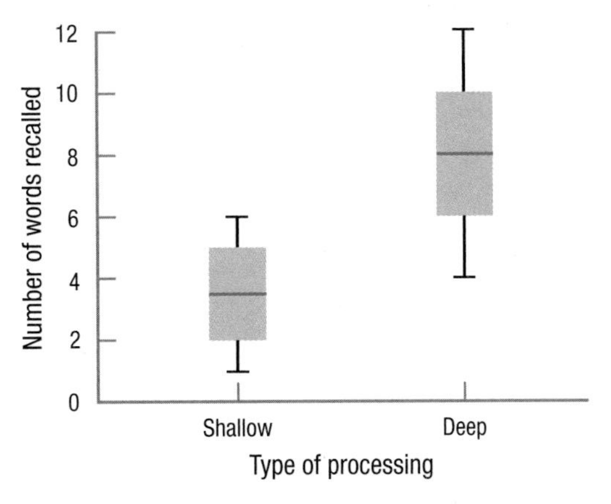

Figure 8.2 Box-and-Whisker Plots Showing the Results of the Depth of Processing Study Note that the middle 50% of cases, represented by the box for the deep processors, recalled more words than the average cases for the shallow processors. This graph appears to show that deep processing leads to better recall than shallow processing. To make sure that the result is a statistically significant one, a researcher would need to complete the appropriate hypothesis test, an independent-samples *t* test.

It certainly appears as if deeper processing leads to better memory, and this was the result that Craik and Tulving found back in 1975. But, in order to see if he replicated their results, Dr. Villanova will need to complete a hypothesis test to see if the difference is a statistically significant one.

Step 1 Pick a Test. The first step in completing a hypothesis test is choosing the correct test. Here, Dr. Villanova is comparing the mean of one population to the mean of another population, so he will use a two-sample *t* test. The two sample sizes are different ($n_{Shallow} = 18$ and $n_{Deep} = 20$), which means they are independent samples (see Table 8.1). In addition, the cases in one sample aren't paired with the cases in the other sample, another indicator of independent samples. As the samples are independent, his planned test is the independent-samples *t* test.

Step 2 Check the Assumptions. Several of the assumptions for the independent-samples *t* test, listed in Table 8.2, are familiar from the single-sample *t* test, but one is new.

TABLE 8.2 Assumptions for the Independent-Samples t Test

	Assumption	Robustness
1. Random samples	Each sample is a random sample from its population.	Robust
2. Independence of observations	Cases within a group are not influenced by other cases within the group.	Not robust
3. Normality	The dependent variable is normally distributed in each population.	Robust, especially if *N* is large
4. Homogeneity of variance	Degree of variability in the two populations is equivalent.	Robust, especially if *N* is large

Note: N is considered large if it is 50 or greater.

The first assumption, random samples, is one seen before. For the depth of processing study, the participants are intro psych students, specifically ones who volunteered to participate in this study. Dr. Villanova may have used random *assignment* to place the participants into a control group and an experimental group, but he doesn't have a random *sample* of participants. Thus, this assumption was violated and this will have an impact on interpreting the results. However, such an assumption is robust, so he can proceed with the *t* test.

With participants being randomly assigned to groups, each case participating by him- or herself, and no one participating twice, the second assumption, that each participant's responding was not influenced by any other participant, was not violated. Dr. Villanova was willing to assume that the third assumption, normality, is not violated. He's willing to make this assumption because memory is a cognitive ability and most psychological variables like this are considered to be normally distributed. In addition, this assumption is robust as long as $N > 30$.[1]

The final assumption is new, homogeneity of variance. This fourth assumption says that the amounts of variability in the two populations should be about equal. Dr. Villanova will assess this by comparing the two sample standard deviations. If the larger standard deviation is not more than twice the smaller standard deviation, then the amounts of variability are considered about equal (Bartz, 1999). For the depth of processing study, the larger standard deviation is 2.74, and the smaller standard deviation is 1.54. The larger standard deviation is not more than twice the smaller standard deviation, so the fourth assumption has not been violated. Dr. Villanova can proceed with the planned independent-samples *t* test.

A Common Question

Q What's the point of the homogeneity of variance assumption? What can I do if it is violated?

A As a step in calculating the independent-samples *t* test, the two sample variances are averaged together. This only gives a meaningful result if the two variances are about the same, homogeneous, to begin with. If the only two siblings in a family are a 192-pound 15-year-old and an 8-pound newborn, it doesn't mean much to report that the average weight of kids in that family is 100 pounds.

If this assumption is violated, there is a way to reduce the degrees of freedom. This correction causes the critical value of *t* to fall further out in the rare zone, making it harder to reject the null hypothesis.

Step 3 List the Hypotheses. Nondirectional, or two-tailed, hypotheses are more common than directional, one-tailed hypotheses, so let's mention those first. Nondirectional hypotheses are used when there is no prediction about which group will have a higher score than the other. In this situation, the null hypothesis will be a negative statement about the two *populations* represented by the two samples. It will say that no difference exists between the mean of one population, μ_1, and the mean of the other population, μ_2.

[1] There are objective ways to assess the normality assumption, but they are beyond the scope of an introductory statistics book like this one. Those who go on to take more classes in statistics will learn these techniques and will be freed from having to assume that variables are normally distributed.

This doesn't mean that the two sample means, M_1 and M_2, will be exactly the same. But, the difference between the sample means should be small enough that it can be explained by sampling error if the two samples were drawn from the same population.

The alternative hypothesis will state that the two *population* means are different. When using a two-tailed test, the alternative hypothesis will just say that the two population means are different from each other; it won't state the direction of the difference. The implication of the alternative hypothesis is that the two *sample* means will be different enough that sampling error is not a likely explanation for the difference.

Written mathematically and using the names of the conditions, the hypotheses are

$$H_0: \mu_{\text{Shallow}} = \mu_{\text{Deep}}$$

$$H_1: \mu_{\text{Shallow}} \neq \mu_{\text{Deep}}$$

Hypotheses like these will always be the null and alternative hypotheses used for a two-tailed independent-samples *t* test.

Dr. Villanova, however, has directional hypotheses and is doing a one-tailed test. Why is he doing a one-tailed test? Because the original study by Craik and Tulving in 1975 found that deep processing worked better than shallow processing, and Dr. Villanova expects that to be the case in his replication. He has made a prediction about the direction of the results in advance of collecting data, so he can do a one-tailed test.

One-tailed tests require a little more thought on the experimenter's part in formulating the hypotheses. With one-tailed tests, it is easier to state the alternative hypothesis first. Dr. Villanova's theory is that deep processing leads to better recall and the alternative hypothesis should reflect this: $H_1: \mu_{\text{Deep}} > \mu_{\text{Shallow}}$.

Once the alternative hypothesis is stated, the null hypothesis is formed by making sure that the two hypotheses are all-inclusive and mutually exclusive. If the alternative hypothesis says that deep processing is better than shallow, then the null hypothesis has to say that shallow processing is as good as, or better than, deep processing. The null hypothesis for the one-tailed independent-samples *t* test would be $H_0: \mu_{\text{Deep}} \leq \mu_{\text{Shallow}}$.

Dr. Villanova, then, would state his hypotheses as

$$H_0: \mu_{\text{Deep}} \leq \mu_{\text{Shallow}}$$

$$H_1: \mu_{\text{Deep}} > \mu_{\text{Shallow}}$$

Step 4 Set the Decision Rule. The critical value of *t*, t_{cv}, is the border between the rare and common zone for the sampling distribution of *t*. t_{cv} is used to set the decision rule and decide whether to reject the null hypothesis for an independent-samples *t* test.

To find the critical value of *t*, Dr. Villanova will use Appendix Table 3. To use this table of critical values of *t*, he needs to know three things: (1) whether he is doing a one-tailed test or a two-tailed test, (2) what alpha level he wants to use, and (3) how many degrees of freedom the test has.

Dr. Villanova predicted in advance of collecting the data that the deep processing participants would do better than the shallow processing participants. And, because he had directional hypotheses, he has a one-tailed test. So, he will be looking at one-tailed values of t_{cv}.

By convention, most researchers are willing to have a 5% chance of making a Type I error and set alpha at .05. (Type I error occurs when one erroneously rejects the null hypothesis.) As Dr. Villanova is replicating previous work, he wants to make sure that if he rejects the null hypothesis, it should have been rejected. He wants to have only a 1% chance of Type I error, so he's setting $\alpha = .01$.

Finally, the degrees of freedom need to be determined. The formula for calculating degrees of freedom (df) for an independent-samples t test is given in Equation 8.1.

Equation 8.1 Formula for Calculating Degrees of Freedom (df) for an Independent-Samples t Test

$$df = N - 2$$

where df = degrees of freedom
N = total number of cases in the two groups

For the depth of processing study, Dr. Villanova calculates $df = 38 - 2 = 36$. Looking in Appendix Table 3 at the row where $df = 36$ and under the column where $\alpha = .01$, one-tailed, we find $t_{cv} = 2.434$.

The t distribution is symmetric, so this value could be either -2.434 or $+2.434$. As Dr. Villanova is doing a one-tailed test, he needs to decide whether his critical value of t will be a positive or a negative number. To do so requires knowing: (1) which mean will be subtracted from which in the numerator of Equation 8.1 when it is used to find the t value; and (2) what the sign of the difference *should* be. The expected sign of the difference is the sign to be associated with the critical value of t for a one-tailed test.

Dr. Villanova has decided that he'll subtract the shallow processing group's mean from the deep processing group's mean. That is, it will be $M_{Deep} - M_{Shallow}$. If his theory is correct, the deep processing group will remember more words than the shallow group, and the difference will have a positive sign. This means that the critical value of t will be positive: 2.434. This value separates the rare zone of the sampling distribution of t from the common zone. The decision rule is shown in Figure 8.3, with the

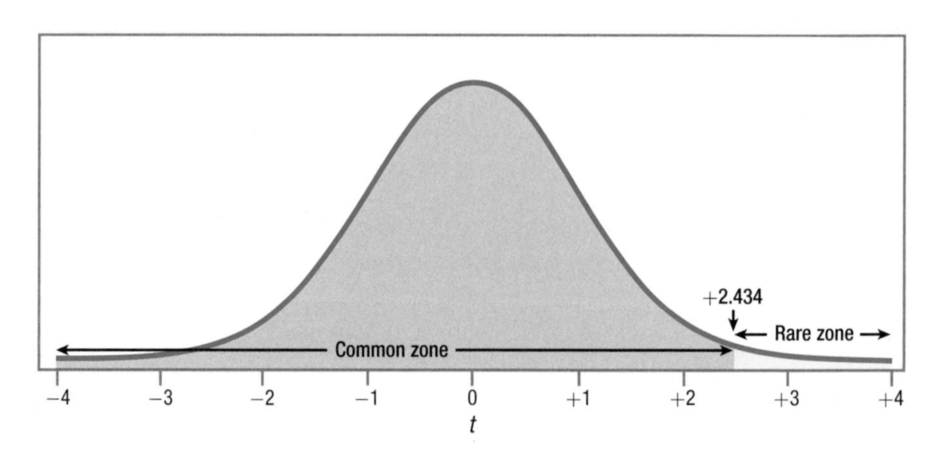

Figure 8.3 The Decision Rule for an Independent-Samples t Test, $df = 36$ For a one-tailed t test with the alpha set at .01 and 36 degrees of freedom, the critical value of t is 2.434.

rare and common zones labeled. Here is how Dr. Villanova states the decision rule for his one-tailed test:

- If $t \geq 2.434$, reject the null hypothesis.
- If $t < 2.434$, fail to reject the null hypothesis.

How would things be different if Dr. Villanova were doing a two-tailed test and had set alpha at .05? The degrees of freedom would still be 36, but t_{cv} would be 2.028 and the decision rule would be:

- If $t \leq -2.028$ *or* if $t \geq 2.028$, reject the null hypothesis.
- If $-2.028 < t < 2.028$, fail to reject the null hypothesis.

This critical value of *t*, 2.028, is closer to zero, the midpoint of the *t* distribution, than is the critical value of *t* that Dr. Villanova is actually using, 2.434. This means it would be easier to reject the null hypothesis if Dr. Villanova were using the two-tailed test with $\alpha = .05$ than with the one-tailed test set at .01 that he is actually using because the rare zone would be larger. He wanted to reduce the likelihood of Type I error, so he has achieved his goal. Table 8.3 summarizes the decision rules for one-tailed and two-tailed two-sample *t* tests.

Step 5 Calculate the Test Statistic. It is now time for Dr. Villanova to calculate the test statistic, *t*. In the last chapter, it was pointed out that the formula for a single-sample *t* test, $t = \frac{M - \mu}{s_M}$, was similar to the *z* score formula, $z = \frac{X - \mu}{\sigma}$. Though it won't look like it by the time we see it, the independent-samples *t* test formula is also like a *z* score formula. The numerator is a deviation score, the deviation of the difference between the two sample means from the difference between the two population means: $(M_1 - M_2) - (\mu_1 - \mu_2)$. Because hypothesis testing proceeds under the assumption that the null hypothesis is true, if we assume $\mu_1 = \mu_2$, we can simplify the numerator to $M_1 - M_2$.

The denominator is a standard deviation, the standard deviation of the sampling distribution of the difference scores. Standard deviations of sampling distributions are called standard errors and this one, called the standard error of the difference, is abbreviated $s_{M_1-M_2}$. Calculating the standard error of the difference proceeds in two steps. First we need to calculate variance for the two samples combined, or pooled, and then we need to adjust this **pooled variance** to turn it into a standard error. Equation 8.2 gives the formula for calculating the pooled variance.

TABLE 8.3 Decision Rules for Independent-Samples t Tests

Two-Tailed Test	One-Tailed Test
If $t \leq -t_{cv}$ *or* if $t \geq t_{cv}$, reject H_0. If $-t_{cv} < t < t_{cv}$, fail to reject H_0.	If $t \geq t_{cv}$, reject H_0. If $t < t_{cv}$, fail to reject H_0. *or* If $t \leq -t_{cv}$, reject H_0. If $t > -t_{cv}$, fail to reject H_0.

Note: *t* is the value of the test statistic, which is calculated in Step 5. t_{cv} is the critical value, which is found in Appendix Table 3. For a one-tailed test, the researcher needs to decide in advance whether the *t* value should be negative or positive in order to reject the null hypothesis.

Equation 8.2 Formula for Calculating the Pooled Variance for an Independent-Samples *t* Test

$$s^2_{\text{Pooled}} = \frac{s_1^2(n_1 - 1) + s_2^2(n_2 - 1)}{df}$$

where s^2_{Pooled} = the pooled variance
n_1 = the sample size for Group (sample) 1
s_1^2 = the variance for Group 1
n_2 = the sample size for Group (sample) 2
s_2^2 = the variance for Group 2
df = the degrees of freedom ($N - 2$)

This equation says the pooled variance is calculated by multiplying each sample variance by 1 less than the number of cases that are in its sample and adding together these products. That sum is then divided by 2 less than the total number of cases, which is the same as the degrees of freedom.

Let's follow Dr. Villanova as he plugs in the values from his depth of processing study into Equation 8.2. Remember, the shallow processing group had 18 cases, with a standard deviation of 1.54; the sample size and standard deviation for the deep processing group were 20 and 2.74, respectively. The equation calls for variances, not standard deviations, but a variance is simply a squared standard deviation, so we are OK. And the equation calls for the degrees of freedom, which we calculated in the previous step as $N - 2 = 38 - 2 = 36$:

$$\begin{aligned} s^2_{\text{Pooled}} &= \frac{s_1{}^2(n_1 - 1) + s_2{}^2(n_2 - 1)}{df} \\ &= \frac{1.54^2(18 - 1) + 2.74^2(20 - 1)}{36} \\ &= 2.3716(17) + 7.5076(19) \\ &= \frac{40.3172 + 142.6444}{36} \\ &= \frac{182.9616}{36} \\ &= 5.0823 \\ &= 5.08 \end{aligned}$$

The pooled variance is 5.08. We can now use the pooled variance to find the standard error of the mean by using Equation 8.3.

Equation 8.3 Formula for the Standard Error of the Mean, $s_{M_1-M_2}$, for an Independent-Samples *t* test

$$s_{M_1-M_2} = \sqrt{s^2_{\text{Pooled}}\left(\frac{N}{n_1 \times n_2}\right)}$$

where $s_{M_1-M_2}$ = the standard error of the difference
s^2_{Pooled} = the pooled variance (from Equation 8.2)
N = the total number of cases
n_1 = the number of cases in Group 1
n_2 = the number of cases in Group 2

This formula says that to find the standard error of the difference, one finds the quotient of the total sample size divided by the two individual sample sizes multiplied together. This quotient is multiplied by the pooled variance. Finally, in a step that is often overlooked, the square root of the product is found. In essence, this is similar to what was done in calculating the standard error of the mean, where the standard deviation was divided by the square root of N.

Let's do the math for the shallow vs. deep processing study:

$$
\begin{aligned}
s_{M_1-M_2} &= \sqrt{s^2_{\text{Pooled}}\left(\frac{N}{n_1 \times n_2}\right)} \\
&= \sqrt{5.08\left(\frac{38}{18 \times 20}\right)} \\
&= \sqrt{5.08\left(\frac{38}{360}\right)} \\
&= \sqrt{5.08 \times 0.1056} \\
&= \sqrt{0.5364} \\
&= .7324 \\
&= .73
\end{aligned}
$$

At this point, the hardest part of calculating an independent-samples t test is over and Dr. Villanova knows that the standard error of the mean is 0.73. All that is left to do is use the formula in Equation 8.4 to find t.

Equation 8.4 Formula for an Independent-Samples t Test

$$t = \frac{M_1 - M_2}{s_{M_1-M_2}}$$

where t = the independent-samples t test value

M_1 = the mean of Group (sample) 1

M_2 = the mean of Group (sample) 2

$s_{M_1-M_2}$ = the standard error of the difference (Equation 8.3)

Equation 8.4 says that an independent-samples t value is calculated by dividing the numerator, the difference between the two sample means, by the denominator, the standard error of the difference. Dr. Villanova, because he is doing a one-tailed test, has already decided that he will be subtracting the shallow processing group's mean, 3.50, from the deep processing group's mean, 8.30:

$$
\begin{aligned}
t &= \frac{M_1 - M_2}{s_{M_1-M_2}} \\
&= \frac{8.30 - 3.50}{0.73} \\
&= \frac{4.8000}{0.73} \\
&= 6.5753 \\
&= 6.58
\end{aligned}
$$

The test statistic *t* that Dr. Villanova calculated for his independent-samples *t* test is 6.58 and Step 5 of the hypothesis test is over. We'll follow Dr. Villanova as he covers Step 6, interpretation, after working through the first five steps with another example.

Worked Example 8.1

For practice with an independent-samples *t* test, let's use an urban example. Dr. Risen, an environmental psychologist, wondered if temperature affected the pace of life. She went to Fifth Avenue in New York City, randomly selected pedestrians who were walking alone, timed how long it took them to walk a block, and converted this into miles per hour (mph).

She did this on two days, one a 20°F day in January and the other a 72°F day in June. Each time, she used the same day of the week and the same hour of the day. She also made sure that on both days there were blue skies and no obstructions, like snow or trash, on the sidewalk. On the cold day she timed 33 people, and on the warm day she timed 28. Her total sample size, *N*, was 61.

The results, displayed in Figure 8.4, show that people walked faster on the cold day ($M = 3.05$ mph, $s = 0.40$ mph) than on the warm day ($M = 2.90$ mph, $s = 0.39$ mph). The results suggest that people pick up their pace when it is cold outside. Let's follow Dr. Risen as she determines if the effect is a statistically significant one. To do so, she'll use the six steps of hypothesis testing. We'll follow her through the first five steps in this section and then tag along for the sixth step, interpretation, in the next part of this chapter.

Step 1 Pick a Test. Two groups of people, those walking on a cold day vs. those walking on a warm day, are being compared in terms of mean walking speed. This calls for a two-sample *t* test. Using Table 8.2, Dr. Risen concludes that the samples are independent—each sample is a random sample from its respective population, the cases in the two samples aren't paired, and the two sample sizes are different. Thus, the appropriate test is the independent-samples *t* test.

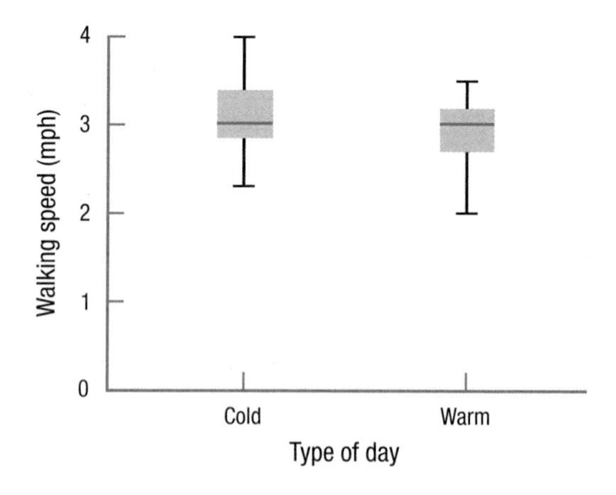

Figure 8.4 Box-and-Whisker Plots Showing the Effect of Temperature on Walking Time The middle 50% of cases (those in the box) look like they are walking a little faster on cold days than on warm days. In order to determine if the effect is a statistically significant one, however, a hypothesis test will need to be completed.

Step 2 Check the Assumptions.

- The random samples assumption is not violated as pedestrians were randomly selected.
- The independence of observations assumption is not violated. Only one case was observed at a time, each person was walking alone, and no person was timed twice. So, no other cases influenced a case.
- The normality assumption is not violated. Dr. Risen is willing to assume that a physical trait, like walking speed, is normally distributed. Plus, her sample size is large, greater than 50, and this assumption is robust if the sample size is large.
- The homogeneity of variance assumption is not violated. The two standard deviations are almost exactly the same ($s_{\text{Cold}} = 0.40$ mph and $s_{\text{Warm}} = 0.39$ mph).

Step 3 List the Hypotheses. Dr. Risen is doing an exploratory study. She's investigating whether temperature affects the pace of life. Hence, her test is two-tailed, and her hypotheses are nondirectional:

$$H_0: \mu_{\text{Cold}} = \mu_{\text{Warm}}$$

$$H_1: \mu_{\text{Cold}} \neq \mu_{\text{Warm}}$$

The null hypothesis says that there is no difference in walking speed in cold weather vs. walking speed in warm weather for the two populations. The alternative hypothesis says that the two population means—walking speed in cold weather vs. walking speed in warm weather—are different.

Step 4 Set the Decision Rule. To set the decision rule, Dr. Risen must find a critical value of *t* in Appendix Table 3. To do so, she needs three pieces of information: (1) whether the test is one-tailed or two-tailed, (2) what alpha level is selected, and (3) how many degrees of freedom there are.

1. The hypotheses were nondirectional, so she's doing a two-tailed test.
2. She's willing to run the standard, 5%, risk of making a Type I error, so $\alpha = .05$.
3. Applying Equation 8.1 to her total sample size, $N = 61$, she calculates degrees of freedom as $df = 61 - 2 = 59$.

Turning to the table of critical values of *t*, she looks for the intersection of the column for a two-tailed test with $\alpha = .05$, the bolded column, and the row for $df = 59$. However, there is no row for $df = 59$. What should she do? She'll follow *The Price Is Right* rule (see Chapter 7) and use the critical value found in the row for the degrees of freedom that are closest to her actual degrees of freedom without going over it. In this instance, that means the row with $df = 55$. The critical value of *t* is ± 2.004. The sampling distribution of *t* with this critical value of *t* is shown in Figure 8.5.

Here's her decision rule:

- If $t \leq -2.004$ *or* if $t \geq 2.004$, reject the null hypothesis.
- If $-2.004 < t < 2.004$, fail to reject the null hypothesis.

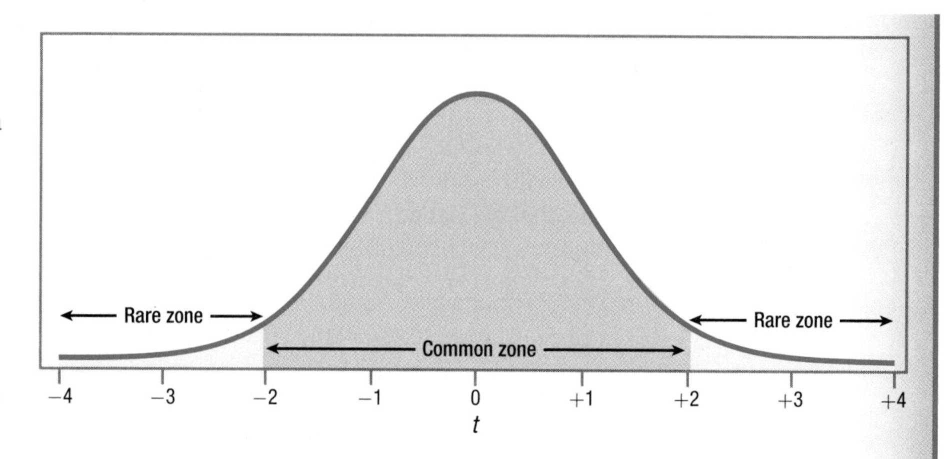

Figure 8.5 The Critical Value of *t* for the Walking Speed Study For a two-tailed *t* test with the alpha set at .05 and 55 degrees of freedom, the critical value of *t* is ±2.004. Note that the critical value associated with $df = 55$ is being used because this is the value closest to, without going over, the actual degrees of freedom of 59.

A Common Question

Q Why do statisticians use *The Price Is Right* rule and, if the actual degrees of freedom aren't in the table of critical values, use the value that is closest without going over the actual value?

A Statisticians like to make it difficult to reject the null hypothesis. Using a critical value associated with a smaller number of degrees of freedom means that the rare zone is smaller and so it is harder to reject the null hypothesis.

Step 5 Calculate the Test Statistic. The first step in finding the *t* value is to use Equation 8.2 to calculate the pooled variance. Once that is done, Dr. Risen can use Equation 8.3 to find the standard error of the difference and Equation 8.4 to calculate the value of the test statistic.

As a reminder, here is the information Dr. Risen has to work with:

- Cold day: $n = 33$, $M = 3.05$ (mph), $s = 0.40$ (mph)
- Warm day: $n = 28$, $M = 2.90$ (mph), $s = 0.39$ (mph)

Let's follow as Dr. Risen plugs the values into Equation 8.2 to find s^2_{Pooled}:

$$
\begin{aligned}
s^2_{\text{Pooled}} &= \frac{s_1^2(n_1 - 1) + s_2^2(n_2 - 1)}{df} \\
&= \frac{0.40^2(33 - 1) + 0.39^2(28 - 1)}{59} \\
&= \frac{0.1600(32) + 0.1521(27)}{59} \\
&= \frac{5.1200 + 4.1067}{59} \\
&= \frac{9.2267}{59} \\
&= 0.1564 \\
&= 0.16
\end{aligned}
$$

The next step is to use Equation 8.3 to find the standard error of the difference:

$$s_{M_1-M_2} = \sqrt{s^2_{\text{Pooled}}\left(\frac{N}{n_1 \times n_2}\right)}$$

$$= \sqrt{0.16\left(\frac{61}{33 \times 28}\right)}$$

$$= \sqrt{0.16\left(\frac{61}{924.0000}\right)}$$

$$= \sqrt{0.16 \times .0660}$$

$$= \sqrt{0.0106}$$

$$= 0.1030$$

$$= 0.10$$

Now that she knows $s_{M_1-M_2} = 0.10$, Dr. Risen can go on to complete Equation 8.4 and find the *t* value:

$$t = \frac{M_1 - M_2}{s_{M_1-M_2}}$$

$$= \frac{3.05 - 2.90}{0.10}$$

$$= \frac{0.1500}{0.10}$$

$$= 1.5000$$

$$= 1.50$$

The value of the test statistic, *t*, is 1.50. This completes Step 5.

A Common Question

Q Is it possible to calculate the standard error of the estimate without computing the pooled variance first?

A With a little algebraic rearranging, almost anything is possible. Here is the combination of Equations 8.2 and 8.3:

$$s_{M_1-M_2} = \sqrt{\left[\frac{s_1^2(n_1 - 1) + s_2^2(n_2 - 1)}{df}\right]\left[\frac{N}{n_1 \times n_2}\right]}$$

Practice Problems 8.2

Apply Your Knowledge

8.04 Previous research has shown that people who have served in the U.S. Armed Forces feel more patriotic about America. A researcher obtains a sample of veterans and a sample of nonveterans, and administers the interval level Sense of Patriotism Scale (SPS). Higher scores on the SPS indicate greater patriotism.

The researcher expects to replicate previous research. Write the researcher's null and alternative hypotheses.

8.05 If $n_1 = 12$, $s_1 = 4$, $n_2 = 16$, and $s_2 = 3$, calculate s^2_{Pooled} and $s_{M_1-M_2}$.

8.06 If $M_1 = 99$, $M_2 = 86$, and $s_{M_1-M_2} = 8.64$, calculate t.

8.3 Interpreting the Independent-Samples *t* Test

The final step in hypothesis testing, Step 6, is interpretation. This will follow the same format for the independent-samples *t* test as for previous hypothesis tests, addressing a series of questions:

1. Was the null hypothesis rejected?
2. How big is the effect?
3. How wide is the confidence interval?

The questions should be answered in order and each one adds additional information. After answering the first question, a researcher will have enough information for a basic interpretation. Answering all three questions, however, gives a deeper understanding of what the results mean and allows a researcher to write a more nuanced interpretation.

Was the Null Hypothesis Rejected?

The null and alternative hypotheses are set up to be all-inclusive and mutually exclusive. If a researcher can reject one hypothesis, he or she will have to accept the other hypothesis. If Dr. Villanova can reject the null hypothesis (that shallow processing is better than or equal to deep processing), he will be forced to accept the alternative hypothesis (that deep processing is better than shallow processing). With a one-tailed test, as this one is, if the null hypothesis is rejected, then the direction of the difference between the populations is known. With a two-tailed test, the researcher would need to examine the sample means in order to tell the direction of the probable population difference.

Here is what we know so far about Dr. Villanova's study on depth of processing:

- Shallow processing: $M = 3.50$, $s = 1.54$, $n = 18$
- Deep processing: $M = 8.30$, $s = 2.74$, $n = 20$
- $df = 36$
- $t_{cv} = 2.434$
- $s^2_{\text{Pooled}} = 5.08$
- $s_{M_1-M_2} = 0.73$
- $t = 6.58$

Dr. Villanova's first move is to plug the observed value of the test statistic, 6.58, into the decision rule generated in Step 4 and decide which statement is true:

- Is $6.58 \geq 2.434$?
- Is $6.58 < 2.434$?

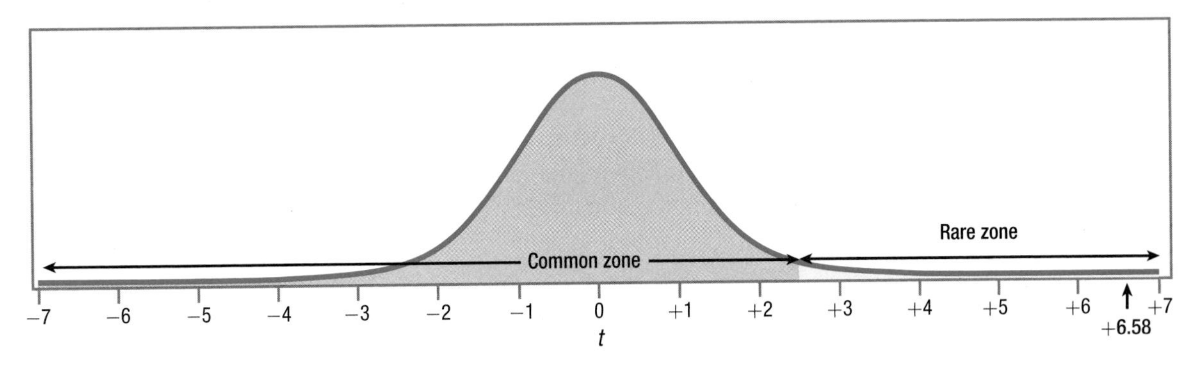

Figure 8.6 The Test Statistic for an Independent-Samples *t* Test for the Depth of Processing Data The test statistic *t*, 6.58, falls in the rare zone, so the null hypothesis is rejected. The alternative hypothesis is accepted. Because this is a one-tailed test, the alternative hypothesis states the direction of the difference.

6.58 is greater than or equal to 2.434, so the first statement is true and Dr. Villanova will reject the null hypothesis (see Figure 8.6) and call the results statistically significant. This means he accepts the alternative hypothesis that the mean number of words recalled by the population of people who use deep processing is greater than the mean number of words recalled by people who use shallow processing.

In APA format, Dr. Villanova would write the results as

$$t(36) = 6.58, p < .01 \text{ (one-tailed)}$$

- The *t* says that he is reporting the results of a *t* test.
- The 36 in parentheses, the degrees of freedom, gives information about how many cases were in the study. Because the degrees of freedom for an independent-samples *t* test is the total sample size minus 2, that means there were a total of 36 + 2, or 38, participants in the study.
- 6.58 is the value of the test statistic that was calculated.
- $p < .01$ indicates two things. The .01 tells that alpha was set at .01 because Dr. Villanova was willing to make a Type I error 1% of the time. The $p < .01$ reveals that the null hypothesis was rejected because the test statistic of 6.58 is a rare occurrence (it happens less than 1% of the time) when the null hypothesis is true.
- Finally, the phrase in parentheses at the end, one-tailed, tells the reader that Dr. Villanova conducted a one-tailed test. As most hypothesis tests are two-tailed, it is only when the test is not two-tailed that this fact is noted.

If Dr. Villanova chose to stop the interpretation at this point, he would have enough information to make a meaningful statement about the results. Here's what he could say:

> In a study comparing deep processing to shallow processing, a statistically significant effect was found [$t(36) = 6.58$, $p < .01$ (one-tailed)]. People who were randomly assigned to use deep processing recalled more words ($M = 8.30$) than did people who used shallow processing ($M = 3.50$).

How Big Is the Effect?

Cohen's *d*

Cohen's d and r^2, the same effect sizes used for the single-sample t test in Chapter 7, will be used for the independent-samples t test to tell how much impact the explanatory variable has on the outcome variable. The same standards can be used to judge the size of the effect for the independent-samples t test as were used for the single-sample t test:

- 0.00 means there is absolutely no effect.
- $d \approx 0.20$ or $r^2 = 1\%$ is a small effect.
- $d \approx 0.50$ or $r^2 = 9\%$ is a medium effect.
- $d \geq 0.80$ or $r^2 \geq 25\%$ is a large effect.

Equation 8.5 shows how to calculate Cohen's d for the independent-samples t test. Note that it makes use of the pooled variance, s^2_{Pooled}, which was 5.08.

Equation 8.5 Formula for Calculating Cohen's *d* for an Independent-Samples t Test

$$d = \frac{M_1 - M_2}{\sqrt{s^2_{\text{Pooled}}}}$$

where d = Cohen's d value
M_1 = the mean for Group (sample) 1
M_2 = the mean for Group (sample) 2
s^2_{Pooled} = the pooled variance (from Equation 8.2)

Here are Dr. Villanova's calculations for the effect size for the depth of processing study. He substitutes in 8.30 as the mean of the deep processing group, 3.50 as the mean of the shallow processing group, and 5.08 as the pooled variance:

$$\begin{aligned} d &= \frac{M_1 - M_2}{\sqrt{s^2_{\text{Pooled}}}} \\ &= \frac{8.30 - 3.50}{\sqrt{5.08}} \\ &= \frac{4.8000}{\sqrt{5.08}} \\ &= \frac{4.8000}{2.2539} \\ &= 2.1296 \\ &= 2.13 \end{aligned}$$

Cohen's d value, 2.13, is greater than 0.80, so Dr. Villanova can consider that the effect of the independent variable (type of processing) on the dependent variable (number of words recalled) is large. Now, in his interpretation, he can note more than the fact that deep processing leads to significantly better recall than shallow processing. He can say that the effect size is large, that how people process information *does* matter in how well they recall information.

r^2

The same formula, Equation 7.4, is used to calculate r^2 for the independent-samples *t* test as was used for the single-sample *t* test. It makes use of two values, *t* and *df*:

$$
\begin{aligned}
r^2 &= \frac{t^2}{t^2 + df} \times 100 \\
&= \frac{6.58^2}{6.58^2 + 36} \times 100 \\
&= \frac{43.2964}{43.2964 + 36} \times 100 \\
&= \frac{43.2964}{79.2964} \times 100 \\
&= 0.5460 \times 100 \\
&= 54.60\,\%
\end{aligned}
$$

r^2, remember, calculates the percentage of variability in the outcome variable that is accounted for by the explanatory variable. Here, r^2 tells how much of the variability in the number of words recalled is accounted for by the group, shallow vs. deep processing, subjects were assigned to. r^2 varies from 0% to 100%; the higher the percentage, the stronger the effect. Here, the effect is quite strong, with over 50% of the variability explained by group status.

How Wide Is the Confidence Interval?

To determine the impact of the independent variable on the dependent variable in the population, a confidence interval is used. For an independent-samples *t* test, a researcher calculates a confidence interval for the difference between population means, the same type of confidence interval calculated for the single-sample *t* test. This confidence interval estimates how close together (or how far apart) the two population means may be. This tells how much of an effect may, or may not, exist in the population.

Though any level of confidence, from greater than 0% to less than 100%, can be used for a confidence interval, the most commonly calculated is a 95% confidence interval. The formula for that is found in Equation 8.6. Two other common confidence intervals are 90% and 99%.

Equation 8.6 Formula for Calculating the 95% Confidence Interval for the Difference Between Population Means

$$95\%\text{CI}\mu_{\text{Diff}} = (M_1 - M_2) \pm (t_{cv} \times s_{M_1 - M_2})$$

where $95\%\text{CI}\mu_{\text{Diff}}$ = the 95% confidence interval for the difference between population means

M_1 = the mean of Group (sample) 1

M_2 = the mean of Group (sample) 2

t_{cv} = the critical value of *t*, two-tailed, $\alpha = .05$, $df = N - 2$ (Appendix Table 3)

$s_{M_1 - M_2}$ = the standard error of the difference (Equation 8.3)

For the depth of processing study, Dr. Villanova is going to calculate the 95% confidence interval. The two sample means are 8.30 and 3.50; the critical value of t, two-tailed, with $\alpha = .05$, and 36 degrees of freedom is 2.028; and the standard error of the difference is 0.73:

$$\begin{aligned} 95\%\text{CI}\mu_{\text{Diff}} &= (M_1 - M_2) \pm (t_{cv} \times s_{M_1-M_2}) \\ &= (8.30 - 3.50) \pm (2.028 \times 0.73) \\ &= 4.8000 \pm 1.4804 \\ &= \text{from } 3.3196 \text{ to } 6.2804 \\ &= [3.32, 6.28] \end{aligned}$$

The 95% confidence interval for the difference between population means ranges from 3.32 to 6.28. In APA format, it would be written as 95% CI [3.32, 6.28]. This confidence interval tells what the effect of the type of processing is on recall in the larger population. It says that the effect probably falls somewhere in the range from deep processing, leading to an average of anywhere from 3.32 to 6.28 more words being recalled over shallow processing.

Just as with the one-sample t test, there are three aspects of the confidence interval to pay attention to: (1) whether it captures zero; (2) how close it is to zero; and (3) how wide it is:

1. If the confidence interval captures zero, then it is plausible that no difference exists between the two population means. Thus, a confidence interval that captures zero occurs when the researcher has failed to reject the null hypothesis, as long as he or she is using a two-tailed test with $\alpha = .05$ *and* as long as he or she is calculating a 95% confidence interval.

2. When the confidence interval comes close to zero, then it is possible that there is little difference between the two population means. When it ranges farther away from zero, then it is possible that the difference between the two populations is more meaningful. In this way, a confidence interval is helpful in thinking about the size of the effect.

3. The width of the confidence interval tells how precisely a researcher can specify the effect in the population. A narrower confidence interval means the researcher can be reasonably certain of the size and meaningfulness of the difference. A wider confidence interval leaves the researcher uncertain of the size and meaningfulness of the difference. In such a situation, it is often reasonable to recommend replicating the study with a larger sample size in order to obtain more precision.

Dr. Villanova's confidence interval ranges from 3.32 to 6.28 for the depth of processing study. With regard to the three points above: (1) The confidence interval doesn't capture zero. It is unlikely that there is no difference between the two population means. Dr. Villanova expected this result as the null hypothesis had been rejected. (2) The low end of the confidence interval, the end that is closer to zero, is 3.32. In Dr. Villanova's opinion, a difference of 3.32 words is still a meaningful difference. Dr. Villanova, who planned and conducted this study, has expertise in this area and with this dependent variable. As a result, his opinion carries some weight. (3) The width of a confidence interval can be calculated by subtracting one side from the other:

$$6.28 - 3.32 = 2.96$$

The confidence interval is almost three words wide. In Dr. Villanova's opinion, based on his expertise, this is a reasonably narrow confidence interval. Thus, it provides a precise-enough estimate of what the population difference is. He feels little need to replicate with a larger sample size to obtain a narrower confidence interval.

A Common Question

Q How are *d*, r^2, and a confidence interval alike? How do they differ?

A *d* and r^2 are officially called effect sizes, but a confidence interval also gives information about how strong the effect of the explanatory variable is. *d* and r^2 reflect the size of the effect as observed in the actual sample; a confidence interval extrapolates the effect to the population. Cohen's *d* is not affected by sample size. If the group means and standard deviations stay the same but the sample size increases, *d* will be unchanged, but the confidence interval will narrow and offer a more precise estimate of the population value. r^2 is inversely affected by sample size—as *N* increases, r^2 decreases.

Putting It All Together

Dr. Villanova has addressed all three of the interpretation questions and is ready to use the information to write an interpretation that explains the results. In the interpretation, he addresses four points:

1. He starts with a brief explanation of the study.
2. He states the main results.
3. He explains what the results mean.
4. He makes suggestions for future research.

There's one more very important thing Dr. Villanova does. He did a lot of calculations—*t*, *d*, r^2, and a confidence interval—in order to understand the results. But, he doesn't feel obligated to report them all just because he calculated them. He limits what he reports in order to give a clear and concise report:

> This study compared how deep processing of words vs. shallow processing of words affected recall on an unexpected memory test. The deep processing group ($M = 8.30$ words, $s = 2.74$ words) recalled more words than the shallow processing group ($M = 3.50$ words, $s = 1.54$ words). This effect was statistically significant [$t(36) = 6.58$, $p < .01$ (one-tailed)] and it is a large effect. Using deep processing leads to markedly better recall when a person is not trying to memorize the words. Whether the effect exists when a person is purposefully trying to learn a list of words should be examined in a subsequent study.

Worked Example 8.2 For practice interpreting results for an independent-samples *t* test, let's return to Dr. Risen's study about how temperature affects the pace of city life. She measured the walking speed for 33 people walking alone on a cold day (20°F) and for 28 people on a warm day (72°F). Here is what is already known:

- For the cold day: $M = 3.05$ mph, $s = 0.40$
- For the warm day: $M = 2.90$ mph, $s = 0.39$
- $df = 59$
- t_{cv}, two-tailed, $\alpha = .05$ is ± 2.004
- $s^2_{\text{Pooled}} = 0.16$
- $s_{M_1 - M_2} = 0.10$
- $t = 1.50$

Was the null hypothesis rejected? The first step is applying the decision rule. Which is true?

- Is $1.50 \leq -2.004$ *or* is $1.50 \geq 2.004$?
- Is $-2.004 < 1.50 < 2.004$?

The second statement is true and the value of the test statistic falls in the common zone, as shown in Figure 8.7. Insufficient evidence exists to reject the null hypothesis, so there is no reason to conclude that temperature affects walking speed. The results are called "not statistically significant." In APA format, the results would be written like this:

$$t(59) = 1.50, p > .05$$

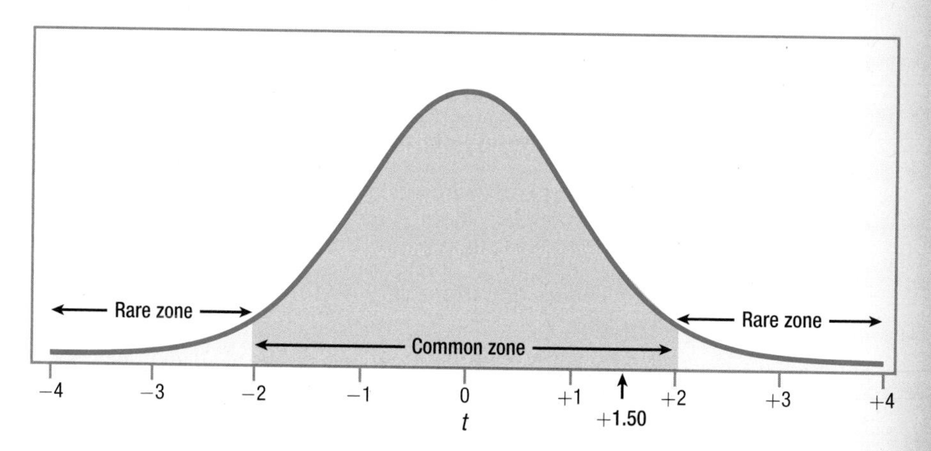

Figure 8.7 The Test Statistic *t* for the Walking Speed Study The test statistic *t*, 1.50, falls in the common zone, so the null hypothesis is not rejected. There is not enough evidence to conclude that the two population means differ.

A Common Question

Q When she looked up the critical value of *t* in Appendix Table 3, Dr. Risen had to use the row with 55 degrees of freedom because there was no row for $df = 59$. But, in reporting the results in APA format, she used $df = 59$, not $df = 55$. Why?

A The degrees of freedom within the parentheses in APA format provide information about how many cases there are, so they should reflect that.

How big is the effect? Using Equation 8.5, Dr. Risen calculated Cohen's d:

$$\begin{aligned} d &= \frac{M_1 - M_2}{\sqrt{s^2_{\text{Pooled}}}} \\ &= \frac{3.05 - 2.90}{\sqrt{0.16}} \\ &= \frac{0.1500}{\sqrt{0.16}} \\ &= \frac{0.1500}{0.4000} \\ &= 0.3750 \\ &= 0.38 \end{aligned}$$

Equation 8.6 is used to calculate r^2:

$$\begin{aligned} r^2 &= \frac{t^2}{t^2 + df} \times 100 \\ &= \frac{1.50^2}{1.50^2 + 59} \times 100 \\ &= \frac{2.2500}{2.2500 + 36} \times 100 \\ &= \frac{2.2500}{38.2500} \times 100 \\ &= 0.0588 \times 100 \\ &= 5.88\,\% \end{aligned}$$

Hypothesis testing says there is not enough evidence to conclude an effect has occurred, but the two effect sizes suggest that a small to moderate effect may be present. Dr. Risen might want to replicate the study with a larger sample size in order to have a better chance of rejecting the null hypothesis and seeing whether temperature does affect the pace of urban life.

How wide is the confidence interval? Applying Equation 8.6, Dr. Risen calculated the 95% confidence interval:

$$\begin{aligned} 95\%\text{CI}\mu_{\text{Diff}} &= (M_1 - M_2) \pm (t_{cv} \times s_{M_1 - M_2}) \\ &= (3.05 - 2.90) \pm (2.004 \times 0.10) \\ &= 0.1500 \pm 0.2004 \\ &= \text{from } -0.0504 \text{ to } 0.3504 \\ &= [-0.05, 0.35] \end{aligned}$$

As expected, the confidence interval, −0.05 to 0.35, captures zero. So, the possibility that there is zero difference between the two population means—walking speed on cold days vs. walking speed on warm days—is plausible. What other in-[illegible]ained in the confidence interval? It tells Dr. Risen that comparing the walking speed of the population of warm-day pedestrians to the population of cold-day pedestrians, the difference may be from 0.05 mph *slower* on cold days to 0.35 mph *faster* on cold days. In other words, from this study she can't tell if, how much, or in what direction the two populations differ.

But, the confidence interval raises the possibility that the null hypothesis is false and pedestrians walk about a third of a mile per hour faster on cold days. In the language of hypothesis testing, she is worried about Type II error, that there may be an effect of temperature on walking speed that she failed to find. To ease her concern, she's going to recommend replicating the study with a larger sample size. This would increase power, making it easier to reject the null hypothesis if it should be rejected, and it would give a narrower confidence interval.

Putting it all together. Here's Dr. Risen's four-point interpretation. She (1) tells what the study was about, (2) gives the main results, (3) explains what the results mean, and (4) makes suggestions for future research:

> This study was conducted to see if temperature affected the pace of life in a city. The walking speed of pedestrians was measured on a cold day (20°F) and on a warm day (72°F). The mean speed on the cold day was 3.05 mph; on the warm day it was 2.90 mph. The difference in speed between the two days was not statistically significant [$t(59) = 1.50$, $p > .05$]. These results do not provide sufficient evidence to conclude that temperature affects the pace of urban life. However, the confidence interval for the difference between population means raised the possibility that there might be a small effect of temperature on walking speed. Therefore, replication of this study with a larger sample size is recommended.

Application Demonstration

One of the most famous studies in psychology was conducted by Festinger and Carlsmith in 1959. In that study, they explored cognitive dissonance, a phenomenon whereby people change their attitudes to make attitudes consistent with behavior. Their study used an independent-samples *t* test to analyze the results. Let's see what they did and then apply this book's interpretative techniques to update Festinger's and Carlsmith's findings.

Festinger and Carlsmith brought male college students into a laboratory and asked them to perform very boring tasks. For example, one task involved a board with 48 pegs. The participant was asked to give each peg a one-quarter clockwise turn, and to keep doing this, peg after peg. After an hour of such boring activity, each participant was asked to help out by telling the next participant that the experiment was enjoyable and a lot of fun. In other words, the participants were asked to lie.

This is where the experimental manipulation took place: 20 participants were paid $1 for lying and 20 were paid [illegible] helping out, each participant was then asked to rate how [illegible] the experiment. The scale used ranged from [illegible] interesting). Obviously, the participants [illegible] tive zone because it was quite boring.

Did how much participants get paid influence their rating? It turned out that it did. The participants who were paid \$1 gave a mean rating in the positive zone (+1.35), and the participants who were paid \$20 gave a mean rating in the negative zone (−0.05). This difference was statistically significant [$t(38) = 2.22$, $p < .05$]. The participants paid less to lie gave statistically significantly higher ratings about their level of interest in the study.

The researchers used cognitive dissonance to explain the results. The participants who had been paid \$20 (\$160 in 2015 dollars) had a ready explanation for why they had lied to the next participant—they had been well paid to do so. The participants who had been paid \$1 had no such easy explanation. The \$1 participants, in the researchers' view, had thoughts like this in their minds: "The task was very boring, but I just told someone it was fun. I wouldn't lie for a dollar, so the task must have been more fun than I thought it was." They reduced the dissonance between their attitude and behavior by changing their attitude and rating the experiment more positively. Festinger's and Carlsmith's final sentence in the study was, "The results strongly corroborate the theory that was tested."

Statistical standards were different 60 years ago. Festinger and Carlsmith reported the results of the *t* test, but they didn't report an effect size or a confidence interval. Fortunately, it is possible to take some of their results and work backward to find that $s^2_{\text{Pooled}} = 3.98$ and $s_{M_1-M_2} = 0.63$. Armed with these values, an effect size and a confidence interval can be calculated.

First, here are the calculations for both effect sizes:

$$\begin{aligned} d &= \frac{M_1 - M_2}{\sqrt{s^2_{\text{Pooled}}}} \\ &= \frac{1.35 - (-0.05)}{\sqrt{3.98}} \\ &= \frac{1.4000}{1.9950} \\ &= 0.7018 \\ &= 0.70 \end{aligned}$$

$$\begin{aligned} r^2 &= \frac{t^2}{t^2 + df} \times 100 \\ &= \frac{2.22^2}{2.22^2 + 38} \times 100 \\ &= \frac{4.9284}{4.9284 + 38} \times 100 \\ &= \frac{4.9284}{42.9284} \times 100 \\ &= 0.1148 \times 100 \\ &= 11.48\,\% \end{aligned}$$

Now, here are the calculations for the confidence interval:

$$\begin{aligned} 95\%\text{CI}\mu_{\text{Diff}} &= (M_1 - M_2) \pm (t_{cv} \times s_{M_1-M_2}) \\ &= (1.35 - (-0.05)) \pm (2.024 \times 0.63) \\ &= 1.4000 \pm 1.2751 \\ &= \text{from } 0.1249 \text{ to } 2.6751 \\ &= [0.12, 2.68] \end{aligned}$$

From d and r^2, it is apparent that the effect size falls in the medium range. From the confidence interval, one could conclude that though an effect exists, it could be a small one. Based on this study, it would be reasonable to conclude that cognitive dissonance has an effect. But, how much of an effect isn't clear. Cognitive dissonance has gone on to be a well-accepted phenomenon in psychology. However, if this early cognitive dissonance study were being reported by a twenty-first-century researcher, he or she shouldn't say that the results "strongly corroborated" the theory.

Practice Problems 8.3

Apply Your Knowledge

8.07 A dermatologist obtained a random sample of people who don't use sunscreen and a random sample of people who use sunscreen of SPF 15 or higher. She examined each person's skin using the Skin Cancer Risk Index, on which higher scores indicate a greater risk of developing skin cancer. For the 31 people in the no sunscreen (control) condition, the mean was 17.00, with a standard deviation of 2.50. For the 36 people in the sunscreen (experimental) condition, the mean was 10.00, with a standard deviation of 2.80. With $\alpha = .05$, two-tailed, and $df = 65$, $t_{cv} = 1.997$. Calculations showed that $s^2_{\text{Pooled}} = 7.11$, $s_{M_1-M_2} = 0.65$, and $t = 10.77$. Use the results to (a) decide whether to reject the null hypothesis, (b) determine if the difference between sample means is statistically significant, (c) decide which population mean is larger than the other, (d) report the results in APA format, (e) calculate Cohen's d and r^2, and (f) report the 95% confidence interval for the difference between population means.

8.08 An exercise physiologist wondered whether doing one's own chores (yard work, cleaning the house, etc.) had any impact on the resting heart rate. (A lower resting heart rate indicates better physical shape.) He wasn't sure, for example, whether chores would function as exercise (which would lower the resting heart rate) or keep people from having time for more strenuous exercise, increasing the heart rate. The 18 people in the control group (doing their own chores) had a mean of 72.00, with a standard deviation of 14.00. The 17 in the experimental group (who paid others to do their chores) had a mean of 76.00, with a standard deviation of 16.00. Here are the results: $\alpha = .05$, two-tailed, $df = 33$, $t_{cv} = 2.035$, $s^2_{\text{Pooled}} = 225.09$, $s_{M_1-M_2} = 5.07$, and $t = 0.79$. Further, $d = 0.27$, $r^2 = 1.86\%$, and the 95% $\text{CI}\mu_{\text{Diff}}$ ranges from -6.32 to 14.32. Use these results to write a four-point interpretation.

SUMMARY

Differentiate between independent samples and paired samples.

- Two-sample *t* tests compare the mean of one sample (e.g., an experimental group) to the mean of another sample (e.g., a control group) in order to conclude whether the population means differ. There are two types of two-sample *t* tests: independent-samples *t* tests and paired-samples *t* tests. With an independent-samples *t* test, how the cases are selected for one sample has no impact on case selection for the other sample. With paired samples, the selection of cases for one sample influences or determines case selection for the other sample.

Conduct the steps for an independent-samples *t* test.

- To conduct an independent-samples *t* test, the assumptions must be checked, the hypotheses generated, and the decision rule formulated. To find *t*, the difference between the sample means is divided by the standard error of the difference.

Interpret an independent-samples *t* test.

- The decision rule is applied to decide if the null hypothesis is rejected. If it is rejected, the researcher concludes that the difference between *sample* means probably represents a difference between *population* means. Next, Cohen's *d* and r^2 are used to calculate effect size and categorize it as small, medium, or large. If the effect size is meaningful and the null hypothesis was not rejected, the researcher should consider the possibility of Type II error. Finally, the researcher should calculate the confidence interval for the difference and interpret it based on whether (1) it captures zero, (2) how close to zero it comes, and (3) how wide it is.

DIY

Put states into two categories on some basis. You could categorize them into whatever groups you wish—for example, states with nice climates vs. those without, or southern states vs. northern states, or states you would like to live in vs. those you would like to avoid. Include about 5–8 states in each group. Pick some outcome variable on which you want to compare the two groups. For example, is the murder rate different in southern states vs. northern states? Do an online search to find the necessary data. Then use an independent-samples *t* test to analyze the data. Report the results in APA format. Don't forget to calculate an effect size.

KEY TERMS

independent samples – the selection of cases for one sample has no impact on the selection of cases for another sample.

independent-samples *t* test – an inferential statistical test used to compare two independent samples on an interval- or ratio-level dependent variable.

paired samples – case selection for one sample is influenced by, depends on, the cases selected for another sample.

pooled variance – the average variance for two samples.

two-samples *t* test – an inferential statistical test used to compare the mean of one sample to the mean of another sample.

CHAPTER EXERCISES

Review Your Knowledge

8.01 To compute either a single-sample z test or a single-sample t test, one must know the population ____.

8.02 Two sample t tests compare the ____ of one sample to the ____ of another sample.

8.03 Two sample t tests use ____ means to draw a conclusion about ____ means.

8.04 A classic experiment might use a two-sample t test to compare a ____ group to an ____ group.

8.05 Two different types of two-sample t tests are the ____ -samples t test and the ____ -samples t test.

8.06 If each sample in a two-sample t test is a random sample from its population, then the test is an ____-samples t test.

8.07 If the selection of cases for one sample determines the cases selected for the other sample, then the samples are ____ samples.

8.08 ____ is the abbreviation for the total sample size in an independent-samples t test; ____ and ____ are the abbreviations for the sizes of the samples in the two groups.

8.09 In order to use an independent-samples t test to analyze data from two samples, the samples have to be ____ and one needs to know the ____ for each sample.

8.10 The nonrobust assumption for an independent-samples t test is ____.

8.11 The ____ assumption for the independent-samples t test is the one that allows a researcher to generalize the results back to the larger population.

8.12 The ____ assumption for the independent-samples t test says that the amount of variability in the two populations is about equal.

8.13 Researchers are often willing, for an independent-samples t test, to assume that the dependent variable is ____.

8.14 One tests the ____ assumption for the independent-samples t test by comparing the ____ of the two samples.

8.15 The hypotheses for an independent-samples t test are either directional or ____ directional.

8.16 The null hypothesis for a two-tailed independent-samples t test, expressed mathematically, is ____.

8.17 The alternative hypothesis for a two-tailed independent-samples t test, expressed mathematically, is ____.

8.18 If the null hypothesis for an independent-samples t test is true, then the observed difference between the sample means is due to ____.

8.19 The critical value of t is the border between the ____ and the ____ zones of the sampling distribution of ____.

8.20 t tests commonly are ____ tailed and have ____ set at .05.

8.21 To calculate the degrees of freedom for an independent-samples t test, subtract ____ from N.

8.22 For a two-tailed test, if t ____ $-t_{cv}$, reject H_0.

8.23 For a two-tailed test, if t ____ t_{cv}, reject H_0.

8.24 When writing the hypotheses for a one-way test, it is easier if one formulates the ____ hypothesis first.

8.25 ____ is the variance of the two samples combined.

8.26 To calculate $s_{M_1-M_2}$, one needs to know the ____ variance, the total sample ____, and the sample ____ of each group individually.

8.27 The numerator in the t equation is the difference between the sample ____.

8.28 If a researcher rejects the ____, the researcher is forced to accept the ____.

8.29 If a researcher reports the results of an independent-samples t test as showing a statistically significant difference, the researcher has ____ the null hypothesis.

8.30 If one rejects the null hypothesis for an independent-samples *t* test, then look at the sample ____ in order to comment on the ____ of the difference.

8.31 The .05 in APA format indicates that there is a ____% chance of a Type I error.

8.32 If the result of an independent-samples *t* test is written as $t(23) = 5.98$, $p < .05$, then *N* was ____.

8.33 In APA format, ____ means one rejected a null hypothesis with alpha set at .05 and ____ means one failed to reject it.

8.34 For an independent-samples *t* test, calculate ____ or ____ to quantify the size of the effect.

8.35 If there is absolutely no effect of the independent variable on the dependent variable, then *d* equals ____.

8.36 A *d* of ≈ ____ is considered a medium effect.

8.37 r^2 calculates the percentage of variability in the ____ that is accounted for by the ____.

8.38 The 95% confidence interval for the difference between population means estimates how ____ or how ____ the difference between the population means might be.

8.39 The 95% confidence interval for the difference between population means *probably* captures the real difference between the ____.

8.40 If the 95% confidence interval for the difference between population means fails to capture zero for a two-tailed test with $\alpha = .05$, then one has ____ the null hypothesis.

8.41 If the 95% confidence interval for the difference between population means comes close to zero, the size of the effect in the population may be ____.

8.42 If the 95% confidence interval for the difference between population means is wide, a reasonable suggestion is to ____ the study with a larger ____.

8.43 If sample size increases but the sample means and standard deviations don't change, then of the three values calculated for the interpretation of the independent-samples *t* test, the one that will not change is ____.

8.44 ____ calculate the size of the effect in the sample; a confidence interval calculates it for the ____.

Apply Your Knowledge

Selecting a test

8.45 A theology professor was curious whether children were as religious as their parents. He obtained a random sample of students at his school and administered an interval-level religiosity scale to them. Using the same scale, he collected information from the same-sex parent for each student. What statistical test should he use to see if there is a difference between a parent's and a child's level of religiosity?

8.46 A demographer working for the U.S. Census Bureau wants to compare salaries for urban vs. rural areas. She gets a sample of psychologists who live in urban areas and a sample of psychologists who live in rural areas. From each, she finds out his or her annual income. What statistical test should she use to see if a difference exists in a psychologist's income as a function of residential status?

8.47 An exercise physiologist classifies people—on the basis of their body mass index, heart rate, and lung capacity—as (a) above average in terms of fitness or (b) below average in terms of fitness. He then directs the same people to walk on a treadmill, individually, at an increasing speed until they can no longer walk. The speed of the treadmill when a person maxes out on walking is the dependent variable. What statistical test should the physiologist use to see if there is a difference in maximum walking speed based on fitness level?

8.48 Some people have white coat hypertension. That is, they grow anxious when a person with a white coat and a stethoscope walks into the examining room to take their blood pressure. As a result, their blood pressure increases.

A family practitioner believes this is quite common. To test her theory, she puts together a random sample of 50 patients and takes two blood pressure measurements, one when each patient first walks into the room and a second, unexpected one after about 15 minutes. What statistical test should she use to see if the two blood pressures differ?

Checking the assumptions

8.49 A developmental psychologist has randomly assigned men to two different groups. After sitting alone at a computer monitor to read a series of stories, each man is asked to rate his level of acceptance of gender typing. (Gender-typed people believe that there are certain roles men should fulfill and certain roles women should fulfill.) Higher scores on the scale indicate higher levels of gender typing; scores on the scale are normally distributed. One group of men, the control group, read gender-typed stories and the other group, the experimental group, read non-gender-typed stories. The researcher found $M_C = 75$, $s_C = 35$, $M_E = 55$, $s_E = 8$. The psychologist is planning to use an independent-samples *t* test to see if the experimental manipulation has had an impact. Check the assumptions and decide if it is OK to proceed with the planned test.

8.50 A clinical psychologist is studying the effects of an experimental medication on depression. He randomly assigns the next 50 patients at his clinic to receive either (a) Prozac or (b) a placebo. Each patient is treated individually. After eight weeks of treatment, each patient completes an interval-level depression scale. The standard deviations for the two groups are similar and the psychologist believes depression level is normally distributed. The psychologist is planning to use an independent-samples *t* test to see if there are differences between the two groups at the eight-week mark. Check the assumptions and decide if it is OK to proceed with the planned test.

Writing hypotheses

8.51 An infectious disease specialist is using an independent-samples *t* test to compare the effectiveness of two treatments for the common cold. (a) Write out H_0 and H_1 and (b) explain what they mean.

8.52 A medical educator is using an independent-samples *t* test to compare the age of physicians who complete the minimum number of continuing education hours per year vs. those who complete extra hours of continuing education. (a) Write out H_0 and H_1 and (b) explain what they mean.

8.53 There is a lot of evidence that fluoride reduces cavities but not all communities add it to their drinking water. A dentist, who expects to replicate this earlier work, classifies randomly selected communities in his state as (1) adding fluoride to their drinking water or (2) not adding fluoride to their drinking water. He then goes to high schools in these communities, inspects the mouths of all high school seniors, and calculates, for each community, the percentage of these students with cavities. He will compare the means for these values between the two types of communities. (a) Write out H_0 and H_1 and (b) explain what they mean.

8.54 The health department physician in a suburban community warned cat owners about a risk of infection. Cats leave their litter boxes and then jump on counters, trailing bacteria behind them. To highlight the greater risk to the health of cat owners than dog owners, the physician went to the homes of cat owners and dog owners, swabbed kitchen counters, cultured the swabs, and counted the number of bacteria that grew. She planned to use a *t* test to compare the mean number of bacteria in cat-owning households vs. dog-owning houses. (a) Write out H_0 and H_1 and (b) explain what they mean.

Finding t_{cv}

8.55 If $n_1 = 223$ and $n_2 = 252$, determine the critical value of *t* for an independent-samples *t* test, two-tailed, $\alpha = .05$.

8.56 If $n_1 = 17$ and $n_2 = 18$, determine the critical value of *t* for an independent-samples *t* test, two-tailed, $\alpha = .05$.

8.57 If $n_1 = 46$ and $n_2 = 46$, determine the critical value of t for an independent-samples t test, two-tailed, $\alpha = .01$.

8.58 If $n_1 = 13$ and $n_2 = 15$, determine the critical value of t for an independent-samples t test, one-tailed, $\alpha = .05$, where the numerator of the t equation is expected to be negative.

Writing the decision rule

8.59 If $t_{cv} = 2.086$, write the decision rule for a two-tailed test for (a) when to reject the null hypothesis and (b) when to fail to reject the null hypothesis.

8.60 If $t_{cv} = 2.396$, write the decision rule for a one-tailed test for (a) when to reject the null hypothesis and (b) when to fail to reject the null hypothesis. (*Hint:* Contemplate the sign of t_{cv} and what that means about what the researcher believes.)

Calculating the pooled variance

8.61 Given $n_1 = 12$, $s_1 = 7.4$, $n_2 = 13$, and $s_2 = 8.2$, calculate s^2_{Pooled}.

8.62 Given $n_1 = 15$, $s_1 = 3.6$, $n_2 = 16$, and $s_2 = 4.3$, calculate s^2_{Pooled}.

Calculating the standard error of the difference

8.63 Given $n_1 = 45$, $n_2 = 58$, and $s^2_{Pooled} = 5.63$, calculate $s_{M_1-M_2}$.

8.64 Given $n_1 = 23$, $n_2 = 19$, and $s^2_{Pooled} = 12.88$, calculate $s_{M_1-M_2}$.

8.65 Given $n_1 = 45$, $n_2 = 58$, $s_1 = 5.98$, and $s_2 = 7.83$, calculate $s_{M_1-M_2}$.

8.66 Given $n_1 = 22$, $n_2 = 28$, $s_1 = 9.58$, and $s_2 = 11.13$, calculate $s_{M_1-M_2}$.

Calculating t

8.67 Given $M_1 = 57$, $M_2 = 68$, and $s_{M_1-M_2} = 2.34$, calculate t.

8.68 Given $M_1 = 5.5$, $M_2 = 4.5$, and $s_{M_1-M_2} = 1.23$, calculate t.

8.69 Given $M_1 = -5$, $s_1 = 4.6$, $n_1 = 72$, $M_2 = -1$, $s_2 = 3.3$, and $n_2 = 60$, calculate t.

8.70 Given $M_1 = 48$, $s_1 = 15.0$, $n_1 = 8$, $M_2 = 52$, $s_2 = 14.2$, and $n_2 = 11$, calculate t.

Deciding whether the null hypothesis was rejected

8.71 Given $M_1 = 98$, $M_2 = 103$, $t_{cv} = 2.060$, $t = 2.060$, and a two-tailed test with $\alpha = .05$, (a) decide whether the null hypothesis was rejected or not, (b) tell whether the difference between sample means is a statistically significant one or not, and (c) make a statement about the direction of the difference between the sample means.

8.72 Given $M_1 = 88$, $M_2 = 83$, $t_{cv} = 2.042$, $t = 2.040$, and a two-tailed test with $\alpha = .05$, (a) decide whether the null hypothesis was rejected or not, (b) tell whether the difference between sample means is a statistically significant one or not, and (c) make a statement about the direction of the difference between the population means.

Using APA format

8.73 Given $N = 23$ and $t = 2.0723$, report the results in APA format for a two-tailed test, $\alpha = .05$.

8.74 Given $N = 35$ and $t = 2.0321$, report the results in APA format for a two-tailed test, $\alpha = .05$.

8.75 Given $N = 10$ and $t = 2.3147$, report the results in APA format for a two-tailed test, $\alpha = .05$.

8.76 Given $N = 73$, $\alpha = .05$, one-tailed, t expected to be negative, and $t = -1.65$, report the results in APA format.

Calculating effect sizes

8.77 Given $M_1 = 12$, $M_2 = 17$, and $s^2_{Pooled} = 4.00$, (a) calculate d and (b) classify the size of the effect.

8.78 Given $M_1 = 88$, $M_2 = 85$, and $s^2_{Pooled} = 81.00$, (a) calculate d and (b) classify the size of the effect.

8.79 Given $t = 9.87$ and $N = 73$, calculate r^2.

8.80 Given $t = 1.34$ and $N = 49$, calculate r^2.

Calculating confidence intervals

8.81 Given $M_1 = 31$, $M_2 = 24$, $s_{M_1-M_2} = 2.88$, and $t_{cv} = 2.045$, (a) calculate the 95% confidence interval for the difference between population means, and (b) based on the confidence interval, decide if the null hypothesis should be rejected for a nondirectional test with $\alpha = .05$.

8.82 Given $M_1 = -13$, $M_2 = -18$, $s_{M_1-M_2} = 1.48$, and $t_{cv} = 2.015$, (a) calculate the 95% confidence interval for the difference between population means, and (b) based on the confidence interval, decide if the null hypothesis should be rejected for a nondirectional test with $\alpha = .05$.

Writing a four-point interpretation

8.83 An elementary education researcher was interested in seeing how the color used to make corrections on students' papers affected their self-esteem. He assembled first graders and asked them to take a third-grade math test. He told the first graders that the test would be very difficult for them and they might not get very many answers right, but he needed their help. After the students were each called into a room to take the test alone, he pretended to grade it. Everyone had 25% of their answers marked wrong. For half the kids, these answers were marked with red ink, and for the other half, the "incorrect" answers were marked with pencil. Each child then took a self-esteem inventory on which higher scores indicate more self-esteem. The 17 red ink (control group) kids had a mean of 23.00 ($s = 5.00$); the 10 pencil (experimental group) kids had a mean score of 29.00 ($s = 5.00$). Given that information and the rest of the results (below), write a paragraph interpreting the results:

- $\alpha = .05$, two-tailed
- $t_{cv} = 2.060$
- $s^2_{Pooled} = 25.00$
- $s_{M_1-M_2} = 1.99$
- $t = 3.02$
- $d = 1.20$
- $r^2 = 26.61\%$
- 95% $CI\mu_{Diff}$ [1.90, 10.10]

8.84 A nutritionist compared the effectiveness of an online diet program to that of an in-person diet program. After three months, she compared the number of pounds of weight lost. The control group (in-person) lost a mean of 18.00 pounds ($s = 14.50$, $n = 16$) and the experimental group (online) lost 16.00 pounds ($s = 13.30$, $n = 21$). Using that and the information below, write a paragraph interpreting the results:

- $\alpha = .05$, two-tailed
- $t_{cv} = 2.030$
- $s^2_{Pooled} = 191.19$
- $s_{M_1-M_2} = 4.59$
- $t = 0.44$
- $d = -0.14$
- $r = 0.54\%$
- 95% $CI\mu_{Diff}$ [−11.32, 7.32]

Completing all six steps of hypothesis testing

8.85 A physician compared the cholesterol levels of a representative sample of Americans who ate an American diet vs. a representative sample of those who followed a Mediterranean diet. Below are the means, standard deviations, and sample sizes for both samples. Though in some studies a Mediterranean diet has been shown to be beneficial, this was one of the first studies on an American population and the physician had made no advance predictions about the outcome.

- American diet (control): $M = 230$, $s = 24$, $n = 36$
- Mediterranean (experimental): $M = 190$, $s = 26$, $n = 36$

8.86 An addictions researcher measured tolerance to alcohol in first-year and fourth-year college students. She gave participants a standard dose of alcohol and then had them walk along a narrow line painted on the floor. The higher the percentage of the distance that they were on the line, the greater their tolerance to alcohol. The researcher expected that the older students would show more tolerance to alcohol. Here is the relevant information:

- 1st year (control): $M = 30$, $s = 12.5$, $n = 20$
- 4th year (experimental): $M = 48$, $s = 14.6$, $n = 16$

Expand Your Knowledge

8.87 A researcher completes an independent-samples *t* test and finds that the probability of two sample means being this far apart, if the null hypothesis is true, is less than .05. Which of the following is true?

a. $\mu_1 = \mu_2$
b. $M_1 \neq \mu_1$
c. There probably is no difference between the two population means.
d. There probably is a difference between the two population means.
e. Sufficient evidence does not exist to draw any conclusion about the population means.

8.88 Which result cannot be true for an independent-samples *t* test?

a. A researcher has rejected the null hypothesis and found $d = 1.50$.
b. A researcher has failed to reject the null hypothesis and found $d = 1.50$.
c. A researcher has rejected the null hypothesis and found $d = 0.10$.
d. A researcher has failed to reject the null hypothesis and found $d = 0.10$.
e. Any of these results can be true.
f. None of these results can be true.

8.89 A consumer group is planning to do 2 two-sample *t* tests. In Test 1, they are going to put together a random sample of items at a jewelry store and compare the prices to a random sample of items at a bookstore, *in order to see which store is more expensive.* In Test 2, they are planning to compare a random sample of textbooks purchased at a campus bookstore to the same books purchased through an online bookseller, *in order to see which store is more expensive.* (a) Determine which test is an independent-samples *t* test and which a paired-samples *t* test. (b) It sounds like each test is answering the same question, "Which store is more expensive?" Rewrite the questions so that they more accurately pose the question that the test answers.

8.90 Explain why the critical value of *t*, one-tailed, $\alpha = .05$ is the same as the critical value of *t*, two-tailed, $\alpha = .10$.

8.91 Margery collected some data from two independent groups and analyzed them with an independent-samples *t* test. No assumptions were violated and she rejected the null hypothesis. Yet, when she calculated a confidence interval for the difference between population means, the confidence interval captured zero. Explain how this is possible.

8.92 Dr. Goddard developed a technique that he thought would increase IQ in adults. He obtained a random sample of 52 adult Americans and randomly assigned half of them to a control group and half to the experimental group. He did nothing to the control group, but he administered his IQ-increasing treatment to the 26 in the experimental group. Afterward, he measured IQs and found that the mean for the control group was 100, while the experimental group had a mean IQ of 102. The standard deviation in both groups was 15. Dr. Goddard found $s_{M_1-M_2} = 4.16$, $t = 0.48$, $r^2 = 0.46\%$, and that the 95% for the difference between population means ranged from –6.36 to 10.36. (a) Did Dr. Goddard reject the null hypothesis? (b) What conclusion should he reach about whether his treatment works to increase IQ? (c) How big is the size of the effect as determined by r^2? (d) What information does the confidence interval give on how sure we are about the impact of the IQ-increasing technique? (e) How worried are you that Dr. Goddard made a Type II error? (f) Do you recommend replicating with a larger sample size?

8.93 Dr. Brigham decided to replicate Dr. Goddard's study (see Exercise 8.92) with a larger sample. She did exactly what Dr. Goddard did, but had 1,002 subjects (501 in the control group and 501 in the experimental group). The means for the two groups were exactly the same, 100 and 102, and both groups again had standard deviations of 15. Dr. Brigham found $s_{M_1-M_2} = 0.95$, $t = 2.11$, $r^2 = 0.44\%$, and the 95% confidence interval ranged from .14 to 3.86. (a) Did Dr. Brigham reject the null hypothesis? (b) What conclusion should she reach about whether Dr. Goddard's treatment increases IQ? (c) How big is the size of the effect as determined by r^2? (d) What information does the confidence interval give on how sure we are about the impact of the IQ-increasing technique? (e) How worried are you that Dr. Brigham made a Type I error?

8.94 Compare your answers for Exercises 8.92 and 8.93. (a) What is the impact of sample size on rejecting the null hypothesis? (b) On conclusions about the effectiveness of treatment? (c) On the size of the effect as determined by r^2? (d) On the confidence interval?

SPSS

Data entry for an independent-samples *t* test in SPSS takes two columns. An example using data for the depth of processing study can be seen in Figure 8.8.

	Depth	Num_recall	va
1	1.00	1.00	
2	1.00	3.00	
3	2.00	11.00	
4	1.00	3.00	
5	1.00	5.00	
6	2.00	6.00	
7	2.00	4.00	
8	2.00	13.00	
9			
10			

Figure 8.8 Data Entry in SPSS for an Independent-Samples *t* Test "Depth" is the independent variable used to classify cases into groups, 1 for shallow processing and 2 for deep processing. The dependent variable is "Num_recall." (Source: SPSS)

The first column, with the variable named "Depth," contains information about each case's status on the independent variable. SPSS calls this the "Grouping Variable" because it is used to assign cases into groups, either into the shallow processing group or the deep processing group. SPSS uses numbers, not words, to classify cases. Here, "1" means the case belongs to the shallow processing group and "2" the deep processing group. Note that all the shallow processing cases don't have to be next to each other. As long as a case has the right group number associated with it, SPSS will correctly classify it.

The second column, with the variable named "Num_recall," contains the dependent variable, how many words were recalled. The first case recalled 1 word, the second 3 words, the third 11 words, and so on.

Figure 8.9 shows where the independent-samples *t* test is located in SPSS, under "Analyze," then "Compare Means," and finally, "Independent-Samples T Test. . . ." When one clicks on Independent-Samples T-Test, the box shown in Figure 8.10 opens up.

In Figure 8.10, the arrow button has already been used to move the dependent variable, Num_recall, into the box for the "Test Variable(s)." The independent variable, Depth, has been moved into the "Grouping Variable" box. Note that the grouping variable now appears as "Depth(? ?)," to indicate that one needs to define the groups.

Figure 8.11 shows the box that opens up when one clicks on the "Define Groups" button in Figure 8.10. A value of 1 for "Group 1" and a value of 2 for "Group 2" were entered. Then click on the "Continue" button, which brings up the box seen in Figure 8.12.

Figure 8.12 is like Figure 8.10, but the grouping variable has been defined. We can tell SPSS to complete the *t* test by clicking the "OK" button.

Figure 8.13 shows all the output that SPSS provides. The first box gives descriptive statistics for the two groups. The SPSS results match Dr. Villanova's closely, though SPSS reports the standard deviations to three decimal places.

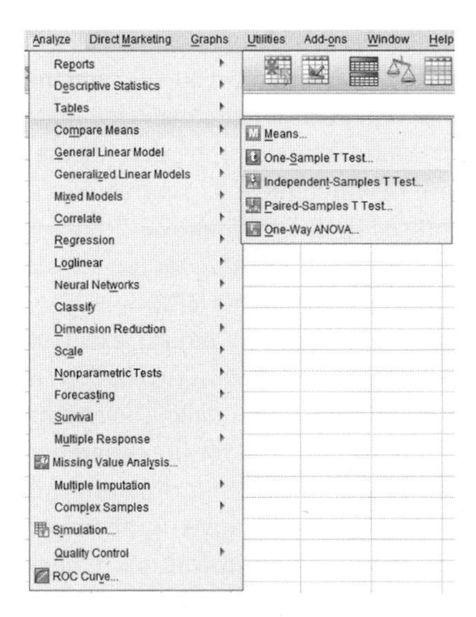

Figure 8.9 Starting an Independent-Samples *t* Test in SPSS The commands for an independent-samples *t* test in SPSS can be found under "Analyze," then "Compare Means." (Source: SPSS)

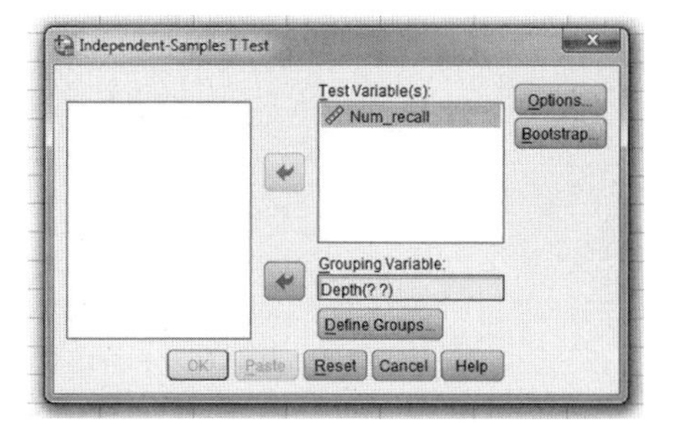

Figure 8.10 Defining Variables as Dependent and Independent in SPSS The dependent variable is called a "Test Variable" in SPSS and the explanatory variable a "Grouping Variable." (Source: SPSS)

Figure 8.11 Defining the Grouping Variable for an Independent-Samples *t* Test in SPSS Once the grouping variable has been defined, SPSS has to be informed which value is for which group. Group 1 represented by the value "1" is the shallow processing group; Group 2 with a value of "2" is the deep processing group. (Source: SPSS)

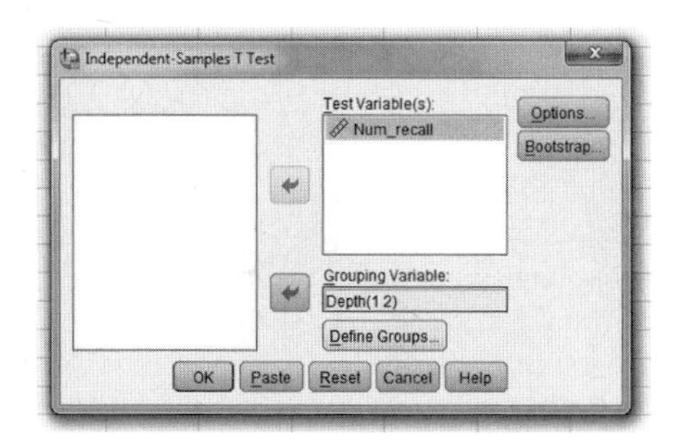

Figure 8.12 Running an Independent-Samples *t* Test in SPSS Once the dependent variable has been set as a test variable and the explanatory variable has been defined as the grouping variable, all that remains is to click on the "OK" button. (Source: SPSS)

Group Statistics

	Depth	N	Mean	Std. Deviation	Std. Error Mean
Num_recall	Shallow	18	3.50	1.543	.364
	Deep	20	8.30	2.736	.612

Independent Samples Test

		Levene's Test for Equality of Variances		t-test for Equality of Means						
									95% Confidence Interval of the Difference	
		F	Sig.	t	df	Sig. (2-tailed)	Mean Difference	Std. Error Difference	Lower	Upper
Num_recall	Equal variances assumed	6.432	.016	-6.558	36	.000	-4.800	.732	-6.284	-3.316
	Equal variances not assumed			-6.744	30.546	.000	-4.800	.712	-6.252	-3.348

Figure 8.13 Printout for an Independent-Samples *t* Test in SPSS SPSS provides a lot of printout. The first table gives descriptive statistics for the two groups and the second the results of the *t* test. Look in the row labeled "Equal variances assumed" to find the results. (Source: SPSS)

The second box of output presents the results of the *t* test. SPSS gives more information than we need. Just pay attention to the first row, the one that says "Equal variances assumed." SPSS reports a negative value for the test statistic, while Dr. Villanova found a positive value. The different sign for the *t* values doesn't matter because it is merely a result of which mean is subtracted from the other. Because SPSS carries more decimal places than Dr. Villanova did, its *t* value (6.558) is more accurate than his (6.58). Similarly, the SPSS value for the standard error of the difference (.732) is more accurate than Dr. Villanova's (.73).

SPSS also reports degrees of freedom (36) and the exact, two-tailed significance level. If this value (here, .000) is less than or equal to .05, then reject the null hypothesis. If the value is greater than .05, then fail to reject the null hypothesis.

SPSS reports the 95% confidence interval for the difference between population means. Because SPSS subtracted the means in a different order than Dr. Villanova did, it reports the confidence interval as negative numbers, from −6.284 to −3.316. (The SPSS confidence interval is also reported with more decimal places.) Don't let the sign become a concern—by referring back to the population means, one can figure out the direction of the confidence interval.

Finally, SPSS does not report Cohen's *d*. And, unfortunately, it does not report the pooled variance so that Cohen's *d* can be calculated by hand. To calculate *d*, go through the first five steps of Equation 8.3 to calculate s^2_{Pooled}.

The Paired-Samples *t* Test

LEARNING OBJECTIVES

- Recognize the different types of dependent samples.
- Calculate and interpret a paired-samples *t* test.

CHAPTER OVERVIEW

This is the third, and final, chapter on *t* tests. Chapter 7 covered a *t* test for comparing a sample mean to a specified value, the single-sample *t* test. Chapter 8 moved to a *t* test for comparing the means of two *independent* samples. Now, in Chapter 9, we add a **paired-samples *t* test**, for comparing the means of two *dependent* samples.

9.1 Paired Samples

When samples are dependent, each case consists of a pair of data points, one data point from each of two samples. In dependent samples, also called paired samples, the data points may be paired in a variety of ways. One pairing, called a **repeated-measures design**, means the same participants provide data at two points in time. (This is also called **longitudinal research** because it follows participants over time or a **pre-post design** as participants are measured on the outcome variable before and after an intervention.) An example of a repeated-measures design would be measuring the level of anxiety in people before and after learning relaxation techniques.

Another type of pairing, called a **within-subjects design,** involves the same participants being measured in two different situations or under two different conditions. For example, a cognitive psychologist might measure how much information people retain when studying in silence and then measure information retention *for the same participants* when they study while listening to music.

Repeated-measures and within-subjects designs involve one sample of cases measured at two points in time or under two conditions. Find this confusing. "Why," they ask, "is it called a *two*-sample test when there is just *one* sample of cases?" Unfortunately, this is statistical terminology that just needs to be learned. To a statistician, each condition in a dependent samples study is considered a "sample."

Repeated-measures and within-subjects designs have a significant advantage over independent-samples designs. These dependent-samples designs control for **individual differences,** attributes that vary from case to case. Because the same participants are in both groups, the researcher can be sure that the two samples are comparable in terms of background characteristics. As a result of this, the researcher can be more confident that any observed difference between the groups on the outcome variable is due to the explanatory variable and not some confounding variable.

In an attempt to derive this benefit, researchers have developed a number of other paired-samples techniques in which different participants are in two conditions. In one, the pairs have some similarity because of some connection, either biological (such as that between two siblings), or formed (such as that between a romantic couple). Another is called *matched pairs.* In **matched pairs**, participants are grouped, by the researcher, into sets of two based on their being similar on potential confounding variables. For example, if a dean were comparing the GPAs of male and female students, she might want to match them, based on IQ, into male–female pairs. That way, a researcher couldn't argue that intelligence was a confounding variable if one sex had a higher GPA.

Because there are so many different types of paired samples, the paired-samples *t* test has more names than any other test in statistics. But, whether it is called a paired-samples *t* test, dependent-samples *t* test, correlated-samples *t* test, related-samples *t* test, matched-pairs *t* test, within-subjects *t* test, or repeated-measures *t* test, it is all the same test.

The wide number of different names reflects how commonly used the paired-samples *t* test is. It is a commonly used test for several reasons. One reason is that many experimental situations are of a pre-post design where the outcome variable is measured before and after the explanatory variable is applied. Another reason is that controlling individual differences makes studies that use paired samples more powerful than studies that use independent samples. In a statistical sense, being more powerful means that the probability of being able to reject the null hypothesis, when it is false, is higher. As a result, a researcher needs a smaller sample size for a paired-samples *t* test than for an independent-samples *t* test. This is a big advantage of paired-samples *t* tests. If a researcher is studying a rare phenomenon or one where participants are hard to come by, a dependent-samples design is the way to go.

If a researcher is studying a rare phenomenon or one where participants are hard to come by, a dependent-samples design is the way to go.

Here is an example of research that used paired samples to investigate how stress affects recovery from a physical wound (Kiecolt-Glaser et al., 1995). One sample, the people who were under stress, consisted of women who were caring for a husband or mother with Alzheimer's disease. Because the researchers believed that age and socioeconomic status might influence physical recovery, they matched each caregiver with a control participant of the same sex, age, and family income who was not a caregiver. So, the participants were matched pairs of women, one a caregiver (the experimental group) and one a control.

Using a dermatology procedure, the researchers made a small wound on each participant's forearm and timed how long it took to heal. The wound took almost 10 days longer on average to heal in the caregivers ($M = 48.7$) than in the controls ($M = 39.3$), and this difference was statistically significant. Why did this difference exist? Well, because the pairs were matched on age and socioeconomic status, it can't be argued that the caregivers were older or poorer. With these confounding variables removed, it seems more plausible that it is the stress of caring for someone who is deteriorating with a chronic illness that affects how quickly one heals from a physical wound.

Now, having seen paired-samples in action and observed their advantages, it's time to learn how to perform a paired-samples *t* test.

9.2 Calculating the Paired-Samples *t* Test

Here are some data that are appropriate for analysis with a paired-samples *t* test. Imagine that a sensory psychologist, Dr. Keim, wanted to examine the effect of humidity on perceived temperature. She obtained six volunteers at her college and tested them, individually, in a temperature- and humidity-controlled chamber. Each participant was tested twice and the tests were separated by 24 hours. For both tests, the temperature in the chamber was set at 76°F. For one test the humidity level was "low," and for the other test the humidity level was "high." In order to avoid any effects due to the order of the tests, which humidity level each participant would experience first was randomly determined. For a test, a participant spent 15 minutes in the chamber, after which he or she was asked what the temperature was inside it. This "perceived temperature" is the study's dependent variable.

The data from the humidity study are shown in Table 9.1, where each case appears on a row and each condition (sample) in a column. Table 9.1 also contains the means and standard deviations for both conditions. Remember, in both conditions the actual temperature was the same, 76°F. The mean perceived temperature on the low-humidity test was 75.00°F, and in the high-humidity condition it was 82.50°F. Between the two test conditions, there was a 7.50°F difference in the means.

TABLE 9.1 The Effect of Humidity Level on Perceived Temperature in °F

Participant	Low-Humidity Test (control condition)	High-Humidity Test (experimental condition)	Difference Score
1	76	81	5.00
2	80	90	10.00
3	78	85	7.00
4	72	82	10.00
5	76	82	6.00
6	68	75	7.00
$M =$	75.00	82.50	7.50
$s =$	4.34	4.93	2.07

The difference score is calculated for each pair (row) of scores by subtracting the value for one test condition from the value for the other test condition. It doesn't matter which value is subtracted from the other, as long as the same order is used consistently. Here, in order to end up with positive numbers, the low-humidity condition is subtracted from the high-humidity condition.

Figure 9.1 is a pair of box-and-whisker plots, showing the median, interquartile range, and minimum/maximum for each condition. When looking at the graph, the difference seems clear—temperature does appear to be perceived as higher when the humidity is higher. But, looks can be deceiving and the sample size is low, so hypothesis testing is necessary to find out if the difference is a statistically significant one or if it can be explained by sampling error.

Step 1 Pick a Test. Dr. Keim, using a within-subjects design, is comparing the means for a sample of people measured in two different conditions, low humidity and high humidity. Remember, when one sample is measured in two different conditions,

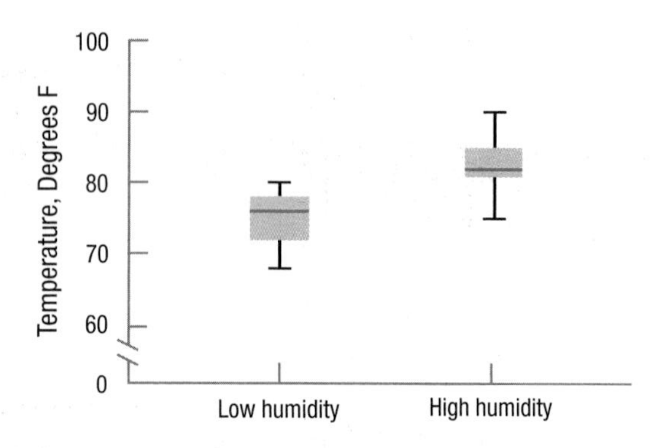

Figure 9.1 Box-and-Whisker Plots Showing Perceived Temperature in Low- and High-Humidity Conditions This graphic appears to show that temperature is perceived as hotter when the humidity is higher. To determine whether the difference between the two conditions is a statistically significant one, a paired-samples t test is needed.

as happens here, statisticians consider this *two* dependent samples. This situation, comparing the means of dependent samples, calls for a paired-samples t test.

Tom **a**nd **H**arry **d**espise **c**rabby **i**nfants

Step 2 Check the Assumptions. There are three assumptions for the paired-samples t test (Table 9.2), and they are familiar. The first assumption is that the sample is a random sample from the population to which the results will be generalized. Dr. Keim would like to be able to generalize her results to people in general, that is, to all the people in the world. She recognizes that she has a convenience sample not a random sample from this population, so this assumption is violated. The random samples assumption is robust, however, so she can continue with the test. She is aware of no one who suggests that Americans and/or college students perceive heat differently from others, so she will be willing to generalize her results more broadly than just to U.S. college students.

The second assumption is that the observations are independent within a sample. Be careful in assessing this assumption: it refers to independence *within* a

TABLE 9.2 Assumptions for the Paired-Samples *t* Test

Random sample: The sample is a random sample from the population.	Robust to violation
Independence of observations: Each case within a group or condition is independent of the other cases in that group or condition.	Not robust to violation
Normality: The population of difference scores is normally distributed.	Robust to violation

sample, not *between* samples. Since the same cases are in both samples, or conditions, *the two samples* are not independent. However, each person in each sample is tested individually and each person is only tested once in each condition. The independence of observations assumption is not violated.

The third assumption, the normality assumption, says that in the larger population *the difference scores* are normally distributed. Look at Table 9.1 and note that there's a third column listing the difference between the value a case had in one condition (high humidity) and its score in the other condition (low humidity). This is called a difference score, abbreviated D, and is needed to calculate a paired-samples t test. The formula for D is shown in Equation 9.1.

Equation 9.1 Formula for Calculating Difference Score, D

$$D = X_1 - X_2$$

where D = the difference score being calculated
X_1 = a case's score in Condition 1
X_2 = a case's score in Condition 2

In calculating D, it matters little which value is subtracted from the other, as long as the same order is followed for all cases. Most people prefer to work with positive numbers, so feel free to decide the order for the subtraction to maximize the number of positive values. Dr. Keim chose to subtract the low-humidity scores from the high-humidity scores in order to have positive difference scores.

For the first case, the person who perceived the low-humidity condition as 76° and the high-humidity condition as 81°F, the difference score is calculated as

$$\begin{aligned} D &= X_{\text{HighHumidity}} - X_{\text{LowHumidity}} \\ &= 81 - 76 \\ &= 5.0000 \\ &= 5.00 \end{aligned}$$

A sample size of 6 is a little small for making decisions about the shape of a parent population. However, based on her previous research, Dr. Keim is willing to assume that the difference scores are normally distributed in the population, so the normality assumption is not violated.

Step 3 List the Hypotheses. The hypotheses, which are statements about populations, are going to be the same as they were for the independent-samples t test. The null hypothesis is going to say that the two population means are the same. The alternative hypothesis will state that the two population means differ, but it won't indicate whether the difference is large or small, positive or negative. These hypotheses are nondirectional or two-tailed, so they don't specify a direction for the difference. Thus, Dr. Keim is testing for two possibilities: (1) low humidity is perceived as hotter than high humidity, or (2) high humidity is perceived as hotter than low humidity. The generic form of two-tailed hypotheses is

$$H_0: \mu_1 = \mu_2$$
$$H_1: \mu_1 \neq \mu_2$$

The specific form of the two hypotheses for this study is

$$H_0: \mu_{\text{LowHumidity}} = \mu_{\text{HighHumidity}}$$
$$H_1: \mu_{\text{LowHumidity}} \neq \mu_{\text{HighHumidity}}$$

Of course, it is possible to have directional hypotheses for a paired-samples *t* test. If Dr. Keim had predicted, in advance, that high humidity would lead to a higher perceived temperature, her hypotheses would have been

$$H_0: \mu_{\text{LowHumidity}} \geq \mu_{\text{HighHumidity}}$$
$$H_1: \mu_{\text{LowHumidity}} < \mu_{\text{HighHumidity}}$$

Step 4 Set the Decision Rule. The decision rule for a paired-samples *t* test is formulated the same way as it was for the independent-samples *t* test. The critical value of *t*, t_{cv}, found in Appendix Table 3, is based on (1) the number of tails for the test, (2) how willing one is to make a Type I error (i.e., the alpha level), and (3) how many degrees of freedom there are. The default option, or the most common form of the paired-samples *t* test, as for other statistical tests, is a two-tailed test with alpha set at .05.

Once the observed value of *t* is calculated (Step 5), it is compared to the critical value in order to decide whether or not to reject the null hypothesis. For a two-tailed test, the general form of the decision rule is:

- If $t \leq -t_{cv}$ or $t \geq t_{cv}$, reject the null hypothesis.
- If $-t_{cv} < t < t_{cv}$, fail to reject the null hypothesis.

Dr. Keim is doing a two-tailed test and is content to use the default alpha level of .05. All that she needs to do is calculate degrees of freedom. Equation 9.2 is the formula for calculating degrees of freedom for a paired-samples *t* test.

Equation 9.2 Degrees of Freedom (*df*) for a Paired-Samples t Test

$$df = N - 1$$

where df = the degrees of freedom
N = the number of *pairs* of cases

Dr. Keim's study involves six *pairs* of cases, so degrees of freedom are calculated as

$$df = 6 - 1$$
$$= 5$$

Look in Appendix 3 and find the intersection of the column for a two-tailed hypothesis test with the alpha set at .05 and the row for $df = 5$. There, the critical value of *t*, ±2.571, is found. Figure 9.2 uses the critical values to mark the rare zone, where the null hypothesis is rejected, and the common zone, where it is not.

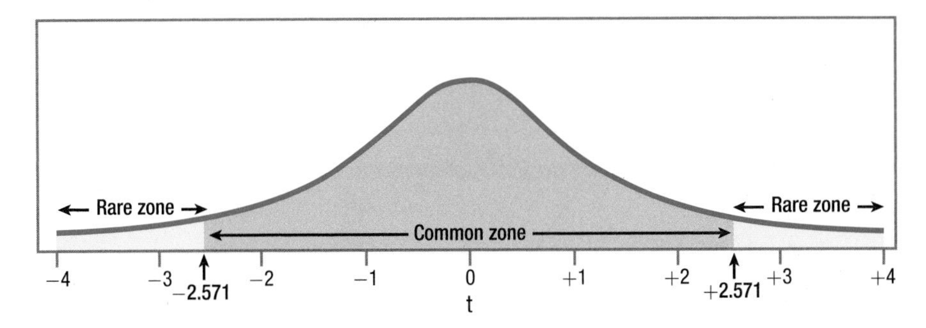

Figure 9.2 Critical Value of t, Two-Tailed, $\alpha = .05$, $df = 5$ This is the sampling distribution of t that would occur if the null hypothesis were true and $\mu_1 = \mu_2$. When the null hypothesis is true, the observed value will fall in the common zone 95% of the time and it will fall in the rare zone only 5% of the time.

Here is the decision rule for Dr. Keim's study:

- If $t \leq -2.571$ *or* $t \geq 2.571$, reject the null hypothesis.
- If $-2.571 < t < 2.571$, fail to reject the null hypothesis.

Step 5 Calculate the Test Statistic. The same general procedure is used for calculating the paired-samples t value as was used for the independent-samples t: divide the difference between the two sample means by the standard error of the difference. This standard error of the difference is the standard error of the difference for difference scores and will be abbreviated s_{M_D} to differentiate it from the other denominators for t tests. The **standard error of the mean difference for difference scores** is the standard deviation of the sampling distribution of difference scores.

Equation 9.3 Formula for the Standard Error of the Mean Difference for Difference Scores (s_{M_D})

$$s_{M_D} = \frac{s_D}{\sqrt{N}}$$

where s_{M_D} = the standard error of the mean difference for difference scores

s_D = the standard deviation (s) of the difference scores

N = the number of pairs of cases

Using Equations 3.6 and 3.7, Dr. Keim has calculated the standard deviation of the difference scores and found $s_D = 2.07$. There are six pairs of cases, so $N = 6$. Plugging these values into Equation 9.3 gives

$$\begin{aligned} s_{M_D} &= \frac{s_D}{\sqrt{N}} \\ &= \frac{2.07}{\sqrt{6}} \\ &= \frac{2.07}{2.4495} \\ &= 0.8451 \\ &= 0.85 \end{aligned}$$

The standard error of the mean difference for the difference scores is 0.85. This value will be used in Equation 9.4, the formula for calculating *t*, the value of the test statistic for the paired-samples *t* test.

Equation 9.4 Formula for Calculating t, the Value of the Test Statistic for a Paired-Samples t Test

$$t = \frac{M_1 - M_2}{s_{M_D}}$$

where t = the value of the test statistic for a paired-samples *t* test
M_1 = the mean of one sample
M_2 = the mean of the other sample
s_{M_D} = standard error of the mean difference for difference scores (Equation 9.3)

Given $s_{M_D} = 0.85$, $M_{\text{LowHumidity}} = 75.00$, and $M_{\text{HighHumidity}} = 82.50$, Dr. Keim is ready to calculate t. Again, it matters little which mean is called M_1 and which is M_2, so Dr. Keim has arranged the calculations to end up with a positive number in the numerator, assuring a positive *t* value:

$$\begin{aligned} t &= \frac{M_{\text{HighHumidity}} - M_{\text{LowHumidity}}}{s_{M_D}} \\ &= \frac{82.50 - 75.00}{0.85} \\ &= \frac{7.5000}{0.85} \\ &= 8.8235 \\ &= 8.82 \end{aligned}$$

Dr. Keim's *t* value is 8.82, and she is done with Step 5. In the next section of the chapter, after more practice with the first five steps, we'll follow Dr. Keim as she applies the decision rule and interprets the results.

Worked Example 9.1

A clinical psychologist, Dr. Althof, was studying the long-term effectiveness of psychodynamic psychotherapy for depression. He obtained a sample of 16 people with moderate to severe depression and assigned each to receive 20 sessions of psychodynamic therapy from a trained therapist. At the end of treatment, he administered a depression scale to each person and determined that the mean level of depression was 14.00. Scores on this depression scale can range from 0 to 50, and higher scores indicate greater depression. Six months later, Dr. Althof tracked down all 16 participants and readministered the depression scale, finding the mean was now 15.00. Figure 9.3 is a pair of box-and-whisker plots showing the depression scores at the two points in time. A 1-point increase on a 50-point scale doesn't sound like very much, but the box-and-whisker plots suggest an increase in the depression level in the six months following the end of treatment. Dr. Althof needs hypothesis testing to find out if the change is a statistically significant one.

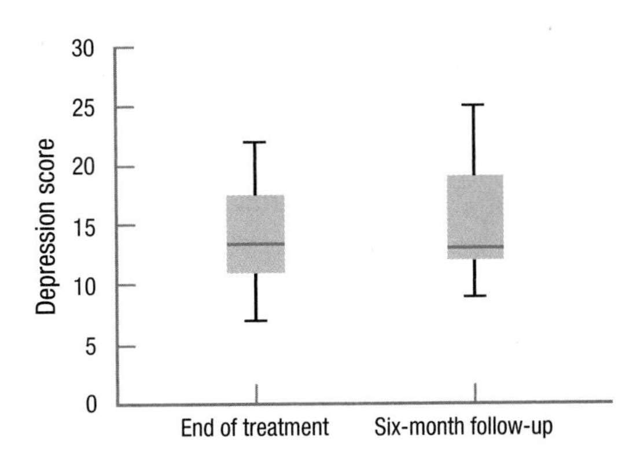

Figure 9.3 Box-and-Whisker Plots Showing Depression Level after Treatment and at Six-Month Follow-Up In this graph, there is some increase in the depression level in the six months following treatment. It will take a paired-samples *t* test to determine if the difference is statistically significant.

Step 1 Pick a Test. Dr. Althof is comparing the mean of a sample at one time to the mean of the sample at a second time. Though it is one sample of people, they are measured at two points in time, making it appropriate for a paired-samples *t* test.

Step 2 Check the Assumptions. The assumptions for the paired-samples *t* test are listed in Table 9.2.

- The random samples assumption is violated. It is not reported that the 16 cases were a random sample from the population of all the people in the world with moderate to severe depression, so it is safe to assume this is not a random sample from that population. When this robust assumption is violated, however, a researcher can still proceed with the test. Dr. Althof just has to be careful about the population to which he generalizes the results.

- The independence of observations assumption is not violated. Each participant received individual therapy, so the participants within a sample didn't influence each other.

- The assumption for the normality of difference scores is not violated. Dr. Althof knows from his review of the literature that depression scores are normally distributed. He is willing to assume that the difference scores (six-month follow-up depression score minus end-of-treatment depression score) will be normally distributed in the larger population.

Step 3 List the Hypotheses. Sometimes the effect of treatment increases over time, but more often the effect of treatment decreases over time, so Dr. Althof was open to both options when he planned the study. As a result, his hypotheses are nondirectional (two-tailed). The null hypothesis states that the two population means (end-of-treatment mean vs. six-month follow-up mean) are the same, and the alternative hypothesis says that the two population means differ:

$$H_0: \mu_{\text{EOT}} = \mu_{\text{6MFU}}$$
$$H_1: \mu_{\text{EOT}} \neq \mu_{\text{6MFU}}$$

Step 4 Set the Decision Rule. The critical value of t depends on the number of tails, alpha, and degrees of freedom. With nondirectional hypotheses, the test is two-tailed. Dr. Althof is comfortable setting the alpha at the usual level, .05, and having a 5% chance of Type I error. Finally, using Equation 9.2, he calculates degrees of freedom for the paired-samples t test:

$$\begin{aligned} df &= N - 1 \\ &= 16 - 1 \\ &= 15 \end{aligned}$$

Consulting Appendix Table 3 in the column for a two-tailed test with an alpha of .05 and the row with 15 degrees of freedom, Dr. Althof finds that the critical value of t is ± 2.131. The common and rare zones for his decision rule are shown in Figure 9.4. The decision rule is:

- If $t \leq -2.131$ *or* $t \geq 2.131$, reject the null hypothesis.
- If $-2.131 < t < 2.131$, fail to reject the null hypothesis.

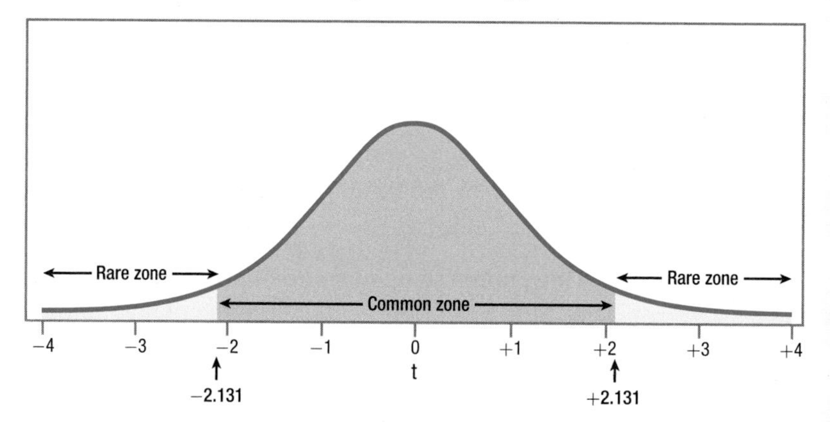

Figure 9.4 Critical Value of t, Two-Tailed, $\alpha = .05$, $df = 15$ This is the sampling distribution of t for 15 degrees of freedom. Compare this to the sampling distribution of t in Figure 9.2, where $df = 5$. Note that this sampling distribution is more peaked, packed more closely around zero. As a result, t_{cv} here, 2.131, is closer to zero, making the rare zone larger. When the rare zone is larger, which occurs when the sample size is larger, it is easier to reject the null hypothesis.

Step 5 Calculate the Test Statistic. The data set, with depression scores at the end of treatment and six months later, is shown in Table 9.3. Also shown are the difference scores and the standard deviation of the difference scores ($s_D = 2.31$).

The first step in the calculations is applying Equation 9.3 to find the standard error of the difference, s_{M_D}:

$$\begin{aligned} s_{M_D} &= \frac{s_D}{\sqrt{N}} \\ &= \frac{2.31}{\sqrt{16}} \\ &= \frac{2.31}{4.0000} \\ &= 0.5775 \\ &= 0.58 \end{aligned}$$

The value for the standard error of the difference, 0.58, is then used in Equation 9.4 to find t. It doesn't matter which mean is subtracted from which, so

TABLE 9.3 Data for Depression Score at the End of Treatment and at Six-Month Follow-Up

Participant	Six-month Follow-up	End of Treatment	Difference Score
1	11	12	−1.00
2	12	8	4.00
3	10	7	3.00
4	19	18	1.00
5	25	21	4.00
6	19	18	1.00
7	13	15	−2.00
8	19	17	2.00
9	13	10	3.00
10	15	12	3.00
11	9	12	−3.00
12	12	10	2.00
13	13	13	0.00
14	16	14	2.00
15	12	15	−3.00
16	22	22	0.00
M	15.00	14.00	1.00
s	4.58	4.37	2.31

Dr. Althof arranged them to end up with a positive value by subtracting the follow-up mean (14.00) from the end-of-treatment mean (15.00):

$$t = \frac{M_{6MFU} - M_{EOT}}{s_{M_D}}$$

$$= \frac{15.00 - 14.00}{0.58}$$

$$= \frac{1.0000}{0.58}$$

$$= 1.7241$$

$$= 1.72$$

Having found $t = 1.72$, Step 5 is complete. All that's left is the interpretation, which we'll turn to in the next section.

Practice Problems 9.1

Apply Your Knowledge

9.01 Given the following pairs of scores, calculate difference scores: 72 and 75; 69 and 45; 42 and 39; 47 and 46; 55 and 61; 50 and 61; 71 and 69; 55 and 69.

9.02 Given $s_D = 8.43$ and $N = 64$, calculate s_{M_D}.

9.03 Given $M_1 = 19.98$, $M_2 = 18.65$, and $s_{M_D} = 2.45$, calculate t.

9.3 Interpreting the Paired-Samples *t* Test

Interpreting a paired-samples *t* test starts by addressing the same questions that were used in interpreting the single-sample *t* test and the independent-samples *t* test:

1. Was the null hypothesis rejected?
2. How big is the effect?
3. How wide is the confidence interval?

And the interpretation ends the same way as well, with a written statement that covers four points:

1. What was the study about?
2. What were its main results?
3. What do these results mean?
4. Are there specific suggestions for future research?

Before Dr. Keim starts the interpretation, let's review her study on the effect of humidity on perceived temperature. She used six participants in a within-subjects design, where each person was tested twice in a 76° room, once at low humidity and once at high humidity. The mean perceived temperature in the low-humidity condition was 75.00°F ($s = 4.34$°F) and 82.50°F ($s = 4.93$°F) in the high-humidity condition. The mean difference score was 7.50 ($s_D = 2.07$°F). For a two-tailed test with $\alpha = .05$ and $df = 5$, t_{cv} was ±2.571°F. Dr. Keim calculated $s_{M_D} = 0.85$°F and $t = 8.82$°F.

Was the Null Hypothesis Rejected?

This first interpretation question can be answered by plugging the observed value of the test statistic, $t = 8.82$, into the decision rule Dr. Keim formulated in Step 4:

- Is $8.82 \leq -2.571$ *or* $8.82 \geq 2.571$?
- Or, is $-2.571 < 8.82 < 2.571$?

The first statement is true because 8.82 is greater than or equal to 2.571. This can be seen in Figure 9.5, where it is clear that the value of the test statistic falls in the rare zone. Dr. Keim has rejected the null hypothesis.

Rejecting the null hypothesis that the two population means are the same leads to accepting the alternative hypothesis that the two population means are different.

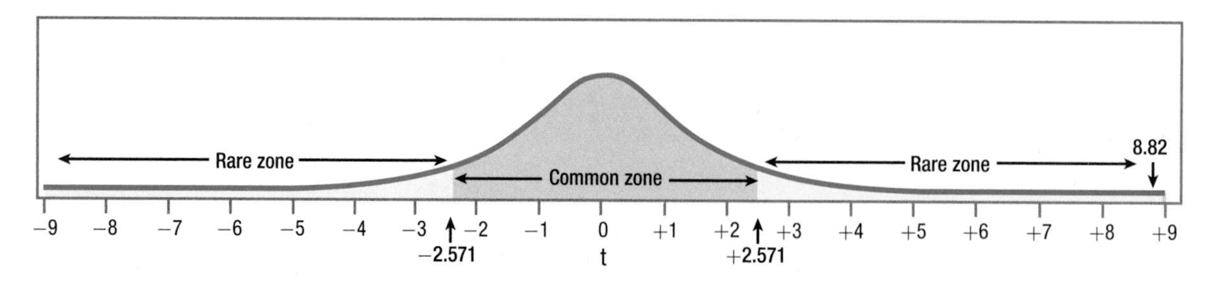

Figure 9.5 Observed Value of *t* from Humidity Study The observed value of *t*, 8.82, falls in the rare zone, so the null hypothesis is rejected.

Dr. Keim can say that a statistically significant difference exists between the perceived temperature in low humidity vs. high humidity.

The Direction of the Difference

Reporting that there is a difference is true, but it is not very useful. The logical follow-up question is, "What is the direction of the difference?" By examining the two sample means, 75.00° for the low-humidity condition vs. 82.50° for the high-humidity condition, Dr. Keim can conclude that 76° feels significantly hotter with high humidity than with low humidity.

A researcher only needs to worry about the direction of the difference when the null hypothesis is rejected. If Dr. Keim had failed to reject the null hypothesis, then there would not have been enough evidence to conclude that a difference exists between the two means, so there would have been no reason to consider the direction of the difference.

APA Format

Reporting the results in APA format for the humidity results, Dr. Keim would write

$$t(5) = 8.82, p < .05$$

For a *t*-test, APA format contains five pieces of information: (1) what test was done, (2) the number of cases, (3) the value of the test statistic, (4) the alpha level chosen, and (5) whether the null hypothesis was rejected.

1. The initial *t* says that this was a *t* test.
2. The number in parentheses, 5, is the degrees of freedom. By adding 1 to the number in the parentheses (i.e., 5 + 1 = 6), one can determine the number of pairs of cases.
3. The number after the equals sign, 8.82, gives the value of the test statistic.
4. The final value, .05, shows that there is a 5% chance of making a Type I error because alpha was set at .05.
5. $p < .05$ is how APA says that the null hypothesis has been rejected. It means that the results are rare, they happen less than 5% of the time, when the null hypothesis is true. (If alpha had been set at something other than .05, say .10, then write $p < .10$ instead.)

Cohen's *d* and r^2 for the Paired-Samples *t* Test

In the last two chapters, for the single-sample *t* test and the independent-samples *t* test, we calculated Cohen's *d* and r^2 in order to quantify how much impact the explanatory variable had on the outcome variable. It is possible to calculate Cohen's *d* and r^2 for a paired-samples *t* test, but doing so is *not* advisable.

The reason that these effect sizes are inappropriate for a paired-samples *t* test is that they measure more than just the impact of the explanatory variable on the dependent variable. Cohen's *d* and r^2 for a paired-samples *t* test also include the effect of individual differences. As a result, they overestimate the size of the effect of the independent variable on the dependent variable. For example, r^2 would be 94% for the humidity data if it were calculated. If true, it would mean that humidity status explains 94% of the variability in perceived temperature. That would be a huge effect for the explanatory variable of humidity level. Unfortunately, it is not true. Here's why.

Look at Table 9.1, which presents the data for each case in each of the two conditions. Notice that there's variability within a condition: not everyone perceives the temperature the same way. As everyone in a condition receives the same treatment, the variability within a condition results from individual differences among the participants. Look at case 6 who perceived the low-humidity condition as the coldest, 68°. Case 6 also gives the lowest rating of the temperature in the high-humidity condition. Case 6 differs from the other individuals in this study in that he or she always feels cold. Case 2, in contrast, differs in the perception of temperature, feeling the warmest in both conditions.

The type of participant, cold-sensitive or heat-sensitive, has a large effect on the temperature the environment is perceived to be. Whether a person is cold-sensitive or heat-sensitive is an individual differences variable. When r^2 or Cohen's d is calculated for a paired-samples t test, they mix in the impact of these individual differences on temperature *with* the impact of the different conditions (high vs. low humidity) on temperature. This means that they give an inflated effect size and are not appropriate to estimate effect size for a paired-samples t test.

So, how does a researcher measure effect size for a paired-samples t test? One way is to calculate the confidence interval for the difference between population means and then use professional expertise—and common sense—to translate it into an effect size. Another approach, not possible until Chapter 12, is to use a test called the repeated-measures ANOVA instead of a paired-samples t test. The repeated-measures ANOVA, unlike the paired-samples t test, separates out the impact of individual differences from the effect of the explanatory variable, allowing the effect of the explanatory variable alone to be assessed.

How Wide Is the Confidence Interval and How Big Is the Effect?

The confidence interval for the difference between population means reveals how small or how large the difference between the population means might be. For the humidity data set, the confidence interval will tell, at a population level, how much hotter the temperature is perceived when the humidity rises. A confidence interval of any percentage could be calculated, but the most common is the 95% confidence interval. Equation 9.5 gives the formula for the 95% confidence interval for the difference between population means for the paired-samples t test.

Equation 9.5 Formula for the 95% Confidence Interval for the Difference Between Population Means for a Paired-Samples t Test

$$95\%\text{CI}\mu_{\text{Diff}} = (M_1 - M_2) \pm (t_{cv} \times s_{M_D})$$

where $95\%\text{CI}\mu_{\text{Diff}}$ = the 95% confidence interval for the difference between population means for a paired-samples t test

M_1 = the mean of one sample

M_2 = the mean of the other sample

t_{cv} = the critical value of t, two-tailed, $\alpha = .05$, $df = N - 1$

s_{M_D} = the standard error of the mean difference for the difference scores (Equation 9.3)

Applying this to her humidity data, Dr. Keim would calculate the 95% confidence interval as follows:

$$\begin{aligned} 95\%\text{CI}\mu_{\text{Diff}} &= (82.50 - 75.00) \pm (2.571 \times 0.85) \\ &= 7.5000 \pm 2.1854 \\ &= \text{from } 5.3146 \text{ to } 9.6854 \\ &= [5.31, 9.69] \end{aligned}$$

Her confidence interval ranges from a lower limit of 5.31° to an upper limit of 9.69°. To make the interpretation more accessible, Dr. Keim has rounded each end to a whole number, making the confidence interval range from 5 to 10 degrees. In interpreting this confidence interval, Dr. Keim will pay attention to three points: (1) whether the confidence interval captures zero, (2) how close it comes to zero, and (3) how wide it is.

1. If a confidence interval captures zero, this means it is possible that the difference between the means of the populations is zero. When such happens, it is plausible that the null hypothesis is true. (This assumes that the confidence interval and the alpha level of the hypothesis test are synchronized.)

2. How close the confidence interval comes to zero provides information about the effect size in the larger population. If both ends of the confidence interval are close to zero, then the effect size is probably small. If both ends of the confidence interval are far from zero, then the effect size is probably large. If one end of the confidence interval is close to zero and the other end is far away, then the researcher will be left unsure of how strong or weak the effect actually is in the population.

3. The width of the confidence interval provides information about how precisely the effect can be specified in the population. Narrower confidence intervals are preferred because they give a more precise sense of the size of the effect in the population. If the confidence interval is wide, the researcher will usually recommend replicating with a larger sample size in order to determine the parameter value more precisely.

For the humidity study, the confidence interval, from 5 to 10 degrees, does not capture zero. This is to be expected as the null hypothesis was rejected. The confidence interval reiterates what is already known—it is unlikely, in the larger population, that there is no difference in how 76° is perceived in low humidity vs. how it is perceived in high humidity. Instead, in the population, the high-humidity condition is probably perceived as 5–10 degrees hotter, on average, than the same temperature at low humidity.

The second step, determining how small and how large the effect may be, takes more thought and expertise for the paired-samples t test than it did for the independent-samples t test. With an independent-samples t test, the researcher could calculate Cohen's d or r^2 and rely on Cohen's standards for small, medium, and large effects. The end of the confidence interval closer to zero (the lower limit) is 5 and the end farther away (the upper limit) is 10. In the larger population, will people perceive the temperature to be a lot hotter or a little hotter if they perceive 76° in conditions of high humidity as 5 degrees hotter than in low humidity? Dr. Keim believes this to be a meaningful effect. She reasons that, on a summer day, there is a noticeable difference in comfort level between being in an air-conditioned room at 71° and one at 76°, so a 5-degree difference is a meaningful one. If a 5-degree difference is meaningful, then a 10-degree difference is even more so. In understanding what a 10-degree difference means, Dr. Keim thinks of how a 76° summer day feels pleasant and an 86° day feels hot.

Finally, she uses Equation 7.5 to calculate the width of the confidence interval. To avoid rounding error, she uses the real limits of the confidence interval, 5.31 to 9.69, not her rounded version of 5 to 10:

$$\begin{aligned} CI_W &= CI_{UL} - CI_{LL} \\ &= 9.69 - 5.31 \\ &= 4.38 \end{aligned}$$

The confidence interval is 4.38° wide. This seems sufficiently narrow to Dr. Keim. She would like to replicate the study in order to make sure the same effect is observed again, and she would like to increase the sample size in order to have a better sample, but she feels no need to replicate with a larger sample size in order to narrow the confidence interval. Given that both a 5-degree difference and a 10-degree difference seem meaningful and that the confidence interval is narrow, Dr. Keim is inclined to focus on the observed difference of 7.50 degrees in her interpretation.

Putting It All Together

Here is Dr. Keim's four-point interpretation in which she states: (1) what the study was about; (2) the main results; (3) what the results mean; and (4) her suggestions for future research. This interpretation is a little longer than previous ones. It takes Dr. Keim two paragraphs to say all that she wants to.

> The impact of humidity on perceived temperature was examined. Using a within-subjects design, six participants judged the temperature after being in a 76° room under conditions of low humidity and high humidity. In the low-humidity condition, they judged the temperature fairly accurately ($M = 75.00$°F), but they judged it as hotter ($M = 82.50$°F) under conditions of high humidity. This 7.50°F difference was statistically significant [$t(5) = 8.82$°F, $p < .05$]. Humidity appears to make people feel hotter. According to this study, people feel almost 8 degrees hotter under conditions of high humidity. This is a noticeable increase in perceived temperature and can move a person from feeling comfortable to being uncomfortable.
>
> There were several limitations to this study that can be rectified in future research. All participants were college students in the United States, people who have experience with heated and cooled environments. It would be interesting to see if the same effect were observed among people with less control over their indoor environments. A second limitation is that only one temperature, 76°, was tested. The effect of humidity on perceived temperature should be observed at both higher and lower temperatures. Finally, there were only six participants in this study. Replicating the study would increase confidence in the robustness of the finding that humidity level affects the perception of temperature.

Worked Example 9.2

For more practice with interpretation, let's return to the study where Dr. Althof followed patients with depression for six months following treatment. At the end of treatment, the mean depression level for the 16 participants was 14.00, and six months later it had climbed to 15.00. The mean difference score was 1.00 ($s_D = 2.31$). Dr. Althof had nondirectional hypotheses and a 5% chance of making a Type I error. t_{cv} was 2.131, s_{M_D} was calculated to be 0.58, and t was 1.72.

Was the null hypothesis rejected? The first step in interpretation is to determine if the null hypothesis is rejected. To do so, Dr. Althof substitutes the observed value of *t*, 1.72, into the decision rule he had generated in Step 4:

- Is $1.72 \leq -2.131$ *or* $1.72 \geq 2.131$?
- Or, is $-2.131 < 1.72 < 2.131$?

The second statement, is true: 1.72 falls between −2.131 and 2.131. Look at Figure 9.6, where it is clear that the value of the test statistic falls in the common zone. Dr. Althof has failed to reject the null hypothesis. There is not enough evidence to conclude that depression level changes, in either a positive or negative direction, in the six months after the end of psychodynamic therapy. In APA format, the results would be written as

$$t(15) = 1.72, p > .05$$

Remember, "$p > .05$" signifies that the result ($t = 1.72$) is an expected, or common, occurrence when the null hypothesis is true. It means that the null hypothesis was not rejected.

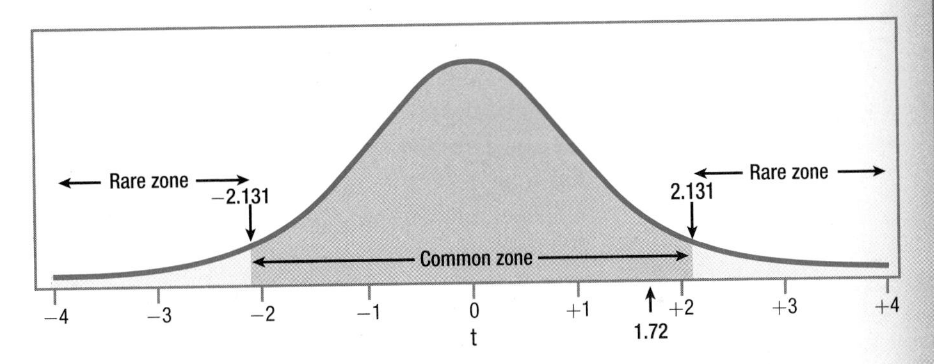

Figure 9.6 Observed Value of *t* from the Depression Follow-Up Study The observed value of *t*, 1.72, falls in the common zone. The null hypothesis is not rejected.

This failure to reject the null hypothesis can be taken as good news. Dr. Althof thinks of it as lack of evidence that relapse to depression occurs within six months of ending psychodynamic treatment for depression. The effectiveness of this treatment seems fairly long-lasting.

How wide is the confidence interval? How big is the effect? The second step in interpretation for a paired-samples *t* test is to use the confidence interval to evaluate the size of the effect. Dr. Althof used Equation 9.5 to calculate the confidence interval:

$$\begin{aligned}95\%\text{CI}\mu_{\text{Diff}} &= (M_1 - M_2) \pm (t_{cv} \times s_{M_D})\\ &= (15.00 - 14.00) \pm (2.131 \times 0.58)\\ &= 1.0000 \pm 1.2360\\ &= \text{from } -0.2360 \text{ to } 2.2360\\ &= [-0.24, 2.24]\end{aligned}$$

For the larger population of people with moderate to severe depression, this confidence interval tells Dr. Althof what the mean difference is in their level of depression from the end of treatment to six months later. It tells him that the mean difference could be anywhere from an average of being 2.24 points *more* depressed six months after treatment to being 0.24 points *less* depressed six months after treatment. And, there is a 5% chance that this interval does not capture the actual mean difference.

There are three points to consider in interpreting a confidence interval: (1) whether it captures zero, (2) how close it comes to zero, and (3) how wide it is.

1. As expected, because Dr. Althof had failed to reject the null hypothesis, zero falls within this interval. This means that the value of zero is a viable option for how much difference occurs in the mean depression level from the end of treatment to six months later. In the larger population, there may be no loss in the effect of treatment in the six months following treatment.
2. One end of the confidence interval (−0.24) falls quite near zero and would be a trivial decrease in depression if it were true. The other end (2.24) doesn't fall far away from zero, considering that the depression scale ranges over 50 points. If the effect in the population were a mean increase of 2.24 depression points on a 50-point scale over six months, that is not much of an effect.
3. The confidence interval is not very wide, being a total of 2.48 points wide, from −0.24 to 2.24. Based on this, Dr. Althof does not feel a great need to replicate the study with a larger sample size. Note, however, that the vast majority of the confidence interval falls on one side of zero. If the sample size had been larger and the confidence interval narrower, then it would have failed to capture zero and Dr. Althof would have rejected the null hypothesis. This makes him worry that a Type II error may have occurred. Perhaps there is a small effect that was not found. For this reason, Dr. Althof is going to suggest replicating with a larger sample size.

This is a good opportunity to talk about the difference between *statistical significance* and *practical significance.* **Statistical significance** indicates that the observed difference between *sample* means is large enough to conclude that there is a difference between *population* means. It doesn't mean that the difference is a meaningful one.

Statistical significance is heavily influenced by sample size: the larger the sample size, the more likely it is that results will be statistically significant. If the sample size were increased in the depression follow-up study from 16 to 23, the results would be statistically significant and the confidence interval would range from 0.004 to 2.00, not capturing zero.

Think about the larger population. In the larger population, if the mean depression score were 14.00 at the end of treatment and 14.004 six months later, then the null hypothesis would be wrong and should be rejected. But a population difference of such a small amount, 0.004, would be so small as to be meaningless. It is of no practical significance. A result is of **practical significance** (also called **clinical significance**) if the size of the effect is large enough to make a real difference. Practical significance means that the explanatory variable has a meaningful impact on the outcome variable. A statistically significant result is no guarantee of a practically significant effect.

A statistically significant result is no guarantee of a practically significant effect.

Judging practical significance requires expertise in and familiarity with a specific area of research. Unless one has experience with a scale, it is hard to know how meaningful a 2-point change on the Beck Depression Inventory. For now, the best option is to use Cohen's small, medium, and large effect sizes as an initial, and very rough, guide to practical and clinical significance.

A Common Question

Q Is it possible for a result to be practically significant but not statistically significant?

A No. When a result is not statistically significant, there is no evidence of a difference between the populations. The observed difference may be unique to the two samples in a study and wouldn't be found again. If an effect doesn't occur consistently, it can't be of practical use.

Putting it all together. Here's what Dr. Althof wrote for an interpretation. He followed the four-point plan for interpretations: (1) stating what the study was about, (2) giving the main results, (3) explaining them, and (4) making suggestions for future research.

A study was conducted that followed 16 people who had received psychodynamic therapy for severe to moderate depression for six months after treatment was complete. Their mean depression level was 14.00 at the end of treatment and 15.00 six months later. This 1-point increase in depression was not a statistically significant change [$t(15) = 1.72$, $p > .05$]. Thus, there is no evidence from this study to conclude that a relapse to depression occurs, for this population of patients, in the six months following treatment with psychodynamic therapy. The effects of psychodynamic therapy seem to be long-lasting.

To increase confidence in the robustness of these results, it would be a good idea to replicate this study. Additionally, it would be a good idea to add another form of therapy for depression to a study in order to see if lack of relapse after treatment ends is unique to psychodynamic therapy.

Application Demonstration

Here are some analyses based on a real data set to end our investigation of the paired-samples *t* test. A student, Kristin Brown, wondered whether there was a difference in cigarette smoking rates between men and women. Kristin found data reporting the rates of smoking for men and for women by state, and we'll use them to select 10 random states and determine if a difference exists in smoking rates between men and women. Figure 9.7 shows the results for the 10 states.

Step 1 Pick a Test. The data are paired together by state—one rate for men and one rate for women. This seems sensible. Individual differences between states (such as whether the state produces tobacco or has active antismoking campaigns) might affect its smoking rate, but should affect both sexes in a state. The men and women are dependent samples, so this situation calls for a paired-samples *t* test.

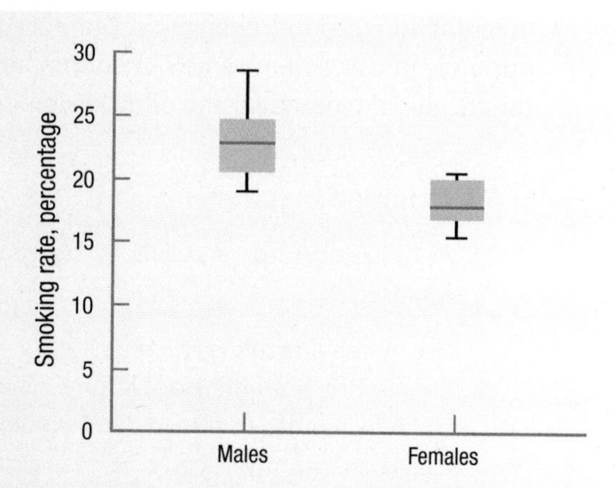

Figure 9.7 Smoking Rates for Males and Females in Sample of 10 States In this graph, it appears as if the smoking rate by state is higher for men than for women. However, to make sure the effect can't be explained by sampling error, hypothesis testing is needed.

Step 2 Check the Assumptions. Table 9.2 lists the assumptions.

- The random samples assumption is not violated. This is a random sample of 10 of the 50 states. The results can be generalized to all 50 states.
- The independence of observations assumption is not violated. The states were measured independently, so cases within a sample don't influence each other. Each state was only in each sample once.
- The normality assumption is not violated. It seems reasonable to assume that there is a normal distribution of difference scores in the larger population. In addition, if this assumption is wrong, it is robust to violations.

None of the assumptions (Table 9.2) was violated, so we can proceed with the paired-samples *t* test.

Step 3 List the Hypotheses. Hypotheses should be generated before any data are collected. The question being addressed is whether there is a difference in the smoking rates of men and women. The direction doesn't matter. This calls for a two-tailed test. The hypotheses are

$$H_0: \mu_{\text{Men}} = \mu_{\text{Women}}$$
$$H_1: \mu_{\text{Men}} \neq \mu_{\text{Women}}$$

Step 4 Set the Decision Rule. For a two-tailed test with alpha set at .05 and 10 – 1 degrees of freedom, the critical value of t is ± 2.262. The decision rule is:

- If $t \leq -2.262$ *or* $t \geq 2.262$, reject the null hypothesis.
- If $-2.262 < t < 2.262$, fail to reject the null hypothesis.

Step 5 Calculate the Test Statistic. Table 9.4 displays the means and standard deviations for men and women, and the difference scores for the 10 states. Using the standard deviation of the difference scores (1.98) and the number of pairs of cases (10), the standard error of the difference is calculated (Equation 9.3):

TABLE 9.4 Summary Data Comparing Smoking Rates of Men and Women

	Men	Women	Difference Score
Sample mean	23.00%	18.23%	4.77%
Sample standard deviation	3.05%	1.69%	1.98%

$$s_{M_D} = \frac{s_D}{\sqrt{N}} = \frac{1.98}{\sqrt{10}} = \frac{1.98}{3.1623} = 0.6261 = 0.63$$

The standard error of the mean difference is used to calculate the t value (Equation 9.4). Remember, it doesn't make a difference which mean is subtracted from the other mean, so subtract the mean for the women (18.23%) from the mean for the men (23.00%) to obtain a positive number:

$$t = \frac{M_1 - M_2}{s_{M_D}} = \frac{23.00 - 18.23}{0.63} = \frac{4.77}{0.63} = 7.5714 = 7.57$$

Step 6 Interpret the Results. *Was the null hypothesis rejected?*

- Is $7.57 \leq -2.262$ *or* $7.57 \geq 2.262$?
- Or, is $-2.262 < 7.57 < 2.262$?

7.57 is greater than or equal to 2.262, so the first statement is true. Figure 9.8 shows that the observed value of t, 7.57, falls in the rare zone, so the null hypothesis is rejected.

The conclusion is that in the larger population of states, there is a difference in the smoking rate between men and women. Examining the sample means, 18.23% for the women and 23.00% for the men, leads to the conclusion that the smoking rate per state is higher for men than it is for women. In APA format, one would report

$$t(9) = 7.57, p < .05$$

How wide is the confidence interval? How big is the effect? To calculate the 95% confidence interval for the difference between population means,

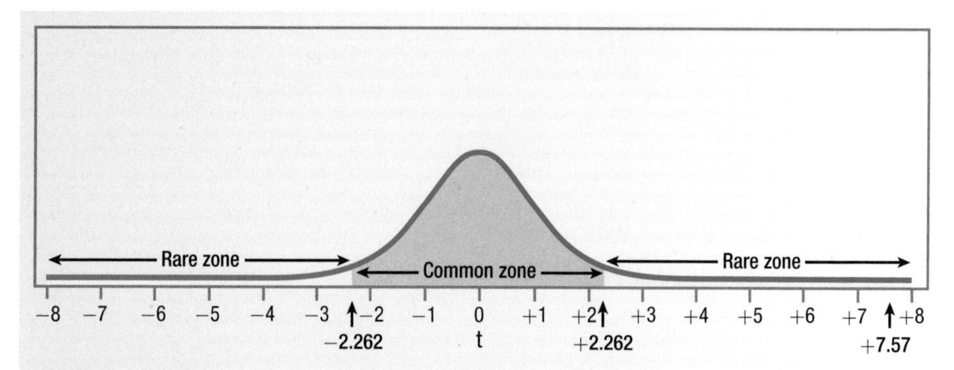

Figure 9.8 **Observed Value of *t* from the Smoking Rate Study** Note that the observed value of *t* falls in the rare zone. This means that the null hypothesis is rejected. It appears that the population mean for men differs from the population mean for women.

Equation 9.5 is used. Arrange the subtraction of one mean (the women at 18.23%) from the other (the men at 23.00%) to end up with a positive value. The other numbers in the equation, 2.262 and 0.63, are the critical value of *t* and the standard error of the difference, respectively:

$$\begin{aligned} 95\%\text{CI}\mu_{\text{Diff}} &= (M_1 - M_2) \pm (t_{cv} \times s_{M_D}) \\ &= (23.00 - 18.23) \pm (2.262 \times 0.63) \\ &= 4.7700 \pm 1.4251 \\ &= \text{from } 3.3449 \text{ to } 6.1951 \\ &= [3.34, 6.20] \end{aligned}$$

The confidence interval says that in the larger population of states, the mean male smoking rate is higher than the mean female smoking rate, by 3.34% to 6.20%.

- *Did the confidence interval capture zero?* As expected, because the null hypothesis was rejected, this confidence interval doesn't capture zero. It is unlikely that the difference in the smoking rates of men and women is zero.
- *How close does the confidence interval come to zero?* The lower end (3.34% difference in smoking rates) is not very close to zero and the upper end (6.20%) is a good distance from zero. It seems reasonable to conclude that the effect is of a moderate size.
- *How wide is the confidence interval?* At 2.86 percentage points, the confidence interval isn't too wide.

Putting it all together. Here's the four-point interpretation:

Data were analyzed from a study comparing the smoking rates of men to those of women for 10 randomly selected states. A mean of 23.00% of men in these states smoked, compared to 18.23% of the women. This 4.77 percentage points higher rate among the men was statistically significant [$t(9) = 7.57, p < .05$] and seems to show a reasonably strong effect of sex on the smoking rate. In the United States, men smoke at a higher rate. In future research, it would be interesting to see if the difference in rates between men and women has changed over time and to examine male/female differences in other countries.

Practice Problems 9.2

Apply Your Knowledge

9.04 Given $N = 46$, $M_1 = 23$, $M_2 = 32$, and $t = 3.67$, (a) write the results in APA format and (b) comment on the direction of the difference. Use $\alpha = .05$, two-tailed.

9.05 Given $N = 20$, $M_1 = 68$, $M_2 = 64$, and $t = 2.01$, (a) write the results in APA format and (b) comment on the direction of the difference. Use $\alpha = .05$, two-tailed.

9.06 Given $M_1 = 55$, $M_2 = 48$, $t_{cv} = 2.023$, and $s_{M_D} = 2.86$, calculate the $95\%\text{CI}\mu_{\text{Diff}}$.

9.07 A tennis instructor compared two homework methods. She took beginning students at her tennis camp and matched them in pairs in terms of their tennis abilities. She then randomly assigned the players to two conditions. Those in the control condition had to practice for half an hour every day with another player. Those in the experimental condition had to practice against a wall for half an hour every day. After two weeks of practice, the instructor measured how often each player could hit targets in different locations on the tennis court. The higher the percentage, the better the player had become. Given the results below, interpret the study:

- $M_C = 48.00$, $M_E = 57.00$, $N = 12$, $s_D = 6.00$
- $s_{M_D} = 1.73$, $t = 5.20$, $95\%\text{CI}\mu_{\text{Diff}} = [5.19, 12.81]$

SUMMARY

Recognize the different types of dependent samples.

- In dependent samples, also called paired samples, cases are pairs of data points. Examples include (a) cases measured at two points in time (repeated measures, longitudinal, or pre-post designs); (b) cases measured in two different conditions (within-subjects designs); or (c) cases matched on some variable, so they are similar (matched-pairs designs).

Calculate and interpret a paired-samples *t* test.

- A paired-samples *t* test is used to compare the means of two dependent samples. Because paired-samples *t* tests control for the effects of individual differences, they have more statistical power, need a smaller sample size to reject H_0, and are useful for studying rare phenomena or those for which cases are hard to come by.

- To conduct a paired-samples *t* test, the assumptions must be met, hypotheses stated, and decision rule set before calculating the actual *t* value.

- In interpreting the results of a paired-samples *t* test, first apply the decision rule to determine if the null hypothesis was rejected, then write the results in APA format. If the null hypothesis was rejected, comment on the direction of the difference between the population means.

- Next, to find the effect size, calculate the 95% confidence interval for the difference between population means. Interpret the confidence interval, paying attention to (a) whether it captures zero, (b) how close it comes to zero, and (c) how wide it is. Do not calculate r^2 or *d*.

- Finally, in a four-point interpretation: (1) explain what the study was about, (2) present the main results, (3) interpret them, and (4) make suggestions for future research.

DIY

Is there a difference in the ages at which men and women get married? Has there been a change in this statistic over time?

Go online or get a newspaper from a recent Sunday and find the wedding section. Note the ages of each bride and groom. Are these paired data?

Check the assumptions and complete a paired-samples *t* test. Is there a difference? In what direction? What do the results tell us?

Want to do some extra work? Go to your school library and have the reference librarian help you find a Sunday newspaper from a decade or two ago. Collect and analyze the data for the ages of the brides and grooms back then. Has a change in the age difference occurred over the years?

KEY TERMS

clinical significance (or practical significance) – the size of the effect is large enough to say the independent variable has a meaningful impact on clinical outcome.

individual differences – attributes that vary from case to case.

longitudinal research (or repeated-measures design) – a study in which the same participants are measured at two or more points in time.

matched pairs – participants are grouped into sets of two based on their being similar on potential confounding variables.

paired-samples t test – hypothesis test used to compare the means of two dependent samples; also known as dependent-samples *t* test, correlated-samples *t* test, related-samples *t* test, matched-pairs *t* test, within-subjects *t* test, or repeated-measures *t* test.

practical significance (or clinical significance) – the size of the effect is large enough to say the independent variable has a meaningful impact on the dependent variable (or the clinical outcome).

pre-post design – participants are measured on the dependent variable before and after an intervention.

repeated-measures design (or longitudinal research) – a study in which the same participants are measured at two or more points in time.

standard error of the mean difference for difference scores – the standard deviation of the sampling distribution of difference scores, abbreviated s_{M_D}; used as the denominator in the paired-samples *t* test equation.

statistical significance – the observed difference between sample means is large enough to conclude that it represents a difference between population means.

within-subjects design – the same participants are measured in two different situations or under two different conditions.

CHAPTER EXERCISES

Review Your Knowledge

9.01 The independent-samples *t* test is used to compare means of two ____ samples; the ____ is used to compare means of two dependent samples.

9.02 The cases in the two samples in a paired-samples *t* test may be paired because they are the same cases measured at two different points in ____ or under two different conditions.

9.03 With matched samples, cases are paired together in order to control for ____ variables.

9.04 Two other names for a paired-samples *t* test are ____ and ____.

9.05 An advantage of a paired-samples *t* test is that it controls for ____.

9.06 An individual difference is an attribute, such as intelligence or weight, that varies from ____.

9.07 Paired-samples *t* tests are more powerful statistically. This means they have a ____ probability of being able to reject the null hypothesis if the null hypothesis was false.

9.08 Paired-samples *t* tests are good to use when ____ are hard to come by.

9.09 Paired-samples *t* tests can be used to determine if a difference between two sample means is statistically significant, or if it could be explained by ____.

9.10 If the random samples assumption for a paired-samples *t* test is not violated, the results can be ____ to the population.

9.11 The independence assumption for a paired-samples *t* test refers to independence ____ a sample, not ____ samples.

9.12 The normality assumption for a paired-samples *t* test refers to the ____ being normally distributed.

9.13 The hypotheses for a paired-samples *t* test are statements about the two ____, not the two ____.

9.14 The null hypothesis for a nondirectional paired-samples *t* test says there is ____ between population means.

9.15 The alternative hypothesis for a two-tailed, paired-samples *t* test says that the difference between population means could be positive or ____, large or ____.

9.16 The decision rule for a paired-samples *t* test compares the ____ value of *t* to the ____ value of *t*.

9.17 The degrees of freedom for a paired-samples *t* test are calculated by subtracting ____ from the number of pairs.

9.18 The default option for a paired-samples *t* test is to set alpha at ____ and to do a ____-tailed test.

9.19 The abbreviation for the standard error of the mean difference scores in a paired-samples *t* test is ____.

9.20 The value of the test statistic for a paired-samples *t* test is obtained by dividing ____ by the standard error of the mean difference of the difference scores.

9.21 It *does/does not* make a difference which sample mean is subtracted from the other mean when calculating a paired-samples *t* value.

9.22 If $t \geq t_{cv}$, one would ____ the null hypothesis.

9.23 If results are written in APA format as p ____ .05, then the researcher has failed to reject the null hypothesis.

9.24 If one fails to reject the null hypothesis, there is no need to comment on the ____ of the difference between the population means.

9.25 The direction of the difference between the population means is determined by comparing the two ____ means.

9.26 If the degrees of freedom for a paired-samples *t* test are 11, then there were ____ pairs of data.

9.27 Cohen's *d*, when calculated for a paired-samples *t* test, includes the effect of ____ as well as the effect of the explanatory variable on the ____.

9.28 r^2 *should/should not* be used as an effect size for a paired-samples *t* test.

9.29 To measure effect size for a paired-samples *t* test, use the ____.

9.30 Whether the 95% confidence interval captures zero for a two-tailed, paired-samples *t* test with alpha set at .05 provides information about whether the null hypothesis is ____.

9.31 If the confidence interval for a paired-samples *t* test does not capture zero, but both ends of it are close to zero, then the size of the effect is probably ____.

9.32 When the confidence interval for a paired-samples t test is ____, one has fairly precisely specified the size of the effect in the population.

Apply Your Knowledge

Picking a test

9.33 A sensory psychologist had participants rate the taste of two coffees, caffeinated and decaffeinated versions of the same brand. Each participant rated both types of coffee. What statistical test should the psychologist use to see if caffeinated and decaffeinated coffees differ in taste?

9.34 A nutritionist wanted to find out if coffee and tea, as served in restaurants, differed in caffeine content. She went to 30 restaurants. In 15 randomly selected restaurants, she ordered coffee; in the other 15 restaurants, she ordered tea. What statistical test should the nutritionist use to see if coffee and tea differ in mean caffeine content?

9.35 A researcher for a health magazine compared the mean caffeine content for a sample of coffees served at coffee houses to the USDA standard for the mean amount of caffeine in a cup of coffee. What statistical test should she use?

9.36 A developmental psychologist wondered if birth order had an impact on academic performance. She found families with two children and compared the mean high school GPA of first-born children to second-born children. What statistical test should she use?

Checking the assumptions

9.37 A group of high school students were in the same math, English, social studies, and science classes. A researcher monitored how much time, during a week, they spent online in school activities (doing homework) vs. how much time they spent online in social activities (on Facebook, chatting, playing multi-player games). The researcher planned to use a paired-samples t test to analyze the data to see if there was a difference in time spent on the two activities. Based on the assumptions, is it OK to proceed with the test?

9.38 Elementary school teachers who did and did not have children of their own were matched in terms of the number of years of experience they had teaching. They were then asked how many minutes of homework a child should complete per night. The researcher planned to use a paired-samples t test to determine if having children of one's own was related to this response. Based on the assumptions for the test, is it OK to proceed?

Writing the hypotheses

9.39 Write the hypotheses for a paired-samples t test for Exercise 9.37.

9.40 Write the hypotheses for a paired-samples t test for Exercise 9.38.

Calculating difference scores

9.41 Calculate difference scores for the following pairs of scores: 5, 10; 7, 3; 6, 8; 4, 3; 7, 8.

9.42 Calculate difference scores for the following pairs of scores: 12, 13; 14, 12; 7, 4; 2, 4; 8, 6.

Calculating degrees of freedom

9.43 A researcher plans to use a paired-samples t test for two dependent samples, each with 10 cases. How many degrees of freedom are there?

9.44 If a consumer researcher compares the mean price of 25 items purchased at one store to the mean price for the same 25 items purchased at a second store, how many degrees of freedom are there?

Finding t_{cv} for a two-tailed test with $\alpha = .05$

9.45 If there are 19 cases in a paired-samples t test, what is t_{cv}?

9.46 If there are 33 pairs of data in a paired-samples t test, what is t_{cv}?

9.47 If $N = 118$ for a paired-samples t test, what is t_{cv}?

9.48 Given 2,012 cases in a paired-samples t test, what is t_{cv}?

Calculating standard error of the mean difference

9.49 Given $s_D = 2.57$ and $N = 17$, calculate s_{M_D}.

9.50 Given $s_D = 12.78$ and $N = 49$, calculate s_{M_D}.

Given s_{M_D}, finding t

9.51 If $M_1 = 18$, $M_2 = 14$, and $s_{M_D} = 5.34$, what is t?

9.52 If $M_1 = -12$, $M_2 = -15$, and $s_{M_D} = 2.81$, what is t?

Calculating t

9.53 Given $M_1 = 25$, $M_2 = 28$, $N = 5$, and $s_D = 7.00$, calculate t.

9.54 Given $M_1 = -7$, $M_2 = -9$, $N = 28$, and $s_D = 2.90$, calculate t.

Implementing the decision rule

9.55 If $t = 2.30$ and $t_{cv} = \pm 2.017$, (a) draw a sampling distribution of t, marking t and t_{cv}, and label the rare and common zones, then (b) report whether or not the null hypothesis was rejected.

9.56 If $t = 1.65$ and $t_{cv} = \pm 2.145$, (a) draw a sampling distribution of t, marking t and t_{cv}, and labeling the rare and common zones, then (b) report whether or not the null hypothesis was rejected.

Writing results in APA format (use $\alpha = .05$, two-tailed)

9.57 Given $N = 5$ and $t = 3.211$, write the results of this paired-samples t test in APA format.

9.58 Given $N = 27$ and $t = 2.033$, write the results of this paired-samples t test in APA format.

9.59 Given $N = 69$ and $t = 1.994$, write the results of this paired-samples t test in APA format.

9.60 Given $N = 181$ and $t = 1.981$, write the results of this paired-samples t test in APA format.

Determining the direction of the difference

9.61 Given these results, comment on the direction of the difference between the population means: $M_1 = 72$, $M_2 = 73$, $t(26) = 2.08$, $p < .05$.

9.62 Given these results, comment on the direction of the difference between the population means: $M_1 = 17$, $M_2 = 24$. $t(35) = 2.01$, $p > .05$.

9.63 Given these results, comment on the direction of the difference between the population means: $M_1 = 50$, $M_2 = 53$, $t(17) = 1.54$, $p > .05$.

9.64 Given these results, comment on the direction of the difference between the population means: $M_1 = 28$, $M_2 = 31$, $t(72) = 7.42$, $p < .05$.

Calculating a confidence interval

9.65 Given the following, calculate the 95% confidence interval for the difference between population means: $M_1 = 108$, $M_2 = 100$, $t_{cv} = 2.052$, and $s_{M_D} = 5.00$.

9.66 Given the following, calculate the 95% confidence interval for the difference between population means: $M_1 = 40$, $M_2 = 50$, $t_{cv} = 2.010$, and $s_{M_D} = 2.44$.

Interpreting confidence intervals (if necessary, assume the test was two-tailed with $\alpha = .05$)

9.67 Based on the confidence interval 18.00 to 27.00, decide if the null hypothesis was rejected for the paired-samples t test.

9.68 Based on the confidence interval −0.40 to 0.50, decide if the null hypothesis was rejected for the paired-samples t test.

Interpreting the results of a paired-samples t test

9.69 A sleep therapist wanted to see if an herbal tea advertised as a sleep aid really worked. He located 46 people with sleep problems and matched them into pairs on the basis of (a) how long they had suffered from insomnia, (b) how long it usually took them to go to sleep at night, (c) how much sleep onset anxiety they experienced, and (d) how suggestible they were. He then randomly assigned one person from each pair to drink the tea at bedtime (the experimental group), while the control group went to sleep as they normally did. He used an EEG to measure the minutes to sleep onset (the fewer the minutes to sleep onset, the better). He found $M_C = 21.20$, $M_E = 19.70$, $s_D = 5.47$, $s_{M_D} = 1.14$, $t = 1.32$, and $95\%CI\mu_{Diff}$ = from −3.86 to 0.86. Write a four-point interpretation.

9.70 A sportswriter was curious if football teams gained more yards rushing (the control condition) or passing (the experimental condition). She randomly selected nine teams and calculated the mean yards gained per game through rushing (101) and through passing (221). s_D was 27.99. She calculated $s_{M_D} = 9.33$, $t = 12.86$, and $95\%CI\mu_{Diff} =$ from 98.49 to 141.51. Write a four-point interpretation.

Completing all six steps of hypothesis testing

9.71 A dermatologist compared a new treatment for athlete's foot (the experimental condition) to the standard treatment (the control condition). He tracked down 30 people with athlete's foot on both feet and, for each participant, randomly assigned one foot to receive the new treatment and the other foot to receive the standard treatment. After three weeks of treatment, he measured the percentage of reduction in symptoms (the larger the number, the better the outcome). He found $M_E = 88$, $M_C = 72$, and $s_D = 8.65$. Analyze and interpret.

9.72 A college president wanted to know how 10-year-after-graduation salaries for academic majors (English, psychology, math, etc.) compared to salaries for career-oriented majors (business, engineering, computer science, etc.). She matched 84 academic majors at her college with 84 career-oriented majors on the basis of SAT scores and GPA. She found $M_{Academic} =$ \$59,250, $M_{Career} =$\$61,000, $s_D = 9{,}500$. Analyze and interpret.

Expand Your Knowledge

9.73 If $N = 24$ and $s_{M_D} = 3.56$, for which situation is the 95% confidence interval for the difference between population means the widest?

a. $M_C = 0$ and $M_E = 2$
b. $M_C = 0$ and $M_E = 5$
c. $M_C = 0$ and $M_E = 10$
d. $M_C = 0$ and $M_E = -2$
e. All confidence intervals are equally wide.
f. Not enough information was presented to answer this question.

9.74 If $M_C = 5$ and $M_E = 10$, for which situation is the 95% confidence interval for the difference between population means the widest?

a. $N = 5$ and $s_{M_D} = 5.28$
b. $N = 10$ and $s_{M_D} = 5.28$
c. $N = 20$ and $s_{M_D} = 5.28$
d. All confidence intervals are equally wide.
e. Not enough information was presented to answer this question.

9.75 If $M_C = 51$ and $M_E = 43$, for which situation is the 95% confidence interval for the difference between population means the widest?

a. $N = 28$ and $s_{M_D} = 2.00$
b. $N = 28$ and $s_{M_D} = 4.00$
c. $N = 28$ and $s_{M_D} = 6.00$
d. $N = 28$ and $s_{M_D} = 9.00$
e. All confidence intervals are equally wide.
f. Not enough information was presented to answer this question.

9.76 If the 95% confidence interval for the difference between population means ranges from 1.00 to 9.00, what is $M_E - M_C$?

9.77 If the 95% confidence interval for the difference between population means ranges from −5.00 to 1.00 and $s_{M_D} = 4.00$, what is t?

9.78 A nurse at a health clinic wanted to see if its ear thermometers and oral thermometers registered the same body temperatures. She selected six healthy staff members and took their temperatures with both thermometers. Apply the six steps of hypothesis testing to the data collected. Is body temperature measured similarly by ear thermometers and oral thermometers?

Body Temperature, Measured in °F

Case	1	2	3	4	5	6
Ear	97.8	98.6	98.9	97.9	99.0	98.2
Oral	97.4	97.9	98.3	97.4	98.1	97.7

SPSS

When entering data for a paired-samples *t* test into the data editor in SPSS, each pair of scores is on a row and the two data points are in separate columns. This means that the data for the humidity study would be entered as shown in Figure 9.9. Note that there is no column for a difference score since SPSS will calculate that internally and automatically.

The command for a paired-samples *t* test, as shown in Figure 9.10, is found under "Analyze" and then "Compare Means."

	low_humid	high_humid
1	76	81
2	80	90
3	78	85
4	72	82
5	76	82
6	68	75

Figure 9.9 SPSS Data Entry for a Paired-Samples *t* Test Note that each case has its own row and that the variables (high_humid and low_humid) are in the columns. (Source: SPSS)

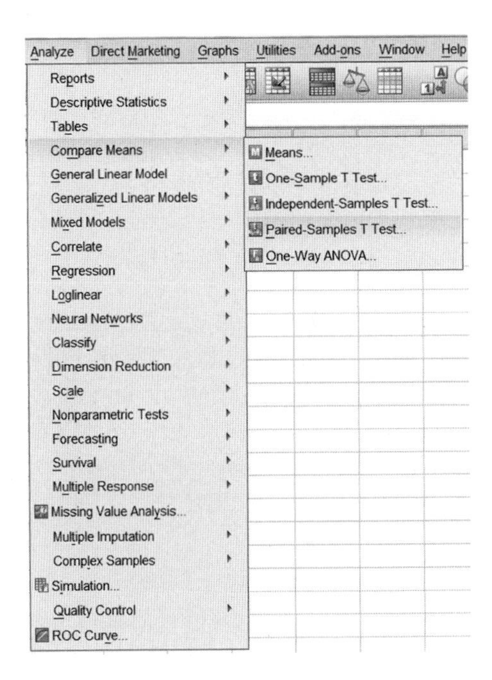

Figure 9.10 SPSS Paired-Samples t Test Command The paired-samples *t* test command in SPSS falls among the "Compare Means" commands. (Source: SPSS)

The commands for completing the paired-samples *t* test in SPSS are shown in Figure 9.11.

The SPSS output for the paired-samples *t* test is shown in Figure 9.12. The first of the three output tables provides descriptive statistics for the two samples. The second table reports the correlation between the two variables. Correlations will be covered in Chapter 13, so ignore this output for now.

The meat of the output appears after the first two tables. SPSS calculated the *t* value as −8.859, whereas in the chapter discussion it is 8.82. The sign is different because SPSS subtracted the two sample means in a different order. SPSS also carries more decimal places in its calculations, so its answer is more exact.

SPSS reports exact significance levels, seen in the last column of the final table. The value for this *t* test is .000. This means that if the null hypothesis is true, a result like the one found here, where $t = -8.859$, is a rare result; it has a probability of less than .001 of occurring. In percentage terms, a result such as this happens less than 0.1% of the time when the null hypothesis is true. The likelihood of a Type I error is very low.

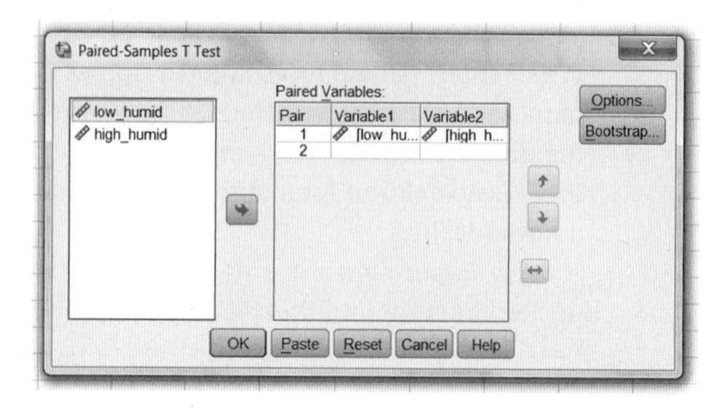

Figure 9.11 Picking the Variables to Analyze with a Paired-Samples *t* Test in SPSS The arrow button was used to move the two variables from the box on the left into the box labeled "Paired Variables." Notice how SPSS puts together the variables as a pair. (Source: SPSS)

Paired Samples Statistics

		Mean	N	Std. Deviation	Std. Error Mean
Pair 1	low_humid	75.00	6	4.336	1.770
	high_humid	82.50	6	4.930	2.012

Paired Samples Correlations

		N	Correlation	Sig.
Pair 1	low_humid & high_humid	6	.908	.012

Paired Samples Test

		Paired Differences					t	df	Sig. (2-tailed)
					95% Confidence Interval of the Difference				
		Mean	Std. Deviation	Std. Error Mean	Lower	Upper			
Pair 1	low_humid - high_humid	-7.500	2.074	.847	-9.676	-5.324	-8.859	5	.000

Figure 9.12 SPSS Output for a Paired-Samples *t* Test The top table provides descriptive statistics for the two samples. The bottom table provides the *t* value, the significance level, and the 95% confidence interval for the difference between the population means. (Ignore the middle table for now.) (Source: SPSS)

For our purposes, if the exact significance value for a test as reported by SPSS is less than or equal to .05, then the null hypothesis is rejected. If the exact significance value for a test is greater than .05, then fail to reject the null hypothesis. Since .001 is less than or equal to .05, the null hypothesis is rejected.

APA format prefers reporting exact significance levels when possible. Here, the exact significance level is reported as .000. APA format would report this as $t(5) = -8.86, p \leq .001$.

The final thing to note is that SPSS calculates the 95% confidence interval for the difference between population means. Here, SPSS reports the difference between population means as ranging from −9.676 to −5.324. Again, the sign is different because of the order in which SPSS subtracted one mean from the other. Otherwise, the numbers calculated by SPSS and in the chapter (5.31 to 9.69) differ slightly because of the number of decimal places carried. Again, SPSS offers the more exact answer because it carries more decimal places.

PART II Test Your Knowledge

These questions are meant to probe your understanding of the material covered in Chapters 6–9. The questions are not in the order of the chapters. Some of them are phrased differently or approach the material from a different direction. A few of them ask you to use the material in ways above and beyond what was covered in the book. This "test" is challenging. But, if you do well on it or puzzle out the answers using the key in the back of the book, you should feel comfortable that you are grasping the material.

1. Using the information below, find $s_{M_1-M_2}$.

Sample 1	Sample 2
$n_1 = 34$	$n_2 = 40$
$M_1 = 26.41$	$M_2 = 32.36$
$s_1 = 3.55$	$s_2 = 5.42$

2. Here is information about a paired-samples t test. Calculate the $95\%CI\mu_{\text{Diff}}$.
   - $M_1 = 48$, $M_2 = 52$
   - $N = 401$, $df = 400$
   - $s_D = 14.53$, $s_{M_D} = 0.73$
   - $t_{cv} = 1.966$, $t = 5.48$

3. Given $N = 18$, $M = 23.42$, $s = 5.82$, and $\mu = 25$, calculate t.

4. Given the information below, calculate s^2_{Pooled}.

	Group 1	Group 2
M	96.86	106.88
s	4.50	6.40
n	12	10

5. Calculate the 95%CI for the difference between population means.

M Group 1 (Control)	3.50
M Group 2 (Experimental)	3.70
N (pairs of cases)	25
s difference scores	0.25
s_{M_D}	0.05

6. Given $N = 36$, $M = 55$, $\mu = 57$, and $\sigma = 12$, calculate z.

7. Derek completes a single-sample z test with 55 cases, where $\mu = 10.6$, $M = 12.9$, and $z = 1.70$. Report the results in APA format.

8. Adele is planning to complete a single-sample t test comparing the GPA of the small number of male psych majors graduating this semester ($n = 10$) to the GPA of the larger number of female psych majors who are graduating ($n = 17$). What df should she use?

9. Meghan used an independent-samples t test to compare the salaries of male professors to those of female professors at her college. To help interpret her results, she calculated a 95%CI for the difference between population means and found that it ranged from \$2,563 to \$42,985. The difference was in favor of the males. What should Meghan suggest for future research?

10. Based on Meghan's results in Question 9, Taylor has decided to replicate the study at her college, but as a one-tailed independent-samples t test, $\alpha = .05$. Her alternative hypothesis is that $\mu_{\text{Men}} > \mu_{\text{Women}}$, and for the numerator of her t test, she uses $M_{\text{Men}} - M_{\text{Women}}$. Don't worry about the exact value of t_{cv}; just draw Taylor's decision rule as a sampling distribution of t in which the rare and common zones are labeled.

11. Michael completed a study. In the interpretation, he expressed a concern that there was an effect, but that his study had failed to find it. Michael had:

_____ rejected the null hypothesis.

_____ failed to reject the null hypothesis.

_____ done an independent-samples t test.

_____ done a paired-samples t test.

_____ found $d = 0$.

_____ a sample size greater than 50.

_____ It is possible to answer the question from the information given, but the correct answer is not listed above.

_____ Based on the information presented above, this question is not answerable.

12. A researcher knows that stimulants are used to treat ADHD. He is curious if caffeine, a stimulant, has an impact on ADHD symptoms. He takes a sample of children who have been diagnosed with ADHD but who are not on medication. One by one, he places each child in a chair to watch a very boring, 15-minute video. The chair has a sensor, so it can measure how many minutes a child remains seated—the fewer the minutes, the worse the child's ADHD symptoms are. A few days later, each child returns to the researcher's office, consumes 16 ounces of Starbucks coffee, and repeats the earlier procedure watching an equally boring video. What statistical test should this researcher use to see if caffeine has an impact on in-seat minutes?

13. An education researcher randomly assigns 36 15-year-olds either to study a driver's manual online or to review a print-out of the same document. She then has each teenager take a multiple-choice test. What statistical test should she use to see if modality of studying, online vs. paper, has an impact on test performance?

14. If $N = 77$, $M_1 = 48$, $M_2 = 44$, and $s_{M_1-M_2} = 2.50$, report the results in APA format.

15. Should this independent-samples t test be conducted? (The dependent variable is minutes spent talking on the phone during the week.)

	Sample 1	Sample 2
M	273	198
s	92	33
n	18	96

16. Should this one-sample z test be completed? (The dependent variable is introversion level.)

$M = 356$; $\mu = 500$; $N = 360$; $s = 24$

17. Stanley compared Druids to Wiccans in terms of spirituality. He used an interval-level measure of spirituality and, other than random samples, violated no assumptions. On the spirituality scale, higher scores mean more spiritual: Druids scored a mean (SD) of 78 (18) and Wiccans 65 (15). In APA format, Stanley wrote: $t(70) = 3.31$, $p < .05$. What should his conclusion be?

_____ In these samples, Druids are more spiritual than Wiccans.

_____ In these samples, Wiccans are more spiritual than Druids.

_____ In the populations, Druids are more spiritual than Wiccans.

_____ In the populations, Wiccans are more spiritual than Druids.

_____ In the populations, Druids are probably more spiritual than Wiccans.

_____ In the populations, Wiccans are probably more spiritual than Druids.

_____ Not enough information was given to answer the question.

_____ Enough information was given, but the correct answer is not listed here.

18. True or False: In order to be able to reject the null hypothesis, power must be $\geq .80$.

19. A public health researcher wanted to test the theory that aluminum is related to Alzheimer's disease (AD). He matched people with AD to people without AD on age, sex, and socioeconomic status. He then found out whether or not each person had used aluminum cookware. Each person received a score for the number of months he or she had prepared meals with such cookware. What statistical test should be used to analyze these data?

20. The researcher from Question 19 was curious if his group of Alzheimer's disease patients could be considered representative of Alzheimer's patients in general. On the Universal Alzheimer's Disease Rating Scale (UADRS), the average score in the Alzheimer's population is 76, with a population standard deviation of 15. (The UADRS is an interval-level scale.) His patients had a mean UADRS score of 72. What statistical test should be used to analyze these data?

21. A storeowner was curious about the effect of music on consumer behavior. He played both happy music and sad music in his shop. Each day he flipped a coin to determine which he should play. He then compared the total dollar value sold in his store on happy days vs. the total sold on sad days. What statistical test should be used to analyze these data?

22. Write a complete interpretation of the following study. Use $\alpha = .05$, two-tailed.

A social psychologist explored what motivates people more—altruism or self-interest. Members of a high school band who were selling LED light bulbs as a fundraiser were randomly assigned to two conditions.

The **altruism** group was told how the money raised was used to support a band program in an impoverished elementary school. It watched a short video of a 10-year-old boy, beaming with joy, as he received a trumpet paid for by the fundraiser's light bulb sales.

The **self-interest** group was told that each person would receive 10% of all the money he or she raised. It watched a short video of a band member, beaming with joy, as he displayed the iPod he bought with his light bulb money.

All band members then sold as many light bulbs as they could. Here are the results:

	Self-interest	Altruism
M	\$78.66	\$106.00
s	19.74	14.79
n	33	39

In addition:

- $df = 70$, $t_{cv} = 1.994$
- $s_{M_1-M_2} = 4.07$; $s_{\text{Pooled}} = 17.23$
- $t = 6.72$
- $d = 1.59$; 95% CI ranges from 19.22 to 35.46
- $r^2 = 40\%$

23. A cognitive psychologist read a study that found that children reared in families who ate organic diets ended up with higher IQs than children from families who did not eat organic food. He thought that if such an effect existed, it could be explained by socioeconomic status (SES). It costs more to buy organic and families that buy organic can afford other advantages for their kids. So, he designed a study:

He found 25 families who were feeding their babies organic food and matched them to 25 non-organic baby food families on the basis of (a) SES, (b) sex and age of the baby, (c) intelligence of the parents, (d) parenting style, and (e) degree of liberalism. Six years later, when the babies were in first grade, he measured the IQs of all 50 kids.

He found that the mean IQ of the "organic" kids was 109 and for the "non-organic" it was 102. In addition:

- $S_{M_D} = 1.40$
- $t_{cv} = 2.064$
- $t = 5.00$
- 95%CI [4.11, 9.84]

Interpret the results. Use $\alpha = .05$, two-tailed.

PART III Analysis of Variance

This section covers a family of tests frequently used in psychology called analysis of variance, abbreviated as ANOVA. ANOVAs are difference tests, like *t* tests. Like *t* tests, they take a variety of forms, depending on the groups being studied. But they are unlike *t* tests in one important way. *t* tests are limited to a maximum of two samples, so they are great for classic studies that have an experimental group and control group. ANOVA can handle three or more groups. So, for example, it can be used with a study where people with headaches are randomly assigned to receive aspirin, acetaminophen, or ibuprofen to see which medication works most quickly.

ANOVAs can also be used for studies with two or more explanatory variables like the one below, in which right-handed people are tested in a driving simulator to see how which hand they have on the wheel and which eye they have on the road affect performance.

	Both Eyes Open	Right Eye Covered	Left Eye Covered
Use both hands	2 eyes, 2 hands	L eye, 2 hands	R eye, 2 hands
Use right hand	2 eyes, R hand	L eye, R hand	R eye, R hand
Use left hand	2 eyes, L hand	L eye, L hand	R eye, L hand

Analyses of variance allow behavioral scientists to ask, and to answer, complex questions, the type of questions that must be posed to understand something as complex as how humans and animals think, feel, and behave.

Between-Subjects, One-Way Analysis of Variance

LEARNING OBJECTIVES

- Explain when ANOVA is used and how it works.
- Complete a between-subjects, one-way ANOVA.
- Interpret the results of a between-subjects, one-way ANOVA.

CHAPTER OVERVIEW

Chapters 8 and 9 covered two-sample *t* tests, tests that compare the means of two groups to see if they have a statistically significant difference. *t* tests are great for experiments with just two groups, like classic experiments with a control group and an experimental group. A *t* test would work well, for example, to see if a medication works better than a placebo in treating an illness. However, experiments that address complex questions often require more than two groups. Finding the best way to treat an illness might involve comparing three different medications and three different doses of each medication, ending up with nine different groups. The statistical technique that compares means when there are more than two groups is called analysis of variance.

10.1 Introduction to Analysis of Variance

Analysis of Variance Terminology

Analysis of variance, called **ANOVA** for short, is a family of statistical tests used for comparing the means of two or more groups. This chapter focuses on **between-subjects, one-way ANOVA**. Between-subjects, one-way ANOVA is an extension of the independent-samples *t* test, so it is used to compare means when there are two or more *independent* samples.

- **Between-subjects** is ANOVA terminology for independent samples.
- **Way** is ANOVA terminology for an explanatory variable. Ways can either be independent variables or grouping variables. (Independent variables are controlled by the experimenter, who assigns subjects to groups. Grouping variables are naturally occurring characteristics used to classify subjects into different groups.)

- A one-way ANOVA has one explanatory variable, either an independent variable or a grouping variable. A two-way ANOVA would have two explanatory variables, a three-way ANOVA would have three, etc.

- Though the explanatory variable in ANOVA is either a grouping variable or an independent variable, it becomes tedious to use both terms. Most statisticians get a bit casual with language and simply call it an independent variable; we'll continue calling them, generically, explanatory variables. Of course, when it is an independent variable, we'll call it that.

- An explanatory variable in ANOVA is also called a **factor,** so a one-way ANOVA can also be called a one-factor ANOVA.

- **Level** is the term in ANOVA for a category of an explanatory variable. The grouping variable sex, for example, has two levels—male and female.

To clarify all this terminology, here's a question where a between-subjects, one-way ANOVA would be used: Is there a difference in artistic ability among right-handed, left-handed, and ambidextrous people? This question could be answered by gathering a sample of people and classifying them into three groups: (1) right-handed people, (2) left-handed people, and (3) ambidextrous people.

Now there are three samples. The samples are independent samples as who is in one sample does not control or determine who is in another sample.

Next, artistic ability is measured with an interval-level scale. Because the dependent variable, artistic ability, is measured at the interval level, the mean level for each of the three groups can be calculated. Means can also be calculated for ratio-level variables, so ANOVA may be used when the outcome variable is measured at the interval or ratio level.

Table 10.1 illustrates this experiment with one cell for each sample. There is one row with three columns. The row represents the explanatory variable of handedness, what in ANOVA terminology is called the *way* or the *factor.* The columns represent the three *levels* of the explanatory variable: right-handed, left-handed, and ambidextrous.

TABLE 10.1 Comparing Artistic Ability in Three Independent Samples

Level 1	Level 2	Level 3
Right-handed	Left-handed	Ambidextrous

A one-way ANOVA has one factor with multiple levels. Here, the factor (explanatory variable) is handedness, and the three levels are right-handed, left-handed, and ambidextrous. Notice how the design is diagrammed as one row with three cells. In diagrams of ANOVA designs, each cell represents a sample. This design has three samples.

Why ANOVA Is Needed

t tests compare means between groups, so why is analysis of variance even needed? ANOVA is needed in order to keep the risk of Type I error at a reasonable level when comparing means of multiple groups. (Remember: Type I error occurs when a researcher erroneously concludes that there is a statistically significant difference.)

To see why this is a problem, consider a study with five conditions—for example, four different medications and a placebo being tested in treating some disease. It would be possible to analyze the data from these five conditions using a series of *t* tests. For example, a researcher could compare the mean of Condition 1 to Condition 2, the mean of Condition 1 to Condition 3, the mean of Condition 1 to Condition 4, and so on as shown in Table 10.2. This would require completing 10 *t* tests.

TABLE 10.2 All Possible Pairs of Five Experimental Groups

1 vs. 2	1 vs. 3	1 vs. 4	1 vs. 5
2 vs. 3	2 vs. 4	2 vs. 5	
3 vs. 4	3 vs. 5		
4 vs. 5			

If a study has five experimental conditions and if each condition is compared to each other, a researcher would need to complete 10 separate *t* tests. If each individual *t* test has a 5% chance of making a Type I error, there's almost a 50% chance that 1 of the 10 would have a Type I error.

It is tedious to do 10 *t* tests, but that is not why ANOVA is preferred. The real problem is that the likelihood of making a Type I error increases as the number of *t* tests increases. Scientists usually set alpha, the probability of a Type I error, at .05, so that they have a 5% chance of making this error. But, with 10 separate tests, the *overall* alpha level rises to be close to 50%. This means that there is close to a 50% chance that 1 of the 10 *t* tests will reach the wrong conclusion, rejecting the null hypothesis when it is true. Those are not good odds. And, the experimenter won't know which, if any, of the statistically significant results is erroneous.

Analysis of variance solves the problem of having a large risk of Type I error, what statisticians call runaway alpha. With ANOVA, one test is completed with a specified chance of Type I error. As with *t* tests, the alpha level, the chance of committing a Type I error, is usually set at .05. One test—the ANOVA—compares all the means at once and determines if any two of the means have a statistically significant difference. If the ANOVA is statistically significant, then the researcher performs what is called a post-hoc test. A **post-hoc test** is a follow-up test, engineered to find out which pairs of means differ while keeping the overall alpha level at a specified level, again, usually .05.

Here's a metaphor for how analysis of variance and post-hoc tests work together. Imagine standing outside of a football stadium on the day of a big game and hearing a loud roar. From the roar, one would know that something significant has happened in the game. That's analysis of variance—it indicates, in general, that something interesting has happened, but it doesn't state specifically what happened. To find out what has happened at the game, one needs to buy a ticket and go into the stadium. Going into the stadium is like doing a post-hoc test. Post-hoc tests are only conducted in analysis of variance when one is sure there is something interesting to be found.

What ANOVA Does

With this background on ANOVA, let's learn why it is called analysis of variance and how it works. ANOVA works by analyzing a set of scores and separating out the different sources of variability in the scores.

To understand this, imagine investigating the effect of alcohol on intoxication. Suppose research participants come to a laboratory where each person consumes one beer. After 20 minutes, researchers measure the level of intoxication by observing the effects of the alcohol on each person's performance on a behavioral task, say, walking a straight line. The higher the intoxication score, the poorer is the ability to walk in a straight line.

Figure 10.1 shows the expected results—not everyone would have exactly the same intoxication score. Rather, there is variability in the scores—with some people acting quite intoxicated, some people not acting at all intoxicated, and most clustered around the average score. Even though everyone received *exactly* the same dose of alcohol, not everyone reacted in *exactly* the same way. This variability *within* a group that receives the same treatment is called **within-group variability.** Within-group variability is primarily caused by individual differences, attributes that vary from case to case. So, how much one weighs, how recently one has eaten, and how much prior experience one has had with alcohol will all affect a person's intoxication score. These are all individual difference factors.

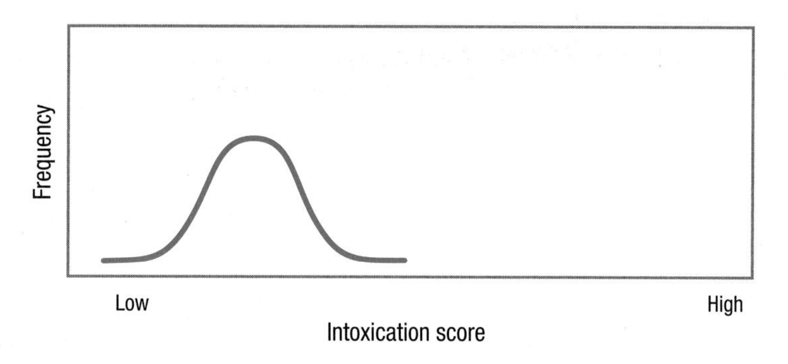

Figure 10.1 Distribution of Intoxication Scores for Participants Who Consumed One Beer Even though all participants received exactly the same dose of alcohol, there was variability in how it affected them. Some people exhibited very little intoxication and other people showed more. This variability within the group is accounted for by individual differences such as sex, weight, time since last meal, and prior experience with alcohol.

Within-group variability can be reduced by making the sample more homogeneous, but the effect of individual differences can't be eliminated entirely. For example, if all participants were men who weighed 175 pounds, had eaten dinner 30 minutes ago, and had been consuming alcohol regularly for over a year, there would still be variability within that group on the intoxication scores.

Within-group variability is one type of variability in analysis of variance. The other type of variability for a one-way ANOVA is *between-group variability.* **Between-group variability** is variability that is due to the different "treatments" that the different groups receive.

Figure 10.2 shows the distribution of intoxication scores for two groups—one group where each participant drank one beer and one group where each participant drank a six-pack. Individual differences explain the variability within a group, but the different doses of alcohol explain the differences *between* groups, why one group is more intoxicated than the other. This is called the **treatment effect** because it refers to the different ways that groups are treated. The treatment effect shows up as an impact on the outcome variable (here, the intoxication score) and is associated

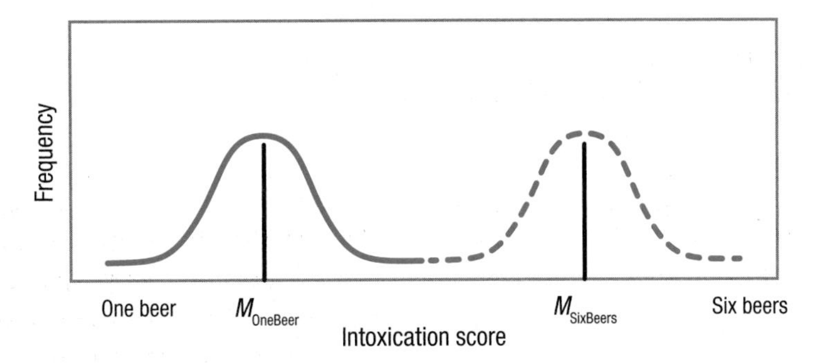

Figure 10.2 Distributions of Intoxication Scores for Participants Who Consumed Different Amounts of Alcohol The curve on the left is the distribution of intoxication scores for participants who consumed one beer. The curve on the right is the distribution of scores for participants who consumed six beers. Note that within each group there is variability in the intoxication level, variability due to individual differences. There is also variability *between* the two groups. The easiest way to see this is to observe that the two curves have different midpoints, with one group having a higher average score than the other.

with the explanatory variable (here, the dose of alcohol, which is controlled by the experimenter).

Between-group variability in one-way ANOVA is made up of two things: the treatment effect and individual differences. We've already covered how treatment plays a role in between-group variability. Now, let's see how individual differences play a role in between-group variability.

Imagine a large group of people randomly divided into two groups. Because of random assignment, the two groups should be fairly similar in terms of sex, weight, time since last meal, experience with alcohol, and so on. Now, each person in each group consumes the same dose of alcohol and is measured for intoxication level. Both groups are similar in terms of their characteristics and receive exactly the same treatment. Will the mean intoxication scores of the two groups be exactly the same? No. Because individual differences exist, the two groups will have slightly different means. Between-group variability is due *both* to individual differences and treatment effect.

How ANOVA Uses Variability

To understand how ANOVA uses within-group variability and between-group variability to see if there is a statistically significant difference, look at the two panels in Figure 10.3. Each panel represents groups that are randomly assigned to receive three different treatments for some illness. Treatment, the explanatory variable, has three levels. The top panel (A) depicts an outcome where there is little impact of treatment on outcome.

The top panel shows little effect of the independent variable because the three means (M_1, M_2, and M_3) are very close to each other. In contrast, the bottom panel (B), where the means (M_4, M_5, and M_6) are far apart, shows that treatment has an impact on outcome because the different treatments lead to dramatically different outcomes.

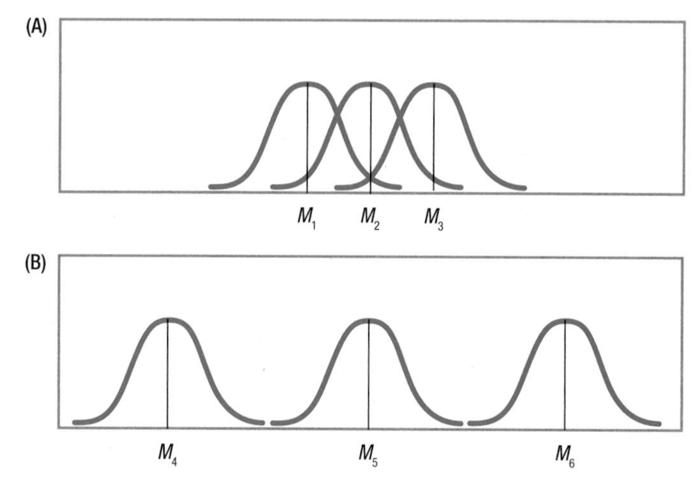

Figure 10.3 Examples of Little Impact of Treatment on Outcome and a Lot of Impact of Treatment on Outcome In the top panel (A), the three different treatments have about the same average effect on outcome. Note that the three means (M_1 to M_3) are very close to each other and that there is a lot of overlap in outcome from group to group. In contrast, the bottom panel (B) shows three treatments with very different effects on outcome. In it, the three means (M_4 to M_6) are further apart from each other and there is no overlap in outcome from group to group.

Analysis of variance could be used to analyze these results. ANOVA would show that the results in the top panel are not statistically significant, while the results in the bottom panel are statistically significant. How does analysis of variance lead to these conclusions?

Look at the means in each panel. Note that there is little variability among the means in the top panel and a lot of variability among the means in the bottom panel. Little variability exists in the top panel as all the means are close to each other. The greater distance between the means in the bottom panel indicates more variability between the means there. In the language of ANOVA, there is more *between-group variability* when the effect of treatment is large (the bottom panel) than when the effect of treatment is small (the top panel).

Figure 10.4 shows how the total variability in the data is partitioned into between-group variability and within-group variability. To decide if the amount of between-group variability is large or small, ANOVA compares it to within-group variability. One-way analysis of variance calculates the ratio of between-group variability to within-group variability. This is called an F ratio, in honor of Sir Ronald Fisher, who developed the procedure.

One-way analysis of variance calculates the ratio of between-group variability to within-group variability.

$$F = \frac{\text{Between Group Variability}}{\text{Within Group Variability}}$$

The F ratio works because within-group variability is made up of individual differences, while between-group variability includes treatment effect *and* individual differences. So, the F ratio could be rewritten as

$$F = \frac{\text{Variability due to treatment effect} + \text{Variability due to individual differences}}{\text{Variability due to individual differences}}$$

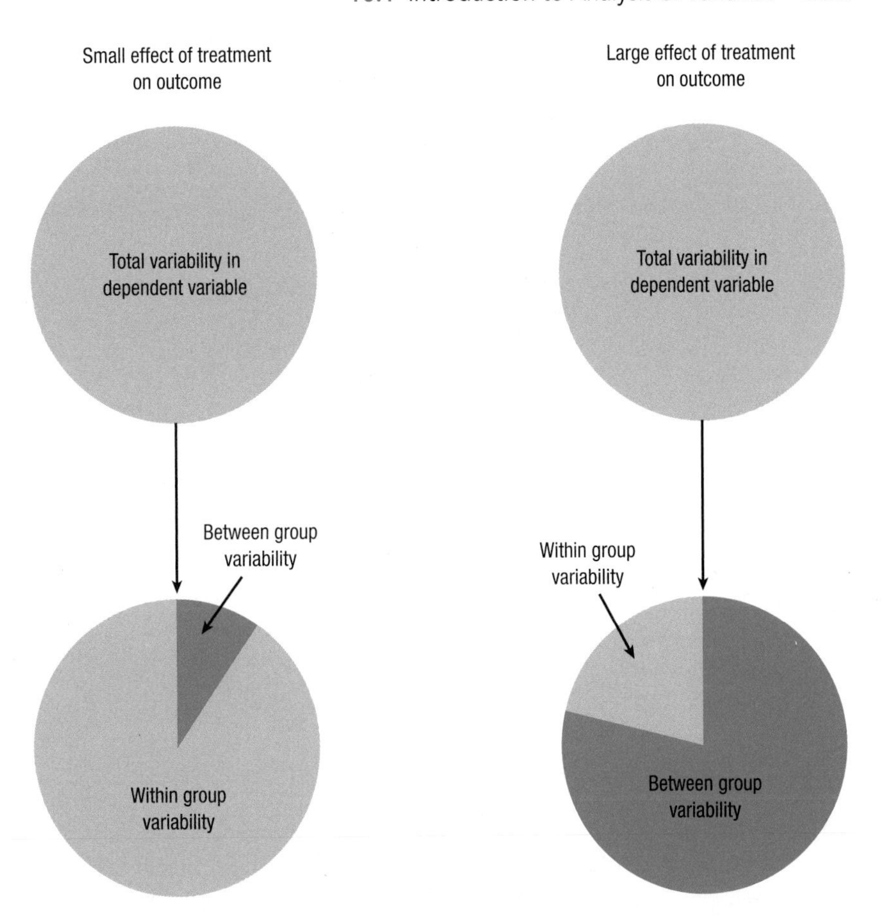

Figure 10.4 Partitioning Variability into Between-Group and Within-Group Variability The panel on the left shows little effect of treatment on outcome as only a small piece of the total variability is explained by differences between groups. In contrast, in the panel on the right, where a large piece of the total variability is accounted for by differences between groups, there is a large effect of treatment on outcome.

Here's what the *F* ratio, also known just as *F*, means:

- If there is no treatment effect, then there is no variability due to treatment and the variability indicated by the numerator of the *F* ratio is due only to individual differences.
- As a result, the *F* ratio has the same numerator (individual differences variability) and denominator (individual differences variability), so it will equal 1.
- As the effect of treatment grows, the numerator becomes larger than the denominator, and the *F* ratio climbs above 1. (Remember, treatment effect refers to the impact of the explanatory variable.)
- As the *F* ratio increases, as it climbs higher above 1, the results are more likely to be statistically significant.
- As variability is never negative, the *F* ratio can't go below 0.

Figure 10.5 gives an example of what the *F* distribution looks like. Note that it starts at 0, has a mode near 1, and tails off to the right. *F* gets bigger when there is more between-group variability than within-group variability.

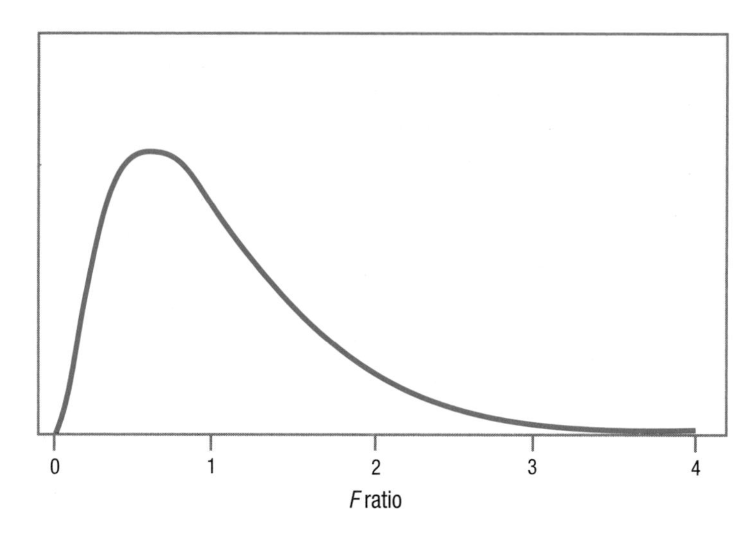

Figure 10.5 An Example of a Sampling Distribution for the *F* Ratio This is an example of a sampling distribution for the *F* ratio, the ratio of between-group variability to within-group variability. Note that *F* can't be lower than zero, that the high point of the curve is close to a value of 1 on the *X*-axis, that the distribution is positively skewed, and that the probability decreases as *F* gets larger than 1.

A Common Question

Q Looking at the ratio of between-group variability to within-group variability is a clever way to see if means differ. Could an ANOVA be used instead of a *t* test when there are just two groups?

A Yes. ANOVA can be used when comparing *two* or more means. In fact, a *t* test is just a variation on ANOVA. If a researcher has two groups, calculates a *t* value, and then squares the *t* value, it will be equal to the *F* ratio obtained by analyzing the same data with an ANOVA.

Practice Problems 10.1

Review Your Knowledge

10.1 What makes up within-group variability?

10.2 What makes up between-group variability?

10.3 An *F* ratio is a ratio of what divided by what?

10.4 When is a post-hoc test used?

10.2 Calculating Between-Subjects, One-Way ANOVA

Imagine that Dr. Chung, a psychologist who studies learning and motivation, designed a study to learn whether rats could discriminate among different types of food and if this discrimination influenced their behavior. He built a large and complex maze and trained 10 rats to run it. A rat would be placed in the start box and had to find its way to the goal box. To motivate the rats, food was placed in the goal box. Dr. Chung trained the rats with three different types of food: a low-calorie food, a normal-calorie food, and a high-calorie food. Each rat received an equal number of trials with each food and the rats were trained until they all ran the maze equally quickly.

Dr. Chung then put all the rats on a diet until they lost 10% of their normal body weight. The purpose of this was to increase their motivation to find food. Up until this point, all the rats had received the same treatment. Now, Dr. Chung implemented his experimental manipulation.

He randomly assigned the rats to three groups based on the type of food they would find in the goal box the next time they ran the maze: low-calorie, medium-calorie, or high-calorie. Each rat, individually, was placed in the maze and allowed to approach the goal box. When the rat got there, however, it found a screen that prevented it from entering. The rat could see and smell the food through the screen, but it couldn't get to the food. The purpose of this was to inform the rat of the type of food that awaited it.

Dr. Chung picked up the rat, removed the screen, and placed the rat in the start box. He then timed, in seconds, how long it took the rat to get to the goal box. This is the dependent variable in his study and the fewer seconds it took the rat to get to the goal box, the faster the rat traveled the maze. Dr. Chung figured that if the different caloric contents of the foods had been recognized by the rats and if hungry rats were more motivated by higher-calorie foods, then the time taken to run the maze should differ among the three groups. The data and the means for the three groups are shown in Table 10.3.

TABLE 10.3 Time to Run Maze (in seconds)

	Low-Calorie Food	Medium-Calorie Food	High-Calorie Food	
	30	28	24	
	31	29	25	
	32	27	26	
		28		Grand
Σ	93.00	112.00	75.00	280.00
n	3	4	3	10
M	31.00	28.00	25.00	28.00
s	1.00	0.82	1.00	2.58

These data represent the dependent variable, the number of seconds it took the 10 rats, randomly assigned to three different conditions, to travel through a maze in order to reach food in the goal box. The independent variable, the caloric density of the food, has three levels: low, medium, and high. The final column, labeled "Grand," provides information for subjects from all groups combined.

The average was 31.00 seconds to get to the low-calorie food, 28.00 seconds for the medium-calorie food, and 25.00 seconds for the high-calorie food (Figure 10.6). It appears as if mean speed increases as calorie content goes up, but the effect is not dramatic. It seems possible that the differences among the three groups may be explained by sampling error. It is also possible that the calorie content of the food in the goal box has had an effect. A statistical test is needed to decide between these two options.

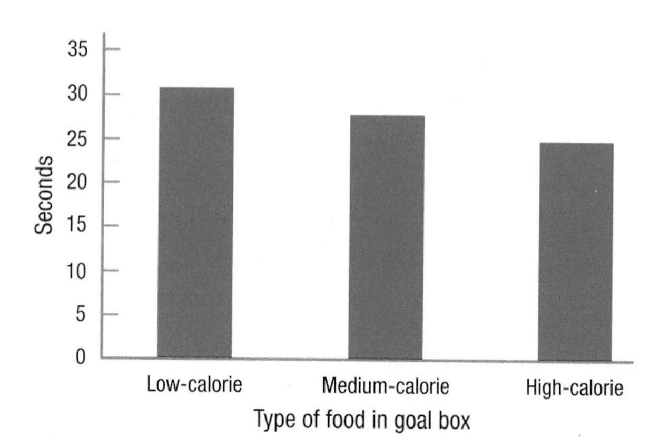

Figure 10.6 The Effect of Calorie Content on Maze Travel Time There appears to be some decrease in the mean time it takes a rat to run the maze (i.e., an increase in speed) as the calorie content of the food in the goal box increases. But, the effect looks modest at best. A between-subjects, one-way ANOVA can be used to see if the effect is a statistically significant one or if the differences are due to sampling error.

Step 1 Pick a Test

The first step of hypothesis testing is picking the right test. This study compares the means of three independent samples. There is one independent variable (calorie content), which has three levels (low, medium, and high). The dependent variable, seconds, is measured at the ratio level, so means can be calculated. This study calls for a between-subjects, one-way ANOVA.

Don't forget: **T**om **a**nd **H**arry **d**espise **c**rabby **i**nfants is a mnemonic to remember the six steps of hypothesis testing.

Step 2 Check the Assumptions

The assumptions for the between-subjects, one-way ANOVA, listed in Table 10.4, are the same as they are for an independent-samples *t* test: (1) each sample should be a random sample from its population, (2) the cases should be independent of each other, (3) the dependent variable should be normally distributed in each population, and (4) each population should have the same degree of variability. The robustness of the assumptions is the same as for *t*. The only nonrobust assumption is the second one, the assumption of independence of observations within each group.

TABLE 10.4 Assumptions for One-Way ANOVA

	Assumption	Robustness
1. Random samples	Each sample is a random sample from its population.	Robust
2. Independence of cases	Each case is not influenced by other cases in the sample.	Not robust
3. Normality	The dependent variable is normally distributed in each population.	Robust, especially if sample size (*N*) is large and the *n*'s are about equal.
4. Homogeneity of variance	The degree of variability in the two populations is equivalent.	Robust, especially if sample size (*N*) is large and the *n*'s are about equal

There's no hard and fast rule as to what a "large" sample size is. Some statisticians say 25 or more, some 35, and some 50.

Here is an evaluation of the assumptions for the maze data:

1. *Random samples.* Though this is an experimental study in which cases were randomly assigned to experimental conditions, the initial sample wasn't a random sample from the population of rats, so this assumption was violated. The assumption is robust, however, so Dr. Chung can proceed. However, he has to be careful about generalizing beyond the specific strain of rats he's working with.
2. *Independence of cases.* Each rat was trained and tested individually. Each rat only provided one data point for the final test. So, this assumption was not violated.
3. *Normality.* It seems reasonable to assume that, within a population of rats, running speed is normally distributed. That is, for rats, there is a mean speed around which most animals cluster and the number of rats with higher and lower speeds tails off in both directions symmetrically.
4. *Homogeneity of variance.* A look at the three standard deviations (1.00, 0.82, and 1.00) shows that they are all about the same. There is no reason to believe that this assumption has been violated.

With no nonrobust assumptions violated, Dr. Chung can proceed with the planned between-subjects, one-way ANOVA.

Step 3 List the Hypotheses

The hypotheses are statements about the populations. In Dr. Chung's study, there are three samples, with each one thought of as having come from a separate population.

His hypotheses will be about the population of rats that finds low-calorie food in the goal box, the population of rats that finds medium-calorie food, and the population of rats that finds high-calorie food.

Look at Figure 10.5, the example of a sampling distribution for an F ratio, and notice two things—there are no negative values on the x-axis and the sampling distribution is positively skewed. It looks like there is only one "tail" to the sampling distribution. This doesn't mean that all ANOVAs are one-tailed; it means just the opposite. Both "tails" are wrapped into this one side, so ANOVA always has nondirectional hypotheses. One doesn't have to decide between a one-tailed or two-tailed test. ANOVA is always two-tailed.

The nondirectional null hypothesis will state that no difference exists in the means of the populations. The generic form of this, where there are k samples, is

$$H_0: \mu_1 = \mu_2 = \ldots = \mu_k$$

For the maze study, with three populations, Dr. Chung's null hypothesis is

$$H_0: \mu_1 = \mu_2 = \mu_3$$

The alternative hypothesis, also nondirectional, says not all population means are equal. It is not written as $\mu_1 \neq \mu_2 \neq \mu_3$ because that doesn't cover options like Populations 1 and 2 having different means but Populations 2 and 3 don't. The alternative hypothesis states that at least two of the population means are different, and maybe all three are. There is no easy way to write this mathematically, so Dr. Chung will write it as

H_1: At least one population mean is different from the others.

Step 4 Set the Decision Rule

This step finds the critical value of F, abbreviated F_{cv}, the value that separates the rare zone from the common zone of the sampling distribution of the F ratio. Figure 10.7 is an example of a sampling distribution of F with the rare and common zones marked.

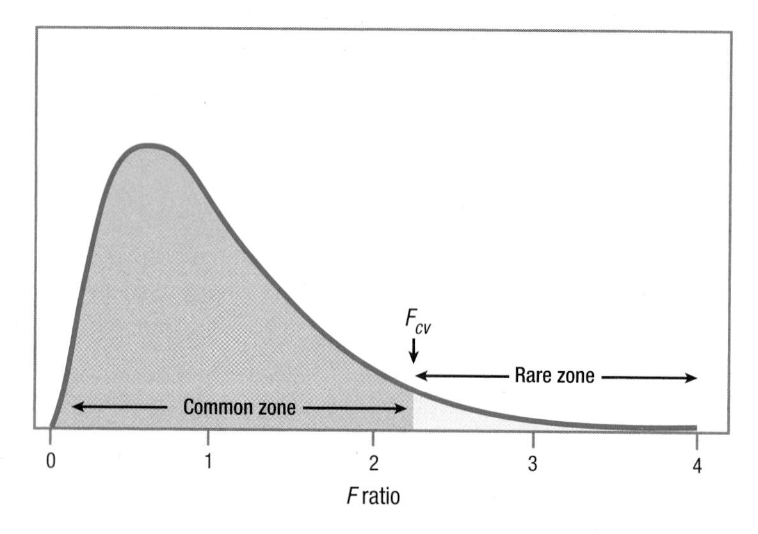

Figure 10.7 Using the Critical Value of F to Mark the Common Zone and the Rare Zone of the Sampling Distribution of F The critical value of F, F_{cv}, marks the boundary between the common zone and the rare zone of the test statistic. Here, 5% of the area under the curve falls in the rare zone that is on and to the right of F_{cv}.

The decision rule will sound familiar as it is similar to what we used for the *z* test and *t* tests: if the calculated value of the test statistic, *F*, falls on the line or in the rare zone, the null hypothesis is rejected.

- If $F \geq F_{cv}$, reject H_0.
- If $F < F_{cv}$, fail to reject H_0.

To obtain the value of F_{cv}, use Appendix Table 4. It provides two F_{cv} tables: one for critical values of *F* with a 5% chance of making a Type I error (i.e., $\alpha = .05$) and one for a 1% chance of making a Type I error ($\alpha = .01$). A portion of the table with the most commonly used alpha level, .05, is shown in Table 10.5.

TABLE 10.5 Part of Appendix Table 4, Critical Values of *F* (F_{cv})

Denominator Degrees of Freedom	Numerator Degrees of Freedom					
	1	2	3	4	5	6
1	161.448	199.500	215.707	224.583	230.162	233.986
2	18.513	19.000	19.164	19.247	19.296	19.330
3	10.128	9.552	9.277	9.117	9.013	8.941
4	7.709	6.944	6.591	6.388	6.256	6.163
5	6.608	5.786	5.409	5.192	5.050	4.950
6	5.987	5.143	4.757	4.534	4.387	4.284
7	5.591	4.737	4.347	4.120	3.972	3.866
8	5.318	4.459	4.066	3.838	3.687	3.581
9	5.117	4.256	3.863	3.633	3.482	3.374
10	4.965	4.103	3.708	3.478	3.326	3.217
11	4.844	3.982	3.587	3.357	3.204	3.095
12	4.747	3.885	3.490	3.259	3.106	2.996
13	4.667	3.806	3.411	3.179	3.025	2.915
14	4.600	3.739	3.344	3.112	2.958	2.848
15	4.543	3.682	3.287	3.056	2.901	2.790

The critical value of *F*, F_{cv}, is found at the intersection of the column for the numerator degrees of freedom and the row for the denominator degrees of freedom.

To determine the critical value of *F*, find the value at the intersection of the *column* for the degrees of freedom in the numerator and the *row* for the degrees of freedom in the denominator. Equation 10.1 shows how to calculate the different degrees of freedom needed for a between-subjects, one-way ANOVA. Recall the formula for *F* ratio from the previous section:

- The numerator degrees of freedom for the *F* ratio, what defines the columns in the table of critical values of *F*, are called between-groups degrees of freedom. This is abbreviated as df_{Between}.
- The denominator degrees of freedom for the *F* ratio, what defines the rows in the table of critical values of *F*, are within-groups degrees of freedom. This is abbreviated as df_{Within}.

- The final degrees of freedom calculated in Equation 10.1 are total degrees of freedom. This is abbreviated as df_{Total} and represents the total number of degrees of freedom in the data. Note that adding together df_{Between} and df_{Within} equals df_{Total}.

Equation 10.1 Degrees of Freedom for Between-Subjects, One-Way ANOVA

$$df_{\text{Between}} = k - 1$$

$$df_{\text{Within}} = N - k$$

$$df_{\text{Total}} = N - 1$$

where df_{Between} = between-groups degrees of freedom (degrees of freedom for the numerator)
df_{Within} = within-groups degrees of freedom (degrees of freedom for the denominator)
df_{Total} = total degrees of freedom
k = the number of groups
N = the total number of cases

For the maze data, there are three groups (low-, medium-, and high-calorie), so $k = 3$. There are a total of 10 participants in the study, so $N = 10$. Given these values, degrees of freedom are calculated as

$$\begin{aligned} df_{\text{Between}} &= k - 1 \\ &= 3 - 1 \\ &= 2 \end{aligned}$$

$$\begin{aligned} df_{\text{Within}} &= N - k \\ &= 10 - 3 \\ &= 7 \end{aligned}$$

$$\begin{aligned} df_{\text{Total}} &= N - 1 \\ &= 10 - 1 \\ &= 9 \end{aligned}$$

To find the critical value of F, Dr. Chung needs numerator degrees of freedom ($df_{\text{Between}} = 2$) and denominator degrees of freedom ($df_{\text{Within}} = 7$). The intersection of the column for 2 degrees of freedom and the row for 7 degrees of freedom leads to a critical value of F of 4.737. That means the decision rule for the maze data can now be written as:

- If $F \geq 4.737$, reject the null hypothesis.
- If $F < 4.737$, fail to reject the null hypothesis.

The sampling distribution of F with degrees of freedom of 2 (numerator) and 7 (denominator) is shown in **Figure 10.8**. The critical value of F, $F_{cv} = 4.737$, is used to separate the rare zone from the common zone. If the observed value of F, the value of the test statistic calculated in Step 5, falls on the line or in the rare zone, the null

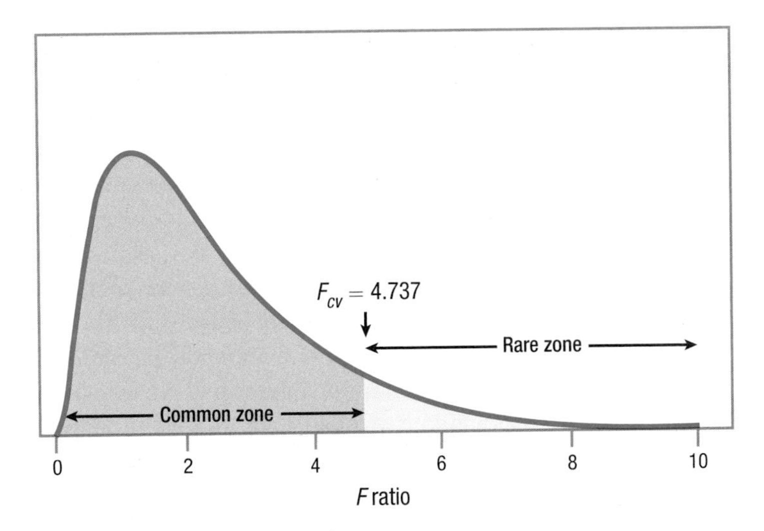

Figure 10.8 The Critical Value of *F* for 2 and 7 Degrees of Freedom The critical value of *F*, F_{cv}, with 2 degrees of freedom in the numerator and 7 degrees of freedom in the denominator, is 4.737. Note that 5% of the *F* distribution falls on or to the right of this point. If the observed value of *F* falls in the rare zone, then the null hypothesis is rejected.

hypothesis is rejected. If it falls in the common zone, Dr. Chung will fail to reject the null hypothesis.

Step 5 Calculate the Test Statistic

To complete the ANOVA, Dr. Chung's next step is to analyze the variability in the maze data. This means taking account of variability both within and between groups:

- Not each rat with the same type of food in the goal box coursed through the maze in the same amount of time, so there is variability within groups.
- The three groups all had different means, so there is variability between groups.

To understand why analysis of variance is called analysis of *variance,* it is useful to review the variance formula from Chapter 3. Equation 3.6 says:

$$s^2 = \frac{\Sigma(X - M)^2}{N - 1}$$

The *numerator* in the variance formula is calculated by following three steps: (1) calculating deviation scores by subtracting the mean from each score, (2) squaring all the deviation scores, and (3) adding up all the squared deviation scores. This numerator, a sum of squared deviation scores, is called a *sum of squares,* abbreviated as *SS*. Calculating sums of squares for the two sources of variability, between-groups and within-groups, is necessary to find the between- and within-group variances that are analyzed in an analysis of variance.

Actually, there are three sums of squares that are calculated. In addition to sum of squares between-groups (SS_{Between}) and sum of squares within groups (SS_{Within}), sum of squares total (SS_{Total}) is also calculated. Sum of squares total represents all the variability in the data. Between-subjects, one-way ANOVA divides SS_{Total} into between-groups sum of squares (SS_{Between}), which measures variability due to treatment, and within-groups sum of squares (SS_{Within}), which measures variability due to individual differences. In other words, $SS_{\text{Total}} = SS_{\text{Between}} + SS_{\text{Within}}$.

SS_{Total} represents all the variability in the scores. Between-subjects, one-way ANOVA divides it into variability due to the different ways the groups are treated, and variability due to individual differences.

Sum of squares total is calculated by treating all the cases as if they belong in one group: the grand mean, the mean of all the scores, is subtracted from each score, these deviation scores are squared, and the squared deviation scores are added up. Voilà, a sum of squares, the numerator for a variance.

Sum of squares between represents the variability between groups. It is calculated by subtracting the grand mean from each group mean, squaring these deviation scores, multiplying each squared deviation score by the number of cases in the group, and then adding them up. The final variance numerator, the one that represents variability within groups, **sum of squares within**, is calculated by taking each score, subtracting from it its group mean, squaring each deviation score, and adding them all up.

The calculations just described are what are called definitional formulas because the calculations explain what the value being calculated is. Unfortunately, definitional formulas often are not straightforward mathematically. In contrast, computational formulas are designed to be easier to use. Equation 10.2 is a computational formula for sum of squares total.

Equation 10.2 Formula for Calculating Sum of Squares Total for Between-Subjects, One-Way ANOVA

$$SS_{\text{Total}} = \Sigma X^2 - \frac{(\Sigma X)^2}{N}$$

where SS_{Total} = total sum of squares
X = raw score
N = the total number of cases

This formula says:

1. Square each score and add them all up.
2. Add up all the scores, square the sum, and divide by the total number of cases.
3. Subtract the result of Step 2 from the result of Step 1.

The easiest way to do this is to take Table 10.3 and add another column to it, one for squared scores. This can be seen in Table 10.6. From the table, we can see that the sum of squared scores for Step 1 is 7,900.00.

TABLE 10.6 Maze Running Time Data from Table 10.3, Squared and Summed, in Preparation for Computing Sums of Squares

	Low-Calorie		Medium-Calorie		High-Calorie		Grand	
	X	X^2	X	X^2	X	X^2	X	X^2
	30	900	28	784	24	576		
	31	961	29	841	25	625		
	32	1,024	27	729	26	676		
			28	784				
Σ	93	2,885	112	3,138	75	1,877	280	7,900
n	3		4		3		10	

Step 2 is next.

$$\frac{(\Sigma X)^2}{N} = \frac{280^2}{10}$$
$$= \frac{78{,}400.00}{10}$$
$$= 7{,}840.00$$

Finally,

$$SS_{\text{Total}} = \Sigma X^2 - \frac{(\Sigma X)^2}{N}$$
$$= 7{,}900.00 - 7{,}840.00$$
$$= 60.00$$

Next, let's use Equation 10.3 to calculate sum of squares between groups.

Equation 10.3 Formula for Calculating Between-Groups Sum of Squares for Between-Subjects, One-Way ANOVA

$$SS_{\text{Between}} = \Sigma\left(\frac{(\Sigma X_{\text{Group}})^2}{n_{\text{Group}}}\right) - \frac{(\Sigma X)^2}{N}$$

where SS_{Between} = between-groups sum of squares
X_{Group} = raw scores for cases in a group
n_{Group} = number of cases in a group
X = raw scores
N = total number of cases

Here's how the formula works:

1. For each group, add up all the scores, square that sum, and divide that square by the number of cases in the group. Add up all these quotients.
2. Add up all the scores, square that sum, and divide that square by the total number of cases.
3. Subtract Step 2 from Step 1.

Again, most of the work has already been done in Table 10.6. Here is Step 1:

$$\Sigma\left(\frac{(\Sigma X_{\text{Group}})^2}{n_{\text{Group}}}\right) = \frac{93^2}{3} + \frac{112^2}{4} + \frac{75^2}{3}$$
$$= \frac{8{,}649.00}{3} + \frac{12{,}544.00}{4} + \frac{5{,}625.00}{3}$$
$$= 2{,}883.00 + 3{,}136.00 + 1{,}875.00$$
$$= 7{,}894.00$$

And, Step 2:

$$\frac{(\Sigma X)^2}{N} = \frac{280^2}{10}$$
$$= \frac{78{,}400.00}{10}$$
$$= 7{,}840.00$$

Finally,

$$SS_{\text{Between}} = \Sigma\left(\frac{(\Sigma X_{\text{Group}})^2}{n_{\text{Group}}}\right) - \frac{(\Sigma X)^2}{N}$$
$$= 7{,}894.00 - 7{,}840.00$$
$$= 54.00$$

The final sum of squares to calculate is the sum of squares within. Equation 10.4 covers this.

Equation 10.4 Formula for Calculating Sum of Squares Within for Between-Subjects, One-Way ANOVA

$$SS_{\text{within}} = \Sigma\left(\Sigma X^2_{\text{Group}} - \frac{(\Sigma X_{\text{Group}})^2}{n_{\text{Group}}}\right)$$

where SS_{Within} = sum of squares within
X_{Group} = raw scores for cases in a group
n_{Group} = number of cases in a group

1. For each group, square each score and add them all up.
2. For each group, add up all the scores, square the sum, and divide by the total number of cases.
3. For each group, subtract the result of Step 2 from the result of Step 1.
4. Add together all the remainders from Step 3.

As with the other sums of squares, Table 10.6 has the components already prepared:

$$SS_{\text{within}} = \Sigma\left(\Sigma X^2_{\text{Group}} - \frac{(\Sigma X_{\text{Group}})^2}{n_{\text{Group}}}\right)$$
$$= \left(2{,}885 - \frac{93^2}{3}\right) + \left(3{,}138 - \frac{112^2}{4}\right) + \left(1{,}877 - \frac{75^2}{3}\right)$$
$$= \left(2{,}885 - \frac{8{,}649.00}{3}\right) + \left(3{,}138 - \frac{12{,}544.00}{4}\right) + \left(1{,}877 - \frac{5{,}625.00}{3}\right)$$
$$= (2{,}885 - 2{,}883.00) + (3{,}138 - 3{,}136.00) + (1{,}877 - 1{,}875.00)$$
$$= 2.00 + 2.00 + 2.00$$
$$= 6.00$$

Earlier in the chapter, it was stated that $SS_{Total} = SS_{Between} + SS_{Within}$. Let's see if that is true. $SS_{Between}$ is 54.00 and SS_{Within} is 6.00. They sum to 60.00, which is what we calculated SS_{Total} to be. This is shown visually in Figure 10.9.

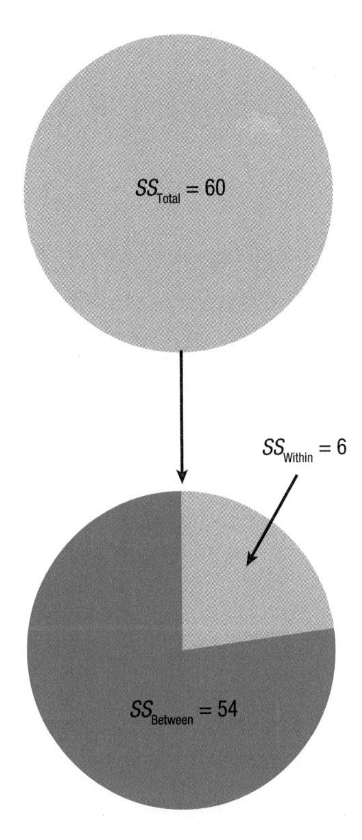

Figure 10.9 Partitioning Total Sum of Squares for Maze Running Time Data SS_{Total} can be broken down into its component parts, $SS_{Between}$ and SS_{Within}. Note that in this example the largest chunk of SS_{Total} is accounted for by $SS_{Between}$.

Now that the degrees of freedom and sums of squares have been calculated, Dr. Chung can go on to calculate the F ratio. By organizing the sums of squares and the degrees of freedom into an ANOVA summary table, the F ratio will almost calculate itself.

Table 10.7 is a template of a summary table for a between-subjects, one-way ANOVA. It is important to note the order in which it is laid out—three rows and five columns. Always arrange a summary table for a between-subjects, one-way ANOVA in exactly this way:

- There is one row for each source of variability:
  - Between-groups variability appears on the top row.
  - Within-groups variability is on the middle row.
  - Total variability goes on the bottom row.

TABLE 10.7 Template for ANOVA Summary Table for One-Way ANOVA

Source of Variability	Sum of Squares	Degrees of Freedom	Mean Square	F ratio
Between groups				
Within groups				
Total				

An ANOVA summary table for a one-way ANOVA is set up with these five columns and these three rows in this order.

- The five columns, in order, tell:
  - The source of variability (between, within, or total).
  - The sum of squares calculated for that source of variability.
  - The degrees of freedom for that source of variability.
  - What is called the "mean square" for that source of variability.
  - The *F* ratio.

Dr. Chung already has enough information to fill in the second and third columns. Mean square, the fourth column, sounds like something new, but it is a variance by another name. To calculate the mean square, one divides the sum of squares in a row by the degrees of freedom for that row.

To see how a mean square is a variance, let's revisit the variance formula from Chapter 3 one more time:

$$s^2 = \frac{\Sigma(X - M)^2}{N - 1}$$

The numerator in this formula is the same as the sum of squares total. And, df_{Total} is $N-1$. Thus, if mean square total were calculated, it would be exactly the same as the variance for all the cases. Analysis of variance really does analyze variances.

The only mean squares needed to complete a between-subjects, one-way ANOVA are between-groups mean square (MS_{Between}) and within-groups mean square (MS_{Within}). As the mean square total is not needed, the convention is not to calculate it and to leave the space blank where mean square total would go.

The formula for calculating the between-groups mean square is given in Equation 10.5.

Equation 10.5 Formula for Between-Groups Mean Square, MS_{Between}, for Between-Subjects, One-Way ANOVA

$$MS_{\text{Between}} = \frac{SS_{\text{Between}}}{df_{\text{Between}}}$$

where MS_{Between} = between-groups mean square
SS_{Between} = between-groups sum of squares
df_{Between} = between-groups degrees of freedom

For the maze data, where $SS_{Between} = 54.00$ and $df_{Between} = 2$, the between-groups mean square would be calculated:

$$MS_{Between} = \frac{SS_{Between}}{df_{Between}}$$
$$= \frac{54.00}{2}$$
$$= 27.00$$

Equation 10.6 offers the formula for the within-groups mean square.

Equation 10.6 Formula for Within-Groups Mean Square, MS_{Within}, for a Between-Subjects, One-Way ANOVA

$$MS_{Within} = \frac{SS_{Within}}{df_{Within}}$$

where MS_{Within} = within-groups mean square
SS_{Within} = within-groups sum of squares
df_{Within} = within-groups degrees of freedom

For the maze data, $SS_{Within} = 6.00$ and $df_{Within} = 7$, so MS_{Within} would be calculated:

$$MS_{Within} = \frac{SS_{Within}}{df_{Within}}$$
$$= \frac{6.00}{7}$$
$$= 0.8571$$
$$= 0.86$$

Once the two mean squares are calculated, it is time to calculate the *F* ratio. *F* is the ratio of variability due to between-group factors and individual differences divided by the variability due to individual differences. *F* is calculated as shown in Equation 10.7.

Equation 10.7 Formula for Calculating *F* for a Between-Subjects, One-Way ANOVA

$$F = \frac{MS_{Between}}{MS_{Within}}$$

where F = the F ratio
$MS_{Between}$ = the between-groups mean square
MS_{Within} = the within-groups mean square

For the maze data, where $MS_{\text{Between}} = 27.00$ and $MS_{\text{Within}} = 0.86$, F is calculated as

$$F = \frac{MS_{\text{Between}}}{MS_{\text{Within}}} = \frac{27.00}{0.86} = 31.3953 = 31.40$$

The complete ANOVA summary table for the maze data is shown in Table 10.8. Table 10.9 is a summary table that provides the organization and formulas for completing a between-subjects, one-way ANOVA.

TABLE 10.8 Completed ANOVA Summary Table for Maze Data

Source of Variability	Sum of Squares	Degrees of Freedom	Mean Square	F ratio
Between groups	54.00	2	27.00	31.40
Within groups	6.00	7	0.86	
Total	60.00	9		

This shows what a complete ANOVA summary table looks like.

TABLE 10.9 How to Complete an ANOVA Summary Table for a Between-Subjects, One-Way ANOVA

Source of Variability	Sum of Squares	Degrees of Freedom	Mean Square	F ratio
Between groups	Equation 10.3	$k - 1$	$\frac{SS_{\text{Between}}}{df_{\text{Between}}}$	$\frac{MS_{\text{Between}}}{MS_{\text{Within}}}$
Within groups	Equation 10.4	$N - k$	$\frac{SS_{\text{Within}}}{df_{\text{Within}}}$	
Total	Equation 10.2	$N - 1$		

This table summarizes all the steps necessary to complete an ANOVA summary table for a between-subjects, one-way ANOVA.

Worked Example 10.1 For practice with a between-subjects, one-way ANOVA, let's explore how effective pain relievers like acetaminophen (Tylenol) and ibuprofen (Advil) are. Dr. Douglas, a sensory psychologist, wanted to see if medications such as these affected the pain threshold (how long it takes to perceive a stimulus as painful). Twelve male undergraduates volunteered for the study, and she randomly assigned them to three groups, one control group and two experimental groups, but tested each person individually. Each participant was given a pill—either a placebo, ibuprofen, or acetaminophen—waited an hour for the medication to take effect, then placed his hand in a bucket of ice water. As this was a test of pain threshold, he was told to remove his hand from the ice water as soon as it became painful. The researcher recorded the elapsed time, in seconds, as shown in Table 10.10. Note that this table has squared and summed scores so that the ANOVA computational formulas can be easily completed.

TABLE 10.10 Results of Pain Threshold Test for Different Pain Relievers (in seconds)

	Placebo		Ibuprofen		Acetaminophen		Grand	
	X	X^2	X	X^2	X	X^2	X	X^2
	15	225	17	289	24	576		
	18	324	24	576	16	256		
	23	529	30	900	26	676		
	14	196	21	441	18	324		
Σ	70	1,274	92	2,206	84	1,832	246	5,312
n	4		4		4		12	
M	17.50		23.00		21.00		20.50	
s	4.04		5.48		4.76		4.95	

These values in the columns labeled X represent how many seconds it took, in the ice water, before a participant removed his hand because it had become painful. This table has been set up with X^2 values and sums so that it is prepared for use in calculating sums of squares.

Figure 10.10 shows what appears to be an effect of the pain relievers on the pain threshold. Compared to the placebo group, those using acetaminophen took longer on average to feel pain, and those given ibuprofen took even longer on average. It is necessary, of course, to conduct a hypothesis test to see if the effect is statistically significant and isn't plausibly accounted for by sampling error.

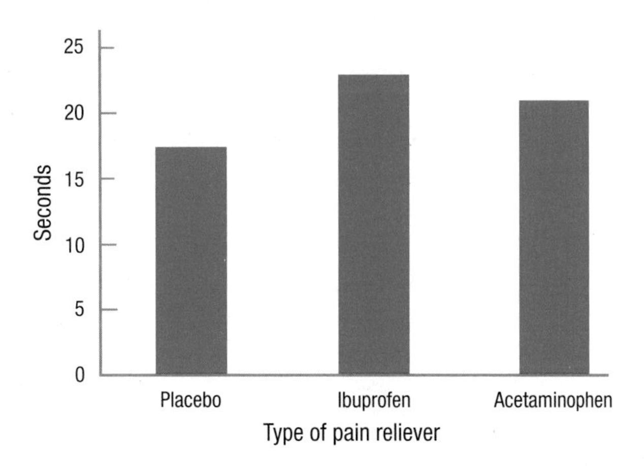

Figure 10.10 Mean Seconds to Mean Pain Threshold for Different Pain Relievers
It appears that ibuprofen, and perhaps acetaminophen, raise the pain threshold compared to placebo. But, a between-subjects, one-way ANOVA has to be conducted to make sure that the effect is statistically significant.

Step 1 Pick a Test

There are a number of conditions that lead to Dr. Douglas's selection of between-subjects, one-way ANOVA as the appropriate test to analyze these data.

- There is one independent variable (factor), the type of drug.
- This factor has three levels (placebo, ibuprofen, and acetaminophen), so there are three samples or groups.

- The cases were randomly assigned to the groups, so the samples are independent samples.
- The dependent variable, number of seconds to pain threshold, is measured at the ratio level so means can be calculated for each group.

Step 2 Check the Assumptions

1. Random samples: The samples are made up of male, college-student volunteers, so they are not random samples from the human population. The random samples assumption is violated, but it is a robust assumption so the researcher can proceed. She'll just have to be careful about the population to which she generalizes the results.
2. Independence of cases: Each participant was in the study only once. Plus, each participant was tested individually, uninfluenced by the other participants. This assumption was not violated.
3. Normality: Researchers are commonly willing to assume that psychological variables, like pain threshold, are normally distributed. So, this assumption is not violated.
4. Homogeneity of variance: All the standard deviations (4.04, 5.48, and 4.76, from Table 10.10) are very similar, so the amount of variability in each population seems about equal.

With no nonrobust assumptions violated, Dr. Douglas can proceed with the planned between-subjects, one-way ANOVA.

Step 3 List the Hypotheses

H_0: $\mu_1 = \mu_2 = \mu_3$.

H_1: At least one population mean is different from the others.

Step 4 Set the Decision Rule

In order to find the critical value of F, Dr. Douglas needs to know the degrees of freedom for the numerator (df_{Between}) and for the denominator (df_{Within}). To do so, she uses Equation 10.1:

$$df_{\text{Between}} = k - 1$$
$$= 3 - 1$$
$$= 2$$

$$df_{\text{Within}} = N - k$$
$$= 12 - 3$$
$$= 9$$

$$df_{\text{Total}} = N - 1$$
$$= 12 - 1$$
$$= 11$$

Looking at the $\alpha = .05$ version of Appendix Table 4—at the intersection of the column for 2 degrees of freedom in the numerator and the row for 9 degrees of freedom in the denominator—she finds F_{cv} is 4.256. Here is the decision rule:

- If $F \geq 4.256$, reject H_0.
- If $F < 4.256$, fail to reject H_0.

Step 5 Calculate the Test Statistic

Dr. Douglas's first step on the path to F is to calculate SS_{Total}, using Equation 10.2 and values from Table 10.10:

$$\begin{aligned} SS_{\text{Total}} &= \Sigma X^2 - \frac{(\Sigma X)^2}{N} \\ &= 5{,}312 - \frac{246^2}{12} \\ &= 5{,}312 - \frac{60{,}516.00}{12} \\ &= 5{,}312 - 5043.00 \\ &= 269.00 \end{aligned}$$

Next, she calculates SS_{Between} following Equation 10.3:

$$\begin{aligned} SS_{\text{Between}} &= \Sigma\left(\frac{(\Sigma X_{\text{Group}})^2}{n_{\text{Group}}}\right) - \frac{(\Sigma X)^2}{N} \\ &= \left(\frac{70^2}{4} + \frac{92^2}{4} + \frac{84^2}{4}\right) - \frac{246^2}{12} \\ &= \left(\frac{4{,}900.00}{4} + \frac{8{,}464.00}{4} + \frac{7{,}056.00}{4}\right) - \frac{60{,}516.00}{12} \\ &= (1{,}225.00 + 2{,}116.00 + 1{,}764.00) - 5{,}043.00 \\ &= 5{,}105.00 - 5{,}043.00 \\ &= 62.00 \end{aligned}$$

Finally, using Equation 10.4, she calculates SS_{Within}:

$$\begin{aligned} SS_{\text{Within}} &= \Sigma\left(\Sigma X^2_{\text{Group}} - \frac{(\Sigma X_{\text{Group}})^2}{n_{\text{Group}}}\right) \\ &= \left(1{,}274 - \frac{70^2}{4}\right) + \left(2{,}206 - \frac{92^2}{4}\right) + \left(1{,}832 - \frac{84^2}{4}\right) \\ &= \left(1{,}274 - \frac{4{,}900.00}{4}\right) + \left(2{,}206 - \frac{8{,}464.00}{4}\right) + \left(1{,}832 - \frac{7{,}056.00}{4}\right) \\ &= (1{,}274 - 1{,}225.00) + (2{,}206 - 2{,}116.00) + (1{,}832 - 1{,}764.00) \\ &= 49.00 + 90.00 + 68.00 \\ &= 207.00 \end{aligned}$$

As a check that she has done the math correctly, she adds together $SS_{Between}$ (62.00) and SS_{Within} (207.00) to make sure that they sum to sum of squares total (269.00). They do. Figure 10.11 shows this visually.

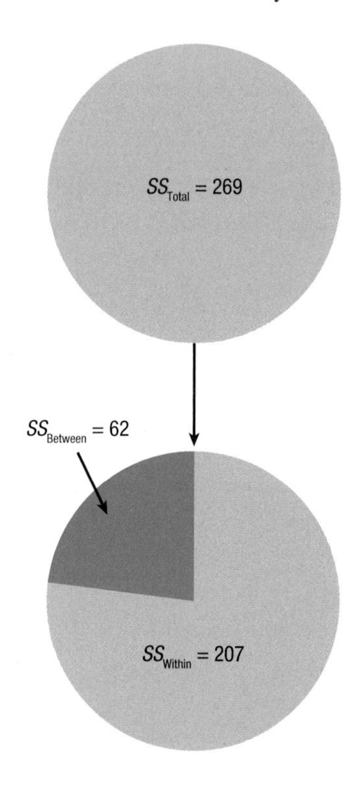

Figure 10.11 Partitioning Total Sum of Squares for Mean Pain Threshold Data SS_{Total} can be broken down into its component parts, $SS_{Between}$ and SS_{Within}. Note that the largest chunk of SS_{Total} is accounted for by SS_{Within}.

The final step is to complete an ANOVA summary table for the pain threshold data. The degrees of freedom and sums of squares already calculated can be used to start to fill in the summary table. Then, following the guidelines in Table 10.9, Dr. Douglas completes the ANOVA summary table (see Table 10.11), finding $F = 1.35$. In the next section, we'll learn how to interpret the results.

TABLE 10.11 ANOVA Summary Table for One-Way ANOVA for Mean Pain Threshold Data

Source of Variability	Sum of Squares	Degrees of Freedom	Mean Square	*F* ratio
Between groups	62.00	2	31.00	1.35
Within groups	207.00	9	23.00	
Total	269.00	11		

This is the completed ANOVA summary table for the study examining the impact of different pain relievers on mean pain threshold.

Practice Problems 10.2

Apply Your Knowledge

10.5 Select the correct test for this example: A gerontologist wants to determine which exercise program, yoga or stretching, leads to greater limberness in elderly people. He uses a ratio-level range-of-motion test to measure limberness and randomly assigns elderly people to receive either eight weeks of yoga or eight weeks of stretching.

10.6 Select the correct test for this example: A human factors psychologist is comparing three different adhesives used in sealing cereal boxes to see which one is easiest to open. He obtained 90 consumers, randomly assigned each participant to open one cereal box, and measured how long it took to open the box. Thirty of the boxes were sealed with Adhesive A, 30 with Adhesive B, and 30 with Adhesive C.

10.7 If the numerator *df* for a between-subjects, one-way ANOVA are 4 and the denominator *df* are 20, what is the decision rule if $\alpha = .05$?

10.8 If a between-subjects, one-way ANOVA has 32 cases randomly assigned to four equally sized groups, what are $df_{Between}$, df_{Within}, and df_{Total}?

10.9 Here are data on an interval-level variable for three independent samples. Prepare this table so that it would be ready for use in computing sums of squares. (Do not compute the sums of squares.)

Group 1	Group 2	Group 3
16	12	13
17	14	15
20		18

10.10 Given the data in the table below, calculate (a) SS_{Total}, (b) $SS_{Between}$, and (c) SS_{Within}.

10.11 Given $df_{Between} = 3$, $df_{Within} = 12$, $df_{Total} = 15$, $SS_{Between} = 716.00$, $SS_{Within} = 228.00$, and $SS_{Total} = 944.00$, complete an ANOVA summary table.

	Group 1		Group 2		Group 3			
	X	X^2	X	X^2	X	X^2		
	16	256	19	361	26	676		
	18	324	20	400	22	484		
	14	196	22	484	25	625	Grand	
	18	324	24	576	27	729	X	X^2
Σ	66.00	1,100.00	85.00	1,821.00	100.00	2,514.00	251.00	5,435.00
n	4		4		4		12	

10.3 Interpreting Between-Subjects, One-Way ANOVA

The interpretation of a between-subjects, one-way ANOVA starts the same way it did for *t* tests, addressing questions of whether the null hypothesis was rejected and how big the effect is. After that, the interpretation answers a different question, where the effect is found.

Why is there a need to worry about where the effect is? This question didn't have to be addressed for a *t* test because in a *t* test just two groups exist. If the results of a *t* test were statistically significant, it was clear that the mean of Group 1 was different from the mean of Group 2. However, there can be more than two groups in an ANOVA. If the results of the ANOVA are statistically significant, then all that is known is that the mean of at least one group is different from at least one other group. If there are three groups, the mean of Group 1 could differ from the mean of Group 2, the mean of Group 1 could differ from the mean of Group 3, or the mean of Group 2 could differ from the mean of Group 3. It is also possible that two of the mean of the three pairs could differ or that all three pairs could differ. This is why it's important to address the question of where the effect is when interpreting a statistically significant ANOVA.

Was the Null Hypothesis Rejected?

Dr. Chung's maze study compared rats that expected to find different foods in the goal box of a maze. Ten rats were assigned to three groups—one that expected to find low-calorie food in the goal box, one that expected medium-calorie food, and one that expected high-calorie food. The rats expecting low-calorie food took 31 seconds from start box to goal box, the rats expecting medium-calorie food took 28 seconds, and the rats expecting high-calorie food took 25 seconds. Dr. Chung had determined $F_{cv} = 4.737$ and formulated the decision rule:

- If $F \geq 4.737$, reject the null hypothesis.
- If $F < 4.737$, fail to reject the null hypothesis.

As shown in the ANOVA summary table, Table 10.8, $F = 31.40$ for the maze data. Applying the decision rule, it is clear that 31.40 is greater than or equal to 4.737, and Figure 10.12 shows how the observed value of F falls in the rare zone. This means that the null hypothesis is rejected and the alternative hypothesis accepted. This leads to the conclusion that at least one of the three groups—rats expecting low-calorie food, rats expecting medium-calorie food, or rats expecting high-calorie food—differs from at least one of the others in terms of the mean time it takes to get to the food. It is not yet clear where the difference lies, so the best interpretation Dr. Chung can make at present is: "The results of the between-subjects, one-way ANOVA were statistically significant [$F(2, 7) = 31.40$, $p < .05$], indicating that there is a statistically significant difference in the time it takes to get from the start box to the goal box depending on the calorie content of the food a rat expects to find."

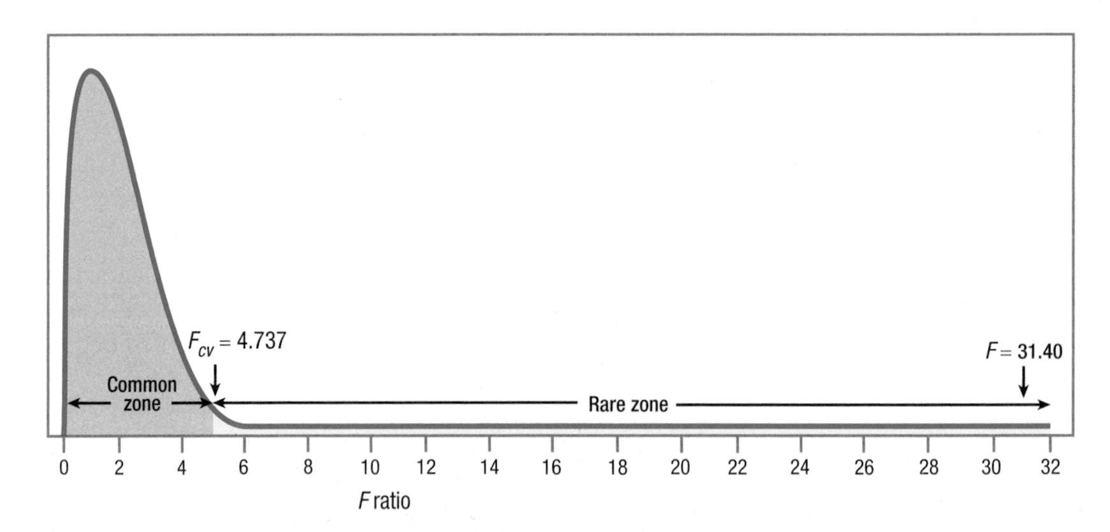

Figure 10.12 Determining Whether the Observed Value of the Test Statistic Falls in the Rare Zone or the Common Zone for the Maze Study The observed value of the test statistic, $F = 31.40$, falls in the rare zone, so the null hypothesis is rejected.

Dr. Chung used APA format in reporting the results of the analysis of variance. APA format provides five pieces of information:

1. F indicates that an F test, an analysis of variance, was used.
2. The numbers in parentheses, 2 and 7, are the degrees of freedom for SS_{Between}, the numerator of the F ratio, and for SS_{Within}, the denominator of the F ratio. By adding these two numbers together and adding 1, it is possible to find the sample size:

$$N = 2 + 7 + 1 = 10$$

3. 31.40 is the observed (calculated) value of the test statistic.
4. .05 reveals that alpha was set at .05 and there is a 5% chance of making a Type I error.
5. $p < .05$ indicates that the null hypothesis was rejected. It means that an F value of 31.40 is a rare occurrence (it happens less than 5% of the time) when the null hypothesis is true.

How Big Is the Effect?

The second question to be addressed in interpreting an ANOVA is how big the *overall* effect is. That is, how much impact does the independent variable have on the dependent variable? For the maze data, this involves asking how much impact the three different types of food (low-, medium-, and high-calorie) have on the time it takes to run the maze.

Once again, r^2 will be used. Equation 10.8 shows how to calculate r^2, the percentage of variability in the outcome variable that is accounted for by group status.

Equation 10.8 Formula for r^2, the Percentage of Variability in the Dependent Variable Accounted for by the Explanatory Variable

$$r^2 = \frac{SS_{\text{Between}}}{SS_{\text{Total}}} \times 100$$

where r^2 = the percentage of variability in the dependent variable that is accounted for by the explanatory variable

SS_{Between} = between-groups sum of squares (Equation 10.3)

SS_{Total} = total sum of squares (Equation 10.2)

This formula says that r^2 is calculated as the between-groups sum of squares divided by the total sum of squares. Then, to turn it into a percentage, the ratio is multiplied by 100. These calculations reveal the percentage of total variability in the scores that is accounted for by the treatment effect. For the maze data, these calculations would lead to the conclusion that $r^2 = 90\%$:

$$\begin{aligned} r^2 &= \frac{SS_{\text{Between}}}{SS_{\text{Total}}} \times 100 \\ &= \frac{54.00}{60.00} \times 100 \\ &= .9000 \times 100 \\ &= 90.00\% \end{aligned}$$

By Cohen's standards, an r^2 of 90% is a huge effect. Figure 10.13 is a visual demonstration of how big an effect this is. Here is what Dr. Chung's interpretation would look like for the maze data now that r^2 is known:

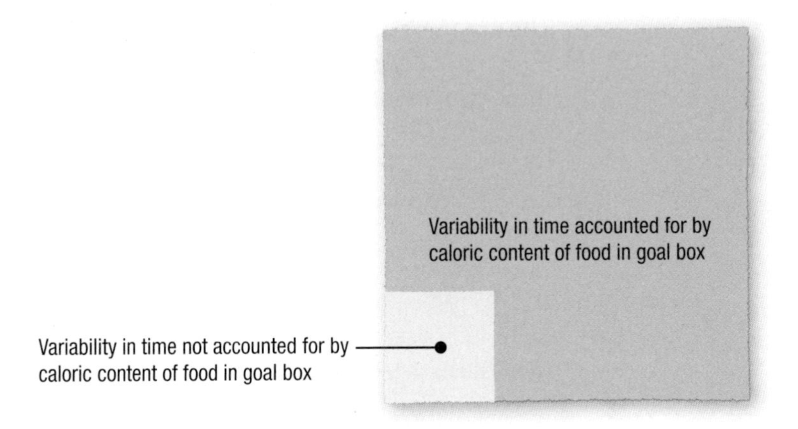

Figure 10.13 Percentage of Variability in Time to Run the Maze Scores That Is Accounted for by Type of Food in Goal Box 90% of the variability in time it takes for rats to run the maze is accounted for by the type of food they expect to find in the goal box. Only 10% of the variability remains, to be accounted for by other factors such as individual differences.

> The results of the one-way ANOVA were statistically significant [$F(2, 7) = 31.40$, $p < .05$], indicating that the time it takes a hungry rat to run a maze is affected by the type of food it expects to find in the goal box. When all the rats are of the same age and from the same strain, as they were here, then the calorie content of the food in the goal box has a very large effect on the speed with which the animal travels the maze. In fact, the type of food in the goal box explains 90% of the variability in running speed.

Here's a heads up for future chapters and for reading results sections in psychology articles—there's another measure of effect size for a between-subjects, one-way ANOVA that is calculated exactly the same way as r^2, but is called something different. It is called "eta squared" and abbreviated η^2. It is calculated the same way as r^2 and provides the same information, how much of the variability in the dependent variable is explained by the independent variable.

A Common Question

Q How is η pronounced?

A η is the lowercase version of the Greek letter eta and there is no consensus on how it should be pronounced. I have heard "eat-uh," "etta," and "ey-tuh" as in "hey." Whichever one you choose, just say it with confidence and you'll be fine.

Where Is the Effect?

So far, Dr. Chung knows that the goal affects running time and how much variability is explained, but he doesn't know where the effect occurs. Is the 3-second difference between the mean of low-calorie time and the mean of medium-calorie time a statistically significant one? Or, does it take the 6-second difference between low-calorie and high-calorie goals to be statistically significant?

Post-hoc tests are meant to be used only after *a statistically significant* F *ratio has been found.*

The *F* ratio, which was statistically significant, reveals that at least one pair of means has a statistically significant difference, but it doesn't tell which one or which ones. For that, a *post-hoc* test is needed. "Post-hoc" is Latin for "after this" and post-hoc tests are meant to be used only *after* a statistically significant *F* ratio has been found. Post-hoc tests are mathematically designed to allow multiple comparisons to be made while keeping alpha at the desired level.

There are a wide variety of post-hoc tests with cool names like the Scheffé, the Newman-Keuls, and the Bonferroni-Dunn. Post-hoc tests vary in terms of what type of error they are more likely to make, a Type I error or a Type II error. No post-hoc test can guarantee avoiding a mistake. Compared to each other, some post-hoc tests are more likely to say that there is a statistically significant difference between means that don't differ (Type I error). Other post-hoc tests are more likely to find no statistically significant difference when a difference exists (Type II error).

The test taught here, the Tukey *HSD*, is the latter type, what is called a conservative test. If it finds a statistically significant difference between a pair of means, then the population means probably really are different. Which is why *HSD* stands for "honestly significant difference." However, there is a cost to this conservatism: the *HSD* has less statistical power and so may overlook a pair of means that are different.

The Tukey *HSD* works by calculating an *HSD* value, which is the minimum difference needed between two means in order for the difference to be considered statistically significant. If the observed difference between a pair of means is greater than or equal to the *HSD* value, then the researcher can conclude that there is a statistically significant difference between the two groups. For example, the mean for the low-calorie group was 31 seconds and for the medium-calorie group it was 28 seconds. There is a 3.00-second difference between the mean of these two groups. If the *HSD* value were, say, 2.50, then the difference between the mean of these two groups would be a statistically significant one. This is the case because the observed difference, 3.00, met or exceeded the *HSD* value of 2.50.

The formula to calculate *HSD* is shown in Equation 10.9. In order to apply that formula, one needs a value called *q*. Values of *q* can be found in Appendix Table 5, a part of which is shown in Table 10.12.

TABLE 10.12 Part of Appendix Table 5, Table of Values of *q* for Use in the Tukey *HSD* Post-Hoc Test

	k		
df	2	3	4
2	6.09	8.33	9.80
3	4.50	5.91	6.83
4	3.93	5.04	5.76
5	3.64	4.60	5.22
6	3.46	4.34	4.90
7	3.34	4.17	4.68
8	3.26	4.04	4.53
9	3.20	3.95	4.42
10	3.15	3.88	4.33

A *q* value is needed to calculate a Tukey *HSD* post-hoc test. The *q* value for a between-subjects, one-way ANOVA is found at the intersection of the column for *k*, the number of groups in the ANOVA, and the row where $df = df_{\text{Within}}$.

There are q tables for $\alpha = .05$ and $\alpha = .01$, for a 5% and a 1% chance of making a Type I error. The q values are found at the intersections of rows and columns.

- The columns represent different values of k, the number of groups being compared in the ANOVA.
- The columns represent within-groups degrees of freedom, df_{Within}.

For the maze data, there are three groups and within-groups degrees of freedom are 7. The q value of 4.17 is found at the intersection of the column where $k = 3$ with the row where $df = 7$.

Equation 10.9 Formula to Calculate Tukey *HSD* Value for a Post-Hoc Test for Between Subjects, One-Way ANOVA

$$HSD = q\sqrt{\frac{MS_{\text{Within}}}{n}}$$

where HSD = value by which, if two means differ, the difference is statistically significant
q = value of q, from Appendix Table 5
$MS_{\text{Within}} = MS_{\text{Within}}$ (from ANOVA summary table)
n = sample size for the smallest group

To calculate *HSD*, one needs to know q, MS_{Within}, and n:

- From Appendix Table 5, it is known that $q = 4.17$.
- Referring back to the ANOVA summary table for the maze data (Table 10.7), $MS_{\text{Within}} = 0.86$.
- The only missing piece is n, the sample size for the smallest group. In Table 10.3, where the maze data were first presented, it can be seen that the smallest group has three cases, so $n = 3$. (Tukey's *HSD* was designed to be used when all sample sizes are equal. When sample sizes are unequal, using the smallest sample size makes the *HSD* value larger, which keeps the test conservative.)

With all the parts in place, the *HSD* value is calculated as

$$\begin{aligned} HSD &= q\sqrt{\frac{MS_{\text{Within}}}{n}} \\ &= 4.17\sqrt{\frac{0.86}{3}} \\ &= 4.17\sqrt{0.2867} \\ &= 4.17 \times 0.5354 \\ &= 2.2326 \\ &= 2.23 \end{aligned}$$

The *HSD* value is 2.23. Any two means that differ by at least 2.23 seconds are honestly significantly different. Here are the three sample means:

- Group 1: Expecting low-calorie food in the goal box = 31.00 seconds
- Group 2: Expecting medium-calorie food in the goal box = 28.00 seconds
- Group 3: Expecting high-calorie food in the goal box = 25.00 seconds

Here are the comparisons:

- Group 1 vs. Group 2: There is a 3.00-second difference between the mean of Group 1 and the mean of Group 2. $3.00 \geq 2.23$, so this is a statistically significant difference.
- Group 1 vs. Group 3: There is a 6.00-second difference between the mean of Group 1 and the mean of Group 3. $6.00 \geq 2.23$, so this is a statistically significant difference.
- Group 2 vs. Group 3: There is a 3.00-second difference between the mean of Group 2 and the mean of Group 3. $3.00 \geq 2.23$, so this is a statistically significant difference.

Once a researcher knows whether a difference is statistically significant, he or she needs to think about the direction of the difference. As with *t* tests, the actual sample means can be used to determine the direction of the difference. Here are Dr. Chung's conclusions:

- Group 1 vs. Group 2: Expecting medium-calorie food compared to low-calorie food leads to rats taking significantly less time to travel through the maze. That is, the medium-calorie rats ran through the maze statistically more quickly on average.
- Group 1 vs. Group 3: There's a significant difference between expecting low-calorie food and expecting medium-calorie food, so it's not surprising that a similar difference exists between expecting low-calorie food and expecting high-calorie food. Rats expecting high-calorie food were statistically faster on average than rats expecting low-calorie food.
- Group 2 vs. Group 3: Rats expecting high-calorie food were statistically faster on average than rats expecting medium-calorie food.

A statistically significant ANOVA reveals that at least one pair of population means differs. Now after the post-hoc test, Dr. Chung knows which pairs differ and what the differences mean. His challenge is to summarize the findings. He could report that the mean of Group 1 differs from the mean of Group 2 and that the mean of Group 1 differs from the mean of Group 3, and so forth, but that would be a focus on the trees, not the forest. Interpretations are more useful if they find a pattern in the results. Dr. Chung's findings could be summarized by saying that rats, this strain of rats at least, appear able to figure out that different foods have different caloric contents and, when hungry, are more motivated to seek nutritionally richer foods.

With all three interpretation questions addressed—Was the null hypothesis rejected? How big is the effect? Where is the effect?—Dr. Chung is ready to write a four-point interpretation that (1) tells what the study was about, (2) presents the main results, (3) explains what the results mean, and (4) makes suggestions for future research:

> In this study, rats were exposed to three different types of food (low-calorie, medium-calorie, and high-calorie) in the goal box while learning a maze. After learning the maze, they were made hungry, randomly assigned to three groups,

and timed while they ran the maze. The three groups expected to find different foods in the goal box: low-calorie, medium-calorie, or high-calorie. There was a statistically significant decrease in mean time, that is, an increase in mean speed, as the caloric content of the food increased [$F(2, 7) = 31.40, p < .05$]. Time decreased from a mean of 31.00 seconds for the rats expecting low-calorie food to 28.00 seconds for those expecting medium-calorie food and to 25.00 seconds for the ones expecting high-calorie food. The effect of caloric content on speed in this study was quite dramatic, accounting for 90% of the variability in speed. This is probably an overestimate of the effect of caloric content as all rats were from the same strain and were the same age.

These results suggest that this strain of rats is able to discriminate the caloric content of food and that, when they are hungry, they are more motivated to seek nutritionally richer food. This motivation makes sense for increasing the likelihood of survival in lean times. An alternative explanation for the results is that the richer-calorie foods tasted better and that rats hurry to food if it tastes good. Future research should be sure to control for food taste. Future research should also employ different strains of rats.

Worked Example 10.2 For practice in interpretation, let's return to Dr. Douglas's data from the study of over-the-counter pain relievers (ibuprofen and acetaminophen) on pain threshold. In that study, 12 male undergraduates were randomly assigned by a sensory psychologist to take a placebo, ibuprofen, or acetaminophen. They then placed their hands in ice water until they felt pain. The participants who received placebos kept their hands in ice water for an average of 17.50 seconds, the ibuprofen participants for 23.00 seconds, and the acetaminophen participants for 21.00 seconds. F_{cv} was 4.256 and the ANOVA summary table is reprinted here (see Table 10.13).

TABLE 10.13 ANOVA Summary Table for One-Way ANOVA for Pain Threshold Data

Source of Variability	Sum of Squares	Degrees of Freedom	Mean Square	*F* ratio
Between groups	62.00	2	31.00	1.35
Within groups	207.00	9	23.00	
Total	269.00	11		

This is the same ANOVA summary table that was printed earlier as Table 10.11.

Was the null hypothesis rejected? From the ANOVA summary table (Table 10.13), it is clear that the observed value of F is 1.35. 1.35 is less than 4.256, F_{cv}, so the null hypothesis was not rejected. Figure 10.14 shows how the observed value of F falls in the common zone. With these data, Dr. Douglas does not have enough evidence to state that any one of these three groups differs from another in terms of the mean pain threshold. In APA format, she would write the results as

$$F(2, 9) = 1.35, p > .05$$

- F tells the reader that an analysis of variance, also known as an F test, was done.

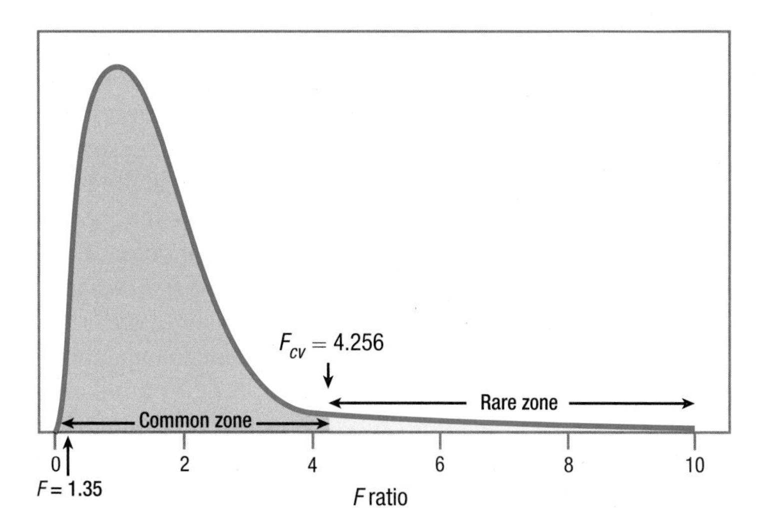

Figure 10.14 Determining Whether the Observed Value of the Test Statistic Falls in the Rare Zone or the Common Zone for the Pain Threshold Study The observed value of the test statistic, $F = 1.35$, falls in the common zone, so we fail to reject the null hypothesis.

- The 2 and 9 in the parentheses are the numerator and denominator degrees of freedom. These are df_{Between} and df_{Within}. If they are added together and 1 is added, that gives N.
- 1.35 is the value of F that Dr. Douglas calculated.
- The .05 is the alpha level that had been selected.
- The $p > .05$ means that a result like $F = 1.35$ is a common one when the null hypothesis is true, as under those conditions it occurs more than 5% of the time. The sensory psychologist has failed to reject the null hypothesis.

How big is the effect? Equation 10.5 is used to calculate the effect size, r^2, the percentage of variability in the dependent variable (seconds to pain threshold) that is accounted for by the independent variable (type of painkiller). To use the equation, two values, SS_{Between} (62.00) and SS_{Total} (269.00), are obtained from the ANOVA summary table:

$$\begin{aligned} r^2 &= \frac{SS_{\text{Between}}}{SS_{\text{Total}}} \times 100 \\ &= \frac{62.00}{269.00} \times 100 \\ &= .2305 \times 100 \\ &= 23.05\% \end{aligned}$$

An r^2 of 23% is close to a large effect, according to Cohen. This gives Dr. Douglas two conflicting pieces of information:

- Not enough evidence exists to say there is an effect.
- However, the effect seems to be large.

One way to reconcile these two conflicting pieces of information is to look at the possibility of a Type II error. Type II errors occur when a researcher fails to find statistically significant evidence of an effect that really does exist. That possibility can't be ruled out here. When interpreting such a situation, the researcher should point out the possibility and suggest that the study be replicated with a larger sample size. Larger sample sizes give tests more power and so increase the likelihood of being able to find an effect when it does exist.

Where is the effect? Post-hoc tests are designed to be conducted only when the null hypothesis is rejected. In this study, the null hypothesis was not rejected. This means there is not enough evidence that an effect exists, so there is no reason to try to find where the effect is.

Putting it all together. Here's Dr. Douglas's four-point interpretation for the pain threshold study. She reveals what the study was about, identifies its main results, interprets them, and makes suggestions for future research.

> Analysis of data for a study comparing the impact of placebo, ibuprofen, and acetaminophen on pain threshold found no evidence that either of the over-the-counter pain relievers had any impact on the threshold for perceiving pain. There was no statistically significant difference in the mean time it took participants to perceive pain when their hands were in ice water [$F(2, 9) = 1.35, p > .05$]. There was some evidence that medication may have an impact, but the small sample size ($N=12$) meant the study did not have enough power to detect it. In order to determine if these over-the-counter pain relievers have an impact on pain threshold, the study should be replicated with a larger sample size.

Practice Problems 10.3

Apply Your Knowledge

10.12 Given $\alpha = .05$, $F_{cv} = 2.690$, $df_{Between} = 4$, $df_{Within} = 30$, and $F = 7.37$, write the results in APA format.

10.13 Given $SS_{Between} = 1{,}827.50$ and $SS_{Total} = 4{,}631.50$, (a) calculate r^2 and (b) comment on the size of the effect.

10.14 Four groups are being compared in a between-subjects, one-way ANOVA. Each group has 10 cases. $MS_{Within} = 77.89$ and $df_{Within} = 36$. The F ratio was statistically significant with $\alpha = .05$.

a. Use this information to complete a post-hoc test comparing $M_1 = 29.00$ and $M_2 = 18.00$; interpret the difference between the two population means.

b. Use this information to complete a post-hoc test comparing $M_1 = 29.00$ and $M_3 = 22.00$; interpret the difference between the two population means.

10.15 An industrial/organizational psychologist wanted to investigate the effect of different management styles on employee performance. She obtained 27 employees at a large company and randomly assigned them, in equal-sized groups, to be supervised by managers who believed in motivating employees by using (a) rewards for good performance, (b) punishment for poor performance, or (c) a mixture of rewards and punishments. No nonrobust assumptions of a between-subjects, one-way ANOVA were violated. One year later, she recorded how much of a raise (as a percentage) each employee received. She believed that employees who did better work would receive bigger raises. The employees supervised by managers who believed in using rewards saw a mean raise of 8%, those supervised by managers who believed in using punishment had a mean raise of 5%, and those supervised by managers who used a mixture of rewards and punishment had a mean raise of 7%.

A between-subjects, one-way ANOVA was used to analyze the results. The results were statistically significant. Given the ANOVA summary table, $r^2 = 30.43\%$ and $HSD = 2.35$, write a four-point interpretation.

Source of Variability	Sum of Squares	Degrees of Freedom	Mean Square	*F* ratio
Between groups	42.00	2	21.00	5.25
Within groups	96.00	24	4.00	
Total	138.00	26		

Application Demonstration

America is a land of movers. If we don't like where we're living, if we think the grass is greener somewhere else, we just pull up our stakes and go. Are salaries for psychologists consistent throughout the United States or do some areas have higher or lower salaries? If so, this information could guide a recent graduate's decision about where to seek employment.

The Census Bureau divides the United States into four regions: the Northeast, Midwest, South, and West. Four states from each region were randomly selected and the mean salary for "clinical, counseling, and school psychologists" was obtained from a Bureau of Labor Statistics website for each state. Salaries were recorded in thousands of dollars; \$79.52, for example, means \$79,520. Recording the values this way keeps the sums of squares from becoming unwieldy numbers in the hundreds of millions. The mean salaries for the four regions (Northeast = \$77.73, Midwest = \$63.55, West = \$68.83, and South = \$61.37) are seen in Figure 10.15.

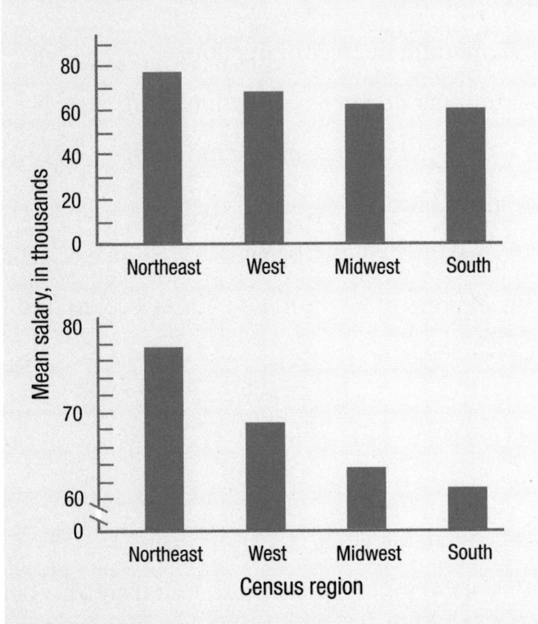

Figure 10.15 Mean Salary for Psychologists by Census Region: Two Different Pictures The mean salary for clinical, counseling, and school psychologists is highest in the Northeast and lowest in the South. Whether the differences among regions look significant or not depends on how the graph is drawn. The two graphs here differ only in the scale on the *Y*-axis.

The top panel in Figure 10.15 shows some slight differences in salary by region. Based on this graph, it seems unlikely that there are any statistically significant differences. The bottom panel in Figure 10.15, where the *y*-axis doesn't start at zero, does make it look like there are significant differences by region. As mentioned in Chapter 2, a graph is a subjective interpretation of the data.

The question of whether there are differences in mean salary is best resolved in an objective manner with a significance test.

Step 1 Pick a Test. There are four samples, they are independent, and the dependent variable is measured at the ratio level, so the test should be a between-subjects, one-way ANOVA.

Step 2 Check the Assumptions.

- The random samples assumption is not violated. Cases were randomly selected from their populations.
- The independence of observations assumption is not violated. Each state (case) is measured only once.
- The normality assumption might be violated. The populations are not large. For example, there are only nine states in the Northeast. With such a small population, it is unlikely to be normally distributed. But, this assumption is robust.
- The homogeneity of variance assumption is not violated. All standard deviations are comparable.

Step 3 List the Hypotheses.

H_0: $\mu_1 = \mu_2 = \mu_3 = \mu_4$.

H_1: At least one population mean differs from the others.

Step 4 Set the Decision Rule. First, find the numerator, denominator, and total degrees of freedom:

$$\text{Numerator degrees of freedom} = df_{\text{Between}} = k - 1 = 4 - 1 = 3$$

$$\text{Denominator degrees of freedom} = df_{\text{Within}} = N - k = 16 - 4 = 12$$

$$\text{Total degrees of freedom} = df_{\text{Total}} = N - 1 = 16 - 1 = 15$$

Next, set the alpha level:

$\alpha = .05$ because willing to have a 5% chance of making a Type I error

Then, use the degrees of freedom and alpha level to find the critical value of F in Appendix Table 4:

$$F_{cv} = 3.490$$

Then, set the decision rule:

- If $F \geq 3.490$, reject H_0.
- If $F < 3.490$, fail to reject H_0.

Step 5 Calculate the Test Statistic. Calculate the between-groups sum of squares using Equation 10.3, the within-groups sum of squares using Equation 10.4, and the total sum of squares using Equation 10.2. To save time, these values have already been calculated:

$$SS_{\text{Between}} = 636.41$$

$$SS_{\text{Within}} = 723.85$$

$$SS_{\text{Total}} = 1{,}360.26$$

Use the degrees of freedom calculated in Step 4 and the sums of squares calculated in Step 4 to fill in the appropriate columns of the ANOVA summary table. Then follow the guidelines in Table 10.7 to complete the ANOVA summary table (Table 10.14).

TABLE 10.14 ANOVA Summary Table for Salaries of Psychologists by Census Regions

Source of Variability	Sum of Squares	Degrees of Freedom	Mean Square	*F* ratio
Between groups	636.41	3	212.14	3.52
Within groups	723.85	12	60.32	
Total	1,360.26	15		

Step 6 Interpret the Results. *Was the null hypothesis rejected?* To determine if the null hypothesis should be rejected, apply the decision rule. Which of these statements is true?

- Is $3.52 \geq 3.490$?
- Is $3.52 < 3.490$?

The first statement is true, so the null hypothesis should be rejected. The results are called statistically significant, the alternative hypothesis is accepted, and it is concluded that at least one population mean differs from at least one other population mean. A post-hoc test will be needed to determine where the differences lie. For now, though, the results can be written in APA format:

$$F(3, 12) = 3.52, p < .05$$

How big is the effect? This question is answered by calculating r^2:

$$r^2 = \frac{SS_{\text{Between}}}{SS_{\text{Total}}} \times 100 = \frac{636.41}{1{,}360.26} \times 100 = .4679 \times 100 = 46.79\%$$

On the basis of Cohen's criteria, the fact that region explains 46.79% of the variability in salary means this is a very large effect.

Where is the effect? Tukey's *HSD* will be used as a post-hoc test to find out which pair, or pairs, of means are statistically significantly different. To calculate the *HSD* value, a researcher must know q, MS_{Within}, and n. q is found in Appendix Table 5 for $\alpha = .05$, in the column for $k = 4$ and the row for $df_{\text{Within}} = 12$. With $q = 4.20$, $MS_{\text{Within}} = 60.32$, and $n = 4$, Equation 10.9 is used to calculate *HSD:*

$$HSD = q\sqrt{\frac{MS_{\text{Within}}}{n}} = 4.20\sqrt{\frac{60.32}{4}} = 4.20\sqrt{15.0800} = 4.20 \times 3.8833$$

$$= 16.3099 = 16.31$$

Any groups that have a mean difference greater than or equal to 16.31 points are statistically significantly different. The largest mean is the Northeast at \$77.73 thousand and the smallest is the South at \$61.37 thousand. Those two are 16.36 points apart and this difference is a statistically significant difference. All other differences are smaller than 16.31 and thus not statistically significant.

The effect seems to be due to the \$16.36 thousand difference between the mean salaries for psychologists in the Northeast and mean salaries in the South. Salaries are statistically significantly higher in the Northeast than they are in the South.

Here is a four-point interpretation:

In this study, a random sample of states from each of the four census regions was taken and the average salary in each state for clinical, counseling, and school psychologists was recorded. The mean salary in the Northeast region was \$77,730, \$63,550 in the Midwest, \$61,370 in the South, and \$68,830 in the West. The overall ANOVA showed a statistically significant effect for the region [$F(3, 12) = 3.52$, $p < .05$] and region accounted for almost 50% of the variability in salary. However, post-hoc testing showed that the only statistically significant difference that existed was between the Northeast and the South—the mean salaries for clinical, counseling, and school psychologists are statistically higher in the Northeast than in the South. The results of this study suggest that it might be worthwhile for a psychologist living in the South to move to the North, but there would be no advantage to moving to the West or Midwest. If this study is replicated, it would be advisable to take into account the cost of living as higher salaries in a region may be mitigated by a higher cost of living.

SUMMARY

Explain when ANOVA is used and how it works.

- Analysis of variance (ANOVA) is a family of tests for comparing the means of three or more groups. Between-subjects, one-way ANOVA is used to compare the means of two or more independent samples when there is just one explanatory variable.
- Between-subjects, one-way ANOVA separates variability in scores into that due to the different ways the groups are treated and that due to variability between cases. If there is more variability between groups due to treatment than within groups due to individual differences, then there is a statistically significant difference among the groups.

Complete a between-subjects, one-way ANOVA.

- To complete an ANOVA, the assumptions must be met and the hypotheses set. The null hypothesis says that all population means are the same and the alternative that at least one population mean differs from the others.

The decision rule compares the observed *F* ratio (variability due to treatment divided by variability due to individual differences) to the critical value of *F*. Calculating an *F* ratio involves finding sums of squares and degrees of freedom for three sources of variability (between group, within group, and total) and organizing that information in an ANOVA summary table.

Interpret the results of a between-subjects, one-way ANOVA.

- If the results are statistically significant, at least one population mean differs from at least one other. Then calculate the size of the effect using r^2 and use post-hoc testing with the Tukey *HSD* to which pair(s) of means differ.
- If results are not statistically significant, then there is not enough evidence to conclude any population means are different. Nonetheless, calculate r^2 in order to consider the possibility of Type II error.

DIY

ANOVA is an extension of the *t* test, so let's extend the DIY from the independent-samples *t* test to the between-subjects, one-way ANOVA. In that DIY, you were asked to compare two groups of states, say, northern states vs. southern states, on some outcome variable, say, murder rate. Now, divide states into three groups, say, northern, middle, and southern states, and compare them on your outcome variable.

KEY TERMS

analysis of variance (ANOVA) – a family of statistical tests for comparing the means of two or more groups.

between-group variability – variability in scores that is primarily due to the different treatments that different groups receive.

between subjects – ANOVA terminology for independent samples.

between-subjects, one-way ANOVA – a statistical test used to compare the means of two or more independent samples when there is just one explanatory variable.

factor – term for an explanatory variable in ANOVA.

level – ANOVA terminology for a category of an explanatory variable.

post-hoc test – a follow-up test to a statistically significant ANOVA, engineered to find out which pairs of means differ while keeping the overall alpha level at the chosen level.

sum of squares between (SS_{Between}) – A sum of the squared deviations scores representing the variability between groups.

sum of squares total (SS_{Total}) – A sum of the squared deviation scores representing the all variability in the scores.

sum of squares within (SS_{Within}) – A sum of the squared deviation scores representing the variability within groups.

treatment effect – the impact of the explanatory variable on the dependent variable.

way – term for an explanatory variable in ANOVA.

within-group variability – variability within a sample of cases, all of which have received the same treatment.

CHAPTER EXERCISES

Review Your Knowledge

10.01 A ____-sample test compares the mean of one group to the mean of another group.

10.02 Analysis of variance is used when comparing the means of ____ or more groups.

10.03 ____ is short for analysis of variance.

10.04 ____, one-way ANOVA compares the means of two or more independent samples.

10.05 Other terms used for explanatory variables in ANOVA are ____ and ____.

10.06 The categories of an explanatory variable in ANOVA are called ____.

10.07 ANOVA keeps the risk of ____ error at a reasonable level.

10.08 The risk of making a Type I error ____ as the number of statistical tests being completed increases.

10.09 A follow-up test in ANOVA is called a ____ test.

10.10 A ____ is used to find out which pairs of means in an ANOVA are statistically significantly different.

10.11 ANOVA works by separating out the different sources of ____ in the scores.

10.12 Variability within a set of scores is called ____ variability.

10.13 Within-group variability is caused by ____.

10.14 Making a sample more homogeneous ____ within-group variability.

10.15 ____ variability is due to the different treatments different groups receive.

10.16 ____ effect is the term used to label the effect of the explanatory variable.

10.17 Between-group variability is caused by ____ as well as treatment effect.

10.18 If treatment has ____ impact on outcome, the sample means are close together.

10.19 When sample means are far apart, the treatment effect is ____.

10.20 When treatment has a large impact, there is a lot of variability between group ____.

10.21 The ratio of between-group variability to within-group variability is called an ____.

10.22 Between-group variability, the numerator in an *F* ratio, is made up of ____ and ____.

10.23 If treatment has no impact on outcome, the *F* ratio should be near ____.

10.24 As treatment has an impact on outcome, the value of the *F* ratio climbs above ____.

10.25 If there are just two groups, $F =$ ____ squared.

10.26 Between-subjects, one-way ANOVA is used when there is one ____ and ____ groups.

10.27 The assumptions for between-subjects, one-way ANOVA are the same as they are for ____.

10.28 For a between-subjects, one-way ANOVA, the cases in each group should be ____ from the population.

10.29 If the cases in the groups for a between-subjects, one-way ANOVA are paired together, a between-subjects, one-way ANOVA *can / cannot* be used.

10.30 Between-subjects, one-way ANOVA assumes that the dependent variable in each population is ____ distributed.

10.31 The assumption that the variability in all groups is about the same is called the ____ assumption.

10.32 The null hypothesis states that there is ____ mean difference between any of the populations.

10.33 The null hypothesis for a between-subjects, one-way ANOVA is always ____ directional.

10.34 The alternative hypothesis says that at least ____ population mean is different from at least ____ other population mean.

10.35 The decision rule for a between-subjects, one-way ANOVA says that if *F* falls in the rare zone of the sampling distribution, the null hypothesis is ____.

10.36 The critical value of *F* depends on the degrees of freedom for the ____ and the degrees of freedom for the ____ of the *F* value.

10.37 The numerator degrees of freedom for a between-subjects, one-way ANOVA *F* ratio is df____.

10.38 The denominator degrees of freedom for a between-subjects, one-way ANOVA *F* ratio is df____.

10.39 In order to calculate total degrees of freedom for a between-subjects, one-way ANOVA, one could add together ____ and ____.

10.40 If $F = F_{cv}$, the null hypothesis is ____.

10.41 ΣX^2 is called a ____.

10.42 Numerators in variance formulas are ____.

10.43 To calculate ____, the grand mean is subtracted from each score, the difference scores are squared, and they are all added up.

10.44 SS_{Total} can be broken down into two components: ____ and ____.

10.45 SS_{Between} isolates the variability in scores that is primarily due to ____.

10.46 $SS_{\text{Total}} - SS_{\text{Between}} =$ ____

10.47 The table used to organize ANOVA results is called an ____.

10.48 The sources of variability in an ANOVA are listed in the ____ column of an ANOVA summary table.

10.49 A ____ is calculated by dividing a sum of squares by its degrees of freedom.

10.50 An *F* ratio in between-subjects, one-way ANOVA is calculated as the ratio of *MS*____ to *MS*____.

10.51 Interpretation of results for a statistically significant ANOVA differs from *t* in that it has to address where the ____ is located.

10.52 One determines if an *F* ratio is statistically significant by comparing *F* to ____.

10.53 If the null hypothesis is rejected, then ____ is accepted.

10.54 The numbers in parentheses when reporting the results of a between-subjects, one-way ANOVA in APA format are ____ and ____.

10.55 If the null hypothesis is not rejected, ANOVA results are reported in APA format as *p* ____ .05.

10.56 ____ is used as the effect size in between-subjects, one-way ANOVA.

10.57 r^2 is calculated as the percentage of total variability in scores that is accounted for by ____ variability.

10.58 The higher the percentage of variability in the dependent variable explained by the explanatory variable, the ____ is the size of the effect.

10.59 The closer r^2 is to ____%, the stronger the effect; the closer r^2 is to ____%, the weaker the effect.

10.60 Cohen considers an r^2 of 9% to be a ____ effect.

10.61 Post-hoc tests are only used if the ANOVA results are ____.

10.62 Some post-hoc tests have a greater likelihood of making a ____ error and others a greater chance of a ____ error.

10.63 The Tukey *HSD* is a conservative test. It is more likely to make a ____ error than a ____ error.

10.64 *HSD* stands for ____.

10.65 If the difference between two means meets or exceeds the *HSD* value, the difference is ____.

10.66 To find a *q* value, one needs to know df_{Within} and ____.

10.67 The *n* in the *HSD* equation represents the sample size for the ____ group.

10.68 The *HSD* test is used to determine which pairs of means have a ____ difference.

Apply Your Knowledge

For the first set of questions, select the appropriate statistical test from (a) single-sample z test; (b) single-sample t test; (c) independent-samples t test; (d) paired-samples t test; (e) between-subjects, one-way ANOVA; or (f) none of the above.

10.69 An adhesives researcher tests the holding power of a wood glue under different humidities. He glues together 100 small pieces of wood and randomly assigns each one to sit for an hour in a room with either 10%, 30%, 50%, 70%, or 90% relative humidity. Each glued piece of wood is then tested to see how many pounds of weight it takes to break the glue bond. The means of the five groups are compared.

10.70 Some parents demand that their children friend them on Facebook and other parents don't. A developmental psychologist wondered if such parental supervision had any impact on behavior. She obtained a sample of first-year college students and classified them as having parents who were or were not Facebook friends. From each student she also learned the number of days, during the first month of college, that he or she had consumed any alcohol. The two means are compared.

10.71 A nutritionist investigates the impact of type of breakfast on mid-morning concentration. He puts together a sample of 50 adults. He feeds them a breakfast and 4 hours later measures their concentration on an interval-level concentration test. On one day he feeds them bacon and eggs, on a second day he feeds them oatmeal, and on a third day he has them skip breakfast. He then calculates the mean concentration score for each condition.

10.72 A social psychologist is interested in how people perceive age-discrepant couples. She gets a random sample of shoppers at a mall, has them read a wedding announcement, and then asks them to predict—on an interval-level scale—the newlyweds' degree of marital happiness after 10 years of marriage. Participants are randomly assigned to groups: one-third read an announcement in which the bride is 25 and the groom 35, one-third read an announcement in which the bride is 35 and the groom 25, and the final third read an announcement in which both bride and groom are 30.

For the next set of questions, determine if assumptions have been violated and if a between-subjects, one-way ANOVA can be completed.

10.73 A sensory psychologist wants to determine if a sensory threshold differs depending on which nostril is used. He obtains 30 introductory psychology student volunteers and measures the absolute threshold, in parts per million, for detecting the scent of peppermint. Participants are randomly assigned to three equal-size groups: (1) scent administered to the left nostril, (2) scent administered to the right nostril, and (3) scent administered to both nostrils. Each participant is tested individually by an experimenter who does not know what the hypothesis is and does not know which nostril is being used.

10.74 A biomedical engineer compared the abrasion resistance—as measured by number of rotations until failure—of three different artificial hips. Manufacturer X makes hip joints out of metal, manufacturer Y out of ceramic, and manufacturer Z out of a metal/ceramic composite. The biomedical engineer gets a random sample of 6 hips from each manufacturer's production line and tests each one individually until failure occurs.

Writing the hypotheses

10.75 Write the null and alternative hypotheses for the between-subjects, one-way ANOVA in Exercise 10.73.

10.76 Write the null and alternative hypotheses for the between-subjects, one-way ANOVA in Exercise 10.74.

Calculating degrees of freedom

10.77 If $N = 48$ and $k = 4$, calculate (a) df_{Total}, (b) $df_{Between}$, and (c) df_{Within}.

10.78 If $N = 120$ and $k = 5$, calculate (a) df_{Total}, (b $df_{Between}$, and (c) df_{Within}.

10.79 If there are four independent samples, each with 15 participants, what are (a) df_{Total}, (b) $df_{Between}$, and (c) df_{Within}?

10.80 If $n_1 = 15$, $n_2 = 12$, and $n_3 = 18$, what are (a) df_{Total}, (b) $df_{Between}$, and (c) df_{Within}?

Finding F_{cv}

10.81 If $df_{Within} = 44$ and $df_{Between} = 2$, what is F_{cv} if $\alpha = .01$?

10.82 If $df_{Between} = 3$ and $df_{Within} = 36$, what is F_{cv} if $\alpha = .05$?

10.83 If $\alpha = .05$, $N = 50$, and $k = 4$, what is F_{cv}?

10.84 If $\alpha = .05$, $N = 80$, and $k = 5$, what is F_{cv}?

Stating the decision rule

10.85 If $df_{Within} = 40$ and $df_{Between} = 2$, what is the decision rule if $\alpha = .05$? Draw a sampling distribution of F and mark the rare and common zones.

10.86 If $df_{Within} = 10$ and $df_{Between} = 3$, what is the decision rule if $\alpha = .05$? Draw a sampling distribution of F and mark the rare and common zones.

Calculating sums of squares

10.87 Prepare the data table for use by computational formulas for sums of squares.

Group 1	Group 2	Group 3
108	100	99
102	105	95
		91

10.88 Prepare the data table for use by computational formulas for sums of squares.

Group 1	Group 2	Group 3
46	54	74
48	58	80

10.89 Given the data in this table, calculate SS_{Total}, $SS_{Between}$, and SS_{Within}.

	Group 1		Group 2		Group 3			
	X	X^2	X	X^2	X	X^2		
	112	12,544	98	9,604	88	7,744		
	104	10,816	90	8,100	85	7,225		
			88	7,744	76	5,776	Grand	
					80		X	X^2
Σ	216.00	23,360.00	276.00	25,448.00	329.00	27,145.00	821.00	75,953.00
n	2		3		4		9	

10.90 Given the data in this table, calculate SS_{Total}, $SS_{Between}$, and SS_{Within}.

	Group 1		Group 2		Group 3			
	X	X^2	X	X^2	X	X^2		
	55	3,025	47	2,209	68	4,624		
	63	3,969	58	3,364	66	4,356		
	72	5,184	63	3,969	62	3,844	Grand	
	59	3,481	48	2,304	73	5,329	X	X^2
Σ	249	15,659	216	11,846	269	18,153	734	45,658
n	4		4		4		12	

10.91 If $SS_{Total} = 98.75$ and $SS_{Between} = 40.33$, what is SS_{Within}?

10.92 If $SS_{Within} = 168.43$ and $SS_{Between} = 764.13$, what is SS_{Total}?

Calculating mean squares

10.93 If $SS_{Between} = 2{,}378.99$ and $df_{Between} = 3$, what is $MS_{Between}$?

10.94 If $SS_{Between} = 138.76$ and $df_{Between} = 4$, what is $MS_{Between}$?

10.95 If $SS_{Within} = 78.95$ and $df_{Within} = 32$, what is MS_{Within}?

10.96 If $SS_{Within} = 452.86$ and $df_{Within} = 102$, what is MS_{Within}?

Calculating F

10.97 If $MS_{Between} = 38.88$ and $MS_{Within} = 17.44$, what is F?

10.98 If $MS_{Within} = 764.55$ and $MS_{Between} = 898.00$, what is F?

Completing an ANOVA summary table

10.99 Complete this ANOVA summary table:

Source of Variability	Sum of Squares	Degrees of Freedom	Mean Square	*F* ratio
Between groups	172.80	2		
Within groups	6,410.80	12		
Total	6,583.60	14		

10.100 Complete this ANOVA summary table:

Source of Variability	Sum of Squares	Degrees of Freedom	Mean Square	*F* ratio
Between groups	47,843.15	3		
Within groups	3,053.88	30		
Total	50,897.03	33		

Deciding if the null hypothesis is rejected

10.101 If $F_{cv} = 3.238$, draw a sampling distribution of F, label the rare and common zones, locate $F = 1.96$, and determine, for this F value, if the null hypothesis should be rejected.

10.102 If $F_{cv} = 2.337$, draw a sampling distribution of F, label the rare and common zones, locate $F = 5.66$, and determine, for this F value, if the null hypothesis should be rejected.

10.103 If $F_{cv} = 3.467$ and $F = 7.64$, is the null hypothesis rejected?

10.104 If $F_{cv} = 2.486$ and $F = 1.98$, is the null hypothesis rejected?

Using APA format

10.105 If $df_{Between} = 3$, $df_{Within} = 17$, $F = 5.34$, and $\alpha = .05$, write the results in APA format. (Use $df_{Between}$ and df_{Within} to find F_{cv} in order to determine if the null hypothesis was rejected.)

10.106 If $df_{Between} = 6$, $df_{Within} = 30$, $F = 2.81$, and $\alpha = .05$, write the results in APA format. (Use $df_{Between}$ and df_{Within} to find F_{cv} in order to determine if the null hypothesis was rejected.)

10.107 Given this ANOVA summary table, write the results in APA format using $\alpha = .05$. (Use $df_{Between}$ and df_{Within} to find F_{cv} in order to determine if the null hypothesis was rejected.)

Source of Variability	Sum of Squares	Degrees of Freedom	Mean Square	F ratio
Between groups	59.98	3	19.99	3.28
Within groups	268.60	44	6.10	
Total	328.58	47		

10.108 Given this ANOVA summary table, write the results in APA format using $\alpha = .05$. (Use $df_{Between}$ and df_{Within} to find F_{cv} in order to determine if the null hypothesis was rejected.)

Source of Variability	Sum of Squares	Degrees of Freedom	Mean Square	F ratio
Between groups	42.34	2	21.17	2.23
Within groups	1,896.22	200	9.48	
Total	1,938.56	202		

Interpreting APA format

10.109 The results of a between-subjects, one-way ANOVA, in APA format, are $F(3, 26) = 4.53$, $p < .05$. What interpretative statement can one make about the differences among the four population means?

10.110 The results of a between-subjects, one-way ANOVA, in APA format, are $F(3, 30) = 2.66$, $p > .05$. What interpretative statement can one make about the differences among the four population means?

Calculating r^2

10.111 If $SS_{Between} = 128.86$ and $SS_{Total} = 413.67$, what is r^2?

10.112 If $SS_{Between} = 17.48$ and $SS_{Total} = 342.88$, what is r^2?

10.113 Using the information in the ANOVA summary table in Exercise 10.107, calculate r^2.

10.114 Using the information in the ANOVA summary table in Exercise 10.108, calculate r^2.

Deciding whether to do a post-hoc test

10.115 Should a post-hoc test be completed for the results in the ANOVA summary table in Exercise 10.107?

10.116 Should a post-hoc test be completed for the results in the ANOVA summary table in Exercise 10.108?

Finding q

10.117 If $k = 4$ and $df_{Within} = 16$, what is q if $\alpha = .01$?

10.118 If $k = 3$ and $df_{Within} = 26$, what is q if $\alpha = .05$?

Calculating HSD

10.119 If $q = 3.55$, $MS_{Within} = 10.44$, and $n = 8$, what is HSD?

10.120 If $q = 4.05$, $MS_{Within} = 6.87$, and $n = 11$, what is HSD?

Interpreting HSD

10.121 If $M_1 = 13.09$, $M_2 = 8.89$, and $HSD = 4.37$, is the difference a statistically significant one? What conclusion would one draw about the direction of the difference between the two population means?

10.122 If $M_1 = 123.65$, $M_2 = 144.56$, and $HSD = 10.64$, is the difference a statistically significant one? What conclusion would one draw about the direction of the difference between the two population means?

10.123 If $M_1 = 67.86$, $M_2 = 53.56$, $M_3 = 61.55$, and $HSD = 8.30$, which pairs of means have statistically significant differences? What is the direction of the differences?

10.124 If $M_1 = 12.55$, $M_2 = 13.74$, $M_3 = 5.49$, and $HSD = 4.44$, which pairs of means have statistically significant differences? What is the direction of the differences?

Completing an interpretation (HSD values are given whether needed or not).

10.125 An addictions researcher was curious about which drug was hardest to quit: alcohol, cigarettes, or heroin. She obtained samples of alcoholics, smokers, and heroin addicts who were in treatment for the second time and found out how long they had remained abstinent, in months, after their first treatment. The mean time to relapse for the 8 alcoholics was 4.63 months; for the 11 smokers, it was 4.91 months; for the 6 heroin addicts, it was 5.17 months. Given the ANOVA summary table below, $r^2 = 0.66\%$, and $HSD = 3.83$, complete a four-point interpretation of the results using $\alpha = .05$.

Source of Variability	Sum of Squares	Degrees of Freedom	Mean Square	F ratio
Between groups	1.02	2	0.51	0.07
Within groups	153.62	22	6.98	
Total	154.64	24		

10.126 A sleep specialist investigated the impact of watching TV and using computers, before bedtime, on sleep onset. He obtained 30 college student volunteers and randomly assigned them to three equally sized groups: (1) work on a computer for 30 minutes before going to bed, (2) watch TV for 30 minutes before going to bed, and (3) don't work on a computer or watch TV before bedtime. He then measured time to sleep onset (in minutes), finding means for the three groups, respectively, of 19.60, 17.40, and 5.30. Given the ANOVA summary table, $r^2 = 65.63\%$, and $HSD = 5.32$, complete a four-point interpretation of the results using $\alpha = .05$.

Source of Variability	Sum of Squares	Degrees of Freedom	Mean Square	F ratio
Between groups	1,185.80	2	592.90	25.78
Within groups	620.90	27	23.00	
Total	1,806.70	29		

For Exercises 10.127–10.128, complete all six steps of hypothesis testing for these data sets. Remember to keep the six steps in the right order.

10.127 A consumer researcher gave consumers a sample shampoo. After using the shampoo, each consumer used an interval-level scale to rate his or her satisfaction with it. Scores could range from 0 to 100, with higher scores indicating greater satisfaction. Consumers didn't know each other and were randomly assigned to three groups: (1) receive a store brand of shampoo in a bottle clearly labeled as such, (2) receive a premium brand of shampoo in the premium brand's bottle, or (3) receive a store brand of shampoo in the premium brand's bottle. Here are the collected data:

Store Brand, Store Bottle	Premium Brand, Premium Bottle	Store Brand, Premium Bottle	
70	85	85	
65	90	80	
65	95	90	
60	90	85	
$M = 65.00$ $s = 4.08$	$M = 90.00$ $s = 4.08$	$M = 85.00$ $s = 4.08$	$M_{Grand} = 80.00$

10.128 An environmental psychologist investigated ways to reduce waste. He randomly assigned office workers in small businesses to three groups: (1) to be in a control group, (2) to receive daily e-mail reminders about the importance of recycling, or (3) to have their current wastebaskets replaced with much smaller wastebaskets. To make sure they didn't influence each other, each worker was selected from a different small business. At the end of the week, the psychologist measured how many pounds of office waste each person had generated. Here are the data:

Control Group	Daily e-mail Reminders	Smaller Wastebaskets	
12	14	6	
14	12	4	
18	10	8	
12			
$M = 14.00$	$M = 12.00$	$M = 6.00$	$M_{Grand} = 11.00$
$s = 2.83$	$s = 2.00$	$s = 2.00$	

Expand Your Knowledge

10.129 Which of the following is in descending order of size?

a. SS_{Total}, SS_{Within}, $SS_{Between}$
b. SS_{Total}, $SS_{Between}$, SS_{Within}
c. SS_{Total}, SS_{Within}, MS_{Within}
d. SS_{Within}, SS_{Total}, MS_{Within}
e. MS_{Within}, SS_{Within}, SS_{Total}
f. None of the above.

10.130 If treatment has an effect, then there is:

a. more variability between groups than in total.
b. more variability within groups than in total.
c. more variability between groups than within groups.
d. more variability within groups than between groups.
e. None of the above.

For Exercises 10.131–10.138, indicate whether what is written could be true or is false.

10.131 $F(3, 18) = 10.98, p < .05$

10.132 $F(3, 25) = 1.98, p < .05$

10.133 $F(4, 20) = -17.89, p < .05$

10.134 $N = 14, k = 3$

10.135 $N = 6, k = 8$

10.136 If $\alpha = .05, q = 1.96$.

10.137 $SS_{Between} = 25, SS_{Within} = 10, SS_{Total} = 35$

10.138 $SS_{Between} = 12.50, SS_{Within} = 12.50, SS_{Total} = 25.00$

SPSS

The data for a between-subjects, one-way ANOVA have to be entered in the SPSS data editor in a specific way. Figure 10.16 contains the data for psychologists' salaries for states in the four census regions. The first column in the data editor tells which state the data come from. The second variable, the column labeled "region," contains information about which group a case is in. Here, a 1 indicates that a case is in the Northeast, a 2 indicates the Midwest, a 3 is for cases in the South, and a 4 is for the West. Note that SPSS does not need cases in a region to be grouped together. The final variable, the column labeled "psych," contains the cases' scores on the dependent variable, the mean salary for the state.

The command to start a one-way ANOVA can be found under "Analyze" and "Compare Means," as shown in Figure 10.17.

Clicking on "One-Way ANOVA..." opens up the commands shown in Figure 10.18. "psych," the dependent variable, has already been moved over to the box labeled "Dependent List." "region," the explanatory variable, is moved over to the box labeled "Factor."

	state	region	psych
1	Alaska	4	72.13
2	Connecticut	1	79.52
3	Delaware	3	66.89
4	Florida	3	68.60
5	Georgia	3	62.76
6	Idaho	4	61.49
7	Indiana	2	64.48
8	Kansas	2	55.96
9	Michigan	2	71.22
10	Mississippi	3	47.21
11	Nevada	4	67.12
12	New York	1	84.09
13	North Dakota	2	62.55
14	Rhode Island	1	82.10
15	Vermont	1	65.21
16	Washington	4	74.59

Figure 10.16 Data Entry in SPSS for a One-Way ANOVA At least two columns are needed to enter data for a one-way ANOVA in SPSS (just as for the independent-samples *t* test). One column, "region," contains information about which group a case belongs to and another column, "psych," has the case's score on the dependent variable. (Source: SPSS)

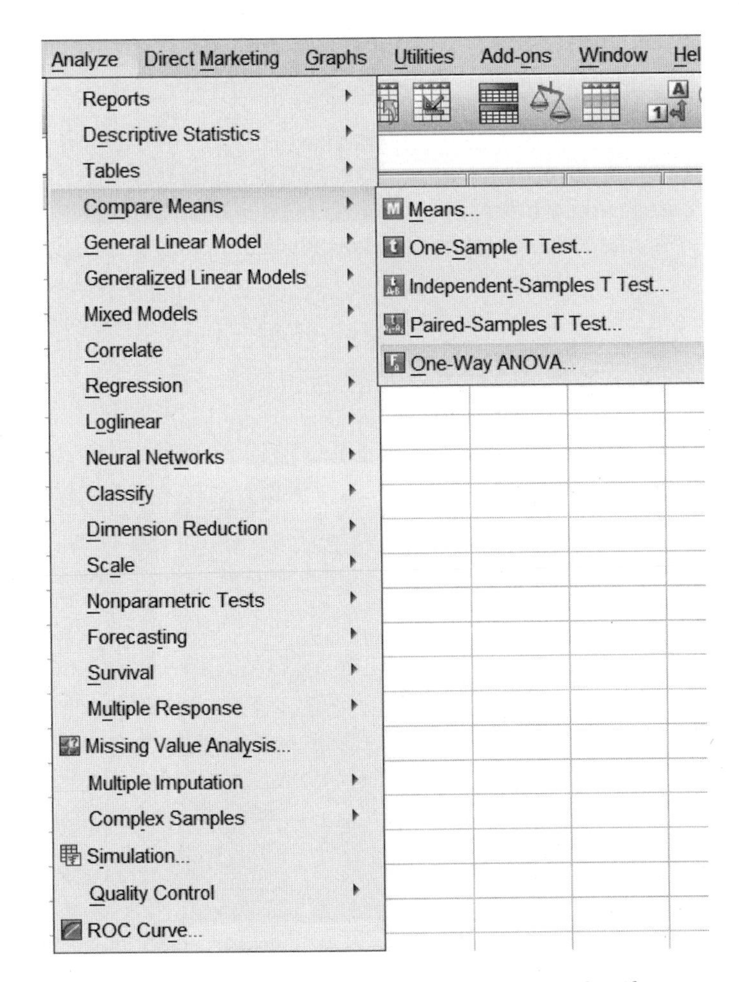

Figure 10.17 Starting a One-Way ANOVA in SPSS The command "One-Way ANOVA. . ." is found under the "Compare Means" heading under "Analyze." (Source: SPSS)

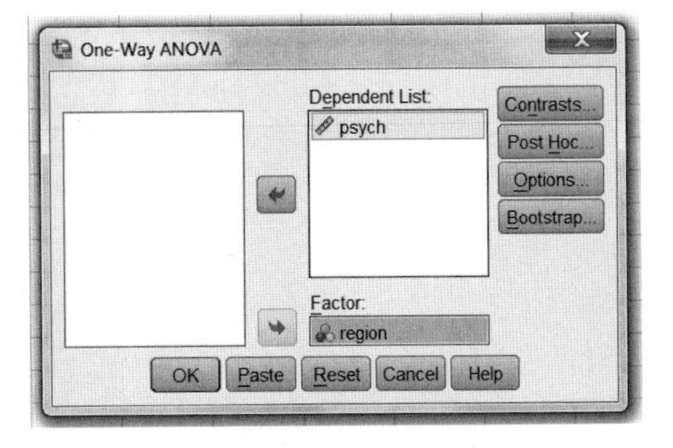

Figure 10.18 Defining the Dependent Variable and Explanatory Variable for a One-Way ANOVA in SPSS Once the dependent variable (psych) has been moved to "Dependent List:" and the explanatory variable (region) to "Factor:", clicking on the "OK" button will start the analysis of variance. (Source: SPSS)

Clicking on "OK" produces the output shown in Figure 10.19. Note that SPSS produces an ANOVA summary table like the one in Table 10.14. The SPSS summary table has an extra column on the right, labeled "Sig." This column reports the exact significance level for the *F* ratio. If the value reported in the "Sig." column is less than or equal to .05, then the results are statistically significant and the results are written in APA format as $p < .05$. If the "Sig." level is greater than .05, then one has failed to reject the null hypothesis and the results are written in APA format as $p > .05$.

ANOVA

psych

	Sum of Squares	df	Mean Square	F	Sig.
Between Groups	636.407	3	212.136	3.517	.049
Within Groups	723.851	12	60.321		
Total	1360.258	15			

Figure 10.19 Example of One-Way ANOVA Output from SPSS SPSS produces an ANOVA summary table. The column labeled "Sig." provides the exact significance level for the *F* ratio. If the "Sig." value is $\leq .05$, the results are written as $p < .05$ in APA format. If the "Sig." value is $> .05$, write the results as $p > .05$. (Source: SPSS)

Here, the exact significance level is .049, which is less than .05, so the null hypothesis is rejected. If the exact significance level is known, as it is here, then it should be reported in APA format. These results would be reported as $F(3, 12) = 3.52$, $p = .049$.

Once a researcher knows that the *F* ratio was statistically significant, he or she can proceed to select a post-hoc test by clicking the "Post Hoc" button seen in Figure 10.17. This opens up the selection of post-hoc tests seen in Figure 10.20. "Tukey," which is the Tukey *HSD* test covered in this chapter, has been selected. If a researcher wants to change the significance level from .05, now is the time to do so.

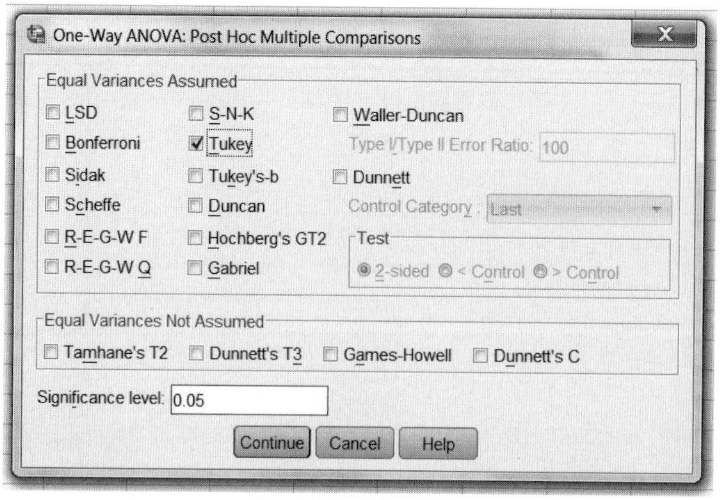

Figure 10.20 Choosing a Post-Hoc Test in SPSS This is the dialog box that opens up when one clicks on "Post Hoc" in Figure 10.18. Note the wide variety of post-hoc tests that are available. "Tukey," which is Tukey's *HSD*, has been selected. (Source: SPSS)

Multiple Comparisons

Dependent Variable: psych

Tukey HSD

(I) region	(J) region	Mean Difference (I-J)	Std. Error	Sig.	95% Confidence Interval	
					Lower Bound	Upper Bound
1 North East	2 Midwest	14.17750	5.49185	.097	-2.1273	30.4823
	3 South	16.36500*	5.49185	.049	.0602	32.6698
	4 West	8.89750	5.49185	.404	-7.4073	25.2023
2 Midwest	1 North East	-14.17750	5.49185	.097	-30.4823	2.1273
	3 South	2.18750	5.49185	.978	-14.1173	18.4923
	4 West	-5.28000	5.49185	.773	-21.5848	11.0248
3 South	1 North East	-16.36500*	5.49185	.049	-32.6698	-.0602
	2 Midwest	-2.18750	5.49185	.978	-18.4923	14.1173
	4 West	-7.46750	5.49185	.546	-23.7723	8.8373
4 West	1 North East	-8.89750	5.49185	.404	-25.2023	7.4073
	2 Midwest	5.28000	5.49185	.773	-11.0248	21.5848
	3 South	7.46750	5.49185	.546	-8.8373	23.7723

* The mean difference is significant at the 0.05 level.

Figure 10.21 Output for Post-Hoc Tests in SPSS Each row in this table compares a pair of means. SPSS is redundant and compares each pair of means twice. For example, the first row (Northeast vs. Midwest) is the same as the fourth row (Midwest vs. Northeast). All that changes is the direction of the mean difference. Also, note the footnote on the table—asterisks that follow the mean difference indicate the difference is a statistically significant one. The exact significance level is found in the column labeled “Sig.” (Source: SPSS)

The output from the *HSD* test is shown in Figure 10.21. SPSS is redundant and has 12 rows in the table when it only needs 6. The first row, for example, compares the Northeast region to the Midwest region, and the fourth row turns this around to compare the Midwest region to the Northeast region. It is exactly the same comparison, though one finds a mean difference of 14.17750 and the other –14.17750. SPSS doesn’t report the *HSD* values, but it does report each mean difference that is statistically significant at the .05 level. It does this, as mentioned in the table footnote, by placing an asterisk after each mean difference that is statistically significant.

One-Way, Repeated-Measures ANOVA

LEARNING OBJECTIVES

- Describe what repeated-measures ANOVA does.
- Complete a one-way, repeated-measures ANOVA.*
- Interpret a one-way, repeated-measures ANOVA.

CHAPTER OVERVIEW

The last chapter introduced between-subjects, one-way ANOVA, an extension of the independent-samples *t* test to situations in which more than two independent samples are being compared. Just as the chapter on independent-samples *t* tests was followed by a chapter on paired-samples *t* tests, the chapter on *between-subjects,* one-way ANOVA is followed by a chapter on *within-subjects,* measures one-way ANOVA. Said another way, the chapter on ANOVA for independent samples is followed by this chapter on ANOVA for dependent samples. Repeated-measures, one-way ANOVA is an extension of paired-samples *t* tests to situations in which there are more than two dependent samples.

11.1 Introduction to Repeated-Measures ANOVA

In this chapter, we learn about *within-subjects* ANOVA, commonly called **repeated-measures ANOVA**. Repeated-measures ANOVA is an extension of the paired-samples *t* test. The paired-samples *t* test can only be used when comparing the means of two dependent samples. But, a repeated-measures ANOVA can be used when there are two or more dependent samples.

There are many different types of experiments that use repeated measures. What they have in common is a connection between the cases in one cell and the cases in the other cells. Repeated measures may involve:

- *The same cases measured at multiple points of time.* For example, to examine how weight changes over college, a group of first-year college students might be weighed at the start and the end of each of the four years of college.

*Note: Formulas for calculating sums of squares appear in an appendix at the end of the chapter.

- *The same cases measured under multiple conditions.* For example, to examine how alcohol affects one's ability to walk a straight line, 21-year-olds might be asked to walk a straight line before drinking anything, after drinking a placebo beer, after drinking one beer, after two, after four, and after six.
- *Different cases, matched so that they are similar in some dimension(s).* For example, a researcher comparing murder rates for cities with high, medium, and low numbers of sunny days per year might be concerned that population density and poverty level could affect the results. To account for these other factors, he could group cities with the same populations and unemployment rates into sets of three: one with a high number of sunshine days, one low on the number of sunshine days, and one in the middle.

In all these examples of dependent samples, we're looking at differences within the same subjects (or matched subjects) over time or across conditions (the treatment). Thus, it is important to make sure that the subjects are arrayed in the same order from cell to cell to cell. If a participant's data point is listed first in one cell, then it should be first in the other cells.

The beauty of analysis of variance is that it partitions the variability in the dependent variable into the components that account for the variability. Between-subjects, one-way ANOVA, covered in the last chapter, distinguishes variability due to how the groups are treated (between-group variability) from variability due to individual differences among subjects (within-group variability). With between-subjects ANOVA, it doesn't matter how subjects in a cell are arrayed. In our example, the three rats that ran the maze to reach low-calorie food ran it in 30, 31, and 32 seconds. And that is how the data showed up in the first cell of Table 10.3. But, the data could just as easily have been listed as 31, 32, and 30 or as 31, 30, and 32. These cases are independent of each other. No matter the order that the results are listed, the results of the between-subjects ANOVA will be the same. This is not true of repeated-measures ANOVA. Because the samples in repeated-measures ANOVA are dependent samples, the order in which they are arranged matters very much. Changing the order, unless it is changed consistently in all cells, will change the outcome.

In repeated-measures ANOVA, the levels of the independent variable, for example, the different times or conditions, make up the columns and the subjects make up the rows. By having each subject on a row by itself, subjects can be treated as a second explanatory variable, as a second factor. Treating subjects as a factor allows repeated-measures ANOVA to measure—to account for—variability due to them. By partitioning out this variability, repeated-measures ANOVA obtains a purer measure of the effect of treatment and is a more statistically powerful test than is between-subjects ANOVA. What does it mean to be a statistically more powerful test? It means that this test gives us a greater likelihood of being able to reject the null hypothesis.

Figure 11.1 shows the similarities and differences between the sources of variability for between-subjects and repeated-measures ANOVA. With between-subjects ANOVA, total variability is partitioned into between and within. For repeated-measures ANOVA, total variability is also partitioned into between and within. But, within-group variability is then broken down further. First, variability due to subjects is removed. Then, the variability left over, labeled "residual," is used as the denominator in the F ratio.

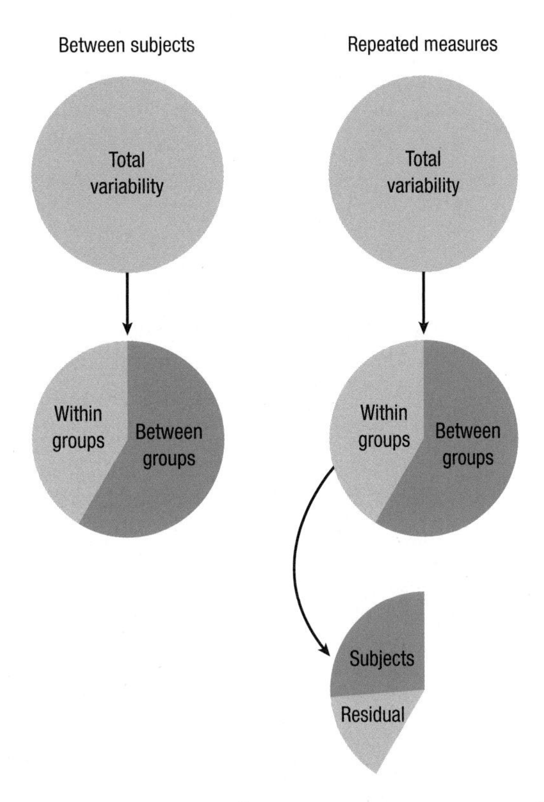

Figure 11.1 Partitioning Variability for One-Way ANOVA Variability in the dependent variable in one-way ANOVA is divided into between-groups and within-groups factors. For between-subjects, one-way ANOVA, the partitioning stops there. For a within-subjects design, the within-groups variability is further divided into that due to subjects and that which still remains.

The statistical power of repeated-measures ANOVA comes from the fact that it partitions out the variability due to subjects. Between-subjects ANOVA divides between-group variability by within-group variability to find the *F* ratio. Repeated-measures ANOVA calculates between-group variability the same way, so it will have the same numerator for the *F* ratio. But the denominator, which has variability due to subjects removed, will be smaller. This means that the *F* ratio will be larger—meaning it is more likely to fall in the rare zone, and therefore more likely to result in a rejection of the null hypothesis than for between-subjects, one-way ANOVA.

To see what this means, let's analyze the same set of data two ways—one way that doesn't partition out the subject variability and one that does. Let's compare two neighboring cities, Erie, PA, and Cleveland, OH, to see if they differ in how hot they get. From weather archives, we have obtained the average high temperature for each city for each month of the year. The data are seen in Figure 11.2. Two things are readily apparent in this graph:

- The high temperatures in the two cities are very similar to each other.
- Both cities share a strong seasonal trend to temperature—it is hottest in the summer and coldest in the winter.

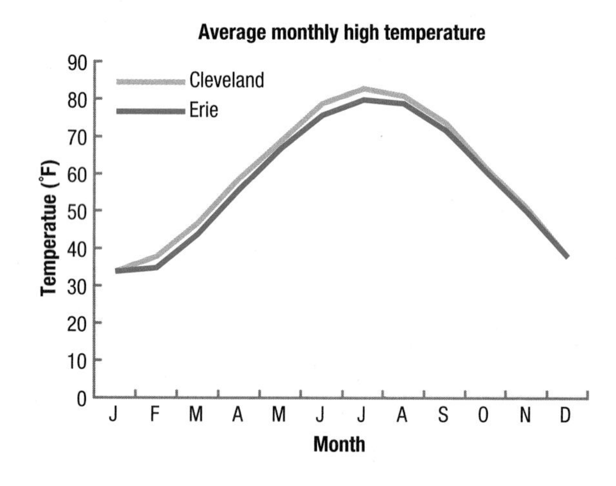

Figure 11.2 Average High Temperature per Month in Cleveland, OH, and Erie, PA Notice how little difference there is in temperature between the two cities each month and how much variability exists from month to month.

The paired-samples t test is appropriate to use to see if there is a statistically significant difference in temperature between the two cities. The t value turns out to be 5.38 and the difference between the two cities is a statistically significant one—Cleveland with a mean high temperature of 59.6°F is statistically significantly hotter than Erie with a mean of 57.7°F. A difference of 1.9°F doesn't sound like much, but it is enough to be statistically significant when analyzed with a paired-samples t test.

The reasonable next question to ask is what the size of the effect is. How much of the variability in the dependent variable of temperature is explained by the grouping variable of city? Judging by Figure 11.2, it shouldn't be much, right? Well, be surprised—72% of the variability in temperature is accounted for by city status! (This book does NOT advocate calculating either d or r^2 for a paired-samples t test. In this one instance, however, and just for this one example, I calculated r^2.)

An r^2 of 72% is a huge effect, a whopping effect. Look at Figure 11.2 again. How can we account for the variability in temperature? Is there more variability in temperature due to differences between the cities each month or is there more variability in temperature due to changes from month to month? Clearly, more variability exists from month to month, which is due to individual differences among our cases, which are months. Regardless of city, months like July and August differ by 40 to 50 degrees F from months like January and February.

Does it make sense that the variability in temperature accounted for by the between-groups factor, city, is 72%? No! And it can't be so. If we just agreed that more of the variability in Figure 11.2 is due to the within-group factor (month) than the between-group factor, and the between-group factor accounted for 72% of the variability, then the within-group factor must explain at least 73% of the variability. That's impossible—we have just accounted for at least 145% of the variability!

So, what is going on? Because a paired-samples t test doesn't partition out the variability due to subjects, it mixes in variability due to individual differences with variability due to the grouping variable when calculating r^2. Thus, r^2 is an overestimate of the percentage of variability due to the grouping variable and should not be used with paired-samples t tests. The same is true for Cohen's d for a paired-samples t test—it also overestimates the size of the effect.

How much of the variability in temperature is really due to the grouping variable of city? To answer that, we need a repeated-measures ANOVA, which treats variability due to the independent variable and that due to the subjects separately. ANOVA, which can be used when comparing *two* or more means, can be used in place of a t test. As we are about to see, there is a distinct advantage to using a repeated-measures ANOVA over a paired-samples t test.

When these temperature data are analyzed with a repeated-measures ANOVA, the difference between the mean temperatures due to location is still a statistically significant one. But now, because the variance due to subjects has been removed, the effect size will be a purer measure of the effect of the independent variable. By removing individual subject variability, the explanatory power of the city where the measurement was taken is reduced from 72% to1%. Yes, 1%. Look at Figure 11.2 again. Doesn't a measure of 1% of the variability due to location reflect reality a lot better than 72%? If you are comparing the means of two dependent samples, use repeated-measures ANOVA, not a paired-samples *t* test .

We'll use the following example throughout the chapter to learn how to complete a repeated-measures ANOVA. Imagine that Dr. King wanted to determine if a new treatment for ADHD was effective. He took four children with ADHD and measured their level of distraction on an interval-level scale where higher scores mean a higher level of distraction. Next, each child received individual behavior therapy for three months, after which the child's level of distraction was assessed again. Finally, to determine if the effects of treatment are long-lasting, each child had his or her level of distraction measured again six months later.

Table 11.1 contains the data, each case in its own row and each phase of treatment in a separate column. The effects of treatment are found in the column means. Figure 11.3 shows that the mean level of distraction is highest before the treatment starts, lowest at the end of treatment, and somewhere in the middle at follow-up. This suggests that treatment has an effect on reducing distraction and that the effect lingers, though in a weakened form, for six months. Of course, Dr. King needs to do a statistical test to determine if the treatment effect is statistically significant or if the differences can be accounted for by sampling error.

TABLE 11.1 Level of Distraction Scores Pre-Treatment, Post-Treatment, and at Six-Month Follow-Up for Four Children with ADHD

	Pre-Treatment	Post-Treatment	Follow-Up
Participant 1	13	10	12
Participant 2	30	20	26
Participant 3	20	13	17
Participant 4	26	17	20
M (*s*)	22.25 (7.41)	15.00 (4.40)	18.75 (5.85)

Looking at the column means, treatment improves the level of distraction from a mean of 22.25 before treatment to a mean of 15.00 after treatment. Further, some improvement lingers for at least six months: after treatment, the distraction score gets worse, but it doesn't return to pre-treatment levels.

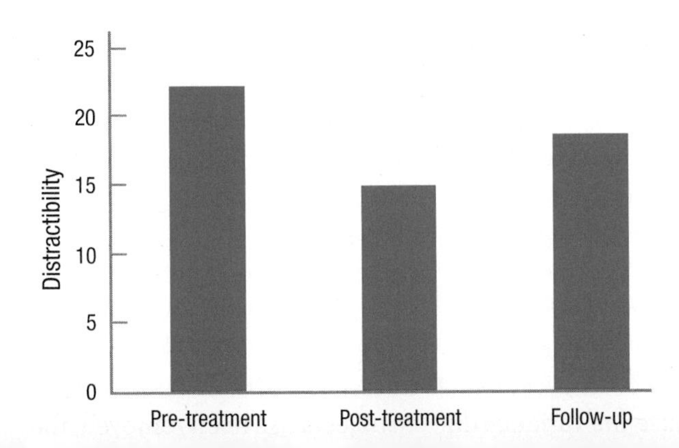

Figure 11.3 ADHD Study Pre-Treatment, Post-Treatment, and Six-Month Follow-Up Distraction Scores The mean level of distraction decreases with treatment and increases when treatment is over. To determine if these changes in level of distraction are statistically significant, a one-way, repeated-measures ANOVA needs to be completed.

The differences between rows are due to individual differences, uncontrolled characteristics that vary from case to case. For example, case 1 has the lowest level of distraction pre-treatment. This could be due to the participant's having a less severe form of ADHD, having eaten a good breakfast on the day of the test, having been taught by a teacher who stressed the skills on the test, or a whole host of other uncontrolled factors.

To reiterate the message of this introduction, the effect of individual differences is intertwined with the effect of treatment. Is the first case's low score at the end of treatment due to the effect of treatment or due to individual differences like a less severe form of ADHD? One-way, repeated-measures ANOVA is statistically powerful because it divides the variability in scores into these two factors. Repeated-measures ANOVA determines (1) how much variability is due to the column effect (the effect of treatment), and (2) how much variability is due to the row effect (individual differences). Let's see it in action in the next section.

Practice Problems 11.1

Review Your Knowledge

11.01 When is repeated-measures ANOVA used?

11.02 Into what two factors does one-way, repeated-measures ANOVA divide variability in a set of scores?

11.03 When comparing the means of two dependent samples, what test should be used? Why?

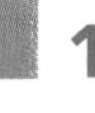

11.2 Calculating One-Way, Repeated-Measures ANOVA

Step 1 Pick a Test

Dr. King's ADHD/distraction study compares the means of three samples, so it calls for an ANOVA to analyze the data. Because there is only one explanatory variable (time), it will be a one-way ANOVA. There are three levels of the explanatory variable—pre-treatment, post-treatment, and follow-up—so there are three groups. This is a longitudinal study in which the same participants are being assessed at three points in time. This means the samples are dependent. The dependent variable (level of distraction) is measured at the interval level, so a mean for each group can be calculated. Comparing the means of three dependent samples calls for a one-way, repeated-measures ANOVA.

Step 2 Check the Assumptions

Assumptions for the one-way, repeated-measures ANOVA are similar to those for the paired-samples *t* test: (1) each sample is a random sample from its population, (2) cases within samples are independent and don't influence each other, and (3) the dependent variable is normally distributed in the population. There is a fourth assumption, which is beyond the scope of this book to assess. It is called the sphericity assumption, and it is a form of the homogeneity of variance assumption: If difference scores were calculated, would variability be the same in each set of difference scores?

Don't forget: **T**om **a**nd **H**arry **d**espise **c**rabby **i**nfants is a mnemonic to remember the six steps of hypothesis testing.

Dr. King makes the following evaluation of the first three assumptions.

1. *Random samples.* There is no evidence that the sample is a random sample of children with ADHD, so this assumption is violated. However, the assumption is robust to violation, so the child psychologist can proceed. He will need to be careful about the population to which the results are generalized.

2. *Independence of observations.* There is no evidence that the cases within a group influence each other, so this assumption is not violated.

3. *Normality.* Dr. King assumes that distraction scores are normally distributed in the larger population of people with ADHD.

Only the random samples assumption is violated (and it's robust to violation), so Dr. King can proceed with the one-way, repeated-measures ANOVA.

Step 3 List the Hypotheses

For a one-way, repeated-measures ANOVA, the null hypothesis states that all the population means are the same. The alternative hypothesis states that at least one population mean is different from at least one other. As with between-subjects one-way ANOVA, the null hypothesis is easy to state using mathematical symbols, but not the alternative hypothesis, so words are used to express the alternative hypothesis. For the ADHD/level of distraction data with three samples, the null and alternative hypotheses are

H_0: $\mu_1 = \mu_2 = \mu_3$.
H_1: At least one population mean is different from at least one of the others.

Step 4 Set the Decision Rule

To set the decision rule with repeated-measures ANOVA, Dr. King needs to find F_{cv}, the critical value of F.

As with the other tests we've covered up to this point, to find F_{cv}, a researcher needs (a) to set the alpha level on the basis of willingness to make a Type I error and (b) to know the degrees of freedom for the numerator and the denominator of the F ratio. These are then used in Appendix Table 4 to find F_{cv}. (One thing a researcher doesn't need to worry about is whether to do a one-tailed or a two-tailed test—ANOVA is always nondirectional.)

In a one-way, repeated-measures ANOVA, the numerator, where the effect of the independent variable is isolated, is called the treatment effect and the denominator is called the residual effect. The formulas for finding degrees of freedom treatment ($df_{\text{Treatment}}$) and degrees of freedom residual (df_{Residual}) for a one-way, repeated-measures ANOVA are given in Equation 11.1. Equation 11.1 also shows how to calculate two other degrees of freedom that will be necessary to complete the ANOVA summary table for repeated-measures ANOVA: (1) degrees of freedom subjects, df_{Subjects}, and (2) degrees of freedom total, df_{Total}.

Equation 11.1 Degrees of Freedom for a One-Way, Repeated-Measures ANOVA

$$df_{\text{Subjects}} = n - 1$$

$$df_{\text{Treatment}} = k - 1$$

$$df_{\text{Residual}} = df_{\text{Subjects}} \times df_{\text{Treatment}}$$

$$df_{\text{Total}} = N - 1$$

where df_{Subjects} = degrees of freedom for variability due to subjects
$df_{\text{Treatment}}$ = degrees of freedom for the treatment effect
df_{Residual} = degrees of freedom for residual variability
df_{Total} = degrees of freedom total
n = sample size per group
k = number of groups
N = total number of observations ($n \times k$)

Dr. King sets alpha at .05. He is willing to run a 5% chance of making a Type I error. To calculate degrees of freedom for the level of distraction/ADHD data, Dr. King needs to know $n = 4$, $k = 3$, and $N = 12$. That is, there are four cases per group and three groups, for a total of 12 cases. These three values are then used in Equation 11.1 to calculate the 4 degrees of freedom:

$$\begin{aligned} df_{\text{Subjects}} &= n - 1 \\ &= 4 - 1 \\ &= 3 \end{aligned}$$

$$\begin{aligned} df_{\text{Treatment}} &= k - 1 \\ &= 3 - 1 \\ &= 2 \end{aligned}$$

$$df_{\text{Residual}} = df_{\text{Subjects}} \times df_{\text{Treatment}}$$
$$= 3 \times 2$$
$$= 6$$

$$df_{\text{Total}} = N - 1$$
$$= 12 - 1$$
$$= 11$$

In this example, note that $3 + 2 + 6 = 11$. For one-way, repeated-measures ANOVA, $df_{\text{Subjects}} + df_{\text{Treatment}} + df_{\text{Residual}} = df_{\text{Total}}$. This means that the total degrees of freedom in a repeated-measures ANOVA are divided into subcomponents for variability due to individual differences (df_{Subjects}), due to the explanatory variable ($df_{\text{Treatment}}$), and remaining variability (df_{Residual}).

To find the critical value of F, the only degrees of freedom needed are $df_{\text{Treatment}}$ and df_{Residual}, the numerator and the denominator, respectively, of the F ratio. For the level of distraction data, those degrees of freedom are 2 and 6. Using Appendix Table 4 with $\alpha = .05$, Dr. King finds the intersection of the column for 2 degrees of freedom and the row with 6 degrees of freedom and arrives at $F_{cv} = 5.143$. Hence, the decision rule for the ADHD/level of distraction data for a one-way, repeated-measures ANOVA is:

- If $F \geq F_{cv}$ of 5.143, reject H_0.
- If $F < F_{cv}$ of 5.143, fail to reject H_0.

Figure 11.4 uses F_{cv} to show the sampling distribution of F with the rare and common zones marked. If the value of F that Dr. King will calculate in the next step falls in the rare zone or on the line that separates the rare zone from the common zone, then (a) the null hypothesis is rejected, (b) the alternative hypothesis is accepted, (c) the results are called statistically significant, and (d) there is reason to believe that at least one difference exists among the population means. If F falls in the common zone, then (a) Dr. King will have failed to reject the null hypothesis, (b) the results will be called not statistically significant, and (c) he concludes there is not enough evidence to conclude any difference exists among any of these population means.

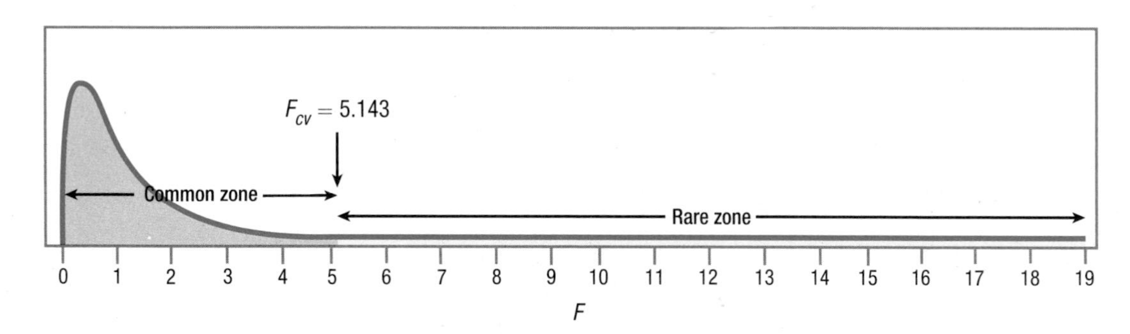

Figure 11.4 Sampling Distribution of *F* for ADHD Distractibility Study This is the sampling distribution of F at the .05 level with 2 and 6 degrees of freedom. If the observed value of F falls in the rare zone, then the null hypothesis is rejected.

Step 5 Calculate the Test Statistic

To calculate the F ratio, a researcher completes an ANOVA summary table. Table 11.2 is a guide to completing the cells in an ANOVA summary table for one-way, repeated-measures ANOVA. Though column labels in this table are identical to the ones for between-subjects, one-way ANOVA, the sources of variability (the rows) are different.

TABLE 11.2 Template for ANOVA Summary Table for One-Way, Repeated-Measures ANOVA

Source of Variability	Sum of Squares	Degrees of Freedom	Mean Square	F ratio
Subjects	See Chapter Appendix	$n - 1$		
Treatment	See Chapter Appendix	$k - 1$	$\frac{SS_{Treatment}}{df_{Treatment}}$	$\frac{MS_{Treatment}}{MS_{Residual}}$
Residual	See Chapter Appendix	$(n - 1)(k - 1)$	$\frac{SS_{Residual}}{df_{Residual}}$	
Total	See Chapter Appendix	$N - 1$		

The sums of squares for subjects, treatment, and residual will be provided in order to complete the ANOVA summary table. n = the number of cases per cell, k = the number of cells, and N = the total number of observations.

- *First column: Source of variability.* In the summary table, the sources of variability are listed in the first column in this order: variability due to subjects, variability due to treatment, residual variability, and total variability. Residual variability is the portion of total variability that is not accounted for by variability due to subjects or treatment.
- *Second column: Sum of squares.* Each source of variability is represented by a sum of squares in the second column. In order to focus on understanding repeated-measures ANOVA and not get slowed down by calculations, the necessary sums of squares—$SS_{Subjects}$, $SS_{Treatment}$, $SS_{Residual}$, and SS_{Total}—will be supplied in this chapter. (For those who wish to learn how to calculate sums of squares, or who have instructors who wish them to learn, formulas are given in an appendix at the end of the chapter.)
- *Third column: Degrees of freedom.* The third column shows the degrees of freedom for each source of variability. The degrees of freedom are calculated through Equation 11.1.
- *Fourth column: Mean squares.* In the fourth column, mean squares for the treatment effect and the residual effect are calculated. These are abbreviated $MS_{Treatment}$ and $MS_{Residual}$. They are calculated by dividing a sum of squares by its degrees of freedom.
- *Fifth column: F ratio.* In the fifth column, the F ratio for the effect of treatment is calculated by dividing the numerator term ($MS_{Treatment}$) by the denominator term ($MS_{Residual}$).

The sums of squares for the ADHD/distraction data follow:

- $SS_{\text{Subjects}} = 308.67$
- $SS_{\text{Treatment}} = 105.17$
- $SS_{\text{Residual}} = 16.83$
- $SS_{\text{Total}} = 430.67$

Then Dr. King retrieves the degrees of freedom he calculated earlier and enters them in the summary table:

- $df_{\text{Subjects}} = 3$
- $df_{\text{Treatment}} = 2$
- $df_{\text{Residual}} = 6$
- $df_{\text{Total}} = 11$

The next step is to calculate the two mean squares, $MS_{\text{Treatment}}$ and MS_{Residual}. To do so, Dr. King divides each sum of squares by its respective degrees of freedom:

$$MS_{\text{Treatment}} = \frac{SS_{\text{Treatment}}}{df_{\text{Treatment}}} = \frac{105.17}{2} = 52.5850 = 52.59$$

$$MS_{\text{Residual}} = \frac{SS_{\text{Residual}}}{df_{\text{Residual}}} = \frac{16.83}{6} = 2.8050 = 2.81$$

Finally, Dr. King divides the mean square for the numerator ($MS_{\text{Treatment}}$) by the mean square for the demoninator (MS_{Residual}) to find the F ratio:

$$F = \frac{MS_{\text{Treatment}}}{MS_{\text{Residual}}} = \frac{52.59}{2.81} = 18.7153 = 18.72$$

At this point, the calculation phase for one-way, repeated-measures ANOVA is over. The complete ANOVA summary table for the ADHD/distraction study is shown in Table 11.3. The next section of the chapter, after more practice with calculation, focuses on interpreting the results.

TABLE 11.3 ANOVA Summary Table for Level of Distraction/ADHD Data

Source of Variability	Sum of Squares	Degrees of Freedom	Mean Square	F ratio
Subjects	308.67	3		
Treatment	105.17	2	52.59	18.72
Residual	16.83	6	2.81	
Total	430.67	11		

Worked Example 11.1 Imagine that Dr. Agosto, an orthopedic surgeon, decided to test the effectiveness of the standard surgical treatment for back pain. She found 30 people with chronic back pain and matched them into groups of three on the basis of age, sex, number of years of pain, and degree of physical impairment. She then randomly assigned members of each group to receive one of three different levels of treatment: (1) standard surgery and physical therapy, (2) sham surgery and physical therapy, or (3) physical therapy alone. (In sham surgery, the patient undergoes anesthesia and has an incision made in his or her back, but is sewn up without any real surgical intervention.) This was a double-blind study for the patients who received surgery—they did not know if they received standard surgery or sham surgery, and neither they did the medical staff who provided follow-up care. After receiving surgery or no surgery, all patients received physical therapy for six months. At this point, the outcome was measured on a quality-of-life scale. Higher scores mean that a person has a better quality of life and is more able to enjoy the regular activities of daily living.

A Common Question

Q Is sham surgery ever really used as a control group?

A Yes. Sham surgery controls for the placebo effect of having a surgical procedure. As long as informed consent is used, there is nothing unethical about using it in a study.

Figure 11.5 shows the outcome of the experiment. Patients in Group 1, real surgery, had a better outcome than the sham surgery patients in Group 2, or the physical therapy patients in Group 3. But, the differences are slight: $M_{RealSurgery} = 51.00$, $M_{ShamSurgery} = 48.00$, and $M_{PhysicalTherapy} = 45.00$. Dr. Agosto suspects that the differences between groups could be explained with sampling error and don't represent an advantage of surgery. She'll need a statistical test to find out.

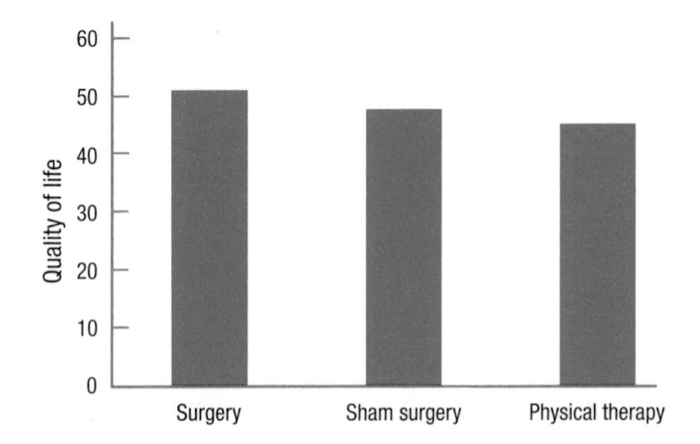

Figure 11.5 Results of Surgery, Sham Surgery, and Physical Therapy on the Quality of Life for People with Chronic Back Pain The surgery patients reported the highest mean for quality of life. However, the degree of improvement seems small and unlikely to be statistically significant. One-way, repeated-measures ANOVA can be used to determine this.

Step 1 Pick a Test

Dr. Agosto is comparing the means of three groups (real surgery vs. sham surgery vs. physical therapy). This calls for an ANOVA. There is only one independent variable (type of treatment), so this is a one-way ANOVA. The cases are made up of matched participants—each one assigned to one type of treatment, so the groups are dependent. That means Dr. Agosto is doing a one-way, repeated-measures ANOVA.

Step 2 Check the Assumptions

- *Random samples.* It was not stated that the 30 participants were a random sample from the population of people with back pain, so it is unlikely they are. Almost certainly the sample is a convenience sample and the random samples assumption has been violated. The random samples assumption is robust, though, so Dr. Agosto can proceed with the one-way, repeated-measures ANOVA. She will just need to be cautious about the population to which she generalizes the results.
- *Independence of observations.* This assumption is about independence within a group, not between groups. Within groups, there is no connection between participants and how one participant performs does not influence the performance of others. Each participant participates only once. This assumption is not violated.
- *Normality.* Dr. Agosto, from her review of the literature, knows that quality-of-life scores are normally distributed when quality of life is measured in the population of people receiving treatment for back pain.

Only the random samples assumption is violated, which is robust, so Dr. Agosto can proceed with the planned one-way, repeated-measures ANOVA.

Step 3 List the Hypotheses

H_0: $\mu_1 = \mu_2 = \mu_3$.

H_1: At least one of the three population means is different from at least one of the others.

Step 4 Set the Decision Rule

To determine the decision rule, Dr. Agosto needs to find the degrees of freedom for the numerator in the F ratio (the treatment effect) and for the denominator (the residual effect). Later on, to complete the ANOVA summary table, she'll need to know the other 2 degrees of freedom—subjects and total—so she might as well use Equation 11.1 to calculate all the degrees of freedom now. To do so, she'll need to know the number of cases per group ($n = 10$), the number of groups ($k = 3$), and the total number of observations ($N = 30$):

$$\begin{aligned} df_{\text{Subjects}} &= n - 1 \\ &= 10 - 1 \\ &= 9 \end{aligned}$$

$$\begin{aligned} df_{\text{Treatment}} &= k - 1 \\ &= 3 - 1 \\ &= 2 \end{aligned}$$

$$
\begin{aligned}
df_{\text{Residual}} &= df_{\text{Subjects}} \times df_{\text{Treatment}} \\
&= 9 \times 2 \\
&= 18
\end{aligned}
$$

$$
\begin{aligned}
df_{\text{Total}} &= N - 1 \\
&= 30 - 1 \\
&= 29
\end{aligned}
$$

Knowing that $df_{\text{Treatment}}$ (the numerator degrees of freedom) = 2 and df_{Residual} (the denominator degrees of freedom) = 18, Dr. Agosto can look in Appendix Table 4 to find the critical value of F. But first, she has to decide how willing she is to make a Type I error.

A Type I error occurs when the null hypothesis is mistakenly rejected. In this study, that would mean concluding there is a treatment effect when such is not the case. Dr. Agosto wants to avoid this error, so she sets alpha at .01, not .05. This means that there is a 1% chance of this error occurring, not a 5% chance.

Using Appendix Table 4, Dr. Agosto finds $F_{cv} = 6.013$. **Figure 11.6** shows the sampling distribution of F with the rare zone marked. The decision rule is:

- If $F \geq 6.013$, reject H_0.
- If $F < 6.013$, fail to reject H_0.

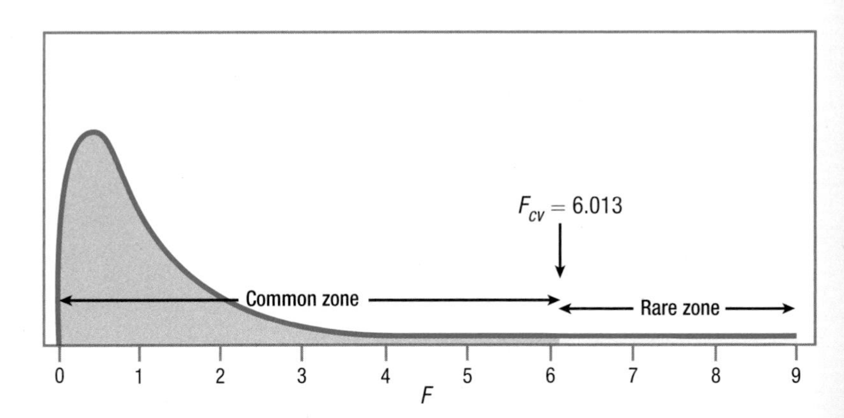

Figure 11.6 Sampling Distribution of F for Effect of Surgery and Physical Therapy on Back Pain F_{cv} of 6.013 is the boundary between the common zone and the rare zone for this sampling distribution of F with 2 and 18 degrees of freedom and $\alpha = .01$. The null hypothesis is not rejected if the results fall in the common zone; it is rejected if results fall on the line or in the rare zone.

Step 5 Calculate the Test Statistic

To calculate the test statistic, Dr. Agosto calculated the sums of squares using the formulas provided in the appendix at the end of this chapter. She calculated:

- $SS_{\text{Subjects}} = 2{,}193.33$
- $SS_{\text{Treatment}} = 180.00$
- $SS_{\text{Residual}} = 50.67$
- $SS_{\text{Total}} = 2{,}424.00$

Next, she uses the degrees of freedom for treatment, 2, and for the residual, 18, to calculate $MS_{\text{Treatment}}$ and MS_{Residual}.

$$MS_{\text{Treatment}} = \frac{SS_{\text{Treatment}}}{df_{\text{Treatment}}} = \frac{180.00}{2} = 90.0000 = 90.00$$

$$MS_{\text{Residual}} = \frac{SS_{\text{Residual}}}{df_{\text{Residual}}} = \frac{50.67}{18} = 2.8150 = 2.82$$

Finally, to find the *F* ratio, the surgeon divides the numerator term ($MS_{\text{Treatment}}$) by the denominator term (MS_{Residual}):

$$\begin{aligned} F &= \frac{MS_{\text{Treatment}}}{MS_{\text{Residual}}} \\ &= \frac{90.00}{2.82} \\ &= 31.9149 \\ &= 31.91 \end{aligned}$$

The complete ANOVA summary table for the back surgery/quality-of-life one-way, repeated-measures ANOVA can be seen in Table 11.4. In the next section, we'll see how Dr. Agosto might interpret her results.

TABLE 11.4 ANOVA Summary Table for Back Surgery/Quality-of-Life Data

Source of Variability	Sum of Squares	Degrees of Freedom	Mean Square	*F* ratio
Subjects	2,193.33	9		
Treatment	180.00	2	90.00	31.91
Residual	50.67	18	2.82	
Total	2,424.00	29		

Practice Problems 11.2

Apply Your Knowledge

11.04 What test should be used to analyze these two studies? Select from the single-sample *z* test; single-sample *t* test; independent-samples *t* test; paired-samples *t* test; between-subjects, one-way ANOVA; and one-way, repeated-measures ANOVA.

a. High school seniors who plan to (i) not go to college, (ii) go to a community college, or (iii) go to a four-year college are matched in terms of intelligence. Ten years later, the three matched groups are compared in terms of mean annual income.

b. A random sample of people with a family history of Alzheimer's disease is compared to a random sample of people from the general population in terms of their mean score on an interval-level measure of fear of dementia.

11.05 List the assumptions for a one-way, repeated-measures ANOVA.

11.06 State the hypotheses for a one-way, repeated-measures ANOVA.

11.07 There's a sample of 12 cases that is measured on an interval-level variable at four points in time. The data will be analyzed with a one-way, repeated-measures ANOVA.

a. What are the numerator degrees of freedom for the *F* ratio?

b. What are the denominator degrees of freedom for the *F* ratio?

11.08 If $df_{\text{Treatment}} = 3$, $df_{\text{Residual}} = 21$, and $\alpha = .05$, what is the decision rule?

11.09 Given $SS_{\text{Subjects}} = 123.00$, $SS_{\text{Treatment}} = 216.00$, $SS_{\text{Residual}} = 410.40$, $SS_{\text{Total}} = 749.40$, $df_{\text{Subjects}} = 19$, $df_{\text{Treatment}} = 3$, $df_{\text{Residual}} = 57$, and $df_{\text{Total}} = 79$, complete a one-way, repeated-measures ANOVA summary table.

11.3 Interpreting One-Way, Repeated-Measures ANOVA

Statistical tests are used to help answer questions and make decisions. It is in the sixth step of hypothesis testing, the interpretation of results, that clear language is used to tell what was found and what it means. In the ADHD/level of distraction study, for example, repeated-measures ANOVA is being used to determine if treatment decreases the level of distraction and if the improvement is maintained when treatment stops.

For a one-way, repeated-measures ANOVA, just as with a between-subjects, one-way ANOVA, there are three tasks involved in interpretation:

1. Determine whether the null hypothesis is rejected.
2. Determine how big the *overall* effect is. That is, how much of the variability in the dependent variable is explained by the independent variable.
3. Determine where the effect is found. All a statistically significant ANOVA reveals is that at least one sample mean differs statistically from at least one other sample mean. Post-hoc tests are needed to find out which pair(s) of means differ and what the direction of the difference is.

Step 6 Interpret the Results

Was the Null Hypothesis Rejected?

As with previous tests, a researcher determines if the null hypothesis was rejected by checking the decision rule generated in Step 4 against the value of the test statistic calculated in Step 5. For the ADHD/level of distraction data, $F_{cv} = 5.143$ and $F = 18.72$. (The F value can be found in the ANOVA summary table, Table 11.3. There will be several other values in the ANOVA summary table needed for interpretation, so it is reprinted here as Table 11.5.)

In Step 4, the decision rule Dr. King generated was:

- If $F \geq 5.143$, reject H_0.
- If $F < 5.143$, fail to reject H_0.

TABLE 11.5 ANOVA Summary Table for Level of Distraction/ADHD Data

Source of Variability	Sum of Squares	Degrees of Freedom	Mean Square	*F* ratio
Subjects	308.67	3		
Treatment	105.17	2	52.59	18.72
Residual	16.83	6	2.81	
Total	430.67	11		

This ANOVA summary table is reprinted here, so values can be taken from it to be used in interpretation.

Because $18.72 \geq 5.143$, the first statement is true and Dr. King rejects the null hypothesis. Figure 11.7 shows that the *F* ratio fell in the rare zone. This means that the results are called statistically significant and the alternative hypothesis is accepted. The psychologist can conclude that at least one of the population means—either the pre-treatment level of distraction, or the post-treatment level of distraction, or the six-month follow-up level of distraction—probably differs from at least one other population mean. It isn't possible yet to talk about the direction of the difference because which pair(s) differs is still unknown. Discussing the direction of the difference needs to wait until post-hoc tests have been completed.

To report the analysis of variance results in APA format, Dr. King would write $F(2, 6) = 18.72, p < .05$. This gives five pieces of information:

1. What test was done (an *F* test).
2. The degrees of freedom for the numerator ($df_{\text{Treatment}} = 2$) and the denominator ($df_{\text{Residual}} = 6$) of the *F* ratio. From these, it is possible to figure out *n*, how many cases there are in each group by using this equation: $n = \frac{df_{\text{Residual}}}{df_{\text{Treatment}}} + 1$. For the ADHD study, this would be $n = \frac{6}{2} + 1 = 3 + 1 = 4$.
3. The observed value of the test statistic (18.72).
4. What alpha level was used (.05). (This indicates how likely a Type I error is.)
5. Whether the null hypothesis was rejected ($p < .05$) or not ($p > .05$).

How Big Is the Effect?

The next step in interpretation involves how big the effect is. This should be calculated whether or not the null hypothesis is rejected.

- If the null hypothesis is rejected, the researcher needs to calculate the size of the effect to quantify how much impact the independent variable has on the dependent variable. In the ADHD/level of distraction example, the question is to what degree time/treatment affects the level of distraction.
- If the null hypothesis is not rejected, it is still important to calculate the effect size. If the effect of the independent variable on the dependent variable is more than small, that alerts the researcher to the possibility of a Type II error. He or she will likely end up suggesting a replication with a larger sample size.

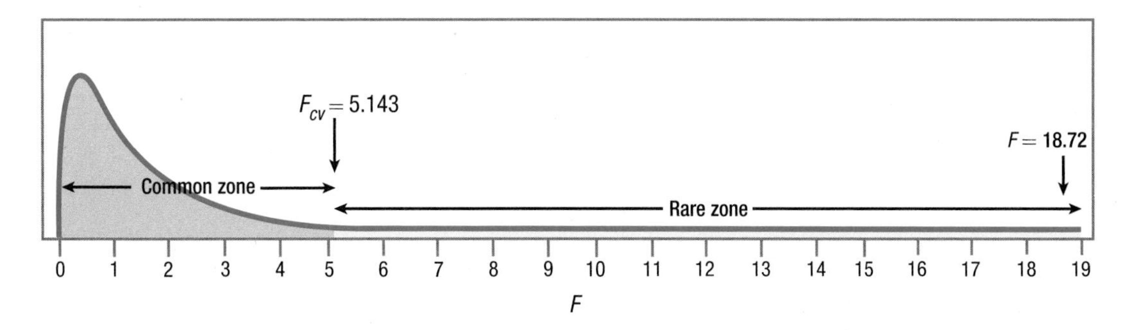

Figure 11.7 Results of ANOVA for Level of Distraction/ADHD Study This sampling distribution shows that the observed *F* value of 18.72 falls in the rare zone. The null hypothesis is rejected and the results are called statistically significant.

Whereas r^2 was used to quantify the size of the treatment effect for a between-subjects, one-way ANOVA, we will use eta squared(η^2) for a one-way, repeated-measures ANOVA. The formula is given in Equation 11.2.

Equation 11.2 Formula for Calculating Eta Squared for the Effect of Treatment for a One-Way, Repeated-Measures ANOVA

$$\eta^2 = \frac{SS_{\text{Treatment}}}{SS_{\text{Total}}} \times 100$$

where η^2 = eta squared for the treatment effect, the percentage of variability in the dependent variable that is explained by the explanatory (treatment) variable

$SS_{\text{Treatment}}$ = sum of squares treatment from the ANOVA summary table

SS_{Total} = sum of squares total from the ANOVA summary table

For the ADHD data, where it is known that $SS_{\text{Treatment}} = 105.17$ and $SS_{\text{Total}} = 430.67$ (see Table 11.5), Dr. King calculated η^2 this way:

$$\eta^2 = \frac{SS_{\text{Treatment}}}{SS_{\text{Total}}} \times 100 = \frac{105.17}{430.67} \times 100 = 0.2442 \times 100 = 24.42\%$$

Eta squared for the treatment effect in the level of distraction data is 24.42%. **Eta squared** tells the percentage of the variability in the dependent variable that is accounted for by the explanatory variable. It can range from 0% (meaning the explanatory variable has no impact on the dependent variable) to 100% (meaning the explanatory variable wholly determines the dependent variable).

For the ADHD/level of distraction data, eta squared shows how well the explanatory variable of treatment status—pre-treatment, post-treatment, or follow-up—predicts outcome as measured by the dependent variable of level of distractibility. If $\eta^2 = 0\%$, then knowing the treatment status doesn't provide any information about a child's level of distraction. If $\eta^2 = 100\%$, then knowing the treatment status precisely predicts a child's level of distraction.

Explaining 100% of the variability never happens and it is all too common to explain only a small percentage of the variability. As a guide to interpreting eta squared, use Cohen's guidelines for r^2:

- $\eta^2 \approx 1\%$ is a small effect.
- $\eta^2 \approx 9\%$ is a medium effect.
- $\eta^2 \approx 25\%$ is a large effect.

Figure 11.8 shows graphically what small, medium, and large effect sizes look like and how much variability they leave unexplained.

As $\eta^2 = 24.42\%$ in the ADHD study, Dr. King can conclude that treatment status has a large effect on the level of distraction. It is now time to find out specifically which group—pre-treatment, post-treatment, or follow-up—has the effect.

Small (≈ 1% of variability explained) | Medium (≈ 9% of variability explained) | Large (≈ 25% of variability explained)

Figure 11.8 Graphic Representations of Cohen's Small, Medium, and Large Effect Sizes Cohen (1988) considers a small effect one that explains about 1% of the variability in a dependent variable. A medium effect explains about 9% and a large effect about 25%. Each figure represents all 100% of the variability that is available to be explained in a dependent variable. The shaded portions represent the percentage of variability that is explained by the different effect sizes and the unshaded portions are left unexplained.

Where Are the Effects and What Is Their Direction?

Once again, the Tukey *HSD* will be used as the post-hoc test. *HSD* stands for "honestly significant difference" and any two sample means that differ by at least the *HSD* value represent two populations that are honestly likely to have different means. The formula for the *HSD* value is given in Equation 11.3. Please remember, post-hoc tests are only performed when the results of the ANOVA are statistically significant!

Equation 11.3 Formula for Calculating Tukey *HSD* Values for Post-hoc Tests for One-Way, Repeated-Measures ANOVA

$$HSD = q\sqrt{\frac{MS_{\text{Residual}}}{n}}$$

where HSD = HSD value being calculated

q = q value from Appendix Table 5, where k = the number of groups and $df = df_{\text{Residual}}$

MS_{Residual} = mean square residual, from the ANOVA summary table

n = number of cases per group

Dr. King needs three values in order to apply Equation 11.3: q, MS_{Residual}, and n.

- To find the q value, Dr. King uses Appendix Table 5. $k = 3$ as there are three groups in the ADHD study, so he looks in that column. The row is based on the residual degrees of freedom, which are 6 for the ADHD data. At the intersection of the column for $k = 3$ and the row for $df = 6$, he finds the q value, $q = 4.34$.
- In the ANOVA summary table (Table 11.3 and Table 11.5), he finds $MS_{\text{Residual}} = 2.81$.
- There are four cases in each group, so $n = 4$.

Given those three values, here are Dr. King's calculations to find the *HSD* value using Equation 11.3:

$$HSD = q\sqrt{\frac{MS_{\text{Residual}}}{n}} = 4.34\sqrt{\frac{2.81}{4}} = 4.34\sqrt{0.7025} = 4.34 \times 0.8382 = 3.6378 = 3.64$$

The Tukey *HSD* value is 3.64. Any pair of sample means for the ADHD/level of distraction study that differs by this much or more has a statistically significant difference. It doesn't matter which mean is subtracted from which in order to compare the difference to the *HSD* value, so Dr. King has arranged the comparisons to give positive numbers. The post-hoc test results are summarized in Table 11.6 and explained in detail below.

TABLE 11.6 Post-Hoc Test Results for Mean Level of Distraction/ADHD Study

	Pre-Treatment	End of Treatment	6-Month Follow-Up
Pre-treatment	—	7.25*	3.50
End of treatment	7.25*	—	3.75*
6-month follow-up	3.50	3.75*	—

The values in the cells represent mean differences between pairs of means. Values followed by an asterisk (*) represent statistically significant differences. The dashes (—) represent comparisons that can't be made. This table contains redundant information as the values below the diagonal are the same as those above.

- Pre-treatment mean level of distraction vs. post-treatment mean level of distraction:
  - $22.25 - 15.00 = 7.25$.
  - $7.25 \geq 3.64$, so Dr. King concludes that there is a statistically significant difference between the level of distraction from pre-treatment to post-treatment.
  - The sample means can be used to comment on the direction of the difference: the *reduction* in level of distraction from pre-treatment to post-treatment is a statistically significant one. Phrased simply: treatment works.
- Pre-treatment mean level of distraction vs. six-month follow-up mean level of distraction:
  - $22.25 - 18.75 = 3.50$.
  - $3.50 < 3.64$, so this difference is not statistically significant. For this population of children with ADHD, there's not enough evidence to conclude that any difference exists in the level of distraction between the start of treatment and six months after treatment. Phrased simply: there's not enough evidence to show that treatment has a long-term impact.
  - However, be aware of two things: (1) the difference was close to being statistically significant, and (2) the sample size was small. These values alert Dr. King to the risk of making a Type II error, so he's going to end up recommending replication with a larger sample size.
- Post-treatment mean level of distraction vs. six-month follow-up mean level of distraction:
  - $18.75 - 15.00 = 3.75$.
  - $3.75 \geq 3.64$, so this difference is statistically significant. For this population of children with ADHD, one can conclude that the level of distraction is statistically

significantly higher six months after treatment than it is at the end of treatment. Phrased simply: some relapse seems to occur when treatment is over.

Putting It All Together

Here is Dr. King's four-point interpretation in which he (1) states what was done, (2) presents the main findings, (3) explains the results, and (4) makes suggestions for future research:

> This study was conducted to evaluate the long-term effectiveness of individual behavior therapy in treating children with ADHD. Four children with ADHD had their levels of distraction assessed before treatment started ($M = 22.25$), at the end of treatment ($M = 15.00$), and again six months later ($M = 18.75$). Using a one-way, repeated-measures ANOVA, there was a large and statistically significant effect on distractibility as a function of when the outcome was assessed [$F(2, 6) = 18.72$, $p < .01$]. Post-hoc tests showed that the mean level of distraction significantly improved from pre-treatment to the end of treatment. Unfortunately, a statistically significant amount of relapse occurred over the next six months, with the mean distractibility level getting worse from the end of treatment to the six-month follow-up. The change in the mean level of distraction from pre-treatment to six-month follow-up was not statistically significant, meaning there was insufficient evidence to show that ADHD symptoms were any better six months after treatment than they were before treatment started. Individual behavior therapy appears effective in treating ADHD, but the results don't remain fully in effect for six months when treatment is over. However, the present study only had four participants and it is possible that the small sample size prevented some effects from being found. To get a better sense of the long-term impact of individual behavior therapy on distractibility, the study should be replicated with a larger sample size.

Worked Example 11.2 For practice interpreting a one-way, repeated-measures ANOVA, let's use the results from the back pain surgery study. In that study, Dr. Agosto, an orthopedic surgeon, took 30 people with chronic back pain and divided them into groups of three that were matched on age, sex, years of pain, and degree of impairment. One of the three people in each group was randomly assigned to receive real surgery, one sham surgery, and one no surgery. Each participant then received physical therapy for six months, at which point his or her quality of life was measured (with higher scores indicating better functioning). The mean was 51.00 for the 10 surgery patients, 48.00 for the sham surgery patients, and 45.00 for the physical therapy only patients. F_{cv} was 3.555 and the ANOVA summary table for the one-way, repeated-measures ANOVA is presented here again as Table 11.7.

TABLE 11.7 Representation of ANOVA Summary Table for Back Surgery/Quality-of-Life Data

Source of Variability	Sum of Squares	Degrees of Freedom	Mean Square	F ratio
Subjects	2,193.33	9		
Treatment	180.00	2	90.00	31.91
Residual	50.67	18	2.82	
Total	2,424.00	29		

Was the Null Hypothesis Rejected?

The decision rule was:

- If $F \geq 3.555$, reject H_0.
- If $F < 3.555$, fail to reject H_0.

The observed value of F (see Table 11.5) is 31.91, which is greater than 3.555. Figure 11.9 shows how the F value landed in the rare zone of the sampling distribution. As a result, the null hypothesis is rejected, the alternative hypothesis is accepted, and the results are called statistically significant. This means it is concluded that at least one of these three treatments is different from at least one other for this population of patients with back pain. In APA format, the results are written $F(2, 18) = 31.91, p < .05$.

The three means (51.00, 48.00, and 45.00) all seemed close to one another and Dr. Agosto had speculated that the ANOVA would not be statistically significant. The fact that the results are statistically significant shows the power of a repeated-measures design. By isolating the variability due to treatment from variability due to individual differences, it is easier to reject the null hypothesis with a repeated-measures design. If one had, erroneously, used a one-way, between-subjects ANOVA to analyze the ADHD data, the results would not have been statistically significant. Repeated-measures tests are more powerful than between-subjects tests.

Repeated-measures tests are more powerful than between-subjects tests.

How Big Is the Effect?

To determine effect size, η^2, Dr. Agosto used Equation 11.2. Eta squared quantifies the percentage of variability in quality of life (the dependent variable) that is accounted for by which treatment cases receive (the independent variable.) To calculate eta squared, a researcher needs to know sum of squares treatment and sum of squares total. Looking in Table 11.7, Dr. Agosto finds $SS_{\text{Treatment}} = 180.00$ and $SS_{\text{Total}} = 2{,}424.00$. Substituting those into Equation 11.2 gives

$$\eta^2 = \frac{SS_{\text{Treatment}}}{SS_{\text{Total}}} \times 100 = \frac{180.00}{2{,}424.00} \times 100 = 0.0743 \times 100 = 7.43\%$$

An η^2 of 7.43% qualifies as a medium effect.

Where Are the Effects and What Is Their Direction?

The ANOVA was statistically significant, so Dr. Agosto will need post-hoc tests to pinpoint where the effect occurred. To use Equation 11.3 to calculate the Tukey *HSD*, she will need to know q, MS_{Residual}, and n.

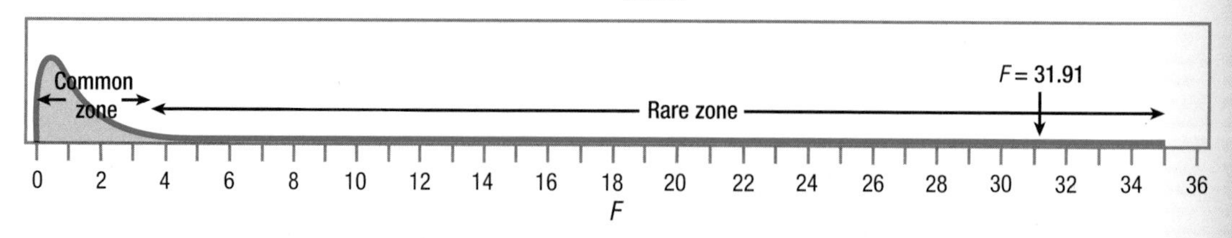

Figure 11.9 ANOVA Results for the Effect of Surgery and Physical Therapy on Back Pain The observed value of F, 31.91, falls in the rare zone of the sampling distribution. The null hypothesis is rejected and the results are called statistically significant. At least one treatment group had an outcome statistically different from at least one other.

- With $k = 3$ and $df_{\text{Residual}} = 18$, $q = 4.70$. (*Note:* Dr. Agosto had set alpha at .01 for the repeated-measures ANOVA to reduce the risk of Type I error. To be consistent, she is using the Appendix 5 q table with $\alpha = .01$.)
- $MS_{\text{Residual}} = 2.82$.
- $n = 10$.

Now she can calculate Tukey *HSD* using Equation 11.3:

$$HSD = q\sqrt{\frac{MS_{\text{Residual}}}{n}} = 4.70\sqrt{\frac{2.82}{10}} = 4.70\sqrt{0.2820} = 4.70 \times 0.5310$$
$$= 2.4957 = 2.50$$

Dr. Agosto has three comparisons to make. Any pairs of means that differ by at least 2.50 points have a statistically significant difference.

- Group 1 vs. Group 2: Real surgery plus physical therapy vs. sham surgery plus physical therapy.
    - $51.00 - 48.00 = 3.00$.
    - $3.00 \geq 2.50$, so the difference is statistically significant. For this population of people with back pain, Dr. Agosto can conclude that people who receive real surgery have a statistically better outcome, in terms of quality of life, than do people who receive sham surgery.
- Group 1 vs. Group 3: Real surgery plus physical therapy vs. physical therapy alone.
    - $51.00 - 45.00 = 6.00$.
    - $6.00 \geq 2.50$, so the difference is statistically significant. For this population of people with back pain, Dr. Agosto can conclude that people who receive real surgery and physical therapy have a statistically better outcome, in terms of quality of life, than do people who receive physical therapy alone.
- Group 2 vs. Group 3: Sham surgery plus physical therapy vs. physical therapy alone.
    - $48.00 - 45.00 = 3.00$.
    - $3.00 \geq 2.50$, so the difference is statistically significant. For this population of people with back pain, Dr. Agosto can conclude that people who receive sham surgery plus physical therapy have a statistically better outcome, in terms of quality of life, than do people who receive physical therapy alone.

There's a lot here to keep track of, so Dr. Agosto uses inequality signs to keep the effects straight: Real surgery > Sham surgery > Physical therapy.

Putting It All Together

Here is Dr. Agosto's four-point interpretation:

I conducted a study investigating how effective surgery is as a treatment for chronic back pain. Thirty people with chronic back pain were matched into groups of three on the basis of age, sex, length of illness, and severity of illness. One member of each group was randomly assigned to receive real surgery, one to receive sham surgery, and one to receive no surgery. Each participant then received six months of physical therapy. Using a one-way, repeated-measures ANOVA, there was a small to modest, statistically significant effect for type of treatment on quality of life [$F(2, 18) = 31.91$, $p < .05$]. People receiving real surgery had a statistically higher quality of life ($M = 51.00$) than people receiving sham surgery ($M = 48.00$), and people receiving sham surgery had a statistically better quality of life than people just receiving physical therapy ($M = 45.00$). These results show that though there is a placebo effect for having surgery for back pain, there is a benefit to receiving surgery that is above and beyond the placebo effect. Future research should investigate which types of back problems are helped by which type of treatment.

Practice Problems 11.3

Apply Your Knowledge

11.10 Given $F_{cv} = 2.866$, $df_{\text{Treatment}} = 3$, $df_{\text{Residual}} = 36$, $F = 4.36$, and $\alpha = .05$, (a) write the results in APA format. (b) Are the results statistically significant?

11.11 Given $SS_{\text{Treatment}} = 35.76$ and $SS_{\text{Total}} = 124.64$, (a) calculate η^2. (b) Is this considered a small, medium, or large effect?

11.12 Given $k = 5$, $n = 16$, $MS_{\text{Residual}} = 12.98$, and $\alpha = .05$, (a) find q and (b) calculate Tukey's *HSD*.

11.13 A software designer was curious as to how comfort with technology varied across generations. He obtained 10 families and brought in a teenager, a parent, and a grandparent from each family. Then he had each of them, individually, install a new piece of software. For each one, he timed how long it took to click on the "Agree to Terms of Installation" button. The mean time for the teens was 1.50 seconds, for the parents 4.50 seconds, and for the grandparents 20.00 seconds. No assumptions were violated and he conducted a one-way, repeated-measures ANOVA with $\alpha = .05$. He calculated $\eta^2 = 94.53\%$ and $HSD = 2.42$. Below is the one-way, repeated-measures ANOVA summary table. Write a four-point interpretation for the results.

Source of Variability	Sum of Squares	Degrees of Freedom	Mean Square	*F* ratio
Subjects	33.03	9		
Treatment	1,971.67	2	985.84	219.08
Residual	81.05	18	4.50	
Total	2,085.75	29		

Application Demonstration

In the most famous example of classical conditioning, a conditioned stimulus (a bell) was repeatedly paired with an unconditioned stimulus (meat) until a dog learned to salivate to the sound of the bell. Salivation to the bell is called a conditioned response. The process of then teaching the dog *not* to salivate to the bell is called extinction. In extinction, the dog hears the bell but doesn't see the meat and eventually the bell no longer elicits salivation. If a researcher then waits a few days after extinction has occurred and rings the bell around the dog again, the dog will salivate to the sound of the bell again. This is called spontaneous recovery.

Let's imagine a researcher who wanted to see if extinction and spontaneous recovery occurred in humans with naturally occurring classical conditioning.

That is, he wanted to see if these phenomena occurred when the conditioned response hadn't been taught in a laboratory, but had naturally developed.

Dr. Brian had noticed that many people salivate when they see a lemon, so he decided to use this conditioned reaction. (Try this thought experiment: Think of someone cutting open a juicy yellow lemon, right in front of you. You watch as she takes a slice of lemon, bites into it, and grimaces a little at the sourness of it. Just thinking of this, do you feel salivation happening in your mouth? That's a conditioned response.)

Dr. Brian brought 35 participants from the experiment participation pool into his lab, one at a time, and measured how many grams of saliva each one produced in 2 minutes (as a baseline). How is salivation measured? Weigh some cotton balls, put them in a mouth for 2 minutes, and then weigh them again. The gain in weight reflects the amount of salivation.

After this, Dr. Brian had the participants smell a lemon for a 2-minute period, during which salivation was measured again. This was the first extinction trial. There was then a 3-minute break, after which the second extinction trial occurred (sniffing a lemon but not tasting it). This continued for seven more extinction trials. Salivation during extinction wasn't measured again until the tenth (and last) extinction trial. The next day, participants came back to the lab and, for the last time, sniffed a lemon for 2 minutes while salivation was measured. This was the spontaneous recovery trial.

The mean grams of saliva for the four trials are shown in Figure 11.10.

- It looks as if there is a conditioned response of salivation to a lemon. The amount of saliva increased from the first baseline measurement to when a lemon is smelled on the first extinction trial.
- It looks as if extinction works. The amount of saliva measured decreases from the first extinction trial to the last trial.
- It looks as if some spontaneous recovery occurs. The amount of saliva increases from the last extinction trial to the next day.

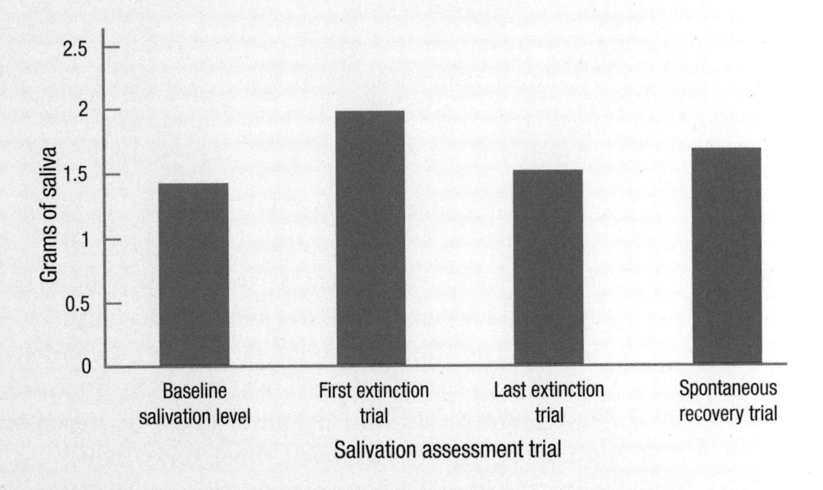

Figure 11.10 Extinction and Spontaneous Recovery of a Conditioned Response in Humans This figure appears to show that there's a conditioned response to smelling a lemon, that this conditioned response can be extinguished, and that spontaneous recovery occurs.

Of course, Dr. Brian can't know if these effects indicate an effect in the larger population until he conducts a statistical test. The first question is, "What test?"

Step 1 Pick a Test. The same participants are measured at four points in time, so the samples are dependent samples. There are more than two groups being compared on a ratio-level variable. Comparing the means of four dependent samples calls for a one-way, repeated-measures ANOVA.

Step 2 Check the Assumptions. As is common in psychological research, the random samples assumption was violated. This did not concern Dr. Brian much, because repeated-measures ANOVA is robust to violations of this assumption and because he has no reason to believe that the response of these participants to a lemon would be much different from anyone else's response to a lemon. The other two assumptions, independence of observations and normality, were not violated. So, Dr. Brian can proceed.

Step 3 List the Hypotheses.

$$H_0: \mu_{\text{Baseline}} = \mu_{\text{Lemon}} = \mu_{\text{Extinction}} = \mu_{\text{SpontaneousRecovery}}.$$

H_1: At least one of the four population means differs from at least one other population mean.

Step 4 Set the Decision Rule. First, calculate the degrees of freedom:

$$df_{\text{Subjects}} = n - 1 = 35 - 1 = 34$$

$$df_{\text{Treatment}} = k - 1 = 4 - 1 = 3$$

$$df_{\text{Residual}} = df_{\text{Subjects}} \times \text{df}_{\text{Treatment}} = 34 \times 3 = 102$$

$$df_{\text{Total}} = N - 1 = 140 - 1 = 139$$

The numerator degrees of freedom, $df_{\text{Numerator}}$, for the F ratio is $df_{\text{Treatment}}$, which is 3, and the denominator degrees of freedom, $df_{\text{Denominator}}$ is df_{Residual}, which is 102. In order to find the critical value of F, F_{cv}, Dr. Brian has to decide how willing he is to make a Type I error. He can accept having a 5% chance of a Type I error, so $\alpha = .05$. Looking in Appendix Table 4 for the column with 3 degrees of freedom and the row with 102 degrees of freedom, Dr. Brian discovers that there is no such row. In such a situation, the row with the degrees of freedom that is closest without going over is used. (Remember, this is called *The Price Is Right* rule.) This is the row with 100 degrees of freedom. $F_{cv} = 2.696$ with $\alpha = .05$. The decision rule is:

- If $F \geq 2.696$, reject H_0.
- If $F < 2.696$, fail to reject H_0.

Step 5 Calculate the Test Statistic. Using the formulas from the appendix to this chapter, Dr. Brian calculated $SS_{\text{Subjects}} = 83.25$, $SS_{\text{Treatment}} = 6.28$, $SS_{\text{Residual}} = 24.35$, and $SS_{\text{Total}} = 113.88$. With the sums of squares and with the degrees of freedom calculated in Step 4, Dr. Brian can complete the ANOVA summary table seen in Table 11.8 by following the template in Table 11.2.

TABLE 11.8 ANOVA Summary Table for Extinction and Spontaneous Recovery of a Naturally Occurring Conditioned Response in Humans

Source of Variability	Sum of Squares	Degrees of Freedom	Mean Square	*F* ratio
Subjects	83.25	34		
Treatment	6.28	3	2.09	8.71
Residual	24.35	102	0.24	
Total	113.88	139		

Step 6 Interpret the Results. *Was the null hypothesis rejected?* The first interpretation question is answered by implementing the decision rule. The observed *F* ratio of 8.71 is greater than or equal to the critical value of *F*, 2.696, so the null hypothesis is rejected. The alternative hypothesis is accepted—it is probably true that at least one of the population means differs from at least one other population mean. Dr. Brian knows that there is a difference in the level of salivation between at least two of these conditions: baseline, the first extinction trial, the last extinction trial, or the spontaneous recovery trial. There may be more than two means that differ. To learn which mean or means differ will have to await post-hoc tests. For now, though, Dr. Brian can write the results in APA format:

$$F(3, 102) = 8.71, p < .05$$

How big is the effect? The second interpretation question for one-way, repeated-measures ANOVA is answered by using Equation 11.2 to calculate η^2. Eta squared provides information about how much of the variability in the amount of salivation is explained by the four different conditions (baseline, first extinction trial, last extinction trial, and spontaneous recovery trial). Below are the calculations, resulting in $\eta^2 = 5.51\%$. Following Cohen's guidelines, Dr. Brian considers explaining 5.51% of the effect in salivation by the treatment condition to be a small to medium effect:

$$\eta^2 = \frac{SS_{\text{Treatment}}}{SS_{\text{Total}}} \times 100 = \frac{6.28}{113.88} \times 100 = .0551 \times 100 = 5.51\%$$

Where are the effects and what is their direction? This is the question Dr. Brian has been waiting to answer. With four groups, there are six possible comparisons, as shown in Table 11.9. In fact, arranging the comparisons this way is very helpful for a researcher thinking about the results, because different comparisons address different questions.

Dr. Brian has three questions he wants to address:

1. Is there a conditioned response of salivation to the sight and smell of a lemon? This is addressed by Comparison 1, which compares baseline salivation to the first time the participants smelled a lemon, which was the first extinction trial.

TABLE 11.9 Possible Comparisons Among the Four Groups in the Lemon Salivation Extinction Study

	First Lemon Extinction Trial	Last Lemon Extinction Trial	Spontaneous Recovery Trial
Baseline Salivation Trial	#1	#2	#3
First Extinction Trial		#4	#5
Last Extinction Trial			#6

When there are four groups, there are six unique paired comparisons.

2. Does extinction occur? This question is broken down into two subquestions. Is there a decrease in salivation from the first extinction trial to the last extinction trial (Comparison 4)? If there is a decrease in salivation, does it go all the way back to baseline levels (Comparison 2)?
3. Does spontaneous recovery occur? This also has two subquestions. Is there more salivation on the spontaneous recovery trial than at the last extinction trial (Comparison 6)? If so, does the increase go all the way back to the level of salivation at the first extinction trial (Comparison 5)?

Note that Comparison 3 is not mentioned. Dr. Brian is not interested in comparing the amount of spontaneous recovery salivation to baseline salivation, so he'll ignore it.

In order to complete the comparisons he has planned, Dr. Brian needs to find the *HSD* value for the Tukey post-hoc test. The first step in using Equation 11.3 is to find the q value in Appendix Table 5. With $k = 4$ and $df_{\text{Residual}} = 102$, $q = 3.74$. (Note that *The Price Is Right* rule was applied.) After he retrieves $MS_{\text{Residual}} = 0.24$ from the ANOVA summary table (Table 11.8) and notes that as he had 35 participants $n = 35$, he can complete Equation 11.3:

$$HSD = q\sqrt{\frac{MS_{\text{Residual}}}{n}}$$

$$= 3.74\sqrt{\frac{0.24}{35}}$$

$$= 3.74\sqrt{.0069}$$

$$= 3.74 \times 0.0831$$

$$= 0.3108$$

$$= 0.31$$

Any pair of means that differs by at least 0.31 grams is a statistically significant difference. Table 11.10 shows the differences between the means, with the statistically significant ones circled. Dr. Brian can now answer his three questions.

TABLE 11.10 Lemon Salivation Study: Differences Between Sample Means

	First Lemon Extinction Trial ($M = 1.99$)	Last Lemon Extinction Trial ($M = 1.53$)	Spontaneous Recovery Trial ($M = 1.68$)
Baseline Salivation Trial ($M = 1.44$)	1.99 – 1.44 = (0.55)	1.53 – 1.44 = 0.09	Not calculated
First Lemon Extinction Trial ($M = 1.99$)		1.99 – 1.53 = (0.46)	1.99 – 1.68 = (0.31)
Last Lemon Extinction Trial ($M = 1.53$)			1.68 – 1.53 = 0.13

Each cell presents the difference between two sample means. Circled differences are statistically significant ones.

1. Is there a conditioned response of salivation to the sight and smell of a lemon? Yes. Smelling a lemon causes a statistically significant increase in mean salivation from baseline.
2. Does extinction occur? Yes. There is a statistically significant decrease in mean salivation from the first extinction trial to the last. In fact, by the last extinction trial, the mean amount of salivation can't be differentiated from baseline salivation, suggesting that extinction was complete. (This is one of the rare instances where nonsignificant results are informative.)
3. Does spontaneous recovery occur? There is not much evidence for spontaneous recovery. There is not statistically significantly more salivation at the spontaneous recovery trial than there was on the last extinction trial.

Putting it all together.

This study investigated whether humans show a naturally occurring response of salivation to the smell of a lemon and whether such a conditioned response can be extinguished and then spontaneously recovered. Thirty-five volunteers measured their baseline salivation and then had 10 extinction trials during which they held and smelled a lemon. Amount of salivation to a lemon was measured on the first and last extinction trials and again, at a spontaneous recovery trial, 24 hours later.

There were statistically significant differences among the means $F(3, 102) = 8.71$, $p < .05$. Post-hoc tests showed that there was a naturally occurring response of salivation to the sight/smell of a lemon and that this response could be extinguished. In this way a naturally occurring conditioned response behaves like an experimentally induced conditioned response. However, the naturally occurring conditioned response did not behave like an experimentally induced conditioned response in terms of spontaneous recovery. Whether naturally occurring conditioned responses really don't show spontaneous recovery, or not a long enough period of time was given for spontaneous recovery is a fruitful area for future research.

DIY

Do you have three grocery stores in your area: maybe a big chain grocery; a smaller, locally owned one; and a discount super-store? Any three will do. Make a list of four or five items you buy regularly—every week or month, such as milk or bread. Go to each store and record the prices of those items. Make sure you find exactly the same items (brand and size/quantity) at each store. Then calculate the sums of squares (following the directions in the appendix to this chapter), complete an ANOVA summary table, and interpret your results. In terms of price, does it matter where you shop?

SUMMARY

Identify what repeated-measures ANOVA does.

- Within-subjects, one-way ANOVA, called repeated-measures ANOVA, is a one-way ANOVA for dependent samples. It divides variability in a set of scores into two factors: (1) variability due to individual differences and (2) variability due to treatment. Thus, repeated-measures ANOVA provides a more pure measure of the effect of treatment than a paired-samples *t* test because it separates out variability due to individual differences.

Complete a one-way, repeated-measures ANOVA.

- Check the assumptions (random samples, independence of observations, normality) and form the null and alternative hypotheses.
- Calculate degrees of freedom, find the critical value of *F*, and set the decision rule.
- Given sums of squares, complete an ANOVA summary table.

Interpret a one-way, repeated-measures ANOVA.

- Determine if the null hypothesis was rejected and write the results in APA format.
- Calculate an effect size, eta squared.
- If the null hypothesis was rejected, complete post-hoc tests (Tukey *HSD*) to determine where effects are and what their direction is.
- Write a four-point interpretation (What was done? What was found? What does it mean? What suggestions are there for future research?).

KEY TERMS

eta squared (η^2) – an effect size that calculates the percentage of variability in the dependent variable accounted for by the independent variable.

repeated-measures ANOVA – a statistical test used to compare three or more dependent samples on an interval- or ratio-level–dependent variable; also called within-subjects ANOVA, dependent-samples ANOVA, or related-samples ANOVA.

CHAPTER EXERCISES

Review Your Knowledge

11.01 Between-subjects, one-way ANOVA extends the ____ to situations in which there are more than two independent samples.

11.02 Within-subjects, one-way ANOVA is like between-subjects, one-way ANOVA but is for ____ samples.

11.03 ____ ANOVA is the common name for within-subjects, one-way ANOVA.

11.04 Repeated-measures ANOVA is used to compare the ____ of ____ or more dependent samples.

11.05 The data used in repeated-measures ANOVA are arranged with cases on ____ and levels of the explanatory variable in ____.

11.06 In repeated-measures ANOVA, the cases should be in the ____ order in each cell.

11.07 Because a repeated-measures ANOVA is more ______ than a between-subjects, one-way ANOVA, there is a ____ likelihood of being able to reject the null hypothesis.

11.08 What is called the between-groups effect in between-subjects, one-way ANOVA is called the _____ effect in repeated-measures ANOVA.

11.09 Because variability due to subjects is removed from within-groups variability in repeated-measures ANOVA, the *numerator/ denominator* of the *F* ratio is smaller than it is in between-subjects, one-way ANOVA.

11.10 If r^2 is calculated for a paired-samples *t* test, it will *over / under*estimate the effect size.

11.11 When comparing the means of two dependent samples, the author advocates using ____, not ____.

11.12 The random samples assumption says that the samples in a repeated-measures ANOVA are ____ samples from the population to which one wishes to generalize the results.

11.13 The random samples assumption is ____ to violation.

11.14 If cases within a sample influence each other's scores on the dependent variable, then the ____ assumption is ____.

11.15 It is often assumed that the dependent variable, if it is of a psychological attribute, is ____ in the ____.

11.16 "All population means are equal." This is the ____ hypothesis.

11.17 $\mu_1 \neq \mu_2 \neq \mu_3$ *is / is not* an accurate statement of the alternative hypothesis.

11.18 The abbreviation for the critical value of *F* is ____.

11.19 The null hypothesis is rejected if *F* is ____ the critical value of *F*.

11.20 The null hypothesis is not rejected if *F* is ____ the critical value of *F*.

11.21 If the results are called statistically significant, then the null hypothesis *was / was not* rejected.

11.22 If the conclusion of a repeated-measures ANOVA is that there is no evidence of a difference among population means, then the null hypothesis *was / was not* rejected.

11.23 If one was forced to accept the alternative hypothesis, then the null hypothesis *was / was not* rejected.

11.24 Values of F_{cv} depend on the numerator and denominator ____.

11.25 *df*____ are the numerator degrees of freedom for a one-way, repeated-measures ANOVA and *df*____ are the denominator degrees of freedom.

11.26 *df*____ and *df*____ are not needed to find F_{cv} for a one-way, repeated-measures ANOVA.

11.27 To apply Equation 11.1 to calculate all 4 degrees of freedom, one needs to know ____, ____, and ____.

11.28 To calculate *N*, one multiplies together ____ and ____.

11.29 If one adds together df_{Subjects}, $df_{\text{Treatment}}$, and df_{Residual}, this equals ____.

11.30 One-way, repeated-measures ANOVA divides df_{Total} into subcomponents for two different sources of ____.

11.31 The effect of individual differences in repeated-measures ANOVA is called the effect of ____ in the ANOVA summary table.

11.32 After df_{Subjects} and $df_{\text{Treatment}}$ are removed from df_{Total}, what is left is called ____.

11.33 If the observed value of *F* falls in the ____ zone of the sampling distribution, the null hypothesis is rejected.

11.34 If *F* falls on the line that separates the rare zone from the common zone, the null hypothesis *is / is not* rejected.

11.35 The first column in an ANOVA summary table lists the sources of ____.

11.36 If one knows SS_{Subjects}, $SS_{\text{Treatment}}$, and SS_{Residual}, one can calculate SS____.

11.37 To calculate a mean square, one divides a ____ by its ____.

11.38 The only mean squares one needs to calculate for a one-way, repeated-measures ANOVA are MS____ and MS____.

11.39 The *F* ratio for one-way, repeated-measures ANOVA is ____ divided by ____.

11.40 The first question to be addressed in an interpretation of a one-way, repeated-measures ANOVA is ____.

11.41 By implementing the ____ from Step 4, one can determine if the results of a statistical test are statistically significant.

11.42 If one knows that the null hypothesis for a one-way, repeated-measures ANOVA was rejected, one *does / does not* know which pairs of sample means had statistically significant differences.

11.43 Results were reported in APA format as $F(2, 12) = 7.89$, $p < .05$. The null hypothesis *was / was not* rejected.

11.44 Results were reported in APA format as $F(2, 22) = 0.89$, $p > .01$. The alpha level was ____.

11.45 If the null hypothesis is not rejected, one *should / should not* calculate an effect size.

11.46 Eta squared provides information about the size of the ____.

11.47 A researcher ends up suggesting replication of study with a larger sample size. The null hypothesis probably *was / was not* rejected.

11.48 If a researcher suggests replication with a larger sample size, then he or she is probably worried about Type ____ error.

11.49 The effect size used with one-way, repeated-measures ANOVA is ____.

11.50 Eta squared tells the percentage of variability in the ____ variable that is explained by the ____ variable.

11.51 Eta squared, like ____, ranges from 0% to100%.

11.52 The closer eta squared is to ____%, the bigger the effect.

11.53 If the treatment effect explained 10% of the variability in a set of scores, Cohen would consider this a ____ effect.

11.54 The ____ is a post-hoc test for use with one-way, repeated-measures ANOVA.

11.55 Post-hoc tests for ANOVA should only be calculated when the null hypothesis is ____.

11.56 To calculate an *HSD* value, one needs to find a ____ value in Appendix Table 5.

11.57 If a pair of sample means differ by less than the *HSD* value, the difference is considered ____ significant.

11.58 If a difference in a post-hoc test is statistically significant, the two sample means can be used to determine the ____ of the difference for the two population means.

Apply Your Knowledge

Selecting the appropriate statistical test. (For 11.59–11.62, select from the single-sample z test; single-sample t test; independent-samples t test; paired-samples t test; between-subjects, one-way ANOVA; and one-way, repeated-measures ANOVA.)

11.59 People who traveled between Philadelphia and New York City by different vehicles (train, car, bus, or plane) were surveyed to see how pleasant the experience had been. Pleasantness was measured on an interval scale. What statistical test should be used to analyze these data to see if different modes of travel were associated with different mean levels of pleasantness?

11.60 First-year college students were surveyed about how much they liked their roommates (a) within five minutes of meeting them, (b) after the first week of classes, and (c) at the end of the semester. An interval measure of liking was used. What statistical test should be used to see if the mean degree of liking changed over the course of the semester?

11.61 When making purchases with cash, some people drop their pennies in the "penny cup" next to the cash register and some don't. A social psychologist wondered if those who did were more altruistic. She obtained a sample of people who dropped their pennies into penny cups and a sample who didn't. To each person, she administered an interval-level measure of altruism. What statistical test should she use to see if the groups differ on the mean level of altruism?

11.62 A recreational therapist knows the U.S. population mean and standard deviation for scores on an interval-level risk-taking scale. She obtains a random sample of people who enjoy riding roller coasters at amusement parks and has them complete the risk-taking scale. What statistical test should she use to see if roller coaster riders differ in the mean level of sensation-seeking from the general population?

Checking the assumptions for a one-way, repeated-measures ANOVA and deciding if it is OK to proceed with the test

11.63 The dean of retention at a college wanted to find out if academic problems or social problems caused students to drop out at her college. She took a random sample of first-year students at her college and, on the basis of SAT scores, matched them into groups of three. One member of each group was then randomly assigned to be in (a) the control group, (b) the academic enhancement group, or (c) the social enhancement group. Nothing was done to the control participants. The academic enhancement participants met together as a group 10 times over the course of the semester, to cover study skills, time management, test anxiety, and so on. The social enhancement participants also met together as a group 10 times over the course of the semester, though the focus was on social skills, dating, alcohol safety, and the like. At the end of the semester, she compared the three group means on the interval-level Adjustment to College Inventory.

a. For each of the three assumptions for a one-way, repeated-measures ANOVA, determine if it was violated.

b. Decide if the dean can proceed with the one-way, repeated-measures ANOVA.

11.64 A clinical psychologist wanted to compare three treatments for generalized anxiety disorder (GAD). She put an ad in the local paper to find people with GAD. She matched them into groups of three, based on severity of symptoms, and randomly assigned each of the matched cases to one of the three treatments. Treatment was administered individually. Outcome was assessed at the end of treatment using an interval-level measure of generalized anxiety.

a. For each of the three assumptions for a one-way, repeated-measures ANOVA, determine if it was violated.

b. Decide if the psychologist can proceed with the one-way, repeated-measures ANOVA.

Stating the hypotheses for a one-way, repeated-measures ANOVA

11.65 **a.** What is the null hypothesis for a one-way, repeated-measures ANOVA?

b. What is the alternative hypothesis for a one-way, repeated-measures ANOVA?

11.66 **a.** What is the null hypothesis for Exercise 11.64?

b. What is the alternative hypothesis for Exercise 11.64?

Calculating degrees of freedom for one-way, repeated-measures ANOVA

11.67 If $k = 3$, $n = 8$, and $N = 24$, what are df_{Subjects}, $df_{\text{Treatment}}$, df_{Residual}, and df_{Total}?

11.68 If $k = 5$, $n = 20$, and $N = 100$, what are df_{Subjects}, $df_{\text{Treatment}}$, df_{Residual}, and df_{Total}?

11.69 If $k = 4$ and $n = 15$, what are df_{Subjects}, $df_{\text{Treatment}}$, df_{Residual}, and df_{Total}?

11.70 If $k = 3$ and $n = 30$, what are df_{Subjects}, $df_{\text{Treatment}}$, df_{Residual}, and df_{Total}?

Setting the decision rule for one-way, repeated-measures ANOVA (assume $\alpha = .05$)

11.71 Given $df_{\text{Treatment}} = 2$ and $df_{\text{Residual}} = 10$, (a) what is F_{cv}? (b) Draw the sampling distribution of F, being sure to label the common zone, the rare zone, and F_{cv}.

11.72 Given $df_{\text{Treatment}} = 3$ and $df_{\text{Residual}} = 36$, (a) what is F_{cv}? (b) Draw the sampling distribution of F, being sure to label the common zone, the rare zone, and F_{cv}.

11.73 If $df_{\text{Treatment}} = 3$ and $df_{\text{Residual}} = 15$, state the decision rule.

11.74 If $df_{\text{Treatment}} = 2$ and $df_{\text{Residual}} = 76$, state the decision rule.

Calculating F

11.75 Given the following, complete an ANOVA summary table for a one-way, repeated-measures ANOVA: $df_{\text{Subjects}} = 11$, $df_{\text{Treatment}} = 2$, $df_{\text{Residual}} = 22$, $df_{\text{Total}} = 35$, $SS_{\text{Subjects}} = 137.50$, $SS_{\text{Treatment}} = 48.48$, $SS_{\text{Residual}} = 115.50$, and $SS_{\text{Total}} = 301.48$.

11.76 Given the following, complete an ANOVA summary table for a one-way, repeated-measures ANOVA: $df_{\text{Subjects}} = 19$, $df_{\text{Treatment}} = 3$, $df_{\text{Residual}} = 57$, $df_{\text{Total}} = 79$, $SS_{\text{Subjects}} = 101.27$, $SS_{\text{Treatment}} = 99.81$, $SS_{\text{Residual}} = 957.48$. and $SS_{\text{Total}} = 1{,}158.56$.

Rejecting the null hypothesis?

11.77 If $F_{cv} = 3.259$ and $F = 3.259$, was H_0 rejected?

11.78 If $F_{cv} = 2.310$ and $F = 1.96$, was H_0 rejected?

11.79 Using the information in this ANOVA summary table, determine if the null hypothesis is rejected if alpha is set at .05:

Source of Variability	Sum of Squares	Degrees of Freedom	Mean Square	F ratio
Subjects	5.00	20		
Treatment	5.00	2	2.50	5.00
Residual	20.00	40	0.50	
Total	30.00	62		

11.80 Using the information in this summary table, determine if the null hypothesis is rejected if alpha is set at .05:

Source of Variability	Sum of Squares	Degrees of Freedom	Mean Square	F ratio
Subjects	12.50	17		
Treatment	5.00	3	1.67	1.78
Residual	48.00	51	0.94	
Total	65.50	71		

Writing results in APA format

11.81 Given $\alpha = .05$, $df_{\text{Treatment}} = 5$, $df_{\text{Residual}} = 45$, $F = 7.84$, and $F_{cv} = 2.579$, (a) write the results in APA format, and (b) state if the results are statistically significant or not.

11.82 Given $\alpha = .05$, $df_{\text{Treatment}} = 3$, $df_{\text{Residual}} = 24$, $F = 2.76$, and $F_{cv} = 3.009$, (a) write the results in APA format, and (b) state if the results are statistically significant or not.

11.83 Given $\alpha = .05$ and the information in this summary table, (a) write the results in APA format, and (b) state if the results are statistically significant or not:

Source of Variability	Sum of Squares	Degrees of Freedom	Mean Square	F ratio
Subjects	88.00	9		
Treatment	60.00	2	30.00	11.24
Residual	48.00	18	2.67	
Total	196.00	29		

11.84 Given $\alpha = .05$ and the information in this summary table, (a) write the results in APA format, and (b) state if the results are statistically significant or not:

Source of Variability	Sum of Squares	Degrees of Freedom	Mean Square	F ratio
Subjects	120.00	21		
Treatment	32.00	2	16.00	1.00
Residual	674.00	42	16.05	
Total	826.00	65		

Interpreting results

11.85 Given the results of a one-way, repeated-measures ANOVA in APA format, $F(2, 40) = 1.52$, $p > .05$, interpret them based only on this.

11.86 Given the results of a one-way, repeated-measures ANOVA in APA format, $F(3, 36) = 12.56$, $p < .05$, interpret them based only on this.

Calculating effect size

11.87 **a.** Given $SS_{Treatment} = 9.89$ and $SS_{Total} = 86.98$, what is η^2?
b. Classify it as small, medium, or large.

11.88 **a.** Given $SS_{Treatment} = 45.55$ and $SS_{Total} =$ 1,893.44, what is η^2?
b. Classify it as small, medium, or large.

11.89 **a.** Given the information in this summary table, what is η^2?
b. Classify it as small, medium, or large.

Source of Variability	Sum of Squares	Degrees of Freedom	Mean Square	F ratio
Subjects	76.00	14		
Treatment	46.00	2	23.00	3.01
Residual	214.00	28	7.64	
Total	336.00	44		

11.90 **a.** Given the information in this summary table, what is η^2?
b. Classify it as small, medium, or large.

Source of Variability	Sum of Squares	Degrees of Freedom	Mean Square	F ratio
Subjects	233.00	16		
Treatment	111.00	2	55.50	5.35
Residual	332.00	32	10.38	
Total	676.00	50		

Worrying about Type II error

11.91 If the results of a one-way, repeated-measures ANOVA are $F(2, 42) = 0.63$, $p > .05$ and if $\eta^2 = 2.12\%$, (a) how worried should the researcher be about having made a Type II error? (b) Should he or she recommend replication with a larger sample size?

11.92 If the results of a one-way, repeated-measures ANOVA are $F(2, 58) = 3.00$, $p > .05$ and if $\eta^2 = 8.11\%$, (a) how worried should the researcher be about having made a Type II error? (b) Should he or she recommend replication with a larger sample size?

Finding q

11.93 If $k = 3$ and $df_{Residual} = 16$, what is q if $\alpha = .01$?

11.94 If $k = 4$ and $df_{Residual} = 87$, what is q if $\alpha = .05$?

Calculating HSD

11.95 If $q = 2.90$, $MS_{Residual} = 12.27$, and $n = 8$, what is *HSD*?

11.96 If $q = 3.70$, $MS_{Residual} = 122.98$, and $n = 30$, what is *HSD*?

Interpreting HSD

11.97 If $M_1 = 12.83$, $M_2 = 14.98$, $M_3 = 8.22$, and $HSD = 3.78$, (a) determine for each possible pair of means if the difference is statistically significant, and (b) comment on the direction of the difference for the populations.

11.98 If $M_1 = 115.54$, $M_2 = 98.98$, $M_3 = 118.22$, and $HSD = 14.78$, (a) determine for each possible pair of means if the difference is statistically significant, and (b) comment on the direction of the difference for the populations.

Writing a complete interpretation

11.99 Researchers from PETA and NIH collaborated on a study to examine the effect, on rats, of being reared in laboratory settings. Sets of 3 rats from 8 litters were randomly assigned to three conditions: (1) being reared in a standard laboratory setting, (2) being reared in an enriched laboratory setting, and (3) being reared in a setting that mimics the wild (e.g., the rats have to forage for their own food; food is occasionally sparse and of poor quality). When the rats reached adulthood, they were given a series of behavioral tasks from which their rat IQ (RIQ) scores were calculated. RIQ scores are interval-level and are scored like human IQ scores. The three means were 95.00 (normal lab), 108.00

(enriched lab), and 94.00 (mimicked wild). No nonrobust assumptions were violated and a one-way, repeated-measures ANOVA was completed with $\alpha = .05$. The results are shown below. F_{cv} was 3.739, η^2 was calculated as 37.14%, and *HSD* was found to be 9.32. Write a four-point interpretation.

Source of Variability	Sum of Squares	Degrees of Freedom	Mean Square	*F* ratio
Subjects	941.33	7		
Treatment	976.00	2	488.00	9.61
Residual	710.67	14	50.76	
Total	2,628.00	23		

11.100 A women's studies professor was curious about the long-term effect, on men, of taking a women's studies class. She thought that it would make them more open-minded and less sexist. She obtained a random sample of 10 first-year male students at her university and assigned them to take Introduction to Women's Studies. She administered a scale that measures sexist beliefs before the class started, again at the end of the semester, and again 10 years later. The scale measures at the interval level and has a mean of 50, with a standard deviation of 10. Scores above 50 indicate a person has more sexist beliefs than average; scores below 50 indicate a person has fewer sexist beliefs than average. Here are the mean scores for her 10 participants: 54.00 (pre-class), 46.00 (post-class), 38.00 (10 years later). No assumptions were violated and a one-way, repeated-measures ANOVA, with the alpha set at .05, was completed. Results are shown below. $F_{cv} = 3.555$, $\eta^2 = 33.21\%$, $HSD = 5.10$. Given this information, complete a four-point interpretation.

Source of Variability	Sum of Squares	Degrees of Freedom	Mean Square	*F* ratio
Subjects	2,215.33	9		
Treatment	1,280.00	2	640.00	32.11
Residual	358.67	18	19.93	
Total	3,854.00	29		

Completing all six steps of hypothesis testing

11.101 A psychologist studying addictions investigated the effectiveness of three different treatments for alcoholism on the number of strong urges to drink alcohol that were experienced at the end of treatment. He compared three treatments: (1) Alcoholics Anonymous, (2) individual psychotherapy, and (3) a medication that is supposed to reduce urges. He matched 30 alcoholics into groups of three based on the severity of their addiction and then randomly assigned them to the three different treatments. (Each person assigned to Alcoholics Anonymous attended a different group.) At the end of treatment, he had the participants keep a diary of how many strong urges to drink alcohol they experienced each day. Here are the results:

	Alcoholics Anonymous	Individual Psychotherapy	Medication
M	8.00	6.00	6.00
s	5.42	4.52	4.37

Given $SS_{\text{Subjects}} = 532.00$, $SS_{\text{Treatment}} = 26.67$, $SS_{\text{Residual}} = 88.00$, and $SS_{\text{Total}} = 646.67$, complete the appropriate statistical test and interpret the results. (Don't forget to follow all six steps of the hypothesis test.)

11.102 A psychologist teaching introductory psychology wanted to demonstrate the effects of mere exposure on liking. On the first day of class, each of her 20 students went to a private booth to view a very abstract piece of modern art. Each was asked to rate how much he or she liked it on a scale ranging from −10 (extremely dislike) to +10 (extremely like). She followed the same procedure once a week for the rest of the semester. At the middle, and again at the end, of the semester, she had the students individually rate their liking of the picture. Here are the results:

	First Class	Mid-Semester	Last Class
M	−6.30	−2.40	1.70
s	2.85	2.35	1.75

Given $SS_{\text{Subjects}} = 176.67$, $SS_{\text{Treatment}} = 640.13$, $SS_{\text{Residual}} = 140.53$, and $SS_{\text{Total}} = 957.33$, complete the appropriate statistical test and interpret the results. (Don't forget to follow all six steps of the hypothesis test.)

Expand Your Knowledge

11.103 If the results of a one-way, repeated-measures ANOVA are $F(3, 45) = 12.56$, $p < .05$, what is N?

11.104 Is it possible, in a one-way, repeated-measures ANOVA, for $df_{\text{Treatment}}$ to be 2 and df_{Residual} to be 75?

11.105 If the results of a one-way, repeated-measures ANOVA are $F(3, 45) = 12.50$, $p < .05$, and $\eta^2 = 28.30$, how concerned should one be about Type II error?

11.106 A group of students had their IQs measured at the start, middle, and end of a school year. The three means, respectively, were 125.5, 127.5, and 130.5. A one-way, repeated-measures ANOVA was used to analyze the data and the results were $F(2, 18) = 0.12$, $p > .05$. Use the summary table below to conduct post-hoc tests as appropriate.

Source of Variability	Sum of Squares	Degrees of Freedom	Mean Square	F ratio
Subjects	4,917.50	9		
Treatment	126.67	2	63.34	0.12
Residual	9,590.00	18	532.78	
Total	14,634.17	29		

11.107 Which HSD value will be larger: one for $\alpha = .01$ or one for .05? Why?

SPSS

Data entry in SPSS for one-way, repeated-measures ANOVA is similar to data entry for a paired-samples t test—each case has a row to itself and the repeated measures each have a column. Figure 11.11 shows the data from the ADHD/level of distraction study in the SPSS Data Editor.

	V1	pre	post	f_u
1	1	13	10	12
2	2	30	20	26
3	3	20	13	17
4	4	26	17	20

Figure 11.11 Data Entry in SPSS for One-Way, Repeated-Measures ANOVA Data for a repeated-measures ANOVA in SPSS are arranged with each case on a separate line and each level of the independent variable in a separate column. (Source: SPSS)

The repeated-measures ANOVA commands are not easy to find in SPSS. They are found under "Analyze" and then "General Linear Model" (see Figure 11.12).

Clicking on "Repeated Measures" opens up a new dialog box, seen in Figure 11.13. "Treatment" is named as the "Within-Subject Factor." There are three levels of treatment, so the "Number of Levels" is "3."

After clicking on the "Add" button (see Figure 11.13), the commands seen in Figure 11.14 open by clicking the "Define" button. Notice that two of the three levels of the "Within-Subjects Variables" are already defined (the variables "pre" and "post"). The third level, "f_u," is highlighted prior to using the arrow key to send it over.

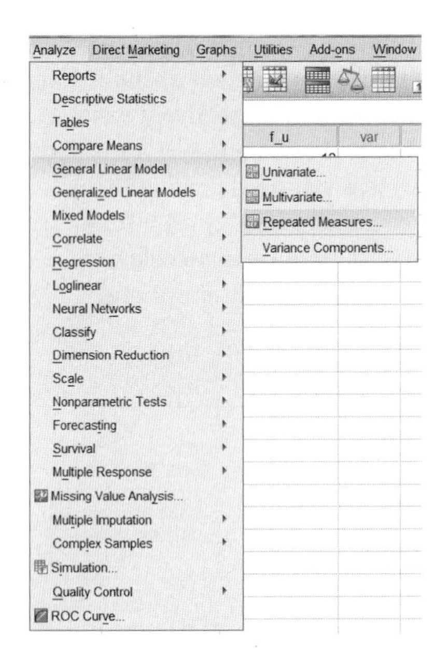

Figure 11.12 Finding Repeated-Measures ANOVA in SPSS The commands for conducting a repeated-measures ANOVA in SPSS are initiated within "General Linear Model." (Source: SPSS)

Figure 11.13 Defining the Within-Subjects Factor in SPSS In this step, the effect is given a name and the number of levels of the independent variable is indicated. (Source: SPSS)

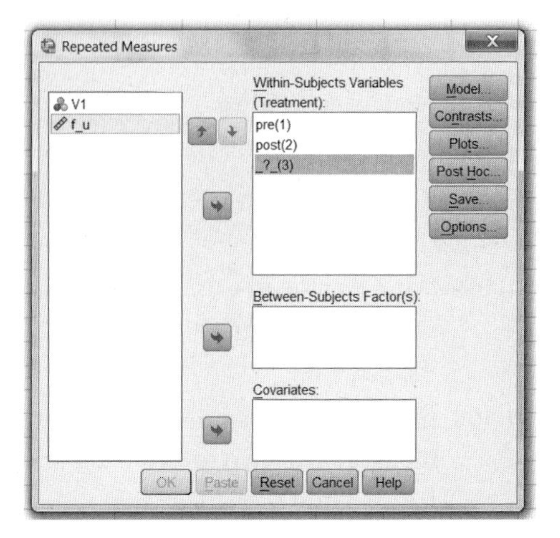

Figure 11.14 Defining the Within-Subjects Variables for a Repeated-Measures ANOVA in SPSS Once the third level of the within-subjects variable, "follow," is moved over to the box where the other two are, the "OK" button will become active and the ANOVA can be completed. (Source: SPSS)

When all three variables have been defined, the "OK" button on the bottom right becomes active. Clicking the OK button causes SPSS to complete the analysis. The printout that SPSS generates for a repeated-measures ANOVA is fairly detailed, but most of the data is not relevant for introductory statistics. Figure 11.15 shows the printout that is most similar to the ANOVA summary table found in Table 11.5.

Tests of Within-Subjects Effects

Measure: MEASURE_1

Source		Type III Sum of Squares	df	Mean Square	F	Sig.
Treatment	Sphericity Assumed	105.167	2	52.583	18.743	.003
	Greenhouse-Geisser	105.167	1.314	80.053	18.743	.011
	Huynh-Feldt	105.167	1.930	54.485	18.743	.003
	Lower-bound	105.167	1.000	105.167	18.743	.023
Error(Treatment)	Sphericity Assumed	16.833	6	2.806		
	Greenhouse-Geisser	16.833	3.941	4.271		
	Huynh-Feldt	16.833	5.791	2.907		
	Lower-bound	16.833	3.000	5.611		

Figure 11.15 ANOVA Summary Table for One-Way, Repeated-Measures ANOVA in SPSS The *F* ratio for the effect of treatment is displayed on the first line, the line for the effect with "Sphericity Assumed." (Source: SPSS)

In this summary table, SPSS reports only two sources of variability, the one called treatment and the one called residual. Here, treatment is called "treatment" because it was labeled that way in Figure 11.13. Residual is called "error." For a basic repeated-measures ANOVA, only pay attention to the first row (of the four rows) for each source, the one labeled "Sphericity Assumed." The first rows are the only lines with integer values for the degrees of freedom. (The slight differences between values calculated in the text and by SPSS are due to the number of decimal places carried.)

Remember, SPSS reports exact significance levels. Here, the significance level for the F ratio of 18.743 is .003. And $.003 \leq .05$, so the results are statistically significant. APA format prefers the use of exact significance levels, so the results would be reported as $F(2, 6) = 18.74, p = .003$.

Appendix

Calculating Sums of Squares for One-Way Repeated-Measures ANOVA

Calculating sums of squares for a one-way repeated-measures ANOVA is quite similar to calculating sums of squares for a between-subjects, one-way ANOVA. Here are the pre-treatment, post-treatment, and follow-up distraction scores, squared and summed just as they were in Chapter 10 for a between-subjects, one-way ANOVA:

	Pre-Treatment		Post-Treatment		Follow-Up			
	X	X^2	X	X^2	X	X^2		
Case 1	13	169	10	100	12	144		
Case 2	30	900	20	400	26	676		
Case 3	20	400	13	169	17	289	Grand	
Case 4	26	676	17	289	20	400	X	X^2
Sum	89.00	2,145.00	60.00	958.00	75.00	1,509.00	224.00	4,612.00
n	4		4		4		12	

Once this is done, the values can be plugged into Equations 10.2, 10.3, and 10.4. SS_{Total} is calculated using Equation 10.2:

$$\begin{aligned} SS_{\text{Total}} &= \Sigma X^2 - \frac{(\Sigma X)^2}{N} \\ &= 4{,}612.00 - \frac{224.00^2}{12} \\ &= 430.67 \end{aligned}$$

Next, we obtain SS_{Between} via Equation 10.3. This is called $SS_{\text{Treatment}}$ for repeated-measures ANOVA.

$$\begin{aligned} SS_{\text{Treatment}} &= \Sigma\left(\frac{(\Sigma X_{\text{Group}})^2}{n_{\text{Group}}}\right) - \frac{(\Sigma X)^2}{N} \\ &= \left(\frac{89^2}{4} + \frac{60^2}{4} + \frac{75^2}{4}\right) - \frac{224^2}{12} \\ &= 105.17 \end{aligned}$$

Then, using Equation 10.4, calculate SS_{Within}:

$$SS_{\text{Within}} = \Sigma\left(\Sigma X^2_{\text{Group}} - \frac{(\Sigma X_{\text{Group}})^2}{n_{\text{Group}}}\right)$$

$$= \left(2145.00 - \frac{89.00^2}{4}\right) + \left(958.00 - \frac{60.00^2}{4}\right) + \left(1509.00 - \frac{75.00^2}{4}\right)$$

$$= 325.50$$

So far, all has been the same as for a between-subjects, one-way ANOVA. Now, it is time to diverge and calculate SS_{Between}. To do so, we need to add two new columns to the data table:

	Pre-Treatment		Post-Treatment		Follow-Up					
	X	X^2	X	X^2	X	X^2			Case Total	T^2/k
Case 1	13	169	10	100	12	144			35	408.33
Case 2	30	900	20	400	26	676			76	1,925.33
Case 3	20	400	13	169	17	289	Grand		50	833.33
Case 4	26	676	17	289	20	400	X	X^2	63	1,323.00
Sum	89.00	2,145.00	60.00	958.00	75.00	1,509.00	224.00	4,612.00	224.00	4,490.00
n	4		4		4		12			

The two new columns are on the far right. The first, labeled "Case Total," is the sum of all the scores for each case. For case 1, this is $T = 13 + 10 + 12$. The next column is that case total squared and then divided by k, the number of conditions. For case 1, this is $35^2/3$. Note that all of these values are summed. SS_{Subjects} is calculated:

$$SS_{\text{Subjects}} = \Sigma\left(\frac{T^2}{k}\right) - \frac{(\Sigma X)^2}{N}$$

$$= 4490.00 - \frac{224.00^2}{12}$$

$$= 308.67$$

Finally, SS_{Residual} is needed. SS_{Residual} is what remains after SS_{Subjects} is removed from SS_{Within}:

$$SS_{\text{Residual}} = SS_{\text{Within}} - SS_{\text{Subjects}}$$

$$= 325.50 - 308.67$$

$$= 16.83$$

Between-Subjects, Two-Way Analysis of Variance

LEARNING OBJECTIVES

- Describe what two-way ANOVA does.
- Complete a between-subjects, two-way ANOVA.*
- Interpret a between-subjects, two-way ANOVA.

CHAPTER OVERVIEW

Chapter 11 introduced one-way ANOVA, a test that allows the examination of the impact of more than two levels of an explanatory variable at one time. Now, this chapter introduces two-way ANOVA, a technique that allows researchers to examine the impact of two explanatory variables at one time. Because each explanatory variable will have at least two levels, as shown in Table 12.1, a two-way ANOVA is used when at least four sample means are being compared.

TABLE 12.1 Design for the Simplest Type of Study to Be Analyzed with a Two-Way Analysis of Variance

	Level 1 of Explanatory Variable 2	Level 2 of Explanatory Variable 2
Level 1 of Explanatory Variable 1	Mean of Cell A	Mean of Cell B
Level 2 of Explanatory Variable 1	Mean of Cell C	Mean of Cell D

Two-way ANOVA is used in studies in which the effects of two explanatory variables are being studied at the same time. As each explanatory variable will have at least two levels, the simplest two-way ANOVA will analyze data from a study, like the one above, in which there are four samples (cells).

12.1 Introduction to Two-Way ANOVA

More complex tools are needed to understand, to take apart, more complex things. A basic tool, like a rock, will get you into a skull and show you that there is grey soft matter in there. But, to see finer details of the brain, a saw and a scalpel are more

*Note: Formulas for calculating sums of squares appear in an appendix at the end of the chapter.

helpful. As tools have evolved, think electron microscopes and MRI machines, so has our understanding of the brain.

Statistical tests are statisticians' tools. Tests like a two-sample t test are simple tools—all they allow a user to do is compare two groups (e.g., What is the impact of 0 vs. 1 drink of alcohol on driving?). One-way ANOVA allows more groups to be compared at once, so it allows more complex questions to be addressed (How does 0 vs. 1 vs. 2 vs. 3 vs. 4 drinks affect driving?). But, the most complex ANOVA questions involve the influence of multiple factors at once (What is the impact on driving of different doses of alcohol for men vs. women, depending on the time elapsed since the last meal?)

A "factor" is ANOVA-speak for an explanatory variable, an independent variable or a grouping variable that is thought to have some impact on the dependent variable. Factorial ANOVA can have two, three, four, even five factors or "ways." (The driving study had three: dose of alcohol, sex, and time since meal.) As the number of ways increases, a factorial ANOVA becomes harder to interpret. In this chapter, we'll limit ourselves to two factors and what is called two-way ANOVA.

To introduce two-way ANOVA, here's an example about the influence of two factors, nature *and* nurture, on personality. Suppose a researcher was studying factors that influence altruism and was interested in *both* how the children were reared (nurture) and what their nervous systems were like (nature). Using adopted children as participants, the researcher classified the children as falling into one of three levels—high, medium, or low—based on how altruistic their adoptive parents were. The adoptive parents reared them, so these three levels of altruism represented the influence of nurture. In addition, the researcher classified the children as being in one of two levels—high or low—based on how altruistic their birth parents were. Birth parents, who provide genetic material, represent the influence of nature. This two-way classification would allow the researcher to study the effect of both nature and nurture at the same time (see Table 12.2).

TABLE 12.2 Design for the Study of the Effect of Nature and Nurture on Altruism

	Adoptive Parents High on Altruism	Adoptive Parents Medium on Altruism	Adoptive Parents Low on Altruism
Birth Parents High on Altruism			
Birth Parents Low on Altruism			

This design is called a 3 × 2 design because there are three levels of one explanatory variable and two levels of the other. (It could also be called a 2 × 3 design.)

Each child who is a participant in the study is classified as belonging in one, and only one, of the six cells.

In Table 12.2, there are two levels of one grouping variable and three levels of the second grouping variable, so this two-way ANOVA would be called a 2 × 3 ANOVA (pronounced "two by three"). (Note that it doesn't matter which number goes first: It's fine to call it a 3 × 2 ANOVA.) Each child in the study would be classified as fitting into only one of the six cells in Table 12.2. This design would allow the researcher to examine the influence of two factors on children's altruism at once.

The type of two-way ANOVA being taught in this chapter is a *between-subjects* ANOVA. **Between-subjects** is an ANOVA term for independent samples. In a between-subjects design, different cases make up the different groups. **Within-subjects** is the ANOVA term for dependent samples. If different participants each rated the taste of a single ice cream (high-fat, low-fat, and no-fat), that would be a between-subjects design. If each taste-tester rated all three of the ice creams, that would be a within-subjects design.

A Common Question

Q Can an ANOVA have more than two ways?

A Absolutely. There can be a three-way ANOVA, a four-way ANOVA, and even more ways than that. The general term for a multiple-way ANOVA is **factorial ANOVA** (ways are also called *factors*). It is rare, however, that a researcher designs an experiment that needs more than a three-way ANOVA because the number of participants required becomes too large and the results become complicated to interpret.

There is an advantage to completing a single two-way ANOVA with two explanatory variables instead of two separate one-way tests, one for each of the explanatory variables. To explore this advantage, imagine a study that compares the final point total in a statistics class based on two explanatory variables: (1) type of instruction (whether students take the class in a classroom or online), and (2) level of math anxiety (whether students are high or low on math anxiety). Each factor has two levels (classroom vs. online, and high math anxiety vs. low math anxiety).

To conduct separate one-way tests, a researcher would need to do two separate studies:

- To examine the effect of type of instruction, the researcher would get a sample of students, assign some to take the course in a classroom, others to take it online, and then compare how much they learned.
- To examine the effect of anxiety, the researcher would put together another sample of students, classify them as high or low on math anxiety, have them all take the same course, and then compare how much they learned.

To conduct the study as a two-way ANOVA means examining both explanatory variables at once. The researcher would obtain *one* sample of students, classify them as high or low on math anxiety, and then assign half of each type to take the course in a classroom and half to take it online. That would give four groups:

- High-anxiety students taking the course in a classroom (Cell A)
- High-anxiety students taking the course online (Cell B)
- Low-anxiety students taking the course in a classroom (Cell C)
- Low-anxiety students taking the course online (Cell D)

The arrangement of the four groups in this study is diagrammed in Table 12.3. Notice how the two ways yield four cells arranged in rows (levels of math anxiety) and columns (where the course is taken). The two explanatory variables are **crossed**, which means that every level of one explanatory variable is paired with every level of the other explanatory variable.

TABLE 12.3 The Four Groups in a Two-Way ANOVA Examining the Impact of Math Anxiety and Type of Instruction on Learning

	Classroom Instruction	Online Instruction
High Math Anxiety	A	B
Low Math Anxiety	C	D

In a two-way ANOVA, there are two explanatory variables, each with at least two levels. Here, there are two levels of the row variable (degree of math anxiety) and two levels of the column variable (type of instruction). This results in four groups, one for each cell of the matrix.

There are two advantages to completing this study as a two-way ANOVA rather than as two one-way ANOVAs. First, it allows the study to be completed and analyzed in one pass, not as two different studies. Second, and more important, the person conducting the study gains more understanding because the variability in the dependent variable can now be divided into *three* effects. By doing the study as a two-way ANOVA, an effect has been gained.

The three effects being studied are:

- Does type of instruction affect learning?
- Is level of math anxiety related to learning?
- Do type of instruction and level of math anxiety *interact* to affect learning?

The first two effects are called **main effects**. Main effects examine the overall impact of an explanatory variable by itself. The third effect, the *interaction* effect, is unique to two-way ANOVA (and factorial ANOVA). An **interaction effect** occurs if the impact of one explanatory variable on the dependent variable depends on the level of the other explanatory variable.

So what exactly is an interaction effect? The easiest way to show what it is, is with a figure. Figure 12.1 gives the outcome for a 2 × 2 ANOVA of type of instruction and math anxiety on performance. When the lines are not parallel, then an interaction exists. That's what is seen in Figure 12.1, which illustrates that the low-math-anxiety people perform equally well whether in the classroom or online, while the high-math-anxiety people do more poorly when taking the class online. The two explanatory variables *interact* to determine the outcome.

When the lines are not parallel, an interaction exists.

A Common Question

Q How nonparallel do the lines have to be for an interaction to occur?

A If the lines are close to parallel, the interaction effect probably is not statistically significant. There should be a reasonable amount of deviation from parallel to speculate that an interaction effect has occurred.

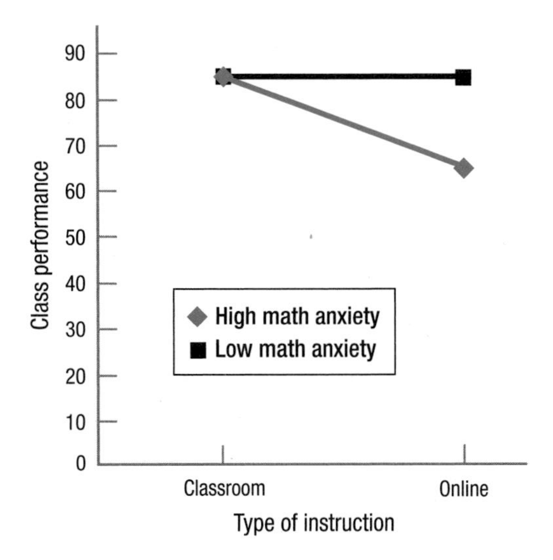

Figure 12.1 Interaction Between Level of Math Anxiety and Type of Instruction on Performance The nonparallel lines indicate that level of math anxiety and type of classroom instruction interact to determine performance. If asked whether classroom instruction or online instruction is better for learning statistics, the answer is that it depends on a student's level of math anxiety.

When an interaction takes place, the answer to the question of whether a main effect exists is, "It depends." Here are questions about the two main effects for the type of instruction/math anxiety study:

- Does type of instruction have an impact on how much one learns?
  - *It depends* on how much math anxiety one has. If a person is low on math anxiety, then he or she will learn equally well whether in the classroom or online. However, if a person is high on math anxiety, then he or she would be wise to take the course in a classroom.
- Is level of math anxiety related to how much one learns?
  - *It depends* on which course one takes. If a person is taking the course in a classroom, then math anxiety doesn't matter. However, if a person is taking the course online, then he or she will probably do better if he or she is low on math anxiety.

When there is an interaction, one needs to be careful about interpreting main effects. To understand why, look at the results for the type of instruction/level of math anxiety study, which are presented in Table 12.4.

Each of the cells has a mean for one of the four groups—a mean of 85 for high-anxiety students in the classroom, a mean of 65 for high-anxiety students online, and so on. Rather than attend to the cell means, pay attention to the means for each row and for each column. Each cell has the same sample size—and that will be true for all examples in this chapter—so the means for the cells in a row or column can be averaged to find the mean for that row or column. The formula for calculating row means and column means is shown in Equation 12.1.

TABLE 12.4 Cell Means, Row Means, and Column Means for the Effect of Level of Math Anxiety and Type of Instruction on Learning

	Classroom Instruction	Online Instruction	
High Math Anxiety	85	65	75.00
Low Math Anxiety	85	85	85.00
	85.00	75.00	

The sample size for each cell is the same, so a row mean or column mean can be calculated by averaging together the cell means in a row or a column. There is an interaction here (see Figure 12.1), so one shouldn't use the row means or column means to describe the main effects for rows or columns.

Equation 12.1 Formulas for Calculating Row Means and Column Means for Two-Way ANOVA When Each Cell Has the Same Sample Size

$$M_{\text{Row}} = \frac{\text{Add up all the cell means in a row}}{\text{The number of cells in the row}}$$

$$M_{\text{Column}} = \frac{\text{Add up all the cell means in a column}}{\text{The number of cells in the column}}$$

where M_{Row} = row mean
M_{Column} = column mean

The researcher would calculate the mean for the high-math-anxiety row as follows:

$$\begin{aligned} M_{\text{RowHigh}} &= \frac{\text{Add up all the means in the row}}{\text{The number of cells in the row}} \\ &= \frac{85 + 65}{2} \\ &= \frac{150.00}{2} \\ &= 75.00 \end{aligned}$$

The researcher would calculate the mean for the low-math-anxiety row as follows:

$$\begin{aligned} M_{\text{RowLow}} &= \frac{\text{Add up all the means in the row}}{\text{The number of cells in the row}} \\ &= \frac{85 + 85}{2} \\ &= \frac{170.00}{2} \\ &= 85.00 \end{aligned}$$

The researcher would calculate the mean for the classroom instruction column as follows:

$$M_{\text{ColumnClass}} = \frac{\text{Add up all the means in the column}}{\text{The number of cells in the column}}$$

$$= \frac{85 + 85}{2}$$

$$= \frac{170.00}{2}$$

$$= 85.00$$

And, finally, here's the mean for the online instruction column:

$$M_{\text{ColumnOnline}} = \frac{\text{Add up all the cell means in a column}}{\text{The number of cells in the column}}$$

$$= \frac{65 + 85}{2}$$

$$= \frac{150.00}{2}$$

$$= 75.00$$

What information do the row and column means provide?

- The row means give information about the main effect of level of anxiety on learning. Comparing the row mean of 75.00 for high-math-anxiety students to the row mean of 85.00 for the low-math-anxiety students suggests that level of anxiety predicts performance: it appears as if the higher the anxiety, the worse the performance. However, we already know, from the interaction effect, that how anxiety affects performance depends on the type of instruction. Though a main effect for type of anxiety exists, interpreting the interaction effect gives a better understanding of the results.

- The column means give information about the main effect of type of instruction on learning. Comparing the column mean of 85.00 for the classroom students to the column mean of 75.00 for the online students suggests that type of instruction affects performance: it seems as if students learn more in the classroom. However, we already know, from the interaction effect, that how type of instruction affects performance depends on a person's level of math anxiety. Again, though a main effect for type of instruction exists, interpreting the interaction seems to do a better job of explaining the results.

In a two-way ANOVA, when there is a statistically significant interaction, statistically significant main effects often play a background role.

This point is important: in a two-way ANOVA, when there is a statistically significant interaction, statistically significant main effects often play a background role.

To help make the difference between main effects and interactions clear, consider an example without an interaction. Suppose a physical therapist had equal numbers of men and women, then measured each person's hand strength. For half of each sex, the physical therapist measured hand strength in the dominant hand. For the other half, she measured hand strength in the nondominant hand.

The graph in Figure 12.2 shows the results—men are stronger than women and the dominant hand is stronger than the nondominant hand. The two lines are parallel, so no interaction exists. In this case, it makes sense to interpret the main effects,

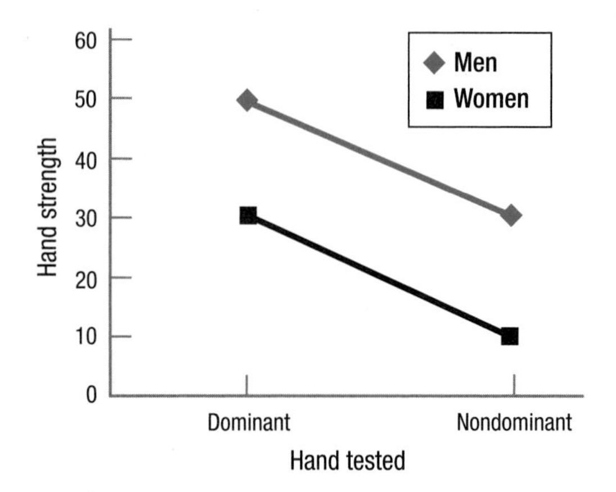

Figure 12.2 Lack of Interaction Between Sex and Tested Hand Strength The lines in this graph are parallel, indicating that sex and hand tested don't interact to determine hand strength. The graph shows a main effect for sex—men are stronger than women—and a main effect for hand tested—dominant hands are stronger than nondominant hands.

which can be seen by examining the row means and the column means found in Table 12.5. There is no "it depends" in answer to the two questions below:

- Is there a difference in hand strength by sex?
  - Yes. Men have more hand strength than women, both in the dominant hand and nondominant hand.
- Is there a difference in strength between dominant and nondominant hands?
  - Yes. Dominant hands are stronger than nondominant hands, both for men and for women.

TABLE 12.5 Effect of Sex and Hand Dominance on Hand Strength

	Dominant Hand	Nondominant Hand	
Men	50	30	40.00
Women	30	10	20.00
	40.00	20.00	

These results have no interaction (see Figure 12.2), so one can use the row means and column means to interpret the main effects for sex and hand tested. The main effects show that men are stronger than women and that dominant hands are stronger than nondominant hands.

Worked Example 12.1 For practice with main effects and interaction effects, let's look at some data on aggression in boys and girls. Suppose a developmental psychologist, Dr. O'Grady, administered an aggression scale to a random sample of 100 boys and 100 girls. Half of each sex had their level of physical aggression measured (getting into fights, throwing sticks and stones, etc.) and half had verbal aggression measured (spreading rumors, telling lies, etc.). On each scale, a higher score means more aggression.

Dr. O'Grady's study compares the means for groups that differ on two independent variables, sex and type of aggression, so it is an example where results are analyzed with a two-way ANOVA. Each way has two levels, so this is called a 2×2 ANOVA and there are four cells. The results from Dr. O'Grady's study can be found in Table 12.6, where each case fits in one, and only one, cell. For example, a boy who answered questions about physical aggressiveness would be in the top left cell.

TABLE 12.6 Measuring Aggression: Effect of Sex and Type of Aggression

	Physical Aggression	Verbal Aggression	
Boys	52	23	37.50
Girls	20	51	35.50
	36.00	37.00	

The graph of these results shows an interaction (see Figure 12.3), so one can't use the row means and column means to demonstrate main effects.

The row means for boys (37.50) and girls (35.50) don't show much of a main effect by sex. Boys are slightly more aggressive on average, but this 2-point difference could easily be due to sampling error. Even if the small difference indicated a difference between population means, it probably wouldn't be a meaningful difference.

The column means tell a similar story—the amount of physical aggression ($M = 36$) and the amount of verbal aggression ($M = 37$) are about equal. Verbal aggression is just slightly higher, but it seems possible that the levels of verbal aggression and physical aggression are the same in the populations and that the difference observed between the samples could easily be due to sampling error. Based on these column means, it seems that there is little difference in the use of the two types of aggression.

For a different view, look at the interaction effect graphed in Figure 12.3.

- Are boys more aggressive than girls?
  - *It depends* on the type of aggression. Boys are more aggressive physically, and girls are more aggressive verbally.
- Is physical aggression as likely to occur as verbal aggression?
  - *It depends* on a person's sex. Physical aggression is more likely to occur in boys and verbal aggression in girls.

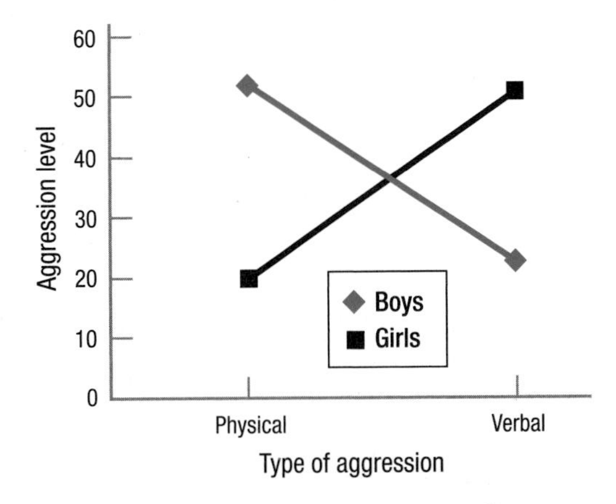

Figure 12.3 Interaction Between Sex and Type of Aggression The nonparallel lines indicate that the two variables, sex and type of aggression, interact. Are boys more aggressive than girls? It depends on the type of aggression. Is physical aggression more common than verbal aggression? It depends on whether one is talking about boys or girls.

It is clear in this example that there is an interaction effect. This is apparent both from the nonparallel lines in Figure 12.3 and from the use of "it depends" in interpreting the results. To reiterate: When an interaction effect exists, be cautious interpreting the main effects.

Practice Problems 12.1

Apply Your Knowledge

12.01 Classify each scenario in terms of the number of ways and the number of levels each way has. For example, classify each scenario as a 2 × 3 design, a 2 × 3 × 5 design, or some other variation.

a. Men and women who are right-handed or left-handed and who use razors with blades vs. electric razors are compared in terms of satisfaction with the smoothness of their shaves.

b. Male and female students studying for traditionally female careers (e.g., nursing) or for traditionally male careers (e.g., engineering) are compared in terms of how androgynous they are.

12.02 Each cell in the matrix below reports the mean of six cases. (a) Calculate the row and column means. (b) Interpret the two main effects.

	Condition 1	Condition 2
Group 1	12.00	18.00
Group 2	3.00	9.00

12.03 Thirty-six cases were randomly divided into four samples. Each person took either a low dose or a high dose of a drug and was queried about either the physical side effects or psychological side effects. Each cell in the matrix below reports the mean number of side effects reported by a sample of nine cases. (a) Graph the cell means. (b) Decide if there is an interaction. (c) Interpret the effect of the two conditions on the two groups.

	Physical Side Effects	Psychological Side Effects
Low Dose	10.00	10.00
High Dose	10.00	18.00

12.2 Calculating a Between-Subjects, Two-Way ANOVA

A two-way ANOVA allows a researcher to see, at one time, the effects of two explanatory variables by themselves (the main effects) and in combination (the interaction). A two-way ANOVA achieves this by partitioning the between-group variability differently. As shown in Figure 12.4, a two-way ANOVA takes the total variability and separates that into within-group variability and between-group variability. It then takes the between-group variability and divides that further into the two main effects, the effect of the row explanatory variable and the effect of the column explanatory variable, and the effect of the interaction of the two main effects. Each of these three effects is tested for statistical significance with its own F ratio.

To learn how to complete the calculations for a between-subjects, two-way ANOVA, here's an example about the effect of caffeine consumption and sleep

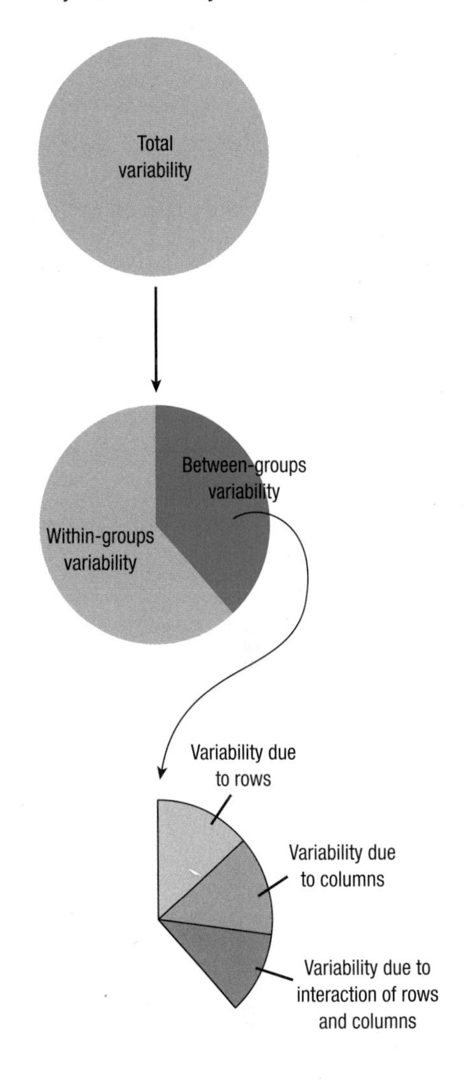

Figure 12.4 Partitioning Variability for Between-Subjects, Two-Way ANOVA This figure shows how the variability for a between-subjects, two-way ANOVA is initially partitioned into between-group variability and within-group variability, as for a between-subjects, one-way ANOVA. The between-group variability is then divided further into three parts: that due to the row variable, the column variable, and the interaction of the two.

deprivation on mental alertness. Imagine that a sleep researcher, Dr. Ballard, obtained 30 participants who were students at his college. He gave each a mental alertness task one hour after waking. (Higher scores on this test indicate more mental alertness.) The participants were randomly assigned into six groups, with five in each group. Half the participants consumed a standard cup of coffee (150 mg of caffeine) 30 minutes after waking and half did not. One third of the participants were allowed a full night's sleep (0-hours sleep deprivation), one third were awakened an hour early (1-hour sleep deprivation), and the final third were awakened two hours early (2-hours sleep deprivation).

The design of the study is shown in **Table 12.7**. Note that the six cells are laid out in a 3×2 format as the two explanatory variables are crossed. The two levels for dose of caffeine are the two rows and the three levels for sleep deprivation are the three columns. Each of the 30 participants is in one and only one cell.

Table 12.7 displays the cell means, row means, and column means, so Dr. Ballard can speculate about main effects. The graph in **Figure 12.5** allows him to consider the presence of an interaction.

Looking at Table 12.7 and Figure 12.5, three effects are apparent: (1) a main effect for caffeine, with those receiving caffeine performing better; (2) a main effect for

TABLE 12.7 Mental Alertness Descriptive Statistics for Caffeine/Sleep Deprivation Data

	0-hours Sleep Deprivation	1-hour Sleep Deprivation	2-hours Sleep Deprivation	
Caffeine	88.00 (3.87)	90.00 (4.69)	86.00 (4.27)	88.00
No Caffeine	80.00 (4.79)	68.00 (4.43)	50.00 (5.07)	66.00
	84.00	79.00	68.00	

Means (and standard deviations) for mental alertness scores are reported for each cell. Row means and column means are also reported.

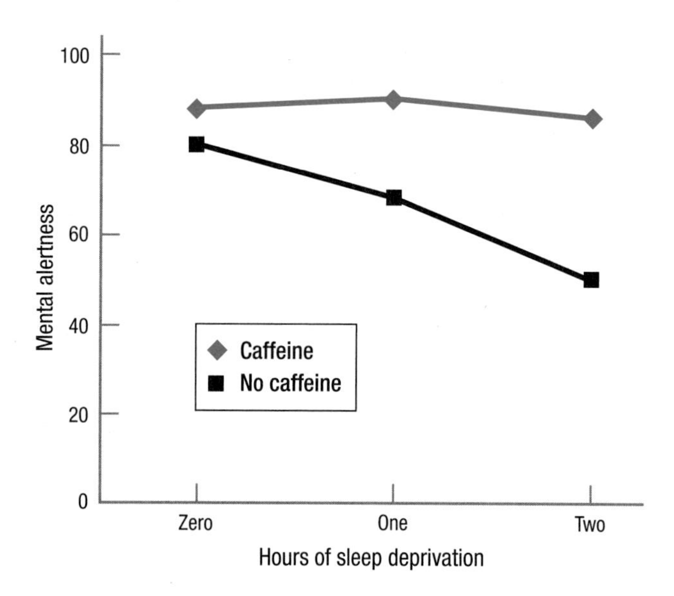

Figure 12.5 Effect of Caffeine Consumption and Sleep Deprivation on Mental Alertness This graph has nonparallel lines. That means there may be an interaction between consumption of caffeine and number of hours of sleep deprivation on mental alertness.

sleep deprivation, with those who were more sleep-deprived performing worse; and (3) an interaction between the two variables. Of course, Dr. Ballard will need to do a statistical test to see if any of these results is statistically significant. And, if the interaction effect is statistically significant, that will take precedence over the main effects.

Conducting a between-subjects, two-way ANOVA requires following the same six steps for hypothesis testing as in previous hypothesis tests.

Step 1 Pick a Test

Dr. Ballard is comparing the means of six groups, formed by the crossing of two independent variables (caffeine consumption and sleep deprivation). The groups are independent samples, so he'll use a between-subjects, two-way ANOVA. Specifically, it is a 2×3 ANOVA, as there are two levels of caffeine consumption (consume caffeine or not consume caffeine) and three levels of sleep deprivation (0, 1, or 2 hours of deprivation).

Tom and Harry despise crabby infants.

Step 2 Check the Assumptions

The assumptions for a two-way ANOVA are the same as they are for a between-subjects, one-way ANOVA: random samples, independence of observations, normality, and homogeneity of variance.

- *Random samples.* Dr. Ballard would love to be able to draw a conclusion from his study about humans in general. But, he doesn't have a random sample of participants from the human population, so the random samples assumption is violated. This is a robust assumption, however, so he can proceed with the two-way ANOVA. He'll need to be careful about generalizing the results.
- *Independence of observations.* Each participant was tested individually, so their results don't influence each other and the independence of samples assumption is not violated. This assumption is not robust, so if it had been violated, the analyses couldn't proceed.
- *Normality.* Dr. Ballard is willing to assume that the dependent variable, mental alertness, is normally distributed in the larger population, so this assumption is not violated. This assumption is robust to violation, especially if the sample size is large.
- *Homogeneity of variance.* All six of the cell standard deviations are about the same (see Table 12.7), which means that the variability in each group is about the same. This assumption is not violated in the study. (And, when the sample size is large, it is a robust assumption.)

For the caffeine/sleep deprivation study, no nonrobust assumptions were violated, so Dr. Ballard can proceed with the ANOVA.

Step 3 List the Hypotheses

For a one-way ANOVA, there is one set of hypotheses. For a two-way ANOVA, with a null hypothesis and an alternative hypothesis for both of the main effects and for the interaction effect, there are three sets of hypotheses.

- For the row main effect, the null hypothesis states that the population means for all levels of the row variable are equal to each other. The alternative hypothesis for the row main effect states that not all population means for the levels of the row variable are the same.
- For the column main effect, the null hypothesis says that the population means for all levels of the column variable are the same. The alternative hypothesis for the column main effect says that not all population means for the levels of the column variable are the same.
- For the interaction effect, the null hypothesis says that there is no interaction effect in the population. This means that the impact of one main effect on the dependent variable is independent of the impact of the other main effect on the dependent variable for all the cells. The alternative hypothesis for the interaction effect says that there is an interaction effect for at least one cell.

For the caffeine/sleep deprivation study, the row variable, caffeine consumption, has two levels, so Dr. Ballard writes the hypotheses:

$$H_{0\text{ Rows}}: \mu_{\text{Row1}} = \mu_{\text{Row2}}$$
$$H_{1\text{ Rows}}: \mu_{\text{Row1}} \neq \mu_{\text{Row2}}$$

There are three levels of the column variable, sleep deprivation, so the column null hypothesis could be written two ways: (1) $\mu_{\text{Column1}} = \mu_{\text{Column2}} = \mu_{\text{Column3}}$ or (2) all column population means are the same.

$H_{0\text{ Columns}}$: All column population means are the same.

$H_{1\text{ Columns}}$: At least one column population mean is different from at least one other column population mean.

There is no easy way to write the hypotheses for the interaction effect symbolically. So, using plain language, Dr. Ballard writes the hypotheses as

$H_{0\text{ Interaction}}$: There is no interactive effect of the two independent variables on the dependent variable in the population.

$H_{1\text{ Interaction}}$: The two independent variables in the population interact to affect the dependent variable in at least one cell.

Step 4 Set the Decision Rules

Just as three sets of hypotheses exist, there will be three decision rules for a two-way ANOVA—one for the row effect, one for the column effect, and one for the interaction effect. Because this is an ANOVA, F ratios will be calculated for each of the three effects (F_{Rows}, F_{Columns}, and $F_{\text{Interaction}}$) and compared to their critical values of F($F_{cv\text{ Rows}}$, $F_{cv\text{ Columns}}$, and $F_{cv\text{ Interaction}}$). If the observed value of F is greater than or equal to F_{cv} for an effect, the null hypothesis is rejected for that effect, the alternative hypothesis accepted, and the effect is called statistically significant.

To find a critical value of F for an F ratio, a researcher needs to decide on the alpha level and know the degrees of freedom for the F ratio. The most commonly used alpha levels for ANOVA are .05 and .01, which correspond to a 5% chance and a 1% chance of making a Type I error. (Type I error occurs when the null hypothesis is erroneously rejected.) Typically, alpha is set at .05. When the consequences of a Type I error are more severe, alpha is set at .01 instead. Dr. Ballard, of course, wants to avoid Type I error, but he can live with a 5% chance of it occurring. So, he sets alpha at .05.

Now he needs to calculate degrees of freedom both for the numerator and the denominator of the F ratios. The degrees of freedom for the numerator will change from F ratio to F ratio, moving from df_{Rows} to df_{Columns} to $df_{\text{Interaction}}$ as a researcher moves from effect to effect. The degrees of freedom for the denominator term for the between-subjects, two-way ANOVA is degrees of freedom within, df_{Within}, and it will remain so across all three of the hypothesis tests. Formulas for calculating all 4 of the degrees of freedom needed to find the critical values of F are shown in Equation 12.2. Table 12.9 (on page 437) shows how to calculate the other 2 degrees of freedom—between groups degrees of freedom and total degrees of freedom—that will be needed to complete the ANOVA summary table.

Equation 12.2 Formulas for Calculating Degrees of Freedom for Between-Subjects, Two-Way ANOVA

$$df_{\text{Rows}} = R - 1$$

$$df_{\text{Columns}} = C - 1$$

$$df_{\text{Interaction}} = df_{\text{Rows}} \times df_{\text{Columns}}$$

$$df_{\text{Within}} = N - (R \times C)$$

where df_{Rows} = degrees of freedom for the row main effect

df_{Columns} = degrees of freedom for the column main effect

$df_{\text{Interaction}}$ = degrees of freedom for the interaction effect

df_{Within} = degrees of freedom for the within-group effect

R = number of rows

C = number of columns

N = total number of cases

For the caffeine/sleep deprivation study, there are two rows. Here are Dr. Ballard's calculations for the degrees of freedom for the row main effect:

$$\begin{aligned} df_{\text{Rows}} &= R - 1 \\ &= 2 - 1 \\ &= 1 \end{aligned}$$

With three columns, his calculations for the degrees of freedom for the column main effect look like this:

$$\begin{aligned} df_{\text{Columns}} &= C - 1 \\ &= 3 - 1 \\ &= 2 \end{aligned}$$

Once df_{Rows} and df_{Columns} are known, it is easy to calculate the numerator degrees of freedom for the interaction effect:

$$\begin{aligned} df_{\text{Interaction}} &= df_{\text{Rows}} \times df_{\text{Columns}} \\ &= 1 \times 2 \\ &= 2 \end{aligned}$$

Finally, checking back and seeing that there were 30 cases, Dr. Ballard calculates the degrees of freedom for the within-group effect:

$$\begin{aligned} df_{\text{Within}} &= N - (R \times C) \\ &= 30 - (2 \times 3) \\ &= 30 - 6 \\ &= 24 \end{aligned}$$

Once all the degrees of freedom have been calculated, they can be used to find the three critical values of F (see Table 12.8). The row main effect has 1 degree of freedom in the numerator, that's df_{Rows}, and 24 in the denominator, df_{Within}. Dr. Ballard looks in Appendix Table 4, at the intersection of the column with 1 degree of freedom and the row with 24 degrees of freedom in the F critical values table for $\alpha = .05$.

TABLE 12.8 Guide to Finding the Critical Value of *F* for Each Effect in a Between-Subjects, Two-Way ANOVA

Effect	Numerator *df*	Denominator *df*
Rows	df_{Rows}	df_{Within}
Columns	df_{Columns}	df_{Within}
Interaction	$df_{\text{Interaction}}$	df_{Within}

The critical values of *F*, F_{cv}, are found in Appendix Table 4 at the intersection of the column for the correct numerator degrees of freedom with the row for the correct denominator degrees of freedom. For a between-subjects, two-way ANOVA, this table reveals what the numerator and denominator degrees of freedom are for each effect.

There, he finds that the critical value of *F*, with $\alpha = .05$, is 4.260 for the row main effect: $F_{cv\ \text{Rows}} = 4.260$.

Next, he finds the critical value of *F* for the column main effect. In Appendix Table 4, he uses the column with 2 degrees of freedom and the row with 24 degrees of freedom to find that F_{cv} for the column effect is 3.403: $F_{cv\ \text{Columns}} = 3.403$. The interaction effect has the same numerator and denominator degrees of freedom as the column effect, 2 and 24. So, the critical value of *F* for the interaction effect is the same as the column effect: $F_{cv\ \text{Interaction}} = 3.403$.

Dr. Ballard can now write the decision rules for the three effects for the caffeine/sleep deprivation study as shown:

- Row main effect (caffeine)
  - If $F_{\text{Rows}} \geq 4.260$, reject $H_{0\ \text{Rows}}$.
  - If $F_{\text{Rows}} < 4.260$, fail to reject $H_{0\ \text{Rows}}$.
- Column main effect (sleep deprivation)
  - If $F_{\text{Columns}} \geq 3.403$, reject $H_{0\ \text{Columns}}$.
  - If $F_{\text{Columns}} < 3.403$, fail to reject $H_{0\ \text{Columns}}$.
- Interaction effect
  - If $F_{\text{Interaction}} \geq 3.403$, reject $H_{0\ \text{Interaction}}$.
  - If $F_{\text{Interaction}} < 3.403$, fail to reject $H_{0\ \text{Interaction}}$.

Step 5 Calculate the Test Statistics

Between-subjects, two-way ANOVA starts the same way between-subjects, one-way ANOVA does. It separates the total variability in a set of scores into between-group variability and within-group variability. Then it goes a step further and separates between-group variability into three subcomponents—variability due to the row effect, variability due to the column effect, and variability due to the interaction effect. Finally, each of these three effects is tested individually with its own *F* ratio.

The sources of variability in the data are separated out by calculating sums of squares, as was done for the one-way ANOVA. The sums of squares that need to be calculated are:

- Sum of squares between groups (SS_{Between})
- Sum of squares rows (SS_{Rows})
- Sum of squares columns (SS_{Columns})
- Sum of squares interaction ($SS_{\text{Interaction}}$)
- Sum of squares within groups (SS_{Within})
- Sum of squares total (SS_{Total})

Remember that a sum of squares in an ANOVA is a sum of squared deviation scores (see Chapter 11). What scores are used and what mean is subtracted from them vary depending on the source of variability being calculated. The sums of squares are arranged in an ANOVA summary table, Table 12.9, along with the formulas to complete the other necessary values. Note that the same columns that were used for a one-way and repeated-measures ANOVA summary table—source of variability, sum of squares, degrees of freedom, mean square, and *F* ratio—are present in the same order in the summary table for the two-way ANOVA. But, the sources of variability are different. Making sure that the correct order is followed for the rows and columns in an ANOVA summary table is key to completing an ANOVA. The three effects that are being tested fall under the between-groups effect. Note how they are indented in our table to show that they are derived from the between-groups effect.

TABLE 12.9 Template for ANOVA Summary Table for Between-Subjects, Two-Way ANOVA

Source of Variability	Sum of Squares	Degrees of Freedom	Mean Square	*F* ratio
Between groups	$SS_{Rows} + SS_{Columns} + SS_{Interaction}$	$df_{Rows} + df_{Columns} + df_{Interaction}$		
Rows	Given	$R - 1$	$\frac{SS_{Rows}}{df_{Rows}}$	$\frac{MS_{Rows}}{MS_{Within}}$
Columns	Given	$C - 1$	$\frac{SS_{Columns}}{df_{Columns}}$	$\frac{MS_{Columns}}{MS_{Within}}$
Interaction	Given	$df_{Rows} \times df_{Columns}$	$\frac{SS_{Interaction}}{df_{Interaction}}$	$\frac{MS_{Interaction}}{MS_{Within}}$
Within groups	Given	$N - (R \times C)$	$\frac{SS_{Within}}{df_{Within}}$	
Total	$SS_{Between} + SS_{Within}$	$N - 1$		

Be sure to label and order the rows and columns in a between-subjects, two-way ANOVA summary table as they are here. Note that some cells are left blank.

Here are the sums of squares for the row effect, the column effect, the interaction effect, and for within-group variability for the caffeine/sleep deprivation study (formulas for calculating sums of squares for a two-way ANOVA are given in an appendix to this chapter):

- $SS_{Rows} = 3{,}630.00$
- $SS_{Between} = 5{,}950.00$
- $SS_{Columns} = 1{,}340.00$
- $SS_{Interaction} = 980.00$
- $SS_{Within} = 748.00$
- $SS_{Total} = 6{,}698.00$

Equation 12.2 has already been used to calculate four degrees of freedom—rows, columns, interaction, and within. Table 12.9 shows how to calculate the other two degrees of freedom—between groups and total. Between-groups variability is broken down into variability for rows, columns, and interaction, so degrees of freedom between groups, df_{Between}, is found by adding up df_{Rows}, df_{Columns}, and $df_{\text{Interaction}}$. For the caffeine/sleep deprivation study, this is

$$\begin{aligned} df_{\text{Between}} &= df_{\text{Rows}} + df_{\text{Columns}} + df_{\text{Interaction}} \\ &= 1 + 2 + 2 \\ &= 5 \end{aligned}$$

As shown in Table 12.9, the degrees of freedom total, df_{Total}, is calculated by subtracting 1 from the total number of subjects:

$$\begin{aligned} df_{\text{Total}} &= N - 1 \\ &= 30 - 1 \\ &= 29 \end{aligned}$$

The next step is to calculate mean squares for rows, columns, interaction, and within groups. Following the instructions in Table 12.9:

$$\begin{aligned} MS_{\text{Rows}} &= \frac{SS_{\text{Rows}}}{df_{\text{Rows}}} \\ &= \frac{3{,}630.00}{1} \\ &= 3{,}630.00 \end{aligned}$$

$$\begin{aligned} MS_{\text{Columns}} &= \frac{SS_{\text{Columns}}}{df_{\text{Columns}}} \\ &= \frac{1{,}340.00}{2} \\ &= 670.00 \end{aligned}$$

$$\begin{aligned} MS_{\text{Interaction}} &= \frac{SS_{\text{Interaction}}}{df_{\text{Interaction}}} \\ &= \frac{980.00}{2} \\ &= 490.00 \end{aligned}$$

$$\begin{aligned} MS_{\text{Within}} &= \frac{SS_{\text{Within}}}{df_{\text{Within}}} \\ &= \frac{748.00}{24} \\ &= 31.1667 \\ &= 31.17 \end{aligned}$$

As directed in Table 12.9, Dr. Ballard finds the three F ratios by dividing the mean squares for the row main effect, the column main effect, and the interaction effect by the mean square for within-group variability:

$$F_{\text{Rows}} = \frac{MS_{\text{Rows}}}{MS_{\text{Within}}} = \frac{3{,}630.00}{31.17} = 116.4581 = 116.46$$

$$F_{\text{Columns}} = \frac{MS_{\text{Columns}}}{MS_{\text{Within}}} = \frac{670.0}{31.17} = 21.4950 = 21.50$$

$$F_{\text{Interaction}} = \frac{MS_{\text{Interaction}}}{MS_{\text{Within}}} = \frac{490.00}{31.17} = 15.7202 = 15.72$$

With all the F ratios calculated, the ANOVA summary table is complete (see Table 12.10). We'll come back to see how the results are interpreted after getting more practice with calculations.

TABLE 12.10 Completed ANOVA Summary Table Showing the Effects of Caffeine and Sleep Deprivation on Mental Alertness

Source of Variability	Sum of Squares	Degrees of Freedom	Mean Square	F ratio
Between groups	5,950.00	5		
Rows	3,630.00	1	3,630.00	116.46
Columns	1,340.00	2	670.00	21.50
Interaction	980.00	2	490.00	15.72
Within groups	748.00	24	31.17	
Total	6,698.00	29		

The effect of caffeine consumption is the row variable, and the effect of sleep deprivation is the column variable.

Worked Example 12.2 The question "What makes relationships work?" fills the covers of supermarket magazines, but it also interests research psychologists. Dr. Larue, a social psychologist, conducted a study to investigate two variables that she believed were associated with relationship satisfaction. The variables were (1) arguing style and (2) type of parental relationship model.

Dr. Larue found college seniors who were in serious relationships and assessed their ability to argue in a positive or negative way. People who argue positively don't become threatened or defensive, don't attack their partners, and help arguments reach a successful conclusion that is satisfactory to both sides. Dr. Larue classified students as positive arguers, mixed arguers, or negative arguers.

Dr. Larue also asked these students to rate the quality of their parents' marriage as being good, average, or bad. With three levels of parental marital quality and three levels of arguing style, there were nine cells in Dr. Larue's design. Dr. Larue randomly selected eight students from each cell. That is, there were eight who perceived their parents' marriages as good and who were positive arguers, eight who rated their parents' marriages as good and who had a mixed arguing style, and so on. With nine cells and eight participants per cell, Dr. Larue had a total of 72 participants.

Dr. Larue then measured the second grouping variable by asking each student to rate his or her satisfaction with his or her current relationship. (Dr. Larue made sure that no one in the study was in a relationship with someone else in the study.) The interval-level satisfaction scores could range from a low of 5 to a high of 35.

Table 12.11 shows the mean (and standard deviation) for each cell, as well as the row means and the column means. Figure 12.6 displays the data graphically. Looking at Figure 12.6 and Table 12.11 together, there are three things to note:

1. It appears that there is a row main effect for arguing style. Mean satisfaction scores decrease as the amount of negative arguing increases, from 27.92 for positive arguers, to 21.25 for mixed arguers, and 16.00 for negative arguers.
2. It appears that there is a column main effect for perceived quality of parental marriage. Mean satisfaction scores decrease as the perception of parental marriage becomes more negative, from 24.42 for those with a positive perception, to 22.08 for those with an average perception, and down to 18.67 for those with a negative perception.

TABLE 12.11 Descriptive Statistics for Relationship Satisfaction Study

	Parents' Marriage Perceived as "Good"	Parents' Marriage Perceived as "Average"	Parents' Marriage Perceived as "Bad"	
Positive Arguing Style	30.50 (3.07)	28.25 (2.43)	25.00 (2.93)	27.92
Mixed Arguing Style	23.75 (3.33)	21.50 (2.45)	18.50 (3.21)	21.25
Negative Arguing Style	19.00 (2.67)	16.50 (2.07)	12.50 (3.21)	16.00
	24.42	22.08	18.67	

Means (and standard deviations) for satisfaction with the current relationship are reported for each cell. Row means and column means are also reported. Satisfaction scores can range from a low of 5 to a high of 35.

3. The three lines are parallel, so there appears to be no interaction between arguing style and quality of parental marriage on relationship satisfaction. Of course, Dr. Larue needs to do a statistical test to see if her observations pan out statistically.

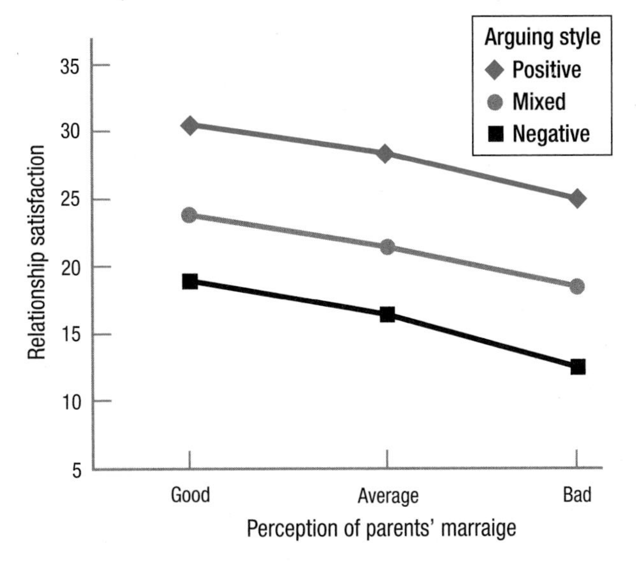

Figure 12.6 Impact of Arguing Style and Perception of Parental Marriage on Relationship Satisfaction This graph shows a main effect of arguing style and a main effect of perception of parents' marriage on relationship satisfaction. The lines are parallel, so there is no interaction between the two grouping variables.

Step 1 Pick a Test. There are two grouping variables (arguing style and parental marriage). Each grouping variable has three levels (positive, mixed, and negative arguing style; good, average, or bad parental marriage). When three levels of arguing style are crossed with three levels of parental marriage, this forms nine groups. Each case is in just one group and each case is not paired with another case, so the groups are all independent. That means this is a between-subjects design. Comparing the means of groups defined by two grouping variables and of groups that are independent samples calls for a between-subjects, two-way ANOVA. Specifically, this is a 3×3 ANOVA.

Step 2 Check the Assumptions.

- *Random samples.* The cases used in this study are random samples from the populations of seniors at Dr. Larue's university who are in relationships and fit in one of the nine cells. This assumption is not violated as long as that is the population to which she wishes to generalize her results.
- *Independence of observations.* No person in the study was in a relationship with anyone else in the study, so this assumption is not violated.
- *Normality.* Dr. Larue is willing to assume that relationship satisfaction is normally distributed within each of the nine populations represented here. Therefore, this assumption is not violated.
- *Homogeneity of variance.* The standard deviations for all nine groups are shown in Table 12.11 and they are all about the same. No standard deviation is twice another, so this assumption is not violated.

Dr. Larue can proceed with the planned statistical test.

Step 3 List the Hypotheses. There are three hypotheses in a two-way ANOVA: one for the row main effect (arguing style), one for the column main effect (parental marriage quality), and one for the interaction effect (arguing style × parental marriage quality).

- Row main effect:

$$H_{0\text{ Rows}}: \mu_{\text{Row1}} = \mu_{\text{Row2}} = \mu_{\text{Row3}}$$

$H_{1\text{ Rows}}$: At least one row population mean is different from at least one other.

- Column main effect:

$$H_{0\text{ Columns}}: \mu_{\text{Column1}} = \mu_{\text{Column2}} = \mu_{\text{Column3}}$$

$H_{1\text{ Columns}}$: At least one column population mean is different from at least one other.

- Interaction effect:

$H_{0\text{ Interaction}}$: There is no interactive effect of the two grouping variables on the dependent variable in the population.

$H_{1\text{ Interaction}}$: The two grouping variables interact to affect the dependent variable in at least one cell in the population.

Step 4 Set the Decision Rules. Setting the decision rules requires deciding on an alpha level and knowing the degrees of freedom for the numerator and the degrees of freedom for the denominator for each of the three effects being tested. Dr. Larue is comfortable having a 5% chance of Type I error, so she sets alpha at .05.

Next, to calculate degrees of freedom, she needs to know R (the number of rows), C (the number of columns), and N (the total number of cases):

- $R = 3$ (There are three levels of arguing style.)
- $C = 3$ (There are three levels of parental marriage quality.)
- $N = 72$ (Each cell has eight cases and there are nine cells: $8 \times 9 = 72$.)

Using Equation 12.2, Dr. Larue calculates the various degrees of freedom:

$$\begin{aligned} df_{\text{Rows}} &= R - 1 \\ &= 3 - 1 \\ &= 2 \end{aligned}$$

$$\begin{aligned} df_{\text{Columns}} &= C - 1 \\ &= 3 - 1 \\ &= 2 \end{aligned}$$

$$\begin{aligned} df_{\text{Interaction}} &= df_{\text{Rows}} \times df_{\text{Columns}} \\ &= 2 \times 2 \\ &= 4 \end{aligned}$$

$$\begin{aligned} df_{\text{Within}} &= N - (R \times C) \\ &= 72 - (3 \times 3) \\ &= 72 - 9 \\ &= 63 \end{aligned}$$

Table 12.8 offers guidance as to which source provides the numerator and denominator degrees of freedom for each effect. The critical value of *F* for the main effect of rows has 2 degrees of freedom in the numerator (df_{Rows}) and 63 degrees of freedom in the denominator (df_{Within}). Looking in Appendix Table 4 at the intersection of the column with 2 degrees of freedom and the row with 63 degrees of freedom, Dr. Larue discovers no such row exists. In these situations, apply *The Price Is Right* rule and use the degrees of freedom value that is closest without going over. Here, that is 60, which makes $F_{cv\ \text{Rows}} = 3.150$.

This study has the same degrees of freedom for the numerator and the denominator for the column main effect, 2 and 63. This means $F_{cv\ \text{Columns}} = 3.150$.

The degrees of freedom for the numerator for the interaction effect ($df_{\text{Interaction}}$) are 4 and the denominator degrees of freedom (df_{Within}) are 63. Using Table 4 in the Appendix and *The Price Is Right* rule, $F_{cv\ \text{Interaction}} = 2.525$.

Here are Dr. Larue's three decision rules:

- Main effect of rows:
  - If $F_{\text{Rows}} \geq 3.150$, reject $H_{0\ \text{Rows}}$.
  - If $F_{\text{Rows}} < 3.150$, fail to reject $H_{0\ \text{Rows}}$.
- Main effect of columns:
  - If $F_{\text{Columns}} \geq 3.150$, reject $H_{0\ \text{Columns}}$.
  - If $F_{\text{Columns}} < 3.150$, fail to reject $H_{0\ \text{Columns}}$.
- Interaction effect:
  - If $F_{\text{Interaction}} \geq 2.525$, reject $H_{0\ \text{Interaction}}$.
  - If $F_{\text{Interaction}} < 2.525$, fail to reject $H_{0\ \text{Interaction}}$.

Step 5 Calculate the Test Statistics. Here are the sums of squares for Dr. Larue's study, derived from the formulas provided in the chapter appendix:

- $SS_{\text{Between}} = 2{,}117.00$
- $SS_{\text{Rows}} = 1{,}712.00$
- $SS_{\text{Columns}} = 401.00$
- $SS_{\text{Interaction}} = 4.00$
- $SS_{\text{Within}} = 469.00$
- $SS_{\text{Total}} = 2{,}586.00$

Dr. Larue has already calculated degrees of freedom for rows, columns, interaction, and within groups (2, 2, 4, and 63, respectively), so she can now calculate degrees of freedom for between groups and total degrees of freedom as directed by Table 12.9:

$$\begin{aligned} df_{\text{Between}} &= df_{\text{Rows}} + df_{\text{Columns}} + df_{\text{Interaction}} \\ &= 2 + 2 + 4 \\ &= 8 \end{aligned}$$

$$\begin{aligned} df_{\text{Total}} &= N - 1 \\ &= 72 - 1 \\ &= 71 \end{aligned}$$

The next step is the calculation of the four mean squares. To do this, Dr. Larue divides each of the four sums of squares by its degrees of freedom:

$$MS_{\text{Rows}} = \frac{SS_{\text{Rows}}}{df_{\text{Rows}}}$$

$$= \frac{1,712.00}{2}$$

$$= 856.00$$

$$MS_{\text{Columns}} = \frac{SS_{\text{Columns}}}{df_{\text{Columns}}}$$

$$= \frac{401.00}{2}$$

$$= 200.50$$

$$MS_{\text{Interaction}} = \frac{SS_{\text{Interaction}}}{df_{\text{Interaction}}}$$

$$= \frac{4.00}{4}$$

$$= 1.00$$

$$MS_{\text{Within}} = \frac{SS_{\text{Within}}}{df_{\text{Within}}}$$

$$= \frac{469.00}{63}$$

$$= 7.4444$$

$$= 7.44$$

Once the four mean squares have been calculated, all that is left is to find the three F ratios. To do this, Dr. Larue divides each of the mean squares—rows, columns, and interaction—by the mean square within groups:

$$F_{\text{Rows}} = \frac{MS_{\text{Rows}}}{MS_{\text{Within}}}$$

$$= \frac{856.00}{7.44}$$

$$= 115.0538$$

$$= 115.05$$

$$F_{\text{Columns}} = \frac{MS_{\text{Columns}}}{MS_{\text{Within}}}$$

$$= \frac{200.50}{7.44}$$

$$= 26.9489$$

$$= 26.95$$

$$F_{\text{Interaction}} = \frac{MS_{\text{Interaction}}}{MS_{\text{Within}}}$$
$$= \frac{1.00}{7.44}$$
$$= 0.1344$$
$$= 0.13$$

The completed ANOVA summary table is shown in Table 12.12. How to interpret the results found in a summary table is the next order of business.

TABLE 12.12 Completed ANOVA Summary Table for Dr. Larue's Relationship Satisfaction Study

Source of Variability	Sum of Squares	Degrees of Freedom	Mean Square	F ratio
Between groups	2,117.00	8		
Rows	1,712.00	2	856.00	115.05
Columns	401.00	2	200.50	26.95
Interaction	4.00	4	1.00	0.13
Within groups	469.00	63	7.44	
Total	2,586.00	71		

Arguing style is the row variable, and perception of parental marriage is the column variable.

Practice Problems 12.2

Apply Your Knowledge

12.04 Read each scenario and decide what statistical test should be used. Select from a single-sample *z* test; single-sample *t* test; independent-samples *t* test; paired-samples *t* test; between-subjects, one-way ANOVA; one-way, repeated-measures ANOVA; and between-subjects, two-way ANOVA.

a. People who are classified as (1) overweight, (2) normal weight, or (3) underweight are randomly assigned to drink either (1) regular soda, (2) diet soda, or (3) water. Thirty minutes later they indicate, on an interval scale, how thirsty they are.

b. Backpacks of elementary school students, middle school students, and high school students are weighed to see if there are differences in how heavy they are.

12.05 List the hypotheses for a between-subjects, two-way ANOVA in which there are four rows and three columns.

12.06 Given $df_{\text{Rows}} = 2$, $df_{\text{Columns}} = 4$, $df_{\text{Interaction}} = 8$, and $df_{\text{Within}} = 165$, list the critical values of *F* for the three *F* ratios for a between-subjects, two-way ANOVA for $\alpha = .05$.

12.07 Given $n = 7$, $R = 2$, $C = 2$, $SS_{\text{Between}} = 650.00$, $SS_{\text{Rows}} = 250.00$, $SS_{\text{Columns}} = 300.00$, $SS_{\text{Interaction}} = 100.00$, $SS_{\text{Within}} = 800.00$, and $SS_{\text{Total}} = 1{,}450.00$, complete an ANOVA summary table for a between-subjects, two-way ANOVA.

12.3 Interpreting a Between-Subjects, Two-Way ANOVA

Interpretation of a statistical test involves stating in plain language what the results mean. The interpretation plan for a two-way ANOVA addresses the same three questions as were addressed for a one-way ANOVA but for more effects: (1) Were the null hypotheses rejected? (2) How large are the effects? (3) Where are the effects and what is their direction?

Let's start with Dr. Ballard's study about the impact of caffeine and sleep deprivation on mental alertness. In this study, 30 college students were randomly assigned to six groups. Before completing a mental alertness task one hour after waking up, half the participants consumed a cup of caffeinated coffee and half didn't. Further, one third of the participants had a full night's sleep, one third were sleep-deprived by one hour, and one third by two hours. The results of the study are shown in Figure 12.5 (see page 432) and Figure 12.7, and in Table 12.13.

- Figure 12.7 graphs the row means for the main effect of caffeine and the column means for the main effect of sleep deprivation.
- Figure 12.5 uses the cell means to show what appears to be an interaction effect.
- Table 12.13 shows the ANOVA summary table, from which Dr. Ballard will need a number of values as he interprets the results.

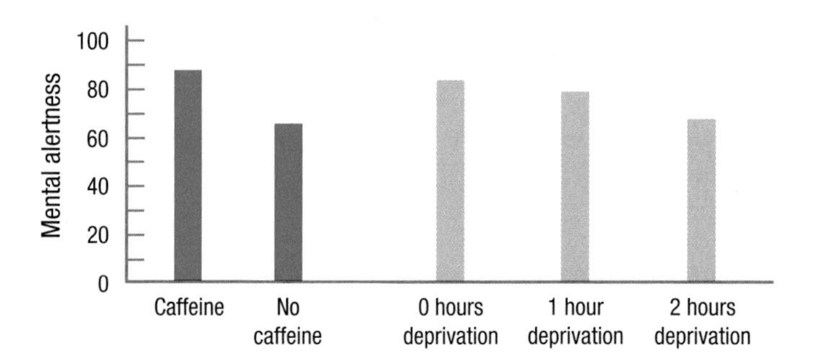

Figure 12.7 Mean Mental Alertness for Row Main Effect (Caffeine Consumption) and Column Main Effect (Sleep Deprivation) The row means and column means appear to show that caffeine improves performance and that sleep deprivation harms it.

TABLE 12.13 ANOVA Summary Table Showing the Effects of Caffeine and Sleep Deprivation on Mental Alertness

Source of Variability	Sum of Squares	Degrees of Freedom	Mean Square	*F* ratio
Between groups	5,950.00	5		
Rows	3,630.00	1	3,630.00	116.46
Columns	1,340.00	2	670.00	21.50
Interaction	980.00	2	490.00	15.72
Within groups	748.00	24	31.17	
Total	6,698.00	29		

Caffeine consumption is the row variable, and sleep deprivation is the column variable.

Were the Null Hypotheses Rejected?

To determine if any of the three null hypotheses was rejected—one for the row main effect, one for the column main effect, and one for the interaction effect—Dr. Ballard needs the decision rules generated in Step 4 and the *F* ratios calculated in Step 5. The critical values of *F* were $F_{cv\ Rows} = 4.260$, $F_{cv\ Columns} = 3.403$, and $F_{cv\ Interaction} = 3.403$. The values of *F* calculated were $F_{Rows} = 116.46$, $F_{Columns} = 21.50$, and $F_{Interaction} = 15.72$ (see Table 12.13). Applying the three decision rules:

- $116.46 \geq 4.260$, so reject $H_{0\ Rows}$, accept $H_{1\ Rows}$, and call the row effect statistically significant.
- $21.50 \geq 3.403$, so reject $H_{0\ Columns}$, accept $H_{1\ Columns}$, and call the column effect statistically significant.
- $15.72 \geq 3.403$, so reject $H_{0\ Interaction}$, accept $H_{1\ Interaction}$, and call the interaction effect statistically significant.

The next step is to write the results in APA format. APA format for the results of an ANOVA means reporting five pieces of information: (1) stating what test was done (an *F* test), (2) indicating the numerator and denominator degrees of freedom for the *F* ratio, (3) reporting the observed value of the test statistic, (4) naming the selected alpha level, and (5) telling whether the observed *F* fell in the rare zone ($p < .05$, i.e., null hypothesis was rejected) or in the common zone ($p > .05$, the null hypothesis was not rejected).

APA format for the main effect of rows is

$$F(1, 24) = 116.46, p < .05$$

For the main effect of columns, APA format is

$$F(2, 24) = 21.50, p < .05$$

For the interaction effect, APA format is

$$F(2, 24) = 15.72, p < .05$$

After determining the status—statistically significant or not—of each *F* ratio, a researcher can begin to interpret the results. It is tempting to start the interpretation with one of the main effects displayed in Figure 12.7, but remember—with a two-way ANOVA and a statistically significant interaction effect, the interaction takes precedence.

When the interaction effect is statistically significant, the null hypothesis that each main effect has an independent effect on the dependent variable is rejected. The alternative hypothesis that the two independent variables interact to affect the dependent variable in at least one cell is accepted. As a result, the main effects are less relevant. To explore further what the interaction means, Dr. Ballard will need to use a post-hoc test to compare individual cell means.

Dr. Ballard can't be sure which mean differences are statistically significant until the post-hoc tests are completed, but inspecting Figure 12.5 gives some sense of the interaction. Figure 12.5 suggests that the amount of sleep deprivation has little effect on mental alertness when people are dosed with caffeine. However, for those who don't receive caffeine, alertness declines as sleep deprivation increases.

What about the statistically significant main effects? Does sleep deprivation affect mental alertness? Does caffeine? Both answers are of the "it depends" variety. Look at Figure 12.5. Whether sleep deprivation affects performance depends on whether one has consumed caffeine. And, whether caffeine affects performance depends on whether a person is sleep-deprived. It doesn't look like the main effects will add much to our interpretation.

How Big Are the Effects?

The same measure of effect, eta squared, is used for two-way ANOVA as was used for repeated-measures ANOVA. For a two-way ANOVA, eta squared can be calculated for each main effect and for the interaction. Eta squared, like r^2, calculates the percentage of variability in the dependent variable that is explained by an explanatory variable or by the interaction of the explanatory variables.

The formulas for calculating eta squared for the row main effect, the column main effect, and the interaction effect are given in Equation 12.3.

Equation 12.3 Formulas for Calculating Eta Squared (η^2) for Row Main Effect, Column Main Effect, and Interaction Effect

$$\eta^2_{\text{Rows}} = \frac{SS_{\text{Rows}}}{SS_{\text{Total}}} \times 100$$

$$\eta^2_{\text{Columns}} = \frac{SS_{\text{Columns}}}{SS_{\text{Total}}} \times 100$$

$$\eta^2_{\text{Interaction}} = \frac{SS_{\text{Interaction}}}{SS_{\text{Total}}} \times 100$$

where η^2_{Rows} = eta squared for the row main effect, the percentage of variability in the dependent variable that is explained by the row explanatory variable

η^2_{Columns} = eta squared for the column main effect, the percentage of variability in the dependent variable that is explained by the column explanatory variable

$\eta^2_{\text{Interaction}}$ = eta squared for the interaction effect, the percentage of variability in the dependent variable that is explained by the interaction between the row explanatory variable and the column explanatory variable

SS_{Rows} = sum of squares rows

SS_{Columns} = sum of squares columns

$SS_{\text{Interaction}}$ = sum of squares interaction

SS_{Total} = sum of squares total

For the caffeine/sleep deprivation study, let's start with η^2 for the statistically significant interaction.

Dr. Ballard calculates eta squared for the interaction effect as follows:

$$\eta^2_{\text{Interaction}} = \frac{SS_{\text{Interaction}}}{SS_{\text{Total}}} \times 100$$

$$= \frac{980.00}{6{,}698.00} \times 100$$

$$= 0.1463 \times 100$$

$$= 14.63\%$$

$$\eta^2_{\text{Columns}} = \frac{SS_{\text{Columns}}}{SS_{\text{Total}}} \times 100$$

$$= \frac{1{,}340.00}{6{,}698.00} \times 100$$

$$= 20.00\%$$

$$\eta^2_{\text{Rows}} = \frac{SS_{\text{Rows}}}{SS_{\text{Total}}} \times 100$$

$$= \frac{3{,}630.00}{6{,}698.00} \times 100$$

$$= 54.20\%$$

The same standards are used for interpreting eta squared as were used for r^2 for one-way ANOVA:

- $\eta^2 \approx 1\%$ is a small effect.
- $\eta^2 \approx 9\%$ is a medium effect.
- $\eta^2 \approx 25\%$ is a large effect.

Even though the rows (caffeine) effect is very large at 54% and the column (sleep deprivation) effect at 20% is stronger than the interaction effect, the focus of the interpretation will be on the interaction. The interaction of the two variables, which explains about 15% of the variability in mental alertness, is a medium effect. Most of the variability that sleep deprivation explains is due to the effect on the no caffeine participants. The line for the caffeine-receiving participants in Figure 12.5 is mostly flat, indicating that degree of sleep deprivation explains little of the variability in mental alertness for these subjects. In contrast, the line for the no caffeine participants is on a downward trajectory as sleep deprivation increases, suggesting that amount of sleep deprivation explains a lot of the variability in mental alertness for these subjects. Does the amount of sleep deprivation explain a lot of the variability in mental alertness? It depends. The main effects are trumped by the interaction effect.

Where Are the Effects, and What Is Their Direction?

Just as with the other ANOVA tests, finding where the effects lie for two-way ANOVA involves the use of post-hoc tests. And, just as with the other ANOVAs, post-hoc tests for two-way ANOVA should be used only when the effect is statistically significant.

- If the row main effect is statistically significant, and if there are three or more levels of the row explanatory variable, then a post-hoc test for the row effect can be used to find which pairs of *row* means differ statistically. (If there are only two row means and the row main effect is statistically significant, then the two existing row means must differ statistically.)
- If the column main effect is statistically significant, and if there are three or more levels of the column explanatory variable, then a post-hoc test for the column effect can be used to find which pairs of *column* means differ statistically. (If there are only two column means and the column main effect is statistically significant, then the two existing column means must differ statistically.)
- If the interaction effect was statistically significant, then a post-hoc test for the interaction effect can be used to find which pair(s) of *cell* means differ statistically.

The post-hoc test for the between-subjects, two-way ANOVA is the same one used for other ANOVAs, the Tukey *HSD*. *HSD*, remember, stands for "honestly significant difference." If a pair of means differs by the *HSD* value or more than the *HSD* value, then the difference is a statistically significant one. The formulas for the calculation of *HSD* values are found in Equation 12.4.

Equation 12.4 Formulas for Calculating Tukey *HSD* Values for Post-Hoc Tests for Between-Subjects, Two-Way ANOVA

$$HSD_{\text{Rows}} = q_{\text{Rows}}\sqrt{\frac{MS_{\text{Within}}}{n_{\text{Rows}}}}$$

$$HSD_{\text{Columns}} = q_{\text{Columns}}\sqrt{\frac{MS_{\text{Within}}}{n_{\text{Columns}}}}$$

$$HSD_{\text{Cells}} = q_{\text{Cells}}\sqrt{\frac{MS_{\text{Within}}}{n_{\text{Cells}}}}$$

where HSD_{Rows} = *HSD* value for the row main effect

q_{Rows} = q value for the row main effect, from Appendix Table 5, where k = number of rows and $df = df_{\text{Within}}$

MS_{Within} = within-groups mean square

n_{Rows} = number of cases in a row

HSD_{Columns} = *HSD* value for the column main effect

q_{Columns} = q value for the column main effect, from Appendix Table 5, where k = the number of columns and $df = df_{\text{Within}}$

n_{Columns} = number of cases in a column

HSD_{Cells} = *HSD* value for the interaction effect

q_{Cells} = q value for the interaction effect, from Appendix Table 5, where k = the number of cells and $df = df_{\text{Within}}$

n_{Cells} = number of cases in a cell

For the caffeine/sleep deprivation data, there is little need to do post-hoc tests for the statistically significant main effects as our focus will be on the interaction. Instead, the *HSD* test will be used to interpret the interaction effect only.

The *HSD* to be calculated will be used to compare cell means—any two cell means that differ by the HSD_{Cells} value have a difference that is large enough to represent a statistically significant difference. And statistically significant sample differences provide evidence for population differences.

Here's what one needs to calculate HSD_{Cells}:

- Determine the alpha level, .05 or .01. Typically, the same alpha level as used in the decision rule for the *F* ratio is utilized. For Dr. Ballard's study, this means $\alpha = .05$.
- To find the q_{Cells} value in Appendix Table 5, know that $k = 6$, because there are six cells, and that $df = 24$, because $df_{\text{Within}} = 24$. The intersection of the column for $k = 6$ and the row for $df = 24$ gives $q_{\text{Cells}} = 4.37$.
- From the ANOVA summary table, note that $MS_{\text{Within}} = 31.17$.
- Each cell has five cases, so $n_{\text{Cells}} = 5$.

Here is Equation 12.4, with those values substituted:

$$\begin{aligned}
HSD_{\text{Cells}} &= q_{\text{Cells}}\sqrt{\frac{MS_{\text{Within}}}{n_{\text{Cells}}}} \\
&= 4.37\sqrt{\frac{31.17}{5}} \\
&= 4.37\sqrt{6.2340} \\
&= 4.37 \times 2.4968 \\
&= 10.9110 \\
&= 10.91
\end{aligned}$$

The HSD_{Cells} value is 10.91 and any two cell sample means that differ by that much or more have a statistically significant difference. Table 12.14 contains all six cell means and there are 15 possible cell-by-cell comparisons (see Table 12.15). Note, in the points below, how Dr. Ballard approaches the comparisons in an organized fashion and indicates the directions of the difference.

- For participants who consumed no caffeine, each increase in sleep deprivation—from 0 hours ($M = 80.00$) to 1 hour ($M = 68.00$), and from 1 hour to 2 hours ($M = 50.00$)—caused a statistically significant decline in mental alertness.
- Consuming caffeine seems to protect against the negative effects of sleep deprivation as there was no statistically significant change in cell means for the caffeine group. (The means for 0, 1, and 2 hours of sleep deprivation were, respectively, 88.00, 90.00, and 86.00.)
- Looking at differences between the caffeine and no-caffeine groups, there is no evidence that caffeine consumption helped performance if no sleep deprivation occurred (means of 88.00 vs. 80.00). But, it did help performance with 1 hour of sleep deprivation (90.00 vs. 68.00) and 2 hours of sleep deprivation (86.00 vs. 50.00).

TABLE 12.14 Mental Alertness Cell Means for Caffeine/Sleep Deprivation Data

	0-hours Sleep Deprivation	1-hour Sleep Deprivation	2-hours Sleep Deprivation
Caffeine	88.00	90.00	86.00
No Caffeine	80.00	68.00	50.00

Any cell means that differ by at least the HSD_{Cells} value of 10.91 are statistically significantly different.

TABLE 12.15 Possible Comparisons of Cell Means for the Caffeine Consumption/Sleep Deprivation Study

	0-hours Sleep Deprivation	1-hour Sleep Deprivation	2-hours Sleep Deprivation
Caffeine	A	B	C
No Caffeine	D	E	F

Possible Cell-to-Cell Comparisons	
A vs.	B, C, D, E, and F
B vs.	C, D, E, and F
C vs.	D, E, and F
D vs.	E and F
E vs.	F

Putting It All Together

Before writing an interpretation of a two-way ANOVA, it is helpful to review the interaction graph, Figure 12.5. The graph shows two things:

- For those who don't consume caffeine, mental alertness decreases as sleep deprivation moves from 0 hours of deprivation to 1 hour and to 2 hours.
- Consuming caffeine keeps mental alertness from deteriorating, at least with 1 or 2 hours of sleep deprivation.

Here's Dr. Ballard's interpretation in which he addresses the following four points: What was done? What was found? What does it mean? What suggestions exist for future research?

> This study explored the effects of caffeine consumption and sleep deprivation on mental alertness. Using a between-subjects design, 30 college students were assigned to six groups and then had their mental alertness tested. Half received caffeine before testing and half didn't; one third had a full night's sleep, one third were awakened 1 hour early, and one third 2 hours early. There was a statistically significant interaction effect of the two variables on mental alertness $F(2, 24) = 15.72$, $p < .05$ as well as statistically significant main effects for caffeine $F(1, 24) = 116.46$, $p < .05$ and sleep deprivation $F(2, 24) = 21.50$, $p < .05$. In general, caffeine

consumption kept mental alertness elevated and, as sleep deprivation increased, performance deteriorated. However, these two variables did not independently affect mental alertness.

The interaction effect was moderately strong and showed that the impact of sleep deprivation on mental alertness depended on whether one consumed caffeine before testing or not. Further, how caffeine affected mental alertness depended on how sleep-deprived one was. For people who did not consume caffeine, increasing sleep deprivation caused a worsening of mental alertness. In contrast, consuming caffeine kept sleep deprivation from affecting mental alertness. This study suggests that a person can compensate for the mental alertness deficit caused by an hour or two of sleep deprivation by drinking a cup of coffee. It would be wise to replicate this study, to see if the effect is found in a different population. If it is, future research should investigate the effect of different doses of caffeine on different amounts of sleep deprivation.

(By the way, the data in this example were made up. Don't put too much faith in caffeine being an effective antidote to sleep deprivation. Sorry.)

Worked Example 12.3 For practice in interpreting two-way ANOVA, a return to Dr. Larue's study of factors affecting relationship satisfaction is in order. In that study, there were three levels of arguing style—positive, mixed, and negative—crossed with three different perceptions of the quality of one's parents' marriage—good, average, and bad. The mean level of relationship satisfaction for each of the nine conditions, measured on a scale ranging from 5 (very low) to 35 (very high), is shown in Table 12.16. The apparent lack of interaction is shown graphically in Figure 12.6. And, Figure 12.8 displays the main effects for arguing style and parental marital quality.

TABLE 12.16 Cell, Row, and Column Means for Relationship Satisfaction for Different Arguing Styles and Different Perceptions of Parental Marriages

	Parents' Marriage Perceived as "Good"	Parents' Marriage Perceived as "Average"	Parents' Marriage Perceived as "Bad"	
Positive Arguing Style	30.50 (3.07)	28.25 (2.43)	25.00 (2.93)	27.92
Mixed Arguing Style	23.75 (3.33)	21.50 (2.45)	18.50 (3.21)	21.25
Negative Arguing Style	19.00 (2.67)	16.50 (2.07)	12.50 (3.21)	16.00
	24.42	22.08	18.67	

The row and column means are suggestive of main effects for the two grouping variables. However, until the absence of an interaction effect is established and until it is determined that the main effects are statistically significant, they cannot be interpreted.

The three critical values of F were $F_{cv\ Rows} = 3.150$, $F_{cv\ Columns} = 3.150$, and $F_{cv\ Interaction} = 2.525$. The ANOVA summary table, which makes a return appearance in Table 12.17, shows that the observed values of F for the three effects were $F_{Rows} = 115.05$, $F_{Columns} = 26.95$, and $F_{Interaction} = 0.13$.

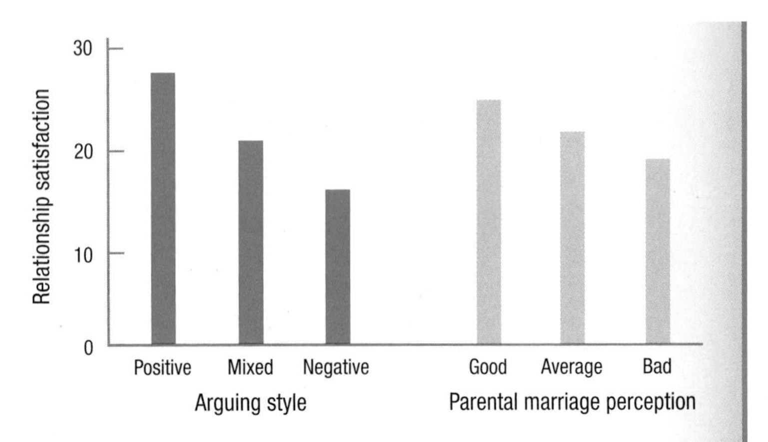

Figure 12.8 Main Effects of Arguing Style and Perception of Parental Marriage on Relationship Satisfaction According to this graph, it looks like there are two main effects: (1) relationship satisfaction decreases as arguing style becomes more negative; (2) relationship satisfaction decreases as perception of quality of parental marriage worsens.

TABLE 12.17 ANOVA Summary Table: Impact of Arguing Style and Perception of Parents' Marriage on Relationship Satisfaction

Source of Variability	Sum of Squares	Degrees of Freedom	Mean Square	*F* ratio
Between groups	2,117.00	8		
Rows	1,712.00	2	856.00	115.05
Columns	401.00	2	200.50	26.95
Interaction	4.00	4	1.00	0.13
Within groups	469.00	63	7.44	
Total	2,586.00	71		

This is the ANOVA summary table for the results of the relationship satisfaction study. Arguing style is the row variable and perception of parental marriage is the column variable.

Were the null hypotheses rejected?

- Row main effect: $115.05 \geq 3.150$, so reject $H_{0\ \text{Rows}}$ and accept $H_{1\ \text{Rows}}$.
    - In APA format: $F(2, 63) = 115.05$, $p < .05$.
    - The row effect, arguing style, is statistically significant.
    - It is reasonable to conclude, in the larger population, that at least one arguing style differs from at least one other in mean relationship satisfaction.
- Column main effect: $26.95 \geq 3.150$, so reject $H_{0\ \text{Columns}}$ and accept $H_{1\ \text{Columns}}$.
    - In APA format: $F(2, 63) = 26.95$, $p < .05$.
    - The column effect, perception of parental marital quality, is statistically significant.
    - It is reasonable to conclude, in the larger population, that at least one level of perceived marital quality differs from at least one other in mean relationship satisfaction.
- Interaction effect: $0.13 < 2.525$, so fail to reject $H_{0\ \text{Interaction}}$.
    - In APA format: $F(4, 63) = 0.13$, $p > .05$.
    - The interaction effect is not statistically significant.
    - There is not enough evidence to conclude, in the larger population, that the effect of arguing style interacts with the effect of perceived parental marital quality to affect relationship satisfaction.

In the caffeine/sleep deprivation study, the interaction effect was statistically significant and, as a result, the statistically significant main effects were ignored. Now, in the relationship satisfaction study, the two main effects are statistically significant and the interaction is not. How does interpretation work with this set of results?

How big are the effects? Effect size is measured by calculating eta squared (Equation 12.3):

$$\eta^2_{\text{Rows}} = \frac{SS_{\text{Rows}}}{SS_{\text{Total}}} \times 100$$

$$= \frac{1{,}712.00}{2{,}586.00} \times 100$$

$$= 0.6620 \times 100$$

$$= 66.20\%$$

$$\eta^2_{\text{Columns}} = \frac{SS_{\text{Columns}}}{SS_{\text{Total}}} \times 100$$

$$= \frac{401.00}{2586.00} \times 100$$

$$= 0.1551 \times 100$$

$$= 15.51\%$$

$$\eta^2_{\text{Interaction}} = \frac{SS_{\text{Interaction}}}{SS_{\text{Total}}} \times 100$$

$$= \frac{4.00}{2586.00} \times 100$$

$$= 0.0015 \times 100$$

$$= 0.15\%$$

Note that eta squared for the not-statistically-significant interaction effect was calculated. Even though the interaction effect was not significant, it is possible for eta squared to be sizable. If that happened, it would serve to alert a researcher to the possibility of Type II error. In the current situation, with the percentage of variability near zero, there is nothing to make Dr. Larue think she missed finding an interaction effect that really exists. From here on out, it is safe to ignore the interaction effect.

Eta squared for the rows effect was about 66% and for the columns effect it was about 16%. Both main effects have an impact on relationship satisfaction, but one more than the other. The impact of arguing style on relationship satisfaction is quite strong. The impact of perceived parental marital quality is smaller but still meaningful. These results suggest that higher levels of relationship satisfaction are associated more with being a positive arguer than with perceiving one's parents' marriage as good, though having a good model of a marriage is associated meaningfully with relationship satisfaction.

Where are the effects, and what is their direction? Now it is time to use Equation 12.4 to conduct some post-hoc tests and find out what is causing the statistically significant effects. Remember, only conduct a post-hoc test when the effect is statistically significant. With the current example, Dr. Larue has no need to find out what caused the interaction effect because there is no evidence that an interaction effect exists.

To apply Equation 12.4, first find the q value from Appendix Table 5. This depends on α (.05), how many means are being compared, and what the degrees of freedom are. For both the row effect and column effect, there are three means, so $k = 3$ in both instances. Both instances have the same degrees of freedom as well: $df_{\text{Within}} = 63$. Turning to Appendix Table 5, there is a column with $k = 3$, but no row for $df = 63$. In these situations, apply *The Price Is Right* rule and use the df value that is closest to 63 without going over. Here, that is $df = 60$. The q value, at the intersection of $k = 3$ and $df = 60$, is 3.40 for $\alpha = .05$.

To apply Equation 12.4, one also needs to know MS_{Within}, which is 7.44, and how many cases are in a row and a column. Each row contains 24 cases, as does each column. All the values are the same for our calculations for HSD_{Rows} and HSD_{Columns}: $q = 3.40$, $MS_{\text{Within}} = 7.44$, and $n = 24$, so both HSD values can be calculated in one pass:

$$
\begin{aligned}
HSD_{\text{Rows}} = HSD_{\text{Columns}} &= q\sqrt{\frac{MS_{\text{Within}}}{n}} \\
&= 3.40\sqrt{\frac{7.44}{24}} \\
&= 3.40\sqrt{.3100} \\
&= 3.40 \times 0.5568 \\
&= 1.8931 \\
&= 1.89
\end{aligned}
$$

With an HSD value of 1.89, any two row means or any two column means that differ by that amount or more are statistically significantly different.

For the row main effect, relationship satisfaction grows statistically significantly worse as there is less positive arguing and more negative arguing:

- Positive arguers ($M = 27.92$) have statistically significantly more relationship satisfaction than do mixed arguers ($M = 21.25$). The difference, 6.67, is greater than the HSD value of 1.89.
- Mixed arguers ($M = 21.25$) have statistically significantly more relationship satisfaction than do negative arguers ($M = 16.00$). The difference, 5.25. is greater than the HSD value.

For the column main effect, relationship satisfaction gets statistically significantly worse as the perception of the parents' marriage worsens:

- Students who rated their parents' marriage as good ($M = 24.42$) rated their own relationship as statistically significantly more satisfying than those who rated their parents' marriage as average ($M = 22.08$). The difference, 2.34, is greater than the HSD value of 1.89.
- Rating one's parents' marriage as average ($M = 22.08$) is associated with a statistically significantly higher score on relationship satisfaction than rating it as bad ($M = 18.67$). The difference, 3.41, is greater than the HSD value.

- *Putting it all together.* Before writing her four-point interpretation, Dr. Larue reviewed the graph displaying the main effects (Figure 12.8), so she had a clear picture of the results in her mind. Here's what she wrote (notice, this is a quasi-experimental study in which nothing is manipulated, so she avoids cause-and-effect language to describe the results):

In this social psychology study, the abilities of two variables, arguing style and perceived quality of parental marriage, to predict relationship satisfaction in college students were examined. Students were classified into three categories of arguing style (positive, negative, or mixed) and with three different perceptions of their parents' marriages (good, average, or bad). Eight students from each of these possible combinations who were in current relationships were randomly selected and completed a survey measuring degree of satisfaction with their current relationship.

Using a between-subjects, two-way ANOVA, both arguing style and parental marital perception had statistically significant effects on relationship satisfaction [respectively, $F(2, 63) = 115.05$, $p < .05$, and $F(2, 63) = 26.95$, $p < .05$]. There was no interaction effect for the two variables $F(4, 63) = 0.13$, $p > .05$.

Arguing style was the stronger predictor of relationship satisfaction. As students' arguing style moved from positive to negative, there was a decrease in relationship satisfaction. The role played by perceived parental marriage, though less powerful, was still meaningful. A more negative view of parents' marriages was associated with lower relationship satisfaction.

This study suggests that the relationships one sees as a child influence one's future relationship satisfaction. The good news is that a more powerful influence on relationship satisfaction is arguing style and a positive arguing style is a skill that can be learned. Future research should examine whether teaching positive arguing skills improves relationship satisfaction.

Practice Problems 12.3

12.08 Given $\alpha = .05$, $df_{Rows} = 3$, $df_{Columns} = 2$, $df_{Interaction} = 6$, $df_{Within} = 60$, $F_{Rows} = 3.25$, $F_{Columns} = 1.22$, and $F_{Interaction} = 0.83$, (a) write each result for this between-subjects, two-way ANOVA in APA format and (b) for each result report whether the effect is statistically significant.

12.09 Given this ANOVA summary table, (a) calculate η^2 for each effect and (b) classify each effect as small, medium, or large. Use $\alpha = .05$.

Source of Variability	Sum of Squares	Degrees of Freedom	Mean Square	F ratio
Between groups	5,357.00	15		
Rows	3,725.00	3	1,241.67	37.63
Columns	1,312.00	3	437.33	13.25
Interaction	320.00	9	35.56	1.08
Within groups	7,392.00	224	33.00	
Total	12,749.00	239		

12.10 Here is a table of cell means for a 3 × 2 between-subjects, two-way ANOVA with seven cases in each cell. Note that the row means and column means have been calculated.

	Column 1	Column 2	Column 3	
Row 1	26.00	30.00	34.00	30.00
Row 2	18.00	22.00	26.00	22.00
	22.00	26.00	30.00	

Here is the between-subjects, two-way ANOVA summary table for these data:

Source of Variability	Sum of Squares	Degrees of Freedom	Mean Square	*F* ratio
Between groups	1,120.00	5	224.00	
Rows	672.00	1	672.00	35.99
Columns	448.00	2	224.00	12.00
Interaction	0.00	2	0.00	0.00
Within groups	672.00	36	18.67	
Total	1,792.00	41		

As appropriate, calculate *HSD* values and comment on the direction of the differences for the effects.

12.11 A kinesiologist wanted to investigate the effect of temperature and humidity on human performance. He found 28 college students and randomly assigned them to four different conditions, during which they were to walk at their normal pace on a treadmill for 60 minutes. He measured how far, in miles, they walked. The conditions varied in temperature and humidity: (1) normal temperature and normal humidity; (2) normal temperature and high humidity; (3) high temperature and normal humidity; (4) high temperature and high humidity. The results looked like this:

	Normal Humidity	High Humidity	
Normal Temperature	3.00 miles	2.80 miles	2.90
High Temperature	2.80 miles	2.00 miles	2.40
	2.90	2.40	

Here is the ANOVA summary table:

Source of Variability	Sum of Squares	Degrees of Freedom	Mean Square	*F* ratio
Between groups	4.13	3		
Rows	1.75	1	1.75	25.00
Columns	1.75	1	1.75	25.00
Interaction	0.63	1	0.63	9.00
Within groups	1.58	24	0.07	
Total	5.71	27		

The critical value of *F* for each effect is 4.260. Eta squared for the row, column, and interaction effects, respectively, are 30.65%, 30.65%, and 11.03%. The *HSD* value for comparing cells is 0.39. Given all this information, write a four-point interpretation.

Application Demonstration

The two examples offered in this chapter so far—Dr. Ballard's caffeine/sleep deprivation study and Dr. Larue's relationship satisfaction study—have both involved statistically significant results. Unfortunately, research doesn't always turn out that way. Here's an example of a two-way ANOVA study in which not one of the three null hypotheses is rejected. How are results interpreted in this situation?

Imagine a sensory psychologist, Dr. Porter, who wanted to explore the threshold for perceiving low-frequency sounds. She tested six men and six women to see how low a sound they could perceive. Half the participants were tested in their left ears and half in their right ears. Sounds were measured in hertz (Hz), and the lower the hertz the better a person's hearing. (Want to see how well you do? Search for "Ultimate Sound Test [10000 Hz–1 Hz]" on YouTube.)

Table 12.18 and Figure 12.9 show the results. There appears to be no interaction and slightly better performance exists (a) for women and (b) for the left ear. Whether the effects are statistically significant or can be explained by sampling error remains to be seen.

TABLE 12.18 Cell, Row, and Column Means for Lowest Hertz Tone Perceived

	Left Ear	Right Ear	
Men	140.00	150.00	145.00
Women	120.00	130.00	125.00
	130.00	140.00	

The lower the mean, the better the hearing. Each cell contains three cases. The row means suggest a small advantage in hearing for women over men. The column means suggest a small advantage for the left ear over the right ear. The lack of interaction between the two variables can be seen in Figure 12.9.

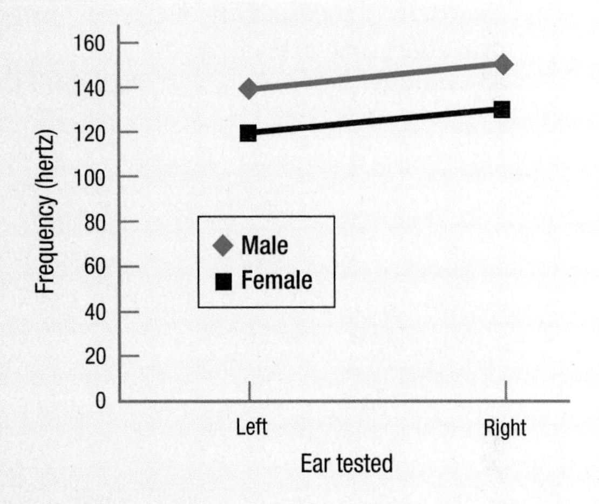

Figure 12.9 Impact of Sex and Ear Tested on Lowest Frequency Perceived
With the lines being parallel, there is no interaction between sex and ear tested. This graph also shows a slightly better performance by women and for the left ear.

The appropriate statistical test to compare the four means from these two independent variables is a two-way ANOVA. The four groups are all independent samples, so the test is a between-subjects, two-way ANOVA, specifically a 2×2 ANOVA.

No assumptions were violated and the ANOVA summary table is presented in Table 12.19. The critical value of F for all three effects was 5.318 and no effect was statistically significant.

- There is not enough evidence to conclude that men and women differ in their ability to hear low-frequency sounds.
- There is not enough evidence to conclude that the right ear differs from the left ear in its ability to hear low-frequency sounds.
- There is not enough evidence to conclude that a person's sex and which ear is tested interact to influence the perception of low-frequency sounds.

TABLE 12.19 Completed ANOVA Summary Table for Low-Frequency Perception Data

Source of Variability	Sum of Squares	Degrees of Freedom	Mean Square	F ratio
Between groups	1,500.00	3		
Rows	1,200.00	1	1,200.00	1.30
Columns	300.00	1	300.00	0.32
Interaction	0.00	1	0.00	0.00
Within groups	7,400.00	8	925.00	
Total	8,900.00	11		

Sex is the row variable, and ear (left vs. right) is the column variable.

With no statistically significant effect, no need exists to do any post-hoc testing. A researcher would be wise, however, to calculate eta squared in order to consider the possibility of a Type II error, the possibility of having failed to find an effect that does occur.

Applying Equation 12.3, Dr. Porter finds

$$\begin{aligned} \eta^2_{\text{Rows}} &= \frac{SS_{\text{Rows}}}{SS_{\text{Total}}} \times 100 \\ &= \frac{1{,}200.00}{8{,}900.00} \times 100 \\ &= 0.1348 \times 100 \\ &= 13.48\% \end{aligned}$$

$$\begin{aligned} \eta^2_{\text{Columns}} &= \frac{SS_{\text{Columns}}}{SS_{\text{Total}}} \times 100 \\ &= \frac{300.00}{8{,}900.00} \times 100 \\ &= 0.0337 \times 100 \\ &= 3.37\% \end{aligned}$$

$$\begin{aligned} \eta^2_{\text{Interaction}} &= \frac{SS_{\text{Interaction}}}{SS_{\text{Total}}} \times 100 \\ &= \frac{0.00}{8{,}900.00} \times 100 \\ &= 0.0000 \times 100 \\ &= 0.00\% \end{aligned}$$

These results show that the row main effect, $\eta^2 = 13.48\%$, is medium. The column main effect, $\eta^2 = 3.37\%$, is small. There is no interaction effect. Not enough evidence exists to say that the row effect of sex—male vs. female—affects the perception of low frequencies, but there is enough of a hint here that Dr. Porter might want to draw attention to it. As always, looking at a picture of the effects, like that shown in Figure 12.9, helps to clarify what the effects were. Here's her four-point interpretation:

A sensory psychologist conducted a study testing the ears (left vs. right) of both men and women to see if one ear or one sex had a lower threshold for low-frequency sounds. According to this study, there was not enough evidence to conclude that either one sex or one ear was better than the other at perceiving low-frequency sounds. However, the results suggested that women ($M = 125.00$ Hz) may have an edge over men ($M = 145.00$ Hz) in perceiving low-frequency sounds. To investigate this, it would be advisable to replicate the study with a larger sample size.

SUMMARY

Describe what two-way ANOVA does.

- Two-way ANOVA measures the impact of two explanatory variables on a dependent variable. It divides the variability in the dependent variable into that due to each explanatory variable separately (the main effects) *and* that due to them together (the interaction effect).
- The interaction effect is an advantage of two-way ANOVA over one-way ANOVA. An interaction effect means that the impact of one explanatory variable depends on the level of the other explanatory variable. When there is a statistically significant interaction effect, the main effects are not interpreted.

Complete a between-subjects, two-way ANOVA.

- Use between-subjects, two-way ANOVA when there are two explanatory variables, each with at least two levels, the samples are independent, and the dependent variable is measured at the interval or ratio level.
- Conducting a two-way ANOVA follows the same procedure as is used for other hypothesis tests. That is, check the assumptions (random samples, independence of observations, normality, and homogeneity of variance); form null and alternative hypotheses for each main effect and for the interaction; calculate degrees of freedom; find the critical value of F; set the decision rule; and, finally, find the value of the test statistic by completing an ANOVA summary table.

Interpret a between-subjects, two-way ANOVA.

- First, determine if the null hypotheses were rejected and write the results in APA format. Then, calculate eta squared as an effect size. If the null hypothesis was rejected, complete post-hoc tests (Tukey *HSD*) to determine where the effects are and what their direction is.
- Finally, write a four-point interpretation (What was done? What was found? What does it mean? What suggestions are there for future research?).

DIY

Two-way ANOVA allows us to examine the impact of two explanatory variables at once. For example, we can look at two sporting events, like swimming, that are structured the same for men and women. Table 12.20 displays the mean times, in seconds, for the top five finishers in the 100-meter freestyle and 100-meter backstroke at the 2012 Summer Olympics.

TABLE 12.20 Mean Times for Top 5 Finishers in Men's and Women's 100-meter Freestyle and Backstroke at the 2012 Summer Olympics

	Freestyle	Backstroke	
Men	48	53	50.5
Women	53	58	55.5
	50.5	55.5	

The graph of this table, Figure 12.10, shows no interaction but the main effects of sex and stroke.

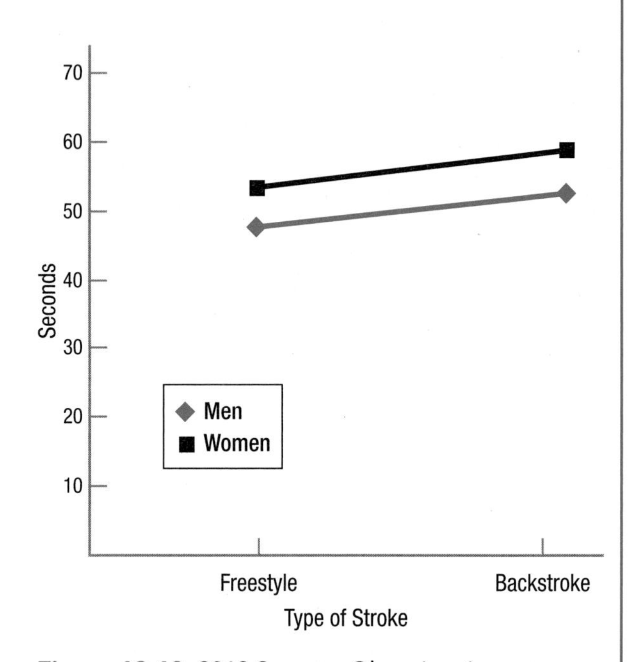

Figure 12.10 2012 Summer Olympics: Average Times of Top 5 Finishers This figure shows the effects of sex and stroke on how long it takes to swim 100 yards.

Figure 12.10 is a graph of the results. What does the graph suggest about the "main effects" of sex and type of stroke as well as the interaction of these two variables? From the graph, it is clear that there is no interaction between sex and stroke and that there are probably main effects of sex and stroke. The main effect of sex, if it is statistically significant, says that no matter the stroke, men swim about five seconds faster over a 100-meter distance. If the main effect of stroke is statistically significant, it would be interpreted as saying that both men and women are about five seconds faster when they cover 100 meters by freestyle rather than backstroke.

Note that the lack of an interaction means that these effects are independent and cumulative. Men are five seconds faster than women and freestyle is five seconds faster than backstroke. Thus, men swimming the freestyle are ten seconds faster than women doing the backstroke.

Now it's your turn. Find some data that vary on two dimensions. Need ideas? See the bulleted examples below. Put the data in a table like the one above. Be sure to label the rows and columns and to calculate marginal values. Graph the results and interpret the graph.

- Sporting events that are structured the same for men and women, at the Olympic, national, collegiate, or high school level
- Sporting events compared across different levels (e.g., high school vs. college)
- Average SAT scores, math vs. reading/writing, for different colleges
- Average salaries, men vs. women, for different professions

KEY TERMS

between-subjects – ANOVA terminology for independent samples.

crossed – a factorial ANOVA in which each level of each explanatory variable occurs with each level of the other explanatory variable.

factorial ANOVA – an analysis of variance in which there is more than one explanatory variable.

interaction effect – situation, in factorial ANOVA, in which the impact of one explanatory variable on the dependent variable depends on the level of another explanatory variable.

main effect – the impact of an explanatory variable, by itself, on the dependent variable.

within-subjects – ANOVA terminology for dependent samples.

CHAPTER EXERCISES

Review Your Knowledge

12.01 ____-way ANOVA examines the impact of two explanatory variables at the same time.

12.02 A 2 × 3 ANOVA could also be called a ____ × ____ ANOVA.

12.03 "Between subjects" means ____ samples and "____ subjects" means dependent samples.

12.04 Two-way ANOVA, three-way ANOVA, and four-way ANOVA are all examples of ____ ANOVA.

12.05 There is an advantage in performing one ____-way ANOVA over two ____-way ANOVAs.

12.06 If every level of one explanatory variable occurs with every level of the other explanatory variable, then the two explanatory variables are said to be ____.

12.07 A two-way ANOVA has two ____ effects and one ____ effect.

12.08 An interaction occurs when the effect of one explanatory variable on the dependent variable ____ on the level of the other explanatory variable.

12.09 If the lines in a graph are ____, then an interaction exists.

12.10 If each cell has the same sample size, the cell means in a row can be averaged to find the ____.

12.11 Row means and column means give information about ____ effects.

12.12 If the interaction effect is statistically significant, ____ interpret the main effects.

12.13 In a between-subjects, two-way ANOVA, the groups are ____ samples.

12.14 The random samples assumption says that the sample is a ____ from the ____ to which one wishes to generalize the results.

12.15 If cases in a cell influence each other's scores on the dependent variable, then the ____ assumption is violated.

12.16 The ____ assumption is tested by looking at cell standard deviations.

12.17 In a two-way ANOVA, there are ____ sets of hypotheses.

12.18 The row's null hypothesis states that the ____ for all levels of the row variable are equal.

12.19 The alternative hypothesis for the columns says that ____ one of the population column means differs from ____ one of the other population column means.

12.20 The alternative hypothesis for the interaction effect says that there is ____ for at least one cell.

12.21 Decision rules exist for the columns main effect, the rows main effect, and the ____.

12.22 The decision rule for the interaction effect involves comparing $F_{Interaction}$ to ____.

12.23 The number of degrees of freedom for the row effect depends on ____.

12.24 The denominator degrees of freedom for all the *F* ratios calculated for a between-subjects, two-way ANOVA is degrees of freedom ____.

12.25 Finding F_{cv} in Appendix Table 4 depends on knowing the numerator ____ and denominator ____ for the *F* ratio.

12.26 If $F_{Interaction} < F_{cv\ Interaction}$, then the null hypothesis is ____.

12.27 Total variance in between-subjects, two-way ANOVA is divided into two factors: ____ and ____.

12.28 In between-subjects, two-way ANOVA, between-group variability is divided into variability due to ____, ____, and ____.

12.29 *SS* is the abbreviation for a sum of ____.

12.30 To calculate a mean square, one divides a ____ by its ____.

12.31 $MS_{Interaction}$ divided by MS_{Within} calculates ____.

12.32 When an ANOVA summary table for a between-subjects, two-way ANOVA is completed, there are ____ *F* ratios.

12.33 If a null hypothesis is rejected, the result is called ____.

12.34 If results, in APA format, are written as $p > .05$, the null hypothesis was ____.

12.35 $p < .05$ means that alpha was set at ____.

12.36 Only interpret a main effect if ____.

12.37 ____ is the effect size used for between-subjects, two-way ANOVA.

12.38 Eta squared is calculated for both ____ effects and for the ____.

12.39 An effect of ≈ ____% is considered a medium effect.

12.40 Eta squared can alert one to the risk of ____ error if the effect was not statistically significant.

12.41 Post-hoc tests should only be used when an effect is ____.

12.42 If the row main effect is statistically significant and there are only two rows, it *is/is not* necessary to do a post-hoc test for the row effect.

12.43 If two cells differ by exactly the *HSD* amount, then the difference is ____.

12.44 The *HSD* value for the interaction effect is used to compare ____ means.

Apply Your Knowledge

Select the right test for the scenario from among a single-sample z test; single-sample t test; independent-samples t test; paired-samples t test; between-subjects, one-way ANOVA; one-way, repeated-measures ANOVA; and between-subjects, two-way ANOVA.

12.45 A dentist randomly assigns people to use one of three different forms of dental hygiene: (1) dental floss, (2) wooden toothpicks, or (3) anti-plaque rinse. After six months he measures, in grams, how much plaque he scrapes off the teeth. What statistical test should he use to see if form of dental hygiene has an impact on the amount of plaque?

12.46 The American Psychological Association wanted to determine if there was a difference in post-graduation success between BA and BS degrees in psychology. It put together a random sample of students with each degree and found out what the annual salary was for their first job after graduation. What statistical test should be used to analyze these data for the effect of type of degree?

12.47 Male and female athletes exercised for an hour on a treadmill. During this time, half of each sex was assigned to drink water and half was assigned to drink a sport beverage. At the end of the hour, lactic acid levels were measured and mean levels were calculated. What statistical test should be used to see how sex and type of beverage affect lactic acid production?

12.48 To study accommodation to the loss of an eye, a sensory psychologist obtained 10 volunteers who agreed to wear an eye patch over one eye for 4 weeks. To test visual ability, the psychologist used a batting test: a pitching machine lobbed 50 pitches to each participant and the psychologist counted how many were hit. This test was conducted (a) before the eye patch was worn, (b) immediately after the eye patch was first put on, (c) 1 week into the study, (d) 2 weeks into the study, (e) 3 weeks into the study, (f) on the last day wearing the eye patch, and (g) 1 week later. What test should be used to see if batting ability changed over the course of the study?

Calculating row and column means

12.49 There are 10 participants in each cell. Each cell is an independent sample. The design is fully crossed. The value in each cell is the cell mean. Calculate (a) row means and (b) column means for this matrix.

	Column 1	Column 2	Column 3
Row 1	25	35	45
Row 2	25	15	5

12.50 There are 17 participants in each cell. Each cell is an independent sample. The design is fully crossed. The value in each cell is the cell mean. Calculate (a) row means and (b) column means for this matrix.

	Column 1	Column 2	Column 3
Row 1	100	110	140
Row 2	140	200	180

Speculating about main effects and interactions

12.51 Given the cell means, row means, and column means below, (a) graph the results to examine if there is an interaction; (b) speculate about which effects would be statistically significant; (c) indicate which effects should be interpreted.

	Column 1	Column 2	
Row 1	17	13	15.00
Row 2	16	12	14.00
	16.50	12.50	

12.52 Given the cell means, row means, and column means below, (a) graph the results to determine if you think there is an interaction; (b) speculate about which effects would be statistically significant; (c) indicate which effects should be interpreted.

	Column 1	Column 2	
Row 1	80	60	70.00
Row 2	60	40	50.00
	70.00	50.00	

Checking the assumptions

12.53 Eighty volunteers who read about a study in a newspaper were randomly assigned to be in the experimental or control group. Within each group, participants were randomly assigned to one of four different conditions. Each participant then participated in the study individually. After this, each participant's status on the interval-level psychological variable was measured. Below are the means (and standard deviations). (a) Evaluate all four assumptions and (b) decide if it is OK to proceed with the test.

	I	II	III	IV
Control	12.34 (2.89)	13.76 (3.89)	14.17 (4.95)	16.88 (6.91)
Experimental	14.88 (3.12)	15.66 (4.18)	16.99 (5.54)	17.33 (6.48)

12.54 A researcher was interested in the impact of different types of day care on children's development. She obtained samples of children who (a) had been watched by a babysitter at home; (b) had attended small, home-based day care; or (c) had gone to a large day-care center. She also classified the children as having received paid care for (1) less than 40 hours a week or (2) 40 or more hours per week. There were eight children in each cell and no siblings were included. She measured each child's adjustment level, on an interval scale, in the first grade. Below are the means (and standard deviations). (a) Evaluate all four assumptions and (b) decide if it is OK to proceed with the test.

	Babysitter	Small Day Care	Large Day Care
<40 Hours	112.66 (14.83)	108.28 (16.76)	106.80 (17.65)
≥40 Hours	114.88 (13.12)	108.71 (15.66)	109.28 (16.77)

Listing hypotheses

12.55 List all the hypotheses for the ANOVA described in Exercise 12.53 for (a) rows, (b) columns, and (c) interaction.

12.56 List all the hypotheses for the ANOVA described in Exercise 12.54 for (a) rows, (b) columns, and (c) interaction.

Calculating degrees of freedom

12.57 If $R = 3$, $C = 4$, and $n = 8$, calculate the degrees of freedom for the following effects: (a) rows, (b) columns, (c) interaction, (d) within groups, (e) between groups, and (f) total.

12.58 If $R = 2$, $C = 3$, and $n = 11$, calculate the degrees of freedom for the following effects: (a) rows, (b) columns, (c) interaction, (d) within groups, (e) between groups, and (f) total.

Finding F_{cv}

12.59 If $\alpha = .05$, $df_{\text{Rows}} = 2$, $df_{\text{Columns}} = 2$, $df_{\text{Interaction}} = 4$, and $df_{\text{Within}} = 36$, find (a) $F_{cv\ \text{Rows}}$, (b) $F_{cv\ \text{Columns}}$, and (c) $F_{cv\ \text{Interaction}}$.

12.60 If $\alpha = .05$, $df_{\text{Rows}} = 3$, $df_{\text{Columns}} = 1$, $df_{\text{Interaction}} = 3$, and $df_{\text{Within}} = 40$, find (a) $F_{cv\ \text{Rows}}$, (b) $F_{cv\ \text{Columns}}$, and (c) $F_{cv\ \text{Interaction}}$.

Writing the decision rule

12.61 If $F_{cv\ \text{Interaction}}$ is 2.668, what is the decision rule for the interaction effect?

12.62 If $F_{cv\ \text{Rows}}$ is 3.295, what is the decision rule for the rows effect?

Calculating mean squares

12.63 If $SS_{\text{Within}} = 168.48$ and $df_{\text{Within}} = 40$, calculate MS_{Within}.

12.64 If $SS_{\text{Interaction}} = 0.60$ and $df_{\text{Interaction}} = 4$, calculate $MS_{\text{Interaction}}$.

Calculating F ratios

12.65 If $MS_{\text{Rows}} = 17.30$ and $MS_{\text{Within}} = 3.33$, what is F_{Rows}?

12.66 If $MS_{\text{Columns}} = 66.54$ and $MS_{\text{Within}} = 78.88$, what is F_{Columns}?

Completing a summary table

12.67 Given $R = 2$, $C = 2$, $n = 5$, $SS_{\text{Between}} = 61.00$, $SS_{\text{Rows}} = 15.00$, $SS_{\text{Columns}} = 12.00$, $SS_{\text{Interaction}} = 34.00$, $SS_{\text{Within}} = 120.00$, and $SS_{\text{Total}} = 181.00$, complete the ANOVA summary table for a between-subjects, two-way ANOVA.

12.68 Given $R = 2$, $C = 4$, $n = 12$, $SS_{\text{Between}} = 225.00$, $SS_{\text{Rows}} = 78.00$, $SS_{\text{Columns}} = 23.00$, $SS_{\text{Interaction}} = 124.00$, $SS_{\text{Within}} = 350.00$, and $SS_{\text{Total}} = 575.00$, complete the ANOVA summary table for a between-subjects, two-way ANOVA.

Applying the decision rule

12.69 If $F_{cv} = 3.443$ and $F = 17.55$, was the null hypothesis rejected?

12.70 If $F_{cv} = 3.191$ and $F = 2.86$, was the null hypothesis rejected?

Writing results in APA format

12.71 If $F_{\text{Interaction}} = 3.70$, $df_{\text{Interaction}} = 2$, $df_{\text{Within}} = 66$, and $F_{cv\ \text{Interaction}} = 3.138$, write the results in APA format. Use $\alpha = .05$.

12.72 If $F_{\text{Rows}} = 2.87$, $df_{\text{Rows}} = 2$, $df_{\text{Within}} = 171$, and $F_{cv\ \text{Rows}} = 3.053$, write the results in APA format. Use $\alpha = .05$.

12.73 If $F_{\text{Columns}} = 2.45$, $df_{\text{Columns}} = 3$, and $df_{\text{Within}} = 168$, write the results in APA format. Use $\alpha = .05$.

12.74 If $F_{\text{Columns}} = 4.00$, $df_{\text{Columns}} = 1$, and $df_{\text{Within}} = 36$, write the results in APA format. Use $\alpha = .05$.

12.75 Given the ANOVA summary table below, write the results in APA format for (a) the rows effect, (b) the columns effect, and (c) the interaction effect. Use $\alpha = .05$.

Source of Variability	Sum of Squares	Degrees of Freedom	Mean Square	F ratio
Between groups	108.56	5		
Rows	45.40	2	22.70	3.21
Columns	7.62	1	7.62	1.08
Interaction	55.54	2	27.77	3.93
Within groups	212.22	30	7.07	
Total	320.78	35		

12.76 Given the ANOVA summary table below, write the results in APA format for (a) the rows effect, (b) the columns effect, and (c) the interaction effect. Use $\alpha = .05$.

Source of Variability	Sum of Squares	Degrees of Freedom	Mean Square	F ratio
Between groups	70.40	7		
Rows	32.34	3	10.78	2.77
Columns	7.62	1	7.62	1.96
Interaction	30.44	3	10.15	2.61
Within groups	342.66	88	3.89	
Total	413.06	95		

Deciding which effects to interpret

12.77 If the null hypothesis is rejected for all three effects (rows, columns, and interaction), which effects should be interpreted?

12.78 If the null hypothesis is rejected for the rows effect and the columns effect but not for the interaction effect, which effects should be interpreted?

12.79 If the null hypothesis is rejected for the interaction effect but not for the rows effect or the columns effect which effects should be interpreted?

12.80 If the null hypothesis is rejected for the rows effect and the interaction effect but not for the columns effect, which effects should be interpreted?

Calculating eta squared

12.81 If $SS_{Rows} = 3.78$ and $SS_{Total} = 47.83$, (a) calculate η^2_{Rows} and (b) classify the effect as small, medium, or large.

12.82 If $SS_{Columns} = 17.91$ and $SS_{Total} = 783.54$, (a) calculate $\eta^2_{Columns}$ and (b) classify the effect as small, medium, or large.

12.83 Given the ANOVA summary table below, calculate η^2 for (a) the rows effect, (b) the columns effect, and (c) the interaction effect.

Source of Variability	Sum of Squares	Degrees of Freedom	Mean Square	F ratio
Between groups	132.00	7		
Rows	33.00	3	11.00	4.56
Columns	54.00	1	54.00	22.41
Interaction	45.00	3	15.00	6.22
Within groups	212.00	88	2.41	
Total	344.00	95		

12.84 Given the ANOVA summary table below, calculate η^2 for (a) the rows effect, (b) the columns effect, and (c) the interaction effect.

Source of Variability	Sum of Squares	Degrees of Freedom	Mean Square	F ratio
Between groups	125.00	5		
Rows	45.00	2	22.50	18.29
Columns	68.00	1	68.00	55.28
Interaction	12.00	2	6.00	4.88
Within groups	140.00	114	1.23	
Total	265.00	119		

Finding q

12.85 Given $R = 3$, $C = 2$, and $n = 8$, find (a) q_{Rows}, (b) $q_{Columns}$, and (c) q_{Cells}.

12.86 Given $R = 3$, $C = 3$, and $n = 11$, find (a) q_{Rows}, (b) $q_{Columns}$, and (c) q_{Cells}.

Calculating HSD

12.87 If $q_{Rows} = 3.49$, $MS_{Within} = 4.56$, and $n_{Rows} = 12$, what is HSD_{Rows}?

12.88 If $q_{Cells} = 2.96$, $MS_{Within} = 12.75$, and $n_{Cells} = 4$, what is HSD_{Cells}?

12.89 If $\alpha = .05$, for which effects should an *HSD* value be calculated?

Source of Variability	Sum of Squares	Degrees of Freedom	Mean Square	*F* ratio
Between groups	324.00	8		
Rows	70.00	2	35.00	28.46
Columns	88.00	2	44.00	35.77
Interaction	166.00	4	41.50	33.74
Within groups	100.00	81	1.23	
Total	424.00	89		

12.90 If $\alpha = .05$, for which effects should an *HSD* value be calculated?

Source of Variability	Sum of Squares	Degrees of Freedom	Mean Square	*F* ratio
Between groups	118.00	8		
Rows	88.00	2	44.00	9.91
Columns	23.00	2	11.50	2.59
Interaction	7.00	4	1.75	0.39
Within groups	600.00	135	4.44	
Total	718.00	143		

Interpreting HSD

12.91 If $M_{\text{Row 1}} = 117.66$, $M_{\text{Row 2}} = 113.63$, $M_{\text{Row 3}} = 128.91$, and $HSD_{\text{Rows}} = 5.89$, determine (a) which rows have a statistically significant difference and (b) the direction of the differences.

12.92 If $M_{\text{Cell 1}} = 55.54$, $M_{\text{Cell 2}} = 48.34$, $M_{\text{Cell 3}} = 36.44$, $M_{\text{Cell 4}} = 59.40$, and $HSD_{\text{Cells}} = 7.83$, determine (a) which cells have a statistically significant difference and (b) the direction of the differences.

Interpreting results

12.93 A consumer psychologist classified shoppers at a grocery store as (a) being males or females, and (b) shopping with or without children. These two variables were crossed and he took a random sample of five shoppers from each of the four cells. Then he gave them the Enjoyment of Shopping Experience Scale (ESES) to be completed for the current shopping experience. The ESES is an interval-level measure, with a mean of 50 indicating neutral feelings about shopping. Scores can range from 20 to 80, with scores above 50 an indication of enjoying shopping; scores below 50 indicate that shopping is not enjoyable. Here are the means:

	Shopping Without Children	Shopping With Children	Row Means
Male Shopper	43.20	56.80	50.00
Female Shopper	57.40	44.80	51.10
Column Means	50.30	50.80	

No nonrobust assumptions were violated and a between-subjects, two-way ANOVA was completed. Here is the ANOVA summary table:

Source of Variability	Sum of Squares	Degrees of Freedom	Mean Square	*F* ratio
Between groups	865.35	3		
Rows	6.05	1	6.05	0.29
Columns	1.25	1	1.25	0.06
Interaction	858.05	1	858.05	41.15
Within groups	333.60	16	20.85	
Total	1,198.95	19		

The researcher calculated $\eta^2_{\text{Rows}} = 0.50\%$, $\eta^2_{\text{Columns}} = 0.10\%$, $\eta^2_{\text{Interaction}} = 71.57\%$, and $HSD_{\text{Cells}} = 8.27$. Complete a four-point interpretation for the results. (*Hint:* Graphing the results will help.)

12.94 A cognitive psychologist decided to investigate whether two beliefs about healthy living had any real impact on performance in college. She took 40, first-semester college student volunteers and randomly assigned half to eat breakfast every day and the other half to skip breakfast every day. This was crossed with another variable. Half the students were randomly assigned to sleep at least 8 hours a night and half were randomly assigned to sleep less than 8 hours a night. At the end of the semester, she recorded each participant's GPA. The results are shown in this table:

	Sleep ≥8 Hours	Sleep <8 Hours	Row Means
Eat Breakfast	3.20	2.72	2.96
Skip Breakfast	3.08	2.77	2.93
Column Means	3.14	2.75	

No nonrobust assumptions for a between-subjects, two-way ANOVA were violated and the ANOVA was completed. Here is the summary table:

Source of Variability	Sum of Squares	Degrees of Freedom	Mean Square	F ratio
Between groups	1.64	3		
Rows	0.01	1	0.01	0.06
Columns	1.56	1	1.56	9.18
Interaction	0.07	1	0.07	0.41
Within groups	6.03	36	0.17	
Total	7.67	39		

The eta squared values for rows, columns, and interaction are, respectively, 0.13%, 20.34%, and 0.91%. HSD_{Cells} is 0.50. Interpret the results. (*Hint:* Graphing the results will help.)

12.95 A developmental psychologist investigated the influence of two crossed variables—exposure to televised violence and type of parental discipline—on teens' acceptance of violence. He obtained a random sample of seniors at the local high school and classified them on two dimensions: (1) whether or not their parents had restricted their television viewing and (2) whether their parents used positive reinforcement or punishment. He took a random sample of six teens from each of the four samples (no siblings were included) and his dependent variable was each teen's score on the interval-level Approval of Violence Scale (AVS). Scores on the AVS range from 0 to 20, with higher scores indicating a greater acceptance of violence as a way to solve disagreements. The cell means (and standard deviations) are shown below. SS_{Rows} was calculated to be 210.04, $SS_{Columns}$ as 222.04, $SS_{Interaction}$ as 15.04, and SS_{Within} as 204.83.

	Restricted TV	Unrestricted TV
Positive Reinforcement	6.00 (2.10)	10.50 (3.62)
Punishment	10.33 (4.08)	18.00 (2.61)

12.96 An industrial organizational psychologist investigated the crossed effects of two variables—being a member of a team sport in high school and extroversion level—on how well professors are liked by their colleagues. The dependent variable was the interval-level Colleague Collegiality Scale (CCS). Scores on the CCS range from 0 to 24; higher scores indicate greater liking by colleagues. From a national and representative sample of college professors, the psychologist randomly selected professors until there were eight in each cell. The table below shows the cell means (and standard deviations). SS_{Rows} was 36.13, $SS_{Columns} = 1.13$, $SS_{Interaction} = 36.13$, and $SS_{Within} = 217.50$.

	Extroverted	Introverted
Played a Team Sport in High School	16.50 (2.51)	14.00 (2.39)
Did Not Play a Team Sport in High School	12.25 (3.62)	14.00 (2.45)

Expand Your Knowledge

12.97 Each cell in the table contains seven cases. (a) Given the cell, row, and column means below, calculate the missing cell means. If it can't be done, say so. (b) Calculate the mean for all 28 cases. If it can't be done, say so.

	Condition 1	Condition 2	Row Means
Control Group	A 4.00	B	8.00
Experimental Group	C	D	7.00
Column Means	6.00	9.00	

12.98 Each cell in this table has the same number of cases, but that number is unknown. (a) Given the cell, row, and column means below, calculate the missing cell means. If it can't be done, say so. (b) Calculate the mean for all the cases. If it can't be done, say so.

	Condition 1	Condition 2	Row Means
Control Group	A	B	6.00
Experimental Group	C 5.00	D	5.50
Column Means	7.00	4.50	

12.99 Each cell in the table has five cases. (a) Given the cell, row, and column means below, calculate the missing cell means. If it can't be done, say so. (b) Calculate the mean for all 30 cases. If it can't be done, say so.

	Condition 1	Condition 2	Condition 3	Row Means
Control Group	A 1.00	B	C	2.00
Experimental Group	D	E	F	6.00
Column Means	2.50	5.50	4.00	

12.100 The means of five independent samples are being compared. Which ANOVA is being used?

a. between-subjects, one-way ANOVA
b. one-way, repeated-measures ANOVA
c. between-subjects, two-way ANOVA
d. (a) or (b)
e. (a) or (c)
f. not enough information provided to decide

12.101 The means of six independent samples are being compared. Which ANOVA is being used?

a. between-subjects, one-way ANOVA
b. one-way, repeated-measures ANOVA
c. between-subjects, two-way ANOVA
d. (a) or (b)
e. (a) or (c)
f. not enough information provided to decide

12.102 Given $R = 3$, $C = 3$, $n = 12$, and the ANOVA summary table below, calculate *HSD* values for the effects as necessary and appropriate.

Source of Variability	Sum of Squares	Degrees of Freedom	Mean Square	*F* ratio
Between groups	184.00	8		
Rows	115.00	2	57.50	9.49
Columns	12.00	2	6.00	0.99
Interaction	57.00	4	14.25	2.35
Within groups	600.00	99	6.06	
Total	784.00	107		

12.103 If $SS_{Rows} = 17.50$, $SS_{Columns} = 13.75$, $SS_{Interaction} = 22.25$, and $SS_{Within} = 44.75$, calculate (a) $SS_{Between}$ and (b) SS_{Total}.

12.104 If $SS_{Rows} = 123.60$, $SS_{Columns} = 80.30$, $SS_{Interaction} = 20.80$, and $SS_{Within} = 168.20$, calculate (a) $SS_{Between}$ and (b) SS_{Total}.

SPSS

- Data to be analyzed for a between-subjects, two-way ANOVA with SPSS have to be entered in the data editor in a certain manner. Each case goes on a separate row and each of the three variables—the two independent variables and the one dependent variable—has a column. Figure 12.11 shows how the caffeine/sleep deprivation data are entered.
- The first column contains the data for the independent variable "Caffeine," which has two levels. Cases who consumed caffeine have a value of 1 and those who did not consume caffeine are given a value of 0.

- The second column, "Sleep," contains information about which amount of sleep deprivation a case has experienced. Those with no sleep deprivation have the value of 0, one hour of sleep deprivation gets a 1, and two hours of sleep deprivation gets a 2.
- The third column, "Alertness," is the case's score on the mental alertness task.

There are two menus that must be accessed to run a between-subjects, two-way ANOVA in SPSS. To start, as shown in Figure 12.12, click on "Analyze," then "General Linear Model," and then "Univariate. . . ."

	Caffeine	Sleep	Alertness
1	1	0	82
2	1	0	85
3	1	0	88
4	1	0	91
5	1	0	94
6	1	1	82
7	1	1	87
8	1	1	90
9	1	1	93
10	1	1	98
11	1	2	79
12	1	2	83
13	1	2	86
14	1	2	89
15	1	2	93
16	0	0	72
17	0	0	76
18	0	0	80

Figure 12.11 Data Entry for Between-Subjects, One-Way ANOVA for SPSS Each case is on a separate row. The values for the two independent variables and for the dependent variable fall in the columns. (Source: SPSS)

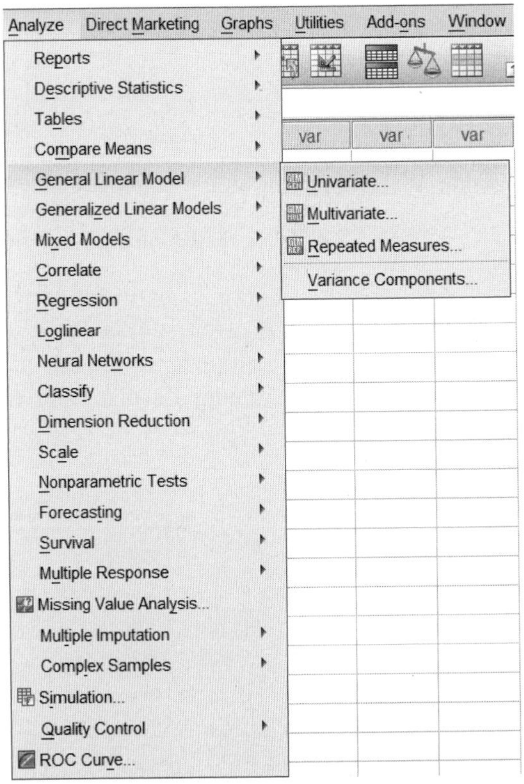

Analyze Direct Marketing Graphs Utilities Add-ons Window
Reports
Descriptive Statistics
Tables
Compare Means
General Linear Model
Generalized Linear Models
Mixed Models
Correlate
Regression
Loglinear
Neural Networks
Classify
Dimension Reduction
Scale
Nonparametric Tests
Forecasting
Survival
Multiple Response
Missing Value Analysis...
Multiple Imputation
Complex Samples
Simulation...
Quality Control
ROC Curve...
var var var
Univariate...
Multivariate...
Repeated Measures...
Variance Components...

Figure 12.12 Initiating a Between-Subjects, One-Way ANOVA in SPSS Between-subjects, one-way ANOVA is found under "General Linear Model" in SPSS. (Source: SPSS)

Clicking on "Univariate. . ." opens up the menu seen in Figure 12.13. Note that the arrow buttons have already been used to send "Alertness" to the dependent variable box and "Caffeine" to the fixed factors box. Once the arrow button is used to send "Sleep" to the fixed factors box, press the "Go" button at the bottom to initiate the calculations.

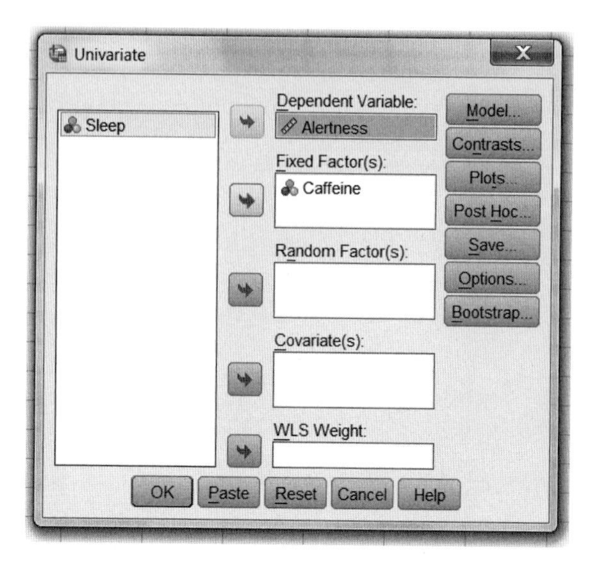

Figure 12.13 Commands to Run a Between-Subjects, One-Way ANOVA in SPSS To run a between-subjects, one-way ANOVA in SPSS, the independent variables are listed as "Fixed Factors" and the dependent variable as the "Dependent Variable." (Source: SPSS)

The results are seen in Figure 12.14. The SPSS ANOVA summary table is arranged a little differently from the one in the text. The columns in the summary table are the same, but there is one additional column labeled "Sig." This gives the exact significance level for an *F* ratio.

Tests of Between-Subjects Effects

Dependent Variable: Alertness

Source	Type III Sum of Squares	df	Mean Square	F	Sig.
Corrected Model	5950.000[a]	5	1190.000	38.182	.000
Intercept	177870.000	1	177870.000	5707.059	.000
Caffeine	3630.000	1	3630.000	116.471	.000
Sleep	1340.000	2	670.000	21.497	.000
Caffeine * Sleep	980.000	2	490.000	15.722	.000
Error	748.000	24	31.167		
Total	184568.000	30			
Corrected Total	6698.000	29			

a. R Squared = .888 (Adjusted R Squared = .865)

Figure 12.14 ANOVA Summary Table Generated by SPSS for a Between-Subjects, One-Way ANOVA What the text calls row, column, and interaction effects are labeled by SPSS with their variable names: "Caffeine," "Sleep," and "Caffeine * Sleep." The asterisk in the interaction effect is an abbreviation for "by," so the interaction is pronounced "Caffeine by sleep." (Source: SPSS)

The rows in the summary table have some different sources of variability. Focus on the three that are labeled "Caffeine," "Sleep," and "Caffeine * Sleep."

- The row labeled "Caffeine" gives the results for the main effect of the caffeine variable.
- The row labeled "Sleep" gives the results for the main effect of the sleep deprivation variable.

- The interaction effect is the one in which the two main effects are connected with an asterisk, "Caffeine * Sleep."
- The row labeled "Error" is what the text calls "Within." This is the denominator term for all three of the F ratios.

The F ratios calculated by SPSS have one more decimal place than those calculated in this book, but otherwise they are the same. The final column in the summary table gives the exact significance level for an effect. If the alpha level is set at .05, then a result is statistically significant as long as the significance level reported in this column is $\leq .05$.

Appendix

Calculating Sums of Squares for Between-Subjects, Two-Way ANOVA

We'll take a few shortcuts in calculating sums of squares for between-subjects, two-way ANOVA. Because $SS_{Total} = SS_{Between} + SS_{Within}$, if one calculates any two, one can figure out the missing sum of squares. And as $SS_{Between} = SS_{Rows} + SS_{Columns} + SS_{Interaction}$, once $SS_{Between}$ is known, if we calculate SS_{Rows} and $SS_{Columns}$, we can figure out $SS_{Interaction}$ by subtraction.

As shown in Table 12.21, the first step is to sum and square the data. These data are the raw data for the caffeine and sleep deprivation study. Of SS_{Total}, SS_{Within}, and $SS_{Between}$, the easiest to calculate are SS_{Total} and $SS_{Between}$. Following Equation 10.2, SS_{Total} is calculated as follows:

$$\begin{aligned} SS_{Total} &= \Sigma X^2 - \frac{(\Sigma X)^2}{N} \\ &= 184{,}568.00 - \frac{2{,}310^2}{30} \\ &= 6{,}698.00 \end{aligned}$$

These data have been summed, squared, and arranged to expedite calculations of sums of squares for a between-subjects, two-way ANOVA.

Using Equation 10.3, $SS_{Between}$ is

$$\begin{aligned} SS_{Between} &= \Sigma\left(\frac{\left(\Sigma X_{Group}\right)^2}{n}\right) - \frac{(\Sigma X)^2}{N} \\ &= \frac{440.00^2}{5} + \frac{450.00^2}{5} + \frac{430.00^2}{5} + \frac{400.00^2}{5} + \frac{340.00^2}{5} \\ &\quad + \frac{250.00^2}{5} - \frac{2{,}310^2}{30} = 5{,}950.00 \end{aligned}$$

Then, SS_{Within} is calculated as

$$\begin{aligned} SS_{Within} &= SS_{Total} - SS_{Between} \\ &= 6{,}698.00 - 5{,}950.00 \\ &= 748.00 \end{aligned}$$

TABLE 12.21 Raw Data Used in Calculating Sums of Squares for a Between-Subjects, Two-Way ANOVA

	Column 1		Column 2		Column 3			
	X	X^2	X	X^2	X	X^2		
	82	6,724.00	82	6,724.00	79	6,241.00		
	85	7,225.00	87	7,569.00	83	6,889.00		
	88	7,744.00	90	8,100.00	86	7,396.00		
	91	8,281.00	93	8,649.00	89	7,921.00		
Row 1	94	8,836.00	98	9,604.00	93	8,649.00	Row total	
n	5		5		5		15	
Σ	440.00	38,810.00	450.00	40,646.00	430.00	37,096.00	1,320.00	116,552.00
	72	5,184.00	62	3,844.00	44	1,936.00		
	76	5,776.00	64	4,096.00	45	2,025.00		
	80	6,400.00	68	4,624.00	50	2,500.00		
	83	6,889.00	72	5,184.00	55	3,025.00		
Row 2	89	7,921.00	74	5,476.00	56	3,136.00	Row total	
n	5		5		5		15	
Σ	400.00	32,170.00	340.00	23,224.00	250.00	12,622.00	990.00	68,016.00
	Column totals						Grand total	
n	10		10		10		30	
Σ	840.00	70,980.00	790.00	63,870.00	680.00	49,718.00	2,310.00	184,568.00

These data have been summed, squared, and arranged to expedite calculations of sums of squares for a between-subjects, two-way ANOVA.

The next step is to compute SS_{Rows} and $SS_{Columns}$:

$$SS_{Rows} = \Sigma\left(\frac{(\Sigma X_{Row})^2}{n_{Row}}\right) - \frac{(\Sigma X)^2}{N}$$

$$= \frac{1{,}320.00^2}{15} + \frac{990.00^2}{15} + \frac{2{,}310^2}{30}$$

$$= 3{,}630.00$$

$$SS_{Cols} = \Sigma\left(\frac{(\Sigma X_{Col})^2}{n_{Col}}\right) - \frac{(\Sigma X)^2}{N}$$

$$= \frac{840.00^2}{10} + \frac{790.00^2}{10} + \frac{680.00^2}{10} - \frac{2{,}310^2}{30}$$

$$= 1{,}340.00$$

The final sum of squares, $SS_{Interaction}$, is calculated by subtraction:

$$SS_{Interaction} = SS_{Between} - SS_{Rows} - SS_{Columns}$$

$$= 5{,}950.00 - 3{,}630.00 - 1{,}340.00$$

$$= 980.00$$

PART III Test Your Knowledge

These questions are meant to probe your understanding of the material covered in Chapters 10–12. The questions are not in the order of the chapters. Some of them are phrased differently or approach the material from a different direction. A few of them ask you to use the material in ways above and beyond what was covered in the book. This "test" is challenging. But, if you do well on it or puzzle out the answers using the key in the back of the book, you should feel comfortable that you are grasping the material.

1. A researcher completed a study with three independent samples, each with four participants. She found $SS_{Total} = 776.00$ and $SS_{Within} = 48.00$. Complete an ANOVA summary table.

2. Here are the descriptive statistics for the groups in a between-subjects, one-way ANOVA:

	Group 1	Group 2	Group 3
n	6	7	8
M	23.00	26.57	36.63
s	4.69	7.28	5.37

And here is the ANOVA summary table:

Source of Variability	Sum of Squares	Degrees of Freedom	Mean Square	F ratio
Between groups	719.36	2	359.68	10.28
Within groups	629.59	18	34.98	
Total	1,348.95	20		

If appropriate to do so, determine which groups have a statistically significant difference in the means. If not appropriate to do so, explain why.

3. Eight hundred participants are randomly assigned to two groups. Half the participants are tested in Environment A and half in Environment B. Prior to testing, different expectations were induced in the participants. Random assignment was used to determine which half of the participants expected a positive outcome and which half expected a negative outcome. Outcome was measured on an interval scale where 100 is neutral, below 100 is poor, and above 100 is good. Here are the means for the 200 people in each cell:

	+ Expect	– Expect
Env A	100	100
Env B	130	70

   a. Calculate marginal means and place them in the appropriate places in the table.
   b. Graph the results.
   c. Answer these two questions:
      1. What is the effect of environment on outcome?
      2. What is the effect of expectation on outcome?

4. Read the ANOVA scenarios below and determine how many "ways" exist in each one. State what each way is (e.g., sex or handedness), indicate whether the samples for the way are independent or dependent, and state the number of levels for the way.

   a. People are assigned to groups based on (1) whether or not they like to exercise and (2) whether they prefer fruits or chips. The groups are compared in terms of weight.

   b. Orchestra members who play string, brass, wind, or percussion instruments are compared on an interval-level measure of perfect pitch.

   c. The same students' political views are measured on an interval-level scale at the end of their first, second, third, and fourth years of college.

d. People with schizophrenia are classified as to whether their symptoms are primarily "positive" or "negative." They are then assigned to receive either standard medications or new experimental medications. Each person's symptom severity is measured after three, six, and nine months of treatment.

e. Psychology majors, English majors, and communications majors are matched in terms of sex and IQ. They are then compared in terms of first-job-after-graduation salary.

5. Design a study in which it would be appropriate to use the tests indicated below to analyze the results. Make sure that you give sufficient details.

a. A between-subjects, one-way ANOVA

b. A one-way, repeated-measures ANOVA

c. A between-subjects, two-way ANOVA

6. Read each scenario below and determine which statistical test should be used to analyze the data to arrive at an answer to the question. Your options are the single-sample z test, the three t tests, and the three ANOVAs. Specify the test being used (i.e., rather than simply indicating "t test," indicate "paired-samples t test"). If the answer calls for a test that has not been covered in the text, say so.

a. The IQs of first-year college students are measured. The IQs for the same students are measured again four years later. Does college make people smarter?

b. The body mass indexes (BMIs) of kindergarten students are measured. The BMIs for the same students are measured 12 years later. Do BMIs change over time?

c. People who suffer from regular headaches are assigned to receive aspirin, ibuprofen, acetaminophen, caffeine, placebo, or a new experimental drug. The subjects track how many minutes it takes for their headaches to go away once they ingest the assigned medication. Is there a difference in the effectiveness of the various treatments?

d. According to the U.S. Hospital Association, every year on July 4th, the average hospital in the United States treats 7.35 ($\sigma = 2.53$) children for fireworks-related injuries. In order to determine whether rural communities are safer for children on July 4, a researcher surveyed a random sample of 37 rural hospitals and found for each one the number of children treated on July 4 for fireworks-related injuries. What statistical test should he use to see if rural communities are safer on July 4?

7. A babysitter once told me that she thought regular babysitting was the best contraceptive in the world. She loved kids and planned to have several someday, but she realized how much responsibility they were and was not going to rush into it. I decided to conduct a study to see if babysitting had any impact on the age at which women had their first child. I found 10 women who had never babysat, 10 who babysat occasionally, and 10 who were frequent babysitters. All participants were college graduates. For each, I determined her age at the birth of her first child. The means for the three groups are: *never babysat* = 23.10 years old at birth of first child; *occasional* = 24.60; *frequent* = 26.90.

Below is the ANOVA summary table:

Source of Variability	Sum of Squares	Degrees of Freedom	Mean Square	F ratio
Between groups	73.27	2	36.64	8.48
Within groups	116.66	27	4.32	
Total	189.93	29		

Interpret the results.

Correlation, Regression, and Nonparametric Statistical Tests

Congratulations! This is the final section of the book and you are just four chapters away from completing your first statistics course. Up to this point, we have discussed a group of hypothesis tests called *difference tests.* These tests, which include the various *t* tests and ANOVAs, allow researchers to determine whether two or more populations differ on an outcome variable.

Now, the final chapters will introduce you to a second group of hypothesis tests called *relationship tests.* These tests allow researchers to analyze whether an association, or correlation, exists between two or more variables.

Chapter 13 introduces the Pearson correlation coefficient, or Pearson *r*, a test used to measure the strength of a relationship between two interval/ratio variables. Chapter 14 then takes the Pearson *r* and uses it in something called linear regression to predict a case's score on the outcome variable from its score on the explanatory variable. Chapter 15 covers nonparametric tests, tests that can be used with ordinal- or nominal-level outcome data.

By the end of Chapter 15, this book will have put more than a dozen statistical tests into your statistical toolbox and taught you the proper use of each one. But, students often stumble when, facing a statistical task on their own, they have to reach into their toolbox and pick the right test. That's where Chapter 16 on selecting the correct test comes in. After completing this final chapter, you should feel confident that when you open your toolbox, you will choose the right test.

The Pearson Correlation Coefficient

LEARNING OBJECTIVES

- Differentiate difference tests from relationship tests.
- Define and describe a relationship.
- Compute a Pearson *r*.
- Interpret the results of a Pearson *r*.
- Take into account the effects of a confounding variable.

CHAPTER OVERVIEW

Previous chapters on hypothesis tests focused on two types of tests—*t* tests and ANOVAs. With these tests, cases are assigned to or classified in groups on the basis of an explanatory variable, and then the outcomes of the groups are compared. These tests are called "difference" tests because they determine whether there is a *difference* in the mean of the dependent variable between the groups. Classic experiments in which an experimental group is compared to a control group to see which has a better mean outcome are examples of difference tests.

In this chapter, we're going to learn about another type of test, what is called a "relationship" test. With relationship tests, there is one group of cases. Each case in the group is measured on two variables to determine if a *relationship*, or an association, exists between the two variables. For example, we could measure a group of college students to determine if there is a relationship between how extroverted they are and how often they date.

13.1 Introduction to the Pearson Correlation Coefficient

There are multiple statistical tests that measure relationships by calculating correlation coefficients. In subsequent chapters, we'll cover the Spearman rank-order correlation coefficient and the chi-square test of independence. But, the focus of this chapter is the head of the relationship test household, the most commonly used relationship test, the Pearson correlation coefficient. Usually, it is referred to as the Pearson *r*, or simply *r*.

Here is what we'll cover in the first part of the chapter as we introduce *r*:

- Defining Pearson *r* and "relationship"
- Exploring what a relationship says about cause and effect
- Seeing how to visualize a relationship
- Learning the difference between weak and strong relationships
- Seeing how *z* scores define the Pearson *r*
- Learning how the Pearson *r* quantifies relationship strength
- Exploring the two directions a relationship can take
- Learning about conditions that affect Pearson *r*

Defining Pearson *r* and Relationship

Measures of association are statistics that quantify the degree of relationship between two variables. If two variables are related, they are said to be *correlated*. This means that the variables vary together systematically, that a change in one variable is associated with a change in the other. For example, alcohol consumption and blood alcohol level vary together systematically—the more alcohol a person consumes, the higher the level of alcohol in his or her blood. Depending on sex and weight as shown in Table 13.1, the relationship between the number of drinks consumed and blood alcohol level is well established.

The Pearson *r* is a specific measure of association. A **Pearson correlation coefficient** quantifies the degree of linear relationship between two interval and/or ratio variables. Because the Pearson *r* uses interval and/or ratio variables, distances between points can be measured, *z* scores calculated, and graphs made. The graphs, called scatterplots, can be examined for the degree to which the relationship between the two variables takes the form of a straight line because the Pearson *r* measures the degree of linear relationship. (We'll discuss nonlinear relationships, and why Pearson *r* is not appropriate to use in those cases, later in this chapter.)

TABLE 13.1 The Relationship Between Number of Drinks Consumed and Estimated Blood Alcohol Percentage

	Body Weight in Pounds					
	Women			Men		
Drinks	120	140	160	140	160	180
0	.00	.00	.00	.00	.00	.00
1	.04	.03	.03	.03	.02	.02
2	.08	.07	.06	.05	.05	.04
3	.11	.10	.09	.08	.07	.06
4	.15	.13	.11	.11	.09	.08
5	.19	.16	.14	.13	.12	.11
6	.23	.19	.17	.16	.14	.13

Blood alcohol level varies systematically with the number of drinks consumed, based on a person's sex and weight.
Source: Pennsylvania Liquor Control Board.

A Common Question

Q If the Pearson *r* only measures association between two interval and/or ratio variables, what is used for ordinal and/or nominal variables?

A If one variable is ordinal and the other is an ordinal, interval, or ratio variable, then association can be measured with a test called the Spearman rank-order correlation coefficient, or Spearman *r* for short. Association between two nominal variables is measured with the chi-square test of independence. Both of these tests are covered in Chapter 15.

Correlation, Causation, and Association

If two variables correlate, the relationship *may* be a cause-and-effect relationship, but it doesn't have to be. Statisticians love to say, "Correlation is not causation." That's because correlation between two variables does not guarantee a cause-and-effect relationship between them. A correlation between two variables only guarantees there is an *association* between the two variables, not that one causes the other.

To understand this, look at Figure 13.1. This type of graph, called a scatterplot, displays the relationship between two variables. For a random sample of 24 states, this graph shows the association between the number of community hospitals in a state (on the *X*-axis) and the number of deaths per year in the state (on the *Y*-axis). Each point represents a state. For example, Delaware, the state at the bottom left of

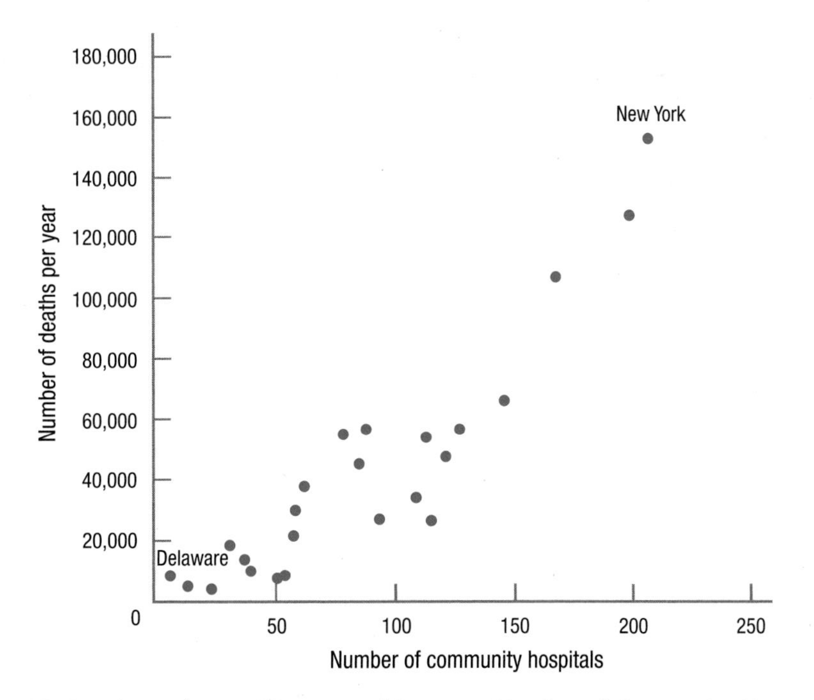

Figure 13.1 Relationship at a State Level Between Number of Community Hospitals and Number of Deaths This scatterplot illustrates the relationship between the number of community hospitals per state and the number of deaths per year in that state. States with more hospitals have more deaths. (Credit: Thanks to Jillian Mrozowski, who obtained these data.)

the graph, has 6 community hospitals and about 7,000 deaths per year; New York, the state at the top right, has 206 community hospitals and about 153,000 deaths.

The scatterplot shows clear evidence of an association between the two variables. The two variables vary together systematically: states like Delaware, with few hospitals have few deaths, and states like New York, with a lot of hospitals, experience a lot of deaths.

The relationship can be read in either direction. So far, this relationship has been viewed as states with more hospitals have more deaths. But, the scatterplot can be viewed just as legitimately as showing that states with more deaths have more hospitals.

If two variables are correlated, it simply means that they vary together systematically. A cause-and-effect relationship may exist between them, but there does not have to be. The number of deaths in a state and the number of community hospitals are correlated, but it doesn't seem plausible that there is a cause-and-effect relationship between them. If such a relationship did exist, then a state could reduce the number of people who die each year by closing down hospitals. That doesn't seem to be a course of action likely to work.

The number of hospitals per state and the number of deaths vary together systematically, but they don't cause each other. Their correlation most likely results because each is associated with a third variable, population. States with more people, like New York, have more of everything than states with fewer people, like Delaware. New York, with a population of almost 20 million, has more pencils, cars, barbers, and murders than Delaware, with under a million residents. New York also has more community hospitals and more deaths, simply because more people live there.

Correlation just means two variables systematically vary together. If the goal of science is to understand cause and effect, it may seem that drawing a conclusion that two variables are associated represents a second-place finish. Not so. Correlation does not *have to* mean causation, but it *may* mean causation. If there is a cause-and-effect relationship between two variables, then they are correlated. An association exists between how much alcohol a person consumes and what his or her blood alcohol level is because consuming alcohol causes the alcohol concentration in the blood to rise.

Correlation may not prove cause and effect, but it can suggest cause. For example, finding that children who watch more violent TV tend to exhibit more aggressive behaviors doesn't prove that TV is the culprit, but it certainly raises such a question and leads to more research. Thus, correlations are not proof of cause and effect, but they often serve as a jumping off point for using experimental techniques to explore cause and effect.

Studies in which the variables in a relationship test are not manipulated by the researcher do not address cause and effect. In a relationship test, it is rare that the explanatory variable is an independent variable and the outcome variable is called a dependent variable. Most commonly, they are just called X and Y and that's what we'll do here. However, if the researcher believes there is an order to the relationship, then the explanatory variable is called the *predictor variable* and the outcome variable is called the *outcome variable*. The language is meant to be straightforward—the variable that comes first, that is thought of as leading to or influencing the other variable, is the **predictor variable**. The one that comes second, and is influenced by the first variable, is the **outcome variable**.

A Common Question

Q Can a correlation give cause-and-effect information?

A There is a difference between correlation, referring to the statistical test, and a correlational study. In the latter, the explanatory variable is not controlled by the experimenter, so a cause-and-effect conclusion cannot be reached. But, it would be possible to design a study where the experimenter manipulates the explanatory variable and where the results would be analyzed with a Pearson *r*. In such a case, a correlation coefficient would give cause-and-effect information.

Visualizing Relationships

The degree to which two variables vary together determines whether the relationship is strong or weak. The strength of the relationship can be visualized in a scatterplot. Figure 13.2 shows the relationship between randomly generated numbers. If *X* values are randomly selected and each *X* is paired with a randomly selected *Y*, then the two do not vary together systematically. Figure 13.2 shows a rectangular scatterplot where there is *no* relationship, what is called a zero correlation, between two variables. Cases with low values on *X* could have low, medium, or high values on *Y*. And, the same is true for cases with medium or high values on *X*.

Though the rectangular scatterplot shows a zero correlation clearly, that is not how statisticians illustrate a zero correlation. Statisticians use a circular scatterplot to demonstrate no relationship between two variables. If both variables are normally distributed, which is one of the assumptions for the Pearson *r*, then the scatterplot has a circular shape when no relationship exists. Figure 13.3 uses the traditional circular shape to illustrate the lack of relationship between two normally distributed variables.

What does a scatterplot look like when there *is* a relationship between two variables and they vary systematically? Figure 13.4 illustrates the relationship between the temperatures of objects measured in both degrees Fahrenheit and Celsius. This figure is an example of the strongest possible linear relationship between two variables, a **perfect relationship.** In a perfect relationship, all the data points fall along a straight line. Figure 13.4 shows that knowing an object's temperature in Fahrenheit tells one exactly what it is in Celsius.

Strength of Relationships

A strong relationship is one in which cases' scores on variable *X* are closely allied with their scores on variable *Y*. When looking at a scatterplot, strength is shown by how well the points fall along a straight line. The more the points fall in a straight line, the stronger the association. Figure 13.5 gives four scatterplots. As the points form less of a line and more of a blob, as they move from a line to an oval or a circle, the relationship grows weaker.

Two situations exist in which the points in a scatterplot form a straight line, but the relationship is not a perfect one. These occur when the line is horizontal or vertical. When the line is horizontal, for example, every value of *X* is paired with the same

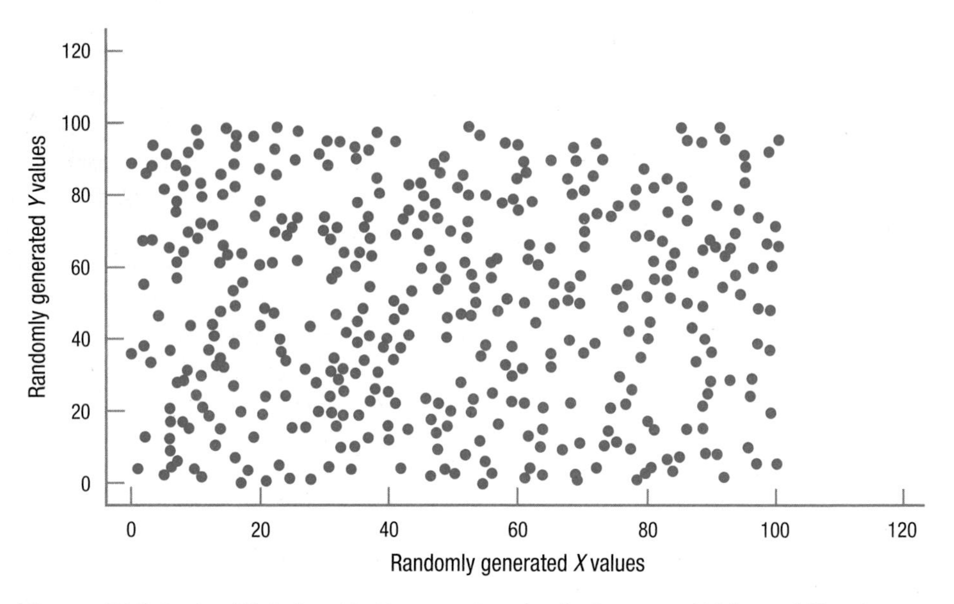

Figure 13.2 Lack of Relationship Between Randomly Generated Values of *X* and Randomly Generated Values of *Y* When the two variables don't vary together systematically, there is no relationship between *X* and *Y*. This is one example of what no relationship looks like.

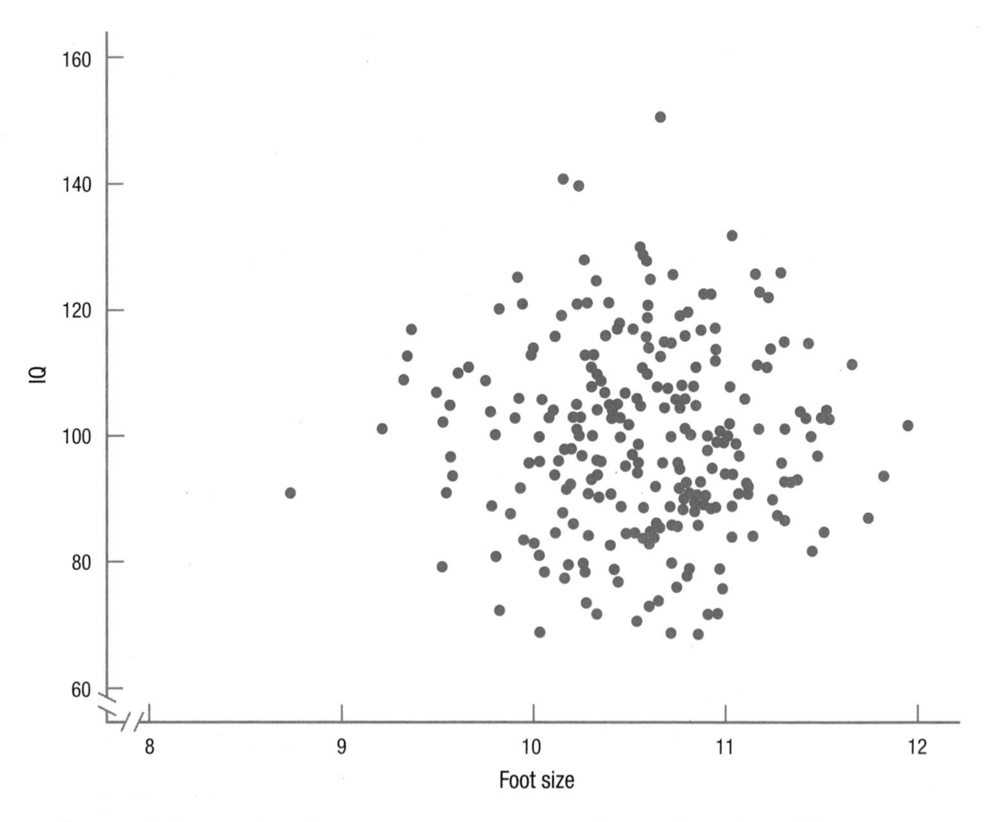

Figure 13.3 Zero Correlation Between Two Normally Distributed Variables If there is no correlation between two normally distributed variables, the scatterplot looks more like a circle than a rectangle. This shows the hypothetical lack of a relationship between foot size and intelligence.

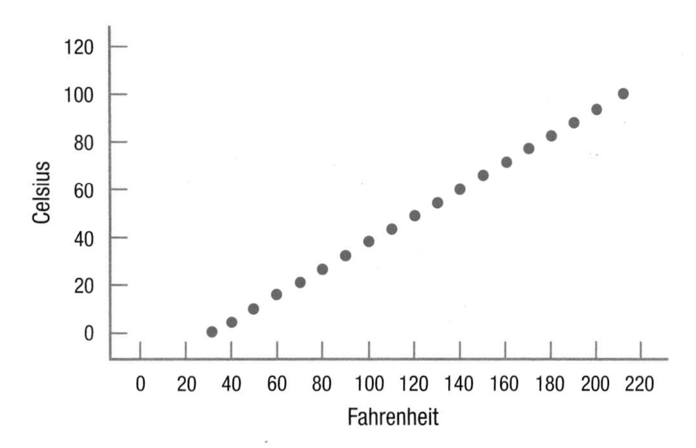

Figure 13.4 A Perfect Linear Relationship Note that all data points fall on a straight line. The relationship between temperature as measured in degrees Fahrenheit and degrees Celsius is a perfect one.

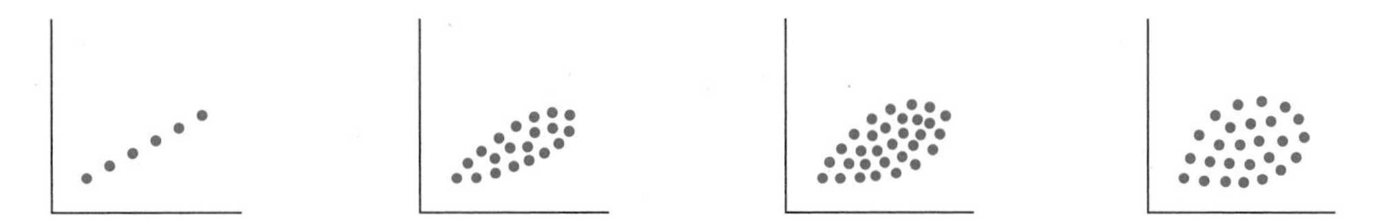

Figure 13.5 Visualizing the Strength of Relationships These four scatterplots differ in the degree to which the points form a straight line or are spread out to form an oval. As the amount of spread increases—as the points fall less along a line—the relationship becomes weaker.

value of *Y*. There is no variability in *Y*. With a vertical line, the opposite is true and there is no variability in *X*. The Pearson *r* formula needs both variables to have variability, and if either variable has none, Pearson *r* can't be calculated. In these situations, *r* is said to be undefined.

Correlations Defined by z Scores

A scatterplot can also be drawn using *z* scores, not raw scores. *z* scores, also called standard scores, were introduced in Chapter 4. They are a transformation of raw scores into scores that reveal how far away from the mean the scores are in standard deviation units. *z* scores can be positive or negative, which means they fall above the mean or below the mean.

Figure 13.6 offers scatterplots for the community hospital/number of deaths data. The panel on the left in 13.6, like Figure 13.1, illustrates the relationship with raw scores, and the panel on the right shows what the scatterplot looks like when the same variables are transformed into *z* scores. Note that transforming from raw scores to *z* scores doesn't change the relationship. The pattern of the dots, the shape, is exactly the same, whether raw scores or *z* scores are being plotted.

Figure 13.7 shows the scatterplot for the relationship between temperature as measured in Fahrenheit and Celsius both with raw scores (the panel on the left) and with *z* scores (the panel on the right). These *z* scores give a new perspective on what

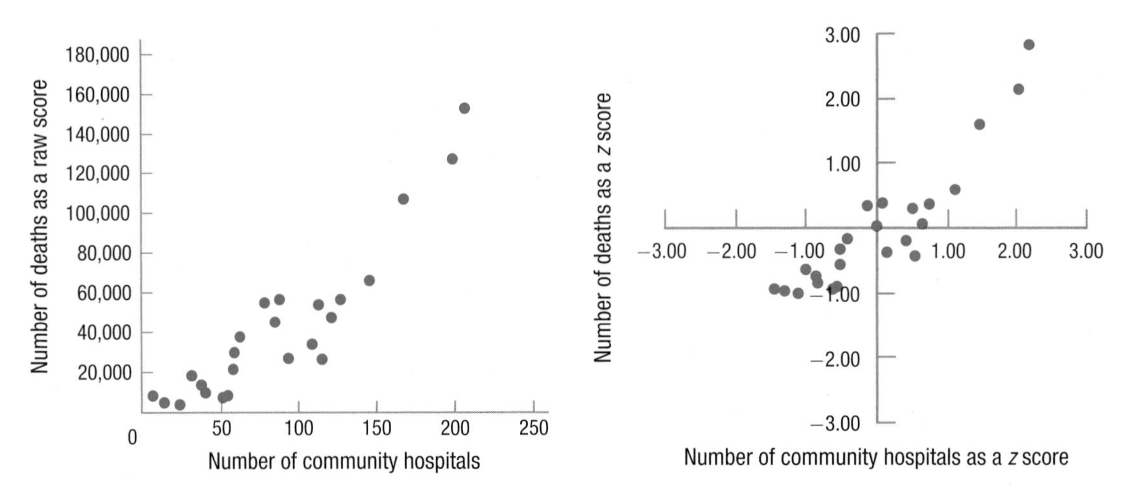

Figure 13.6 Number of Community Hospitals and Number of Deaths: Raw Scores and z Scores When raw scores are transformed into *z* scores, the scales on the axes change, but not the shape of the scatterplot. The dots form the same pattern in both of these scatterplots.

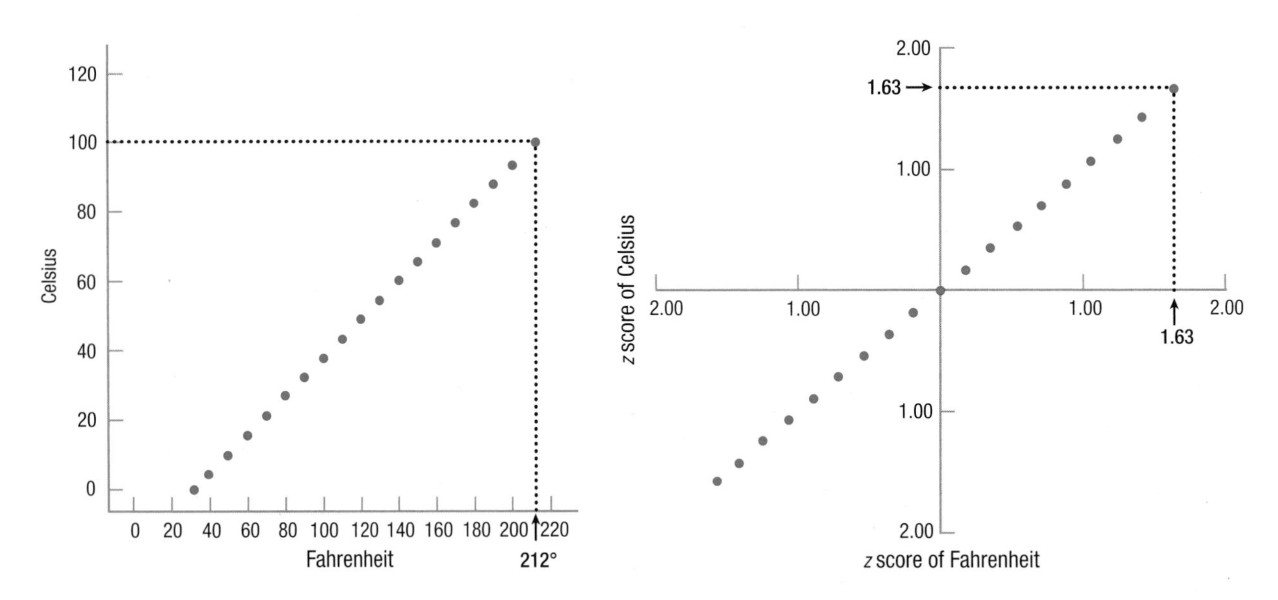

Figure 13.7 A Perfect Relationship: Raw Scores vs. z Scores With a perfect relationship, a case's *z* score on the *X* variable has exactly the same value as it does on the *Y* variable. In the panel on the left, the data point for boiling water occurs at the intersection of 212°F and 100°C. In the panel on the right, the data point is at the intersection of *z* scores of 1.63 on both axes.

a correlation means. Look at the dot on the upper right, which is for boiling water. In the panel on the left, the *X* value for this point is 212° (Fahrenheit) and the *Y* value is 100° (Celsius). To a person who doesn't know much about Fahrenheit and Celsius, those scores don't sound similar as they are 112 points apart. In the panel on the right, the z scores for this point are $z_X = 1.63$ and $z_Y = 1.63$. That is, the boiling point of water is exactly as far above the mean when measured in Fahrenheit as when it is measured in Celsius. That's what a perfect correlation means.

As correlations get weaker, the similarity between z_X and z_Y lessens. The community hospital/number of deaths data show a strong correlation, but not a perfect one. The point on the upper right is New York with 206 hospitals and 153,000 deaths. Converted, those are z scores of 2.2 and 2.8. The two z scores are not the same, but they are in the same ballpark, both far above the mean. As correlations grow weaker and weaker, the similarity between the two z scores for each data point becomes less and less.

In fact, the similarity–dissimilarity of the pairs of z scores is one way of calculating what a Pearson correlation coefficient is. This formula, called the definitional formula, calculates a Pearson r as the average of multiplied-together pairs of z scores. Each case's raw score on X is transformed into a z score and its raw score on Y is transformed into a z score. The two z scores for each case are multiplied together and the mean of these products is calculated. The mean of all such multiplied-together scores is a Pearson correlation coefficient. Thus, the stronger the correlation, the more similar—on average—the z scores are.

Here is a quick example to illustrate the benefit of the z score method. Many years ago, the United States Postal Service ran an ad showing that they, compared to the United Parcel Service and FedEx, had the largest fleet of trucks. The totals were, respectively, 200,000, 130,000, and 35,000. The USPS also had the lowest price for overnight delivery among the three shippers—respectively, \$3, \$6, and \$12. As the scatterplot in Figure 13.8 makes clear, this is a strong relationship: more trucks is associated with lower cost.

Table 13.2 shows the data, both as raw scores and transformed into z scores. Note how much clearer the relationship between the two variables is when expressed in z scores. The number of trucks the United States Postal Service has is as far above the mean, $z = 0.95$, as its price is below the mean, $z = -0.87$. This, the strength of the relationship between variable X expressed as a z score and variable Y expressed as a z score, is what the Pearson r expresses.

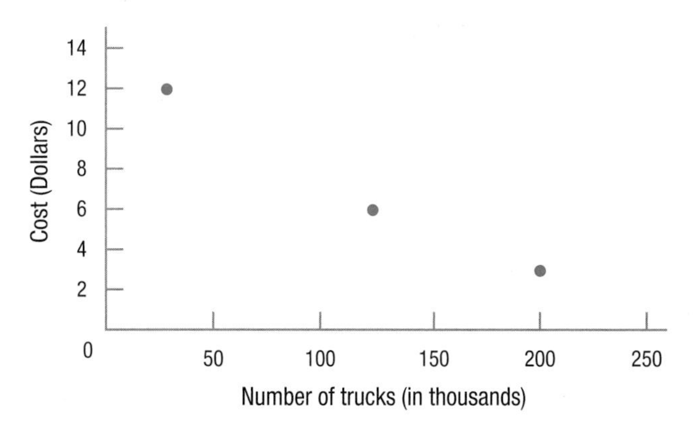

Figure 13.8 Relationship Between Fleet Size and Cost of Overnight Letter This scatterplot shows a strong relationship between how many trucks a delivery service has and the amount it charges for overnight delivery.

TABLE 13.2 Size of Delivery Service Fleet and Cost of Overnight Delivery

	Number of Trucks	Overnight Delivery Cost	*z* Score (trucks)	*z* Score (cost)
USPS	200,000	$3	0.95	−0.87
UPS	130,000	$6	0.10	−0.22
FedEx	35,000	$12	−1.05	1.09

Notice how much clearer the relationship is between fleet size and cost when expressed as *z* scores. The *z* scores show that each delivery service is as far above the mean on one variable as it is below on the other.

Quantifying Relationships

A correlation coefficient is a number that summarizes the strength of the relationship between two variables.

Scatterplots are great for visualizing relationships, but their interpretation is subjective. For example, can the owner of a baseball team buy his or her way to the World Series? Figure 13.9 is a scatterplot showing the relationship between the payrolls of Major League Baseball teams and their winning percentages. How strong is the association? Just by looking at the graph, it is difficult to tell how strong the relationship between these two variables is. Deciding how strong a relationship is on the basis of a scatterplot is subjective and different people can have different—and valid—opinions. One person might interpret the points on Figure 13.9 as roughly linear and conclude that the payroll is strongly related to a team's performance. Another may see the circular clump of scores in the middle of the graph and conclude that there doesn't appear to be much of a relationship.

The Pearson *r* gets around this problem of how specific the prediction is by calculating a statistic called a *correlation coefficient*. A **correlation coefficient** is a number that summarizes the strength of the linear relationship between two variables.

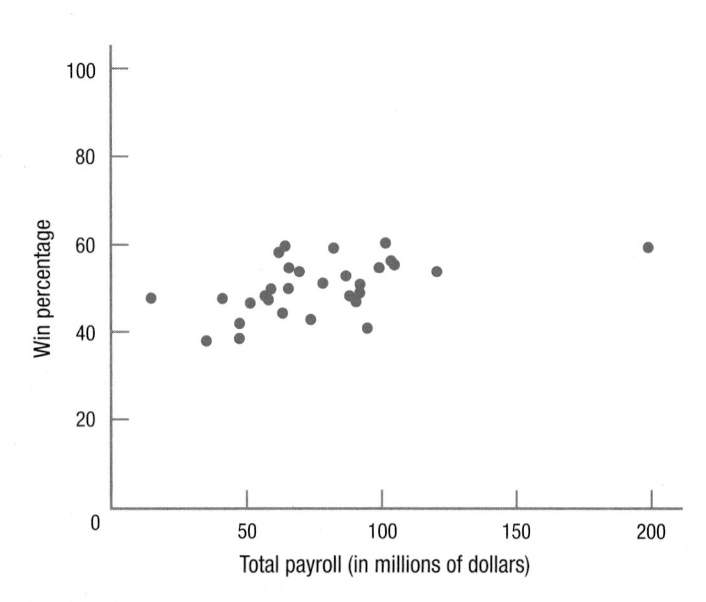

Figure 13.9 Relationship Between Team Payroll and Team Success in Major League Baseball This scatterplot shows the relationship between the payrolls of Major League Baseball teams and their winning percentages.

For the Pearson r, the correlation coefficient is abbreviated as r (think of r as being short for "relationship"). For a Pearson correlation, r is a value that ranges from −1.00 to +1.00. An r value of zero is less than a weak relationship; it means that there is *no* relationship between the two variables. Figure 13.2, where the dots form a rectangle, and Figure 13.3, where the dots form a circle, have Pearson r values of zero.

As the r value moves further away from zero, in either a negative or positive direction, it represents a stronger relationship between X and Y. For example, an r of −.60 represents a stronger relationship than an r of .30. Pearson r values of −1.00 and 1.00, though they differ in sign, both represent perfect relationships.

Direction of the Relationship

Though Pearson r values of −1.00 and 1.00 both represent perfect relationships, the two indicate different types of relationships because their signs differ. The sign of a Pearson r, either positive or negative, gives information about the *direction* of the relationship. There are two options for direction: positive or negative.

- **Positive relationships.** Positive r's are found for what are called direct relationships. **Direct relationships** have scatterplots where the points tend to fall along a line moving from the bottom on the left, up and to the right. This means that cases with low scores on the X variable tend to have low scores on the Y variable and those with high scores on the X variable will tend to have high scores on the Y variable. For example, within a community there's a positive relationship between the size of a house and how much it costs. In general, smaller houses cost less money and bigger houses cost more money. This positive relationship is illustrated in Figure 13.10.

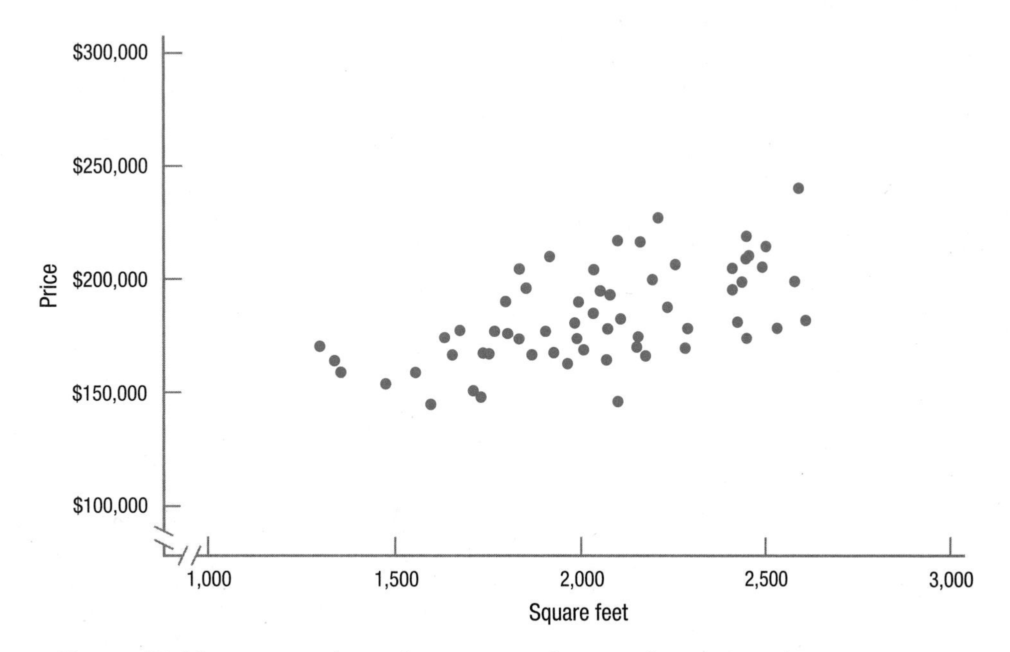

Figure 13.10 Positive Relationship Between the Size of a House and Its Cost In a positive relationship, the data points in a scatterplot fall along a diagonal line that moves up and to the right. In this scatterplot, as the size of the house increases, the price generally does as well.

- **Negative relationships.** Negative *r*'s are also called inverse relationships. **Inverse relationships** have scatterplots where the points fall along a line moving from the top left, down and to the right. This means that cases with low scores on *X* tend to have high scores on *Y* and cases with high scores on the *X* variable will tend to have low scores on the *Y* variable. For example, Figure 13.11 shows a relationship between a car's horsepower and its fuel economy. The relationship is a negative one—in general, lower horsepower means better fuel economy and higher horsepower means worse fuel economy.

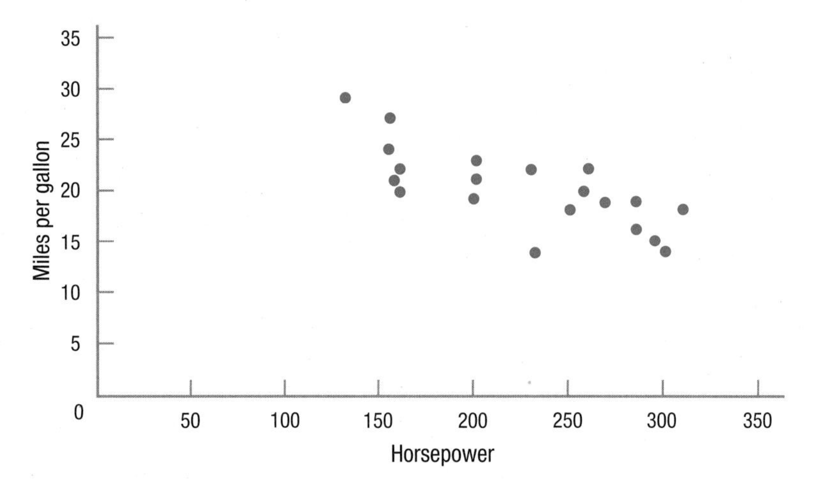

Figure 13.11 Negative Relationship Between Horsepower and Fuel Economy In a negative relationship, points form a line moving downward and to the right. This means that as one variable increases, the other decreases. Here, as the horsepower of cars goes up, miles per gallon go down. (Credit: Thanks to Griffon Olon who collected these data.)

Conditions Affecting Pearson *r*

There are three conditions that affect a Pearson *r*: (1) nonlinearity, (2) outliers, and (3) restriction of range.

Nonlinearity

The Pearson *r* is used to measure the degree of *linear* relationship between two variables. A linear relationship exists when the points in the scatterplot for the relationship between *X* and *Y* fall along a *straight* line. Figure 13.12 illustrates a nonlinear association, what is called the Yerkes-Dodson law. The Yerkes-Dodson law states that arousal affects performance in an orderly way—low arousal leads to poor performance, moderate arousal leads to optimal performance, and high arousal impairs performance. This scatterplot shows a relationship between *X* and *Y*, but the points in the scatterplot don't fall on a straight line. Using a Pearson correlation to quantify this relationship would provide a value close to zero, meaning there is little *linear* relationship between the two variables. The Pearson *r* only measures how much *linear* relationship exists between two variables. If the relationship appears nonlinear, do not use a Pearson *r* to measure it.

The Pearson r only measures how much linear *relationship exists between two variables.*

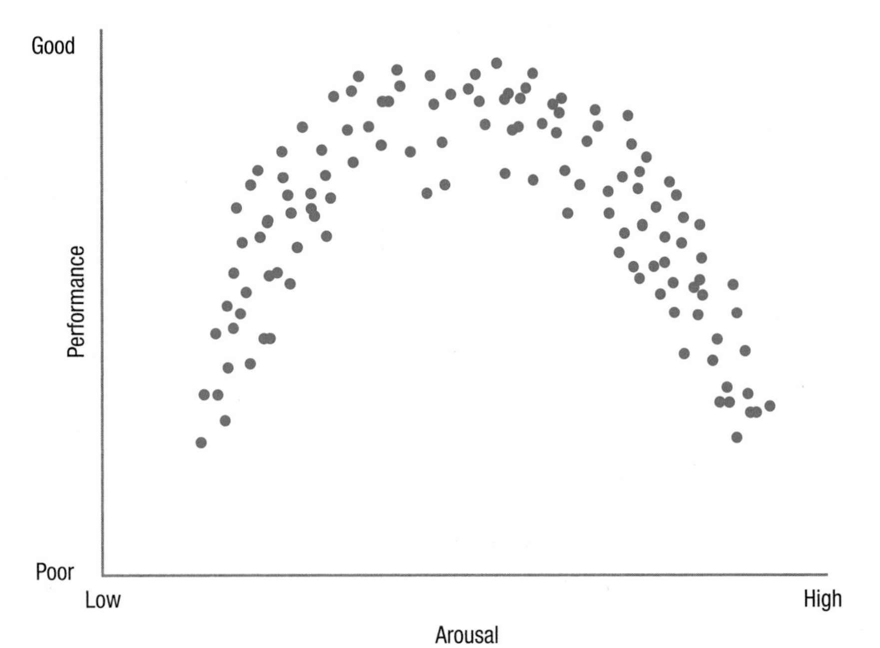

Figure 13.12 Example of a Curvilinear Relationship There is a relationship between X and Y in this scatterplot, but it is curvilinear, not linear. Because the relationship is not a straight-line relationship, the Pearson r would be near zero. An r of zero doesn't mean there is no relationship. It just means that no linear relationship exists.

Outliers

Outliers are cases with extreme values, values that fall far away from the values that other cases have. One outlier in a data set can dramatically change the value of a correlation. The left panel in Figure 13.13 shows a scatterplot for a small data set. The points form a rough circle and the correlation between X and Y is zero.

The right panel in Figure 13.13 adds one data point, an outlier with extreme values on both X and Y. This data point has an X value and a Y value that are dramatically higher than any other case. As a result of adding this one case, the correlation between X and Y changes from $r = .00$ to $r = .63$.

Outliers inflate the strength of a relationship between two variables. Because outliers may exist, it is always a good idea to create and inspect a scatterplot before calculating a correlation coefficient.

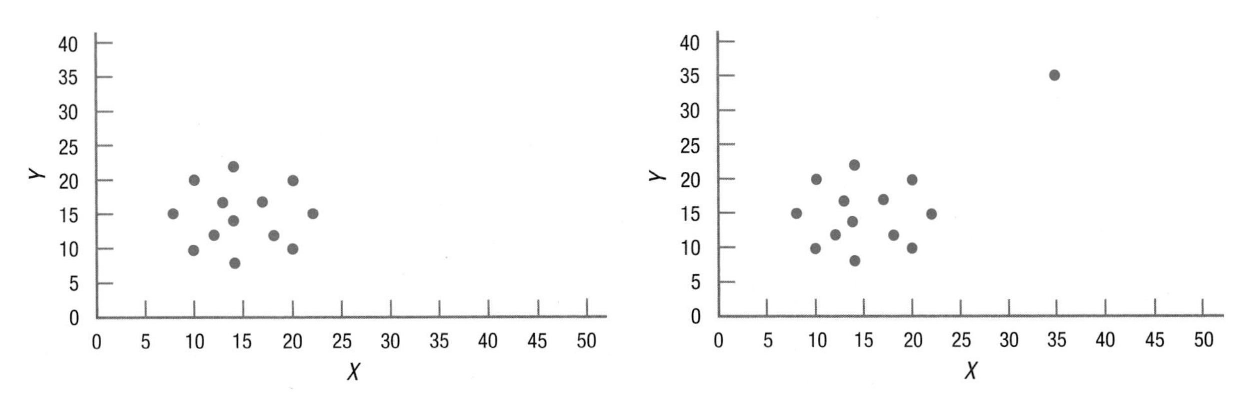

Figure 13.13 Effect of an Outlier on a Pearson Correlation The correlation in the left scatterplot is zero. In the right scatterplot, the correlation is .63 thanks to the addition of one case. The additional case is an outlier, with extreme values on X and Y.

Restriction of Range

While outliers inflate the value of a correlation, restriction of range tends to deflate the value of a correlation. This means that a restricted range could lead a researcher to conclude that there is less of a relationship between two variables than actually exists.

What is restriction of range? Let's consider an unrestricted range first. Look at the left panel of Figure 13.14, a scatterplot showing the hypothetical relationship between IQ and GPA in a sample of high school students. There is a full (unrestricted) range of IQ scores, ranging from 70 to 130. And, there is a full (unrestricted) range of GPA, from 0 to 4. Judging by the noncircular shape of the scatterplot, the correlation between these two variables is strong.

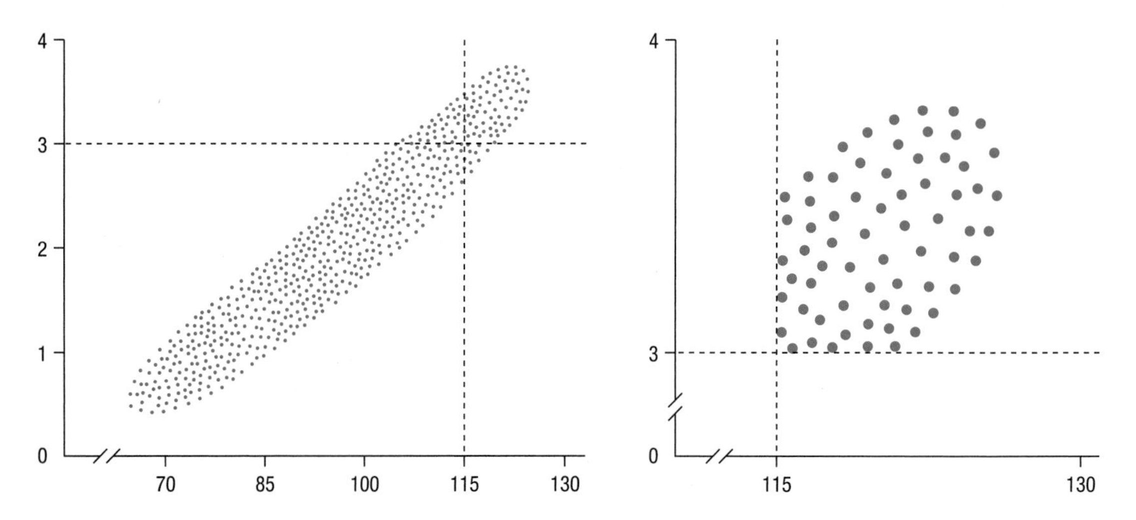

Figure 13.14 Effect of Restriction of Range on Correlation Coefficients The left scatterplot shows the strong relationship between IQ and GPA for a large sample of high school students. The right scatterplot depicts the relationship when the analysis is restricted to people with IQs of 115 or higher *and* GPAs of 3 or higher. Note how the shape of this sample with restricted ranges on both variables is more circular, which means the relationship between IQ and GPA will be much weaker. A restricted range deflates the value of a correlation.

Suppose a researcher decided to restrict her examination of the relationship between IQ and GPA to students likely to be accepted at Ivy League colleges. Thus, she restricted her sample to students with IQs above 115 *and* GPAs above 3.00. The two lines in the top panel are the cut-off values for her sample and the few cases that fall in the upper right quadrant are the new sample.

This subsample has a restricted range on *both* the *X* variable and the *Y* variable. The new sample is shown in a scatterplot all by itself, in the right panel of Figure 13.14. What is the shape of this scatterplot? The points fall roughly in a circle, meaning that little correlation will be found between *X* and *Y*. Restricting the range of one or both variables in a correlation deflates its value.

Practice Problems 13.1

Apply Your Knowledge

13.01 Make a scatterplot for these data:

X	Y	X	Y
100	110	80	95
90	85	100	95
85	95	110	115
90	95	85	80
80	85	90	105
110	125		

Use the nine scatterplots below to answer Problems 13.02–13.07.

13.02 Which figure or figures have a linear relationship?

13.03 Which figure or figures represent no relationship?

13.04 Which figure or figures have a weak linear relationship?

13.05 Which figure or figures have a strong, but not perfect, linear relationship?

13.06 Which figure or figures have a perfect linear relationship?

13.07 Which figure or figures have a negative linear relationship?

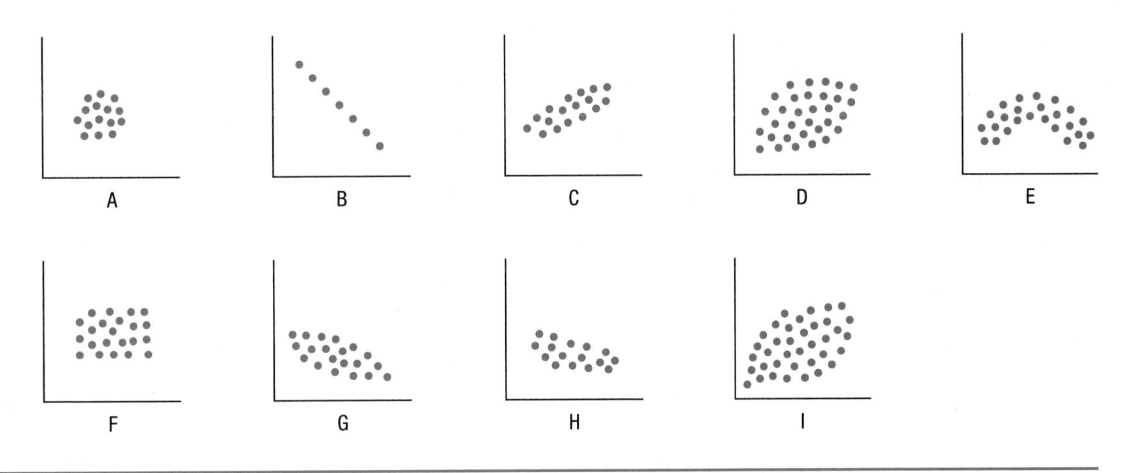

13.2 Calculating the Pearson Correlation Coefficient

To learn how to calculate a Pearson r, imagine some data collected by a marriage therapist. Dr. Paik was interested in seeing whether there was a relationship between gender role flexibility and marital satisfaction. Gender role flexibility refers to the ability to express both male and female traits. Dr. Paik wanted to find out if women's marital satisfaction correlated with how gender role flexible their husbands were.

To measure gender role flexibility, he used the Role Flexibility Test (RFT). The RFT is scored on an interval level and scores range from 0 to 40. Higher scores mean more role flexibility. To measure marital satisfaction, he asked the women to grade their husbands A–F on a number of dimensions. He averaged these grades together, and then he calculated a marital "GPA." Just like an academic GPA, a marital GPA ranges from 0 to 4, with 0 = F and 4 = A.

Dr. Paik obtained a random sample of eight heterosexual married couples from his city and measured two characteristics for each couple (see the data in Table 13.3). *X* was the husband's level of gender role flexibility, and *Y* was the wife's rating of marital satisfaction. Dr. Paik's question was simple. He did not specify a direction, direct or inverse, for the relationship, he just wanted to know: Are the two variables related?

TABLE 13.3 Data for Dr. Paik's Study of the Relationship Between Gender Role Flexibility and Marital Satisfaction

Case	*X* Gender Role Flexibility	*Y* Marital Satisfaction
1.	8	.8
2.	15	2.0
3.	22	1.5
4.	31	2.3
5.	35	1.5
6.	38	3.3
7.	15	1.5
8.	36	3.1
$M =$	25.00	2.00
$s =$	11.49	0.86

The husband's gender role flexibility test score is *X*. The wife's marital satisfaction score is *Y*.

Step 1 Pick a Test

The first step in hypothesis testing involves picking the correct statistical test. Here, one group of cases ($N = 8$) is used to examine the linear relationship between two interval/ratio level variables. This calls for a Pearson *r*.

Step 2 Check the Assumptions

The Pearson *r* has four assumptions that should be met in order to proceed with the test. Three of the assumptions are familiar (random sample, independence of observations, and normality). One assumption, linearity, is new.

- *Random sample.* This assumption says that the sample in which the correlation is being calculated is a random sample from the population to which the results will be generalized. As with *t* tests and ANOVAs, the random sample assumption is robust to violation. With the marital satisfaction study, the random sample assumption is not violated. This is one of those rare times there is a random sample. The researcher can generalize the results of the study to the whole city. However, the sample size is small, so replication should be on Dr. Paik's mind already.

"**T**om **a**nd **H**arry **d**espise **c**rabby **i**nfants" is the mnemonic for the six steps of hypothesis testing: (1) Pick a **test**, (2) check the **assumptions**, (3) list the **hypotheses**, (4) set the **decision rule**, (5) **calculate** the test statistic, and (6) **interpret** the results.

- *Independence of observations.* This assumption says each case in the sample is independent of every other case, that the cases don't influence each other. This assumption is not robust, so a researcher shouldn't use the Pearson r if it is violated. Even though the cases are couples, the independence of observations assumption is not violated for the marital satisfaction study because each couple was measured individually and was only in the sample once.
- *Normality.* This assumption says that each of the variables in a Pearson r, both X and Y, is normally distributed in the population. This assumption is robust to violation, especially if the sample size is large. The normality assumption is not violated for the marital satisfaction data. It seems reasonable to assume, in the larger population of the city, that both gender role flexibility and marital satisfaction are normally distributed.
- *Linearity.* This assumption is a new one. It says that the relationship between the two variables in the population is a linear one. The linearity assumption is tested by making a scatterplot for the data. If the dots in the scatterplot fall on a curve, as they do in Figure 13.12, then this assumption is violated. If the dots form an irregular shape, like that seen in Figure 13.13, or fall in a straight line, then the linearity assumption is not violated. This assumption is not robust, so if it is violated, the test should not proceed. The scatterplot for Dr. Paik's data appears in **Figure 13.15**, which shows the points falling roughly along a straight line that moves up and to the right. The shape is not curvilinear, so the linearity assumption is not violated for the marital satisfaction data.

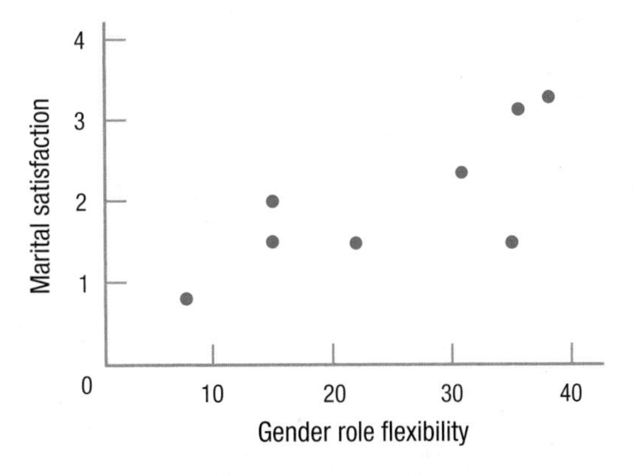

Figure 13.15 Relationship Between Male Gender Role Flexibility and Female Marital Satisfaction The data points in this scatterplot for the most part fall along a straight line, so the linearity assumption is not violated.

With no assumptions violated, Dr. Paik can proceed to the next step.

Step 3 List the Hypotheses

The null and alternative hypotheses are statements about populations, not samples. For the population value of a correlation, the Greek letter rho, ρ, will be used as the abbreviation. Though ρ looks like the letter "p," it is the Greek letter "r" and is pronounced "row." ρ represents the *population* value of a correlation, while r is the value of a correlation when it is calculated for the data in a *sample.* The correlation coefficient Dr. Paik is calculating for the sample of eight cases in the marital satisfaction study is an r. If the correlation coefficient were calculated for all the married couples in the city, it would be a ρ, a population value.

For a nondirectional or two-tailed test, the null hypothesis states that there is no relationship, in the population, between X and Y. This is expressed by saying that ρ equals zero:

$$H_0: \rho = 0$$

The alternative hypothesis says there is *some* relationship between X and Y in the population. It might be a strong relationship (near 1.00), a very weak relationship (say, .01), or anywhere between those extremes. It might be a positive relationship or it might be a negative relationship. It just is something other than a zero-relationship. So, the alternative hypothesis is

$$H_1: \rho \neq 0$$

If a researcher has a directional hypothesis, then he or she would specify what the expected direction of the relationship is before any data are collected. This direction is then stated as the alternative hypothesis. For example, if the researcher expected a direct relationship between X and Y, that is, a correlation coefficient with a positive sign, then the alternative hypothesis would be $\rho > 0$; for an expected inverse relation, the alternative hypothesis would read $\rho < 0$. The null hypothesis is then set so that the two hypotheses together are all-inclusive and mutually exclusive. If the alternative hypothesis is $\rho > 0$, then a null hypothesis of $\rho \leq 0$ would satisfy those criteria.

Step 4 Set the Decision Rule

The sampling distribution of r is based on the population value of the correlation being zero. If the null hypothesis is nondirectional, then the sampling distribution of r would look like Figure 13.16. Note the following points about the sampling distribution:

- All the values of r fall from -1.00 to 1.00.
- The distribution is centered on zero, and $r = .00$ has the highest frequency.
- The distribution is symmetric. One is just as likely to draw a random sample that has a negative r as a positive r.
- As one moves away from $r = .00$, the frequencies become smaller.

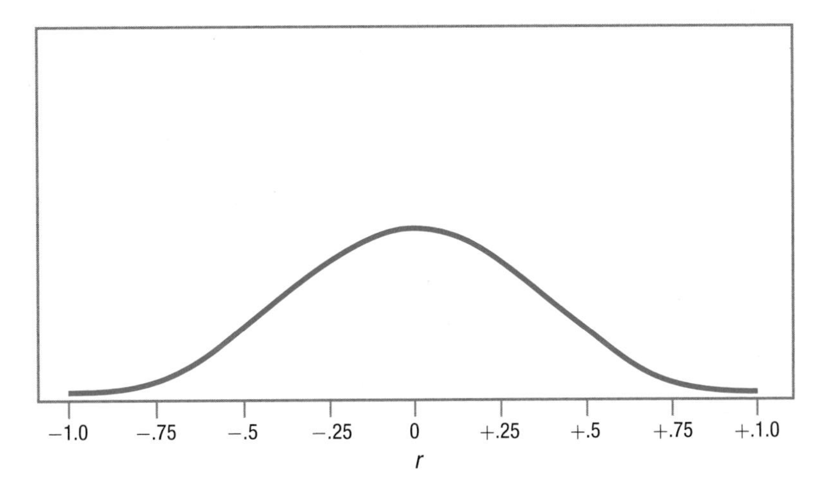

Figure 13.16 Example of a Sampling Distribution of Pearson *r* Centered at zero, *r* is symmetrical and ranges from −1.00 to 1.00. Note that the frequencies decrease as *r* moves away from zero.

As with the sampling distribution for *t* and *F*, this sampling distribution can be divided into common and rare zones:

- The common zone is the middle section of the sampling distribution, the section centered around zero where it would be common to find the *r* value for a sample when the null hypothesis is true.
- The rare zone is the extreme sections of the sampling distribution. If the null hypothesis is true, it is rare to find a sample *r* value that falls in this section.

The decision rule for the Pearson *r* is similar to the other tests:

- If the value of *r* calculated for the sample falls in the rare zone, reject the null hypothesis.
- If the value of *r* calculated for the sample falls in the common zone, fail to reject the null hypothesis.

The boundary between the common zone and the rare zone is called the critical value of *r*, abbreviated r_{cv}. Most commonly, the rare zone is set to be 5% of the sampling distribution. Phrased in the language of statistics, this means alpha (α) is set at .05, giving a 5% chance of a Type I error.

Besides deciding what alpha level to use, a researcher has to decide whether to do a one-tailed or two-tailed test. With a one-tailed test, the hypotheses are directional, so the entire rare zone is placed in one tail of the sampling distribution. A directional alternative hypothesis makes a statement about the direction of ρ, such as $\rho > 0$ or $\rho < 0$. Directional hypotheses need to be stated before data are collected.

With a two-tailed test, the hypotheses are nondirectional, so the rare zone is split evenly between the two tails. Two-tailed tests are more common. For example, the hypotheses for the marital satisfaction study are nondirectional, with the alternative hypothesis simply stating $\rho \neq 0$. The hypotheses are nondirectional because Dr. Paik hasn't predicted whether higher gender role flexibility scores or lower gender role flexibility scores are associated with more marital satisfaction.

Two-tailed tests with $\alpha = .05$ are the "default" option for the Pearson correlation coefficient. This is what Dr. Paik plans to use. When there is a two-tailed test and $\alpha = .05$, the middle 95% of the sampling distribution is the common zone and the rare zone is made up of the 2.5% on the far left of the sampling distribution *and* the 2.5% on the far right. The sampling distribution for the marital satisfaction study, with the rare and common zones marked, is shown in Figure 13.17.

Critical values of r are listed in Appendix Table 6, a portion of which is in Table 13.4. Finding the critical value of r follows the same process as finding a critical value of t:

- First, select a column to use based on whether it is a one-tailed test or a two-tailed test *and* what the level of alpha is.

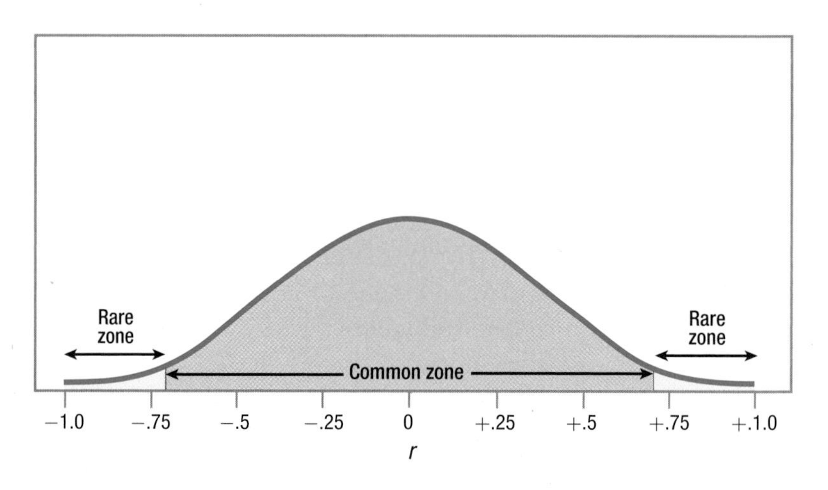

Figure 13.17 Common Zone and Rare Zones for the Sampling Distribution of r for Dr. Paik's Marital Satisfaction Study The critical value of r is ±.707. If the observed value of r falls in the common zone of the sampling distribution, one will fail to reject the null hypothesis. If r falls on either of the lines or in either rare zone, reject the null hypothesis.

TABLE 13.4 Section of Appendix Table 6, Table of Critical Values of r

df	$\alpha = .10$, two-tailed *or* $\alpha = .05$, one-tailed	$\alpha = .05$, two-tailed *or* $\alpha = .025$, one-tailed
1	.988	**.997**
2	.900	**.950**
3	.805	**.878**
4	.729	**.811**
5	.669	**.754**
6	.621	**.707**
7	.582	**.666**
8	.549	**.632**
9	.521	**.602**
10	.497	**.576**

The critical value of r, r_{cv}, is found at the intersection of the column for the selected alpha level and number of tails and the row for the degrees of freedom.

- Then, find the row that contains r_{cv} by selecting the row with the correct degrees of freedom. (If there is no row for the degrees of freedom one needs, follow *The Price Is Right* rule and use the row for the degrees of freedom that is closest without going over.)

Equation 13.1 shows how to calculate degrees of freedom for a Pearson correlation coefficient.

Equation 13.1 Formula for Calculating Degrees of Freedom (*df*) for a Pearson Correlation Coefficient

$$df = N - 2$$

where df = degrees of freedom
N = number of cases in the sample

For the marital satisfaction study, there are eight cases in the sample, so $N = 8$. Applying Equation 13.1, Dr. Paik determines $df = 6$:

$$\begin{aligned} df &= N - 2 \\ &= 8 - 2 \\ &= 6 \end{aligned}$$

Looking in the table of critical values of r, the intersection of the row where $df = 6$ and the column where $\alpha = .05$, two-tailed, we determine that the critical value of r is .707. There is no sign attached to this critical value. Dr. Paik has opted for a two-tailed test and the sampling distribution is symmetric, so $r_{cv} = \pm.707$. The decision rule can now be written for the marital satisfaction study:

- If $r \leq -.707$ *or* if $r \geq .707$, reject H_0.
- If $-.707 < r < .707$, fail to reject H_0.

Step 5 Calculate the Test Statistic

Equation 13.2 is the computational formula for calculating the Pearson r. Computational formulas are mathematically easier than definitional formulas. Equation 13.2 may look complex, but it isn't difficult when broken down into chunks.

Equation 13.2 Formula for Calculating *r*, the Pearson Correlation Coefficient

$$r = \frac{\Sigma[(X - M_X)(Y - M_Y)]}{\sqrt{SS_X SS_Y}}$$

where r = Pearson correlation coefficient
X = a case's score on variable X
M_X = mean score on variable X
Y = a case's score on variable Y
M_Y = mean score on variable Y
SS_X = sum of the squared deviation scores for variable X
SS_Y = sum of the squared deviation scores for variable Y

The easiest way to complete Equation 13.2 is in three pieces: one for the numerator, one for the denominator, and one to finish the division.

The Numerator. There are four steps in calculating the numerator:

1. Take each X score and subtract the mean of the X scores from it. The resulting values are the X deviation scores.
2. Take each Y score and subtract the mean of the Y scores from it. The resulting values are the Y deviation scores.
3. For each case, multiply the two deviation scores together.
4. Add up all the multiplied pairs of deviation scores. This sum is the numerator in Equation 13.2.

The Denominator. There are six steps in calculating the denominator:

1. Take each X deviation score and square it.
2. Add up all the squared X deviation scores. This results in the sum of squares for X, abbreviated SS_X.
3. Take each Y deviation score and square it.
4. Add up all the squared Y deviation scores. This results in the sum of squares for Y, abbreviated as SS_Y.
5. Multiply SS_X from Step 2 by SS_Y from Step 4.
6. The square root of the product found in Step 5 is the denominator in Equation 13.2.

Final Calculations. The last step in calculating Pearson r is to take the numerator and divide it by the denominator, yielding the r value.

Doing the Calculations. The marital satisfaction data are re-presented in Table 13.5. The first two columns contain the X value and the Y value for each case (the same

TABLE 13.5 Calculating Pearson r for Dr. Paik's Marital Satisfaction Study

	1	2	3	4	5	6	7
	X	Y	$X - M_X$	$Y - M_Y$	$(X - M_X)(Y - M_Y)$	$(X - M_X)^2$	$(Y - M_Y)^2$
	8	.8	−17.00	−1.20	20.40	289.00	1.44
	15	2.0	−10.00	0.00	0.00	100.00	0.00
	22	1.5	−3.00	−0.50	1.50	9.00	0.25
	31	2.3	6.00	0.30	1.80	36.00	0.09
	35	1.5	10.00	−0.50	−5.00	100.00	0.25
	38	3.3	13.00	1.30	16.90	169.00	1.69
	15	1.5	−10.00	−0.50	5.00	100.00	0.25
	36	3.1	11.00	1.10	12.10	121.00	1.21
M	25.00	2.00			$\Sigma = 52.70$	$\Sigma = SS_X = 924.00$	$\Sigma = SS_Y = 5.18$

The original data can be found in column 1 (gender role flexibility) and column 2 (marital GPA). Columns 3–7, as outlined in the text, are steps in the calculation of the Pearson r.

information as in Table 13.3). But, Table 13.5 contains five new columns (numbered as columns 3–7), each of which will be explained below as it is used.

To find Pearson *r*, calculate the numerator first:

1. Column 3, labeled "$X - M_X$," is used to calculate deviation scores for the *X* variable. It is completed by subtracting the mean for the *X* scores, which is 25.00 here, from each *X* score. The first case has an *X* score of 8.00, so the deviation score is $8.00 - 25.00 = -17.00$. To complete column 3, repeat this step for each of the *X* scores.
2. Column 4, labeled "$Y - M_Y$," calculates deviation scores for the *Y* variable. The same action is taken here for each *Y* value that was done for each *X* value in column 3. That is, the mean for *Y* is subtracted from each *Y* score. For example, the first case has a *Y* score of 0.8 and M_Y is 2.00, so the deviation score is $0.8 - 2.00 = -1.20$. To complete column 4, repeat this step for each of the *Y* scores.
3. In column 5, labeled "$(X - M_X)(Y - M_Y)$," the two deviation scores for each case are multiplied together. For the first case, which is in the first row, $-17.00 \times -1.20 = 20.40$. For each row, column 3 is multiplied by column 4.
4. Finally, add up all the deviation scores that were multiplied together in column 5. At the bottom of column 5, it says $\Sigma = 52.70$. That sum of the multiplied-together deviation scores is the numerator Dr. Paik will use to calculate *r*.

The next step to find Pearson *r* is calculating the denominator:

1. In column 6, labeled "$(X - M_X)^2$," square the *X* deviation scores from column 3. The *X* deviation score for the first case was −17.00, so $-17.00^2 = 289.00$. Repeat this step for each of the scores in column 3.
2. Once all the *X* deviation scores have been squared, add them all up at the bottom of column 6. $SS_X = 924.00$.
3. In column 7, labeled "$(Y - M_Y)^2$," square the *Y* deviation scores from column 4. For example, for the first case, $-1.20^2 = 1.44$. Repeat this step for each of the scores in column 4.
4. Next, add up the squared *Y* deviation scores and write the value at the bottom of column 7. $SS_Y = 5.18$.
5. Then, multiply together the two sums of squared deviation scores, 924.00 (Step 2) and 5.18 (Step 4): $924.00 \times 5.18 = 4{,}786.32$.
6. Next, to find the denominator, take the square root of the product that was calculated in Step 5: $\sqrt{4{,}786.32} = 69.1832 = 69.18$.

Now that the numerator of Equation 13.2 (52.70) and the denominator of Equation 13.2 (69.18) are known, Dr. Paik can calculate *r*:

$$r = \frac{\Sigma[(X - M_X)(Y - M_Y)]}{\sqrt{SS_X SS_Y}}$$

$$= \frac{52.70}{69.18}$$

$$= .7618$$

$$= .76$$

Dr. Paik now knows that $r = .76$. His next task is to interpret this correlation. We'll turn to that after one more experience going through the first five steps of hypothesis testing for a Pearson *r*.

A Common Question

Q Why is *r* reported as .76, not 0.76?

A APA format says not to use a zero before the decimal point for correlations.

Worked Example 13.1 Imagine that a developmental psychologist, Dr. Solomon, wanted to determine if there were a relationship between the age at which children started to walk and their intelligence at age 16. She went to a pediatrician's office and randomly selected 10 charts of 16-year-old girls. In the charts, she found the age (in months) at which each girl started walking and then she gave each girl a standard IQ test.

In this instance, there is a clear order to the variables—age of walking comes first and IQ is measured later. This is an example of a relationship test with a predictor variable (age) and an outcome variable (IQ). When the variables can be classified as predictor and outcome, the convention is to make the predictor variable *X* and the outcome variable *Y*. That is how the data are shown in Table 13.6, with age of walking as *X* and IQ as *Y*. Remember, correlations measure association. The fact that one variable is used to predict the other should not be taken to mean that the one causes the other.

TABLE 13.6 Data for Dr. Solomon's Study Exploring the Relationship Between Age of First Walking and IQ at Age 16

	Age of Walking (in months)	Intelligence at Age 16
1.	9	115
2.	15	100
3.	11	90
4.	14	95
5.	13	115
6.	10	100
7.	12	110
8.	18	90
9.	17	105
10.	16	90
M	13.50	101.00
s	3.03	9.94

A Pearson *r* will be used to see if age of first walking, the *X* variable, predicts IQ, the *Y* variable.

Step 1 Pick a Test. There is one group of cases in which two interval/ratio level variables were measured in order to see if a relationship exists between them. This calls for a Pearson *r*.

Step 2 Check the Assumptions.

- *Random sample.* The random sample assumption is not violated. The sample is from one pediatric practice, though, so the results shouldn't be generalized beyond that practice.
- *Independence of observations.* Each case is in the sample only once. There's no reason to think that the cases influence each other in terms of age of first walking or IQ. The independence of observations assumption is not violated.
- *Normality.* Researchers consider IQ to be normally distributed. It seems reasonable to consider age at first walking normally distributed as well. This assumption is not violated.
- *Linearity.* There is no obvious curvilinear relationship in the scatterplot (Figure 13.18), so this assumption is not violated.

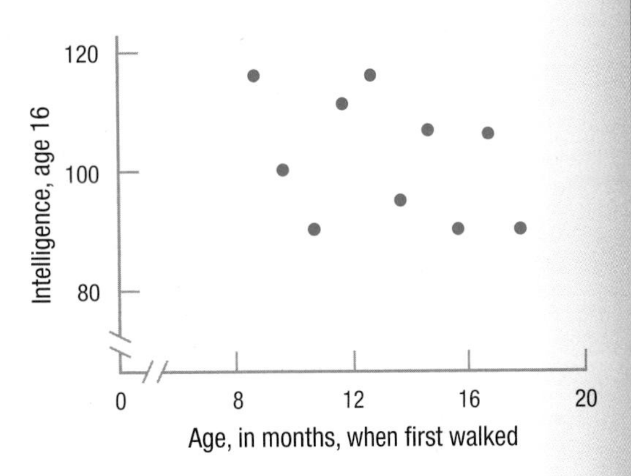

Figure 13.18 Dr. Solomon's Data Showing the Relationship Between Age of First Walking and Intelligence at Age 16 There is no obvious curvilinear relationship in this scatterplot, so the linearity assumption is not violated.

With no assumptions violated, Dr. Solomon can proceed with the Pearson *r*.

Step 3 List the Hypotheses. Dr. Solomon has not made a prediction about the direction of the relationship, so her hypotheses are nondirectional or two-tailed:

$$H_0: \rho = 0$$
$$H_1: \rho \neq 0$$

Step 4 Set the Decision Rule. With $N = 10$, the degrees of freedom are calculated with Equation 13.1:

$$\begin{aligned} df &= N - 2 \\ &= 10 - 2 \\ &= 8 \end{aligned}$$

Dr. Solomon is examining whether there is a relationship between age of first walking and IQ, either positive or negative, so the test is two-tailed. She's willing to have a 5% chance of a Type I error, so $\alpha = .05$. Next, she consults the table of critical values of r, Appendix Table 6. The intersection of the row with 8 degrees of freedom and the column for $\alpha = .05$, two-tailed gives $r_{cv} = \pm.632$. The decision rule is:

- If $r \leq -.632$ *or* if $r \geq .632$, reject H_0.
- If $-.632 < r < .632$, fail to reject H_0.

Figure 13.19 shows how Dr. Solomon sketched the decision rule as a sampling distribution of r, marking the rare and common zones.

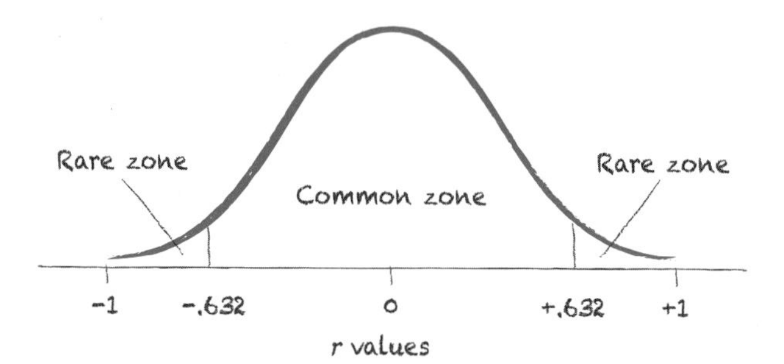

Figure 13.19 Decision Rule for Dr. Solomon's IQ/Walking Study This sketch of a sampling distribution of r uses the critical value of r, $\pm.632$, to mark the rare and common zones.

Step 5 Calculate the Test Statistic. Table 13.7 shows the deviation scores for age of first walking (X) and IQ (Y), the deviation scores that were multiplied together, and the squared deviation scores.

- For the numerator in Equation 13.2, add up the deviation scores that were multiplied together. The bottom of column 5 gives the result: -120.00.
- The sums of the squared deviation scores are 82.50 (SS_X in column 6) and 890.00 (SS_Y in column 7):
- Multiply these together: $82.50 \times 890.00 = 73{,}425.00$.
- Take the square root of this product to find the denominator: $\sqrt{73{,}425.00} = 270.97$.

TABLE 13.7 Calculating Pearson r for Dr. Solomon's Age of Walking/IQ Study

	1	2	3	4	5	6	7
	X	Y	$X - M_X$	$Y - M_Y$	$(X - M_X)(Y - M_Y)$	$(X - M_X)^2$	$(Y - M_Y)^2$
	9	115	−4.50	14.00	−63.00	20.25	196.00
	15	100	1.50	−1.00	−1.50	2.25	1.00
	11	90	−2.50	−11.00	27.50	6.25	121.00
	14	95	0.50	−6.00	−3.00	0.25	36.00
	13	115	−0.50	14.00	−7.00	0.25	196.00
	10	100	−3.50	−1.00	3.50	12.25	1.00
	12	110	−1.50	9.00	−13.50	2.25	81.00
	18	90	4.50	−11.00	−49.50	20.25	121.00
	17	105	3.50	4.00	14.00	12.25	16.00
	16	90	2.50	−11.00	−27.50	6.25	121.00
M	13.50	101.00			$\Sigma = -120.00$	$\Sigma = SS_X = 82.50$	$\Sigma = SS_Y = 890.00$

The calculations in this table lead through the steps necessary to obtain the numerator and denominator for calculating a Pearson correlation coefficient.

Now that the numerator (−120.00) and the denominator (270.97) are known, the calculation of Pearson r is straightforward:

$$r = \frac{\Sigma[(X - M_X)(Y - M_Y)]}{\sqrt{SS_X SS_Y}}$$

$$= \frac{-120.00}{270.97}$$

$$= -.4429$$

$$= -.44$$

Dr. Solomon now knows that $r = -.44$. Her next step will be to interpret the results.

Practice Problems 13.2

Apply Your Knowledge

13.08 Read each scenario and decide if the data can be analyzed with a Pearson r.

a. A researcher from a facial tissue manufacturer rates the severity of people's colds on a 15-point interval scale and measures how many tissues they use in a 24-hour period. She wants to determine if there is a relationship between the severity of a cold and tissue use.

b. A high school counselor obtains each student's class rank and IQ score. She wants to know if a relationship exists between class rank and IQ.

c. People who use Apple computers and those who use Windows-based computers are measured on an interval level of creativity. Is there a relationship between the type of computer a person uses and his or her level of creativity?

13.09 A dietician goes to a mall on a Sunday afternoon and finds people shopping by themselves who are willing to complete questionnaires about weekly food consumption and weekly exercise. She wants to see if any relationship exists between caloric consumption and caloric expenditure. She is planning to use a Pearson r. (a) Check as many assumptions as possible and (b) decide if she can proceed.

13.10 A cosmetician develops a theory that the longer a man's hair is, the more tattoos and piercings he is likely to have. She obtains a sample of 37 men, measures how long their hair is in millimeters, and counts how many tattoos and piercings each has.

a. List her hypotheses.

b. With $\alpha = .05$, list her decision rules.

13.11 Given the following values for X and Y, complete the table:

	Col. 1	Col. 2	Col. 3	Col. 4	Col. 5	Col. 6	Col. 7
	X	Y	$X - M_X$	$Y - M_Y$	$(X - M_X)(Y - M_Y)$	$(X - M_X)^2$	$(Y - M_Y)^2$
	10	20					
	14	24					
	9	37					
M	11.00	27.00			$\Sigma =$	$\Sigma = SS_X =$	$\Sigma = SS_Y =$

13.12 Given $\Sigma[(X - M_X)(Y - M_Y)] = 80.00$, $SS_X = 200.00$, and $SS_Y = 90.00$, what is r?

13.3 Interpreting the Pearson Correlation Coefficient

Step 6 Interpret the Results

To interpret a Pearson correlation coefficient, there are three questions to be addressed. The answers to these three questions provide the raw material from which the researcher selects the most salient pieces to build a four-point interpretation. The three questions are the same ones used in the interpretation of an independent-samples *t* test:

- Was the null hypothesis rejected?
- How big is the effect?
- How wide is the confidence interval?

After those three questions have been covered for the Pearson *r*, it is time to add more nuance and depth to the interpretation with a fourth question, "Did this test have adequate power?" As we learned in Chapter 6, power is the likelihood of being able to reject the null hypothesis when it should be rejected. Having adequate power is usually raised as a concern when the null hypothesis is not rejected. So, we will save our exploration of power until the next worked example, when Dr. Solomon interprets the Pearson *r* for the relationship between age of walking and intelligence. Be sure to read the worked example—it covers new ground.

Until then, we will follow Dr. Paik as he interprets the results from his marital satisfaction study. In that study, he obtained a random sample of eight couples from the city where he lives. For each couple, he measured the husband's degree of gender role flexibility and the wife's level of marital satisfaction in order to see if the two variables were related.

Was the Null Hypothesis Rejected?

To determine if the null hypothesis is rejected, Dr. Paik will use the decision rule from Step 4. For the marital satisfaction study, the observed value of *r* was calculated, in Step 5, to be .76. In Step 4, the critical value of *r* was found to be $\pm.707$. Dr. Paik has to decide:

- Is $.76 \leq -.707$ *or* is $.76 \geq .707$?
- Is $-.707 < .76 < .707$?

The second part of the first statement, $.76 \geq .707$, is true, so the null hypothesis is rejected. Figure 13.20 shows how the results fall in the rare zone. The null hypothesis is rejected and the results are called statistically significant. Dr. Paik can say that this Pearson *r* of .76 is statistically different from zero. In APA format, he would write

$$r(6) = .76, p < .05$$

APA format for a Pearson *r* contains five pieces of information: (1) what test is being done, (2) how many cases there are, (3) the observed value of the test statistic, (4) what alpha level was selected, and (5) whether the null hypothesis was rejected.

1. The *r* tells that the test was a Pearson *r*.

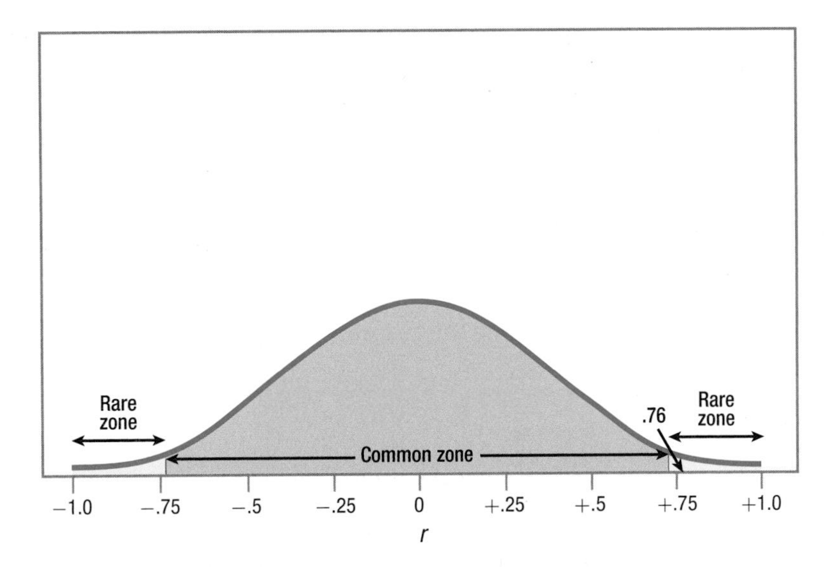

Figure 13.20 Implementing the Decision Rule for Dr. Paik's Marital Satisfaction Study
The observed value of *r*, .76, falls in the rare zone of the sampling distribution. It is an unusual result if the null hypothesis is true, so the null hypothesis is rejected.

2. The 6 in the parentheses is the degrees of freedom for the test. Degrees of freedom for a Pearson correlation coefficient are calculated by subtracting 2 from the sample size. Thus, 2 can be added to the degrees of freedom to determine the sample size, which was 8.
3. The .76 is the value of the test statistic found in the sample.
4. The .05 shows what level alpha was set and that a 5% chance of making a Type I error exists.
5. Finally, $p < .05$ reveals that the results landed in the rare zone, and the null hypothesis is rejected.

What conclusion can Dr. Paik draw so far? He has rejected the null hypothesis, which means he is forced to accept the alternative hypothesis, that there is a relationship between these two variables. Being "forced" to accept the alternative hypothesis is not a hardship. Remember, it is almost always the alternative hypothesis that the researcher believes to be true. Dr. Paik should bear in mind that there is a 5% chance that his decision to reject the null hypothesis is wrong. He should state that the alternative hypothesis is *probably* true.

If the alternative hypothesis is true, it doesn't tell much besides the fact that the population value of the correlation (ρ) between *X* and *Y* is not zero. Phrased without the negative, Dr. Paik can conclude that some relationship exists between the two variables in the population, but he doesn't know how much. At this point, how strong ρ is remains unknown, but it is possible to comment on its direction, whether the relationship is direct or inverse:

- If the null hypothesis is rejected and the sign of the *r* is positive, then a researcher would conclude that the association between *X* and *Y* is a positive one. This means that, in general, cases with high scores on *X* have high scores on *Y*, and cases with low scores on *X* have low scores on *Y*. This is a direct relationship.

- If the null hypothesis is rejected and the sign of the r is negative, then a researcher would conclude that the association between X and Y is a negative one. This means that, in general, cases with high scores on X will have low scores on Y, and cases with low scores on X will have high scores on Y. This is an inverse relationship.
- If the null hypothesis is not rejected, then r is not statistically different from zero. This means sufficient evidence does not exist to conclude that there is a relationship between X and Y. One can't say that the relationship isn't a zero relationship, and one can't assert that it is a zero relationship. The researcher is left in limbo.

At this point, Dr. Paik's interpretation would read something like this:

> The relationship between a husband's gender role flexibility and a wife's level of marital satisfaction was examined in a random sample of married couples in a city. The relationship was positive and statistically significant [$r(6) = .76, p < .05$]. For couples in this city, there is a direct relationship between these two variables: higher levels of female marital satisfaction are associated with higher levels of male gender role flexibility.

How Big Is the Effect?

It is a good idea to quantify the size of the effect whether the observed value of r was statistically significant or not:

- If the Pearson r was statistically significant, an effect probably exists in the population. The question is how large is the effect? How strong is the relationship between the two variables? Calculating an effect size will answer this question.
- If the Pearson r was not statistically significant, then not enough evidence is available to conclude that an effect exists. However, a Type II error might have been made, and there was a failure to find an effect when one exists. When the null hypothesis was not rejected, if the effect size is meaningful, then the researcher should raise a concern about Type II error.

The effect size used for r has a formal name, the **coefficient of determination,** but it is commonly called r^2. This is the same effect size that was calculated for the between-subjects, one-way ANOVA.

$\boldsymbol{r^2}$ tells the percentage of variability in one variable that is accounted for by the other variable. The amount of variability that can be explained ranges from 0% to 100%. The closer the size of the effect is to 100%, the stronger it is. The closer the size of the effect is to 0%, the weaker it is.

r^2 can be thought of as indicating how much overlap occurs between what the two variables measure. If $r^2 = 100\%$, then the two variables overlap 100% and measure exactly the same thing (though they are on different scales). This occurs if $r = 1.00$ or $r = -1.00$. For example, the correlation between temperatures measured in Fahrenheit and Celsius is 1.00 and $r^2 = 100\%$. Fahrenheit and Celsius measure the same thing. Saying something's temperature is 212° in Fahrenheit is exactly the same as saying it is 100° Celsius.

An r^2 of 0% would occur if there were no relationship between X and Y, if r were zero. This means that no overlap, none at all, exists between the two variables. An r^2

near zero means the effect is very weak; near 100% means the effect is very strong. Very high r^2 values are rare. Much more common are low to mid-level values. Cohen (1988) provides standards for judging effect sizes for r^2:

- Small $\approx$1%
- Medium $\approx$9%
- Large $\approx$25%

The formula for calculating r^2 is shown in Equation 13.3. To calculate r^2, take the Pearson r, square it, and multiply that number by 100 to turn it into a percentage.

Equation 13.3 Formula for Calculating r^2, the Percentage of Variability in One Variable That Is Accounted for by the Other Variable

$$r^2 = (r)^2 \times 100$$

where r^2 = percentage of variability in one variable that is accounted for by the other variable

r = Pearson r

For the marital satisfaction study, $r = .76$. Here are the calculations for r^2:

$$\begin{aligned} r^2 &= (r)^2 \times 100 \\ &= .76^2 \times 100 \\ &= .5776 \times 100 \\ &= 57.76\% \end{aligned}$$

The two variables in this correlational study, husbands' gender role flexibility and wives' marital satisfaction, explain almost 58% of the variability in each other. This is a large effect, indicating that the two variables are strongly correlated with each other. Remember, correlation is not causation, so a researcher needs to be careful about phrasing the results. The order in which the variables are listed has implications for the conclusion about the order of the relationship. There's a difference between "The husbands' gender role flexibility explains 58% of the variability in the wives' marital satisfaction" and "The wives' marital satisfaction explains 58% of the variability in the husbands' gender role flexibility." The variable that is listed first tends to be perceived as the influential one. Unless one wants to suggest an order to the relationship, it is better to say something like the following: "The two variables, husbands' gender role flexibility and wives' marital satisfaction, explain 58% of the variability in each other."

How Wide Is the Confidence Interval?

Most of the interpreting done so far has been based on r, the correlation coefficient calculated for the data in the *sample*. The task of inferential statistics is to use the sample to draw a conclusion about the larger *population* of cases. A confidence interval allows a researcher to use a sample statistic to estimate the range within which the population parameter probably falls. A confidence interval for Pearson r makes a statement about ρ, the population correlation coefficient, based on r, the sample correlation coefficient.

Calculating the 95% confidence interval for ρ, 95% CIρ, is a three-step procedure:

1. First, transform the observed *r* value into a *z* score, z_r. This is called a Fisher's *r* to *z* transformation. Appendix Table 7 makes the transformation easy.
2. Then, use z_r along with Equations 13.4 and 13.5 to calculate the confidence interval.
3. This confidence interval is in z_r format, not *r* format. So, the final step is to use Appendix Table 8 to transform the confidence interval back into *r* value format.

Step 1 Transform *r* to z_r

Appendix Table 7, shown in Table 13.8, is used to transform *r* to z_r. The transformation is necessary because the sampling distribution of *r* becomes less normally distributed as ρ deviates more from zero.

TABLE 13.8 Appendix Table 7: Transformation Table for Fisher's *r* to *z*

	.00	.01	.02	.03	.04	.05	.06	.07	.08	.09
.0	0.00	0.01	0.02	0.03	0.04	0.05	0.06	0.07	0.08	0.09
.1	0.10	0.11	0.12	0.13	0.14	0.15	0.16	0.17	0.18	0.19
.2	0.20	0.21	0.22	0.23	0.24	0.26	0.27	0.28	0.29	0.30
.3	0.31	0.32	0.33	0.34	0.35	0.37	0.38	0.39	0.40	0.41
.4	0.42	0.44	0.45	0.46	0.47	0.48	0.50	0.51	0.52	0.54
.5	0.55	0.56	0.58	0.59	0.60	0.62	0.63	0.65	0.66	0.68
.6	0.69	0.71	0.73	0.74	0.76	0.78	0.79	0.81	0.83	0.85
.7	0.87	0.89	0.91	0.93	0.95	0.97	1.00	1.02	1.05	1.07
.8	1.10	1.13	1.16	1.19	1.22	1.26	1.29	1.33	1.38	1.42
.9	1.47	1.53	1.59	1.66	1.74	1.83	1.95	2.09	2.30	2.65

In this table, the rows represent the first digit of an *r* value and the columns represent the second digit of an *r* value. For example, $r = .76$ is broken down into the row for .7 and the column for .06. The z_r value is found at the intersection of the row and column. An *r* value of .76 becomes a z_r of 1.00. (Be sure to maintain the sign associated with the *r*. If *r* had been −.76, z_r would have been −1.00.)

The rows in Appendix Table 7 represent the first digit of a two-digit *r* value and the columns represent the second digit of the two-digit *r* value. For example, an *r* of .76 is broken down into .7 for the row and .06 for the column. The numbers at the intersections of the row and column are the *r* value transformed into a *z* value. Table 13.7 demonstrates this for the Pearson *r* for the marital satisfaction study where $r = .76$. The intersection of the row for *r*'s that start with .7 and the column for *r*'s that end in .06 gives $z_r = 1.00$. The original *r*, .76, was positive, so the z_r value is also positive. The *r* of .76 is transformed into a z_r of 1.00.

Step 2 Calculate the 95% Confidence Interval for the *z* Value

The formula for calculating the 95% confidence interval around the z_r value is given in Equation 13.4.

Equation 13.4 95% Confidence Interval for z_r

$$95\%CI_{z_r} = z_r \pm 1.96s_r$$

where $95\%CI_{z_r}$ = 95% confidence interval for ρ, expressed in z_r units

z_r = Pearson r transformed into z format, using Appendix Table 7

s_r = standard error of r (Equation 13.5)

This formula says that 1.96 standard errors of r, abbreviated s_r, are added to and subtracted from z_r. So before calculating 95%CIρ, s_r needs to be calculated using Equation 13.5.

Equation 13.5 Standard Error of r

$$s_r = \frac{1}{\sqrt{N-3}}$$

where s_r = standard error of r

N = number of pairs of cases used in calculating the Pearson r

For Dr. Paik's marital satisfaction study, where $N = 8$, s_r is calculated like this:

$$\begin{aligned} s_r &= \frac{1}{\sqrt{N-3}} \\ &= \frac{1}{\sqrt{8-3}} \\ &= \frac{1}{\sqrt{5}} \\ &= \frac{1}{2.2361} \\ &= 0.4472 \\ &= 0.45 \end{aligned}$$

A Common Question

Q Are a 1 and a 3 always used in Equation 13.4 to calculate s_r?

A Yes. They are constants.

The standard error of r from Equation 13.5 can now be used in Equation 13.4 to calculate the 95% confidence interval for ρ in z_r units:

$$\begin{aligned} 95\%CI_{z_r} &= z_r \pm 1.96s_r \\ &= 1.00 \pm (1.96 \times 0.45) \\ &= 1.00 \pm 0.8820 \\ &= \text{from } 0.1180 \text{ to } 1.8820 \\ &= \text{from } 0.12 \text{ to } 1.88 \end{aligned}$$

The 95% confidence interval for ρ, expressed in z_r units, ranges from 0.12 to 1.88. All that is left is to transform the confidence interval back into *r* units.

Step 3 Transform the Confidence Interval from z_r Units to *r* Values

Transforming from z_r units to *r* units uses Appendix Table 8, a reversal of Appendix Table 7. Appendix Table 8 is shown in Table 13.9.

TABLE 13.9 Appendix Table 8: Transformation Table for Fisher *z* to *r*

	.00	.01	.02	.03	.04	.05	.06	.07	.08	.09
0.0	.00	.01	.02	.03	.04	.05	.06	.07	.08	.09
0.1	.10	.11	.12	.13	.14	.15	.16	.17	.18	.19
0.2	.20	.21	.22	.23	.24	.24	.25	.26	.27	.28
0.3	.29	.30	.31	.32	.33	.34	.35	.35	.36	.37
0.4	.38	.39	.40	.41	.41	.42	.43	.44	.45	.45
0.5	.46	.47	.48	.49	.49	.50	.51	.52	.52	.53
0.6	.54	.54	.55	.56	.56	.57	.58	.58	.59	.60
0.7	.60	.61	.62	.62	.63	.64	.64	.65	.65	.66
0.8	.66	.67	.68	.68	.69	.69	.70	.70	.71	.71
0.9	.72	.72	.73	.73	.74	.74	.74	.75	.75	.76
1.0	.76	.77	.77	.77	.78	.78	.79	.79	.79	.80
1.1	.80	.80	.81	.81	.81	.82	.82	.82	.83	.83
1.2	.83	.84	.84	.84	.85	.85	.85	.85	.86	.86
1.3	.86	.86	.87	.87	.87	.87	.88	.88	.88	.88
1.4	.89	.89	.89	.89	.89	.90	.90	.90	.90	.90
1.5	.91	.91	.91	.91	.91	.91	.92	.92	.92	.92
1.6	.92	.92	.92	.93	.93	.93	.93	.93	.93	.93
1.7	.94	.94	.94	.94	.94	.94	.94	.94	.94	.95
1.8	.95	.95	.95	.95	.95	.95	.95	.95	.95	.96
1.9	.96	.96	.96	.96	.96	.96	.96	.96	.96	.96
2.0	.96	.96	.97	.97	.97	.97	.97	.97	.97	.97
2.1	.97	.97	.97	.97	.97	.97	.97	.97	.97	.98
2.2	.98	.98	.98	.98	.98	.98	.98	.98	.98	.98
2.3	.98	.98	.98	.98	.98	.98	.98	.98	.98	.98
2.4	.98	.98	.98	.98	.98	.99	.99	.99	.99	.99
2.5	.99	.99	.99	.99	.99	.99	.99	.99	.99	.99
2.6	.99	.99	.99	.99	.99	.99	.99	.99	.99	.99
2.7	.99	.99	.99	.99	.99	.99	.99	.99	.99	.99
2.8	.99	.99	.99	.99	.99	.99	.99	.99	.99	.99
2.9	.99	.99	.99	.99	.99	.99	.99	.99	.99	.99
3.0	1.00	1.00	1.00	1.00	1.00	1.00	1.00	1.00	1.00	1.00

The intersection of the row for the first two digits of *z* (1.8) with the column for the third digit of *z* (.08) gives the *r* value into which the z_r value is transformed. $z_r = 1.88$ is transformed into $r = .95$. Be sure to maintain the sign associated with the *z* value.

Appendix Table 8 is set up the same way as Appendix Table 7. Each row covers the first decimal place of a z_r value and each column is the second decimal place of a z_r value. At the intersection of the row and column, the z_r value is transformed back into an r value. Of course, be sure to maintain the sign. If the z_r was a negative number, then the r value is a negative value as well.

The first z_r value for the marital satisfaction study, the one at the lower end of the confidence interval, was 0.12. Looking at the intersection of the 0.1 row with the .02 column in the z to r table, Dr. Paik finds $r = .12$. The second z_r value for the marital satisfaction study was 1.88. The intersection of the 1.8 row with the .08 column gives $r = .95$. The 95% confidence interval for ρ, the population value of the relationship between male gender role flexibility and female marital satisfaction, ranges from .12 to .95 when expressed in Pearson correlation coefficient units.

The observed r in the sample was .76, which is a strong and positive correlation. But, the confidence interval indicates uncertainty about the strength of the relationship between gender role flexibility and marital satisfaction in the population. ρ could be (1) an almost perfect correlation coefficient of .95, or (2) a correlation of .12 that is much closer to zero, or (3) a correlation anywhere between those two extremes. And, there's a fourth option, a 5% chance that this confidence interval doesn't capture ρ. A very wide confidence interval, like this one ranging from .12 to .95, tells a researcher that there is a lot of uncertainty about how strong the correlation is in the underlying population.

To interpret a confidence interval for ρ, pay attention to three factors:

1. Whether the confidence interval captures zero. Capturing zero means it is possible that there is no relationship in the population between the two variables. In other words, it is possible that $\rho = 0$, just like the null hypothesis said. The confidence interval should capture zero when the null hypothesis is not rejected.

2. How close the confidence interval comes to zero. The closer an end of the confidence interval comes to zero, the weaker the relationship may be in the population.

3. How wide the confidence interval is. The wider the confidence interval is, the less sure a researcher is of the population value of the correlation coefficient.

For Dr. Paik's marital satisfaction study, the results were statistically significant and the null hypothesis was rejected. As a result, the confidence interval for ρ should not capture zero and that's exactly what happened. There's little reason, beyond the possibility of a Type I error, to think $\rho = 0$. The whole confidence interval, from .12 to .95, is on the positive side of zero. This leaves the researcher confident that the relationship between gender role flexibility and marital satisfaction is a direct one.

The low end of the confidence interval is .12. Using Equation 13.4 to calculate r^2 for this correlation coefficient, Dr. Paik finds $r^2 = 1.44\%$. This is a small effect, which tells Dr. Paik that the size of the effect in the overall population might also be small. On the other hand, if $\rho = .95$, the other end of the confidence interval, then $r^2 = 90.25\%$, which represents a huge effect.

Dr. Paik is bothered by the width of the confidence interval. It ranges from near 0 (.12) to near 1 (.95). That's a wide range. This lack of precision in the confidence interval makes it unclear how much association exists between role flexibility and marital satisfaction in the population. The relationship in the population from which this sample came could be near zero or near perfect.

The reason the confidence interval is so wide is that the sample size was small. If N had been 80 (not 8), then s_r would have been 0.11 (not 0.45), and the 95% confidence interval would have been a much more precise .65 to .84, with the low end much further away from zero. Such a confidence interval would inspire more confidence that, as a result of this study, something was known about the relationship between these two variables in the larger population.

Putting It All Together

Before writing the interpretation, Dr. Paik reviews the scatterplot made in Step 2 when evaluating the assumptions. Seen in Figure 13.14, it is a graphic display of the relationship between gender role flexibility (X) and marital satisfaction (Y). Scatterplots help one think about the strength and direction of relationships. The scatterplot here shows a direct and strong relationship—higher levels of gender role flexibility are associated with more marital satisfaction. Below is Dr. Paik's four-point interpretation in which he (1) tells what the study was about, (2) indicates its main results, (3) explains what they mean, and (4) makes suggestions for future research:

> In this study, the relationship between a husband's flexibility in his gender role and a wife's degree of marital satisfaction was assessed in a random sample of eight married couples from one city. There was a statistically significant, direct relationship [$r(6) = .76$, $p < .05$] between the two variables: couples with a high score on one variable tended to have a high score on the other. Unfortunately, the sample size in the present study was small, making it impossible to say how strong the relationship between these two variables is in the larger population. The study should be replicated with a larger sample size in order to better determine the strength of the relationship.
>
> In addition, this is a correlational study, so it does not address whether (1) more gender flexible husbands lead to more satisfied wives, or (2) wives who are more satisfied give their husbands the leeway to be more gender flexible, or (3) that a third variable—such as age, education level, or socioeconomic status—could influence both marital satisfaction and gender role flexibility. Future research should attempt to determine the order of the relationship.

Worked Example 13.2 For practice interpreting a Pearson correlation coefficient, let's return to Dr. Solomon's study that investigated the relationship between the age at which children begin walking and their later IQ. In that study, ten 16-year-old girls were randomly selected from a pediatrician's practice. They were given IQ tests and their charts were examined to determine the age at which they started to walk. Dr. Solomon used a Pearson correlation coefficient to examine the relationship between the two variables and found $r = -.44$.

Was the null hypothesis rejected? The null hypothesis said that there was no relationship between the two variables. With $df = 8$, $\alpha = .05$, and a two-tailed test, the critical value of r was $\pm.632$. Here is the decision rule:

- If $-.44 \leq -.632$ *or* if $-.44 \geq .632$, reject H_0.
- If $-.632 < -.44 < .632$, fail to reject H_0.

The second statement is true, so Dr. Solomon has failed to reject the null hypothesis. As shown in Figure 13.21, the observed *r* of −.44 falls between −.632 and .632, which is in the common zone of the sampling distribution of *r*. That means −.44 is a result that will commonly occur if the null hypothesis is true. There is no reason to reject the null hypothesis.

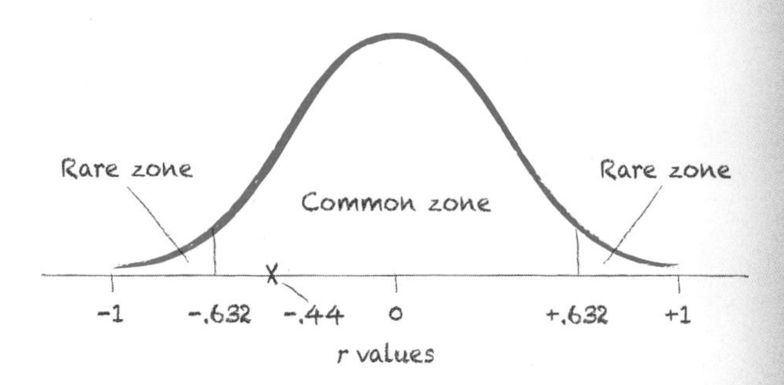

Figure 13.21 Implementing the Decision Rule for Dr. Solomon's Age of Walking/IQ Study The observed *r* of −.44 falls in the common zone of the sampling distribution of *r*. It is a common result when the null hypothesis is true, so the null hypothesis is not rejected.

With the observed *r* in the common zone, the null hypothesis is not rejected, and the results are called "not statistically significant." Dr. Solomon has to state that the observed *r* of −.44 is not statistically different from zero. When she calculates the confidence interval for ρ, she should find that it captures zero.

When a researcher fails to reject the null hypothesis, he or she can't speculate about the direction of the relationship because there isn't enough evidence to say one exists.

The null hypothesis wasn't rejected, so Dr. Solomon can't draw a conclusion about the direction of the relationship. The observed *r*, −.44, may be negative, but she can't say that there is a negative relationship between age at first walking and IQ in the population. That is, she can't conclude that people who start walking at a younger age have higher IQs at age 16. When a researcher fails to reject the null hypothesis, he or she can't speculate about the direction of the relationship because there isn't enough evidence to say a relationship exists.

In APA format, Dr. Solomon reports the results as

$$r(8) = -.44,\ p > .05$$

- The "*r*" tells what test was done, a Pearson *r*.
- The "8" is the degrees of freedom. Adding 2 to it gives the number of participants: 10.
- "−.44" is the observed value of the test statistic, Pearson *r*.
- ".05" indicates the alpha level.
- "$p > .05$" reveals that results like this occur more than 5% of the time when the null hypothesis is true. That is, the null hypothesis was not rejected.

How big is the effect? Though Dr. Solomon failed to reject the null hypothesis, she should still calculate an effect size to help her consider the possibility of Type II error.

Using Equation 13.3, she calculates the effect size, r^2:

$$\begin{aligned} r^2 &= (r)^2 \times 100 \\ &= -.44^2 \times 100 \\ &= 0.1936 \times 100 \\ &= 19.36\% \end{aligned}$$

An r^2 of 19.36% is more than a medium effect, according to Cohen (1988). As a result, Dr. Solomon is worried that a Type II error may have been made and that she missed finding a relationship between age of walking and IQ. She is going to recommend replicating the study with a larger sample size in order to have a better chance of finding the effect if it exists.

How wide is the confidence interval? Calculating the confidence interval for ρ is a three-step process.

Step 1 Convert r to z_r. To do this, Dr. Solomon uses Appendix Table 7. Given $r = -.44$, she finds the z_r transformation for this at the intersection of the row for .4 and the column for .04. The r was negative, so the z_r should be, too: $z_r = -0.47$.

Step 2 Calculate the Confidence Interval in z_r Units. The first thing Dr. Solomon does in this step is use Equation 13.5 to calculate s_r. The only variable to plug into the equation is the sample size, which is 10:

$$\begin{aligned} s_r &= \frac{1}{\sqrt{N-3}} \\ &= \frac{1}{\sqrt{10-3}} \\ &= \frac{1}{\sqrt{7}} \\ &= \frac{1}{2.6458} \\ &= 0.3780 \\ &= 0.38 \end{aligned}$$

Now that she knows $s_r = 0.38$ and $z_r = -0.47$, she can use Equation 13.4 to calculate the confidence interval in z_r units:

$$\begin{aligned} 95\%\text{CI}_{z_r} &= z_r \pm 1.96 s_r \\ &= -0.47 \pm (1.96 \times 0.38) \\ &= -0.47 \pm 0.7448 \\ &= \text{from } -1.2148 \text{ to } 0.2748 \\ &= [-1.21, 0.27] \end{aligned}$$

Step 3 Convert the Confidence Interval Back into r Units. The confidence interval, in z_r units, ranges from -1.21 to 0.27. In order to interpret the confidence interval, Dr. Solomon needs to convert it back to Pearson correlation coefficient units using Appendix Table 8. She will maintain the sign of the values, positive or negative, in the conversion process.

- At the intersection of the row for 1.2 and the column for .01, she transforms a z_r of -1.21 into $r = -.84$.
- At the intersection of the row for 0.2 and the column for .07, she transforms a z_r of .27 into $r = .26$.

- Her 95% confidence interval for the population value of the correlation between age at first walking and IQ at age 16 is [−.84, .26].

There are several things Dr. Solomon can note about this confidence interval:

- The confidence interval captures zero. This was expected because the null hypothesis was not rejected. The confidence interval capturing zero means it is possible, in the population, that there is a zero relationship between age at first walking and IQ.
- The confidence interval doesn't just come close to zero, it includes zero. This means that the relationship might not just be small: it might be nonexistent.
- The confidence interval is very wide. It goes all the way from far below zero, −.84, to a modest distance above zero, .26. With such a wide range, the population value is uncertain. Is the population value positive, zero, or negative? Is it a weak relationship or a strong one? With a range from −.84 to .26, one just can't tell. The study should be replicated with a larger sample size in order to obtain a more precise estimate of ρ.

Step 4 Was There Adequate Power? How Likely Was a Type II Error? Dr. Solomon is already concerned that a Type II error might have been made. That is, she's concerned the study might've been underpowered. Remember, power is the probability that a null hypothesis that should be rejected is rejected. Statisticians like power to be at least .80 to be considered adequate. That is, if the null hypothesis should be rejected, they want an 80% chance of being able to do so.

Power is the flipside to Type II error, it is heads to error's tails. If the null hypothesis should be rejected, then there are only two options: it is rejected or it is not rejected. If there is an 80% chance of rejecting it, then there's a 20% chance of failing to reject it. Beta, β, is the probability of Type II error. As discussed in Chapter 6, power + β = 1.00, so if power = .80, then β = .20.

Appendix Table 9 makes it easy to find power for a Pearson r. Power depends on the number of cases and the strength of the correlation. The rows in the table are for different sample sizes and the columns offer different correlations. Dr. Solomon had 10 cases and found $r = -.44$. For both rows and columns, follow the *Price Is Right* rule and select the option that is closest without going over. The sign of the correlation does not matter. So, Dr. Solomon looks at the intersection of the row with 10 cases and the column for $r = .40$, where she finds power = .20. As she would like power to be .80 and it is only .20, this study is quite underpowered.

With power of .20, β = .80. If there really is a relationship between age of walking and intelligence and Dr. Solomon tries to find it with only 10 cases, then there's an 80% chance the results will not find evidence of a relationship. Those are not very good odds.

Appendix Table 9 can also be used to see how many cases Dr. Solomon would need to have an 80% chance of finding a relationship if the relationship had the strength of $r = .40$. She would simply go down the column until she reaches power of .80 and then look to the left to see how many cases would be needed. Dr. Solomon should have used 48 cases.

Putting it all together. Finally, Dr. Solomon is ready to interpret results:

> This study investigated if a relationship existed between the age at which children started walking and their IQs at age 16. Using a random sample of 10 girls from a pediatrician's practice, there wasn't enough evidence to conclude that the age at which a child begins to walk provides any information about intelligence [$r(8) = -.44$, $p > .05$]. However, the sample size was small, so a relationship may exist that the researcher failed to find. Future research should replicate the study with a sample size of at least 48 participants to increase the chances of finding a relationship if one does exist. The sample should also include boys and be drawn from multiple sites.

A Common Question

Q A lot of interpretations in this book suggest replication with a larger sample size in order to get a better idea if an effect exists and/or how strong it is. Is this really a common recommendation in studies?

A It's not as common as in this book. In order to make the math easy in the examples, small sample sizes are used. Most real studies use larger sample sizes, making it easier to reject the null hypothesis when it is false and yielding narrower confidence intervals.

Practice Problems 13.3

Apply Your Knowledge

13.13 If $N = 22$, $\alpha = .05$, $r_{cv} = .444$, and $r = .63$, (a) state whether the null hypothesis was rejected, (b) indicate whether the results are statistically significant, and (c) report the results in APA format.

13.14 (a) If $r = -.34$, what is r^2? (b) Is this a small, medium, or large effect?

13.15 Given $N = 12$ and $r = .40$, calculate the 95% confidence interval for ρ.

13.16 If power $= .72$, what is β?

13.17 Given $N = 33$ and $r = .37$, what is power?

13.18 If $r(19) = .36$, $p > .05$, how likely is Type II error?

13.19 If a researcher thinks $r = .30$, how many cases would he need to have in order to have power of .80? Of .95?

13.20 A clinical psychologist randomly sampled 402 parents-to-be in the United States and administered to each couple an interval-level measure of marital harmony. She tracked down all of the children 18 years later and to each child administered an interval-level measure of mental health. Both scales were scored so that higher scores indicated more marital harmony or more mental health. She found $r(400) = .38$, $p < .05$. Given 95% CI [.29, 36] and $r^2 = 14.44\%$, write a four-point interpretation.

13.4 Calculating a Partial Correlation

We started this chapter with a scatterplot showing that there was an association between the number of community hospitals per state and the number of deaths per state. That correlation did not mean that having hospitals caused the deaths. Rather, it seemed more plausible that a confounding variable, the population of a state, caused both the number of hospitals and the number of deaths. Now, we are going to learn about a technique called *partial correlation* that can be used to quantify objectively if and how a confounding variable exerts influence.

A confounding variable is a third variable, let's call it Z, that is not measured and/or not controlled, that correlates with both X and Y, and that potentially explains why X and Y are correlated. Here is an example. Suppose we give $10 each to ten 21-year-old college students on a Friday night and ask them to rendezvous with us at midnight. When they return, we find out how much money they spent and measure how well they can walk a straight line. We find a relationship—the more money a student has spent, the worse he or she is at walking the line. Clearly, a third variable, the amount of alcohol a student has purchased, would explain both why his or her funds are depleted and why his or her performance is impaired. If we took into account each student's alcohol consumption, the correlation between money spent and performance would be explained.

A **partial correlation** mathematically removes the influence of a third variable on a correlation. It is a useful technique that moves a correlational study a little bit in the direction of an experimental study. Partial correlations do not allow cause-and-effect conclusions to be drawn, but they can make potential cause-and-effect conclusions less or more plausible. In the student study above, if the results turned out as described, it would be less plausible to conclude that spending money is the cause of impaired performance.

How can a partial correlation make a cause-and-effect conclusion more plausible? Suppose a researcher finds a relationship between the number of years a person has smoked cigarettes and the degree of impairment of lung function. This researcher claims that the impairment is caused by smoking. Another researcher points out that people who have been smoking longer are usually older, and posits that it is age that causes the impaired lung function. If the effect of age is removed and the correlation between years of smoking and impaired lung function remains strong, then the notion that it is the number of years of smoking that leads to impaired lung function remains plausible.

The abbreviation for a partial correlation is r_{XY-Z}, the correlation between X and Y minus the influence of Z. The formula for a partial correlation, shown in Equation 13.6, makes use of three pieces of information: the correlation between X and Y, the correlation between X and Z, and the correlation between Z and Y.

Equation 13.6 Partial Correlation of X with Y, Controlling for Z

$$r_{XY-Z} = \frac{r_{XY} - (r_{XZ} \times r_{YZ})}{\sqrt{(1 - r_{XZ}^2)(1 - r_{YZ}^2)}}$$

where r_{XY-Z} = the partial correlation of X and Y controlling for the influence of Z

r_{XY} = the correlation between X and Y

r_{XZ} = the correlation between X and Z

r_{YZ} = the correlation between Y and Z

Let's use a partial correlation to control for the effect of population on the correlation between the number of community hospitals per state and the number of deaths per state. The scatterplot in Figure 13.1 at the start of this chapter shows a strong and direct relationship. Not surprisingly, the correlation is strong and significant: $r(23) = .89$, $p < .05$. But, the correlation between the number of hospitals (X) and the state population (Z) was .92, and for the number of deaths (Y) and the population, it was .98.

Using $r_{XY} = .89$, $r_{XZ} = .92$, and $r_{YZ} = .98$, we can complete Equation 13.6:

$$
\begin{aligned}
r_{XY-Z} &= \frac{r_{XY} - (r_{XZ} \times r_{YZ})}{\sqrt{(1 - r_{XZ}^2)(1 - r_{YZ}^2)}} \\
&= \frac{.89 - (.92 \times .98)}{\sqrt{(1 - .92^2)(1 - .98^2)}} \\
&= \frac{.89 - .9016}{\sqrt{1 - .9604)(1 - .8464)}} \\
&= \frac{-.0116}{\sqrt{.0396 \times .1536}} \\
&= \frac{-.0116}{\sqrt{.0061}} \\
&= \frac{-.0116}{.0781} \\
&= -.1485 \\
&= -.15
\end{aligned}
$$

The correlation of the number of hospitals with the number of deaths was .89. But, when population was controlled for, the correlation fell to $-.15$. It went from statistically significant to failing to reject the null hypothesis. This means that there is not enough evidence to suggest the number of community hospitals has any causal impact on the number of deaths in a given state—the apparent association between the two variables can be explained by the states' populations.

A Common Question

Q How many degrees of freedom does a partial correlation have?

A *df* for a partial correlation is $N - 3$. For the community hospital study, there were 25 states, so $df = 25 - 3 = 22$. The critical value of r is $\pm .396$; thus, the observed r of $-.15$ falls in the common zone. The results would be reported as $r(22) = -.15$, $p > .05$.

Worked Example 13.3 A researcher gathered a random sample of 160 students at her college and measured their physical and mental health on two dimensions: the distance that they are able to run in 30 minutes and happiness, as indicated on a survey. She theorized, from a biopsychosocial perspective, that if there is a mind–body connection, then a relationship should exist between physical health and psychological health.

She found in her sample that the correlation between the two variables was .37 [$r(158) = .37$, $p < .05$], suggesting that a correlation does indeed exist between physical health and mental health.

She tells a colleague, an obesity researcher, about these findings, and the colleague suggests that the observed relationship could be explained by the body mass index (BMI)—people with high BMIs are in poorer mental and physical health. Luckily, the biopsychosocial researcher had recorded the heights and weights of her subjects, so she calculates BMIs and finds that the correlation between physical health and BMI is .42; for BMI and emotional health, it is .18. Using these values, she calculates a partial correlation:

$$\begin{aligned} r_{XY-Z} &= \frac{r_{XY} - (r_{XZ} \times r_{YZ})}{\sqrt{(1 - r_{XZ}^2)(1 - r_{YZ}^2)}} \\ &= \frac{.37 - (.42 \times .18)}{\sqrt{(1 - .42^2)(1 - .18^2)}} \\ &= \frac{.37 - .0756}{\sqrt{(1 - .1764)(1 - .0324)}} \\ &= \frac{.2944}{\sqrt{.8236 \times .9676}} \\ &= \frac{.2944}{\sqrt{.7969}} \\ &= \frac{.2944}{.8927} \\ &= .3298 \\ &= .33 \end{aligned}$$

The partial correlation of .33 is reduced slightly from .37, but it is still statistically significant: $r(157) = .33$, $p < .05$. The relationship between physical and emotional health is not explained by BMI.

Practice Problems 13.4

13.21 Using $N = 54$, $r_{XY} = .53$, $r_{XZ} = .46$, and $r_{YZ} = .25$, calculate $r_{XY\text{-}Z}$.

13.22 Using $N = 27$, $r_{XY} = .35$, $r_{XZ} = .18$, and $r_{YZ} = .17$, calculate $r_{XY\text{-}Z}$.

Application Demonstration

To see how the Pearson *r* is used in real research, let's examine a study that involves both psychology and political science. A researcher wanted to know if living in a nation with better government services was associated with greater happiness. He used the average happiness ratings reported by citizens of 130 nations (Ott, 2011). Happiness ratings by individuals could range from 0 ("Right now I am living the worst possible life") to 10 ("Right now I am living the best possible life").

The individual ratings were averaged together to yield a score for a country. The country with the greatest average happiness was Denmark ($M = 8.00$) and the lowest average happiness was Togo ($M = 3.24$); the United States was near the top ($M = 7.26$).

The researcher also developed a measure of government services for each of the 130 nations. Nations that provided better public services had better regulations, experienced less corruption, etc., received higher scores on government services.

The scatterplot showing the relationship between the two variables can be seen in Figure 13.22. The r value for these data is .75, and the results are statistically significant [$r(128) = .75, p < .05$]. There is a positive relationship between the two variables in these countries—better government services are associated with happier citizens. Squaring the correlation gives the effect size, $r^2 = 56.25\%$. According to Cohen (1988), this is a large effect.

The 95% confidence interval for ρ ranges from .66 to .82. Note that (1) the confidence interval doesn't capture zero, (2) the bottom end (.66) is far from zero, and (3) the confidence interval is quite narrow. One can conclude that there is

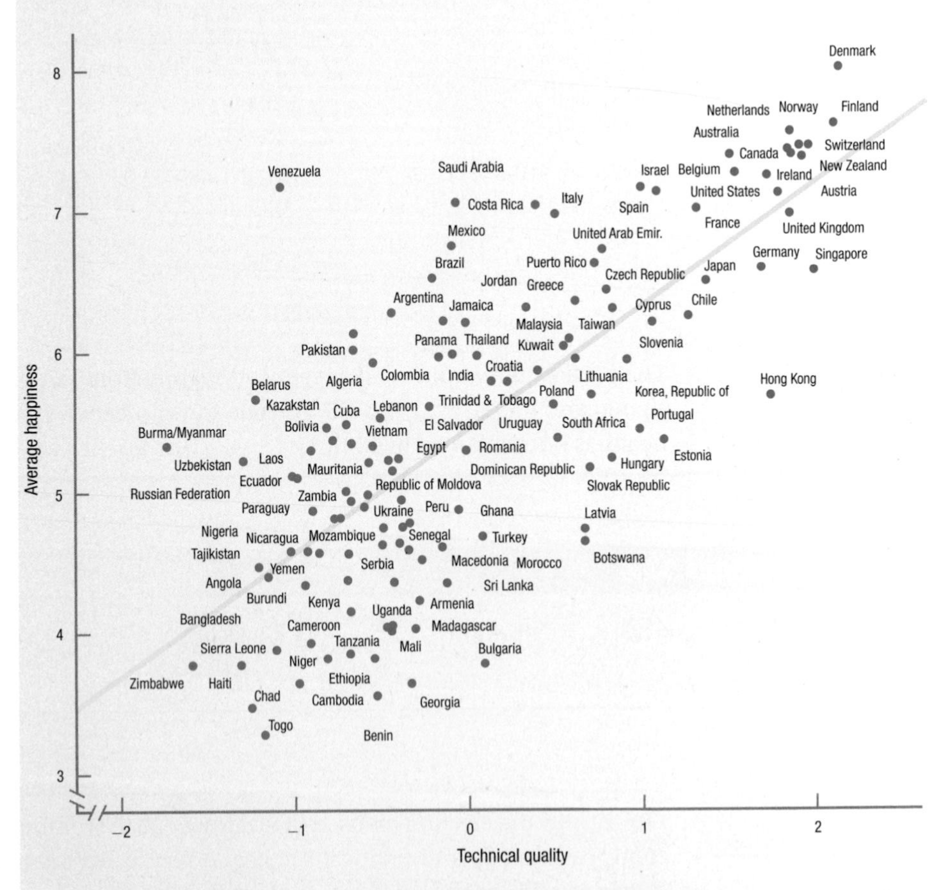

Figure 13.22 Relationship Between the Quality of a Nation's Governments and the Happiness of the People This scatterplot suggests a positive relationship between the quality of government and the happiness of the people. Data from Ott, 2011.

probably a strong, positive correlation between these two variables in the larger population of countries.

The researcher thought of government services as predicting, or leading to, happiness. But, do better government services *cause* more happiness? This is a correlational study where nothing is manipulated. All that one can be sure of is that there is an *association* between better government services (X) and more happiness (Y). Correlation is not causation, so we shouldn't draw conclusions about cause and effect.

When two variables are correlated, three possible explanations exist for the correlation:

1. X causes Y.
2. Y causes X.
3. Some other variable, Z, causes X and Y.

For the happiness/government services data, all three explanations are plausible. First, it is possible that X causes Y, that better government services lead to more happiness. This seems sensible—if government provides better services to its citizens, then their lives should be better and they should be happier.

The second explanation, Y causes X, also seems plausible. Happier citizens could cause better government. It seems possible that happier people will make their government function better because they are more likely to vote, more willing to serve on committees, less likely to disobey laws, and so on.

The third explanation—that Z causes both X and Y—is plausible if one can think of a third variable, Z, that has an impact on both X and Y. Wealth is a third variable that springs to mind. Nations with more wealth might have better government services because they can afford to spend more money on providing them. And, nations with more wealth might have happier citizens because, well, because money does help buy happiness. So, perhaps wealth causes *both* better government services and more happiness.

Untangling cause and effect in a correlational study is a challenge for any researcher. Looking at the scatterplot (Figure 13.22) and the .75 r value, there is clearly an association between the two variables. Just as clearly, the relationship is a direct one. But, how can a researcher put the results in context and explain what they mean? Here's an interpretation that addresses, even embraces, the uncertainty.

A researcher examined the relationship between the quality of government services and happiness in 130 nations around the world. He found that there was a strong, positive association between these two variables [$r(128) = .75, p < .05$]. This means that governments with better services had happier citizens. One explanation for the observed relationship is that countries with better government services provide an environment that leads to more happiness among their citizens. Improving government then should raise up citizens.

But, this study is correlational and other explanations are possible. Perhaps the relationship goes in the other direction. That is, happy citizens lead to better government because happy citizens are more likely to be involved civically and to be concerned for, and supportive of, the welfare of others.

There is a third explanation for the correlation between these two variables, an explanation that also seems plausible. Another variable may exist, called a confounding variable, that causes *both* good government and citizen happiness. Examples of potential confounding variables are how wealthy a nation is and how industrialized it is. Nations with more wealth, for example, can afford to provide more government services. And, for individual people, maybe money does buy some happiness. Thus, it seems reasonable to suggest replicating this study with confounding variables being measured so their influence can be assessed.

SUMMARY

Differentiate difference tests from relationship tests.

- Difference tests—like *z*, *t*, and *F* tests—test for differences between groups on a dependent variable. Relationship tests determine if a relationship exists between two variables measured in one group of cases.

Define and describe a relationship.

- If two variables are correlated, they vary together systematically. The relationship may be one of cause and effect, but correlation does not mean causation: it may just mean association.
- Scatterplots display a relationship graphically. The scatterplot will look the same if it is made for *z* scores or raw scores. The strength of a Pearson correlation coefficient, which measures the degree of linear relationship between two interval and/or ratio variables, is shown by how well the points form a line. As the shape made by the points moves from a line to an oval, the relationship grows weaker. Scatterplots should be inspected for three conditions that can affect correlations: nonlinearity, outliers, and restriction of range.
- A Pearson *r* is a number that summarizes the strength and direction of the linear relationship. It is defined as the average of the pairs of *z* scores for each case multiplied together. *r* ranges from -1.00 to $+1.00$ with $r = .00$ meaning no relationship and an *r* of 1.00 or -1.00 indicating a perfect relationship. Positive *r*'s are called direct relationships (larger values of *X* are associated with larger values of *Y*), and negative *r*'s are called inverse relationships (larger values of *X* are associated with smaller values of *Y*).

Compute a Pearson *r*.

- Pearson *r* is used to examine the linear relationship between two interval/ratio-level variables. The assumptions for it must be met and hypotheses, either directional or nondirectional, should be set. The decision rule depends on the hypotheses, the researcher's willingness to run the risk of making a Type I error, and the number of cases.

Interpret the results of a Pearson *r*.

- The interpretation of a Pearson *r* is based on the answers to three questions: (1) Was the null hypothesis rejected? (2) How big is the effect? (3) How wide is the confidence interval? The effect size, r^2, gives a sense of the size of the effect in the sample, while the confidence interval offers a sense of ρ, the population value of the correlation. In addition, if the null hypothesis is not rejected, one can see how likely Type II error is and estimate how many cases one would need to have adequate power.

Take into account the effect of a confounding variable.

- A partial correlation is a way, mathematically, to see how much a relationship between two variables is influenced by a third variable.

KEY TERMS

coefficient of determination – formal name for the effect size r^2.

correlation coefficient – a statistic that summarizes, in a single number, the strength of a relationship between two variables.

direct relationship – a relationship in which high scores on X are associated with high scores on Y. Also called a positive relationship.

inverse relationship – a relationship in which high scores on X are associated with low scores on Y. Also called a negative relationship.

negative relationship – a relationship in which high scores on X are associated with low scores on Y.

outcome variable – the variable in a relationship test, Y, that is predicted from the other variable, X. Sometimes called the dependent variable.

partial correlation – a correlation between two variables from which the influence of a third variable has been mathematically removed.

Pearson correlation coefficient – a statistical test that measures the degree of linear relationship between two interval/ratio-level variables.

perfect relationship – a relationship between two variables in which the value of one can be exactly predicted from the other.

positive relationship – a relationship in which high scores on X are associated with high scores on Y.

predictor variable – the variable in a relationship test, X, that is used to predict the other variable, Y. Sometimes called the independent variable.

r^2 – an effect size that reveals the percentage of variability in one variable that is accounted for by the other variable.

DIY

Is there a relationship between how long a person's foot is, in inches, and how tall, in inches, he or she is. Connect with ten friends, ask them their heights, and then measure how long their feet are. Which variable is X and which is Y? Why did you make that decision? Compute a Pearson r. Is there a relationship? Is it strong? (Be sure to keep track of who is in your sample and who isn't. That answer will be important in Chapter 14's DIY.)

CHAPTER EXERCISES

Review Your Knowledge

13.01 Relationship tests look for a relationship between ____ variables in ____ group of cases.

13.02 A Pearson r measures the degree of ____ relationship between two ____ variables.

13.03 Measures of ____ quantify the degree of relationship between two variables.

13.04 If there's a relationship between two variables, they are said to be ____.

13.05 If two variables are correlated, they vary together ____.

13.06 A Pearson r measures how much the points in a scatterplot fall in a ____ line.

13.07 If two variables are correlated, the relationship between them may be a ____ relationship.

13.08 Statisticians love to say that correlation ____ causation.

13.09 ____ are used to visualize the relationship between two interval/ratio variables.

13.10 If there is a cause-and-effect relationship between two variables, then they are ____. But, if two variables are correlated, there is not necessarily a ____ relationship between them.

13.11 The two variables in a correlation are usually labeled as ____ and ____.

13.12 The predictor variable in a correlation is the one that is viewed as ____ the other variable.

13.13 The dots fall in a ____ for a scatterplot showing no linear relationship between two normally distributed variables.

13.14 The strength of relationship, in a scatterplot, is determined by how well the points form a ____.

13.15 If the points in a scatterplot form a horizontal line, then r is ____.

13.16 Making a scatterplot using z scores instead of raw scores *does / does not* change the shape of the scatterplot.

13.17 In a perfect correlation, if a case has a z score of .75 for X, then its z score on Y is ____.

13.18 By the definitional formula for Pearson r, r is the ____ of multiplied-together pairs of ____.

13.19 A correlation coefficient is a number that ____ the strength of the relationship between ____.

13.20 If there is no linear relationship, then $r =$ ____.

13.21 If $r = .45$, the relationship is ____ than if $r = .70$.

13.22 An r of ____ or ____ indicates a perfect linear relationship.

13.23 If r ranges from .01 to 1.00, the direction of the relationship is ____.

13.24 X and Y have an inverse relationship. As scores on X go up, scores on Y go ____.

13.25 An ____ is a score that falls far away from other scores.

13.26 An outlier can ____ the value of a correlation.

13.27 If the range of a variable is ____, then the full range of the variable is present in the sample.

13.28 A deflated value of a correlation can be caused by a ____.

13.29 The assumption that the sample is a random sample *can / cannot* be violated.

13.30 The linearity assumption is tested by making a ____.

13.31 The population value of a correlation is represented by the Greek letter ____.

13.32 The null and alternative hypotheses are statements about the ____ value of the correlation.

13.33 A nondirectional null hypothesis says there is ____ relationship in the population between X and Y.

13.34 The alternative hypothesis for a two-tailed test says there is ____ relationship in the population between X and Y, but it doesn't state the ____ of the relationship.

13.35 If a researcher is doing a one-tailed test, first determine the ____ hypothesis.

13.36 If the alternative hypothesis is $\rho < 0$, then the researcher believes the relationship between X and Y is *direct / inverse*.

13.37 If the null hypothesis for a two-tailed test is true, then the sampling distribution of r ranges from ____ to ____ and is centered on ____.

13.38 If the hypotheses are nondirectional, then one is completing a ____-tailed test.

13.39 By setting alpha at .05, one is willing to make a Type ____ error 5% of the time.

13.40 Finding r_{cv} depends on the number of tails in the test, alpha, and ____.

13.41 If $r \geq r_{cv}$, then ____ the null hypothesis.

13.42 APA says a correlation should be reported as ____, not 0.45.

13.43 When getting ready to interpret Pearson r results, it is helpful to review the ____.

13.44 The first question addressed in interpretation is ____.

13.45 To decide if the null hypothesis is rejected, implement the ____ developed in Step 4 of the hypothesis-testing procedure.

13.46 If the null hypothesis is not rejected, the results are said to be ____ significant.

13.47 If APA format says $r(8) = -.23$, $p > .05$, then $N =$ ____.

13.48 If APA format says $r(8) = -.23$, $p > .05$, then the results were ____ significant.

13.49 If the null hypothesis is rejected and r is $-.23$, then one can conclude that there is a ____ relationship between X and Y, also known as an ____ relationship.

13.50 If the relationship between X and Y is a direct one, then cases with high scores on X are likely to have ____ scores on Y.

13.51 If the null hypothesis is not rejected, ____ evidence exists to say that there is a ____ between X and Y.

13.52 If a result is not statistically significant, then there is a possibility of a ____ error.

13.53 ____ reveals the percentage of variability in one variable that is accounted for by the other variable.

13.54 As r^2 gets closer to ____%, the effect is considered stronger.

13.55 If a relationship is perfect, r^2 is ____%.

13.56 Cohen considers an r^2 of ____% or greater to be a large effect.

13.57 The confidence interval for a Pearson correlation coefficient gives the range within which ____ likely falls.

13.58 When transforming a confidence interval from z_r units to r units, make sure to maintain the ____ of the z_r units.

13.59 If the confidence interval captures zero, the null hypothesis *was / was not* rejected.

13.60 If the confidence interval is wide, one is likely to recommend ____ of the study with a larger ____.

13.61 Power is the probability of ______ the null hypothesis when the null hypothesis *should / should not* be rejected.

13.62 Statisticians like power to be at least ______.

13.63 If power = .46, the probability of Type II error = ______.

13.64 To have an 80% chance of rejecting the null hypothesis if $r = .60$, one should have about _____ cases.

13.65 Correlation is not causation. A technique for dealing with a confounding variable is called _____.

13.66 A partial correlation mathematically _________ the effect of a third variable.

Apply Your Knowledge

Making scatterplots

13.67 Given these data, make a scatterplot:

Case	1	2	3	4	5	6	7	8	9	10	11	12	13	14	15
X	5	8	12	16	19	22	26	31	12	25	18	6	30	22	9
Y	5	10	13	12	9	4	5	11	13	6	10	6	8	6	8

13.68 Given these data, make a scatterplot:

Case	1	2	3	4	5	6	7	8	9	10	11	12	13	14	15
X	44	46	48	47	49	50	48	49	52	52	53	54	54	55	56
Y	29	25	28	34	26	30	32	36	27	37	24	31	34	25	30

Select the appropriate test from among a single-sample z test; single-sample t test; independent-samples t test; paired-samples t test; between-subjects, one-way ANOVA; one-way, repeated-measures ANOVA; between-subjects, two-way ANOVA; and Pearson r.

13.69 "People who wear hats in the wintertime are compared to those who don't wear hats in terms of how many days they suffer from a headcold. Does wearing a hat make a difference?" What test should be used to decide?

13.70 "People are measured, on an interval scale, to determine how fair-skinned they are. A dermatologist then examines them to see how many suspicious moles they have in order to determine if fairness of skin relates to the number of suspicious moles." What statistical test should she use to answer this question?

13.71 "A behavioral therapist had patients with spider phobias rate the level of their fear on an interval-level scale. He then asked the spider phobics to enter a room with a spider in a cage and come as close to the spider as they felt comfortable. He measured the distance in feet. Do people with more self-reported fear stay a greater distance away?" What statistical test should the therapist use to find out?

13.72 "A kindergarten teacher classifies students as coloring inside the lines or coloring outside the lines. Years later, he uses police records to determine how many driving violations each former student has had. Is there a relationship between following the rules in kindergarten and following them on the highway?" What statistical test should be used?

Checking assumptions

13.73 A philosopher wanted to see if curiosity about the meaning of life had any correlation, either positive or negative, with the age of college students and planned to use a Pearson r on her data. She obtained a random sample of 250 undergraduate students at the college where she taught and administered the Degree of Curiosity About the Meaning of Life Scale (an interval-level scale) to each one. Here is the scatterplot of her results. (a) Check the assumptions and (b) decide if she can proceed with the planned Pearson r.

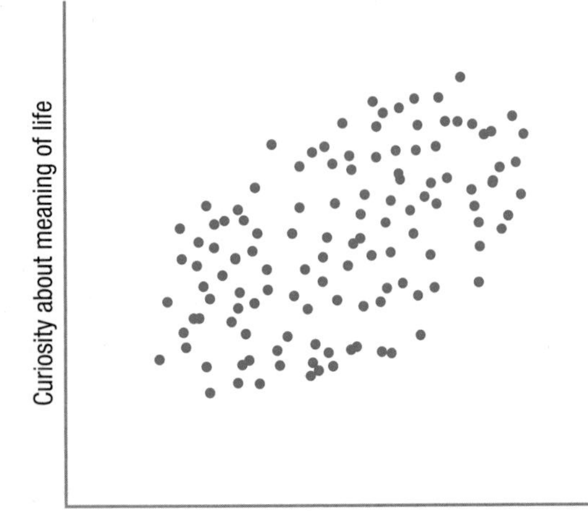

13.74 A psychodynamic psychologist used an interval-level scale to measure the degree of Oedipal conflict in a large sample of elementary age boys. He then had each one, by himself, play a target-shooting game where the boy had equal opportunity to shoot at male and female targets. The psychologist calculated the percentage of time the male target was shot at and considered that a measure of degree of hostility toward father figures. He wanted to use a Pearson r to examine the relationship, if any and if positive or negative, between Oedipal conflict and hostility toward father figures. Here is the scatterplot of the results. (a) Check the assumptions and (b) decide if he can proceed with the planned Pearson r.

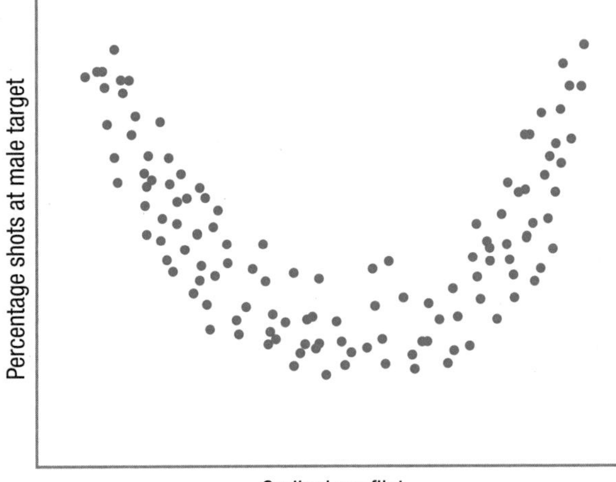

Listing hypotheses

13.75 Assuming that a Pearson r can be used to analyze the data in Exercise 13.74, list the hypotheses.

13.76 A sleep specialist believes that the more caffeine a person consumes per day, the more episodes of restless sleep that person experiences. What are her null and alternative hypotheses?

Calculating degrees of freedom

13.77 What is df for a Pearson r if $N = 63$?

13.78 What is df for a Pearson r if $N = 212$?

Finding r_{cv}

13.79 Given a two-tailed test with $\alpha = .05$ and $df = 7$, what is r_{cv}?

13.80 Given a two-tailed test with $\alpha = .05$ and $df = 27$, what is r_{cv}?

13.81 Given a one-tailed test, $\alpha = .05$, $N = 33$, and $H_1: \rho > 0$, what is r_{cv}?

13.82 Given a one-tailed test, $\alpha = .05$, $N = 33$, and $H_1: \rho < 0$, what is r_{cv}?

Setting the decision rule

13.83 If $r_{cv} = \pm.438$, what is the decision rule?

13.84 If $r_{cv} = +.554$, what is the decision rule?

13.85 If $r_{cv} = \pm.200$, draw the sampling distribution of r and label the rare and common zones.

13.86 If $r_{cv} = \pm.404$, draw the sampling distribution of r and label the rare and common zones.

Calculating deviation scores

13.87 Given $M_X = 9.00$ and $M_Y = 5.00$, calculate deviation scores for X and Y.

X	Y
12	6
8	3
7	6

13.88 Given $M_X = 16.33$ and $M_Y = 14.00$, calculate deviation scores for X and Y.

X	Y
18	7
8	11
23	24

Multiplying deviation scores

13.89 (a) Multiply together the pairs of deviation scores and (b) sum the products.

X	Y	X Deviation Scores	Y Deviation Scores
10	65	−2.33	11.67
18	55	5.67	1.67
9	40	−3.33	−13.33
$M = 12.33$	$M = 53.33$		

13.90 (a) Multiply together the pairs of deviation scores and (b) sum the products.

X	Y	X Deviation Scores	Y Deviation Scores
540	115	36.67	11.33
510	88	6.67	−15.67
460	108	−43.33	4.33
$M = 503.33$	$M = 103.67$		

Squaring deviation scores

13.91 (a) Square the deviation scores and (b) sum the squared scores.

X	Y	X Deviation Scores	Y Deviation Scores	Product of Deviation Scores	Squared X Deviation Scores	Squared Y Deviation Scores
21	110	−4.33	10.67	−46.20		
38	90	12.67	−9.33	−118.21		
17	98	−8.33	−1.33	11.08		
$M = 25.33$	$M = 99.33$			$\Sigma = -153.33$		

13.92 (a) Square the deviation scores and (b) sum the squared scores.

X	Y	X Deviation Scores	Y Deviation Scores	Product of Deviation Scores	Squared X Deviation Scores	Squared Y Deviation Scores
230	55	−10.00	−8.33	83.30		
210	68	−30.00	4.67	−140.10		
280	67	40.00	3.67	146.80		
$M = 240.00$	$M = 63.33$			$\Sigma = 90.00$		

Multiplying sums of squares

13.93 (a) Multiply together the two sums of the squared deviation scores and (b) find the square root of that product.

X	Y	X Deviation Scores	Y Deviation Scores	Product of Deviation Scores	Squared X Deviation Scores	Squared Y Deviation Scores
88	23	1.67	−5.00	−8.35	2.79	25.00
93	28	6.67	0.00	0.00	44.49	0.00
78	33	−8.33	5.00	−41.65	69.39	25.00
$M = 86.33$	$M = 28.00$			$\Sigma = -50.00$	$\Sigma = 116.67$	$\Sigma = 50.00$

13.94 (a) Multiply together the two sums of the squared deviation scores and (b) find the square root of that product.

X	Y	X Deviation Scores	Y Deviation Scores	Product of Deviation Scores	Squared X Deviation Scores	Squared Y Deviation Scores
780	560	−62.67	−33.33	2,088.79	3,927.53	1,110.89
848	740	5.33	146.67	781.75	28.41	21,512.09
900	480	57.33	−113.33	−6,497.21	3,286.73	12,843.69
$M = 842.67$	$M = 593.33$			$\Sigma = -3{,}626.67$	$\Sigma = 7{,}242.67$	$\Sigma = 35{,}466.67$

Calculating r

13.95 Given $\Sigma[(X - M_X)(Y - M_Y)] = 255.70$ and $\sqrt{SS_X SS_Y} = 337.29$, calculate r.

13.96 Given $\Sigma[(X - M_X)(Y - M_Y)] = -564.32$ and $\sqrt{SS_X SS_Y} = 15{,}342.55$, calculate r.

13.97 Given the information in this table, calculate r:

X	Y	X Deviation Scores	Y Deviation Scores	Product of Deviation Scores	Squared X Deviation Scores	Squared Y Deviation Scores
7.6	5.6	−0.90	0.11	−0.10	0.81	0.01
6.8	7.4	−1.70	1.91	−3.25	2.89	3.65
9.0	4.8	0.50	−0.69	−0.35	0.25	0.48
9.0	4.0	0.50	−1.49	−0.75	0.25	2.22
11.0	3.4	2.50	−2.09	−5.23	6.25	4.37
8.0	3.5	−0.50	−1.99	1.00	0.25	3.96
7.6	7.8	−0.90	2.31	−2.08	0.81	5.34
8.0	6.6	−0.50	1.11	−0.56	0.25	1.23
8.7	7.8	0.20	2.31	0.46	0.04	5.34
9.3	4.0	0.80	−1.49	−1.19	0.64	2.22
$M = 8.50$	$M = 5.49$			$\Sigma = -12.03$	$\Sigma = 12.44$	$\Sigma = 28.81$

13.98 Given the information in this table, calculate r:

X	Y	X Deviation Scores	Y Deviation Scores	Product of Deviation Scores	Squared X Deviation Scores	Squared Y Deviation Scores
−2	−3	−1.80	−2.60	4.68	3.24	6.76
−2	2	−1.80	2.40	−4.32	3.24	5.76
5	−5	5.20	−4.60	−23.92	27.04	21.16
−6	3	−5.80	3.40	−19.72	33.64	11.56
2	−6	2.20	−5.60	−12.32	4.84	31.36
−2	0	−1.80	0.40	−0.72	3.24	0.16
0	1	0.20	1.40	0.28	0.04	1.96
−1	−1	−0.80	−0.60	0.48	0.64	0.36
2	2	2.20	2.40	5.28	4.84	5.76
2	3	2.20	3.40	7.48	4.84	11.56
$M = -0.20$	$M = -0.40$			$\Sigma = -42.80$	$\Sigma = 85.60$	$\Sigma = 96.40$

Implementing the decision rule for a two-tailed test

13.99 $r_{cv} = \pm.217$ and $r = -.17$. State (a) whether the null hypothesis was rejected, and (b) whether or not the results are statistically significant.

13.100 $r_{cv} = \pm.301$ and $r = -.31$. State (a) whether the null hypothesis was rejected, and (b) whether or not the results are statistically significant.

Writing APA format with $\alpha = .05$, two-tailed

13.101 Given $df = 20$ and $r = -.67$, write the results in APA format.

13.102 Given $df = 87$ and $r = -.18$, write the results in APA format.

Interpreting APA format

13.103 If $r(18) = -.50$, $p < .05$, (a) was the null hypothesis rejected, and (b) were the results statistically significant?

13.104 If $r(80) = -.18$, $p > .05$, (a) was the null hypothesis rejected, and (b) were the results statistically significant?

13.105 If $r(24) = .36$, $p > .05$, (a) what conclusion can be drawn about the existence of a relationship between X and Y in the population? (b) Comment on the direction of the relationship in the population.

13.106 If $r(30) = .40$, $p < .05$, (a) what conclusion can be drawn about the existence of a relationship between X and Y in the population? (b) Comment on the direction of the relationship in the population.

Calculating r^2

13.107 If $r = .32$, (a) calculate r squared and (b) classify the effect as small, medium, or large.

13.108 If $r = -.05$, (a) calculate r squared and (b) classify the effect as small, medium, or large.

Transforming r to z_r

13.109 If $r = -.46$, what is z_r?

13.110 If $r = .33$, what is z_r?

Calculating s_r

13.111 If $N = 81$, what is s_r?

13.112 If $N = 36$, what is s_r?

Finding the confidence interval in z_r units

13.113 If $z_r = -0.34$ and $s_r = .24$, what is the 95% confidence interval for ρ in z_r units?

13.114 If $z_r = 0.87$ and $s_r = .08$, what is the 95% confidence interval for ρ in z_r units?

13.115 If $r = .54$ and $N = 45$, what is the 95% confidence interval for ρ in z_r units?

13.116 If $r = -.28$ and $N = 144$, what is the 95% confidence interval for ρ in z_r units?

Transforming a confidence interval back into r units

13.117 If a 95% confidence interval for ρ in z_r units ranges from -0.85 to -0.41, what is it in r units?

13.118 If a 95% confidence interval for ρ in z_r units ranges from -1.52 to -0.26, what is it in r units?

Interpreting confidence intervals

13.119 A 95% confidence interval for ρ ranges from $-.09$ to .93.

a. Was the null hypothesis rejected?
b. Were the results statistically significant?
c. Could the relationship between X and Y in the population be weak?
d. How sure is one about the population value of the correlation?

13.120 A 95% confidence interval for ρ ranges from .61 to .70.

a. Was the null hypothesis rejected?
b. Were the results statistically significant?
c. Could the relationship between X and Y in the population be weak?
d. How sure is one about the population value of the correlation?

Calculating power, β, and N

13.121 If $r = .66$ and $N = 72$, what is its power and what is β?

13.122 If $r = .45$ and $N = 17$, what is its power and what is β?

13.123 How many cases should one have to have in order to have adequate power if $r = .28$.

13.124 How many cases should one have to have in order to have adequate power if $r = .57$.

Completing an interpretation

13.125 A kinesiologist and a psychologist collaborated on a study to investigate the relationship between exercise and mental health. They obtained a random sample of 1,002 American men. They administered an interval-level mental health scale to each man and measured how many minutes of aerobic exercise each got per week. (On both measures, higher scores are better.) No assumptions were violated for the Pearson r and they found $r(1{,}000) = .36$, $p < .05$. r^2 was 12.96% and the 95% confidence interval ranged from .31 to .41. Complete a four-point interpretation.

13.126 An economist wanted to investigate the relationship between the personality variable of conscientiousness and saving money. She obtained a random sample of 502 white-collar workers. She had each worker complete a personality inventory on which higher scores indicated more conscientiousness. She also asked each worker to report the percentage of income saved each month. No assumptions were violated for the Pearson r and she found $r(500) = .11$, $p < .05$. r^2 was 1.21% and the 95% confidence interval ranged from .03 to .19. Complete a four-point interpretation.

Calculating a partial correlation

13.127 Using $N = 32$, $r_{XY} = .28$, $r_{XZ} = .20$, and $r_{YZ} = .22$, calculate $r_{XY\text{-}Z}$.

13.128 Using $N = 108$, $r_{XY} = .63$, $r_{XZ} = .66$, and $r_{YZ} = .45$, calculate $r_{XY\text{-}Z}$.

Interpreting partial correlations

13.129 Dr. Jones investigates the relationship between intelligence and number of absences from school in high school students. She finds a statistically significant relationship of $-.33$: those with higher IQs have fewer absences. She then partials out socioeconomic status and the correlation becomes $-.13$. Because her sample size was large, the correlation is still statistically significant. Interpret the results.

13.130 Dr. Atropos examines the correlation between years of sun exposure and healthiness of the skin. He finds a statistically significant correlation of $-.44$: the more years of sun exposure, the less healthy the skin. Thinking that more educated people might be more aware of the risk of melanoma, he partials out years of education and finds $r_{XY-Z} = -.41$. Interpret the results.

Completing all six steps of hypothesis testing

13.131 A dentist wanted to determine if a relationship existed between childhood fluoride exposure and cavities. She took a sample of adults in her practice and counted how many cavities each person had in his or her permanent teeth. She also determined how many years of childhood each person was exposed to tap water with fluoride. The minimum value on this variable was 0 and the maximum was 18. Here are the data she collected:

	Case 1	Case 2	Case 3	Case 4	Case 5	Case 6
X (years of fluoride)	0	18	2	12	3	10
Y (number of cavities)	10	1	7	3	5	4

13.132 A sociologist wanted to see if there was a relationship between a family's educational status and the eliteness of the college that their oldest child attended. She measured educational status by counting how many years of education beyond high school the parents had received. In addition, she measured the eliteness of the school by its yearly tuition, in thousands (e.g., 5 = \$5,000). She obtained a random sample of 10 families.

	1	2	3	4	5	6	7	8	9	10
Years post-HS education	0	7	8	8	4	5	12	17	8	2
Yearly tuition	12	26	33	18	20	7	15	38	41	5

Expand Your Knowledge

13.133 A researcher has completed a study and found $r(28) = -.23$, $p > .05$. What should she conclude about the direction of the relationship between X and Y in the population?

13.134 Dr. Smith had a sample of 19 cases, violated no assumptions, and calculated $r = .46$. She was conducting a two-tailed test with $\alpha = .05$. $r^2 = 21.16\%$ and the 95% CI ranges from .01 to .76. What type of error does he need to worry about?

a. Type I
b. Type II
c. Both Type I and Type II
d. Neither Type I nor Type II
e. Type II because it is a two-tailed test. It would be Type I if the test was one-tailed.
f. There's not enough information to tell.

13.135 Dr. Jones had 92 cases, violated no assumptions, conducted a two-tailed test with $\alpha = .05$, and calculated $r = -1.26$. Which are accurate descriptions of the relationship between X and Y? (Select as many as are appropriate.)

a. It is an inverse relationship.
b. It is a direct relationship.
c. It is a curvilinear relationship.
d. It is a linear relationship.
e. It is not significantly different from zero.
f. It is significantly different from zero
g. There was some error in calculating r.

13.136 The owner of a hair salon is curious about the relationship between the amount of hair cut and the amount of money earned by the stylists. At the end of a randomly selected day, she weighed the amount of hair around the chairs of her eight stylists and counted their earnings. Below are the data she collected. Analyze them using the six-step hypothesis-testing procedure.

	1	2	3	4	5	6	7	8
Pounds of hair	1	10	2	8	4	7	5	6
Earnings	195	223	150	165	104	130	60	100

13.137 How many cases should one have in order to have adequate power to use a Pearson r to find a small effect? Medium effect? Large effect?

13.138 Rebekah used an independent-samples t test to compare a control group to an experimental group. She had 284 cases and found $t = 4.38$. Was her study underpowered?

SPSS

The data for a Pearson correlation coefficient in SPSS should be arranged with each variable in a column and each case in a row. Figure 13.23 shows how the data for the eight cases in Dr. Paik's marital satisfaction study would look.

V1	Role_Flex	Marital_Stat
1	8	.8
2	15	2.0
3	22	1.5
4	31	2.3
5	35	1.5
6	38	3.3
7	15	1.5
8	36	3.1

Figure 13.23 Data Entry for a Pearson *r* in SPSS Each variable has a separate column and each case has a separate row. (Source: SPSS)

The commands for a Pearson *r* in SPSS are found under "Analyze," then "Correlate," and finally "Bivariate," as shown in Figure 13.24.

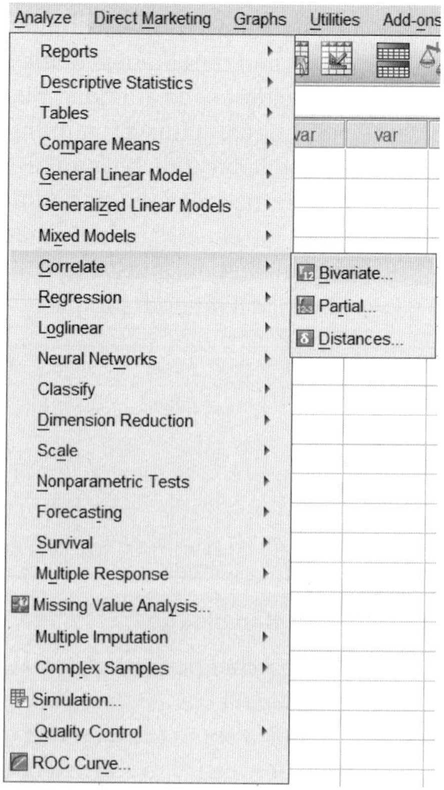

Analyze Direct Marketing Graphs Utilities Add-ons
Reports
Descriptive Statistics
Tables
Compare Means
General Linear Model
Generalized Linear Models
Mixed Models
Correlate
Regression
Loglinear
Neural Networks
Classify
Dimension Reduction
Scale
Nonparametric Tests
Forecasting
Survival
Multiple Response
Missing Value Analysis...
Multiple Imputation
Complex Samples
Simulation...
Quality Control
ROC Curve...
Bivariate...
Partial...
Distances...
var
var

Figure 13.24 Starting Pearson *r* in SPSS The command in SPSS for a Pearson *r* can be found under the "Bivariate" command. (Source: SPSS)

Clicking on "Bivariate" opens the menu seen in Figure 13.25. Highlight each of the variables to correlate and use the arrow button to move them over to the box labeled "Variables." In Figure 13.25, note that the SPSS default options include Pearson *r*, a two-tailed test, and flagging significant correlations.

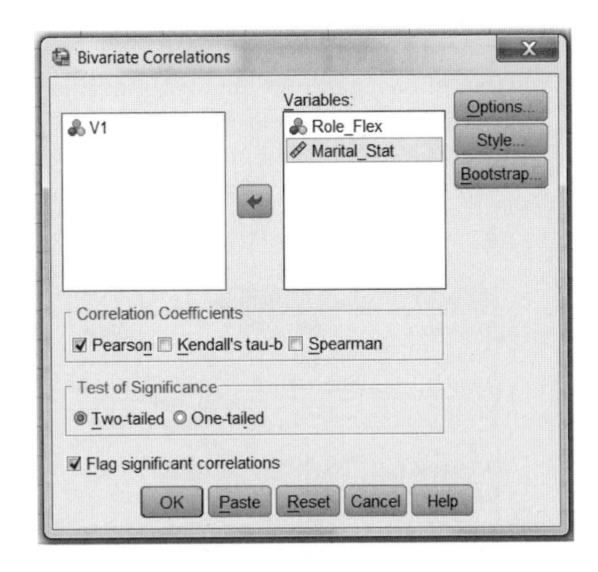

Figure 13.25 Commands for Pearson *r* in SPSS To move the variables-to-be-correlated over to the box labeled "Variables," highlight them and use the arrow button. Once they are in the "Variables" box, use the OK button to start the analysis. (Source: SPSS)

The SPSS output is shown in Figure 13.26. SPSS does each correlation twice (X with Y and Y with X) and displays the results in what is called a correlation matrix. In the text, the correlation between role flexibility and marital satisfaction was found to be .76. SPSS carries more decimal places, so it reports $r = .762$.

Correlations

		Role_Flex	Marital_Stat
Role_Flex	Pearson Correlation	1	.762*
	Sig. (2-tailed)		.028
	N	8	8
Marital_Stat	Pearson Correlation	.762*	1
	Sig. (2-tailed)	.028	
	N	8	8

*. Correlation is significant at the 0.05 level (2-tailed).

Figure 13.26 Output for a Pearson *r* in SPSS The value of the correlation is .762. The asterisk to the right of the correlation indicates, as the note under the table states, that the result is statistically significant at the .05 alpha level, two-tailed. (Source: SPSS)

Underneath the value of the correlation is the number .028. This is the significance level for the correlation. If the value is less than or equal to .05, which .028 is, then the results are statistically significant. Want some help deciding if the results are statistically significant? Note the asterisk to the right of .762. Because the option to flag significant correlations was left on, SPSS places an asterisk next to correlations that are significant at the $\alpha = .05$ level, two-tailed.

For an example of partial correlation, return to the correlation between number of hospitals and number of deaths when controlling for the population value. We will use a data set with values on these three variables from 25 states, as shown in Figure 13.27. The commands for a partial correlation in SPSS are found under "Analyze," then "Correlate," and finally "Partial," as shown in Figure 13.28.

State	Deaths	Hospitals	POP
Delaware	7,000	6	830,364
Georgia	66,000	146	8,829,383
Idaho	10,000	39	1,393,262
Indiana	54,000	113	6,237,569
Iowa	27,000	115	2,954,451
Massachu...	55,000	78	6,416,505
Mississippi	28,000	93	2,902,966
Montana	8,000	54	926,865
Nevada	18,000	30	2,334,771
New Mexico	14,000	37	1,903,289
New York	153,000	206	19,227,088
Ohio	106,000	166	11,459,011
Oklahoma	34,000	109	3,523,553
Oregon	30,000	58	3,594,586
Pennsylvania	128,000	197	12,406,292
South Caro...	37,000	62	4,198,068
South Dak...	7,000	51	770,883
Tennessee	56,000	127	5,900,962
Texas	153,000	418	22,490,022
Vermont	5,000	14	621,394
Virginia	57,000	88	7,459,827
Washington	45,000	85	6,203,788
West Virgi...	21,000	57	1,815,354
Wisconsin	46,000	121	5,509,026
Wyoming	4,000	24	506,525

Figure 13.27 Data Entry for a Partial Correlation in SPSS There are four variables: state, deaths, hospitals, and population. (Source: SPSS)

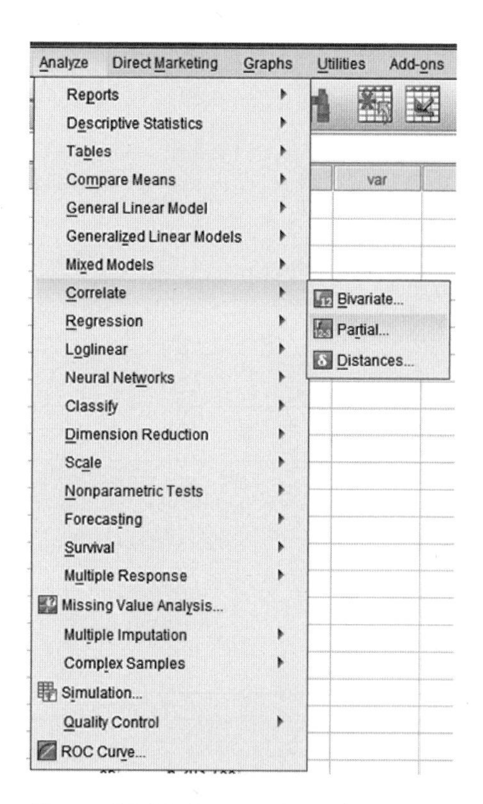

Figure 13.28 Starting Partial Correlation in SPSS The command in SPSS for a partial correlation can be found under the "Partial" command. (Source: SPSS)

Clicking on "Partial" opens the menu seen in Figure 13.29. The two variables we want to correlate, hospitals and deaths, have been moved into the "Variables" box. The variable we want to control for, "population," has been moved into the "Controlling for" box. In Figure 13.29, note that the SPSS default options include a two-tailed test and the option to display significant correlations.

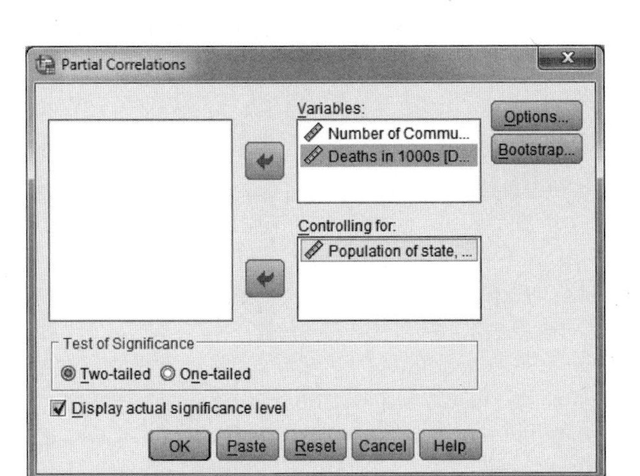

Figure 13.29 Commands for Partial Correlation in SPSS To move the variables-to-be-correlated over to the box labeled "Variables," highlight them and use the arrow button. To move the variable-to-be-controlled over to the box labeled "Controlled for," highlight it and use the arrow button. Once all variables are in place, use the OK button to start the analysis. (Source: SPSS)

Correlations

Control Variables			Number of Community Hospitals	Deaths in 1000s
Population of state, 2004	Number of Community Hospitals	Correlation	1.000	-.197
		Significance (2-tailed)	.	.357
		df	0	22
	Deaths in 1000s	Correlation	-.197	1.000
		Significance (2-tailed)	.357	.
		df	22	0

Figure 13.30 Output for a Partial Correlation in SPSS The value of the correlation is –.197. The partial correlation is not significant. Therefore, we conclude that at the state level, there is not a significant correlation between hospitals and deaths when controlling for the number of people in the population. (Source: SPSS)

The SPSS output is shown in Figure 13.30. SPSS does each partial correlation twice (X with Y and Y with X) and displays the results in what is called a correlation matrix. In the chapter, the correlation between hospitals and deaths controlling for population was $-.15$. Due to differences in rounding, SPSS reports $r = -.197$ which is nonsignificant as indicated by the significance value of .357, which exceeds .05.

Simple and Multiple Linear Regression

LEARNING OBJECTIVES

- Calculate and apply a linear regression equation.
- Measure uncertainty in regression predictions.
- Describe how multiple regression works.

CHAPTER OVERVIEW

Psychology, the study of behavior and mental processes, has four goals. The first goal is to *describe* behavior and mental processes, and the second goal is to *understand* what causes them. If the causes are known, then the third goal is *predicting* how an organism will behave, think, or feel. Accurate predictions help with the fourth goal, *influencing* how an organism behaves, thinks, or feels.

The difference between the second goal (understanding) and third goal (predicting) is the same as the difference between correlation (Chapter 13) and regression (this chapter). Correlation is about finding associations between variables, about understanding how one variable relates to another variable. This chapter, on regression, looks at the procedure for predicting one variable from the other variable. The procedure is called linear regression, and it is used to make statements like, "A person with 18 years of education is predicted to earn an annual salary of $72,000."

14.1 Simple Linear Regression

In **linear regression**, one or more predictor variables are used to predict cases' scores on an outcome variable. For example:

- If a person has X level of depression, what will be his or her level of depression after 12 sessions of cognitive-behavioral therapy?
- If we reduce truancy by X amount, how much will the high school graduation rate improve?
- If a child is bullied at age X, how will that affect her self-esteem?
- What effect does the height of the mother, the height of the father, and the annual family income have on the height of a child?

In **simple linear regression,** one predictor variable, X, is used to predict Y, the outcome variable.

Simple linear regression uses the Pearson r to develop an equation, called a regression equation, to predict Y from X. All the predictions made by a regression equation won't be perfectly accurate, but using a regression equation helps one to arrive at better decisions overall. Of course, making predictions only makes sense if there is evidence that a relationship exists between X and Y. That means simple linear regression should only be used with a statistically significant Pearson r.

Using a Regression Line for Prediction

To see how linear regression works, let's start with a straightforward example. Figure 14.1 shows a perfect correlation ($r = 1.00$) between temperature measured in Fahrenheit and in Celsius. All six data points in Figure 14.1 fall on a straight line.

For these six data points, their values on X (Fahrenheit) and Y (Celsius) are known. For example, the point on the bottom left of the scatterplot has a Fahrenheit value of 32° and a Celsius value of 0°. These six are known, but what about all the other possible Fahrenheit values? If an object's temperature is measured and found to be 86° Fahrenheit, what would it be in Celsius?

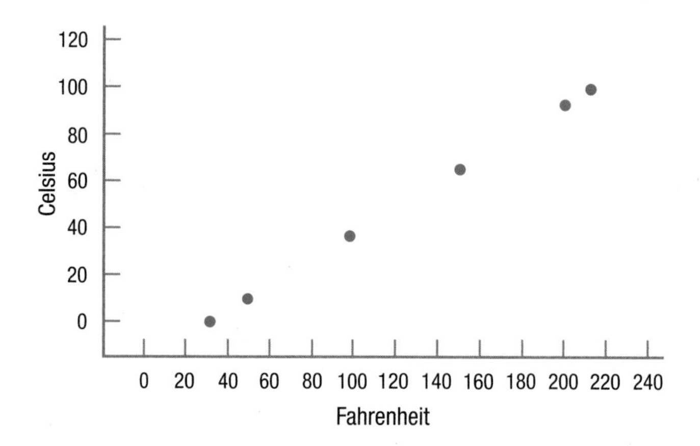

Figure 14.1 Temperature Measured in Fahrenheit and Celsius In a perfect relationship, all points in a scatterplot fall on a straight line.

Figure 14.2 shows how to estimate Y for a given value of X, like 86°. In Figure 14.2, the six points have been connected with a line. This line, called the **regression line,** allows one to find a Y value for any X value. Here's how to do it:

- Draw a vertical line from 86° on the X-axis up to the diagonal line.
- Draw a horizontal line over to the Y-axis from the point on the diagonal line.
- Estimate the value of Y where the horizontal line intersects the Y-axis, say, 30°.
- Thus, the predicted value of Y is approximately 30°.

Before moving on to the next example, let's add some terminology. The six data points in Figure 14.1 have X scores and Y scores. In the case above, we had an X score (86°F), but no Y score. The Y score that was found, 30°C, is a predicted or estimated value. A predicted value of Y has a special name, ***Y* prime**, abbreviated Y'.

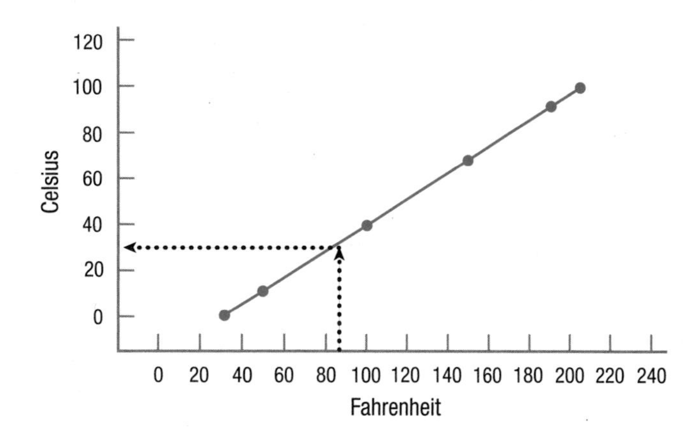

Figure 14.2 Regression Line for Predicting Celsius from Fahrenheit The regression line inserted in this scatterplot makes it easy to predict an object's temperature in Celsius if we know its temperature in Fahrenheit. An object that is 86°F would be about 30°C.

($\hat{Y}$, called "Y hat," is also commonly used as an abbreviation for the predicted value of *Y*.)

In Figure 14.1, all the data points fall on a line, so it is clear where to place the regression line. It is less clear what to do in a situation like that found in Figure 14.3, which displays Dr. Paik's marital satisfaction data from Chapter 13. In that study, a marital therapist randomly selected eight couples and found a statistically significant, positive relationship between the husband's gender role flexibility and the wife's marital satisfaction [$r(6) = .76$, $p < .05$].

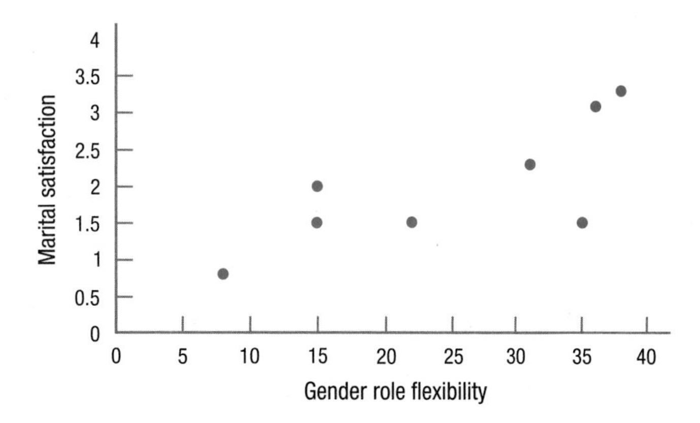

Figure 14.3 Relationship Between Gender Role Flexibility and Marital Satisfaction Though there is a strong ($r = .76$) relationship between the two variables in this scatterplot, it is not clear where the best place is to draw a regression line for predicting *Y* from *X*.

The relationship between gender role flexibility and marital satisfaction is a strong one. And a look at Figure 14.3 shows that it is a linear relationship. But where the best place would be to draw the regression line is not clear. Figure 14.4 illustrates the marital satisfaction data with three different potential lines (labeled I, II, and III). Which one is the best regression line? Which one is the worst?

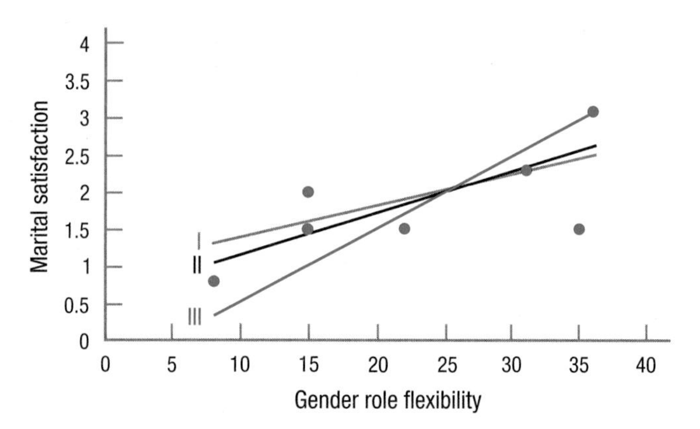

Figure 14.4 Three Potential Regression Lines for Predicting Marital Satisfaction Which of these three lines best "fits" the data in this scatterplot? By what criterion should one decide? Statisticians use the "least squares" criterion, which means the best-fitting line is the one that, overall, minimizes the discrepancies between actual *Y* scores and predicted *Y* scores.

How to Judge Whether Prediction Is Good

To learn how statisticians decide which line is the best, imagine that a memory test is given to all the students at a college. Scores can range from 1.00 to 19.00, and the mean score is calculated to be 10.00. Further, the scores are normally distributed with a standard deviation of 3.00. A frequency distribution of the memory scores is shown in Figure 14.5.

Now, imagine 12 students are randomly selected from this college and a contest is held to guess what their scores on the memory test are. Any prediction from 1.00 to 19.00 is fair and there is a substantial prize for guessing correctly. Nothing is known about the selected students: not their GPAs, years in school, histories of head trauma, or any other fact. As a result, anyone making a prediction is guessing blindly. What to

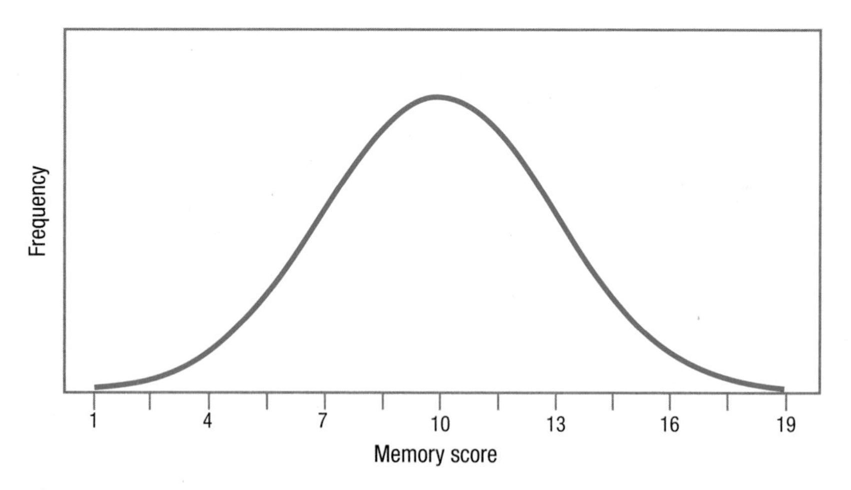

Figure 14.5 Frequency Distribution of Memory Scores If memory scores are normally distributed with a mean of 10, then the most commonly occurring score is 10. If asked to guess what a randomly selected person's memory score is, one is more likely to be right by guessing the mean than any other value.

do? A statistician would say, "Guess the mean for each one." That is, make the same guess, 10.00, twelve times in a row. This is the best strategy for two reasons:

1. First, there is a greater chance of being right guessing the mean than guessing any other value. Look at Figure 14.5—the score at the midpoint, the mean, occurs with the greatest frequency.

2. Second, the errors will be smaller, on average, if the mean is guessed for each person. Here's how to think of this. The scores on the memory test range from 1.00 to 19.00, so the most one can be off by guessing the mean (10.00) is 9.00 points. Guessing any other value increases the potential size of the error. For example, if the guess was 14.50 and the student's score was 2.00, then the guess would be off by 12.50 points.

The best prediction is the one that yields the smallest errors between predicted outcomes and actual outcomes.

The second point, about minimizing errors, is important because it is how statisticians judge prediction. The best prediction is the one that yields the smallest errors between predicted outcomes and actual outcomes. In fact, minimizing errors is how the regression line is defined—it is the best-fitting straight line by the least squares criterion. The **least squares criterion** means that the prediction errors are squared and the best-fitting line is the one that has the smallest sum of squared errors. Why are we concerned with squared values? Let's return to Dr. Paik's study.

Figure 14.6 shows the scatterplot for the marital satisfaction data with line II from Figure 14.4. In Figure 14.6, double-headed arrows are used to mark the distance for the eight cases from their actual values (the dots) to the line. These distances represent errors in prediction: the distance from Y (the wives' real satisfaction scores) to Y' (their predicted satisfaction scores) is the error in prediction. Sometimes the errors are small, as for points A, B, and E. Sometimes the errors are large, as for point F.

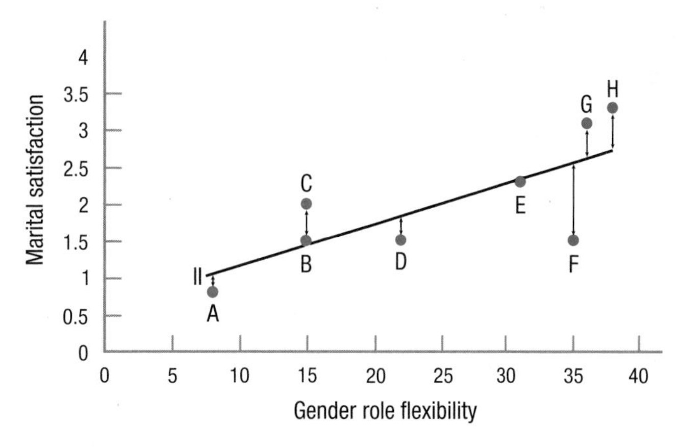

Figure 14.6 Errors in Prediction This figure compares the cases' actual marital satisfaction scores to their predicted scores for line II from Figure 14.4. The double-headed arrows from the actual cases to the line show the sizes of the errors. Notice that for some cases (e.g., A, B, and E), the errors are small and the predicted values are close to the actual values. For other cases, like F, the error is large and the predicted value is far from the actual value.

Look at the top panel in Table 14.1. The top panel shows the Y scores for the eight marital satisfaction cases. The first column gives the actual Y value for each case and the next three columns give the predicted Y scores, one for each of the three lines in Figure 14.4. The first row is for case A, where the wife's marital satisfaction score

TABLE 14.1 Three Prediction Lines for Dr. Paik's Marital Satisfaction Data

	Y: Actual Marital Satisfaction Score	Y': Marital Satisfaction Score Predicted by Line I	Y': Marital Satisfaction Score Predicted by Line II	Y': Marital Satisfaction Score Predicted by Line III
A	0.80	1.32	1.03	0.30
B	1.50	1.60	1.43	1.00
C	2.00	1.60	1.43	1.00
D	1.50	1.88	1.83	1.70
E	2.30	2.24	2.34	2.60
F	1.50	2.40	2.57	3.00
G	3.10	2.44	2.62	3.10
H	3.30	2.52	2.74	3.30

	$Y - Y'$ for Line I	$Y - Y'$ for Line II	$Y - Y'$ for Line III
A	−0.52	−0.23	0.50
B	−0.10	0.07	0.50
C	0.40	0.57	1.00
D	−0.38	−0.33	−0.20
E	0.06	−0.04	−0.30
F	−0.90	−1.07	−1.50
G	0.66	0.48	0.00
H	0.78	0.56	0.00
	$\Sigma = 0.00$	$\Sigma = 0.00$	$\Sigma = 0.00$

	$(Y - Y')^2$ for Line I	$(Y - Y')^2$ for Line II	$(Y - Y')^2$ for Line III
A	0.27	0.05	0.25
B	0.01	0.00	0.25
C	0.16	0.32	1.00
D	0.14	0.11	0.04
E	0.00	0.00	0.09
F	0.81	1.14	2.25
G	0.44	0.23	0.00
H	0.61	0.32	0.00
	$\Sigma = 2.44$	$\Sigma = 2.17$	$\Sigma = 3.88$

This set of tables examines three different prediction lines for Dr. Paik's marital satisfaction data. Each row in the top panel shows the actual marital satisfaction scores for the eight cases discussed in Chapter 13 and the scores predicted by each of the linear equations. The bottom left panel shows the residual scores, $Y - Y'$, for each linear equation. For each line, the residual scores sum to zero. The bottom right panel displays the squared residual scores. It is by the least squares criterion that the best-fitting line is selected. By this criterion, line II is considered the best-fitting of these three lines because it minimizes the sum of the squared residual scores.

is 0.80. Line I predicts her level of marital satisfaction to be higher, 1.32; line II also predicts high with $Y' = 1.03$; line III underestimates her satisfaction with a predicted value of 0.30.

The bottom left panel in Table 14.1 shows the differences between the actual scores and the predicted scores. These values are sizes of the errors. They are what is left over after Y' is subtracted out, so they are called **residuals**. For example, case A in the first row has a Y' for line I that is off by −0.52 points, for line II off by −0.23 points, and for line III off by 0.50 points. Notice that each column is a mixture of positive and negative residuals, of overestimates and underestimates. For these three lines, the residuals for each column sum to zero, meaning that the positive and negative errors balance each other out. Thus, comparing the sums of the error scores does not make one of these lines stand out over the others. So, how can one differentiate these three lines?

The answer is to square the residual scores. As a result, the squared error scores are all positive (see the bottom panel of Table 14.1) and sum to a positive number when added together. The squared error scores sum to 2.44 for line I, 2.17 for line II, and 3.88 for line III. Linear regression uses the least squares criterion, which minimizes the sum of the squared errors, so we can now conclude that line II is the best-fitting line of these three and line III is the worst-fitting line.

Line II is the best-fitting of these three lines, but is it the best-fitting line out of all other possible lines? The regression formula we are about to learn determines the equation for the best-fitting line.

The Linear Regression Equation

Most students remember the formula for a straight line from algebra. The abbreviations may have been different, but it looked something like this:

$$Y = bX + a$$

In this equation, Y is the value being calculated; b is the slope of the line; X is the value for which Y is being calculated; and a is the point where the line intersects the Y-axis, the Y-intercept.

The regression line equation, Equation 14.1, is similar, but it calculates Y', the predicted value of Y, not Y.

Equation 14.1 Formula for Calculating a Regression Line

$$Y' = bX + a$$

where Y' = predicted value of Y
b = slope of the regression line (Equation 14.2)
X = value of X for which one wants to find Y'
a = Y-intercept of the regression line (Equation 14.3)

In order to apply the regression line formula, three factors need to be known: (1) the X value for which one wants to predict a Y value; (2) the slope, b; and (3) the Y-intercept, a. The first of these, X, does not need to be calculated. It will either be given to or determined by the researcher. But the other two values, the slope and the Y-intercept, need to be calculated in order to apply Equation 14.1.

Understanding Slope

Slope represents the tilt of the line. It tells how much up or down change in Y is predicted for each 1-unit change in X. This is often called "rise over run."

- If the slope is positive, then the line is moving up and to the right. (The slope is positive for direct relationships where increases on one variable are associated with increases on the other variable.)
- If the slope is negative, then the line is moving down and to the right. (The slope is negative for inverse relationships. In an inverse relationship, increases in X are associated with decreases in Y.)
- If the slope is zero, then the line is horizontal.

Here's the formula for calculating the slope.

Equation 14.2 Formula for the Slope, *b*, of the Regression Line

$$b = r\left(\frac{s_Y}{s_X}\right)$$

where b = slope of the regression line
r = observed correlation between X and Y
s_Y = standard deviation of the Y scores
s_X = standard deviation of the X scores

For Dr. Paik's marital satisfaction study, $r = .76$, $s_Y = 0.86$, and $s_X = 11.49$. He would calculate the slope as follows:

$$\begin{aligned} b &= r\left(\frac{s_Y}{s_X}\right) \\ &= .76\left(\frac{0.86}{11.49}\right) \\ &= .76 \times 0.0748 \\ &= 0.0568 \\ &= 0.06 \end{aligned}$$

The slope, 0.06, is positive. This was expected because the correlation coefficient, .76, was positive. The value of the slope, 0.06, means that, on average, for every 1-point increase in a husband's level of gender role flexibility, there is a predicted increase of 0.06 points in the wife's level of marital satisfaction. It also means that a 1-point *decrease* in gender role flexibility is associated with a 0.06-point *decrease* in marital satisfaction.

It is important to note the careful use of language here. Correlational designs give information about association, not cause and effect. Dr. Paik is careful *not* to say that a 1-point increase in gender role flexibility causes a 0.06-point increase in marital satisfaction.

Understanding the Y-Intercept

The slope was calculated first because it is needed to calculate *a*, the *Y*-intercept. The ***Y*-intercept** indicates the spot where the regression line would pass through the *Y*-axis. It gives information about the "altitude" of the line, how high or low it is:

- If the *Y*-intercept is positive, the line passes through the *Y*-axis above zero.
- If the *Y*-intercept is negative, the line passes through the *Y*-axis below zero.
- If the *Y*-intercept is zero, the line passes through the *Y*-axis at zero.
- The bigger the absolute value of the *Y*-intercept, the further away from zero the intercept passes through the *Y*-axis.

Here is the formula for calculating the *Y*-intercept.

Equation 14.3 Formula for the Y-Intercept, *a*, for the Regression Line

$$a = M_Y - bM_X$$

where a = *Y*-intercept for the regression line
M_Y = mean of the *Y* scores
b = slope of the regression line (Equation 14.2)
M_X = mean of the *X* scores

Dr. Paik has already calculated the slope and found it to be 0.0568, which he rounded to $b = 0.06$. Consulting his data, he finds $M_Y = 2.00$ and $M_X = 25.00$. Using these values, the *Y*-intercept is calculated as follows:

$$\begin{aligned} a &= M_Y - bM_X \\ &= 2.00 - (0.0568 \times 25.00) \\ &= 2.00 - (1.4200) \\ &= 0.5800 \\ &= 0.58 \end{aligned}$$

(*Note:* Because very precise numbers are needed for an example to work later in the chapter, this equation uses a value of the slope to four decimal places, $b = 0.0568$.)

The *Y*-intercept, the spot where the regression line would intersect the *Y*-axis, is 0.58. Now that the slope, $b = 0.06$, and the *Y*-intercept, $a = 0.58$, are known, Dr. Paik can complete the regression formula, Equation 14.1:

$$\begin{aligned} Y' &= bX + a \\ &= 0.06X + 0.58 \end{aligned}$$

Predicting Y

Here is how a researcher could apply the formula and use it to draw the regression line. Dr. Paik needs to select an *X* value for which to predict a *Y* score. He must select a value that is within the range used to develop the regression formula. So, he selects a gender role flexibility score of 30 and substitutes that for *X* in Equation 14.1:

$$\begin{aligned} Y' &= bX + a \\ &= (0.06 \times 30) + 0.58 \\ &= 1.8000 + 0.58 \\ &= 2.3800 \\ &= 2.38 \end{aligned}$$

Dr. Paik has just predicted that a man with a gender role flexibility score of 30 will have a partner who rates her level of marital satisfaction as 2.38. Given that marital satisfaction is rated on a 4-point scale like GPA, this means she's predicted to rate her marriage at the C+ level.

Drawing the Regression Line

Putting a regression line into a scatterplot helps to highlight the relationship between the two variables. Any two points can be connected with a straight line, so the regression line can be drawn once two points are known. All Dr. Paik needs is the two points.

The regression equation is meant to make predictions for the range of values it was based on. So, Dr. Paik will find Y' for the lowest X value (8), and Y' for the largest (38). (Again, because precise numbers are needed for an example later in the chapter, a four decimal place version of slope $b = 0.0568$ instead of $b = .06$ will be used.)

$$
\begin{aligned}
Y' &= bX + a \\
&= (0.0568 \times 8) + 0.58 \\
&= 0.4544 + 0.58 \\
&= 1.0344 \\
&= 1.03
\end{aligned}
$$

$$
\begin{aligned}
Y' &= bX + a \\
&= (0.0568 \times 38) + 0.58 \\
&= 2.1584 + 0.58 \\
&= 2.7384 \\
&= 2.74
\end{aligned}
$$

Dr. Paik now knows two points that anchor the line: (8, 1.03) and (38, 2.74). Figure 14.7 shows the scatterplot with the two points marked and a line drawn through them. Yes, this is the same as line II in Figure 14.4.

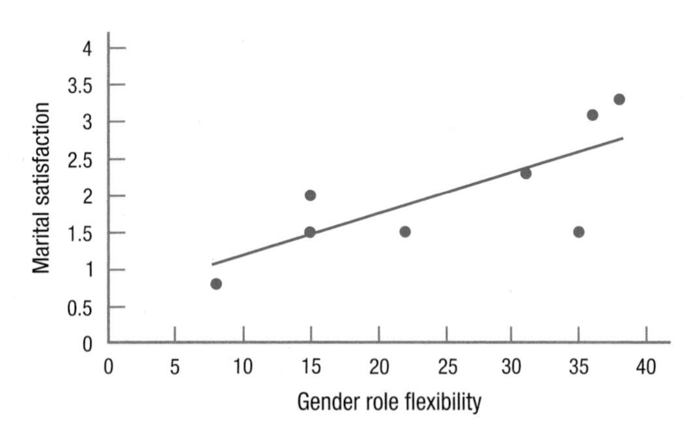

Figure 14.7 Regression Line for Predicting Marital Satisfaction Finding Y' for the two points at two ends of the range of X scores allows a researcher to draw a regression line. Remember, the regression line should only be used to predict Y' for the range of X scores used to derive the regression equation.

Worked Example 14.1 For another example of developing a regression equation from start to finish, imagine the following: A large and representative sample of cigarette smokers ($N = 2{,}500$) was obtained in order to see if there were a relationship between how much a person smoked and his or her physical health. To measure amount of smoking, each person reported how many years he or she had been smoking cigarettes. The mean, M_X, was 22 years, with a standard deviation of 9. As a measure of physical health, each person's lung function was measured. This was reported as a percent of the predicted normal level, so lower scores mean worse functioning. A lung function score of 100 would mean that the person's lung capacity was normal for his or her age and sex. A score of 50 would mean that the smoker's level of lung function was only 50% of what was expected for a person of the same age and sex. The mean level of function was 76%, with a standard deviation of 13. The average person had been smoking for 22 years and had lungs functioning at 76% of what was expected.

Not surprisingly, the relationship between years of smoking and degree of lung function was negative and strong: [$r(2{,}498) = -.68$, $p < .05$]. As years of smoking went up, the percent of normal lung function went down (see the scatterplot in Figure 14.8).

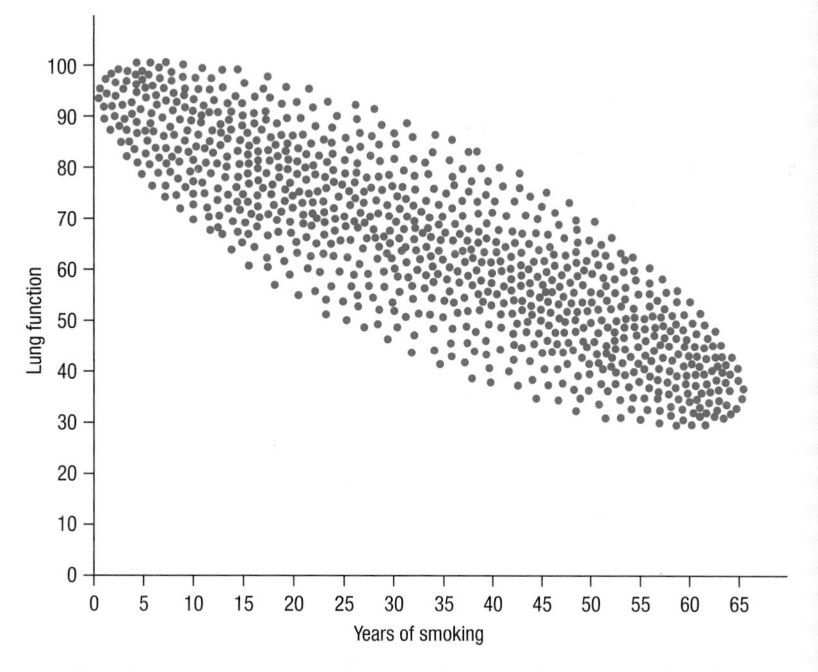

Figure 14.8 Relationship Between Years of Cigarette Smoking and Lung Function In this scatterplot, the relationship between years of smoking and loss of lung function is strong, statistically significant, and inverse.

To calculate the slope (b) for the regression line, plug $r = -.68$, $s_X = 9$, and $s_Y = 13$ into Equation 14.2:

$$\begin{aligned} b &= r\left(\frac{s_Y}{s_X}\right) \\ &= -.68\left(\frac{13}{9}\right) \\ &= -.68 \times 1.4444 \\ &= -0.9822 \\ &= -0.98 \end{aligned}$$

For the regression line for the years of smoking/lung function data, the slope is $b = -0.98$. A slope of *negative* 0.98 means that for every 1-point *increase* in X, there's a predicted *decrease* of 0.98 points in Y. To put this in the context of this example, every year of smoking is associated with an additional decrease of 0.98 percentage points from normal lung function. In this way, a slope can be a meaningful tool for interpreting regression.

To calculate the Y-intercept, one needs the slope, which was just found to be -0.98, and the two means, M_X and M_Y. The predictor variable is years of smoking so $M_X = 22$; the predicted variable is percent of normal function and $M_Y = 76$. These values can be plugged into Equation 14.3 to find the Y-intercept:

$$
\begin{aligned}
a &= M_Y - bM_X \\
&= 76 - (-0.98 \times 22) \\
&= 76 - (-21.5600) \\
&= 76 + (21.5600) \\
&= 97.5600 \\
&= 97.56
\end{aligned}
$$

The Y-intercept for the regression line, which is the predicted value of Y when $X = 0$, is 97.56. Given $b = -0.98$ and $a = 97.56$, it is now possible to complete the regression equation, Equation 14.1:

$$
\begin{aligned}
Y' &= bX + a \\
&= -0.98X + 97.56
\end{aligned}
$$

This is the regression equation. Regression equations are used to predict a Y score for a case from its score on X. In the present instance, it can be used to predict a smoker's lung function (Y') based on how many years he or she has been smoking (X). Let's see it in action and predict the percentage of normal lung function for a person who has been smoking for eight years. Or, phrased mathematically, if $X = 8$, what is Y'? Applying Equation 14.1 to answer that question, it is predicted that a person who has been smoking for eight years will have lungs that function at 89.72% of normal capacity:

$$
\begin{aligned}
Y' &= bX + a \\
&= -0.98X + 97.56 \\
&= (-0.98 \times 8) + 97.56 \\
&= -7.8400 + 97.56 \\
&= 89.7200 \\
&= 89.72
\end{aligned}
$$

Now let's draw the regression line. The regression line should only span the range of existing X values. Look at the scatterplot in Figure 14.8 and see that the X values range from 1 to 65. Below, Y' scores for these two X values are calculated and they are used to draw the regression line seen in Figure 14.9 from (1, 96.58) to (65, 33.86):

$$\begin{aligned} Y' &= bX + a \\ &= -0.98X + 97.56 \\ &= (-0.98 \times 1) + 97.56 \\ &= -0.9800 + 97.56 \\ &= 96.5800 \\ &= 96.58 \end{aligned}$$

$$\begin{aligned} Y' &= bX + a \\ &= -0.98X + 97.56 \\ &= (-0.98 \times 65) + 97.56 \\ &= -63.7000 + 97.56 \\ &= 33.8600 \\ &= 33.86 \end{aligned}$$

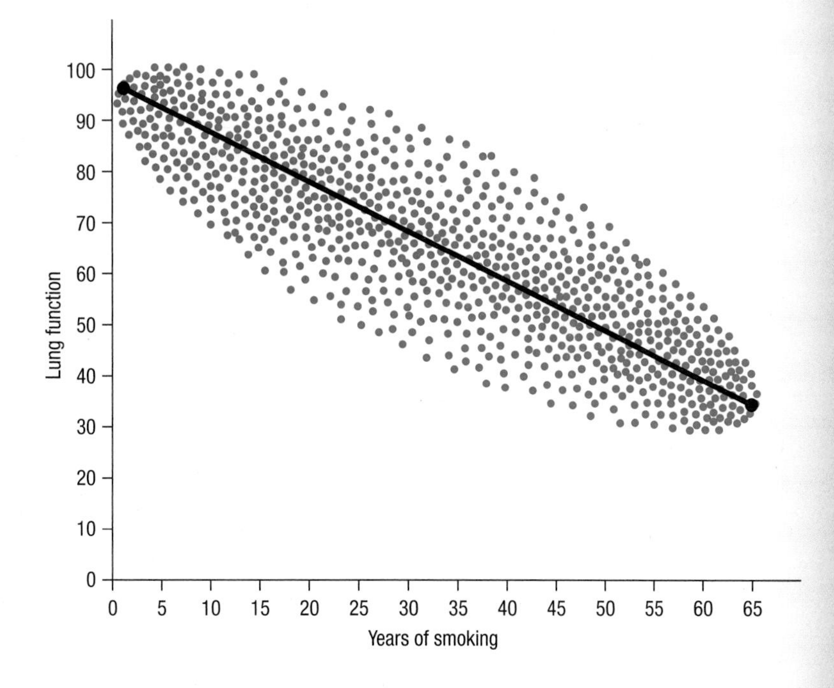

Figure 14.9 Regression Line for Years of Cigarette Smoking and Lung Function Study The regression line for this data set has a negative slope because the relationship is inverse—as the years of smoking go up, lung function goes down.

A Common Question

Q Predicting that a person who has smoked for eight years will have lungs that function at 89.72% of normal capacity sounds quite exact. Is such a precise prediction accurate?

A No. A prediction like 89.72% is a point estimate. An interval estimate, which gives a range within which Y' probably falls, is a better way to go. We will discuss such an interval, called a predication interval, in the next section, though calculating the prediction interval is beyond the scope of this text.

Practice Problems 14.1

Apply Your Knowledge

14.01 Given $r = -.37$, $s_X = 12.88$, and $s_Y = 9.33$, find the slope.

14.02 Given $M_X = 10.65$, $M_Y = 45.64$, and $b = 4.54$, find the Y-intercept.

14.03 Given a slope of 1.80 and a Y-intercept of -12.42, form a regression equation.

14.04 Given $Y' = 1.50X + 4.50$, (a) find the predicted values of Y for $X = 2$ and $X = 12$, and (b) draw the regression line.

14.05 Given the regression line to the right, estimate Y' if $X = 40$.

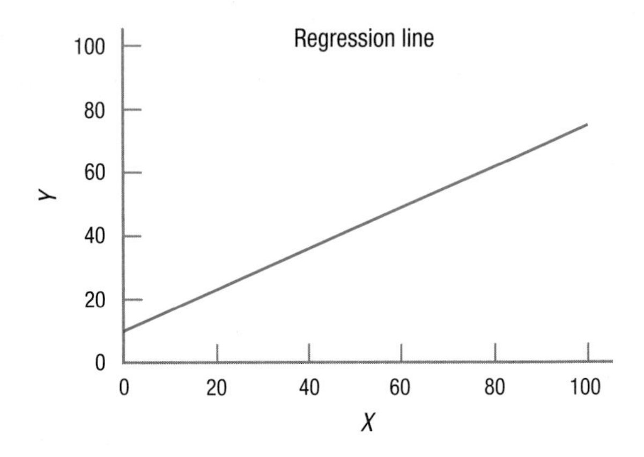

14.2 Errors in Regression

The goal of regression is to be able to predict Y values for X values. And the more accurately this can be done, the better. That's why researchers need a way to measure how much error occurs in prediction. As we'll soon learn, the measure for error in prediction is called the **standard error of the estimate**. Here's how it works.

With a perfect correlation, where all the data points fall on a line, each X value is associated with one and only one Y value. The story is more complex when r is not perfect. Look back at Figure 14.9, which shows the scatterplot and regression line for the $-.68$ correlation between years of smoking and loss of lung function. Using the regression equation, it was predicted that a person with eight years of smoking would have 89.72% of normal lung function. And that exact same prediction, 89.72%, would be made for every person who had been smoking for eight years. Whether the person was male or female, whether the person exercised regularly or not, whether the person smoked five cigarettes a day or two packs—all those things don't matter for the purposes of this estimate. If a person had been smoking for eight years, his or her lung capacity would be estimated at 89.72%.

But from the scatterplot in Figure 14.9, it is apparent that people who smoke for eight years have lung function scores that range from about 70% all the way up to 100%. There is variability in the scores. How much the actual scores deviate from the predicted scores is a measure of the degree of error in prediction. The more deviation there is—the larger the error is—the less sure a researcher is of the accuracy of the prediction. The statistic that summarizes how much error exists is called the standard error of the estimate.

Let's use Dr. Paik's marital satisfaction data set, with only eight cases, to calculate the standard error of the estimate. The first two columns in Table 14.2 contain his data set and the third column the predicted Y values for each of the X values. Case A, for example, has a gender role flexibility score of 8 and a marital satisfaction score of 0.8. Using the linear regression equation, its predicted marital satisfaction

TABLE 14.2 Actual, Predicted, and Residual Scores

	X: Gender Role Flexibility Score	Y: Marital Satisfaction Score	Y′: Predicted Marital Satisfaction Score	Y − Y′: Residual Score
A	8	0.8	1.03	−0.23
B	15	1.5	1.43	0.07
C	15	2	1.43	0.57
D	22	1.5	1.83	−0.33
E	31	2.3	2.34	−0.04
F	35	1.5	2.57	−1.07
G	36	3.1	2.62	0.48
H	38	3.3	2.74	0.56
	$M = 25.00$ $s = 11.49$	$M = 2.00$ $s = 0.86$		$s = 0.56$

This table shows two things. First, the standard deviation of the residual scores is the definitional formula for the standard error of the estimate. Second, it shows how the variability in *Y* scores can be broken down into two components—that which can be explained by *X*, *Y*′, and that which is left unexplained, the residual score.

score is 1.03. The final column is labeled residual scores. It shows the deviation of the predicted score from the actual score, calculated as $Y - Y'$. The predicted *Y* score for case A was off by 0.23 points and, because of the direction of the difference, is reported as −0.23. This difference is a measure of how wrong the predicted score is, so it is a measure of error. Case B, for example, where the predicted score was off from the actual score by 0.07 points, had less error in its predicted score than case C, where $Y - Y'$ was off by 0.57 points.

The last column contains deviation scores (error scores), which sum to zero. So, how can the average amount of error be represented? With a standard deviation! And that is what a standard error of the estimate is, the standard deviation of the residual scores. As can be seen at the bottom of the column, for Dr. Paik's data, the standard deviation of the residual scores, the standard error of the estimate, is 0.56.

The definitional formula requires that we calculate the standard error of the estimate as the standard deviation of the residual scores. Equation 14.4 gives the easier-to-use computational formula, a formula that can be used as long as one knows *r* and s_Y.

Equation 14.4 Formula for the Standard Error of the Estimate

$$s_{Y-Y'} = s_Y \sqrt{1 - r^2}$$

where $s_{Y-Y'}$ = standard error of the estimate
s_Y = standard deviation of the *Y* scores
r = the Pearson *r* value

This equation says that the standard error of the estimate may be calculated by (1) squaring the correlation coefficient, (2) subtracting the square from 1, (3) taking the square root of the difference, and (4) multiplying this square root by the standard deviation of the Y scores. Here are the calculations for Dr. Paik's data. The value about to be calculated, $s_{Y-Y'} = 0.56$, is exactly the same value that was found as the standard deviation of the difference scores in Table 14.2:

$$\begin{aligned} s_{Y-Y'} &= s_Y\sqrt{1 - r^2} \\ &= 0.86\sqrt{1 - .76^2} \\ &= 0.86\sqrt{1 - .5776} \\ &= 0.86\sqrt{.4224} \\ &= 0.86 \times .6499 \\ &= 0.5589 \\ &= 0.56 \end{aligned}$$

What does a standard error of the estimate of 0.56 mean? Loosely, one can think of standard error of the estimate as the average residual score, the average difference between the actual Y scores and the predicted Y scores. Is 0.56 a lot of error? It depends on the possible range of scores. Here the variable being predicted is marital satisfaction, which is measured on a scale ranging from 0 to 4. Being off by 0.56 points, on average, on a 4-point scale means being off, on average, by 14%. That's not good.

Want a concrete example? Suppose Neil goes to a county fair and stops at an "I'll Guess Your Weight" booth. The carny guesses Neil's weight as 150 pounds. But, if he's off by 14%, Neil could weigh 171 pounds and the carny underestimated his weight. The error could go the other way as well. The carny could have overestimated Neil's weight. Maybe Neil only weighs 129 pounds, which is off from 150 by 14% in the opposite direction.

This range, from 129 pounds to 171 pounds, gives the general idea of what a prediction interval is. A **prediction interval** gives a range within which there is some certainty that a case's real Y score falls. The calculation of the interval is based on the estimated Y score and the standard error of the estimate. The smaller the standard error of the estimate, the narrower the prediction interval and the better the prediction.

Worked Example 14.2

For another example of calculating the standard error of the estimate, a return to the cigarette smoking and lung function study is in order. In that study, 2,500 smokers reported how many years they had been smoking ($M = 22$, $s = 9$) and had their lung function measured as a percentage of normal ($M = 76$, $s = 13$). There was a strong and statistically significant inverse relationship, $r = -.68$: the longer people smoked, the lower their lung function. After a regression equation was developed, it was used to predict that a person who had been smoking for eight years would have lungs functioning at 89.72% of normal capacity. How much confidence should we have that this estimate is accurate?

The way to answer this is by calculating $s_{Y-Y'}$ using Equation 14.4:

$$\begin{aligned} s_{Y-Y'} &= s_Y\sqrt{1 - r^2} \\ &= 13\sqrt{1 - (-.68^2)} \\ &= 13\sqrt{1 - .4624} \\ &= 13\sqrt{.5376} \\ &= 13 \times .7332 \\ &= 9.5316 \\ &= 9.53 \end{aligned}$$

This standard error of the estimate of 9.53 means that the actual *Y* scores and *Y′* scores for the 2,500 people in the sample differed by almost 10 points, on average, on a 100-point scale. That seems like a fair amount of error. This suggests that predictions based on this regression equation aren't very accurate.

A Common Question

Q So far, both examples have had standard errors of the estimate that are large, suggesting prediction is not very good. What does it take to have a small standard error of the estimate?

A As r grows larger and s_Y becomes smaller, $s_{Y-Y'}$ gets smaller.

Practice Problems 14.2

Apply Your Knowledge

14.06 Given $r = .42$ and $s_Y = 5.64$, find $s_{Y-Y'}$.

14.07 Dr. Binet developed a regression equation to predict adult IQ from childhood language abilities. IQ can range from 55 to 145. The standard error of the estimate for the regression equation is 14. Is that a large error of the estimate?

14.3 Multiple Regression

Here's a thought experiment. In which scenario could one more accurately predict a student's GPA?

A. Knowing how many hours the student spends on schoolwork each week

B. Knowing how many hours the student spends on schoolwork each week *plus* his or her high school GPA, his or her IQ, and how much alcohol the student consumes each week

Most people believe the additional information in Scenario B is relevant to predicting academic performance and they are correct. In Scenario B, the prediction should be more accurate because more factors are taken into account.

The difference between Scenario A and Scenario B is the difference between simple regression and multiple regression. The previous section focused on simple linear regression. **Simple regression** uses just one predictor variable to calculate Y'. **Multiple regression** uses several predictor variables to calculate Y'. If the different X variables have different influences on the outcome variable being predicted, when they are combined, they will do a better job of prediction than any one variable by itself.

r^2, the percentage of variability in the outcome variable that is accounted for by the predictor variable(s), is called R^2 in multiple regression. Better prediction means a higher percentage of variability is accounted for with multiple regression than with simple regression. In this way, multiple regression is a more powerful technique than simple regression.

Deriving a multiple regression equation is beyond the scope of this text. But, here is an example to show how it works. Every year, colleges have many more applicants than they can admit. Part of the admissions process involves deciding which applicants can do college-level work. Multiple regression plays a role in predicting which applicants will fare well in college.

The College Board, the folks who created the SAT, provide a service to colleges that it calls ACES, the Admitted Class Evaluation Service. ACES uses admissions information from a first-year class to develop a multiple regression equation to predict first-year GPA. Once this equation is developed, the college can apply it in subsequent years to applicants to predict what their GPAs will be. The college can decide whom to admit, objectively, on the basis of predicted GPA.

The College Board offers a sample ACES report on its website (collegeboard.com). Using hypothetical data, the Board examines how well four variables—SAT reading subtest scores, SAT writing subtest scores, SAT math subtest scores, and high school class rank—predict first-year GPA for about a thousand students at one college.* [High school class rank is transformed to range from 100 (the best student) to 0 (the worst student).] Here are the Pearson r correlation coefficients for each of these variables predicting GPA by itself:

- SAT reading test, $r = .42$
- SAT writing test, $r = .42$
- SAT math test, $r = .39$
- High school class rank, $r = .52$

These r's are all fairly close to each other in terms of size. The r with the strongest correlation with GPA, meaning the one that is the strongest predictor, is high school class rank. There is certainly some overlap in what these four variables measure and how well they predict GPA. For example, general intelligence level plays a role in all four scores and intelligence plays a role in determining GPA. But, each of the four variables also measures something unique. For example, part of how well one does on the math test is not a result of one's general level of intelligence or the reading and writing skills that help on any test. But, to some degree, performance on a math test is determined by specialized math skills. And, to some degree, these same specialized math skills play a role in some of the courses that determine the GPA. Multiple regression adds together the unique predictive power of each variable. As a result, multiple regression usually accounts for a bigger percentage of the variability in the outcome variable than is accounted for by any single variable.

* Note: This example is based on the three-section SAT in use prior to 2016. Beginning March 2016, the SAT includes only two sections, Reading/Writing and Math.

When the four College Board variables are combined together to predict GPA in a multiple regression, the correlation climbs to $R = .57$. (The abbreviation for the correlation coefficient for multiple regression is R, not r.) This doesn't sound like much of an increase from the .52 correlation between class rank and GPA. But, it is. The percentage of variance explained changes from 27.04% to 32.49%. Predicting an extra 5 percentage points of variability is very worthwhile.

The multiple regression equation the College Board develops for a college can be used to predict GPA from SAT scores and high school rank for a potential student. Their equation is a more complex version of the linear regression equation from earlier in this chapter. The equation has "weights" for each of the predictor variables. The weights are like the slope in the linear regression equation. And there is a constant that is like the Y-intercept. When all of this information is put together, it makes for a long equation. Here is how estimated GPA, GPA', would be calculated:

$$GPA' = (SAT_{ReadingScore} \times Weight_{ReadingScore}) + (SAT_{WritingScore} \times Weight_{WritingScore}) + (SAT_{MathScore} \times Weight_{MathScore}) + (HSRank \times Weight_{HSRank}) + Constant$$

Here are the four weights and the constant for the College Board example:

- Reading weight = 0.0012
- Writing weight = 0.0013
- Math weight = 0.0006
- HS rank weight = 0.0029
- Constant = 0.7821

If an applicant were good at reading (SAT score = 600), not so good at writing (SAT score = 450), very good at math (SAT score = 760), and had a very good class rank (90), then her predicted GPA would be

$$\begin{aligned} GPA' &= (SAT_{ReadingScore} \times Weight_{ReadingScore}) + (SAT_{WritingScore} \times Weight_{WritingScore}) + \\ &\quad (SAT_{MathScore} \times Weight_{MathScore}) + (HSRank \times Weight_{HSRank}) + Constant \\ &= (600 \times 0.0012) + (450 \times 0.0013) + (760 \times 0.0006) + (90 \times 0.0029) + 0.7821 \\ &= 0.7200 + 0.5850 + 0.4560 + 0.2610 + 0.7821 = 2.8041 = 2.80 \end{aligned}$$

A person with those SAT scores and class rank would be predicted to end up with a GPA of 2.80 at the end of her first year.

The multiple regression equation is built from cases where the first-year GPA is known. The equation can be used to predict first-year GPA for these students. As a result, the students have both actual and predicted GPAs, and it is possible to see how well the predicted GPA predicts the actual GPA. The correlation between the two is .46. Multiple regression makes objective predictions that minimize errors in prediction *overall.* In that sense, it makes better decisions. But, unless $R = 1.00$, it doesn't make perfect predictions.

Worked Example 14.3

Many Americans are trying to lose weight, either by counting calories or by using Weight Watchers®. One of the two Weight Watchers plans counts points, not calories. In its system, foods are assigned a point value based on a mysterious combination of how much protein, carbohydrates, fat, and fiber the food contains. The number of calories in the food is not part of the equation. With the point system, a cup of lettuce is worth 0.3 points and a McDonald's Quarter Pounder is worth 13.4 points.

Imagine that a nutritionist, Dr. Feldman, wanted to crack the secret equation that Weight Watchers uses and figure out how these four variables—protein, carbs, fat, and fiber—are combined to generate a point score. This calls for multiple regression.

First, Dr. Feldman draws a random sample of foods and for each one he finds out how much protein, carbs, fat, and fiber the food contains. He also consults the Weight Watchers Web site and locates the point value for each food. Armed with these four predictor variables (protein, carbs, fat, and fiber) and the one outcome variable (points), he uses SPSS to find the multiple regression equation. The equation that calculates *Points′*, the estimated number of points, is

$$\begin{aligned} Points' &= (Grams_{\text{Protein}} \times Weight_{\text{Protein}}) + (Grams_{\text{Carbs}} \times Weight_{\text{Carbs}}) + (Grams_{\text{Fat}} \times \\ &\quad Weight_{\text{Fat}}) + (Grams_{\text{Fiber}} \times Weight_{\text{Fiber}}) + Constant \\ &= (Grams_{\text{Protein}} \times 0.074) + (Grams_{\text{Carbs}} \times 0.096) + (Grams_{\text{Fat}} \times 0.279) + \\ &\quad (Grams_{\text{Fiber}} \times -0.101) + 0.112 \end{aligned}$$

Note that three of the weights are positive, but the weight for fiber is negative. This reveals that fiber plays a different role in determining points than do the other three variables. As the levels of proteins, carbohydrates, and fats in a food go up, so does the point value for the food. However, as the amount of fiber in a food goes up, the point value goes down.

An important use of a regression equation is to predict a value for a new case. Suppose Dr. Feldman is about to eat a BLT and wants to know how many points it is worth. From the menu, he learns that the sandwich has 15 grams of protein, 28 grams of carbohydrates, 17 grams of fat, and 3 grams of fiber. Here's how he would calculate its points:

$$\begin{aligned} Points\ BLT' &= (Grams_{\text{Protein}} \times 0.074) + (Grams_{\text{Carbs}} \times 0.096) + (Grams_{\text{Fat}} \times 0.279) \\ &\quad + (Grams_{\text{Fiber}} \times -0.101) + 0.112 = (15 \times 0.074) + (28 \times 0.096) \\ &\quad + (17 \times 0.279) + (3 \times -0.101) + 0.112 \\ &= 1.1100 + 2.6880 + 4.7430 - 0.3030 + 0.112 = 8.3500 = 8.35 \end{aligned}$$

Dr. Feldman has now estimated (predicted) that eating a BLT at lunch will use up 8.35 of a person's daily point allowance.

Practice Problems 14.3

Review Your Knowledge

14.08 Explain why multiple regression explains a larger percentage of variability in the predicted variable than does simple regression.

Apply Your Knowledge

14.09 A multiple regression equation has a constant of 55.12, a weight of 13.17 for variable 1, and a weight of 4.55 for variable 2. If a case has a score of 12 on variable 1 and a score of 33 on variable 2, what is Y'?

Application Demonstration

Let's see multiple regression in action. In this cost-conscious era, hospitals try to save money by reducing the length of stay of their patients. It would be beneficial to a hospital if it could predict a patient's length of stay at the time of admission. If so, therapeutic resources could be directed to the patients predicted to be in the hospital for a long time, in order to help them get better more quickly.

Some researchers turned their attention to predicting length of stay for patients admitted to a large, metropolitan psychiatric hospital (Huntley, Cho, Christman, & Csernansky, 1998). In a six-month period, almost 800 patients were admitted to the facility and they spent an average of 16.3 days in the hospital. The hospital database contained a lot of information about each patient, including each patient's sex, age, primary and secondary diagnoses, number of prior admissions, and legal status. The researchers combined these variables using multiple regression to see if length of stay could be predicted.

There turned out to be five variables that played a statistically significant role in predicting a patient's length of stay: (1) a primary diagnosis of schizophrenia, (2) the number of previous admissions, (3) a primary diagnosis of a mood disorder, (4) age, and (5) an alcohol or drug problem as a secondary diagnosis. These variables can be thought of as reflecting difficult cases. For example, someone with five previous psychiatric admissions probably has a more severe problem, one that may take longer to treat, than a patient for whom this admission is the first hospitalization.

Together, these five variables predicted 17% of the variance in length of stay. This may not sound like much, but Cohen (1988) would call it a medium effect. Is it enough to be useful?

So far, what these researchers did is not unusual. But, now their work took an interesting direction. They used their regression equation to calculate the predicted length of stay for each patient. As a result, there were two pieces of data for each patient—the actual length of stay and the predicted length of stay. The researchers then added a third variable for each patient—the psychiatrist in charge of the patient's care. There were 12 psychiatrists at this hospital and newly admitted patients were assigned to their care on a rotating basis. In essence, patients were randomly assigned to psychiatrists.

If patients are randomly assigned to psychiatrists and if all psychiatrists provide equivalent care, then the mean length of stay should be roughly the same for each psychiatrist. The blue bars in Figure 14.10 show the mean length of stay of the patients for each of the psychiatrists—it ranges from less than 10 days (Psychiatrist 1) to more than 25 days (Psychiatrist 12). Either differences in the effectiveness of the psychiatrists exist or some psychiatrists had more or less than their fair share of hard-to-treat patients.

How could one tell if a psychiatrist were assigned difficult or easy patients? Difficult patients should have a longer predicted length of stay. So, calculating the mean predicted length of stay for each psychiatrist should answer that question. The grey bars in Figure 14.10 show the mean predicted length of stay corresponding to each psychiatrist.

Look at Psychiatrist 1. Earlier, when the focus was only on the blue bars showing actual length of stay, he appeared to be doing a good job because his patients had the shortest length of stay. Now, looking at the grey bar, it is apparent that

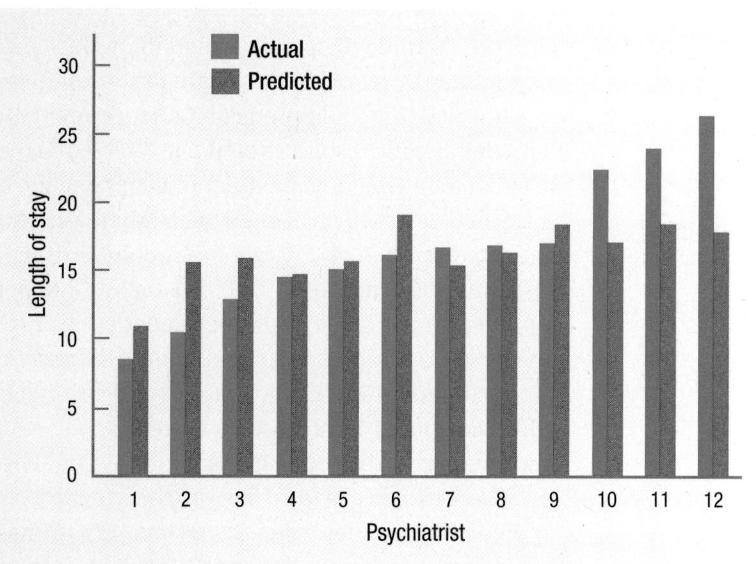

Figure 14.10 Actual and Predicted Length of Stay for Patients Treated by Different Psychiatrists Multiple regression was used to calculate the predicted length of stay for each patient. Comparing the two means allows one to speculate about a psychiatrist's skills. (Data from Huntley, Cho, Christman, & Csernansky, 1998.)

this psychiatrist was assigned the healthiest patients. No wonder he discharged them quickly.

The two bars, one the actual length of stay and the other predicted by multiple regression, allow us to think about this psychiatrist's performance in a more complex fashion. Psychiatrist 1's patients have a mean length of stay around 8 days but are predicted to need approximately 11 days. Why the difference? There are three likely explanations, two of which involve his skills and one that involves multiple regression. First, perhaps he is a phenomenal psychiatrist and cures people quickly. It could happen. Second, maybe he is a terrible psychiatrist who can't assess patients' progress and discharges them before they are ready. That could happen, too. Third, maybe there are errors in prediction. Maybe this is especially true for the healthier cases and their lengths of stay are overestimated.

Whatever the explanation turns out to be, this study shows how multiple regression is used in psychology and how predictions are made and utilized. Regression, either simple or multiple, is a useful tool that helps researchers understand their results in more detail.

SUMMARY

Calculate and apply a linear regression equation for a Pearson correlation coefficient.

- Linear regression predicts a value of Y, Y', for X when there is a statistically significant relationship between X and Y. The prediction equation uses the slope and Y-intercept to generate a regression line, the best-fitting line that minimizes the errors between Y and Y'. Slope indicates how much change in Y is predicted for each 1-unit change in X, and the Y-intercept tells where the line passes through the Y-axis.

- As r approaches zero, the regression line becomes horizontal and predicted Y values approach M_Y. When $r = 0$, then X doesn't predict Y and the best prediction that can be made for Y' is M_Y.

Measure uncertainty in regression predictions.

- Error in prediction is the difference between the actual score, Y, and the predicted score, Y'.
- The average amount of error is summarized in a statistic called the standard error of the estimate, which is the standard deviation of the residual scores.

Describe how multiple regression works.

- Simple regression uses a single predictor variable to predict Y'; multiple regression uses two or more predictor variables. By combining the unique predictive ability of multiple predictor variables, multiple regression accounts for more variability in the outcome variable.

KEY TERMS

least squares criterion – prediction errors are squared and the best-fitting regression line is the one that has the smallest sum of squared errors.

linear regression – a predictor variable is used to predict a case's score on another variable and the prediction equation takes the form of a straight line.

multiple linear regression – prediction in which multiple predictor variables are combined to predict an outcome variable.

prediction interval – a range around Y' within which there is some certainty that a case's real value of Y falls.

regression line – the best-fitting straight line for predicting Y from X.

residual – the difference between an actual score and a predicted score; the size of the error in prediction.

simple linear regression – prediction in which Y' is predicted from a single predictor variable.

slope – the tilt of the line; rise over run; how much up or down change in Y is predicted for each 1-unit change in X.

standard error of the estimate – the standard deviation of the residual scores, a measure of error in regression.

Y-intercept – the spot where the regression line would pass through the Y-axis.

Y prime – the value of Y predicted from X by a regression equation; Y'.

DIY

In the DIY of Chapter 13, you calculated the correlation between foot size and height. Now, take that same correlation coefficient and generate the regression equation to predict height from foot size. When you have arrived at the equation, use it to calculate Y' for the students on whom the equation was based. For each of the cases, calculate residual scores. Do they sum to zero? Now, find the standard deviation of the residual scores. Then, use Equation 14.4 to calculate the standard error of the estimate. Is that the same value you calculated for the standard deviation?

Want more fun? Select 10 new cases and use the regression equation to calculate Y' scores for them. Will the regression equation be as accurate for them as it was for the original group? Investigate this by calculating residual scores and finding their standard deviation. Is it larger or smaller than the first standard deviation? Why?

CHAPTER EXERCISES

Review Your Knowledge

14.01 The correlation chapter was about understanding ____ between variables; this chapter, on regression, is about ____ one ____ from another ____.

14.02 In linear regression, the ____ variable is X and Y is the ____ variable.

14.03 It is reasonable to do linear regression if the correlation between X and Y is ____.

14.04 ____ is the abbreviation for a predicted value of Y.

14.05 If blindly guessing a person's score, the best guess is the ____.

14.06 Statisticians use the ____ criterion to judge the regression line.

14.07 On the basis of least squares, the best-fitting line is the one that ____ the sum of the ____.

14.08 The difference between a case's actual Y score and its predicted Y score is called a ____.

14.09 A residual score is a measure of ____.

14.10 In the linear regression equation, b is the ____ and a is the ____.

14.11 Y' in the regression equation is ____ and X is the ____.

14.12 A ____ slope means the line is moving down and to the right.

14.13 If the correlation is positive, the slope of the regression line is ____.

14.14 If a slope is -0.50, then for every 1-point increase in X, there is a 0.5-point ____ in Y.

14.15 The spot where the regression line would pass through the ____ is called the Y-intercept.

14.16 Predictions of Y from X should only be made for X values that fall within the range of ____ used to develop the regression equation.

14.17 If $X = 17$ and $Y' = 29.37$, then the value 29.37 is a ____ estimate.

14.18 An ____ estimate is better than a ____ estimate.

14.19 If one uses a regression equation to predict Y' from X, and several cases have the same X value, then each time Y' is predicted for these cases, Y' will be *the same / different.*

14.20 The standard deviation of the residual scores is called the ____.

14.21 The standard error of the estimate can be thought of as the average ____ in prediction.

14.22 A prediction interval gives the range within which it is likely that a case's Y / Y' value falls.

14.23 The standard error of the estimate has an impact on the ____ of a prediction interval.

14.24 Simple regression uses ____ predictor variable to predict Y'; multiple regression uses ____.

14.25 Comparing multiple regression to simple regression, ____ usually accounts for a higher percentage of variability in Y than does ____.

14.26 Multiple regression is often used in college ____ decisions.

Apply Your Knowledge

Using regression lines to predict Y

14.27 Given this regression line, predict Y for an X value of 30:

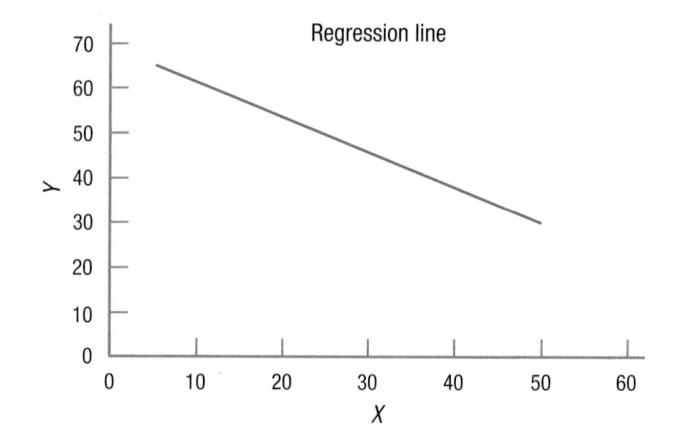

14.28 Given this regression line, predict Y for $X = 25$:

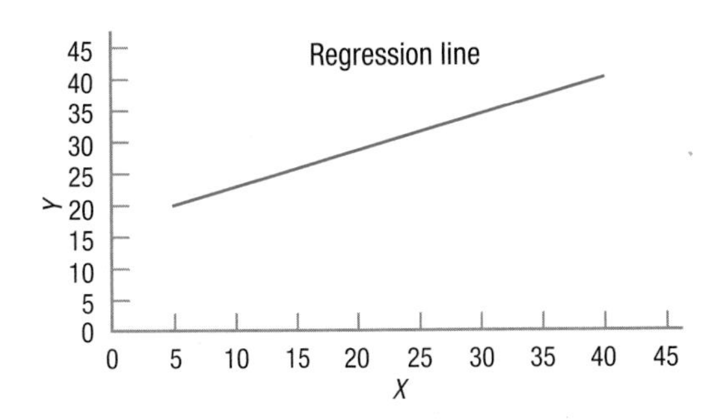

Calculating slope

14.29 If $r = -.47$, $s_Y = 3.65$, and $s_X = 9.66$, what is b?

14.30 If $r = .28$, $s_Y = 0.34$, and $s_X = 0.28$, what is b?

Interpreting slope

14.31 An automotive magazine used the price of gas in cents per gallon to predict the number of miles families drove on their summer vacations. The slope of the regression line was −35. Use the slope to interpret the impact of gas prices on vacation driving.

14.32 An exercise physiologist used the number of hours of TV watched per week at age 30 to predict the number of pounds gained over the next 10 years. The slope of the regression line was 1.25. Use the slope to interpret the impact of watching TV on weight gain.

Calculating the Y-intercept

14.33 If $b = 4.33$, $M_X = 5.00$, and $M_Y = 17.50$, what is a?

14.34 If $b = -2.45$, $M_X = 53.45$, and $M_Y = 112.23$, what is a?

Forming a regression equation

14.35 Given $b = 12.98$ and $a = -5.00$, write a regression equation.

14.36 Given $b = -0.68$ and $a = 7.50$, write a regression equation.

14.37 Given $r = -.24$, $M_X = 55.00$, $s_X = 11.00$, $M_Y = 25.00$, and $s_Y = 3.98$, write a regression equation.

14.38 Given $r = .33$, $M_X = 2.50$, $s_X = 1.50$, $M_Y = 112.50$, and $s_Y = 21.50$, write a regression equation.

Predicting Y

14.39 Find Y' if $X = 25$ for $Y' = 0.37X + 15$.

14.40 Find Y' if $X = -10$ for $Y' = 13X + 88$.

Drawing a regression line

14.41 (a) Given the endpoints of (10, 20) and (80, 50), draw a regression line. (b) What is the range of X values for which Y' can be calculated?

14.42 (a) Given the endpoints of (0, 70) and (50, 0), draw a regression line. (b) What is the range of X values for which Y' can be calculated?

14.43 Here is a regression equation: $Y' = 2.50X - 12.50$. If X values can range from 30 to 70, draw the regression line.

14.44 Here is a regression equation: $Y' = -1.10X + 115$. If X values can range from 70 to 130, draw the regression line.

Calculating residual scores

14.45 If $X = 75.45$, $Y = 12.96$, and $Y' = 13.43$, what is the residual score?

14.46 If $X = 24.77$, $Y = 33.43$, and $Y' = 31.22$, what is the residual score?

Calculating standard error of the estimate

14.47 If $r = .28$ and $s_Y = 10.55$, what is the standard error of the estimate?

14.48 If $r = .20$ and $s_Y = 30.42$, what is the standard error of the estimate?

Interpreting standard error of the estimate

14.49 Dr. Lansing is using high school GPA to predict combined SAT scores. Combined SAT scores from the two SAT subtests can range from 400 to 1600. He calculated the standard error of the estimate as 40. Interpret this error of the estimate.

14.50 Dr. Pallas is predicting severity of depression in adulthood from a childhood behavior checklist. The severity of depression scale ranges from 0 to 50 and the standard error of the estimate is 2.25. Interpret this error of the estimate.

Calculating Y' in multiple regression

14.51 A multiple regression equation has a constant of 5.74, a weight of 3.42 for variable 1, and a weight of −0.76 for variable 2. If a case has a score of 56.66 on variable 1 and a score of 88.99 on variable 2, what is Y'?

14.52 A multiple regression equation has a constant of 25.12, a weight of −4.55 for variable 1, a weight of −8.86 for variable 2, and a weight of 10.76 for variable 3. If a case has a score of 7.33 on variable 1, a score of 12.20 on variable 2, and a score of 18.85 on variable 3, what is Y'?

Expand Your Knowledge

14.53 Jeff is an above-average golfer. He decides to change sports and take up ping pong. If there is a strong, positive correlation between golf ability and ping pong ability, predict how he'll perform as a ping pong player?

a. Excellent
b. Above average
c. Average
d. Below average
e. Terrible
f. Not enough information given to reach a conclusion.
g. Ping pong and golf are different sports and one can't be predicted from the other.

14.54 Sue is an above-average golfer. She decides to change sports and take up archery. If a strong, negative correlation exists between golf ability and archery ability, predict how she'll perform as an archer?

a. Excellent
b. Above average
c. Average
d. Below average
e. Terrible
f. Not enough information given to reach a conclusion.
g. Archery and golf are different sports and one can't be predicted from the other.

14.55 David is an above-average golfer. He decides to change sports and take up wrestling. If there is a no correlation between golf ability and wrestling ability, predict how he'll rate as a wrestler?

a. Excellent
b. Above average
c. Average
d. Below average
e. Terrible
f. Not enough information given to reach a conclusion.

14.56 Shemekia applied to a college that uses multiple regression to select students. This college only considers students whose predicted first-year GPA is 3.0 or higher. Shemekia's predicted GPA was 2.9, and she did not get accepted. The correlation between the predictor variable and GPA is .30 and the standard deviation for GPA is .40. Based on this, what argument could Shemekia make to the college for why she should be considered?

14.57 Continue with Exercise 14.56. The college responds to Shemekia. Based on the same information, what argument could the college make for why she shouldn't be offered admission?

14.58 If Y' and $s_{Y-Y'}$ are known, make an educated guess as to what the 95% prediction interval would be.

14.59 There is a graph with a regression line for predicting Y. Could it be used to predict X?

SPSS

SPSS does calculate linear regression. We'll use Dr. Paik's marital satisfaction data to show how it works. Figure 14.11 illustrates how the data are arranged with each variable (role flexibility and marital satisfaction) in its own column and each case in its own row.

Role_Flex	Marital_Stat
8	.8
15	2.0
22	1.5
31	2.3
35	1.5
38	3.3
15	1.5
36	3.1

Figure 14.11 Data Entry for Linear Regression in SPSS Each variable is in a separate column and each case appears on its own row. (Source: SPSS)

Figure 14.12 shows how to access the linear regression commands in SPSS. Click on "Analyze," then "Regression," and finally "Linear." SPSS calls the predictor variable "independent" and the outcome variable "dependent."

Clicking on "Linear" opens up the commands seen in Figure 14.13. Notice that *Y*, the dependent variable "Marital_Sat," has been moved over to be the

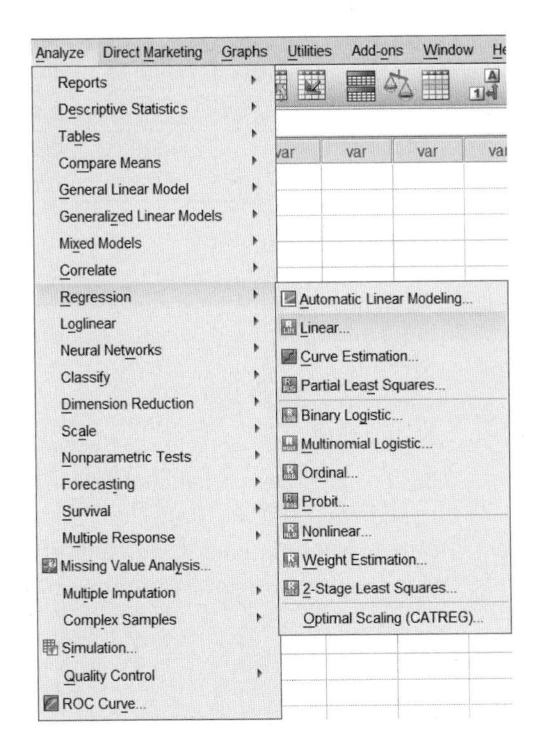

Figure 14.12 Starting Regression Line Analysis in SPSS Linear regression is started in SPSS by clicking on "Analyze," then "Regression," and "Linear." (Source: SPSS)

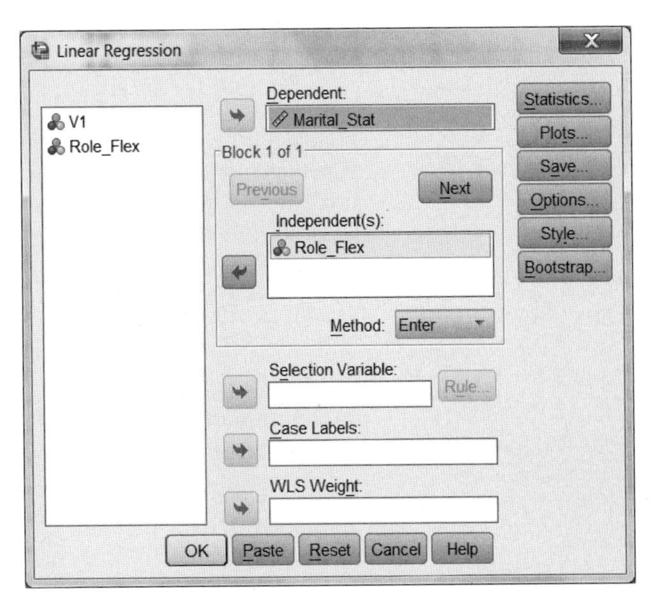

Figure 14.13 Selecting Variables for Regression Analysis in SPSS The predicted variable—here, "Marital_Sat"—is the dependent variable and the predictor variable—here, "Role_Flex"— is an independent variable. (Source: SPSS)

dependent variable. Also, X, the independent variable "Role_Flex," has been moved over to be an independent variable. Clicking on "OK" in the lower right starts the calculations.

The output, of which SPSS generates a lot, is shown in Figure 14.14. First, look in the table named "Model Summary." There, it says "R = .762." R is the value of the correlation coefficient. (R is the abbreviation for a multiple regression. We've borrowed it to do a simple regression and SPSS can't tell.)

Model Summary

Model	R	R Square	Adjusted R Square	Std. Error of the Estimate
1	.762[a]	.580	.510	.6020

a. Predictors: (Constant), Role_Flex

ANOVA[a]

Model		Sum of Squares	df	Mean Square	F	Sig.
1	Regression	3.006	1	3.006	8.294	.028[b]
	Residual	2.174	6	.362		
	Total	5.180	7			

a. Dependent Variable: Marital_Stat

b. Predictors: (Constant), Role_Flex

Coefficients[a]

Model		Unstandardized Coefficients		Standardized Coefficients	t	Sig.
		B	Std. Error	Beta		
1	(Constant)	.574	.539		1.065	.328
	Role_Flex	.057	.020	.762	2.880	.028

a. Dependent Variable: Marital_Stat

Figure 14.14 SPSS Regression Output In this SPSS output, look at the last table, "Coefficients," where the slope, .057, is listed as B, the unstandardized coefficient for the predictor variable, and the Y-intercept, .574, is the unstandardized B coefficient for the constant. (Source: SPSS)

SPSS carries many more decimal places than this text does, so its answers will differ from the answers here. The slope of the line, which was calculated as 0.06 in the text, is 0.057 in SPSS and is found next to the predictor variable, "Role_Flex" under the "B" column under "Unstandardized coefficients." The Y-intercept, which was calculated as 0.50 in the text, is 0.574 in SPSS and is found in the same column in the row labeled "(Constant)."

Nonparametric Statistical Tests: Chi-Square

LEARNING OBJECTIVES

- Differentiate parametric tests from nonparametric tests.
- Calculate and interpret a chi-square goodness-of-fit test.
- Calculate and interpret a chi-square test of independence.
- Know when to use a Spearman rank-order correlation coefficient and a Mann–Whitney *U* test.

CHAPTER OVERVIEW

So far, all the tests covered in this text have had two things in common. One is that no matter the test—whether for a *z*, *t*, *F*, or *r*—the outcome variable has always been measured at the interval or ratio level. The other is that for each test, the outcome variable was supposed to be normally distributed in the population.

But, sometimes a researcher takes on a study where the outcome variable is ordinal or nominal. Sometimes, the outcome variable isn't normally distributed. Tests for these situations, what are called nonparametric tests, are the subject of this chapter.

15.1 Introduction to Nonparametric Tests

z, *t*, *F*, and *r* all come from a family of tests called *parametric tests.* **Parametric tests** should only be used when assumptions about the population, about the parameters, are met. In contrast, **nonparametric tests** don't have to meet the same assumptions. There are two situations where nonparametric tests are used:

1. If the outcome variable is nominal or ordinal, then a nonparametric test is planned from the outset.
2. If a researcher is planning to use a parametric test, but a nonrobust assumption is violated, then the researcher can "fall back" to a nonparametric test.

Nonparametric tests are desirable because they are less restricted by assumptions. However, this comes at a cost—nonparametric statistical tests are usually less powerful than parametric tests. This is a problem because a test with less power is less likely to succeed in rejecting the null hypothesis, the usual goal of hypothesis testing. Also, nonparametric tests work with nominal- or ordinal-level data, not interval/ratio. Nominal- and ordinal-level numbers contain less information than do interval- and ratio-level numbers and this can make it harder to find an effect.

To see how less information means less power, imagine a medication that is only slightly effective in reducing fever. This small effect would be more evident if the temperature were measured to a hundredth of a degree than simply measuring whether or not a person has a fever. The more precisely a researcher can measure the outcome, the greater the ability to find an effect.

Statisticians prefer parametric tests because of their greater power. But, when parametric tests can't be used, when their assumptions are not met, or when the outcome variable is nominal or ordinal, it is time for a nonparametric test. Our primary focus in this chapter will be two different versions of the most commonly used nonparametric test, the chi-square. (The "chi" in chi-square is abbreviated with an uppercase Greek letter, χ, pronounced "kai," so chi-square is written χ^2.) Then, at the end of the chapter, the nonparametric version of a Pearson *r* and nonparametric version of an independent-samples *t* test are explored.

15.2 The Chi-Square Goodness-of-Fit Test

The **chi-square goodness-of-fit test** is a nonparametric, single-sample test that can be used with a nominal dependent variable or a higher-level variable treated as categorical, such as turning actual scores on an exam into letter grades. This type of test is called a **single-sample test** because it compares the results from a single sample to a specific value, usually a population value. The chi-square goodness-of-fit test is the nonparametric version of the single-sample *z* test and single-sample *t* test.

Here's an example of how the chi-square goodness-of-fit test might be used. Suppose a university administrator wants to survey a sample of students to get their opinion about a planned decrease in the intramural sports program. The administrator wants to make sure that the sample represents the university population, particularly in terms of gender. His sample is 59% female. From the registrar, he learns that 52% of the students at the university are female. A chi-square goodness-of-fit test could be used to compare the percentage of women in the sample to the percentage of women in the population. This allows the researcher to determine if obtaining a sample that is 59% female is a common occurrence if the population is 52% female. If it is a common occurrence, then he'll decide that the sample may be representative of the population on this variable and he'll be more likely to trust the results. If it is an uncommon occurrence, then the sample may be an odd one and its results unrepresentative.

To learn how to calculate and interpret the chi-square goodness-of-fit test, imagine this example from a small community where teenagers believe that the local police single them out more often than adult drivers for traffic stops. To investigate this, a traffic researcher, Dr. Koenig, randomly selected 72 tickets from all the tickets issued during a calendar year. As the age of the driver was recorded on each ticket, Dr. Koenig determined that 11 of the tickets went to teen drivers and 61 went to adults. The results are displayed in two cells, shown in Table 15.1.

TABLE 15.1 Ages of 72 Ticketed Drivers

Teenagers	Adults
11	61

11 of the 72 ticketed drivers (15.28%) were teenagers. 61 of the 72 ticketed drivers (84.72%) were adults.

Teenagers received 11 of the 72 traffic tickets (15.28%) in the sample. Dr. Koenig found, from the Department of Motor Vehicles, that 8% of licensed drivers in the population of the town are teenagers. She reasoned that if teens were treated the same as adults, 8% of the traffic tickets should go to teens. Yet, the teens received more than 15% of the tickets. Do teens get more than their fair share of traffic tickets? Or, can the difference between what is expected (8%) and what is observed (15.28%) be explained by sampling error?

"**T**om **and** **H**arry **d**espise **c**rabby **i**nfants" is the mnemonic for the six steps of hypothesis testing for a chi-square goodness-of-fit test: (1) Pick the **test,** (2) check the **assumptions,** (3) list the **hypotheses,** (4) set the **decision rule,** (5) **calculate** the test statistic, and (6) **interpret** the results.

Step 1 Pick a Test

This scenario calls for a single-sample test because a sample value, 15.28%, is being compared to a specific value, 8%. It may seem as if there are two samples (a sample of teens and a sample of adults), but it is just one sample (a sample of ticketed drivers) where the age status of each case is being measured (teen or adult) as the outcome variable. The outcome variable (the category of who gets a ticket, teen or adult) is a nominal-level variable, so the appropriate test is a chi-square goodness-of-fit test. Goodness-of-fit is abbreviated GOF.

Step 2 Check the Assumptions

The chi-square goodness-of-fit test has three assumptions:

1. *Random sample.* The sample should be a random sample from the population. This is a robust assumption, so it can be—and usually is—violated. As with parametric tests, if the random sample assumption is violated, then the researcher needs to be careful about generalizing the results from the sample. For the ticket data, there is a random sample of tickets, so the first assumption is not violated. But, the sample is only from tickets given in this town, so generalizability is limited.
2. *Independence of observations.* The cases in the sample should be independent of each other. This means that the observations don't influence each other. This assumption is not robust, so one can't proceed with the planned test if it is violated. With the ticket data, random sampling was used, so the cases within the sample aren't connected to each other. In addition, no case was in the sample more than once. The independence of observations assumption is not violated.
3. *Expected frequencies.* In order to conduct a chi-square test, all cells must have expected frequencies of at least 5. The chi-square test only works if there are enough cases in each cell. We will cover the method for calculating expected frequencies later in this chapter. Just know that the expected frequencies assumption is not robust, so the chi-square can't be calculated if each cell doesn't have enough cases. (For this example, expected frequencies will be large enough and this assumption isn't violated.)

Step 3 List the Hypotheses

The hypotheses for the chi-square goodness-of-fit test are easier to express in words than in mathematical symbols:

- The null hypothesis states that the proportion of each category in the population matches specified values. This means that since the sample comes from the population, the proportions in the sample should match the specified values. But, as sampling error exists, one shouldn't expect the sample to match the specified values exactly.
- The alternative hypothesis says that the distribution of the characteristic in the population is different from the specified values. The alternative hypothesis means that the difference between the percentages in the sample and the specified values is too large to be explained by sampling error.

H_0: In the population, the percentage of tickets received by teenage drivers = 8%.

H_1: In the population, the percentage of tickets received by teenage drivers ≠ 8%.

Step 4 Set the Decision Rule

The decision rule for a chi-square goodness-of-fit test follows the same format as the decision rules for *z*, *t*, *F*, and *r*:

- Find the critical value of chi-square, χ^2_{cv}, using a table of critical values of chi-square (Appendix Table 9).
    - The critical value is the value that separates the rare zone from the common zone in the sampling distribution of chi-square.
    - The sampling distribution of chi-square is the distribution of chi-square values that would occur if the null hypothesis were true.
- Compare χ^2, the observed value of the chi-square statistic to χ^2_{cv}.
    - If $\chi^2 \geq \chi^2_{cv}$, then the observed value of chi-square falls in the rare zone, and reject the null hypothesis.
    - If $\chi^2 < \chi^2_{cv}$, then the observed value of chi-square falls in the common zone, so fail to reject the null hypothesis.

A portion of Appendix Table 9, the table of critical values of chi-square, is shown in Table 15.2. There are three characteristics to note about the table of critical values of chi-square:

1. The chi-square is always a two-tailed test. Because chi-square is calculated using squared values, it is always positive. If the results are statistically significant, the researcher will need to look at the direction of the difference in the sample to figure out the direction of the difference in the population.
2. There are three alpha levels: .01, .05, and .10. The alpha level most commonly used, the column where $\alpha = .05$, appears in bold. That alpha level represents a 5% chance of making a Type I error. (Type I error occurs when the null hypothesis is rejected by mistake.)
3. The critical value of chi-square also depends on how many degrees of freedom there are. Each row in the table of critical values represents a different number of

degrees of freedom. The critical value of chi-square is found at the intersection of the column for the desired alpha level with the row for the correct number of degrees of freedom.

TABLE 15.2 Critical Values of Chi-Square (Appendix Table 9)

	Alpha Level		
df	.10	.05	.01
1	2.706	**3.841**	10.828
2	4.605	**5.991**	13.816
3	6.251	**7.815**	16.266
4	7.779	**9.488**	18.467
5	9.236	**11.070**	20.515
6	10.645	**12.592**	22.458
7	12.017	**14.067**	24.322

The critical value of χ^2 is found at the intersection of the row with the correct number of degrees of freedom and the column with the desired alpha level. The most commonly used alpha level is $\alpha = .05$, so those values are in bold.

The next step is to find the degrees of freedom, which can be calculated using Equation 15.1.

Equation 15.1 Degrees of Freedom for a Chi-Square Goodness-of-fit Test

$$df = k - 1$$

where df = degrees of freedom
k = number of categories

Equation 15.1 says to calculate degrees of freedom (df) for a chi-square goodness-of-fit test as the number of categories, k, minus 1. For the ticket data, the variable is age of drivers and it has two categories—adults and teens. This can be seen in the two cells in Table 15.1. With $k = 2$, degrees of freedom is calculated as follows:

$$\begin{aligned} df &= k - 1 \\ &= 2 - 1 \\ &= 1 \end{aligned}$$

Dr. Koenig is willing to have a 5% chance of making a Type I error, so she sets alpha at .05. Looking in the table of critical values of chi-square at the intersection of the row with $\alpha = .05$ and the row with $df = 1$, she finds $\chi^2_{cv} = 3.841$. The sampling distribution of chi-square, with the rare and common zones marked, is seen in Figure 15.1.

Now that χ^2_{cv} is known, the decision rule can be written:

- If $\chi^2 \geq 3.841$, reject H_0.
- If $\chi^2 < 3.841$, fail to reject H_0.

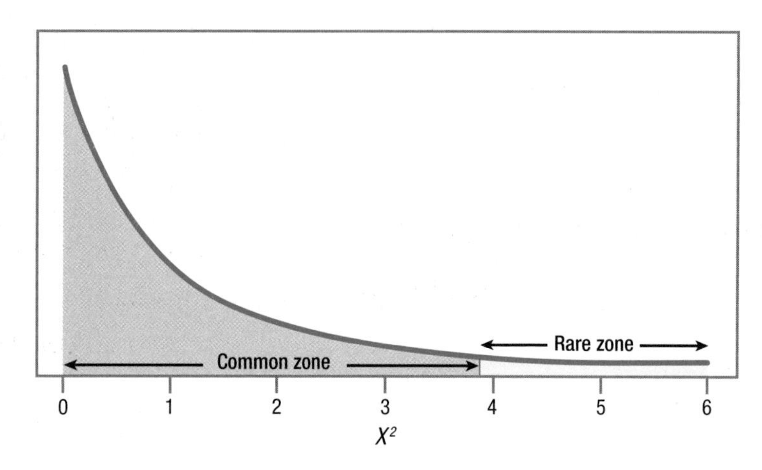

Figure 15.1 Sampling Distribution of Chi-Square with 1 Degree of Freedom When $df = 1$, the critical value of chi-square is 3.841 if $\alpha = .05$.

Step 5 Calculate the Test Statistic

Calculating the test statistic, χ^2, involves comparing the *observed* frequency of each category of the dependent variable in the sample to the frequency that would be *expected* if the null hypothesis were true:

- If the differences between observed and expected are small enough to be accounted for by sampling error, then the χ^2 value will be small, it will land in the common zone of the sampling distribution, and the null hypothesis won't be rejected.
- If the differences between observed and expected are too large to be accounted for by sampling error, then the χ^2 value will be large, it will land in the rare zone, and the null hypothesis will be rejected.

In order to calculate chi-square, one needs to know the observed frequency for each category and the expected frequency for each category. In Chapter 2, f was introduced as the abbreviation for frequency. Now f_{Observed} will be used as the abbreviation for observed frequency and f_{Expected} for expected frequency.

The observed frequencies for each cell (category) of the dependent variable in the sample are already known—there were 11 teens with tickets and 61 adults with tickets. Those are the observed frequencies and Table 15.3 organizes them with each observed frequency in a cell.

The next step is to calculate the expected frequencies for each cell. It is already known that 8% of drivers are teens. The total percentage of drivers is 100%. So, 100% – 8%, or 92%, of drivers must be adults. These two percentages, 8% for teens

TABLE 15.3 Observed Frequencies for the Ages of Ticketed Drivers ($N = 72$)

	Teenagers	Adults
Observed Frequency	11	61

In the random sample of 72 traffic tickets, 11 were issued to teenagers and 61 to adults.

and 92% for adults, are the expected *percentages*, abbreviated $\%_{Expected}$, for the two cells. If the null hypothesis is true, one would expect that 8% of tickets would go to teens and 92% to adults. These expected percentages are used to calculate the expected frequencies, as shown in Equation 15.2.

Equation 15.2 Formula for Calculating Expected Frequency

$$f_{Expected} = \frac{\%_{Expected} \times N}{100}$$

where $f_{Expected}$ = expected frequency for a cell/category
$\%_{Expected}$ = expected percentage for a cell/category
N = total number of cases in the sample

To calculate the expected frequency of traffic tickets for teenagers in this sample where $\%_{Expected} = 8\%$ and $N = 72$, Dr. Koenig uses Equation 15.2:

$$\begin{aligned} f_{Expected} &= \frac{\%_{Expected} \times N}{100} \\ &= \frac{8 \times 72}{100} \\ &= \frac{576.0000}{100} \\ &= 5.7600 \\ &= 5.76 \end{aligned}$$

This means that if the null hypothesis were true and teens received their fair share of the tickets, which is 8%, one would expect 5.76 of these 72 tickets to have been issued to teenagers. Don't be bothered by the fact that there are fractional tickets. Expected frequencies don't have to be whole numbers.

Equation 15.2 is also used by Dr. Koenig to calculate the expected frequency for adult tickets. The sample size, *N*, is still 72, but the expected percentage is now 92%:

$$\begin{aligned} f_{Expected} &= \frac{\%_{Expected} \times N}{100} \\ &= \frac{92 \times 72}{100} \\ &= \frac{6{,}624.0000}{100} \\ &= 66.2400 \\ &= 66.24 \end{aligned}$$

If the null hypothesis were true, one would expect the vast majority of the 72 tickets in the sample, 66.24, to go to adults. Only 5.76 are expected to go to teens. Table 15.4 shows the observed frequencies and the expected frequencies. Note that all expected frequencies are greater than 5, so the third assumption was not violated.

TABLE 15.4 Observed Frequencies and Expected Frequencies for the Ages of Ticketed Drivers

	Teenagers	Adults	
Observed Frequency	11	61	$N = 72$
Expected Frequency	5.76	66.24	$\Sigma = 72.00$

The sum of the expected frequencies is the same as the sample size.

Notice something interesting in Table 15.4. If the expected frequencies for all the categories are added up, the total is the same as the original sample size: $5.76 + 66.24 = 72.00$. This will always be the case and is a good way to check that the math was done correctly in calculating expected frequencies.

A Common Question

Q Do I have to use Equation 15.2 to calculate expected frequencies for each cell?

A No. You can take advantage of the fact that the expected frequencies add up to *N* and only calculate expected frequencies for as many categories as there are degrees of freedom. Then subtract the calculated frequencies from *N* to find the missing frequency.

Now that the observed frequencies and expected frequencies are known, they will be used to calculate the test statistic, chi-square. The formula for chi-square is shown in Equation 15.3.

Equation 15.3 Formula for Calculating Chi-Square (χ^2)

$$\chi^2 = \Sigma \frac{(f_{\text{Observed}} - f_{\text{Expected}})^2}{f_{\text{Expected}}}$$

where χ^2 = chi-square value

f_{Observed} = observed frequency for a cell/category

f_{Expected} = expected frequency for a cell/category

To use Equation 15.3 to calculate chi-square, follow these four steps:

1. For each cell/category, subtract the expected frequency from the observed frequency.
2. Square each difference.
3. Divide each squared difference by its respective expected frequency to yield a quotient.
4. Sum all the quotients to obtain the chi-square value.

Here are the calculations for the ticket data, where it is found that $\chi^2 = 5.18$:

$$\begin{aligned}
\chi^2 &= \Sigma\frac{(f_{\text{Observed}} - f_{\text{Expected}})^2}{f_{\text{Expected}}} \\
&= \frac{(11-5.76)^2}{5.76} + \frac{(61-66.24)^2}{66.24} \\
&= \frac{5.2400^2}{5.76} + \frac{-5.2400^2}{66.24} \\
&= \frac{27.4576}{5.76} + \frac{27.4576}{66.24} \\
&= 4.7669 + 0.4145 \\
&= 5.1814 \\
&= 5.18
\end{aligned}$$

Step 6 Interpret the Results

In interpreting the results of a chi-square goodness-of-fit test, there are two questions to be addressed: (1) Was the null hypothesis rejected? (2) If so, what is the direction of the results?

Was the Null Hypothesis Rejected?

To answer this question, refer back to the decision rule generated in Step 4. Decide which of the two decision rules is true:

- Is $5.18 \geq 3.841$? If so, reject H_0 and call the results statistically significant.
- Is $5.18 < 3.841$? If so, fail to reject H_0 and call the results not statistically significant.

With the ticket data, the chi-square value (5.18) is greater than the critical value (3.841), so the first statement is true and the null hypothesis is rejected. This means the difference between the percentages observed in the sample and those found in the population is statistically significant. The differences are too large to be explained by sampling error, so we conclude that the distribution of the dependent variable in the population differs from the specified value in the null hypothesis. Here, this means that the percentage of tickets issued to teens in the population is probably not 8%.

Writing results in APA format for chi-square calls for six pieces of information: (1) what test was done, (2) how many degrees of freedom there were, (3) what the sample size was, (4) what the value of the test statistic was, (5) what alpha level was selected, and (6) whether the null hypothesis was rejected. For the traffic ticket data, Dr. Koenig would write

$$\chi^2(1, N = 72) = 5.18, p < .05$$

1. χ^2 reveals that the test statistic is a chi-square value.
2. The 1 in the parentheses states the degrees of freedom.
3. $N = 72$ gives the sample size.
4. 5.18 is the value of the test statistic that was calculated.

5. .05 indicates that alpha was set at the .05 level.

6. $p < .05$ says that the null hypothesis was rejected and that the value of the test statistic (5.18) is a rare one if the null hypothesis is true.

What Is the Direction of the Results?

If the results were statistically significant, the researcher needs to comment on the direction of the difference. This can be done by comparing what was observed to what was expected. If the results were not statistically significant, then there's not enough evidence to say a difference exists so there's no need to worry about the direction of the difference.

With the traffic ticket data, there are just two categories, so it is easy to tell the direction of the difference. Teenagers account for 8% of the drivers, but they received 15.28% of the tickets. The researcher can say that 15.28% is statistically higher than 8%. Dr. Koenig can conclude that teens in this town get statistically significantly more tickets than expected for the number of teens who are drivers.

Putting It All Together

A four-point interpretation (What was done? What was found? What does it mean? What suggestions are there for future research?) can be completed for a chi-square goodness-of-fit test. Here's an interpretation for the traffic ticket data:

> A traffic researcher analyzed a sample of traffic tickets from a town in order to determine if teen drivers received proportionally more tickets than adult drivers. 8% of the drivers in the town were teens, but 15.28% of traffic tickets issued went to teens. The difference was statistically significant [$\chi^2(1, N = 72) = 5.18$, $p < .05$]. Teenage drivers in this town get more than their fair share of traffic tickets, almost twice as many as expected. Future research should extend the study to other municipalities. It would also be worthwhile to explore whether the over-ticketing of teen drivers is deserved because they are worse drivers or because they are being unfairly targeted by the police.

Worked Example 15.1

Here's another example for a chi-square goodness-of-fit test. Courtney had not done well in intro psych and tried to make the case that the instructor, Dr. Wald, was an unfairly harsh grader. As evidence, she counted the number of A's, B's, C's, D's, and F's given as final grades for the 84 students in Dr. Wald's two sections of intro psych (Table 15.5). Courtney then found out, from the registrar, the percentages of A's, B's, C's, D's, and F's given in all other sections of intro psych that semester. In this table, D's and F's are combined into one category for reasons that will be explained shortly. This distribution, the information from the registrar, is shown in gray in Figure 15.2. Courtney then transformed the distribution of grades for Dr. Wald's two sections of intro psych into percentages, shown in dark blue in Figure 15.2. As Courtney complained to the school dean, "It is clear in my graph—Professor Wald gives fewer A's and B's than the other instructors, but more C's and more D's and F's. Dr. Wald is a harsh and unfair grader." Is he? To try to clear his name, Dr. Wald is going to need a statistical test.

TABLE 15.5 Observed Distribution of Final Grades in Dr. Wald's Sections of Intro Psych and Percentages of Different Grades Awarded in All Sections of Intro Psych

	A	B	C	D or F	
Number of grades awarded in Dr. Wald's sections	10	24	41	9	$N = 84$
Percentage of grades awarded in all sections of intro psych	19%	33%	42%	6%	$\Sigma = 100\%$

Grades of D and F are counted together because there are so few of them. This will serve to keep the expected frequency for this cell above 5.

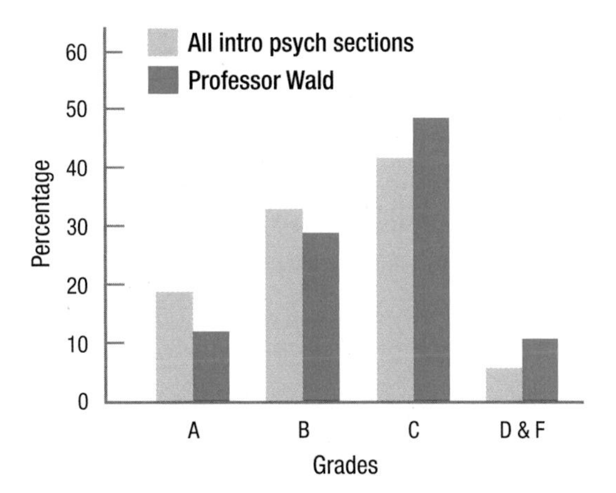

Figure 15.2 Final Grades in Professor Wald's Intro Psych Classes Compared to All Intro Psych Instructors Professor Wald gives fewer high grades and more low grades than do the other instructors. Is the difference a statistically significant one? Or, is it explained by normal variability in sampling?

Step 1 Pick a Test. The question being asked is whether the distribution of grades in Dr. Wald's class is different from the distribution in the larger population. In other words, is the difference between his classes and those of the other professors small enough to be due to the random error associated with sampling? A chi-square goodness-of-fit test addresses this question. Here, letter grades, which are ordinal, are treated as categorical. The appropriateness of this will become clear when two grade categories are merged into one.

Step 2 Check the Assumptions. The three assumptions for a chi-square goodness-of-fit test are (1) random sample, (2) independence of observations, and (3) adequate expected frequencies:

1. *Random sample.* With this example, the importance of the random sample assumption is crystal clear. Comparing this sample of grades from two sections to the larger population only makes sense if these two sections have students similar to the students in the other sections. If, for example, both of Dr. Wald's sections were 8 A.M. sections, it is possible that the students who register for an early morning section differ from those who register for classes that meet at a more reasonable time.

The ideal scenario, from an experimenter's standpoint, would be that students indicate a desire to take intro psych and then are randomly assigned to sections. But, of course, that didn't happen, so the random samples assumption is violated. The objective of the assumption, though, is to study a sample that is representative of the population. When the random samples assumption is violated, a researcher can proceed with the study if he or she makes the case that the sample is representative. As there was nothing unusual about his two sections, Dr. Wald is willing to continue with the chi-square goodness-of-fit test.

2. *Independence of observations.* The independence of observations assumption is not violated as each student was in only one section.
3. *Adequate expected frequencies.* All cells must have expected frequencies of at least 5. As will be seen later, each cell has an expected frequency greater than 5, so this assumption is not violated. To achieve this, Dr. Wald put together the small number of D's and F's in one cell.

Step 3 List the Hypotheses. The null hypothesis will state that the distribution of outcomes in the population is the same as is specified. The alternative hypothesis will state that the distribution of outcomes in the population is different from the specified values. The specified values, as percentages, are found in Table 15.5.

H_0: In the population, the distribution of A's, B's, C's, and D's/F's is, respectively, 19%, 33%, 42%, and 6%.

H_1: In the population, the distribution of A's, B's, C's, and D's/F's is not, respectively, 19%, 33%, 42%, and 6%.

Step 4 Set the Decision Rule. Finding the critical value of chi-square depends on what is set as an acceptable risk of Type I error and the number of degrees of freedom. Dr. Wald follows the convention for Type I error and uses $\alpha = .05$. Degrees of freedom are calculated with Equation 15.1, which involves knowing k, the number of categories of outcome that are possible. There are four categories for grades (A, B, C, and D or F), so $k = 4$. To calculate degrees of freedom, apply Equation 15.1:

$$\begin{aligned} df &= k - 1 \\ &= 4 - 1 \\ &= 3 \end{aligned}$$

Looking in Appendix Table 9, the table of critical values of χ^2, at the intersection of the column for $\alpha = .05$ and the row for $df = 3$, one finds that $\chi^2_{cv} = 7.815$. Here is the decision rule:

- If $\chi^2 \geq 7.815$, reject H_0.
- If $\chi^2 < 7.815$, fail to reject H_0.

Step 5 Calculate the Test Statistic. To calculate the chi-square value, a researcher needs to know the observed and expected frequencies. The observed frequencies were shown in Table 15.5, which also displays the percentages for the expected frequencies.

To calculate the expected frequency for A's, use Equation 15.2, where $\%_{\text{Expected}} = 19\%$ and $N = 84$:

$$
\begin{aligned}
f_{\text{Expected}} &= \frac{\%_{\text{Expected}} \times N}{100} \\
&= \frac{19 \times 84}{100} \\
&= \frac{1{,}596.0000}{100} \\
&= 15.9600 \\
&= 15.96
\end{aligned}
$$

(The value 19% can be found in Table 15.5.)

Dr. Wald gave only 10 A's, but if his grade distribution was exactly the same as that found in the entire population, he would have distributed 15.96 A's. The expected frequency for B's is

$$f_{\text{Expected}} = \frac{\%_{\text{Expected}} \times N}{100} = \frac{33 \times 84}{100} = 27.72$$

The expected frequency for C is

$$f_{\text{Expected}} = \frac{\%_{\text{Expected}} \times N}{100} = \frac{42 \times 84}{100} = 35.28$$

To find the expected frequency for the last cell, Dr. Wald takes advantage of the fact that there are 3 degrees of freedom, and once the expected frequencies for three cells are known, the fourth can be determined. The sum of the expected frequencies for all the cells will be the same as N, which in this case is 84. So, Dr. Wald subtracts the expected frequencies for the A, B, and C cells from 84, to find the remainder, which is the expected frequency for the D or F cell:

$$84 - 15.96 - 27.72 - 35.28 = 5.04$$

Table 15.6 shows the observed frequencies and the expected frequencies for the four categories. To calculate the chi-square statistic, Dr. Wald uses Equation 15.3:

$$
\begin{aligned}
\chi^2 &= \Sigma \frac{\left(f_{\text{Observed}} - f_{\text{Expected}}\right)^2}{f_{\text{Expected}}} \\
&= \frac{(10 - 15.96)^2}{15.96} + \frac{(24 - 27.72)^2}{27.72} + \frac{(41 - 35.28)^2}{35.28} + \frac{(9 - 5.04)^2}{5.04} \\
&= \frac{-5.9600^2}{15.96} + \frac{-3.72^2}{27.72} + \frac{5.7200^2}{35.28} + \frac{3.96^2}{5.04} \\
&= \frac{35.5216}{15.96} + \frac{13.8384}{27.72} + \frac{32.7184}{35.28} + \frac{15.6816}{5.04} \\
&= 2.2257 + 0.4992 + 0.9274 + 3.1114 \\
&= 6.7637 \\
&= 6.76
\end{aligned}
$$

For these data, $\chi^2 = 6.76$.

TABLE 15.6 Observed Frequencies and Expected Frequencies for Students in Two Sections of Intro Psych

	A	B	C	D or F	
$f_{Observed}$	10	24	41	9	$\Sigma = 84$
$f_{Expected}$	15.96	27.72	35.28	5.04	$\Sigma = 84.00$

Once the observed frequencies and expected frequencies are known for each category, Equation 15.3 can be used to calculate the chi-square value.

Step 6 Interpret the Results. To decide if the null hypothesis is rejected, Dr. Wald uses the decision rule:

- Is $6.76 \geq 7.815$? If so, reject the null hypothesis and call the results statistically significant.
- If $6.76 < 7.815$? If so, fail to reject the null hypothesis and the results are called not statistically significant.

The second statement is true: 6.76 is less than 7.815. As seen in Figure 15.3, the observed value of chi-square, 6.76, falls in the common zone. Dr. Wald has failed to reject the null hypothesis. There is not enough evidence to conclude that the percentages of different grades in the population differ from what was indicated by the null hypothesis. With a sample size of 84, the discrepancy of the observed frequencies from the expected frequencies was small enough that sampling error could account for it. There is insufficient evidence to conclude that Dr. Wald grades more harshly than do the other intro psych faculty.

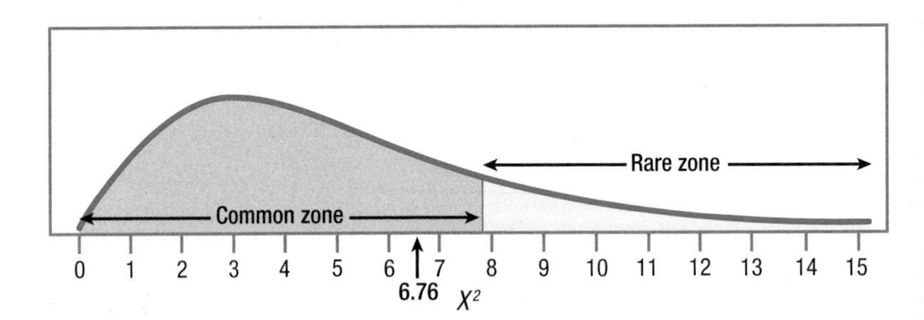

Figure 15.3 Is Dr. Wald Harsh and Unfair? Comparing an Observed Chi-Square Value to the Expected Distribution of Chi-Square Values The observed chi-square value, 6.76, falls in the common zone, meaning insufficient evidence exists to justify rejecting the null hypothesis. There is not enough evidence to conclude that this professor is harsh and unfair.

The results in Figure 15.2 show C's, and D's and F's, turning up more often in Dr. Wald's grades than among the grades of the other instructors. So, it may not seem correct to conclude that insufficient evidence exists to say that Dr. Wald gives more low grades. However, remember that hypothesis testing works like the American legal system: finding a defendant not guilty is different from saying the defendant is innocent. Saying that the evidence is insufficient is not the same as saying Dr. Wald gives the same grades as the other instructors.

In APA format, the results would be reported as

$$\chi^2(3, N = 84) = 6.76, p > .05$$

- χ^2 says that the test is a chi-square test.
- 3 is the degrees of freedom.
- $N = 84$ gives the sample size.
- 6.76 is the calculated value of chi-square.
- .05 indicates that alpha was set at the .05 level.
- $p > .05$ says that the null hypothesis was not rejected. The value of the test statistic (6.76) is a common one. It happens more than 5% of the time when the null hypothesis is true.

Here is a four-point interpretation for these results:

A psychology professor gave more C's, D's, and F's than the other instructors. A chi-square goodness-of-fit test was used to see if he was a harsher grader than the others. The results showed that his grade distribution was not statistically different [$\chi^2(3, N = 84) = 6.76, p > .05$]. This means there is not enough evidence to say that this professor grades differently from the other instructors. If one were to replicate this study, it would be desirable to use a larger sample and to measure characteristics like sex and GPA in order to make sure that the students in his classes are similar to students in the other classes.

Practice Problems 15.1

15.01 People in the world are classified in terms of their natural hair color: black, blonde, brown, red, auburn, chestnut, grey, or white. A hair salon owner keeps track of the natural color of her clients. If she wants to use a chi-square goodness-of-fit test to compare the distribution of hair colors in her salon to the worldwide distribution, how many degrees of freedom will she have?

15.02 If $\alpha = .01$ and $df = 2$, what is χ^2_{cv}?

15.03 Current worldwide estimates are that 51.69% of births are boys and 48.31% are girls. A demographer in the United States obtained a random sample of 5,873 births from all 50 states, determined the sex of each child, and planned to use a chi-square goodness-of-fit test to compare the sex of children born in the United States to the world rate. What are his expected frequencies?

15.04 A researcher is conducting a chi-square goodness-of-fit test on a variable that has two categories. The observed frequencies are 75 for Cell A and 82 for Cell B. The expected frequencies are 62.80 for Cell A and 94.20 for Cell B. Calculate χ^2.

15.05 If $N = 73$, $df = 1$, $\alpha = .05$, and $\chi^2 = 4.72$, write the results in APA format.

15.3 Calculating the Chi-Square Test of Independence

There's a second chi-square test to learn about, the *chi-square test of independence*. Like the chi-square goodness-of-fit test, it uses nominal data (or higher-level data treated as categorical), but it is not a single-sample test. The **chi-square test of**

independence is a nonparametric test that answers the question of whether two or more samples of cases differ on some nominal-level variable. For example, comparing a sample of boys to a sample of girls to see if the nominal variable of having an eating disorder differs between the two sexes calls for a chi-square test of independence. The chi-square test of independence is also known as the chi-square test of association or the chi-square test for contingency tables.

This chi-square test is called a test of "independence" because it answers the question of whether two variables are related to each other or are independent. The chi-square test of independence functions like a difference test ("Do the sexes differ in the prevalence of eating disorders?") and like a relationship test ("Is there a relationship between sex and the presence of an eating disorder?"). This shows that though relationship tests and difference tests may look different, at their core they are the same.

To learn how to conduct a hypothesis test with a chi-square test of independence, imagine the following example: An educational psychologist, Dr. Pradesh, wants to explore whether a student reads a textbook before or after class has an impact on how well he or she does in that class. Dr. Pradesh puts together a random sample of 50 students from introductory psychology classes at her university, then randomly assigns 26 to read the textbook chapters before class and the other 24 to read them after the lectures. At the end of the semester, the students' grades are classified as high (A or B) or low (C, D, or F). The question she asks can be phrased as a relationship question ("Is there a *relationship* between when one reads the text and how well one does in the class?") or a difference question ("Does class performance *differ* depending on when one reads the text?"). These two questions are really the same and both are appropriate for a chi-square test of independence.

Table 15.7 shows the results of Dr. Pradesh's study. The matrix in Table 15.7 is called a **contingency table**, because it shows how the values of the cases on the dependent variable depend on the category of the independent variable. This table illustrates the degree to which students' grades are *contingent on* when they read the text. It is also called a cross-tabulation table because it indicates how the levels of one variable intersect with the levels of the other variable.

There are four cells in this table and each student fits in only one cell. Following convention, the independent variable (when the text is read) is the row variable and

TABLE 15.7 Observed Frequencies for Read the Text Before Lecture vs. Read the Text After Lecture Study

	High Grade	Low Grade	
Read text before lecture	A 20	B 6	26
Read text after lecture	C 12	D 12	24
	32	18	$N = 50$

This contingency table has four cells (A through D), one for each combination of the independent variable (when the text is read) and the dependent variable (grade). The numbers in the far right column (24 and 26) are the frequencies for the rows and the numbers in the bottom row (32 and 18) are the frequencies for the columns.

the dependent variable (high or low grade) is the column variable. Here is what each of the four cells tallies:

- Cell A counts the students who were in the "read before" group and received a high grade.
- Cell B counts the students who were in the "read before" group and received a low grade.
- Cell C counts the students who were in the "read after" group and received a high grade.
- Cell D counts the students who were in the "read after" group and received a low grade.

Contingency tables allow researchers to calculate and compare percentages. 20/26 students who read the text before class (76.92%,) received high grades compared to only 12/24 who read the text after class (50.00%). This difference, 76.92% vs. 50.00%, suggests that reading the text before class makes it more likely one will receive a high grade. However, is the difference statistically significant? To answer that question, Dr. Pradesh needs to use a hypothesis test.

Step 1 Pick a Test

This scenario involves comparing two groups (read text before class vs. read text after class) to see if a difference exists for a variable used to categorize people as good performers vs. poor performers. This calls for a chi-square test of independence.

Step 2 Check the Assumptions

The three assumptions for the chi-square test of independence are the same as for the chi-square goodness-of-fit test:

1. *Random samples.* The samples should be random samples from their populations. The random samples assumption is a robust assumption. If it is violated, one can proceed with the chi-square test of independence, but must be careful about the population to which the results are generalized. In the example, the participants in the read before class vs. read after class study are a random sample from introductory psychology classes who are then randomly assigned to the two groups. The random samples assumption was not violated and the results can be generalized to intro psych classes at this university.
2. *Independence of observations.* The cases in the sample should be independent of each other. That is, the same case can't be in the sample twice. The independence of observations assumption is not robust, so one can't proceed with the planned test if it is violated. In the example, who was in which group was up to chance thanks to random sampling and random assignment, and each student fits in only one cell. The independence of observations assumption was not violated.
3. *Expected frequencies.* All cells must have expected frequencies of at least 5. This assumption is not robust, so the chi-square can't be calculated if the expected frequencies in the cells are small. Until expected frequencies are calculated, this assumption can't be evaluated.

Step 3 List the Hypotheses

The chi-square test of independence tests to see if two variables are independent. Two variables are independent of each other when no relationship exists between them. So, the null hypothesis for a chi-square test of independence states that the two variables are independent of each other *in the population*. The alternative hypothesis for a chi-square test of independence states that, in the population, the two variables are not independent of each other. Notice several things about the alternative hypothesis:

- It doesn't say what direction the relationship goes, whether one variable has a positive or negative impact on the other.
- It doesn't say whether the relationship between the two variables is small or large.
- It just says there is something other than a zero relationship between the two variables in the population.

The simplest way to express the null and alternative hypotheses for the chi-square test of independence is the same way they were expressed for the Pearson r, using ρ, the abbreviation for the population value of a correlation, to say whether there is a relationship between the variables:

$$H_0: \rho = 0$$
$$H_1: \rho \neq 0$$

Step 4 Set the Decision Rule

Setting the decision rule for a chi-square test depends on the alpha level and the number of degrees of freedom (df). For the read before class vs. read after class experiment, Dr. Pradesh was willing to have a 5% chance of making a Type I error, so she set $\alpha = .05$.

Determining df for a chi-square test of independence depends on how many rows (R) and how many columns (C) are in the contingency table. Look back to Table 15.7:

- There are two rows, one for the students who read the text before class and one for the students who read the text after class, so $R = 2$.
- There are two columns, one for students with high grades and one for students with low grades, so $C = 2$.

Equation 15.4 uses R (the number of rows in the contingency table) and C (the number of columns in the contingency table) to calculate the degrees of freedom for a chi-square test of independence.

Equation 15.4 Formula for Degrees of Freedom (df) for a Chi-Square Test of Independence

$$df = (R - 1) \times (C - 1)$$

where df = degrees of freedom
R = number of rows in the contingency table
C = number of columns in the contingency table

For the textbook reading data, there is 1 degree of freedom:

$$\begin{aligned} df &= (R - 1) \times (C - 1) \\ &= (2 - 1) \times (2 - 1) \\ &= 1 \times 1 \\ &= 1 \end{aligned}$$

Now that the degrees of freedom are known, Dr. Pradesh can find the critical value of chi-square in Appendix Table 9. The intersection of the column where $\alpha = .05$ and the row where $df = 1$ shows that $\chi^2_{cv} = 3.841$. If the value of chi-square calculated for the data in the sample is greater than or equal to this critical value, then the results will fall in the rare zone, the null hypothesis is rejected, and the results are called statistically significant. If χ^2 is less than χ^2_{cv}, then χ^2 falls in the common zone, the null hypothesis is not rejected, and the results are called not statistically significant. Here is the decision rule:

- If $\chi^2 \geq 3.841$, reject H_0.
- If $\chi^2 < 3.841$, fail to reject H_0.

Step 5 Calculate the Test Statistic

χ^2 for a chi-square test of independence is calculated with the same formula (Equation 15.3) used to calculate a chi-square goodness-of-fit test. Here it is again.

Equation 15.3 Formula for Calculating Chi-Square (χ^2)

$$\chi^2 = \Sigma \frac{(f_{\text{Observed}} - f_{\text{Expected}})^2}{f_{\text{Expected}}}$$

where χ^2 = chi-square value
f_{Observed} = observed frequency for a category
f_{Expected} = expected frequency for a category

To apply the formula, two values are needed for each cell: (1) the observed frequency, f_{Observed}, and (2) the expected frequency, f_{Expected}. The contingency table in Table 15.7 gives the observed frequencies for the cells, so those are known. Finding the expected frequencies takes Equation 15.5.

Equation 15.5 Formula for Calculating Cell Expected Frequencies (f_{Expected})

$$f_{\text{Expected}} = \frac{N_{\text{Row}} \times N_{\text{Column}}}{N}$$

where f_{Expected} = the expected frequency for a cell
N_{Row} = number of cases in the row with that cell
N_{Column} = number of cases in the column with that cell
N = total number of cases in the contingency table

This formula says that the expected frequency for a cell is found by multiplying together the N for the row that contains the cell by the N for the column that contains the cell. This product is then divided by the total sample size.

Dr. Pradesh applies Equation 15.5 to the read before class vs. read after class data. From Table 15.7, the following is known:

- For the first row, $N_{\text{Row}} = 26$. For the second row, $N_{\text{Row}} = 24$.
- For the first column, $N_{\text{Column}} = 32$. For the second column, $N_{\text{Column}} = 18$.
- $N = 50$.

Using Equation 15.5 to calculate the expected frequency for Cell A finds:

$$\begin{aligned} f_{\text{Expected}A} &= \frac{N_{\text{Row}} \times N_{\text{Column}}}{N} \\ &= \frac{26 \times 32}{50} \\ &= \frac{832.0000}{50} \\ &= 16.6400 \\ &= 16.64 \end{aligned}$$

Continuing to use Equation 15.5, the expected frequencies for the other three cells are found in a similar way.

$$\begin{aligned} f_{\text{Expected}B} &= \frac{N_{\text{Row}} \times N_{\text{Column}}}{N} \\ &= \frac{26 \times 18}{50} \\ &= \frac{468.0000}{50} \\ &= 9.3600 \\ &= 9.36 \end{aligned}$$

$$\begin{aligned} f_{\text{Expected}C} &= \frac{N_{\text{Row}} \times N_{\text{Column}}}{N} \\ &= \frac{24 \times 32}{50} \\ &= \frac{768.0000}{50} \\ &= 15.3600 \\ &= 15.36 \end{aligned}$$

$$\begin{aligned} f_{\text{Expected}D} &= \frac{N_{\text{Row}} \times N_{\text{Column}}}{N} \\ &= \frac{24 \times 18}{50} \\ &= \frac{432.0000}{50} \\ &= 8.6400 \\ &= 8.64 \end{aligned}$$

The four expected frequencies are shown in Table 15.8. There are five things to note:

1. Expected frequencies don't have to be whole numbers. Don't be bothered by saying, for example, that 16.64 cases are expected to fall in the read before/get high grades cell.

2. The row frequencies for the *expected* frequencies (Table 15.8) are exactly the same as those found for the *observed* frequencies (Table 15.7).

3. Similarly, the column frequencies for the expected frequencies are the same as the column frequencies for the observed frequencies.

4. Finally, the total number of cases in the expected frequency cells is the same as the total number of cases in the observed frequency cells.

5. All the expected frequencies were at least 5, so Dr. Pradesh now knows that the third assumption was not violated.

TABLE 15.8 Expected Frequencies for the Read Text Before Lecture vs. Read Text After Lecture Study

	High Grade	Low Grade	
Read text before lecture	A 16.64	B 9.36	26
Read text after lecture	C 15.36	D 8.64	24
	32	18	$N = 50$

Equation 15.16 is used to calculate the expected frequencies for a contingency table. Note that the row frequencies and column frequencies found in this table are the same as those in Table 15.7, the contingency table showing observed frequencies for these data.

A Common Question

Q Does one have to use Equation 15.5 to calculate an expected frequency for each cell?

A No. Equation 15.5 is only needed to calculate expected frequencies for as many cells as there are degrees of freedom. For the read before class vs. read after class study, where $df = 1$, once the first cell is known, the other three can be figured out. If the frequency for the first row is 26 and $f_{Expected} = 16.64$ for Cell A, then Cell B must have an expected frequency of $26 - 16.64$, or 9.36.

Now that the expected frequencies have been found, Equation 15.3 may be used to calculate χ^2. Using the observed frequencies for the read before class vs. read after class data (see Table 15.7) and the expected frequencies (see Table 15.8), the formula finds $\chi^2 = 3.93$:

$$\chi^2 = \Sigma \frac{(f_{Observed} - f_{Expected})^2}{f_{Expected}}$$

$$= \frac{(20 - 16.64)^2}{16.64} + \frac{(6 - 9.36)^2}{9.36} + \frac{(12 - 15.36)^2}{15.36} + \frac{(12 - 8.64)^2}{8.64}$$

$$= \frac{3.3600^2}{16.64} + \frac{-3.3600^2}{9.36} + \frac{-3.3600^2}{15.36} + \frac{3.3600^2}{8.64}$$

$$= \frac{11.2896}{16.64} + \frac{11.2896}{9.36} + \frac{11.2896}{15.36} + \frac{11.2896}{8.64}$$

$$= 0.6785 + 1.2062 + 0.7350 + 1.3067$$

$$= 3.9264$$

$$= 3.93$$

The value of the test statistic, χ^2, for the text-reading data is 3.93. After some more practice with the first five steps, we'll move on to the sixth step of hypothesis testing, interpretation.

Worked Example 15.2 Imagine an elementary school teacher, Mr. Conaway, who is a follower of Carl Rogers and believes that unconditional positive regard leads to psychological health and positive behavior. He obtains a simple random sample of 40 children from other teachers' classrooms at his school and, through home observation, he categorizes each child as receiving unconditional positive regard (1) frequently, (2) sometimes, or (3) rarely. He then has each child's teacher classify the child as a behavior problem or not.

The contingency table for the relationship between the two variables—the grouping variable of frequency of unconditional positive regard and the dependent variable of being a behavior problem—is shown in Table 15.9:

- Five of the 15 frequent recipients of unconditional positive regard (33.33%) were behavior problems.

TABLE 15.9 Frequencies for the Relationship Between Frequency of Unconditional Positive Regard and Being a Behavior Problem, with Row Frequencies and Column Frequencies Added

	Behavior Problem	Not a Behavior Problem	
Frequently receives unconditional positive regard	A 5	B 10	15
Sometimes receives unconditional positive regard	C 5	D 8	13
Rarely receives unconditional positive regard	E 8	F 4	12
	18	22	$N = 40$

This contingency table appears to show that the likelihood of being a behavior problem increases for those students who receive unconditional positive regard less frequently. A chi-square test of independence will be necessary to determine if the effect is a statistically significant one.

- Five of the 13 sometimes recipients of unconditional positive regard (38.46%) were behavior problems.
- Eight of the 12 students who rarely received unconditional positive regard (66.67%) were behavior problems.

These results are graphed in Figure 15.4, which shows that the likelihood of being a behavior problem increases as the frequency of receiving unconditional positive regard decreases. To determine if this is a real effect—or can be explained by sampling error—Mr. Conaway will need a statistical test.

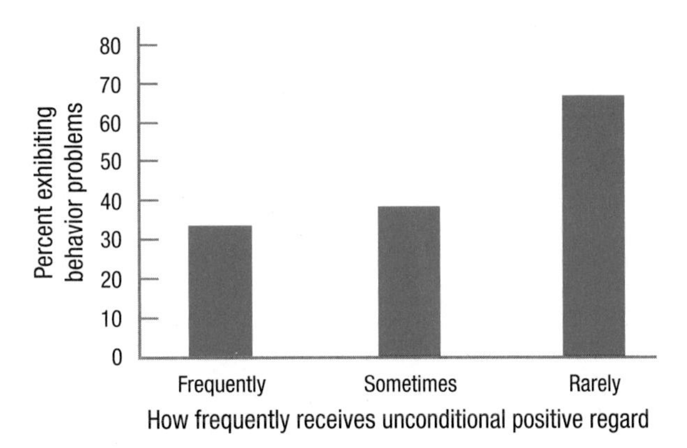

Figure 15.4 Relationship Between Being a Behavior Problem in School and Frequency of Receiving Unconditional Positive Regard These data show that the likelihood of being a behavior problem goes up as the frequency of receiving unconditional positive regard goes down. Whether the effect is a statistically significant one or can be explained by sampling error will require the completion of a chi-square test of independence.

Step 1 Pick a Test. A chi-square test of independence is appropriate to determine if there is a relationship between three categories of one variable (frequency of receiving unconditional positive regard) and two categories of another variable (behavior problem or not a behavior problem) in a sample of elementary school children.

Step 2 Check the Assumptions. There are three assumptions for the chi-square test of independence:

1. *Random samples.* The cases come from a number of classrooms at the school, but it is not stated that the sample is a random sample. The random samples assumption is violated, but it is robust so the teacher can still proceed with the chi-square test of independence. However, he should be careful about generalizing the results beyond this sample.
2. *Independence of observations.* Each child is in the sample only once, so this assumption is not violated.
3. *Adequate expected frequencies.* All cells must have expected frequencies of at least 5. Until the expected frequencies are calculated, one can't be sure that this assumption is met. If it turns out not to be met, one could collapse some

categories and, for example, the frequent recipients of unconditional positive regard could be compared to a combined group of the sometimes and rare recipients of unconditional positive regard.

Step 3 List the Hypotheses. The hypotheses are listed below. The null hypothesis says that no relation exists in the population between the amount of unconditional positive regard one receives and the likelihood of being a behavior problem. The alternative hypothesis states there is some relationship in the population.

$$H_0: \rho = 0$$

$$H_1: \rho \neq 0$$

Step 4 Set the Decision Rule. To set the decision rule, a researcher needs to know the alpha and degrees of freedom. Mr. Conaway has set alpha at .05. He uses Equation 15.4 to calculate the degrees of freedom:

$$\begin{aligned} df &= (R - 1) \times (C - 1) \\ &= (3 - 1) \times (2 - 1) \\ &= 2 \times 1 \\ &= 2 \end{aligned}$$

Looking in the table of critical values of chi-square, Appendix Table 9, at the intersection of the column for $\alpha = .05$ and the row for $df = 2$, he finds $\chi^2_{cv} = 5.991$. The decision rule is:

- If $\chi^2 \geq 5.991$, reject H_0.
- If $\chi^2 < 5.991$, fail to reject H_0.

Step 5 Calculate the Test Statistic. The first step in calculating a chi-square test for independence (χ^2) is using Equation 15.5 to calculate expected frequencies. This contingency table has six cells and it is a bit tedious to perform Equation 15.5 six times. So, Mr. Conaway will use a bit of logic and just use Equation 15.5 twice. Why twice? Because the contingency table has 2 degrees of freedom, and once two cells are known, the rest can be calculated. The two to be calculated have to be chosen a little carefully, so Mr. Conaway has picked Cells A and C.

$$\begin{aligned} f_{\text{Expected}A} &= \frac{N_{\text{Row}} \times N_{\text{Column}}}{N} \\ &= \frac{18 \times 15}{40} \\ &= \frac{270.0000}{40} \\ &= 6.7500 \\ &= 6.75 \end{aligned}$$

$$f_{\text{Expected}C} = \frac{N_{\text{Row}} \times N_{\text{Column}}}{N} = \frac{18 \times 13}{40} = \frac{234.0000}{40} = 5.8500 = 5.85$$

Table 15.10 contains the results so far. Now comes the logic.

- If the first row has a total of 15 cases and Cell A has 6.75 of them, then Cell B contains 15 – 6.75 cases, or 8.25 cases.
- The same logic applies to Cell D: 13 – 5.85 = 7.15 cases.
- To find the expected frequency for Cell E, look at the total number of cases in the column, 18. Subtracting the expected frequencies for Cells A and C from this yields the expected frequency of 5.40 for Cell E.
- Finally, for Cell F, subtract Cell E (5.40) from 12, the total number of cases in the row to find 6.60.

TABLE 15.10 Expected Frequencies for Two Cells of the Contingency Table for the Unconditional Positive Regard Frequency/Behavior Problem Data

	Behavior Problem	Not a Behavior Problem	
Frequently receives unconditional positive regard	A 6.75	B	15
Sometimes receives unconditional positive regard	C 5.85	D	13
Rarely receives unconditional positive regard	E	F	12
	18	22	$N = 40$

The row frequencies and column frequencies are set so the expected frequencies for the other cells in this contingency table can be determined once the expected frequencies for two of the cells are known.

The expected frequencies for all six cells are shown in Table 15.11. Note that all the expected frequencies are greater than 5. Mr. Conaway can finally determine that the remaining assumption, that all expected frequencies were large enough, was not violated. The expected frequencies, in conjunction with the observed frequencies (Table 15.9), are needed to calculate the value of the test statistic using Equation 15.3:

$$\chi^2 = \Sigma \frac{\left(f_{\text{Observed}} - f_{\text{Expected}}\right)^2}{f_{\text{Expected}}}$$

$$= \frac{(5-6.75)^2}{6.75} + \frac{(10-8.25)^2}{8.25} + \frac{(5-5.85)^2}{5.85} + \frac{(8-7.15)^2}{7.15} + \frac{(8-5.40)^2}{5.40} + \frac{(4-6.60)^2}{6.60}$$

$$= \frac{-1.75000^2}{6.75} + \frac{1.75^2}{8.25} + \frac{-0.85^2}{5.85} + \frac{0.85^2}{7.15} + \frac{2.60^2}{5.40} + \frac{-2.60^2}{6.60}$$

$$= \frac{3.0625}{6.75} + \frac{3.0625}{8.25} + \frac{0.7225}{5.85} + \frac{0.7225}{7.15} + \frac{6.7600}{5.40} + \frac{6.7600}{6.60}$$

$$= 0.4537 + 0.3712 + 0.1235 + 0.1010 + 1.2519 + 1.0242$$

$$= 3.3255$$

$$= 3.33$$

TABLE 15.11 Expected Cell Frequencies for the Contingency Table for the Unconditional Positive Regard Frequency/Behavior Problem Data

	Behavior Problem	Not a Behavior Problem	
Frequently receives unconditional positive regard	A 6.75	B 8.25	15
Sometimes receives unconditional positive regard	C 5.85	D 7.15	13
Rarely receives unconditional positive regard	E 5.40	F 6.60	12
	18	22	$N = 40$

This table contains each cell's expected frequency. Note that the row frequencies and column frequencies here are the same as in Table 15.9, the contingency table with observed frequencies.

The chi-square test of independence value for the unconditional positive regard study is 3.33. Now it is time to learn how to interpret the results of a chi-square test of independence.

Practice Problems 15.2

15.06 Given a chi-square test of independence where the explanatory variable has four categories and the dependent variable has three, (a) How many degrees of freedom are there? (b) If $\alpha = .05$, what is χ^2_{cv}?

15.07 Given the cell frequencies below, calculate (a) the row totals and (b) the column totals.

5	10	12
5	8	10

15.08 The row and column totals are shown below. Calculate the expected frequencies for the cells.

A	B	39
C	D	49
28	60	

15.09 Given the information below, calculate χ^2.

$f_{Observed} = 50$ $f_{Expected} = 47.22$	$f_{Observed} = 36$ $f_{Expected} = 38.78$
$f_{Observed} = 45$ $f_{Expected} = 47.77$	$f_{Observed} = 42$ $f_{Expected} = 39.23$

15.4 Interpreting the Chi-Square Test of Independence

Now that χ^2 has been calculated for a chi-square test of independence, it is time to move on to the sixth step of hypothesis testing and interpret the results. To interpret a chi-square test of independence, there are three questions to answer: (1) Was the null hypothesis rejected? (2) If so, what is the direction of the difference? (3) How big is the effect?

Let's use the study by the educational psychologist Dr. Pradesh to see how to answer those questions for a chi-square test of independence. Dr. Pradesh's study compared the grades, high vs. low, for 50 students who were randomly assigned to read the text either before or after class.

Step 6 Interpret the Results

Was the Null Hypothesis Rejected?

For Dr. Pradesh's study, alpha was set at .05, there was 1 degree of freedom, and the critical value of chi-square was 3.841. Dr. Pradesh calculated $\chi^2 = 3.93$. Applying the decision rule $3.93 \geq 3.841$, so the null hypothesis is rejected, the alternative hypothesis is accepted, and the results are considered statistically significant. The results can be phrased two ways:

1. There is a relationship between when a student reads the text and what grade he or she receives in the class.
2. A statistically significant difference exists in grades between the students who read the text before class vs. those who read the text after class.

What Is the Direction of the Difference?

Both statements above are true, but neither is satisfying because neither reveals the direction of the difference, whether it is better to read the text before class or after class. With Dr. Pradesh's study, it is easy to determine the direction of the difference because there are only two groups.

By examining the results in the samples, Dr. Pradesh can draw a conclusion about the direction for the populations. Twenty of the 26 students in the read-before-class group (76.92%) received high grades compared to 12 of the 24 students (50.00%) of the read-after-class group. So, Dr. Pradesh concludes, for the larger population of introductory psychology students at this university, that completing the textbook readings before class leads to a better grade outcome than completing the reading after class.

APA Format

The results are reported in APA format for the chi-square test of independence the same way they are for a chi-square goodness-of-fit test. Here are the six pieces of information for the read before class vs. read after class study:

$$\chi^2(1, N = 50) = 3.93, p < .05$$

- χ^2 indicates that the test statistic was a chi-square.
- 1 gives the degrees of freedom.
- $N = 50$ tells how many cases there were.
- 3.93 is the value of the test statistic.
- .05 provides information about the alpha level selected.
- And, $p < .05$ means that the null hypothesis was rejected.

How Big Is the Effect?

Determining the size of the effect for a chi-square test of independence involves transforming the chi-square value into Cramer's *V*, a statistic that is a lot like a Pearson *r*:

- Cramer's *V* ranges from 0 to 1.
- As Cramer's *V* gets closer to 1, the effect is stronger.
- As Cramer's *V* gets closer to 0, the effect is weaker.

The formula for calculating Cramer's *V* is shown in Equation 15.7.

Equation 15.7 Formula for Calculating Cramer's *V*

$$V = \sqrt{\frac{\chi^2}{N \times df_{RC}}}$$

where V = Cramer's V
χ^2 = chi-square value, calculated via Equation 15.3
N = total number of cases in the contingency table
df_{RC} = $(R - 1)$ or $(C - 1)$, whichever is smaller
(R = number of rows in the contingency table;
C = number of columns in the contingency table)

Here's how to calculate *V* for the read before class vs. read after class study. It is already known that $\chi^2 = 3.93$ and $N = 50$. To calculate *V* for this equation, one needs to know the number of rows (2) and the number of columns (2) in the contingency table. The degrees of freedom for Equation 15.7 are either $R - 1$, which is $2 - 1$, or $C - 1$, which is also $2 - 1$, whichever is smaller. Both are the same, so $df_{RC} = 1$.

$$V = \sqrt{\frac{\chi^2}{N \times df_{RC}}}$$

$$= \sqrt{\frac{3.93}{50 \times 1}}$$

$$= \sqrt{\frac{3.93}{50}}$$

$$= \sqrt{0.0786}$$

$$= .2804$$

$$= .28$$

Cramer's *V* for the text-reading data is .28. When there is only 1 degree of freedom, as here, Cramer's *V* is equivalent to a Pearson *r*. So, Dr. Pradesh could say that there is a .28 correlation between the independent variable (when the text is read) and the dependent variable (high grade or low grade). This is a medium-strength correlation. Guidelines for interpreting Cramer's *V* are given in Table 15.12.

- When $df_{RC} = 1$, Cramer's *V* is interpreted like a Pearson *r* in terms of what is a small, medium, and large effect size.
- As df_{RC} increases, the criteria for considering an effect size meaningful become more lenient.

For the read before class vs. read after class study, the effect, $V = .28$, would be considered a medium one.

TABLE 15.12 Guidelines for Interpreting Cramer's *V*

	Small Effect Size	Medium Effect Size	Large Effect Size
$df_{RC} = 1$	.10	.30	.50
$df_{RC} = 2$	.07	.21	.35
$df_{RC} = 3$	.06	.17	.29
$df_{RC} = 4$	.05	.15	.25
$df_{RC} = 5$	.05	.13	.22

These effect sizes are from Cohen (1988). Note that when $df_{RC} = 1$, the effect size for Cramer's *V* is the same as it is for a Pearson *r*. As degrees of freedom increase, a smaller *V* is interpreted as a larger effect.

Putting It All Together

Here is Dr. Pradesh's four-point interpretation. In it she (1) tells what the study was about, (2) gives the results, (3) explains what they mean, and (4) makes suggestions for future research.

> This study investigated whether completing readings before or after class has an impact on the grade a student receives. Fifty students in introductory psychology classes were randomly assigned to complete the readings before class or after it. Reading the text before attending class resulted in a significantly higher percentage of students receiving high grades: 77% vs. 50%, $\chi^2(1, N = 50) = 3.93$, $p < .05$. The impact on grades of when the reading is completed is more than small, but less than large. To increase the generalizability of results and certainty about the size of the effect, the study should be replicated in courses other than psychology and at other universities.

(Though the results do sound reasonable, please be aware that these data were manufactured for this example.)

Worked Example 15.3 For an example demonstrating how to interpret the chi-square test of independence results that are not statistically significant, a return to Mr. Conaway's study of the relationship between the frequency of receiving unconditional positive regard and being a behavior problem is in order. In that study, Mr. Conaway, an elementary school teacher, "measured" 40 children. The grouping variable was how frequently the child received unconditional positive regard (frequently, sometimes, or rarely) and the dependent variable was whether the child's teacher rated him or her as a behavior problem (yes or no). Alpha was set at .05, there were 2 degrees of freedom, and the critical value of chi-square was 5.991. Using Equation 15.3, Mr. Conaway found $\chi^2 = 3.33$.

Was the null hypothesis rejected? Here is the decision rule:

- If $\chi^2 \geq 5.991$, reject H_0.
- If $\chi^2 < 5.991$, fail to reject H_0.

3.33 < 5.991, so the results fall in the common zone (see Figure 15.5), the null hypothesis is not rejected, and the results are called "not statistically significant." There is not enough evidence to conclude that a relationship exists between the frequency with which elementary school children received unconditional positive regard at home and their being a behavior problem. Alternatively, one could say that there is not enough evidence to conclude that the rate of behavior problems differs among children who receive unconditional positive regard frequently, sometimes, or rarely. That is, the conclusion could be stated in either relationship test terms or difference test terms.

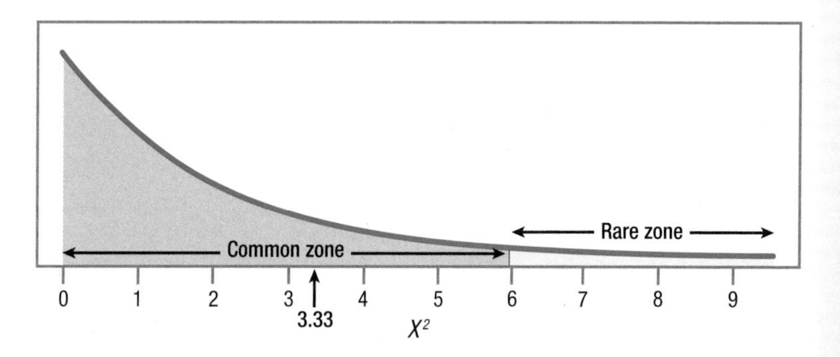

Figure 15.5 Chi-Square Results for the Effect of Frequency of Unconditional Positive Regard on Being a Behavior Problem The observed value of χ^2, 3.33, falls in the common zone of the sampling distribution of chi-square with $df = 2$ and $\alpha = .05$. This means that there is insufficient evidence to justify rejecting the null hypothesis.

Direction of the difference. Because not enough evidence is available to say that a difference exists, there is no need to comment on the direction of the difference.

APA format. In APA format, these results would be written as

$$\chi^2(2, N = 40) = 3.33, p > .05$$

- χ^2 indicates that the statistic calculated is a chi-square value.
- 2 indicates that there are 2 degrees of freedom.
- $N = 40$ tells the sample size.
- 3.33 is the chi-square value found in the sample.
- .05 indicates that alpha was set at .05.
- $p > .05$ indicates that the results fell in the common zone, that the null hypothesis was not rejected.

How big is the effect? Cramer's *V*, the effect size for a chi-square test of independence, should be calculated even when results are not statistically significant. Doing so allows a researcher to evaluate the likelihood of a Type II error. To calculate *V*, one needs the observed chi-square value, sample size, and degrees of freedom (Equation 15.7). For his data, Mr. Conaway already knows $\chi^2 = 3.33$ and $N = 40$. To find df_{RC}, he needs to subtract 1 from the number of rows or 1 from the number of columns, whichever is smaller. There are fewer columns (two) than rows (three), so $df_{RC} = 2 - 1 = 1$. Next, he plugs the numbers into Equation 15.7 to calculate V:

$$V = \sqrt{\frac{\chi^2}{N \times df_{RC}}}$$

$$= \sqrt{\frac{3.33}{40 \times 1}}$$

$$= \sqrt{\frac{3.33}{40}}$$

$$= \sqrt{.0833}$$

$$= .2886$$

$$= .29$$

According to Table 15.13, $V = .29$ is a medium effect size. This medium effect conflicts with the fact that there is insufficient evidence for an effect. In such situations, the first explanation that comes to mind is the possibility of a Type II error. Perhaps there really is an effect and, due to sampling error, Mr. Conaway failed to find it. This could have happened because the study was underpowered. To determine if an effect does exist, the study should be replicated with a larger sample size.

Putting it all together. Here is Mr. Conaway's four-point interpretation:

> Data from 40 elementary school students were collected to see if the frequency with which they received unconditional positive regard at home was related to their being considered a behavior problem at school. Though the likelihood of being a behavior problem increased as the likelihood of receiving unconditional positive regard decreased, there was not sufficient evidence to conclude that a relationship exists between these two variables [$\chi^2(2, N = 40) = 3.33, p > .05$]. Failure to find an effect could have been due to the small sample size. In order to have adequate power to determine if a relationship exists, it would be a good idea to replicate this study with a larger sample size.

Practice Problems 15.3

15.10 If $\chi^2 = 4.76$, $df = 1$, $N = 36$, and $\alpha = .05$, write the results in APA format.

15.11 If $N = 45$, $\chi^2 = 6.78$, and $df_{RC} = 2$, calculate V.

15.12 A human factors psychologist randomly assigned volunteers to drive on a twisty mountain road in a driving simulator. The control group drove in a normal car with a dashboard. The experimental group drove in a car with a heads-up display, where the dashboard information was projected onto the windshield. The dependent variable was whether the group crashed the car or not. Given the contingency table below, $\chi^2(1, N = 180) = 5.27$, $p < .05$, and Cramer's $V = .17$, write a four-point interpretation:

	Good Outcome: Did Not Crash	Bad Outcome: Crashed
Control Group	42	43
Experimental Group	63	32

15.5 Other Nonparametric Tests

In this brief section, two other nonparametric tests are introduced: the Spearman rank-order correlation coefficient and the Mann–Whitney *U*.

The Spearman Rank-Order Correlation Coefficient

The **Spearman rank-order correlation coefficient** (abbreviated Spearman *r* or r_s) is the nonparametric version of the Pearson correlation coefficient. The Spearman *r* examines the relationship between (1) two ordinal-level variables, (2) one ordinal-level variable and one interval/ratio-level variable, or (3) as a fallback option with two interval/ratio-level variables when the assumptions for the Pearson *r* have been violated.

One can think of the Spearman *r* as a Pearson *r* in which the correlation is calculated for data that are converted to ranks. This means that a researcher can directly apply the hypothesis-testing steps about hypotheses, decision rules, calculation, and interpretation from the Pearson *r* to the Spearman *r*—as long as he or she remembers that ordinal, or ranked, data are being used.

For an example of how the Spearman *r* works, imagine a psychologist who examined the relationship between romantic attractiveness and academic success in high school students. His measure of romantic attractiveness was an interval scale that ranged from 0 (low) to 10 (high). Class rank, an ordinal measure, was the measure of academic success. He obtained both of these measures on a random sample of 24 seniors, and Panel A in Figure 15.6 shows the relationship between them. It appears as if students who are doing less well academically are more romantically attractive.

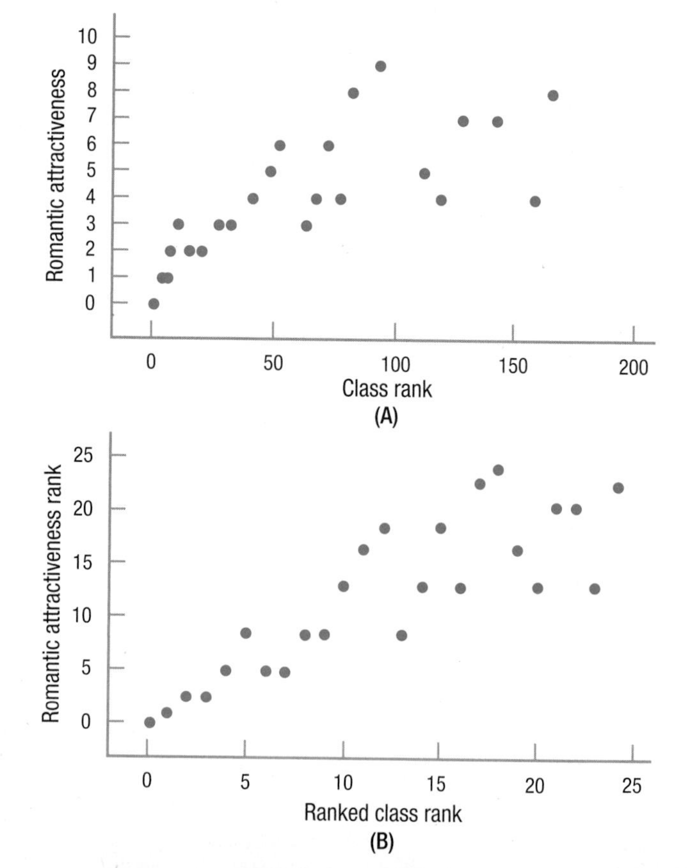

Figure 15.6 Relationship Between Academic Success and Romantic Attractiveness for Unranked Data and Ranked Data Both scatterplots show the relationship between a student's class rank and his or her desirability as a romantic partner. In the top graph (A), the naturally occurring range is used for each variable. In the bottom graph (B), the variables have been converted to ranks. For example, in Panel A, the *X*-axis ranges from 0 to 200, which is sufficient to capture the person with the highest class rank, 165. In Panel B, when the class ranks are ranked from 1 to 24, the person with a rank of 165 now has a score of 24 and the *X*-axis only has to go up to 25.

Class rank is an ordinal-level variable, so a Pearson *r* is not appropriate to use to analyze these data. Instead, a Spearman *r* is the appropriate test. To conduct the Spearman *r*, each variable is assigned a rank from 1 to 24 and the rankings are correlated. Why is each variable assigned a rank from 1 to 24? Because there are 24 cases in the sample. The scatterplot for the ranked data is displayed in Panel B of Figure 15.6, and it shows a very similar pattern to Panel A.

Look at class rank on the *X*-axis in the top panel of Figure 15.6. For these 24 students, it ranges from 1 to 165 and the *X*-axis goes all the way up to 200 to accommodate this. Then look at the *X*-axis in the bottom panel—instead of ranging from 0 to 180, it ranges from 0 to 25. That's because the Spearman *r* takes class rank, which was already an ordinal-level variable, and turns it into pure ranks. The student with the worst class rank in this sample, who had a class rank of 165, now has a score of 24, which indicates the worst class rank in the sample.

Similarly, the *Y*-axis variable, romantic attractiveness, is converted into ranks. In Panel A, the scores ranged from 0 to 10. In Panel B, when the scores are ranked, the highest possible score (for the most attractive person) is now 24, not 10.

The Spearman *r* is just like the Pearson *r*, except that it correlates ranks, not raw scores. The difference between the two can be seen in the axes of the two scatterplots in Figure 15.6. In addition, look closely at the shape of the two scatterplots in Figure 15.6—they are similar, but not exactly the same. Converting raw scores to ranks can change the shape. Sometimes, as here, it makes the relationship appear stronger. Other times it makes the relationship appear weaker. However, making the relationship stronger or weaker isn't the objective. The objective is to use the appropriate statistical test for the data. If correlating two ordinal variables, one ordinal variable with one interval or ratio variable, or two interval and/or ratio variables where it's not possible to proceed with a Pearson *r*, then the Spearman rank-order correlation coefficient is the appropriate test.

The Mann–Whitney *U* Test

The Mann–Whitney *U* test is the nonparametric equivalent of the independent-samples *t* test. It is used to compare two independent populations when the dependent variable is measured at the ordinal level, or as a fallback test when nonrobust assumptions for an independent-samples *t* test have been violated. Because it uses an ordinal outcome variable, the Mann–Whitney *U* test determines if the median of one population is significantly different from the other.

The Mann–Whitney *U* works by combining the two independent samples and then assigning a rank to each case based on its standing in the pooled group. The two samples are then separated, and the sum of the ranks for each of the samples is calculated and used to calculate *U*. If the two populations have similar scores on the dependent variable, then each sample will have a mixture of high ranks and low ranks. However, if the two populations differ, then one will have more high ranks than the other. When this happens, the *U* value will reflect the difference in ranks between the samples, and the null hypothesis of no difference in the medians will be rejected.

Here's an example where the difference between the two samples is very obvious. Table 15.13 shows the speeds, in miles per hour, of the five fastest tennis serves by men and the five fastest serves by women. Clearly, the fastest male tennis serves are faster than the fastest female tennis serves. Let's see how the Mann–Whitney *U* test would show this.

TABLE 15.13 Five Fastest Tennis Serves by Men and by Women

Men	Women
163	129
156	128
155	126
155	126
153	125

These are the speeds, in miles per hour, of the five fastest serves by men and the five fastest serves by women. (From Wikipedia, the source of all contemporary knowledge.)

TABLE 15.14 Fastest Tennis Serves by Men and Women, Combined and Ranked

Speed	Sex	Rank
163	Male	1
156	Male	2
155	Male	3.5
155	Male	3.5
153	Male	5
129	Female	6
128	Female	7
126	Female	8.5
126	Female	8.5
125	Female	10

The data from the two samples in Table 15.13 are combined here in order. It is clear that the five men have faster serves than the five women. And, the sum of the ranks for the two sexes, 15 vs. 40, reflects this.

In Table 15.14, the two samples are combined into one group and the cases are ordered from fastest to slowest. The rank of 1 is given to the fastest serve and the rank of 10 to the slowest serve. There is no overlap in the ranks assigned to men and women. Ranks 1 to 5 belong to the men and ranks 6 through 10 to the women. The sum of the men's ranks, 15, and the sum of the women's ranks, 40, are very different numbers. And, the U value that the Mann–Whitney U test would calculate for these data would reflect as much.

That's how the Mann–Whitney U test works. It combines the two samples into one group, assigns ranks to the whole group, then puts the cases back into their original samples, and compares the ranks. If the ranks are evenly distributed between the two samples, then there is no evidence that the populations differ. But if one sample has a lot of high ranks and the other a lot of low ranks, then the difference between the populations is statistically significant.

Practice Problems 15.4

For these practice questions, select the appropriate statistical test from the following: Pearson r, Spearman r, independent-samples t test, or Mann–Whitney U test.

15.13 A psychologist wanted to examine the relationship, in adolescents, between days of drug use in the past year and IQ score. Days of drug use turned out to be very positively skewed.

15.14 The psychologist then divided the adolescents into two groups, those with no days of drug use over the past year and those with one or more days. She wanted to compare the two groups in terms of IQ.

15.15 A sports economist wanted to know if countries that invested more in their sports programs did better in the Olympics. He divided countries into those that competed in 10 or more sports in the summer Olympics and those that competed in fewer than 10 sports. For each country, he found its medal count rank, and he divided this by the number of sports in which the country competed to obtain his dependent variable, rank per events entered.

Application Demonstration

To see chi-square in action, let's analyze some data from a real-life study. A group of researchers was interested in determining if survival after surgery was influenced by a person's nutritional status before surgery. They took almost 400 patients about to undergo surgery for kidney cancer and classified them as malnourished or not. Surgery was then performed and the researchers followed the patients for three years to see how many lived and how many died (Morgan et al., 2011).

Step 1 Pick a Test. Though the researchers used more complex statistical techniques, we'll apply a chi-square test of independence to the data:

- The grouping variable, nutritional status, is nominal with two categories: (1) malnourished or (2) not malnourished.
- The dependent variable, survival three years later, is nominal with two categories: (1) alive or (2) dead.

Table 15.15 displays the results as a contingency table. The table shows cell frequencies, as well as row percentages and column percentages. Eighty-five of the 369 patients (23%) were classified as malnourished and 76 of the 369 patients (21%) died within three years.

TABLE 15.15 Contingency Table for the Effect of Nutritional Status on Mortality

	Alive	Dead	
Malnourished	A 50 (58.82%)	B 35 (41.18%)	85 (23.04%)
Not Malnourished	C 243 (85.56%)	D 41 (14.44%)	284 (76.96%)
	293 (79.40%)	76 (20.60%)	$N = 369$

The percentages in parentheses are row percentages and column percentages. The percentages in the cells are for that row only. The mortality rate for the malnourished patients (41%) was almost 3 times higher than for the patients who were not malnourished (14%). Data from Morgan et al., 2011.

The percentages in the cells suggest that being malnourished prior to surgery diminishes one's chance of surviving for three years after surgery: 35 of the 85 malnourished patients died—that's 41%—compared to only 41 of the 284 not malnourished patients, or 14%. In other words, there appears to be a relationship between malnourishment status and mortality status. But, is this a statistically significant relationship? To answer that question about these nominal-level variables, a chi-square test of independence is required.

Step 2 Check the Assumptions.

- *Random samples.* The sample was not a random sample, leaving uncertainty about the population to which the results can be generalized. The sample is large; it might be representative of patients with kidney cancer at this hospital.

Whether these patients are like patients at other hospitals or in other states and countries is unknown. The random samples assumption is robust to violation, so it's OK to proceed with the planned chi-square test of independence.

- *Independence of observations.* No evidence exists that the patients influenced each other. There's no reason to believe any patients were in the study twice. The independence of observations assumption does not appear to have been violated.
- *Adequate expected frequencies.* All cells must have expected frequencies of at least 5. This assumption can't be assessed until the expected frequencies are calculated.

Step 3 List the Hypotheses. The null hypothesis says that malnutrition status and survival are independent and the alternative hypothesis states that they have some relationship:

$$H_0: \rho = 0$$
$$H_1: \rho \neq 0$$

Step 4 Set the Decision Rule. The alpha level is set at .05, as usual. The degrees of freedom are calculated using Equation 15.4:

$$\begin{aligned} df &= (R - 1) \times (C - 1) \\ &= (2 - 1) \times (2 - 1) \\ &= 1 \times 1 \\ &= 1 \end{aligned}$$

Use Appendix Table 9 to find the critical value of chi-square, $\chi^2_{cv} = 3.841$. Here is the decision rule:

- If $\chi^2 \geq 3.841$, reject H_0 and say the results are statistically significant.
- If $\chi^2 < 3.841$, fail to reject H_0 and say the results are not statistically significant.

Step 5 Calculate the Test Statistic. To save time, the expected frequencies for the cells are shown in Table 15.16. Equation 15.5 was used to calculate the expected frequency for Cell A and then logic was utilized to find the expected frequencies for the other three cells. None of the expected frequencies is less than 5, so it's now known that the third assumption was not violated.

TABLE 15.16 Expected Frequencies for the Effect of Nutritional Status on Mortality

	Alive	Dead	
Malnourished	A 67.50	B 17.50	85
Not Malnourished	C 225.50	D 58.50	284
	293	76	$N = 369$

Equation 15.6 was used to find the expected frequency for Cell A. There is 1 degree of freedom in this contingency table, so once one cell frequency is known, the other cell frequencies can be determined from the column frequencies and row frequencies.

Using Equation 15.3, the chi-square value is calculated as 28.63:

$$\chi^2 = \sum \frac{\left(f_{\text{Observed}} - f_{\text{Expected}}\right)^2}{f_{\text{Expected}}}$$

$$= \frac{(50-67.50)^2}{67.50} + \frac{(35-17.50)^2}{17.50} + \frac{(243-225.50)^2}{225.50} + \frac{(41-58.50)^2}{58.50}$$

$$\text{Step 1} = \frac{-17.5000^2}{67.50} + \frac{17.5000^2}{17.50} + \frac{17.5000^2}{225.50} + \frac{-17.5000^2}{58.50}$$

$$\text{Step 2} = \frac{306.2500}{67.50} + \frac{306.2500}{17.50} + \frac{306.2500}{225.50} + \frac{306.2500}{58.50}$$

$$\text{Step 3} = 4.5370 + 17.5000 + 1.3581 + 5.2350$$

$$\text{Step 4} = 28.6301$$

$$= 28.63$$

Step 6 Interpret the Results. The calculated value of chi-square (28.63) is greater than the critical value (3.841), so the null hypothesis is rejected and the results are written in APA format as $\chi^2(1, N = 369) = 28.63, p < .05$. A statistically significant difference exists between the mortality rate for malnourished patients and patients who were not malnourished. There are only two groups, so we know the direction of the difference: the mortality rate is higher for malnourished patients than for patients who were not malnourished, 41% vs. 14%.

Using Equation 15.7, Cramer's V is calculated as .28:

$$V = \sqrt{\frac{\chi^2}{N \times c}}$$

$$= \sqrt{\frac{28.63}{369 \times 1}}$$

$$= \sqrt{\frac{28.63}{369}}$$

$$= \sqrt{0.776}$$

$$= 0.2786$$

$$= .28$$

By Cohen's standards (1988), this is a medium effect. However, Cohen's standards are meant as guidelines. A change in the mortality rate from 14% to 41%—almost a threefold increase—seems like a strong effect, especially as death is the outcome.

Putting it all together.

Data were analyzed from a study in which researchers compared the mortality rate for kidney cancer patients who were and were not malnourished prior to surgery. Three years after surgery, 41% of the malnourished patients had died compared to only 14% of the patients who were not malnourished. This almost threefold increase in mortality was a statistically significant difference [$\chi^2(1, N = 369) = 28.63, p < .05$]. The effect seems to be a strong one. For kidney cancer patients undergoing surgery, being malnourished prior to surgery is associated with almost a threefold increase in the risk of dying in the next three years. In future studies, it would be important to assess the reason for the malnourishment. If it resulted from having a more aggressive form of cancer, it might be the cancer, not the malnourishment, that led to the increase in mortality.

DIY

What can grocery shopping tell us about gender equality? Do men do more shopping than women? Are women more likely than men to be accompanied by children?

Make a data collection sheet like the one below. Then go to a grocery store during the day and put a tally mark in Cell A for every woman you see shopping with children. A mark goes in Cell B for every woman shopping by herself. Similarly, use Cell C to keep track of men shopping with children and D for men shopping alone.

Add together Cells A and B, and Cells C and D. Use a chi-square goodness-of-fit test with expected percentages of 50% and 50% to see if men and women share shopping responsibility equally. Do the results say something about contemporary gender roles?

	Shopping with Children	Shopping without Children
Female Shoppers	A	B
Male Shoppers	C	D

Next, let's see if one sex is more likely to shop with children. To do so, use the values in all four cells as the observed frequencies for a chi-square test of independence.

SUMMARY

Differentiate parametric tests from nonparametric tests.

- Parametric tests have interval or ratio outcome variables that must be normally distributed. Nonparametric tests can be done on nominal or ordinal outcome variables and don't require a normal distribution. Less restricted by assumptions, nonparametric tests often have less power than parametric tests.

Calculate and interpret a chi-square goodness-of-fit test.

- The chi-square goodness-of-fit test is a nonparametric, single-sample test for use with a nominal outcome variable. It compares the observed frequencies to the expected frequencies to calculate a chi-square value. Interpretation concerns whether the null hypothesis was rejected and, if so, what the direction of the difference was.

Calculate and interpret a chi-square test of independence.

- A chi-square test of independence determines whether two or more populations differ on a nominal-level outcome variable. Data for it are arrayed in a contingency table, which shows the degree to which the outcome variable is determined by the explanatory variable. The chi-square value is calculated from the discrepancy between the observed frequencies and the expected frequencies. In the interpretation, three questions are addressed: (1) whether the null hypothesis was rejected, (2) what the direction of the difference is, and (3) how big the effect is.

Know when to use a Spearman rank-order correlation coefficient and a Mann–Whitney *U* test.

- The Spearman rank-order correlation coefficient, used with ordinal-level data, is the nonparametric alternative to Pearson *r*, the correlation coefficient used with interval/ratio-level data. The Mann–Whitney *U* test, used with an ordinal outcome variable, is the nonparametric alternative to an independent-samples *t* test.

KEY TERMS

chi-square goodness-of-fit test – a nonparametric, single-sample test used to compare the distribution of a categorical (nominal- or ordinal-level) outcome variable in a sample to a known population value.

chi-square test of independence – a nonparametric test used to determine whether two or more populations of cases differ on a categorical (nominal- or ordinal-level) outcome variable.

contingency table – a table showing the degree to which a case's value on the outcome variable depends on its category on the explanatory variable.

Mann–Whitney U test – a nonparametric test used to compare two independent samples on an ordinal-level outcome variable that utilizes ranking.

nonparametric test – a statistical test for use with nominal- or ordinal-level outcome variables, and for which assumptions about the shape of the population don't have to be met.

parametric test – a statistical test for use with interval- or ratio-level outcome variables, and for which assumptions about the shape of the population must be met.

single-sample test – a statistical test used to compare the results in a sample to a known population value.

Spearman rank-order correlation coefficient – a nonparametric test that examines the relationship between two ordinal-level variables or one ordinal and an interval/ratio variable.

CHAPTER EXERCISES

Review Your Knowledge

15.01 The statistical tests in previous chapters all had ____-level or ____-level outcome variables.

15.02 ____ tests are for nominal-level or ____-level outcome variables.

15.03 The independent-samples *t* test is an example of a ____ test.

15.04 Parametric tests assume that the outcome variable is ____ in the population.

15.05 If a researcher violates a nonrobust assumption for a parametric test, then a nonparametric test can be used as a ____ test.

15.06 Nonparametric tests are less restricted by ____ than are ____ tests.

15.07 Nonparametric tests usually have less ____ than ____ tests.

15.08 The chi-square goodness-of-fit test is a ____-sample test used with a ____ outcome variable.

15.09 The chi-square goodness-of-fit test sees if the difference between what is ____ and what is ____ can be explained by sampling error.

15.10 The abbreviation for chi-square is ____, where the Greek letter is pronounced ____.

15.11 The sample in a chi-square goodness-of-fit test should be a ____ from the population.

15.12 There should be at least ____ cases in each cell in a chi-square goodness-of-fit test.

15.13 The chi-square goodness-of-fit test compares the distribution of the outcome variable in the ____ to the specified distribution in the ____.

15.14 The critical value of chi-square is abbreviated ____.

15.15 If the ____ lands in the ____ of the sampling distribution of chi-square, then the null hypothesis is rejected.

15.16 The degrees of freedom for a chi-square goodness-of-fit test depend on the number of ____ in the study.

15.17 If the differences between the observed frequencies and the expected frequencies in a chi-square goodness-of-fit test are small, the null hypothesis is ____.

15.18 It is *impossible / possible* for expected frequencies in a chi-square goodness-of-fit test to be fractional numbers.

15.19 The expected frequencies in a chi-square goodness-of-fit test add up to ____.

15.20 The first step in calculating a chi-square is to ____ the expected frequencies from the ____.

15.21 In APA format for a chi-square goodness-of-fit test, the first number inside the parentheses represents the ____.

15.22 In APA format, if the results are written "$p > .05$," this means the results fell in the ____ zone.

15.23 The direction of the results for a chi-square goodness-of-fit test needs to be determined if the results are ____.

15.24 The chi-square test of independence differs from the chi-square goodness-of-fit test in terms of the number of ____ in the test.

15.25 The chi-square test of independence can be conceptualized as a ____ test or a ____ test.

15.26 To conduct a chi-square test of independence, construct a ____ table that cross-tabulates the values of the ____ with the levels of the ____.

15.27 In a contingency table, each ____ is placed in one, and only one, ____.

15.28 The assumptions for the chi-square test of independence are the same as for the ____.

15.29 The null hypothesis for the chi-square test of independence states that the ____ and the ____ are ____.

15.30 The hypotheses for a chi-square test of independence are stated the same way as they are for a ____.

15.31 The degrees of freedom for a chi-square test of independence depend on the number of ____ and the number of ____.

15.32 The formula for the chi-square test of independence *is similar to / not similar to* the formula for the chi-square goodness-of-fit test.

15.33 The number of cases in the rows in a chi-square test of independence add up to ____%.

15.34 The sum of the expected frequencies for the rows in a chi-square test of independence are the same as the sum of the observed frequencies for the ____.

15.35 In a Spearman rank-order correlation coefficient, the ____ of the two variables are correlated.

15.36 The ____ is the nonparametric alternative to the independent-samples *t* test.

15.37 The Mann–Whitney *U* test takes two independent ____, combines them into one ____, and then assigns a ____ to each case in the combined group.

15.38 The Mann–Whitney *U* test compares the ranks of the cases in one ____ to the ____ of the cases in the other sample.

Apply Your Knowledge

Pick the correct test from the following: the chi-square goodness-of-fit test, chi-square test of independence, Spearman rank-order correlation coefficient, and Mann–Whitney U test.

15.39 A high school principal developed a theory that caffeinated sodas cause more burping than decaffeinated sodas. She obtained a large sample of students and randomly assigned them to drink sodas with or without caffeine.

She then waited 15 minutes and classified each student as having burped or not having burped during that time. What test should she use?

15.40 The principal also kept track of how many burps each student produced. For both the caffeinated and de-caffeinated groups, the number of burps was extremely positively skewed. What statistical test should she use to see if the number of burps differs between the two groups?

15.41 A social psychologist assigned ranks to all the second-grade boys in a school based on how much the other boys liked them. He then did the same thing for all the boys again, but based the ranks on how much the girls liked them. What test should the psychologist conduct to see if there's an association between how boys and how girls view boys?

15.42 A statistics teacher looked out at her class of 32 students and noted that 21 of them were female. Assuming there are equal numbers of males and females in the world, what test should this teacher use to see if her class is overpopulated with women?

Checking the assumptions for a chi-square goodness-of-fit test or a chi-square test of independence

15.43 A sociologist planned to study patterns of criminality in small towns. She drew a random sample of 50 small towns from all across America. Before conducting her study, she wanted to make sure that her sample was representative of small towns. From the FBI, she learned that 2% of small towns have experienced a murder over the past 10 years. Can she use a chi-square goodness-of-fit test to compare the percentage of small towns that had experienced a murder to the national percentage?

15.44 An addictions researcher wants to see if male and female alcoholics differ in the type of alcohol they consume. She goes to a large alcohol detox facility, gets a sample of men and a sample of women, and checks each person's chart to find the beverage of choice. She classifies the beverages as (a) wine, (b) beer, or (c) hard liquor. The table that follows shows the expected frequencies. Can the researcher use a chi-square test of independence to analyze her data?

	Beer	Wine	Hard Liquor
Men	9.07	8.53	6.40
Women	7.93	7.47	5.60

Stating the hypotheses for chi-square tests

15.45 State the null and alternative hypotheses for Exercise 15.43.

15.46 State the null and alternative hypotheses for Exercise 15.44.

Finding degrees of freedom for chi-square goodness-of-fit tests

15.47 Given this matrix, how many degrees of freedom are there for this chi-square goodness-of-fit test?

15.48 Given this matrix, how many degrees of freedom are there for this chi-square goodness-of-fit test?

15.49 Given this matrix, how many degrees of freedom are there for this chi-square test of independence?

15.50 Given this matrix, how many degrees of freedom are there for this chi-square test of independence?

Finding χ^2_{cv} and setting the decision rule. Use $\alpha = .05$.

15.51 (a) If $df = 2$, what is χ^2_{cv}? (b) What is the decision rule?

15.52 (a) If $df = 4$, what is χ^2_{cv}? (b) What is the decision rule?

Calculating expected frequencies for a chi-square goodness-of-fit test

15.53 Given $N = 97$ and the expected percentages below, find the expected frequencies:

A 63%	B 37%

15.54 Given $N = 182$ and the expected percentages below, find the expected frequencies:

A 14%	B 38%	C 48%

Calculating χ^2 for a chi-square goodness-of-fit test

15.55 Given the information in the following matrix, calculate χ^2:

f_{Observed}	9	14	18	$\Sigma = 41$
f_{Expected}	6.97	15.58	18.45	$\Sigma = 41.00$

15.56 Given the information in the following matrix, calculate χ^2:

f_{Observed}	28	23	20	$\Sigma = 71$
f_{Expected}	33.37	19.98	17.65	$\Sigma = 71.00$

15.57 Given the information in the following matrix, calculate χ^2:

f_{Observed}	40	50
$f_{\text{\% Expected}}$	35%	65%

15.58 Given the information in the following matrix, calculate χ^2:

f_{Observed}	22	37
$f_{\text{\% Expected}}$	42%	58%

Using APA format with $\alpha = .05$

15.59 If $df = 1$, $N = 43$, $\chi^2 = 4.81$, and $\chi^2_{cv} = 3.841$, write the results in APA format.

15.60 If $df = 3$, $N = 78$, $\chi^2 = 5.99$, and $\chi^2_{cv} = 7.815$, write the results in APA format.

15.61 If $df = 4$, $N = 55$, and $\chi^2 = 9.49$, write the results in APA format.

15.62 If $df = 5$, $N = 234$, and $\chi^2 = 3.85$, write the results in APA format.

Determining the direction of the difference for a statistically significant chi-square goodness-of-fit test

15.63 Given the information below, determine the direction of the results:

	Category I	Category II
f_{Observed}	24	26
f_{Expected}	8.77	41.23

15.64 Given the information below, determine the direction of the results:

	Category I	Category II
f_{Observed}	37	23
f_{Expected}	49.23	10.77

Interpreting chi-square goodness-of-fit test

15.65 Dr. Lowry investigated whether fortunetellers have extrasensory perception. He obtained a random sample of 124 fortunetellers. For each one, a card was randomly selected from a well-shuffled 52-card deck and placed face down. Then the fortuneteller had to determine if the card was red or black. For each fortuneteller, Dr. Lowry recorded if he or she guessed correctly. If that person had no ESP ability, then he or she had a 50% chance of guessing correctly. Dr. Lowry found that 72 of the 124 fortunetellers (58.06%) correctly guessed the color of the card and 52 (41.94%) were wrong. The table below shows the observed and expected frequencies. The chi-square goodness-of-fit test results were $\chi^2(1, N = 124) = 3.23$, $p > .05$. Write a four-point interpretation.

	Guessed Correctly	Category Incorrectly
f_{Observed}	72	52
f_{Expected}	62.00	62.00

15.66 According to the U.S. government, 68% of American adults weigh more than they should. A dietician, Dr. Christiansen, maintained a theory that drinking 8 or more glasses of water a day was associated with not being overweight. She obtained a random sample of 580 American adults who consumed 8 or more glasses of water a day, weighed each adult, and classified each one as overweight or not. The table below shows the observed frequencies and the expected frequencies. The chi-square goodness-of-fit test results were $\chi^2(1, N = 580) = 15.62, p < .05$. Write a four-point interpretation.

	Not Overweight	Overweight
$f_{Observed}$	230	350
$f_{Expected}$	185.60	394.40

Completing all six steps of hypothesis testing for a chi-square goodness-of-fit test

15.67 A consumer psychologist, Dr. Wessells, obtained a random sample of 880 people who had strong preferences as to what cola they preferred. He gave each one, individually, a taste test in which each participant tasted four different colas and then had to pick his or her favorite. Dr. Wessells wondered whether consumers would be able to tell the difference. If they couldn't, they would have a 25% chance of being right and a 75% chance of being wrong. It turned out that 279 participants correctly chose their preferred cola and 601 chose incorrectly. Use hypothesis testing to decide if all colas taste the same, even to people who have strong preferences.

15.68 Dr. Constantinople, a cancer physician, wondered whether wearing a hat protected one against skin cancer. From prior research, she knew that about 1% of adults develop skin cancer every year. She obtained a random sample of 2,040 adults who wore hats and followed them for a year. During that year, 14 developed skin cancer and 2,026 did not. Use hypothesis testing to determine if wearing a hat is associated with a change in the risk of developing skin cancer.

Generating contingency tables

15.69 A researcher compared people in their 20s, 40s, and 60s in terms of whether they supported gay marriage or not. Support was measured as a "yes" or "no." Create a labeled contingency table that could be used to cross-tabulate the results.

15.70 An experimental group and a control group were compared in terms of whether they recovered from an illness in 5 or fewer days, 6 to 10 days, or more than 10 days. Make a labeled contingency table that could be used to cross-tabulate the results.

Finding the number of cases per row and column for a chi-square test of independence

15.71 Given the contingency table below, find the number of cases for each row and column:

	Outcome A	Outcome B
Group I	34	86
Group II	67	113

15.72 Given the contingency table below, find the number of cases for each row and column:

	Outcome A	Outcome B
Group I	28	12
Group II	32	18

Calculating expected frequencies

15.73 Given $N = 72$ and the information in the following contingency table, calculate the expected frequencies for the cells:

	Outcome 1	Outcome 2	
Group I	A	B	34
Group II	C	D	38
	30	42	

15.74 Given $N = 123$ and the information in the following contingency table, calculate the expected frequencies for the cells:

	Outcome 1	Outcome 2	
Group I	A	B	34
Group II	C	D	32
Group III	E	F	57
	27	96	

Calculating χ^2

15.75 Given the information below, calculate χ^2:

	Observed Frequencies	
	Outcome A	Outcome B
Group I	28	12
Group II	12	12

	Expected Frequencies	
	Outcome A	Outcome B
Group I	25.00	15.00
Group II	15.00	9.00

15.76 Given the information below, calculate χ^2:

	Observed Frequencies	
	Outcome A	Outcome B
Group I	35	55
Group II	67	40

	Expected Frequencies	
	Outcome A	Outcome B
Group I	46.60	43.40
Group II	55.40	51.60

15.77 Given the information below, calculate χ^2:

	Observed Frequencies	
	Outcome A	Outcome B
Group I	14	21
Group II	18	18

15.78 Given the information below, calculate χ^2:

	Observed Frequencies	
	Outcome A	Outcome B
Group I	6	18
Group II	30	10

Types of error

15.79 Given $\chi^2(1, N = 45) = 4.92$, $p < .05$, which type of error does the researcher need to worry about?

15.80 Given $\chi^2(1, N = 88) = 3.45$, $p > .05$, which type of error does the researcher need to worry about?

Determining the direction of the difference

15.81 The chi-square yielded statistically significant results. Given the information below, determine the direction of the difference:

	Observed Frequencies	
	Outcome A	Outcome B
Group I	6	18
Group II	30	10

	Expected Frequencies	
	Outcome A	Outcome B
Group I	13.50	10.50
Group II	22.50	17.50

15.82 The chi-square yielded statistically significant results. Given the information below, determine the direction of the difference:

	Observed Frequencies	
	Outcome A	Outcome B
Group I	12	19
Group II	17	9

	Expected Frequencies	
	Outcome A	Outcome B
Group I	15.77	15.23
Group II	13.23	12.77

Calculating and interpreting Cramer's V

15.83 Given $\chi^2(3, N = 78) = 8.92$, $p < .05$ where $R = 4$ and $C = 2$, (a) calculate V; (b) classify the effect as small, medium, or large; and (c) determine if the researcher should worry about Type II error.

15.84 Given $\chi^2(1, N = 102) = 3.53$, $p > .05$ where $R = 1$ and $C = 1$, (a) calculate V; (b) classify the effect as small, medium, or large; and (c) determine if the researcher should worry about Type II error.

Interpreting a chi-square test of independence

15.85 Tennis players are more likely to win points when they are serving. A tennis journalist wondered whether the server's advantage diminished as a rally lasted longer. She randomly selected 209 points from all the matches played at the major tennis tournaments during a year. For each point, she recorded whether the rally was short (the ball went back and forth no more than twice before the point was decided) or the rally was long (the ball went back and forth more than two times). She also recorded whether the server won the point or not. The table below shows the results. She found $\chi^2(1, N = 209) = 3.93$, $p < .05$ and calculated Cramer's $V = .14$. Write a four-point interpretation.

	Observed Frequencies	
	Server Wins	Server Loses
Short rally	66	41
Long rally	49	53

	Expected Frequencies	
	Server Wins	Server Loses
Short rally	58.88	48.12
Long rally	56.12	45.88

15.86 A college dean noticed that political science seemed to attract more male majors than psychology. She wondered if her observation were true. So, she obtained a random sample from each major and recorded the sex of each student. The table below shows her results. The dean found $\chi^2(1, N = 55) = 3.66$, $p > .05$; $V = .26$. Write a four-point interpretation.

	Observed Frequencies	
	Female	Male
Psychology	23	13
Political science	7	12

	Expected Frequencies	
	Female	Male
Psychology	19.64	16.36
Political science	10.36	8.64

Completing all six steps of hypothesis testing

15.87 A surgeon decided to compare sutures and staples in closing incisions. When he finished operating on a patient, he flipped a coin in order to determine if he would use sutures or staples to close the wound. Six months later, he called each patient and found out whether he or she was satisfied, yes vs. no, with how the scar looked. The results appear below. Use hypothesis testing to determine if there's a difference in satisfaction between the two techniques.

	Observed Frequencies	
	Not Satisfied	Satisfied
Sutures	8	13
Staples	9	12

15.88 A political scientist developed a theory that after an election, supporters of the losing candidate removed the bumper stickers from their cars faster than did supporters of the winning candidate. The day before a presidential election, he randomly selected parking lots, and at each selected parking lot, he randomly selected one car with a bumper sticker and recorded which candidate it supported. The day after the election, he followed the same procedure with a new sample of randomly selected parking lots. For both days, he then classified the bumper stickers as supporting the winning or losing candidate. Below are the results. Use hypothesis testing to see if a difference exists between how winners and losers behave.

	Observed Frequencies	
	Winner	Loser
Before	34	32
After	28	10

Expand Your Knowledge

15.89 Complete the following contingency table. If it can't be done, explain why.

12		20
		30
26	24	

15.90 Complete the following contingency table. If it can't be done, explain why.

9		7	24
			17
3		2	10
18	18	15	

15.91 Complete the following contingency table. If it can't be done, explain why.

9		18
		36
		42
46	50	

15.92 Is it possible, for the contingency table below, where $N = 24$, for f_{Observed} to equal f_{Expected} for each cell?

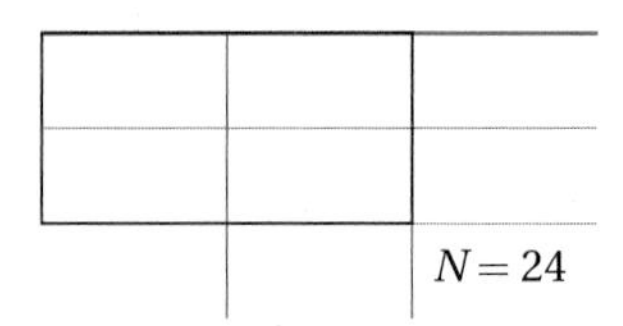

15.93 If $f_{\text{Observed}} = f_{\text{Expected}}$ for every cell, what does that mean for the chi-square value and for the null hypothesis?

15.94 Is there any value for degrees of freedom from 1 to 13 that cannot occur for a chi-square test of independence?

SPSS

Data for a chi-square test of independence in SPSS are entered with each variable in a separate column and each case in a row. Figure 15.7 shows the data for the text-reading study. The second column, text, has a 1 if the student read the text before class and a 2 if the text was read after class. In the third column, grade, a 1 means a high grade (A or B) and a 2 means a low grade (C, D, or F).

	V1	text	grade
1	1	1	2
2	2	2	2
3	3	2	2
4	4	2	2
5	5	2	2
6	6	1	1
7	7	1	1
8	8	1	2
9	9	1	1
10	10	1	1
11	11	1	1
12	12	1	1
13	13	1	1

Figure 15.7 Data Entry for Chi-Square Test of Independence in SPSS Data for a chi-square test of independence are entered in SPSS with each variable in a separate column. (Source: SPSS)

In SPSS, the chi-square test of independence is called "Crosstabs," which is short for a cross-tabulation (contingency) table. Figure 15.8 shows that Crosstabs is located under "Analyze" and "Descriptive Statistics."

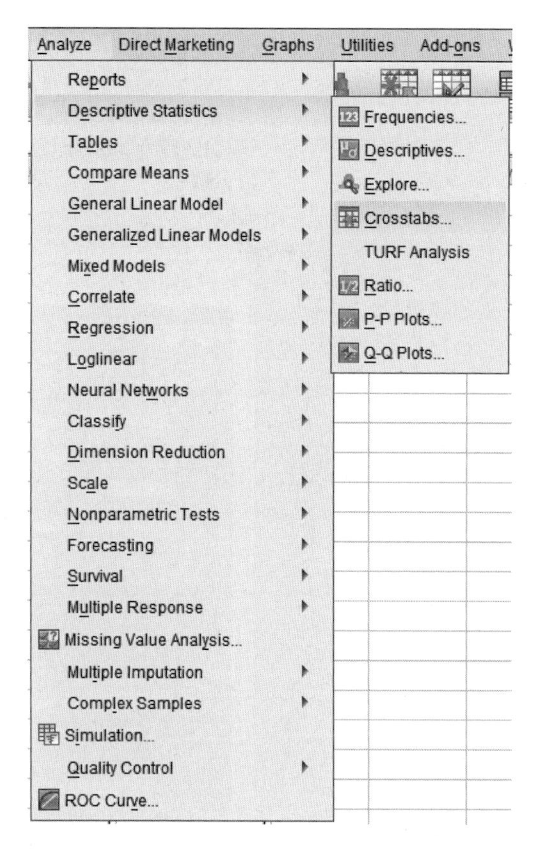

Figure 15.8 The Command for Crosstabs in SPSS "Crosstabs" is the SPSS command for a chi-square test of independence. It is found in the menu under "Analyze" and "Descriptive Statistics." (Source: SPSS)

Clicking on the Crosstabs command opens up the menu seen in Figure 15.9. Note that the variable "text" has already been sent over to be the row variable and the variable "grade" is getting ready to be sent to be the column variable. Be consistent—make the explanatory variable the row variable and the dependent variable the column variable.

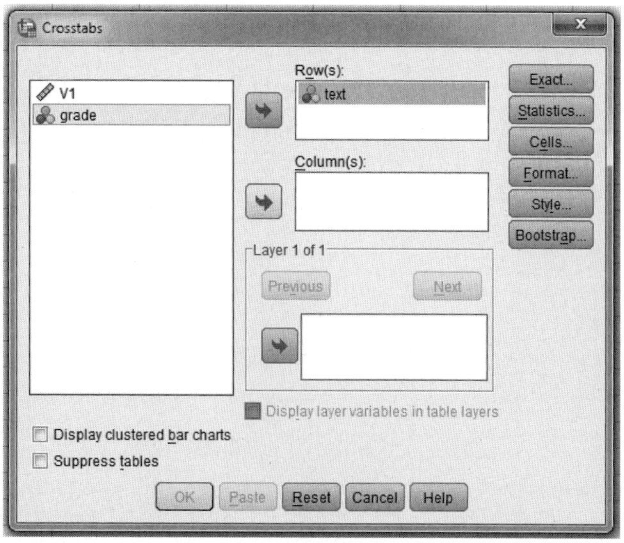

Figure 15.9 Menu for the Crosstabs Command Once this menu is active, one variable is designated as the row variable and a second variable as the column variable. (Source: SPSS)

Clicking on the "Statistics" button in the upper-right-hand corner of Figure 15.9 opens the menu seen in Figure 15.10. Note that the boxes for "Chi-Square" and "Phi and Cramer's V" have been checked. Then click on the "Continue" button in Figure 15.10 and the "OK" button in Figure 15.9.

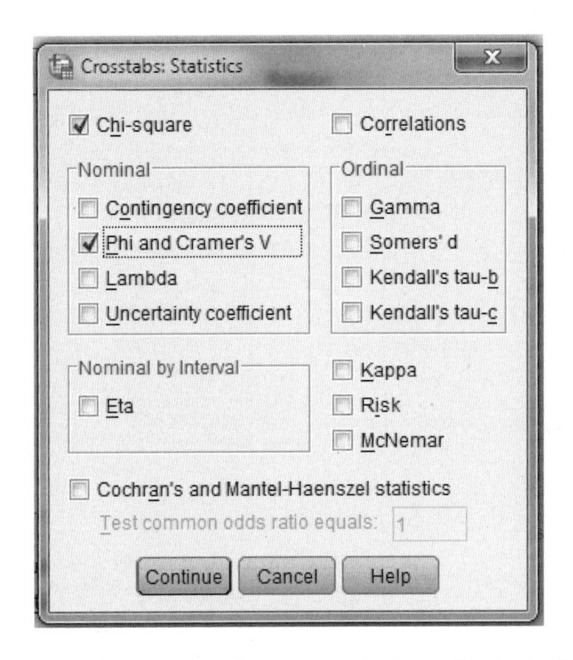

Figure 15.10 Selecting Chi-square and Cramer's *V* Once the desired statistics have been selected, click on the "Continue" button. (Source: SPSS)

Once the statistics have been selected, clicking on the OK button on the cross-tabs menu generates the output seen in Figure 15.11. The first box in the output (A) is the contingency table. The second box (B) gives the chi-square value, the degrees of freedom, and the exact significance level. SPSS calls chi-square "Pearson chi-square" and reports it on the first row as 3.926. Note that the subscript "a" means that all cells had expected frequencies greater than 5. The degrees of freedom are reported as 1 and the significance level as .048. If alpha is set at .05, then the results are statistically significant as long as the significance level is $\leq .05$. Finally, Cramer's V is reported in the third box of the printout (C) as equal to .280.

text * grade Crosstabulation

Count

		grade		Total
		1	2	
text	1	20	6	26
	2	12	12	24
Total		32	18	50

Chi-Square Tests

	Value	df	Asymp. Sig. (2-sided)	Exact Sig. (2-sided)	Exact Sig. (1-sided)
Pearson Chi-Square	3.926[a]	1	.048		
Continuity Correction[b]	2.845	1	.092		
Likelihood Ratio	3.980	1	.046		
Fisher's Exact Test				.077	.045
Linear-by-Linear Association	3.848	1	.050		
N of Valid Cases	50				

a. 0 cells (0.0%) have expected count less than 5. The minimum expected count is 8.64.

b. Computed only for a 2x2 table

Symmetric Measures

		Value	Approx. Sig.
Nominal by Nominal	Phi	.280	.048
	Cramer's V	.280	.048
N of Valid Cases		50	

Figure 15.11 Output for Crosstabs in SPSS The chi-square value, here called "Pearson Chi-Square," can be found on the first line of the second box. Cramer's *V* is found on the second line of the third box. (Source: SPSS)

Selecting the Right Statistical Test

LEARNING OBJECTIVES

- Define the different tasks of statistics.
- Select the correct statistical test.

CHAPTER OVERVIEW

Most people have a toolbox with a few tools in it. There's probably a hammer, a screwdriver, and a pair of pliers. Most people know that if a screw has to be tightened, a pair of pliers won't be much use, so they reach for the screwdriver. Statistical tests are tools—pliers, screwdrivers, and hammers—designed for different research jobs. This book has filled your statistical toolbox with a set of tools for the most commonly encountered statistical jobs. Now it is time to learn how to pick the right statistical tool for the particular research job at hand.

16.1 Review of Statistical Tasks

The tool that the statistician selects depends on the task, and there are three different types of tasks:

1. Summarizing a set of data
2. Deciding if a group of cases differs from another group of cases on some underlying parameter
3. Determining if two variables vary together systematically

The first type of task involves what are called *descriptive statistics*, statistics used to describe a set of observations. Reporting that the average length of stay in a psychiatric hospital is two weeks is an example of a descriptive statistic.

The other two tasks for statisticians involve *difference tests* and *relationship tests*. Difference tests look for differences in some underlying parameter among groups of cases. For example, reporting that it takes longer, on average, to treat someone for depression than for anxiety would be the result of a difference test. Relationship tests see if variables correlate. Finding that the longer a person has suffered from depression tends to increase the difficulty in treating it would be the result of a relationship test.

Chapter 1 covered three different types of studies—experimental designs, quasi-experimental designs, and correlational designs. Table 16.1 summarizes their similarities and differences. Knowing what type of study was conducted is important because different types of studies call for different statistics:

- *Descriptive studies* call for the use of descriptive statistics.
- *Experimental designs* and *quasi-experimental designs* divide participants into groups, so they are analyzed with difference tests.
- *Correlational designs,* which measure two variables in a single sample of participants, call for relationship tests.

TABLE 16.1 Comparing Correlational, Experimental, and Quasi-Experimental Studies

	Correlational	Experimental	Quasi-Experimental
Is the explanatory variable manipulated/ controlled?	No	Yes	No
Cases are . . .	Measured for naturally occurring variability on both variables	Assigned to groups by an experimenter using random assignment	Classified in groups on the basis of naturally occurring status
The study can draw a firm conclusion about cause and effect.	No	Yes	No
Do you need to worry whether confounding variables exist?	Yes	No	Yes
Question being asked by the study	Is there a relationship between the two variables?	Do the different populations possess different amounts of the dependent variable?	Do the different populations possess different amounts of the dependent variable?
Advantage of study	A correlational design can study conditions that can't be manipulated.	An experimental study can draw conclusions about cause and effect.	A quasi-experimental design can study conditions that can't be manipulated.

The flowchart in Figure 16.1 shows how to determine whether a study is experimental, quasi-experimental, or correlational, as well as whether it is descriptive.

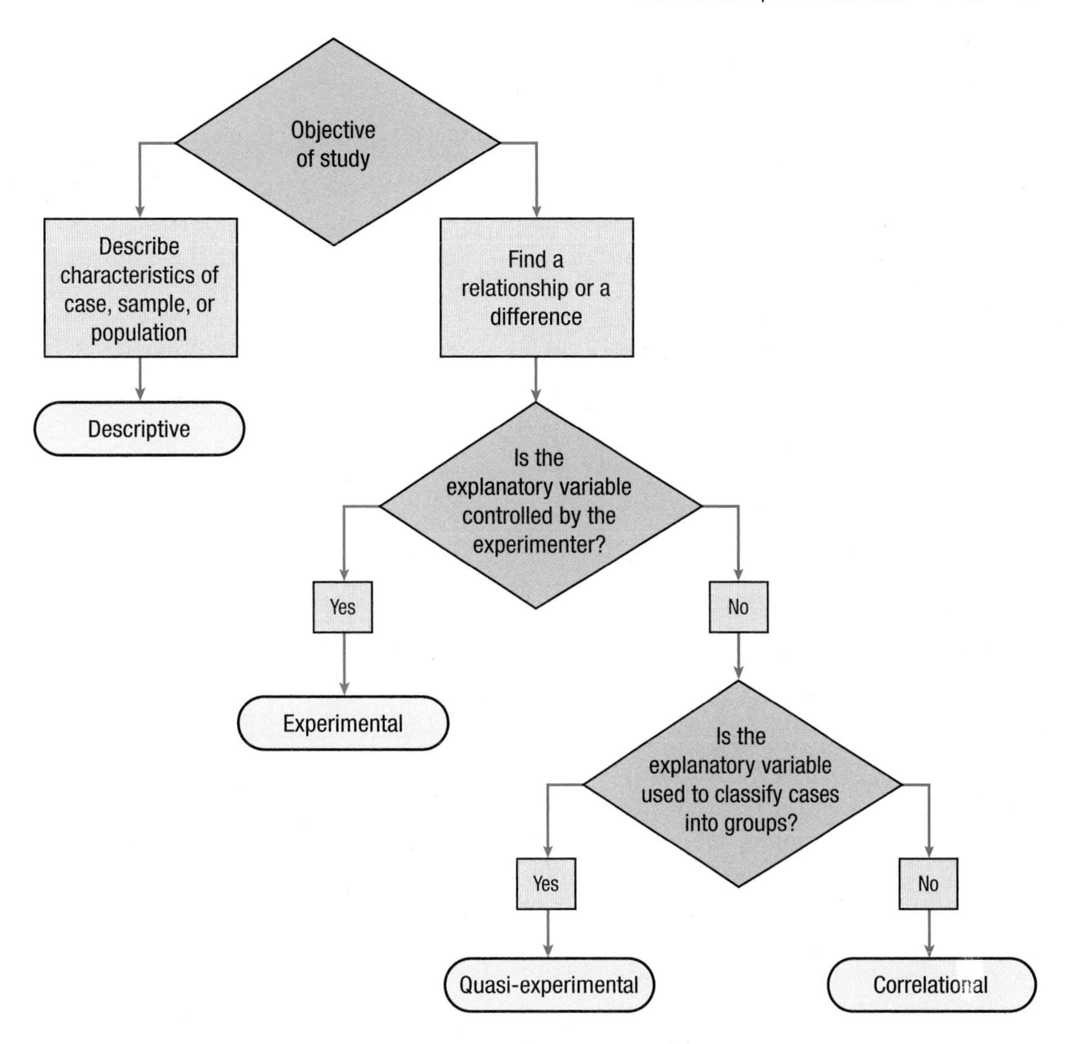

Figure 16.1 How to Choose: Type of Study This flowchart can be used to determine whether a study is descriptive, correlational, experimental, or quasi-experimental. The type of study depends on whether the explanatory variable is manipulated by the experimenter and, if the explanatory variable is not manipulated, whether cases are classified into groups.

The flowchart in Figure 16.1 may involve making a decision about the explanatory variable, often called the independent variable, in a study. Studies often have two variables, an explanatory variable and an outcome variable, commonly referred to, respectively, as the independent variable and the dependent variable. It is important to be able to tell which is which. The explanatory variable is the one used to assign or classify cases into groups or to predict the outcome, and the outcome variable is the one where the outcome of the study is measured.

16.2 Descriptive Statistics

The most basic task in statistics is taking a set of numbers and summarizing them in some way. Statistical techniques that do this are called **descriptive statistics** because these techniques are used to describe a set of cases.

The flowchart in Figure 16.2 walks through how to choose the correct descriptive statistic. Choosing the correct descriptive statistic depends on three things:

1. At what level the variable is measured: nominal, ordinal, or interval/ratio
2. What is being described: an individual case or a sample of cases
3. What is to be described: central tendency or variability

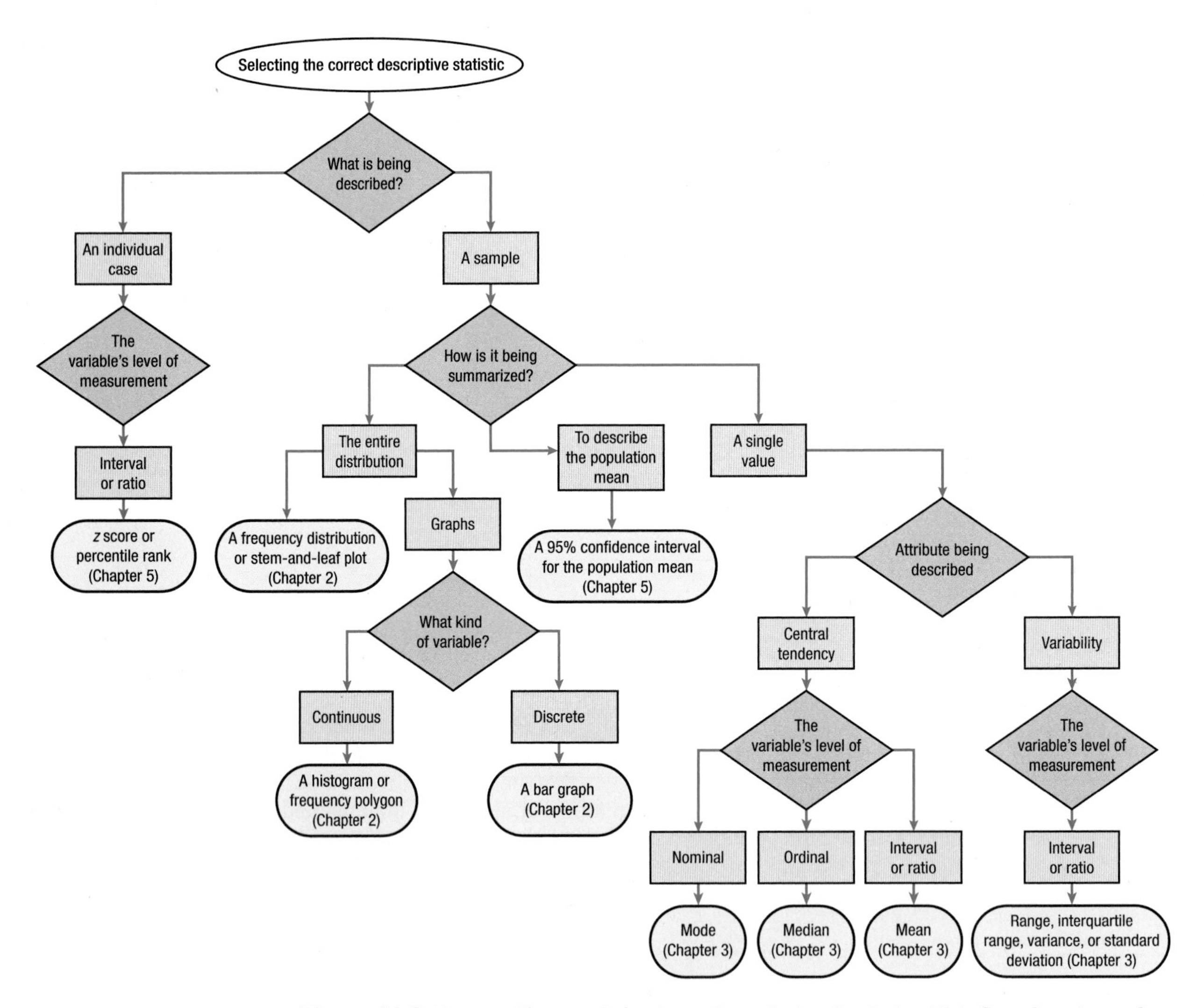

Figure 16.2 How to Choose: Selecting a Descriptive Statistic This flowchart is used to select the correct descriptive statistic, depending on (1) whether one is trying to describe an individual case, a sample, or a population; (2) what attribute one wants to describe (central tendency or variability); and (3) the variable's level of measurement (nominal, ordinal, interval, or ratio).

Knowing the level of measurement for the variable—nominal, ordinal, interval, or ratio—is important in statistics (see Chapter 1). Table 16.2 shows the distinctions among the four levels of measurement, and the flowchart in Figure 16.3 leads one through the steps for determining the level of measurement of a variable.

TABLE 16.2 Level of Measurement: Information Contained in Numbers

	Same/Different	Direction of Difference (more/less)	Amount of Distance (equality of units)	Proportion (absolute zero point)
Nominal	✓			
Ordinal	✓	✓		
Interval	✓	✓	✓	
Ratio	✓	✓	✓	✓

As levels of measurement become more complex, numbers contain more information.

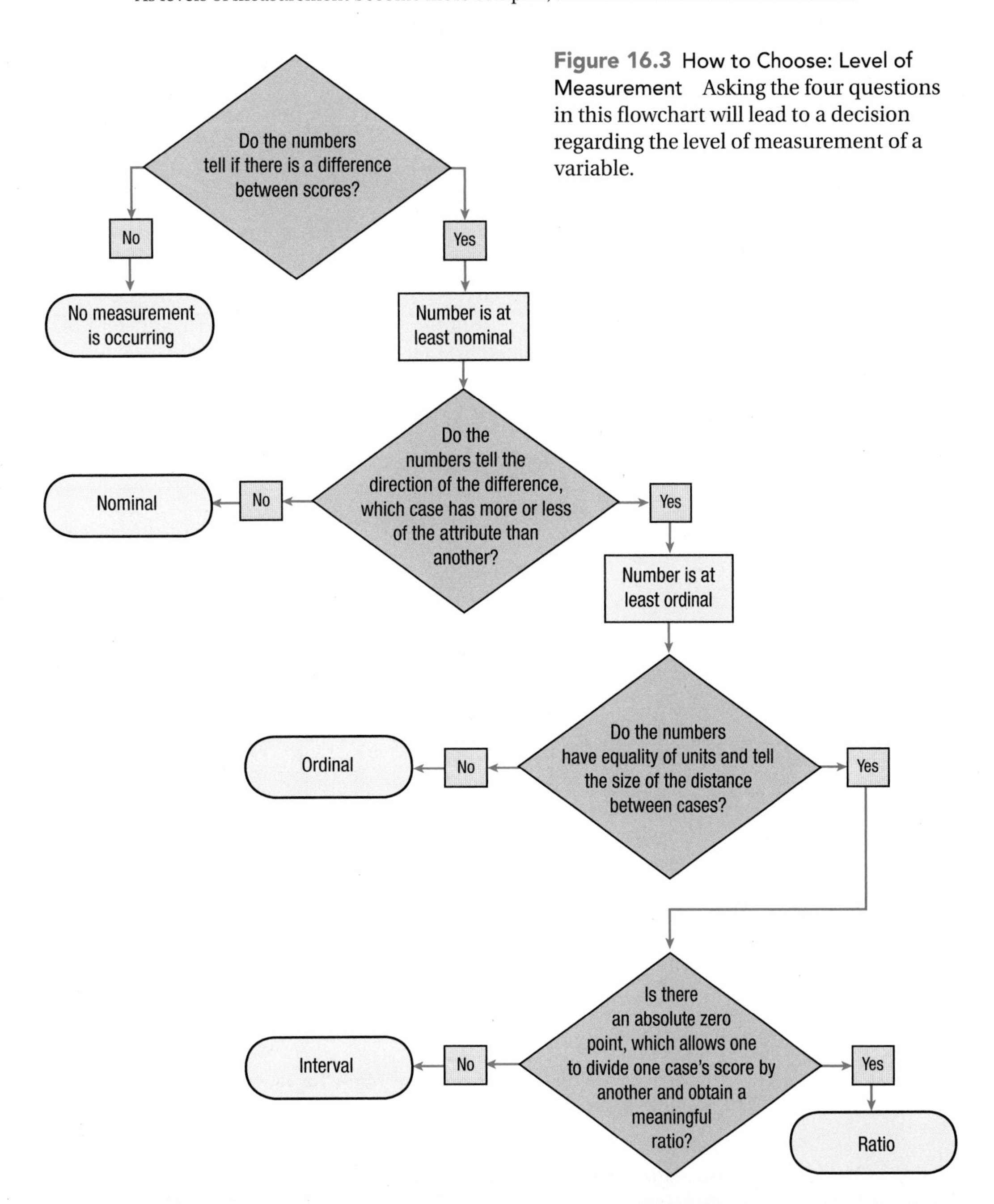

Figure 16.3 How to Choose: Level of Measurement Asking the four questions in this flowchart will lead to a decision regarding the level of measurement of a variable.

Describing Individual Cases

Sometimes the objective is to describe a single case. If the characteristic being measured for an individual case is measured at the interval or ratio level, it can be left as a raw score or transformed. Transformed scores have an advantage because they put scores into context. This book covered two types of transformed scores in Chapter 4:

- z scores reveal (1) whether a score falls above or below the mean, and (2) how far away from the mean it falls.
- Percentile ranks tell the percentage of scores in the distribution that fall at or below a score.

For example, a person with an IQ score of 115 could have his or her score transformed into a z score of 1.00 or a percentile rank of 84.13, both of which provide more information than the raw score.

Describing a Sample of Cases

When a set of scores is a sample from a population, it can be summarized (1) by presenting the entire distribution or (2) with a single value that describes some aspect of the sample. As was covered in Chapter 2, the entire distribution can be summarized in a table (i.e., a frequency distribution or stem-and-leaf plot) or as a graph. Graphs have a visual advantage over tables. Which graph can be used (histogram, frequency polygon, or bar graph) depends on whether the variable is continuous or discrete. Figure 16.4 is a flowchart for determining if a variable is continuous or discrete.

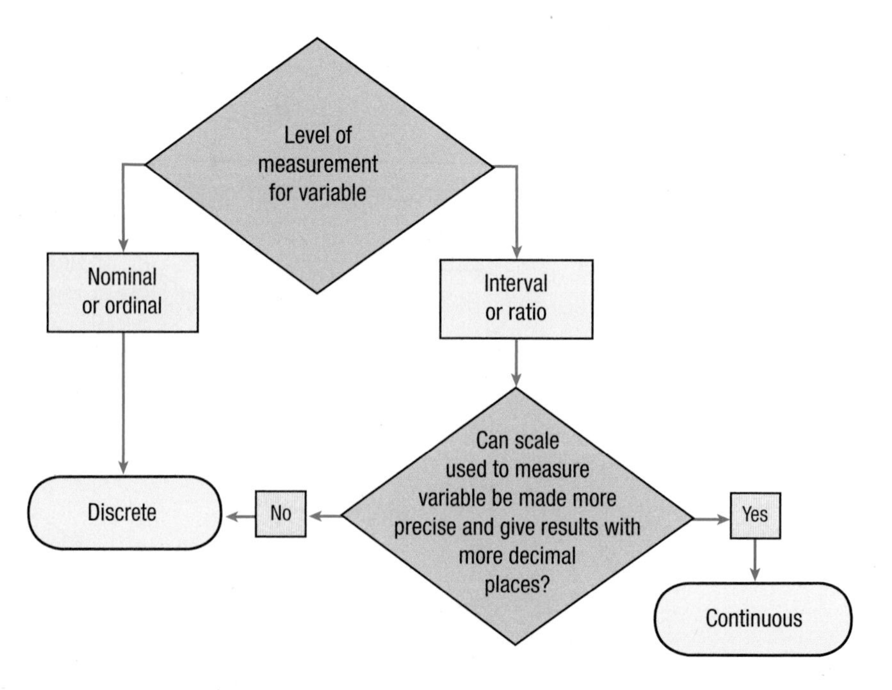

Figure 16.4 How to Choose: Continuous Numbers vs. Discrete Numbers To determine if the variable is continuous or discrete, decide whether the variable is (1) nominal or ordinal, or (2) interval or ratio. Then follow the flowchart.

When using a descriptive statistic to describe a sample of cases, a researcher can describe the central tendency or the variability in a set of cases (see Chapter 3). Usually, both are described. For example, a researcher might report both the average age of the participants *and* the range of ages.

A measure of central tendency for a variable is chosen depending on its level of measurement (see Table 16.3). But other factors—such as the shape of the distribution (see Chapter 3)—also need to be taken into account when selecting a measure of central tendency.

- To describe central tendency in a sample for a variable measured at the interval or ratio level, the go-to option is the mean (M).
- Central tendency for an ordinal-level variable is best measured with the median (Mdn).
- Central tendency for a nominal variable can only be measured with a mode.

TABLE 16.3 How to Choose: Which Measure of Central Tendency for Which Level of Measurement

	Measure of Central Tendency		
Level of Measurement	Mode	Median	Mean
Nominal	✓		
Ordinal	✓	✓	
Interval or ratio	✓	✓	✓

Not all measures of central tendency can be used with all levels of measurement. When more than one measure of central tendency may be used, choose the one that utilizes more of the information available in the numbers. If considering a mean, be sure to check the shape of the data set for skewness and modality.

All of the descriptive statistics for variability (see Chapter 3) are used for interval-level or ratio-level variables. These statistics include the range, interquartile range (IQR), variance (s^2), and standard deviation (s). The standard deviation is the most commonly used measure of variability. Different abbreviations are used for variance and standard deviation depending on whether one is referring to a population value or sample value. A lowercase Greek sigma, σ, is used for population standard deviations and s for sample standard deviations.

Just like seeing a mountain from several sides gives a better idea of what it looks like, using multiple perspectives—multiple descriptive statistics—gives a better sense of a set of data. Here are multiple ways a teacher might report how well her class had done on a test. Each descriptive statistic by itself answers a specific question. Taken together, they provide a detailed and comprehensive view of class performance:

- The teacher might use a frequency distribution, grouping scores into A's, B's, and so on to provide an overview of the class performance.
- A single score like the mean would summarize the class' performance, giving some sense of how easy (or difficult) the test was and how well (or how poorly) the class as a whole had done.

- The standard deviation would indicate how much variability existed in the scores, how tightly they are packed around the mean.
- The range would tell how badly the worst student had performed and how well the best student had performed. It would also reveal how much distance separated the best and the worst students.
- If a student came in to talk about his or her performance, the teacher might transform that student's score into a percentile rank in order to compare that student's performance to the rest of the class.

Sometimes the results from a sample are used to describe a population. The population standard deviation, σ, is an example of this. σ is a point estimate, a single value. Chapter 5 covered confidence intervals, which give a range within which it is likely that the population value falls. If one has a sample measured on an interval- or ratio-level variable and wants to estimate what the average value is in the larger population, a confidence interval for the mean is the way to go.

Practice Problems 16.1

16.01 Read each scenario and decide whether the study is descriptive, experimental, quasi-experimental, or correlational.

a. An ethologist attaches motion sensors to house cats and to wild cats so that he can measure the amount of time that each type of cat spends resting/sleeping each day. He's curious if wild cats spend more, less, or the same amount of time sleeping/resting as do house cats.

b. A physical fitness instructor is curious about how much time people spend at the gym. At his gym, people have to swipe in and swipe out. He collects data for a week, finding that the amount of time ranges from less than a minute to more than 9 hours, with a median time of 77 minutes.

c. An industrial/organizational psychologist obtained a random sample of employees at a large company. Among the questions she asked them was how good they thought the company was. She found that as the ages of the workers increased, so did their ratings of the "goodness" of the company.

16.02 Read the scenarios below and for each determine the level of measurement for the variable.

a. People who were not Chinese and who could not read Chinese viewed samples of Chinese writing written by Chinese people of different ages. They were asked to judge whether the passage was written by a pre-schooler (1), an elementary school–age child (2), a middle schooler (3), a high schooler (4), a young adult (5), a middle-aged person (6), or a senior citizen (7). The numbers in parentheses are the scores assigned to each answer.

b. Researchers wanted to find out if people could judge what sex a person was from a distance. They had participants watch videos of people walking toward them from 200 yards away and classify the subjects as –1 (male), 0 (can't tell), or 1 (female).

c. Some personality psychologists wanted to measure "manliness." They showed women pictures of men and had them rate the images as –1 (negative levels of manliness), 0 (neutral level of manliness), 1 (a small degree of manliness), 2 (more than a little, but less than average level of manliness), 3 (average level of manliness), 4 (somewhat above average in manliness), 5 (far above average in manliness), and 6 (off the top of the chart in terms of manliness).

16.03 Read each scenario and then select the appropriate descriptive statistic.

a. John attends a large university and is one of 450 people in introductory psychology. The last exam posed 73 questions, and John answered 62 correctly. John goes to see the professor because he would like to know how his performance compares to others in the same class.
b. The physical fitness instructor in Practice Problem 16.01(b) above wants to display the whole range of amounts of time people spend in the gym, as well as how "popular" the different amounts of time are.
c. The researchers in Practice Problem 16.02(b) above want to report what the average perceived sex is.
d. A hospital clothing manufacturer wants to re-size its one-size-fits-all hospital gowns now that Americans have grown larger. The manufacturer measures the weights of a sample of hospital patients from all 50 states and wishes to know the average weight of all Americans.

16.3 Hypothesis Tests I: Difference Tests

Let's move from selecting the correct descriptive statistic to selecting the right difference test. Difference tests are the most popular type of hypothesis test used in psychology. *t* tests and ANOVAs are examples of difference tests. **Difference tests** are used when cases are sorted into groups that are defined on the basis of an explanatory variable. They are used to answer the question of whether groups differ on some outcome variable. For example, whether people with depression or people with anxiety are more impaired by their disorder would be answered by a difference test.

Difference tests are the go-to tests for experimental and quasi-experimental designs. In experimental designs, the researcher controls the assignment of participants to groups on the basis of the independent variable. With quasi-experimental designs, participants are classified as being in groups on the basis of their naturally occurring status on the grouping variable. In either event, a difference test is used to compare the groups in terms of the dependent variable.

The flowchart in Figure 16.5 shows how to choose the correct difference test to answer a research question. Picking the right test depends on (1) the number of groups of cases, (2) the number of explanatory variables, (3) the type of samples—independent or dependent, and (4) the level of measurement for the dependent variable.

Whether a study is experimental or quasi-experimental, the first decision point involves how many groups of cases there are: one, two, or three or more groups. If just one group exists, then the sample is being compared to a specified value or a population value. The specific test to be used in making that comparison is determined by the dependent variable's level of measurement.

If there is more than one group, it is important to determine whether the comparison is between just two groups or among three or more groups. This determination leads to the next decision point. When the comparison is between two groups, the next decision point is whether the groups are independent samples or dependent samples (see Chapters 8 and 9 on *t* tests).

Table 16.4 presents guidelines for determining the type of samples—independent or dependent. Once the type of samples has been determined for a two-sample test, the specific difference test to use depends on the level of measurement for the dependent variable.

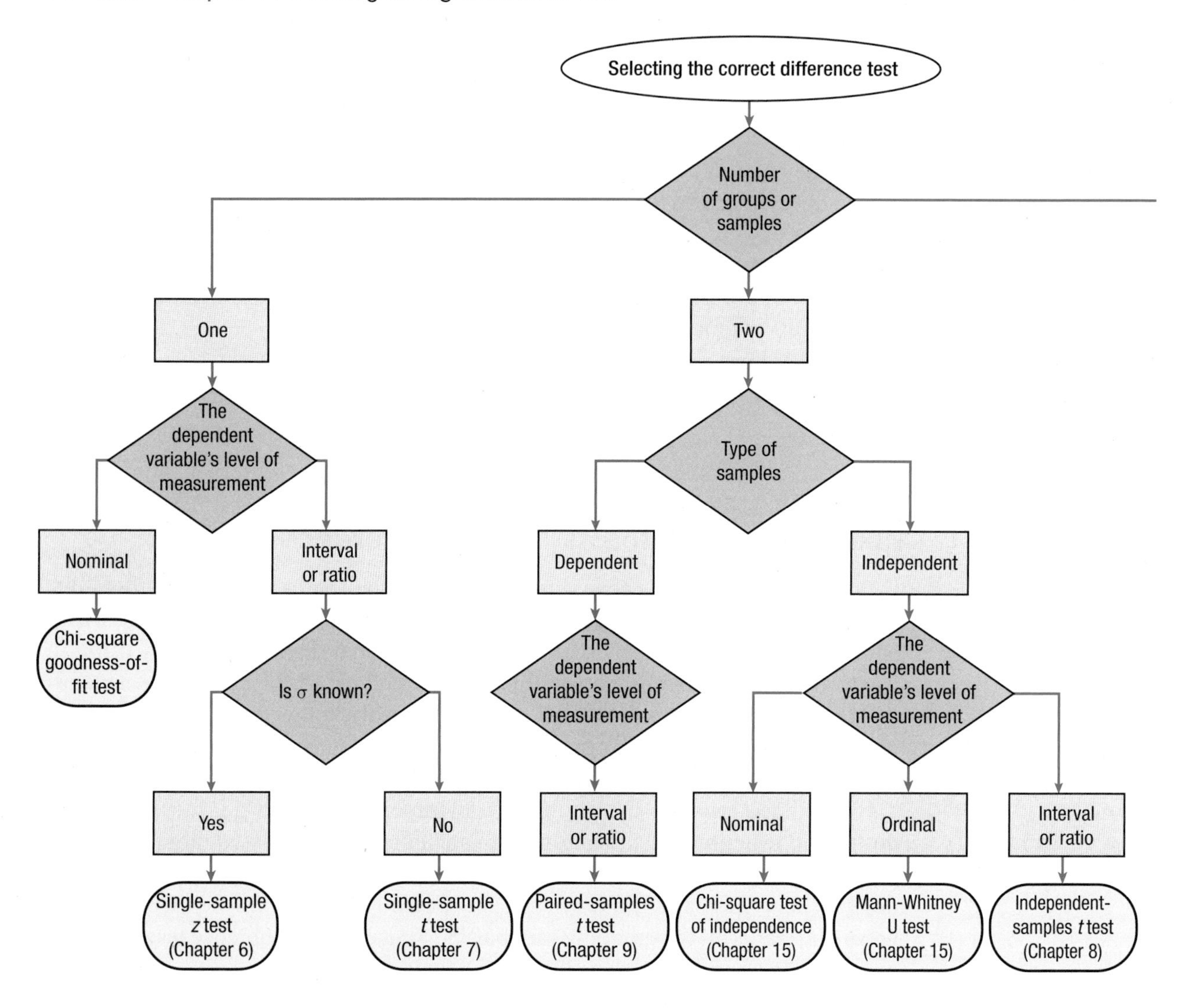
Selecting the correct difference test
Number of groups or samples
One
Two
The dependent variable's level of measurement
Nominal
Interval or ratio
Chi-square goodness-of-fit test
Is σ known?
Yes
No
Single-sample z test (Chapter 6)
Single-sample t test (Chapter 7)
Type of samples
Dependent
Independent
The dependent variable's level of measurement
Interval or ratio
Paired-samples t test (Chapter 9)
The dependent variable's level of measurement
Nominal
Ordinal
Interval or ratio
Chi-square test of independence (Chapter 15)
Mann-Whitney U test (Chapter 15)
Independent-samples t test (Chapter 8)

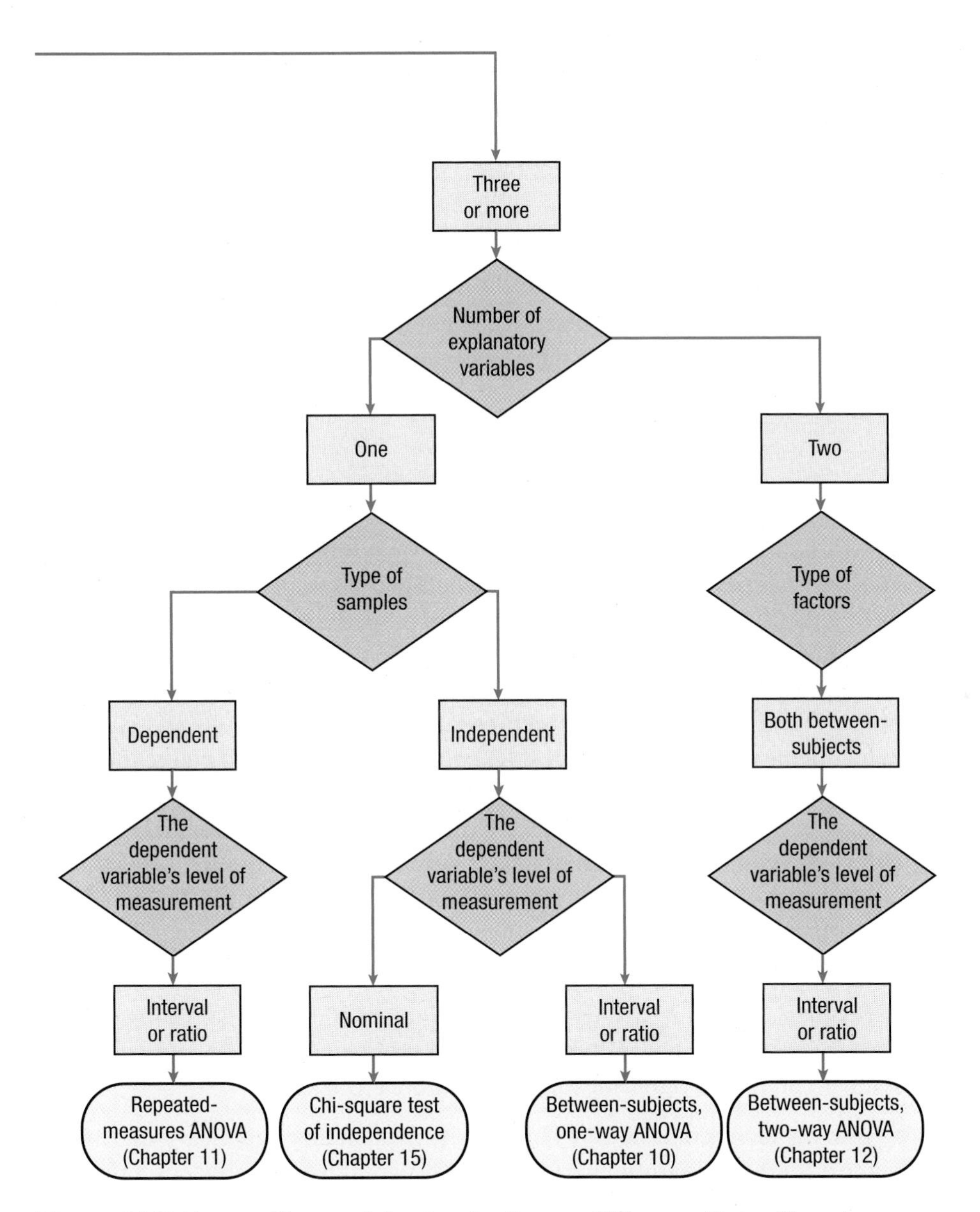

Figure 16.5 How to Choose: Selecting the Correct Difference Test Choosing a difference test depends on the number of samples, the type of samples, the dependent variable's level of measurement, and the number of explanatory variables.

TABLE 16.4 How to Choose: Guidelines for Determining If Samples Are Independent Samples or Dependent Samples

- Is each sample a random sample from its respective population?
  - If yes, independent samples.
  - If no, then there is no information about the type of samples.
- Is the size of each sample different? (That is, does $n_1 \neq n_2$?)
  - If yes, independent samples.
  - If no, then there is no information about the type of samples.
- Do the samples consist of the same cases measured at more than one point in time or in more than one condition?
  - If yes, dependent (paired) samples.
  - If no, then there is no information about the type of samples.
- Does the selection of cases for one sample determine the selection of cases for the other sample? Are cases paired together in some way?
  - If yes, dependent (paired) samples.
  - If no, independent samples.

When there are three or more samples, the flowchart branches out depending on (1) the number of explanatory variables, (2) whether the samples are independent or dependent, and (3) the dependent variable's level of measurement.

To see how the flowchart works, imagine a clinical psychologist who wanted to see how the loss of a parent before age 6 affected a child's adult level of psychological functioning. She obtained a large and representative sample of adults and classified them as belonging to one of four groups: (1) neither parent had died before the child, who is now an adult, was 6 years old; (2) the mother died before the child was 6 years old; (3) the father died before the child was 6 years old; or (4) both parents died before the child was 6 years old. To measure adult psychological functioning, she used an interval-level measure of psychological resilience. What statistical test should she use to find out if the mean level of psychological resilience differs among these four groups? The difference test flowchart, Figure 16.5, will help determine the answer.

- The first decision point involves how many groups of cases exist. There was one sample of cases, which was divided into four groups, so proceed down the flowchart path on the far right side for three or more groups.
- The next decision point involves how many explanatory variables exist. In this situation, there is one grouping variable, early parental death, that has four levels (none died, mother died, father died, both died).
- The next choice point involves the type of samples, dependent or independent. In this example, each case is classified as belonging to only one group and cases are not paired with each other. Who is classified as being in one group has no impact on who is classified as being in one of the other groups. The samples are independent.
- The final question to be addressed involves the level of measurement of the dependent variable. The dependent variable, psychological resilience, is measured at the interval level, which leads to the selection of a between-subjects, one-way ANOVA as the appropriate statistical test.

A Common Question

Q What should be done if there is no test listed in the flowchart? For example, what happens if the dependent variable for the early parental loss study was ordinal, not interval?

A The flowcharts here only choose among the statistical tests taught in this book. To find the correct test for a different situation, see the SPSS section at the end of this chapter, or consult a more advanced text.

Worked Example 16.1 For practice choosing the correct difference test, imagine an economist who wanted to investigate whether parents invest equally in their children's education depending on the children's birth order. From around the United States, he obtained a sample of families with two (and only two) children in which both children had attended college. His dependent variable was how much money the parents contributed to each child's college education. Is there a difference in mean parental funding for first-born vs. second-born children?

- There are two samples, a sample of first-born children and a sample of second-born children.
- The children in the two samples are paired together—they come from the same family—so the samples are dependent samples.
- The dependent variable, dollars spent, is a ratio-level variable.

Following the difference test flowchart, Figure 16.5, leads to selecting the paired-samples *t* test as the correct test to analyze the results.

Practice Problems 16.2

Select the correct statistical test.

16.04 A public health researcher wanted to know if hand washing or hand sanitizer was more effective in removing germs. She randomly assigned visitors at a hospital to wash their hands with soap and water or to use hand sanitizer. She then cultured the visitors' hands to see how many bacterial colonies there were.

16.05 People who wanted to lose weight were randomly assigned to drink 1 cup, 2 cups, or 3 cups of green tea each day. After three months, each person was weighed to determine how many pounds he or she had lost.

16.06 People from countries with spicy cuisines were matched in terms of age and sex with people from countries with bland cuisines. All participants completed the Willingness to Engage in Risky Behavior Scale. Are people who eat spicy food more likely to engage in risky behavior?

16.4 Hypothesis Tests II: Relationship Tests

Having covered the selection of descriptive statistics and difference tests, it is time to turn to relationship tests. **Relationship tests** are meant to examine whether a correlation exists between two variables. Relationship tests are used when there is one sample of cases and each case is measured on two variables in order to see if the variables vary together systematically. For example, one might measure both attractiveness and number of dates in college students to see if the two are associated. The flowchart for selecting the correct relationship test is shown in Figure 16.6.

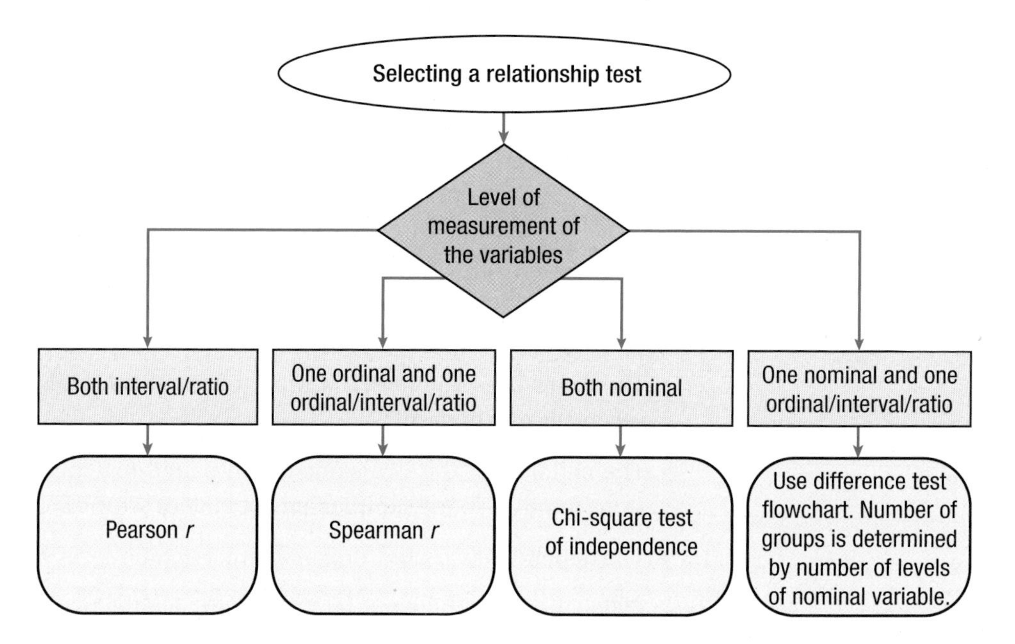

Figure 16.6 How to Choose: Selecting the Correct Relationship Test The selection of the correct relationship test depends on the level of measurement of the variables. If one variable is nominal and the other is not, then a difference test is used.

The decision about which relationship test to use in a given situation is determined by the level of measurement for each of the two variables:

- If both variables are measured at the interval or ratio level, then the Pearson *r* is used (see Chapter 13).
- If one variable is ordinal and the other is ordinal or interval or ratio, use a Spearman *r* (see Chapter 15).
- If both variables are nominal-level, then the chi-square test of independence is used (see Chapter 15).
- When one of the variables is nominal and the other is not, the relationship question can be conceptualized as a difference question. To do so, treat the nominal variable as the explanatory variable and use it to classify cases into groups. Then, the difference test flowchart (Figure 16.5) can be used to determine the appropriate test.

For an example of how to select the correct relationship test, imagine an animal behaviorist who first trained rats to run a maze to find food in a goal box and then measured how many days it took for the behavior to cease (for the rats to stop running to the goal box) once food was no longer being placed in the goal box. She wondered if smarter rats, those who learned the maze more quickly, were also quicker in learning a reward was no longer present. Is there a relationship between the times it takes to learn these two tasks?

This calls for a relationship test. To select the appropriate test, use the flowchart in Figure 16.6. To use that flowchart, the level of measurement must be known for each of the two variables—number of trials to learn the maze and number of days to extinction. Using the flowchart in Figure 16.2, both variables are classified as being at the ratio level. Turning back to the relationship test flowchart, Figure 16.6, when both variables are measured at the interval or ratio level, the Pearson *r* is the test to use.

Let's complete one more relationship test example, one in which one variable is nominal and the other isn't. Imagine a researcher who has a random sample of male college students and a random sample of female college students, and who wonders if there is a relationship between sex (male or female) and GPA. To determine what statistical test should be used, go to the relationship test flowchart, Figure 16.6. Notice that one variable is nominal (sex) and the other variable is interval (GPA). This combination of variables prompts us to exit the relationship test flowchart and move to the difference test flowchart. The relationship question is now phrased, "Is there a difference between male and female students in terms of mean GPA?"

In the difference test flowchart, the nominal-level variable, sex, is treated as the explanatory variable, and the interval/ratio variable, GPA, as the dependent variable. Here are the answers to the choice points in Figure 16.5 for the sex/GPA test decision:

- The explanatory variable has two levels, male and female, so two groups are being compared.
- Each sample was a random sample from its population, so the two samples are independent.
- The dependent variable, GPA, is measured at the interval level.

These choice points lead to the selection of an independent-samples *t* test as the appropriate test to analyze the data to determine if a relationship exists between sex and mean GPA. A difference test can answer a relationship question.

Worked Example 16.2

A criminologist developed a theory that elementary school teachers can quickly tell which kids are going to be a problem in their classroom. After only two weeks of school, he had asked teachers to make a judgment for each child: whether he or she "would be trouble in the classroom." A dozen years later, as the same kids are ready to graduate from high school, he tracked down all their records and determined which ones had been arrested, spent time in jail, dropped out of school, and so on. Anyone for whom one of these events had occurred was classified as "having gotten into trouble in life." What test should the criminologist use to see if an association exists between a prediction made by a first-grade teacher, after only two weeks of observation, and getting into serious trouble over the next 12 years?

This means asking if there is a relationship between the two variables, so the place to start is with the flowchart in Figure 16.6. The first question concerns the level of measurement of the variables. Both variables are nominal. Thus, the appropriate statistical test is the chi-square test of independence.

Practice Problems 16.3

Select the correct statistical test.

16.07 A hair salon owner watches people walk by his shop, notes whether they are male or female, and classifies them as having long hair or short hair. Does a relationship exist between sex and hair length?

16.08 Is there a relationship, for adults, between how many text messages they send per week and how many minutes they talk on their phones?

16.09 An education researcher wants to see if class rank in high school is associated with class rank in college.

Application Demonstration

To see how selecting the correct statistical test works in real life, here are three classic studies in psychology: one that uses descriptive statistics, one that uses difference tests, and one that uses relationship tests.

Obedience to Authority

Perhaps the most famous study in psychology was reported by Stanley Milgram in 1963. Called a "behavioral study in obedience," Milgram investigated whether normal people would behave in inhumane ways just because they were following orders.

In the study, 40 males, from age 20 to 50, served as participants and were called "teachers." The teachers were asked to test another participant, called the "learner," to see how well he had memorized a list of words. For every word the learner got wrong, the teacher was asked to punish him with a shock. And, after delivering a shock, the teacher was to adjust the shock generator to deliver a higher level of shock as the next punishment. The shock generator had 30 levels, ranging from 15 to 450 volts, and the levels were labeled from "Slight Shock" to "Danger: Severe Shock." At 300 volts, the learner stopped responding—presumably, he was now unconscious or dead from the shocks—and the teacher was instructed to treat no response as a wrong answer, to administer a shock, and to advance to the next question. (By the way, the learner was a confederate of the experimenter, offered incorrect answers according to a script, and received no shocks.)

This is all that there was to Milgram's experiment. It included no experimental group vs. control group aspect. All he included was an experimental group. Milgram's question was simple: How many participants would continue to give shocks all the way up to 450 volts? And his answer—a descriptive statistic that 26 of 40 normal men, 65%, were willing to shock a man to death in a psychology experiment simply because a researcher in a lab coat asked them to—shocked a nation. Sometimes the simplest statistic is the most powerful.

Television and Aggression

Leonard Eron was one of the first to study the relationship between children watching violence on TV and behaving aggressively in real life. In 1972, along with some colleagues, he presented data on children followed for 10 years, from 9 to 19 years old.

When the children were 9 years old, the researchers measured the number of violent shows the children liked to watch. At the same time, Eron asked the children to answer questions about their peers. From this, he was able to develop a measure for each child as to how aggressive he or she was, as rated by peers. Similar measures were obtained when the kids were 19 years old.

Eron used Pearson correlation coefficients to analyze the results. Here are the highlights of his findings:

- There was a positive, statistically significant correlation of .21 between violent TV watching at age 9 and rating of aggressiveness by peers at age 9.
- The correlation between watching violent TV at age 9 and rating of aggressiveness by peers 10 years later was also positive and statistically significant. But, the relationship was a stronger one: $r = .31$.
- The correlation between violent TV watching at age 19 and aggression at age 19 was not statistically different from zero.

Taken together, these correlations suggest that early viewing of violent TV is more strongly related to aggression 10 years later than it is to aggression as a child. These relationship tests suggest that there is a critical period in childhood during which images from television may have an impact on personality. Eron's research has led to thousands of other studies about the effects of violence on television. Simple correlations can make powerful points.

Language and Memory

Elizabeth Loftus, a cognitive psychologist, is one of the people most responsible for eyewitness testimony no longer being held in high esteem. One of her early studies, published with John Palmer in 1974, used difference tests to make this point.

In that study, participants were shown movies of car crashes and asked to estimate how fast the cars were traveling when the crash occurred. Different participants were asked slightly different questions, ranging from how fast the cars were going when they *contacted* each other to how fast they were going when they *smashed into* each other. There were five different options: smashed, collided, bumped, hit, and contacted. Loftus's research question was whether the different words would elicit different estimates of speed.

And, that was what she found. Asked how fast cars were traveling when they "contacted" each other, the mean response was 32 mph. Each word that suggested more speed was associated with an increase of perceived speed by about 2 mph, until cars that "smashed into" each other were going almost 41 mph. The analysis of variance that Loftus used to analyze these results showed that the results were statistically significant. Here's a *difference* test that made a difference—it showed that perceptions can be manipulated by a person asking about them.

DIY

Many people have pet theories about the world. A friend of mine in graduate school, many years ago, believed that no matter what she bought at the grocery store, the cost always turned out to be about $15 per bag. I believe that the day after a presidential election, many more supporters of the loser have removed bumper stickers from their cars than have supporters of the winner.

Do you have a pet theory? If not, develop one. What statistical test would you use to answer it?

SUMMARY

Define the different tasks of statistics.

- Statistical tests are tools, appropriate for specific tasks. Tasks include summarizing a set of data with descriptive statistics, using difference tests to compare groups in experimental and quasi-experimental designs, and using relationship tests to see if two variables vary together systematically (relationship tests).

Select the correct statistical test.

- Descriptive statistics summarize a set of numbers to describe an individual case or a sample of cases in terms of central tendency and/or variability. Difference tests are used when cases are sorted into groups on the basis of an explanatory variable. Relationship tests are used when there is one sample of cases and each case is measured on two variables to see if the variables covary.

KEY TERMS

descriptive statistics – statistics used to describe a set of observations.

difference tests – statistical tests that look for differences among groups of cases.

relationship tests – statistical tests that determine if two variables in a group of cases covary.

CHAPTER EXERCISES

Review Your Knowledge

16.01 Descriptive statistics ____ a set of observations.

16.02 To determine if one group differs from another, the type of test to use is a ____ test.

16.03 Relationship tests are used to see if ____ vary together systematically.

16.04 If cases are divided into groups, the researcher is probably conducting a ____ test.

16.05 The ____ is thought of as the causal agent.

16.06 The two characteristics of a data set most commonly described are ____ and ____.

16.07 Scores are transformed in order to put them into ____.

16.08 In order to choose the correct ____ to display a variable's frequency, it matters if the variable is discrete or continuous.

16.09 The ____ of the distribution of the data, as well as the variable's ____, determines what measure of central tendency should be chosen.

16.10 This book covered no measures of variability for ____ or ____ variables.

16.11 Level of measurement *is* / *is not* important in deciding what difference test to use.

16.12 Two variables are measured on each case in one sample of cases. This sounds like a ____ test is being planned.

Apply Your Knowledge

Determining the type of study

For Exercises 16.13–16.20, determine if the study is descriptive, experimental, quasi-experimental, or correlational.

16.13 A restaurant chain surveys its customers to learn demographic characteristics such as age in years, marital status, and annual income.

16.14 A developmental psychologist has a teacher rate the boys in his class in terms of how helpful they are. He then classifies the boys as to whether they were ever Boy Scouts or not, and compares the helpfulness of the two groups.

16.15 A clinical psychologist obtained a large sample of adults. Each adult kept track of all the food and beverages he or she consumed for a week. From this, the psychologist calculated the average grams of caffeine consumed per day. The psychologist believed that caffeine, a stimulant, might be used to self-medicate for symptoms of ADHD. So, she also had all participants complete an ADHD symptom inventory. She wanted to see if people with more ADHD symptoms consumed more caffeine.

16.16 An economist wondered if job satisfaction was associated with salary. She obtained a sample of first-year teachers from all around the country and found each one's salary, adjusted for the local cost of living. She also had the teachers complete a job satisfaction survey.

16.17 A cognitive psychologist wondered if the type of computer one used affected how one thought. She put together a sample of kids, randomly divided them into two groups, and gave Apple Macs to one group and Windows PCs to the other group. A year later, she administered the Smith & Jones Measure of Creative and Divergent Thinking to each child.

16.18 A real estate company wondered why some of its agents were more successful than others. The company classified its agents as high-volume producers, medium-volume producers, and low-volume producers. All agents took a personality test and the company compared the three groups on the personality traits measured.

16.19 A milk-processing plant had strict standards regarding how much milk went into each container. A quart container, for example, was supposed to contain no less than 31.5 ounces and no more than 32.5 ounces. Every day, quality control randomly sampled 100 quart containers, precisely measured their contents, and prepared a graph showing the distribution of amount per container.

16.20 A researcher on superstition decided to investigate if walking under a ladder really brought about bad luck. He took 100 people to his laboratory, one at a time. As each person entered, he flipped a coin. If it turned up heads, he had the person walk under a ladder; tails, the person walked next to the ladder. A week later, all the participants returned and reported if anything unlucky had happened to them during the week.

Determining a variable's level of measurement

For Exercises 16.21–16.26, determine (a) whether the variable is measured at the nominal, ordinal, interval, or ratio level, and (b) whether it is continuous or discrete.

16.21 Ounces of milk, as measured in Exercise 16.19

16.22 Whether anything unlucky or not happened, as measured in Exercise16.20

16.23 Marital status

16.24 Level of anxiety before a test, as measured 0–4 on a scale marked 0 = none, 1 = slight, 2 = moderate, 3 = considerable, and 4 = extreme

16.25 Heights of randomly selected pieces of land as measured by meters above or below sea level

16.26 Number of light bulbs in a house

Selecting the right descriptive statistic

For Exercises 16.27–16.30, select the appropriate statistical technique from among the options in Figure 16.2.

16.27 A measure of variability in IQ scores for students in law school

16.28 The most common marital status among 50-year-old men living in Rhode Island

16.29 A graph showing the number of high-, medium-, and low-volume producers from the real estate company in Exercise 16.18

16.30 The average salary for a sample of first-year teachers in the United States

Selecting the correct difference test

For Exercises 16.31–16.38, select the appropriate statistical technique from among those in Figure 16.5.

16.31 10% of the U.S. population is left-handed. Several recent presidents (Barack Obama, Bill Clinton, George H. W. Bush) have been left-handed. This caused a cognitive psychologist to wonder if presidents were representative of the U.S. population in terms of handedness. What test should she use?

16.32 A demographer was curious if an age difference existed between men and women who were getting married for the first time. From wedding announcements in local papers throughout the United States, he randomly sampled 250 couples for whom it was the first marriage for both the bride and the groom. From the announcements, he obtained their ages. What statistical test should he use to compare the ages of husbands and wives?

16.33 In some educational programs, there are a lot of options. One nursing program, for example, allows its students to decide between (a) traditional, semester-long courses; (b) online courses; (c) weekend-only courses; or (d) courses that meet for 8 hours a day, 6 days in a row. The dean of the school wanted to find out if the type of course content had any impact on how much was learned. At the end of its programs, she administered a 50-question, multiple-choice test about basic nursing facts. What test should she use to see if the groups differ on the mean percentage of questions answered correctly?

16.34 Students often gain a few pounds when they move away to college and start eating institutional food. A nutritionist weighed first-year students at the start of the semester, right before Thanksgiving, on their return to school in January, and at the end of the second semester. What statistical test should he use to see if their mean weight changes over time?

16.35 An ecologist obtained a random sample of 85 small rural towns and a random sample of 72 small suburban towns. For each town, he calculated pounds of trash per resident per year that went to the landfill. What statistical test should he use to see if rural and suburban residents differ in the average amount of trash they generate every year?

16.36 A dentist wanted to examine if milk consumption as a child had an impact on the need for dentures by age 50. He classified his patients as having consumed a lot of milk as a child (yes vs. no) and as to whether they wore dentures by age 50 (yes vs. no). What statistical test should he use to determine this?

16.37 The same dentist wanted to find out if his practice was unusual in terms of how many cavities his patients had. He learned that the average American has 11.23 decayed, filled, or missing permanent teeth. In his dental practice, the mean was 8.76, with a standard deviation of 3.42. What statistical test should he use?

16.38 A textbook publisher calculated the percentage of deadlines met by each of its authors. It then classified the authors as having made money for the company, broken even, or lost money for the company. What statistical test should the company use to see if the three types of authors differ in the degree to which they met their deadlines.

Selecting the right relationship test

For Exercises 16.39–16.42, select the appropriate statistical technique from among those in Figure 16.6.

16.39 The dentist from Exercises 16.36 and 16.37 has a new question to contemplate. He wondered if people get most of their cavities when they are young or if they occur consistently over time. To investigate this, he decided to examine the relationship between the ages of his patients and how many of their permanent teeth were decayed, filled, or missing. What test should he use?

16.40 An anti-doping agency decided to investigate if an athlete's eliteness was related to the likelihood of doping. It unexpectedly drug-tested the top 10 finishers in all the track and field events at a large meet. From this, it calculated the percentage of first-place finishers who tested positive, the percentage of second-place finishers who tested positive, and so on. The agency then wanted to examine the relationship between the place in which an athlete finished and the likelihood of testing positive. What test should the agency use?

16.41 A member of a bicycle club wondered if a relationship existed between how expensive a person's bicycle was and how fast that person rode. He found riders with bikes in two categories. One group rode bikes that were classified by a cycling magazine as "decent bikes for not too much money." The others rode bikes classified as "amazing, but even Bill Gates would think twice about spending this much for a bike." The club member then asked each person to ride solo on a track and time his or her fastest lap. What test should the club member use to see if there's a relationship between the cost of the bike and the speed of the rider?

16.42 Some people like cats and some like dogs. Cats are generally thought of as more independent and aloof. A personality psychologist wondered if a relationship existed between the personality trait of introversion/extroversion and the type of pet a person preferred. He obtained a sample of cat owners and a sample of dog owners. Each pet owner took a personality test and was classified as an introvert or an extrovert. What statistical test should be used to see if the two variables are related?

Expand Your Knowledge

In this series of questions, there are no headings that tell which flowchart to use. Plus, some gaps occur in the flowchart. For example, if the scenario has two dependent samples with an ordinal-level dependent variable, there is no such test in the flowchart. Tests exist for such situations; they just weren't covered in this book. So, if a scenario in this series calls for a "missing" test, just write, "No such test in book" as your answer.

16.43 A neonatologist is curious to find out what the long-term effect of premature birth is on mental development. He goes to a large school district, assembles a sample of sixth graders, and determines each student's IQ score. From each student's parents, he finds out if the student was full-term or premature at birth. What statistical test should he complete to figure out if there is a relationship between prematurity and IQ?

16.44 A summer camp director wanted to find out if campers' degree of homesickness changed over time. She took a sample of them at her camp and at three points in time—the beginning, middle, and end of camp—administered the interval-level Homesickness Inventory. What statistical test should she choose to analyze her data?

16.45 A high school principal claimed that academic talent was evenly distributed among the social strata at his high school. To test this, he classified the students as jocks, cheerleaders, techies, stoners, band geeks, and so on and compared class rank among the different groups. What statistical test should he use?

16.46 A health-care researcher obtained a sample of respondents from throughout the United States and asked each one to indicate how worried he or she is about the avian flu on an interval-level scale. Before reporting the results for how worried Americans are about avian flu, the researcher wants to make sure the sample may be representative of the United States in terms of geographic distribution, based on the U.S. Census Bureau's division of the United States into four regions (Northeast, Midwest, South, and West). What statistical test should he use to answer this question?

16.47 A medical school dean wondered whether students who did better in her medical school became better physicians. To test this, she gathered a sample of medical students in their final year and determined their GPAs. She also administered to each student the Differential Diagnosis Test (DDT). The DDT presents students with 50 diagnostic dilemmas. Each dilemma is considered equivalent, and scores on the DDT can range from 0 (no questions answered correctly) to 50 (all answered correctly). What statistical test should the dean use to answer her question?

16.48 Graduates of a particular college were classified as having (a) liberal arts degrees or (b) professional degrees. They were also classified as (1) planning to pursue graduate education or (2) not going on for graduate education. When the graduates returned for their 25th reunion, they completed an interval-level, life satisfaction scale. Is life satisfaction influenced by type of degree and/or graduate study? What statistical test will answer this question?

16.49 A sample of students at a college were asked to classify themselves as (1) Asian; (2) Black; (3) Hispanic; (4) Native American; (5) White; or (6) other. Each person was also asked whether his or her family would consider it acceptable (yes or no) if he or she married someone from a different racial category. What statistical test should be used to see if the degree of acceptability differs among racial/ethnic categories?

16.50 The very last question in this book is a challenging one. It is rare that a researcher does a study that involves just one statistical analysis. Usually, there are multiple dependent variables and a branching series of questions and analyses. Here is an example. An infectious disease specialist believes she has developed a vaccine for the common cold. In order to test its efficacy, she designs a study. As a first step, potential participants call in and provide demographic information (age, sex, race) and complete an interval-level measure of general physical health. After this, they learn more about the study, and about 40% of those who originally called decide that they wish to take part in the study.

The actual study is a double-blind study in which the volunteers will be inoculated either with the active vaccine or a placebo. At the end of the first week, each participant completes a side effect checklist, which counts how many side effects each person has experienced.

For the next 12 months, the research team will keep track of each participant and whether he or she contracts a cold. If a participant contracts a cold, the research team will note how many days have elapsed from the vaccine until the cold. In addition, the researchers will assess the severity of the cold in terms of two dimensions: (1) measuring how many days the cold lasts, and (2) having the cold sufferers report how congested they felt on an ordinal scale.

For each scenario below, decide which statistical test or tests should be used to answer the research question. Be aware—many of these questions involve multiple dependent variables and multiple analyses.

a. The researcher wants to know if those who volunteered for the study differ from those who chose not to volunteer.

b. The researcher wants to know if her sample of volunteers is demographically similar to the U.S. population.

c. Did the active vaccine have more side effects than the placebo?

d. The primary measure of the vaccine's effectiveness is whether or not people in the experimental group, compared to the control group, did or did not contract a cold during the year after the vaccine. How should she answer this question?

e. Secondary measures of the vaccine's effectiveness were, for those who contracted colds, how long the cold lasted and how sick they were. What tests should the researcher do to answer these questions?

f. For the control group participants, the researcher was interested in the relationship between their general physical health and contracting a cold. How can this question be answered?

g. For the control group participants who contracted a cold, the researcher was interested in the relationship between general physical health and the severity of the cold. How should she answer this question?

SPSS

SPSS, like most serious software programs, has a variety of help options. One help option is Statistics Coach, which leads users through a series of questions to the selection of a statistical test. Let's see how it works, using as an example the study in which the effect of early parental loss is measured years later in adults. There were four independent samples (no loss, mother died, father died, both died), and the dependent variable was measured at the interval level.

Figure 16.7 shows the help options available within SPSS. Statistics Coach is on the fifth line down. Clicking on it opens up the screen shown in Figure 16.8, where SPSS asks, "What do you want to do?" There are seven possible answers to the question. Option 1 and option 2 involve descriptive statistics, option 4 applies to difference tests, and option 5 relationship tests. The other three options lead to more advanced statistics.

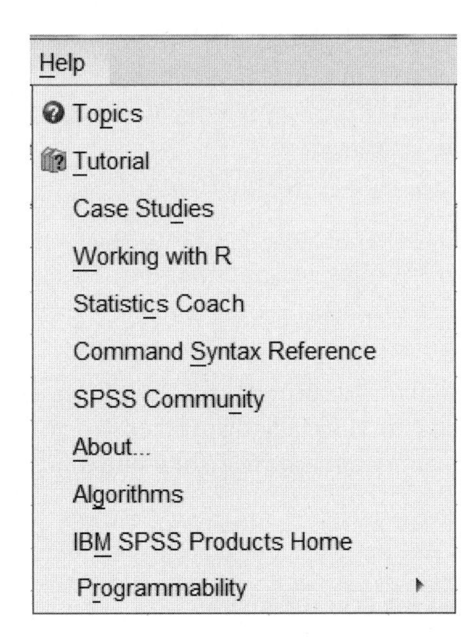

Figure 16.7 The SPSS Help Menu Statistics Coach leads an SPSS user through a series of questions to select an appropriate statistical test. (Source: SPSS)

Statistics Coach >

Statistics Coach

What do you want to do?

Summarize, describe, or present data

Look at variance and distribution of data

Create OLAP report cubes

Compare groups for significant differences

Identify significant relationships between variables

Identify groups of similar cases

Identify groups of similar variables

Figure 16.8 Initiating the Statistics Coach The first step is to decide what one wants to do statistically. (Source: SPSS)

Clicking on the option for difference tests leads to the query, seen in Figure 16.9, about the kind of data. SPSS calls interval and ratio data "scale" data. Here, it is checking how the cases are divided into groups, on the basis of the level of measurement for the variable. Click on the option for "scale numeric data divided into groups."

The next question, seen in Figure 16.10, concerns how many groups there are. This study involves three or more groups.

Statistics Coach > Statistics Coach
Compare groups for significant differences

What kind of data do you want to compare?

Data in categories (nominal, ordinal)

Scale numeric data divided into groups

Figure 16.9 Determining Level of Measurement In this step, SPSS asks one to choose whether one's data are nominal/ordinal or interval/ratio. (Source: SPSS)

Statistics Coach > Statistics Coach > Compare groups for significant differences
Scale numeric data divided into groups

How many groups or variables do you want to compare?

One group or variable compared to a known value

Two groups or variables

Three or more groups

Figure 16.10 Deciding How Many Groups One Has In this step, one decides whether to do a single-sample test, a two-sample test, or a multiple-sample test. (Source: SPSS)

Clicking on three or more groups brings up the next screen, Figure 16.11, which inquires how many "grouping" or "factor" variables exist. These are what the text calls independent variables, and there is only one of them.

Statistics Coach > Statistics Coach > Compare groups for significant differences > Scale numeric data divided into groups
Three or more groups

How many grouping (factor) variables do you have?

One (e.g, revenue for three groups defined by region)

Two or more (e.g., revenue for groups defined by division within each region)

Figure 16.11 Verifying the Number of Independent Variables In this step, the decision is whether to do a one-way test or a multiway test. (Source: SPSS)

Clicking on one grouping variable brings up the next-to-last screen, Figure 16.12, which asks whether the data are normally distributed (i.e., a parametric test), not normally distributed (i.e., a nonparametric test), or if one wants to check.

Statistics Coach > Statistics Coach > Compare groups for significant differences > Scale numeric data divided into groups > Three or more groups
One (e.g, revenue for three groups defined by region)

Which test do you want?

Test that assumes data are normally distributed within groups

Test that does not assume data are normally distributed

Check normality of data

Figure 16.12 Parametric or Nonparametric? In this step, one decides whether a parametric or nonparametric test should be done. (Source: SPSS)

Selecting normally distributed leads to a decision about what test should be done, a one-way ANOVA. As can be seen in Figure 16.13, SPSS also provides information about where the commands are located.

Statistics Coach > Statistics Coach > Compare groups for significant differences > Scale numeric data divided into groups > Three or more groups > One (e.g. revenue for three groups defined by region)

Test that assumes data are normally distributed within groups

To Obtain a One-Way Analysis of Variance

This feature requires the Statistics Base option.

1. From the menus choose:

   Analyze > Compare Means > One-Way ANOVA...

2. Select one or more scale, numeric test variables for the Dependent List.
3. Select a numeric grouping variable (a variable that divides cases into three or more groups) for the Factor.

The factor (grouping) variable must be numeric. If your grouping variable is a string (alphanumeric) variable, use Automatic Recode on the Transform menu to convert the string values to integers.

Related information:
One-Way ANOVA

Figure 16.13 Selecting the Test Not only does SPSS select a test, but it also provides guidance about where to find the commands and what commands to select. (Source: SPSS)

PART IV Test Your Knowledge

These questions are meant to probe your understanding of the material covered in Chapters 13–16. The questions are not in the order of the chapters. Some of them are phrased differently or approach the material from a different direction. A few of them ask you to use the material in ways above and beyond what was covered in the book. This "test" is challenging. But, if you do well on it or puzzle out the answers using the key in the back of the book, you should feel comfortable that you are grasping the material.

1. Given $r(38) = -.44$, $p < .05$, calculate the 95% CI for ρ.

2. Given the following information, find Y':

   $N = 132$, $r = .68$, $M_X = 60$, $s_x = 16$, $M_Y = 55$, $s_y = 17$, and $X = 50$

3. A professor obtains data from a random sample of his former students. The following scatterplot shows the relationship between each student's grade on the first test and his or her grade on the final.

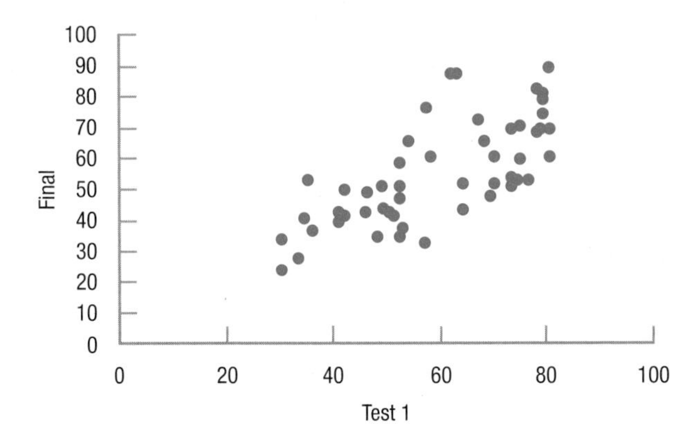

   a. Should the professor calculate a Pearson r? Why or why not?

   b. What type of relationship is shown: direct, indirect, inverse, obverse, curvilinear, or near zero?

   c. What is the predictor variable? What is the criterion variable?

   d. The range of scores for which predictions can be made, roughly, is:

   _____ There is no limitation on the range.
   _____ 0 to 100
   _____ 10 to 90
   _____ 20 to 80
   _____ 30 to 70
   _____ 30 to 80
   _____ 30 to 90
   _____ 25 to 90

4. Read each of the following scenarios and decide which statistical test should be used to analyze the data to answer the question. Be specific: simply indicating t test, ANOVA, or chi-square is not sufficient, as there are multiple versions of these tests. If the answer calls for a test which has not been covered in this text, indicate that.

   a. The level of social anxiety of kindergarten students is measured. The social anxiety level for the same students is measured 8 years later. Does degree of social anxiety at kindergarten predict degree of social anxiety 8 years later?

   b. The level of social anxiety of kindergarten students is measured. The social anxiety level for the same students is measured 8 years later. Does degree of social anxiety change over time?

   c. People with a fear of needles were assigned to receive either relaxation therapy, anti-anxiety medication, or a placebo. They then went to get their blood drawn and measured how many minutes it took for their heart rate to return to normal. Is there a difference in the effectiveness of the various treatments as measured by how long it takes heart rate to return to normal?

   d. People with a fear of needles were assigned to receive either relaxation therapy, anti-anxiety medication, or a placebo. They then went to get their blood drawn and measured how many minutes it took for their heart rate to return to normal. Is there a relationship between how long it takes the heart rate to normalize and the type of treatment received?

e. Candidates for a job were randomly assigned to be interviewed by either the supervisor or by employees. All candidates considered by the majority of interviewers to be of good fit for the job were hired. Six months later, the new hires were rated as being good employees or not good employees. What test should be used to analyze these data to see if supervisors or employees are better predictors of job success?

5. Given the data below, calculate r:

X	Y
93	8
48	11
18	17
$M = 53.00$	$M = 12.00$
$s = 37.75$	$s = 4.58$

6. Answer each of the following questions. (*Sample answer:* An independent-samples t test is used to compare the means of two independent groups.)

a. When should a single sample t-test be used?

b. When should a Spearman r be used?

c. When should a Mann–Whitney U be used?

d. When should a repeated-measures ANOVA be used?

7. There has been a lot of talk about how a Mediterranean diet (which is high in fruits, vegetables, and olives and low in dairy and meat) is good for physical health. A dietician decided to investigate its impact on mental health. He obtained a large and representative sample of Americans 55 years old and gave them the Mediterranean Diet Scale (MDS). The MDS is an interval-level test that measures the degree, from 0% to 100%, to which a person follows the Mediterranean diet. He then waited 15 years and, using the STMIS (Short-Term Memory Impairment Scale), measured the degree of short-term memory impairment being experienced by the now 70-year-old participants. Scores on the interval-level STMIS can range from 0 (no impairment of STM) to 100 (total STM impairment). No assumptions were violated for the Pearson r. The dietician found $r(2{,}498) = -.14$. He calculated the 95% CI as ranging from $-.18$ to $-.10$. Interpret these results.

8. The principal of the elementary school in a small town in Ohio was looking at the roster for the incoming kindergarten class and noticed something unusual—of the 125 students in the class, 81 of them were boys. The principal expected that half the class should be boys and wouldn't have been surprised if it had been 53% or even 55%, but 65% seemed odd. Use a statistical technique to determine if the class is abnormal in terms of how many boys it includes.

9. Many parents use the hand-on-the-forehead test, rather than a thermometer, to decide if their child has a fever. A pediatrician wondered how accurate this method was. He attended several PTA meetings at elementary schools in the area and brought a life-size doll whose forehead temperature could be rapidly switched from 98.6°F to 100.6°F. He invited parents to feel the doll's forehead and indicate whether or not it had a fever. Each parent was randomly assigned to touch either a "healthy" doll (98.6°F) or a doll with a "fever" of 100.6°F. The results are presented in the table below. Use a statistical technique to determine if these parents are accurate in determining whether the doll has a fever. How accurate, or inaccurate, are they?

	Say "Fever"	Say "No fever"
100.6°	610	423
98.6°	433	634

10. The correlation between X and Y is .47. If $X = 73$, $M_X = 128$, $s_x = 12$, $M_Y = 42$, and $s_y = 6$, what is Y'?

APPENDIX A

Statistical Tables

Table 1: *z* Scores

Area under the normal curve

z score	A Below +*z* Above −*z*	B From mean to *z* or to −*z*	C Above +*z* Below −*z*	*z* score	A Below +*z* Above −*z*	B From mean to *z* or to −*z*	C Above +*z* Below −*z*
0.00	50.00%	0.00%	50.00%	0.25	59.87%	9.87%	40.13%
0.01	50.40%	0.40%	49.60%	0.26	60.26%	10.26%	39.74%
0.02	50.80%	0.80%	49.20%	0.27	60.64%	10.64%	39.36%
0.03	51.20%	1.20%	48.80%	0.28	61.03%	11.03%	38.97%
0.04	51.60%	1.60%	48.40%	0.29	61.41%	11.41%	38.59%
0.05	51.99%	1.99%	48.01%	0.30	61.79%	11.79%	38.21%
0.06	52.39%	2.39%	47.61%	0.31	62.17%	12.17%	37.83%
0.07	52.79%	2.79%	47.21%	0.32	62.55%	12.55%	37.45%
0.08	53.19%	3.19%	46.81%	0.33	62.93%	12.93%	37.07%
0.09	53.59%	3.59%	46.41%	0.34	63.31%	13.31%	36.69%
0.10	53.98%	3.98%	46.02%	0.35	63.68%	13.68%	36.32%
0.11	54.38%	4.38%	45.62%	0.36	64.06%	14.06%	35.94%
0.12	54.78%	4.78%	45.22%	0.37	64.43%	14.43%	35.57%
0.13	55.17%	5.17%	44.83%	0.38	64.80%	14.80%	35.20%
0.14	55.57%	5.57%	44.43%	0.39	65.17%	15.17%	34.83%
0.15	55.96%	5.96%	44.04%	0.40	65.54%	15.54%	34.46%
0.16	56.36%	6.36%	43.64%	0.41	65.91%	15.91%	34.09%
0.17	56.75%	6.75%	43.25%	0.42	66.28%	16.28%	33.72%
0.18	57.14%	7.14%	42.86%	0.43	66.64%	16.64%	33.36%
0.19	57.53%	7.53%	42.47%	0.44	67.00%	17.00%	33.00%
0.20	57.93%	7.93%	42.07%	0.45	67.36%	17.36%	32.64%
0.21	58.32%	8.32%	41.68%	0.46	67.72%	17.72%	32.28%
0.22	58.71%	8.71%	41.29%	0.47	68.08%	18.08%	31.92%
0.23	59.10%	9.10%	40.90%	0.48	68.44%	18.44%	31.56%
0.24	59.48%	9.48%	40.52%	0.49	68.79%	18.79%	31.21%

able 1: z Scores (continued)

z score	A Below +z Above −z	B From mean to z or to −z	C Above +z Below −z	z score	A Below +z Above −z	B From mean to z or to −z	C Above +z Below −z
0.50	69.15%	19.15%	30.85%	0.87	80.78%	30.78%	19.22%
0.51	69.50%	19.50%	30.50%	0.88	81.06%	31.06%	18.94%
0.52	69.85%	19.85%	30.15%	0.89	81.33%	31.33%	18.67%
0.53	70.19%	20.19%	29.81%	0.90	81.59%	31.59%	18.41%
0.54	70.54%	20.54%	29.46%	0.91	81.86%	31.86%	18.14%
0.55	70.88%	20.88%	29.12%	0.92	82.12%	32.12%	17.88%
0.56	71.23%	21.23%	28.77%	0.93	82.38%	32.38%	17.62%
0.57	71.57%	21.57%	28.43%	0.94	82.64%	32.64%	17.36%
0.58	71.90%	21.90%	28.10%	0.95	82.89%	32.89%	17.11%
0.59	72.24%	22.24%	27.76%	0.96	83.15%	33.15%	16.85%
0.60	72.57%	22.57%	27.43%	0.97	83.40%	33.40%	16.60%
0.61	72.91%	22.91%	27.09%	0.98	83.65%	33.65%	16.35%
0.62	73.24%	23.24%	26.76%	0.99	83.89%	33.89%	16.11%
0.63	73.57%	23.57%	26.43%	1.00	84.13%	34.13%	15.87%
0.64	73.89%	23.89%	26.11%	1.01	84.38%	34.38%	15.62%
0.65	74.22%	24.22%	25.78%	1.02	84.61%	34.61%	15.39%
0.66	74.54%	24.54%	25.46%	1.03	84.85%	34.85%	15.15%
0.67	74.86%	24.86%	25.14%	1.04	85.08%	35.08%	14.92%
0.68	75.17%	25.17%	24.83%	1.05	85.31%	35.31%	14.69%
0.69	75.49%	25.49%	24.51%	1.06	85.54%	35.54%	14.46%
0.70	75.80%	25.80%	24.20%	1.07	85.77%	35.77%	14.23%
0.71	76.11%	26.11%	23.89%	1.08	85.99%	35.99%	14.01%
0.72	76.42%	26.42%	23.58%	1.09	86.21%	36.21%	13.79%
0.73	76.73%	26.73%	23.27%	1.10	86.43%	36.43%	13.57%
0.74	77.04%	27.04%	22.96%	1.11	86.65%	36.65%	13.35%
0.75	77.34%	27.34%	22.66%	1.12	86.86%	36.86%	13.14%
0.76	77.64%	27.64%	22.36%	1.13	87.08%	37.08%	12.92%
0.77	77.94%	27.94%	22.06%	1.14	87.29%	37.29%	12.71%
0.78	78.23%	28.23%	21.77%	1.15	87.49%	37.49%	12.51%
0.79	78.52%	28.52%	21.48%	1.16	87.70%	37.70%	12.30%
0.80	78.81%	28.81%	21.19%	1.17	87.90%	37.90%	12.10%
0.81	79.10%	29.10%	20.90%	1.18	88.10%	38.10%	11.90%
0.82	79.39%	29.39%	20.61%	1.19	88.30%	38.30%	11.70%
0.83	79.67%	29.67%	20.33%	1.20	88.49%	38.49%	11.51%
0.84	79.95%	29.95%	20.05%	1.21	88.69%	38.69%	11.31%
0.85	80.23%	30.23%	19.77%	1.22	88.88%	38.88%	11.12%
0.86	80.51%	30.51%	19.49%	1.23	89.07%	39.07%	10.93%

z score	A Below +z Above −z	B From mean to z or to −z	C Above +z Below −z	z score	A Below +z Above −z	B From mean to z or to −z	C Above +z Below −z
1.24	89.25%	39.25%	10.75%	**1.61**	94.63%	44.63%	5.37%
1.25	89.44%	39.44%	10.56%	**1.62**	94.74%	44.74%	5.26%
1.26	89.62%	39.62%	10.38%	**1.63**	94.84%	44.84%	5.16%
1.27	89.80%	39.80%	10.20%	**1.64**	94.95%	44.95%	5.05%
1.28	89.97%	39.97%	10.03%	**1.65**	95.05%	45.05%	4.95%
1.29	90.15%	40.15%	9.85%	**1.66**	95.15%	45.15%	4.85%
1.30	90.32%	40.32%	9.68%	**1.67**	95.25%	45.25%	4.75%
1.31	90.49%	40.49%	9.51%	**1.68**	95.35%	45.35%	4.65%
1.32	90.66%	40.66%	9.34%	**1.69**	95.45%	45.45%	4.55%
1.33	90.82%	40.82%	9.18%	**1.70**	95.54%	45.54%	4.46%
1.34	90.99%	40.99%	9.01%	**1.71**	95.64%	45.64%	4.36%
1.35	91.15%	41.15%	8.85%	**1.72**	95.73%	45.73%	4.27%
1.36	91.31%	41.31%	8.69%	**1.73**	95.82%	45.82%	4.18%
1.37	91.47%	41.47%	8.53%	**1.74**	95.91%	45.91%	4.09%
1.38	91.62%	41.62%	8.38%	**1.75**	95.99%	45.99%	4.01%
1.39	91.77%	41.77%	8.23%	**1.76**	96.08%	46.08%	3.92%
1.40	91.92%	41.92%	8.08%	**1.77**	96.16%	46.16%	3.84%
1.41	92.07%	42.07%	7.93%	**1.78**	96.25%	46.25%	3.75%
1.42	92.22%	42.22%	7.78%	**1.79**	96.33%	46.33%	3.67%
1.43	92.36%	42.36%	7.64%	**1.80**	96.41%	46.41%	3.59%
1.44	92.51%	42.51%	7.49%	**1.81**	96.49%	46.49%	3.51%
1.45	92.65%	42.65%	7.35%	**1.82**	96.56%	46.56%	3.44%
1.46	92.79%	42.79%	7.21%	**1.83**	96.64%	46.64%	3.36%
1.47	92.92%	42.92%	7.08%	**1.84**	96.71%	46.71%	3.29%
1.48	93.06%	43.06%	6.94%	**1.85**	96.78%	46.78%	3.22%
1.49	93.19%	43.19%	6.81%	**1.86**	96.86%	46.86%	3.14%
1.50	93.32%	43.32%	6.68%	**1.87**	96.93%	46.93%	3.07%
1.51	93.45%	43.45%	6.55%	**1.88**	96.99%	46.99%	3.01%
1.52	93.57%	43.57%	6.43%	**1.89**	97.06%	47.06%	2.94%
1.53	93.70%	43.70%	6.30%	**1.90**	97.13%	47.13%	2.87%
1.54	93.82%	43.82%	6.18%	**1.91**	97.19%	47.19%	2.81%
1.55	93.94%	43.94%	6.06%	**1.92**	97.26%	47.26%	2.74%
1.56	94.06%	44.06%	5.94%	**1.93**	97.32%	47.32%	2.68%
1.57	94.18%	44.18%	5.82%	**1.94**	97.38%	47.38%	2.62%
1.58	94.29%	44.29%	5.71%	**1.95**	97.44%	47.44%	2.56%
1.59	94.41%	44.41%	5.59%	**1.96**	97.50%	47.50%	2.50%
1.60	94.52%	44.52%	5.48%	**1.97**	97.56%	47.56%	2.44%

Table 1: z Scores (continued)

z score	A Below +z Above −z	B From mean to z or to −z	C Above +z Below −z	z score	A Below +z Above −z	B From mean to z or to −z	C Above +z Below −z
1.98	97.61%	47.61%	2.39%	2.35	99.06%	49.06%	0.94%
1.99	97.67%	47.67%	2.33%	2.36	99.09%	49.09%	0.91%
2.00	97.72%	47.72%	2.28%	2.37	99.11%	49.11%	0.89%
2.01	97.78%	47.78%	2.22%	2.38	99.13%	49.13%	0.87%
2.02	97.83%	47.83%	2.17%	2.39	99.16%	49.16%	0.84%
2.03	97.88%	47.88%	2.12%	2.40	99.18%	49.18%	0.82%
2.04	97.93%	47.93%	2.07%	2.41	99.20%	49.20%	0.80%
2.05	97.98%	47.98%	2.02%	2.42	99.22%	49.22%	0.78%
2.06	98.03%	48.03%	1.97%	2.43	99.25%	49.25%	0.75%
2.07	98.08%	48.08%	1.92%	2.44	99.27%	49.27%	0.73%
2.08	98.12%	48.12%	1.88%	2.45	99.29%	49.29%	0.71%
2.09	98.17%	48.17%	1.83%	2.46	99.31%	49.31%	0.69%
2.10	98.21%	48.21%	1.79%	2.47	99.32%	49.32%	0.68%
2.11	98.26%	48.26%	1.74%	2.48	99.34%	49.34%	0.66%
2.12	98.30%	48.30%	1.70%	2.49	99.36%	49.36%	0.64%
2.13	98.34%	48.34%	1.66%	2.50	99.38%	49.38%	0.62%
2.14	98.38%	48.38%	1.62%	2.51	99.40%	49.40%	0.60%
2.15	98.42%	48.42%	1.58%	2.52	99.41%	49.41%	0.59%
2.16	98.46%	48.46%	1.54%	2.53	99.43%	49.43%	0.57%
2.17	98.50%	48.50%	1.50%	2.54	99.45%	49.45%	0.55%
2.18	98.54%	48.54%	1.46%	2.55	99.46%	49.46%	0.54%
2.19	98.57%	48.57%	1.43%	2.56	99.48%	49.48%	0.52%
2.20	98.61%	48.61%	1.39%	2.57	99.49%	49.49%	0.51%
2.21	98.64%	48.64%	1.36%	2.58	99.51%	49.51%	0.49%
2.22	98.68%	48.68%	1.32%	2.59	99.52%	49.52%	0.48%
2.23	98.71%	48.71%	1.29%	2.60	99.53%	49.53%	0.47%
2.24	98.75%	48.75%	1.25%	2.61	99.55%	49.55%	0.45%
2.25	98.78%	48.78%	1.22%	2.62	99.56%	49.56%	0.44%
2.26	98.81%	48.81%	1.19%	2.63	99.57%	49.57%	0.43%
2.27	98.84%	48.84%	1.16%	2.64	99.59%	49.59%	0.41%
2.28	98.87%	48.87%	1.13%	2.65	99.60%	49.60%	0.40%
2.29	98.90%	48.90%	1.10%	2.66	99.61%	49.61%	0.39%
2.30	98.93%	48.93%	1.07%	2.67	99.62%	49.62%	0.38%
2.31	98.96%	48.96%	1.04%	2.68	99.63%	49.63%	0.37%
2.32	98.98%	48.98%	1.02%	2.69	99.64%	49.64%	0.36%
2.33	99.01%	49.01%	0.99%	2.70	99.65%	49.65%	0.35%
2.34	99.04%	49.04%	0.96%	2.71	99.66%	49.66%	0.34%

Table 2: Random Number Table

	A	B	C	D	E	F
1	8607	1887	5432	2039	5502	3174
2	5574	4576	5273	8582	1424	9439
3	5515	8367	6317	6974	3452	2639
4	0296	8870	3197	4853	4434	1571
5	0149	1919	8684	9082	0335	6276
6	8211	4653	2421	8635	8388	2544
7	2848	7715	5620	2649	7561	0766
8	3007	3419	4373	6721	2428	1532
9	2221	4703	7265	4061	9277	0900
10	8670	0480	0672	8572	9597	7785
11	5475	9133	5481	7966	8873	3147
12	7294	5418	1795	5198	4946	1615
13	3498	1061	4566	0370	3225	8464
14	3186	8239	8706	8345	2373	4830
15	7037	6540	9220	6516	0370	8777
16	3953	0689	3746	6861	4949	3386
17	2136	4209	8825	2571	6623	7126
18	3761	0535	4566	6536	9985	5070
19	8048	6079	2496	9461	2638	9390
20	4275	6909	5832	9159	4191	1325
21	2092	1191	7593	4784	9688	0476
22	8545	4468	8530	8935	1195	6530
23	4863	2618	9081	7876	8383	0235
24	8354	8405	4918	4851	6941	4597
25	8010	5343	3199	3236	6898	2562
26	5158	9039	8902	0905	2472	070[illegible]
27	9943	8717	2530	0421	1351	792[illegible]
28	0540	9804	7933	2358	8892	46[illegible]
29	9761	0723	3059	5386	5249	8[illegible]0
30	1276	9555	5058	5119	1543	2[illegible]66
31	3625	7693	3127	5576	4385	0356
32	7543	6216	2586	6012	7964	9972
33	5065	2734	6829	6362	6208	1577
34	7910	8629	7253	3425	4733	9927
35	9080	8616	4977	0703	3784	2608
36	4142	9849	7180	0053	6437	4342
37	5633	2804	4612	0386	2020	8173
38	5371	2571	1339	4213	1945	4844
39	1440	2099	5031	6049	5047	6239
40	6914	5610	2821	8760	5634	7445

Table 3: Critical Values of *t*

df	α = .05, one-tailed -or- α = .10, two-tailed	α = .025, one-tailed -or- α = .05, two-tailed	α = .01, one-tailed -or- α = .02, two-tailed	α = .005, one-tailed -or- α =.01, two-tailed
1	6.314	12.706	31.821	63.657
2	2.920	4.303	6.965	9.925
3	2.353	3.182	4.541	5.841
4	2.132	2.776	3.747	4.604
5	2.015	2.571	3.365	4.032
6	1.943	2.447	3.143	3.707
7	1.895	2.365	2.998	3.499
8	1.860	2.306	2.896	3.355
9	1.833	2.262	2.821	3.250
10	1.812	2.228	2.764	3.169
11	1.796	2.201	2.718	3.106
12	1.782	2.179	2.681	3.055
13	1.771	2.160	2.650	3.012
14	1.761	2.145	2.624	2.977
15	1.753	2.131	2.602	2.947
16	1.746	2.120	2.583	2.921
17	1.740	2.110	2.567	2.898
18	1.734	2.101	2.552	2.878
19	1.729	2.093	2.539	2.861
20	1.725	2.086	2.528	2.845
21	1.721	2.080	2.518	2.831
22	1.717	2.074	2.508	2.819
23	1.714	2.069	2.500	2.807
24	1.711	2.064	2.492	2.797
25	1.708	2.060	2.485	2.787
26	1.706	2.056	2.479	2.779
27	1.703	2.052	2.473	2.771
28	1.701	2.048	2.467	2.763
29	1.699	2.045	2.462	2.756
30	1.697	2.042	2.457	2.750
31	1.696	2.040	2.453	2.744
32	1.694	2.037	2.449	2.738
33	1.692	2.035	2.445	2.733
34	1.691	2.032	2.441	2.728
35	1.690	2.030	2.438	2.724
36	1.688	2.028	2.434	2.719
37	1.687	2.026	2.431	2.715
38	1.686	2.024	2.429	2.712
39	1.685	2.023	2.426	2.708

df	α = .05, one-tailed -or- α = .10, two-tailed	α = .025, one-tailed -or- α = .05, two-tailed	α = .01, one-tailed -or- α = .02, two-tailed	α = .005, one-tailed -or- α =.01, two-tailed
40	1.684	**2.021**	2.423	2.704
41	1.683	**2.020**	2.421	2.701
42	1.682	**2.018**	2.418	2.698
43	1.681	**2.017**	2.416	2.695
44	1.680	**2.015**	2.414	2.692
45	1.679	**2.014**	2.412	2.690
46	1.679	**2.013**	2.410	2.687
47	1.678	**2.012**	2.408	2.685
48	1.677	**2.011**	2.407	2.682
49	1.677	**2.010**	2.405	2.680
50	1.676	**2.009**	2.403	2.678
55	1.673	**2.004**	2.396	2.668
60	1.671	**2.000**	2.390	2.660
65	1.669	**1.997**	2.385	2.654
70	1.667	**1.994**	2.381	2.648
75	1.665	**1.992**	2.377	2.643
80	1.664	**1.990**	2.374	2.639
85	1.663	**1.988**	2.371	2.635
90	1.662	**1.987**	2.368	2.632
95	1.661	**1.985**	2.366	2.629
100	1.660	**1.984**	2.364	2.626
120	1.658	**1.980**	2.358	2.617
140	1.656	**1.977**	2.353	2.611
160	1.654	**1.975**	2.350	2.607
180	1.653	**1.973**	2.347	2.603
200	1.653	**1.972**	2.345	2.601
250	1.651	**1.969**	2.341	2.596
300	1.650	**1.968**	2.339	2.592
350	1.649	**1.967**	2.337	2.590
400	1.649	**1.966**	2.336	2.588
450	1.648	**1.965**	2.335	2.587
500	1.648	**1.965**	2.334	2.586
600	1.647	**1.964**	2.333	2.584
700	1.647	**1.963**	2.332	2.583
800	1.647	**1.963**	2.331	2.582
900	1.647	**1.963**	2.330	2.581
1000	1.646	**1.962**	2.330	2.581
∞	1.645	**1.960**	2.326	2.576

Table 4a: Critical Values of F, $\alpha = .05$

Denominator degrees of freedom	Numerator degrees of freedom 1	2	3	4	5	6	7
1	161.448	199.500	215.707	224.583	230.162	233.986	236.768
2	18.513	19.000	19.164	19.247	19.296	19.330	19.353
3	10.128	9.552	9.277	9.117	9.013	8.941	8.887
4	7.709	6.944	6.591	6.388	6.256	6.163	6.094
5	6.608	5.786	5.409	5.192	5.050	4.950	4.876
6	5.987	5.143	4.757	4.534	4.387	4.284	4.207
7	5.591	4.737	4.347	4.120	3.972	3.866	3.787
8	5.318	4.459	4.066	3.838	3.687	3.581	3.500
9	5.117	4.256	3.863	3.633	3.482	3.374	3.293
10	4.965	4.103	3.708	3.478	3.326	3.217	3.135
11	4.844	3.982	3.587	3.357	3.204	3.095	3.012
12	4.747	3.885	3.490	3.259	3.106	2.996	2.913
13	4.667	3.806	3.411	3.179	3.025	2.915	2.832
14	4.600	3.739	3.344	3.112	2.958	2.848	2.764
15	4.543	3.682	3.287	3.056	2.901	2.790	2.707
16	4.494	3.634	3.239	3.007	2.852	2.741	2.657
17	4.451	3.592	3.197	2.965	2.810	2.699	2.614
18	4.414	3.555	3.160	2.928	2.773	2.661	2.577
19	4.381	3.522	3.127	2.895	2.740	2.628	2.544
20	4.351	3.493	3.098	2.866	2.711	2.599	2.514
21	4.325	3.467	3.072	2.840	2.685	2.573	2.488
22	4.301	3.443	3.049	2.817	2.661	2.549	2.464
23	4.279	3.422	3.028	2.796	2.640	2.528	2.442
24	4.260	3.403	3.009	2.776	2.621	2.508	2.423
25	4.242	3.385	2.991	2.759	2.603	2.490	2.405
26	4.225	3.369	2.975	2.743	2.587	2.474	2.388
27	4.210	3.354	2.960	2.728	2.572	2.459	2.373
28	4.196	3.340	2.947	2.714	2.558	2.445	2.359
29	4.183	3.328	2.934	2.701	2.545	2.432	2.346
30	4.171	3.316	2.922	2.690	2.534	2.421	2.334
31	4.160	3.305	2.911	2.679	2.523	2.409	2.323
32	4.149	3.295	2.901	2.668	2.512	2.399	2.313
33	4.139	3.285	2.892	2.659	2.503	2.389	2.303
34	4.130	3.276	2.883	2.650	2.494	2.380	2.294
35	4.121	3.267	2.874	2.641	2.485	2.372	2.285
36	4.113	3.259	2.866	2.634	2.477	2.364	2.277
37	4.105	3.252	2.859	2.626	2.470	2.356	2.270
38	4.098	3.245	2.852	2.619	2.463	2.349	2.262

Numerator degrees of freedom								
8	9	10	11	12	13	14	15	16
238.883	240.543	241.882	242.983	243.906	244.690	245.364	245.950	246.464
19.371	19.385	19.396	19.405	19.413	19.419	19.424	19.429	19.433
8.845	8.812	8.786	8.763	8.745	8.729	8.715	8.703	8.692
6.041	5.999	5.964	5.936	5.912	5.891	5.873	5.858	5.844
4.818	4.772	4.735	4.704	4.678	4.655	4.636	4.619	4.604
4.147	4.099	4.060	4.027	4.000	3.976	3.956	3.938	3.922
3.726	3.677	3.637	3.603	3.575	3.550	3.529	3.511	3.494
3.438	3.388	3.347	3.313	3.284	3.259	3.237	3.218	3.202
3.230	3.179	3.137	3.102	3.073	3.048	3.025	3.006	2.989
3.072	3.020	2.978	2.943	2.913	2.887	2.865	2.845	2.828
2.948	2.896	2.854	2.818	2.788	2.761	2.739	2.719	2.701
2.849	2.796	2.753	2.717	2.687	2.660	2.637	2.617	2.599
2.767	2.714	2.671	2.635	2.604	2.577	2.554	2.533	2.515
2.699	2.646	2.602	2.565	2.534	2.507	2.484	2.463	2.445
2.641	2.588	2.544	2.507	2.475	2.448	2.424	2.403	2.385
2.591	2.538	2.494	2.456	2.425	2.397	2.373	2.352	2.333
2.548	2.494	2.450	2.413	2.381	2.353	2.329	2.308	2.289
2.510	2.456	2.412	2.374	2.342	2.314	2.290	2.269	2.250
2.477	2.423	2.378	2.340	2.308	2.280	2.256	2.234	2.215
2.447	2.393	2.348	2.310	2.278	2.250	2.225	2.203	2.184
2.420	2.366	2.321	2.283	2.250	2.222	2.197	2.176	2.156
2.397	2.342	2.297	2.259	2.226	2.198	2.173	2.151	2.131
2.375	2.320	2.275	2.236	2.204	2.175	2.150	2.128	2.109
2.355	2.300	2.255	2.216	2.183	2.155	2.130	2.108	2.088
2.337	2.282	2.236	2.198	2.165	2.136	2.111	2.089	2.069
2.321	2.265	2.220	2.181	2.148	2.119	2.094	2.072	2.052
2.305	2.250	2.204	2.166	2.132	2.103	2.078	2.056	2.036
2.291	2.236	2.190	2.151	2.118	2.089	2.064	2.041	2.021
2.278	2.223	2.177	2.138	2.104	2.075	2.050	2.027	2.007
2.266	2.211	2.165	2.126	2.092	2.063	2.037	2.015	1.995
2.255	2.199	2.153	2.114	2.080	2.051	2.026	2.003	1.983
2.244	2.189	2.142	2.103	2.070	2.040	2.015	1.992	1.972
2.235	2.179	2.133	2.093	2.060	2.030	2.004	1.982	1.961
2.225	2.170	2.123	2.084	2.050	2.021	1.995	1.972	1.952
2.217	2.161	2.114	2.075	2.041	2.012	1.986	1.963	1.942
2.209	2.153	2.106	2.067	2.033	2.003	1.977	1.954	1.934
2.201	2.145	2.098	2.059	2.025	1.995	1.969	1.946	1.926
2.194	2.138	2.091	2.051	2.017	1.988	1.962	1.939	1.918

Table 4a: Critical Values of F, $\alpha = .05$ (continued)

Denominator degrees of freedom	Numerator degrees of freedom						
	1	2	3	4	5	6	7
39	4.091	3.238	2.845	2.612	2.456	2.342	2.255
40	4.085	3.232	2.839	2.606	2.449	2.336	2.249
41	4.079	3.226	2.833	2.600	2.443	2.330	2.243
42	4.073	3.220	2.827	2.594	2.438	2.324	2.237
43	4.067	3.214	2.822	2.589	2.432	2.318	2.232
44	4.062	3.209	2.816	2.584	2.427	2.313	2.226
45	4.057	3.204	2.812	2.579	2.422	2.308	2.221
46	4.052	3.200	2.807	2.574	2.417	2.304	2.216
47	4.047	3.195	2.802	2.570	2.413	2.299	2.212
48	4.043	3.191	2.798	2.565	2.409	2.295	2.207
49	4.038	3.187	2.794	2.561	2.404	2.290	2.203
50	4.034	3.183	2.790	2.557	2.400	2.286	2.199
55	4.016	3.165	2.773	2.540	2.383	2.269	2.181
60	4.001	3.150	2.758	2.525	2.368	2.254	2.167
65	3.989	3.138	2.746	2.513	2.356	2.242	2.154
70	3.978	3.128	2.736	2.503	2.346	2.231	2.143
75	3.968	3.119	2.727	2.494	2.337	2.222	2.134
80	3.960	3.111	2.719	2.486	2.329	2.214	2.126
85	3.953	3.104	2.712	2.479	2.322	2.207	2.119
90	3.947	3.098	2.706	2.473	2.316	2.201	2.113
95	3.941	3.092	2.700	2.467	2.310	2.196	2.108
100	3.936	3.087	2.696	2.463	2.305	2.191	2.103
120	3.920	3.072	2.680	2.447	2.290	2.175	2.087
140	3.909	3.061	2.669	2.436	2.279	2.164	2.076
160	3.900	3.053	2.661	2.428	2.271	2.156	2.067
180	3.894	3.046	2.655	2.422	2.264	2.149	2.061
200	3.888	3.041	2.650	2.417	2.259	2.144	2.056
250	3.879	3.032	2.641	2.408	2.250	2.135	2.046
300	3.873	3.026	2.635	2.402	2.244	2.129	2.040
350	3.868	3.022	2.630	2.397	2.240	2.125	2.036
400	3.865	3.018	2.627	2.394	2.237	2.121	2.032
450	3.862	3.016	2.625	2.392	2.234	2.119	2.030
500	3.860	3.014	2.623	2.390	2.232	2.117	2.028
600	3.857	3.011	2.620	2.387	2.229	2.114	2.025
700	3.855	3.009	2.618	2.385	2.227	2.112	2.023
800	3.853	3.007	2.616	2.383	2.225	2.110	2.021
900	3.852	3.006	2.615	2.382	2.224	2.109	2.020
1000	3.851	3.005	2.614	2.381	2.223	2.108	2.019
∞	3.841	2.996	2.605	2.372	2.214	2.099	2.010

Numerator degrees of freedom								
8	9	10	11	12	13	14	15	16
2.187	2.131	2.084	2.044	2.010	1.981	1.954	1.931	1.911
2.180	2.124	2.077	2.038	2.003	1.974	1.948	1.924	1.904
2.174	2.118	2.071	2.031	1.997	1.967	1.941	1.918	1.897
2.168	2.112	2.065	2.025	1.991	1.961	1.935	1.912	1.891
2.163	2.106	2.059	2.020	1.985	1.955	1.929	1.906	1.885
2.157	2.101	2.054	2.014	1.980	1.950	1.924	1.900	1.879
2.152	2.096	2.049	2.009	1.974	1.945	1.918	1.895	1.874
2.147	2.091	2.044	2.004	1.969	1.940	1.913	1.890	1.869
2.143	2.086	2.039	1.999	1.965	1.935	1.908	1.885	1.864
2.138	2.082	2.035	1.995	1.960	1.930	1.904	1.880	1.859
2.134	2.077	2.030	1.990	1.956	1.926	1.899	1.876	1.855
2.130	2.073	2.026	1.986	1.952	1.921	1.895	1.871	1.850
2.112	2.055	2.008	1.968	1.933	1.903	1.876	1.852	1.831
2.097	2.040	1.993	1.952	1.917	1.887	1.860	1.836	1.815
2.084	2.027	1.980	1.939	1.904	1.874	1.847	1.823	1.802
2.074	2.017	1.969	1.928	1.893	1.863	1.836	1.812	1.790
2.064	2.007	1.959	1.919	1.884	1.853	1.826	1.802	1.780
2.056	1.999	1.951	1.910	1.875	1.845	1.817	1.793	1.772
2.049	1.992	1.944	1.903	1.868	1.837	1.810	1.786	1.764
2.043	1.986	1.938	1.897	1.861	1.830	1.803	1.779	1.757
2.037	1.980	1.932	1.891	1.856	1.825	1.797	1.773	1.751
2.032	1.975	1.927	1.886	1.850	1.819	1.792	1.768	1.746
2.016	1.959	1.910	1.869	1.834	1.803	1.775	1.750	1.728
2.005	1.947	1.899	1.858	1.822	1.791	1.763	1.738	1.716
1.997	1.939	1.890	1.849	1.813	1.782	1.754	1.729	1.707
1.990	1.932	1.884	1.842	1.806	1.775	1.747	1.722	1.700
1.985	1.927	1.878	1.837	1.801	1.769	1.742	1.717	1.694
1.976	1.917	1.869	1.827	1.791	1.759	1.732	1.707	1.684
1.969	1.911	1.862	1.821	1.785	1.753	1.725	1.700	1.677
1.965	1.907	1.858	1.816	1.780	1.748	1.720	1.695	1.672
1.962	1.903	1.854	1.813	1.776	1.745	1.717	1.691	1.669
1.959	1.901	1.852	1.810	1.774	1.742	1.714	1.689	1.666
1.957	1.899	1.850	1.808	1.772	1.740	1.712	1.686	1.664
1.954	1.895	1.846	1.805	1.768	1.736	1.708	1.683	1.660
1.952	1.893	1.844	1.802	1.766	1.734	1.706	1.681	1.658
1.950	1.892	1.843	1.801	1.764	1.732	1.704	1.679	1.656
1.949	1.890	1.841	1.799	1.763	1.731	1.703	1.678	1.655
1.948	1.889	1.840	1.798	1.762	1.730	1.702	1.676	1.654
1.938	1.880	1.831	1.789	1.752	1.720	1.692	1.666	1.644

Table 4b: Critical Values of F, $\alpha = .01$

Denominator degrees of freedom	Numerator degrees of freedom						
	1	2	3	4	5	6	7
1	4052.18	4999.50	5403.35	5624.58	5763.65	5858.99	5928.36
2	98.503	99.000	99.166	99.249	99.299	99.333	99.356
3	34.116	30.817	29.457	28.710	28.237	27.911	27.672
4	21.198	18.000	16.694	15.977	15.522	15.207	14.976
5	16.258	13.274	12.060	11.392	10.967	10.672	10.456
6	13.745	10.925	9.780	9.148	8.746	8.466	8.260
7	12.246	9.547	8.451	7.847	7.460	7.191	6.993
8	11.259	8.649	7.591	7.006	6.632	6.371	6.178
9	10.561	8.022	6.992	6.422	6.057	5.802	5.613
10	10.044	7.559	6.552	5.994	5.636	5.386	5.200
11	9.646	7.206	6.217	5.668	5.316	5.069	4.886
12	9.330	6.927	5.953	5.412	5.064	4.821	4.640
13	9.074	6.701	5.739	5.205	4.862	4.620	4.441
14	8.862	6.515	5.564	5.035	4.695	4.456	4.278
15	8.683	6.359	5.417	4.893	4.556	4.318	4.142
16	8.531	6.226	5.292	4.773	4.437	4.202	4.026
17	8.400	6.112	5.185	4.669	4.336	4.102	3.927
18	8.285	6.013	5.092	4.579	4.248	4.015	3.841
19	8.185	5.926	5.010	4.500	4.171	3.939	3.765
20	8.096	5.849	4.938	4.431	4.103	3.871	3.699
21	8.017	5.780	4.874	4.369	4.042	3.812	3.640
22	7.945	5.719	4.817	4.313	3.988	3.758	3.587
23	7.881	5.664	4.765	4.264	3.939	3.710	3.539
24	7.823	5.614	4.718	4.218	3.895	3.667	3.496
25	7.770	5.568	4.675	4.177	3.855	3.627	3.457
26	7.721	5.526	4.637	4.140	3.818	3.591	3.421
27	7.677	5.488	4.601	4.106	3.785	3.558	3.388
28	7.636	5.453	4.568	4.074	3.754	3.528	3.358
29	7.598	5.420	4.538	4.045	3.725	3.499	3.330
30	7.562	5.390	4.510	4.018	3.699	3.473	3.304
31	7.530	5.362	4.484	3.993	3.675	3.449	3.281
32	7.499	5.336	4.459	3.969	3.652	3.427	3.258
33	7.471	5.312	4.437	3.948	3.630	3.406	3.238
34	7.444	5.289	4.416	3.927	3.611	3.386	3.218
35	7.419	5.268	4.396	3.908	3.592	3.368	3.200
36	7.396	5.248	4.377	3.890	3.574	3.351	3.183
37	7.373	5.229	4.360	3.873	3.558	3.334	3.167
38	7.353	5.211	4.343	3.858	3.542	3.319	3.152

Numerator degrees of freedom								
8	9	10	11	12	13	14	15	16
5981.07	6022.47	6055.85	6083.32	6106.32	6125.86	6142.67	6157.28	6170.10
99.374	99.388	99.399	99.408	99.416	99.422	99.428	99.433	99.437
27.489	27.345	27.229	27.133	27.052	26.983	26.924	26.872	26.827
14.799	14.659	14.546	14.452	14.374	14.307	14.249	14.198	14.154
10.289	10.158	10.051	9.963	9.888	9.825	9.770	9.722	9.680
8.102	7.976	7.874	7.790	7.718	7.657	7.605	7.559	7.519
6.840	6.719	6.620	6.538	6.469	6.410	6.359	6.314	6.275
6.029	5.911	5.814	5.734	5.667	5.609	5.559	5.515	5.477
5.467	5.351	5.257	5.178	5.111	5.055	5.005	4.962	4.924
5.057	4.942	4.849	4.772	4.706	4.650	4.601	4.558	4.520
4.744	4.632	4.539	4.462	4.397	4.342	4.293	4.251	4.213
4.499	4.388	4.296	4.220	4.155	4.100	4.052	4.010	3.972
4.302	4.191	4.100	4.025	3.960	3.905	3.857	3.815	3.778
4.140	4.030	3.939	3.864	3.800	3.745	3.698	3.656	3.619
4.004	3.895	3.805	3.730	3.666	3.612	3.564	3.522	3.485
3.890	3.780	3.691	3.616	3.553	3.498	3.451	3.409	3.372
3.791	3.682	3.593	3.519	3.455	3.401	3.353	3.312	3.275
3.705	3.597	3.508	3.434	3.371	3.316	3.269	3.227	3.190
3.631	3.523	3.434	3.360	3.297	3.242	3.195	3.153	3.116
3.564	3.457	3.368	3.294	3.231	3.177	3.130	3.088	3.051
3.506	3.398	3.310	3.236	3.173	3.119	3.072	3.030	2.993
3.453	3.346	3.258	3.184	3.121	3.067	3.019	2.978	2.941
3.406	3.299	3.211	3.137	3.074	3.020	2.973	2.931	2.894
3.363	3.256	3.168	3.094	3.032	2.977	2.930	2.889	2.852
3.324	3.217	3.129	3.056	2.993	2.939	2.892	2.850	2.813
3.288	3.182	3.094	3.021	2.958	2.904	2.857	2.815	2.778
3.256	3.149	3.062	2.988	2.926	2.871	2.824	2.783	2.746
3.226	3.120	3.032	2.959	2.896	2.842	2.795	2.753	2.716
3.198	3.092	3.005	2.931	2.868	2.814	2.767	2.726	2.689
3.173	3.067	2.979	2.906	2.843	2.789	2.742	2.700	2.663
3.149	3.043	2.955	2.882	2.820	2.765	2.718	2.677	2.640
3.127	3.021	2.934	2.860	2.798	2.744	2.696	2.655	2.618
3.106	3.000	2.913	2.840	2.777	2.723	2.676	2.634	2.597
3.087	2.981	2.894	2.821	2.758	2.704	2.657	2.615	2.578
3.069	2.963	2.876	2.803	2.740	2.686	2.639	2.597	2.560
3.052	2.946	2.859	2.786	2.723	2.669	2.622	2.580	2.543
3.036	2.930	2.843	2.770	2.707	2.653	2.606	2.564	2.527
3.021	2.915	2.828	2.755	2.692	2.638	2.591	2.549	2.512

Table 4b: Critical Values of F, $\alpha = .01$ (continued)

Denominator degrees of freedom	Numerator degrees of freedom						
	1	2	3	4	5	6	7
39	7.333	5.194	4.327	3.843	3.528	3.305	3.137
40	7.314	5.179	4.313	3.828	3.514	3.291	3.124
41	7.296	5.163	4.299	3.815	3.501	3.278	3.111
42	7.280	5.149	4.285	3.802	3.488	3.266	3.099
43	7.264	5.136	4.273	3.790	3.476	3.254	3.087
44	7.248	5.123	4.261	3.778	3.465	3.243	3.076
45	7.234	5.110	4.249	3.767	3.454	3.232	3.066
46	7.220	5.099	4.238	3.757	3.444	3.222	3.056
47	7.207	5.087	4.228	3.747	3.434	3.213	3.046
48	7.194	5.077	4.218	3.737	3.425	3.204	3.037
49	7.182	5.066	4.208	3.728	3.416	3.195	3.028
50	7.171	5.057	4.199	3.720	3.408	3.186	3.020
55	7.119	5.013	4.159	3.681	3.370	3.149	2.983
60	7.077	4.977	4.126	3.649	3.339	3.119	2.953
65	7.042	4.947	4.098	3.622	3.313	3.093	2.928
70	7.011	4.922	4.074	3.600	3.291	3.071	2.906
75	6.985	4.900	4.054	3.580	3.272	3.052	2.887
80	6.963	4.881	4.036	3.563	3.255	3.036	2.871
85	6.943	4.864	4.021	3.548	3.240	3.022	2.857
90	6.925	4.849	4.007	3.535	3.228	3.009	2.845
95	6.909	4.836	3.995	3.523	3.216	2.998	2.833
100	6.895	4.824	3.984	3.513	3.206	2.988	2.823
120	6.851	4.787	3.949	3.480	3.174	2.956	2.792
140	6.819	4.760	3.925	3.456	3.151	2.933	2.769
160	6.796	4.740	3.906	3.439	3.134	2.917	2.753
180	6.778	4.725	3.892	3.425	3.120	2.904	2.740
200	6.763	4.713	3.881	3.414	3.110	2.893	2.730
250	6.737	4.691	3.861	3.395	3.091	2.875	2.711
300	6.720	4.677	3.848	3.382	3.079	2.862	2.699
350	6.708	4.666	3.838	3.373	3.070	2.854	2.691
400	6.699	4.659	3.831	3.366	3.063	2.847	2.684
450	6.692	4.653	3.825	3.361	3.058	2.842	2.679
500	6.686	4.648	3.821	3.357	3.054	2.838	2.675
600	6.677	4.641	3.814	3.351	3.048	2.832	2.669
700	6.671	4.636	3.810	3.346	3.043	2.828	2.665
800	6.667	4.632	3.806	3.343	3.040	2.825	2.662
900	6.663	4.629	3.803	3.340	3.038	2.822	2.659
1000	6.660	4.626	3.801	3.338	3.036	2.820	2.657
∞	6.635	4.605	3.782	3.319	3.017	2.802	2.639

Numerator degrees of freedom								
8	9	10	11	12	13	14	15	16
3.006	2.901	2.814	2.741	2.678	2.624	2.577	2.535	2.498
2.993	2.888	2.801	2.727	2.665	2.611	2.563	2.522	2.484
2.980	2.875	2.788	2.715	2.652	2.598	2.551	2.509	2.472
2.968	2.863	2.776	2.703	2.640	2.586	2.539	2.497	2.460
2.957	2.851	2.764	2.691	2.629	2.575	2.527	2.485	2.448
2.946	2.840	2.754	2.680	2.618	2.564	2.516	2.475	2.437
2.935	2.830	2.743	2.670	2.608	2.553	2.506	2.464	2.427
2.925	2.820	2.733	2.660	2.598	2.544	2.496	2.454	2.417
2.916	2.811	2.724	2.651	2.588	2.534	2.487	2.445	2.408
2.907	2.802	2.715	2.642	2.579	2.525	2.478	2.436	2.399
2.898	2.793	2.706	2.633	2.571	2.517	2.469	2.427	2.390
2.890	2.785	2.698	2.625	2.562	2.508	2.461	2.419	2.382
2.853	2.748	2.662	2.589	2.526	2.472	2.424	2.382	2.345
2.823	2.718	2.632	2.559	2.496	2.442	2.394	2.352	2.315
2.798	2.693	2.607	2.534	2.471	2.417	2.369	2.327	2.289
2.777	2.672	2.585	2.512	2.450	2.395	2.348	2.306	2.268
2.758	2.653	2.567	2.494	2.431	2.377	2.329	2.287	2.249
2.742	2.637	2.551	2.478	2.415	2.361	2.313	2.271	2.233
2.728	2.623	2.537	2.464	2.401	2.347	2.299	2.257	2.219
2.715	2.611	2.524	2.451	2.389	2.334	2.286	2.244	2.206
2.704	2.600	2.513	2.440	2.378	2.323	2.275	2.233	2.195
2.694	2.590	2.503	2.430	2.368	2.313	2.265	2.223	2.185
2.663	2.559	2.472	2.399	2.336	2.282	2.234	2.192	2.154
2.641	2.536	2.450	2.377	2.314	2.260	2.212	2.169	2.131
2.624	2.520	2.434	2.360	2.298	2.243	2.195	2.153	2.114
2.611	2.507	2.421	2.348	2.285	2.230	2.182	2.140	2.102
2.601	2.497	2.411	2.338	2.275	2.220	2.172	2.129	2.091
2.583	2.479	2.392	2.319	2.257	2.202	2.154	2.111	2.073
2.571	2.467	2.380	2.307	2.244	2.190	2.142	2.099	2.061
2.562	2.458	2.372	2.299	2.236	2.181	2.133	2.090	2.052
2.556	2.452	2.365	2.292	2.229	2.175	2.126	2.084	2.045
2.551	2.447	2.360	2.287	2.224	2.170	2.121	2.079	2.040
2.547	2.443	2.356	2.283	2.220	2.166	2.117	2.075	2.036
2.541	2.437	2.351	2.277	2.214	2.160	2.111	2.069	2.030
2.537	2.433	2.346	2.273	2.210	2.155	2.107	2.064	2.026
2.533	2.429	2.343	2.270	2.207	2.152	2.104	2.061	2.023
2.531	2.427	2.341	2.267	2.204	2.150	2.101	2.058	2.020
2.529	2.425	2.339	2.265	2.203	2.148	2.099	2.056	2.018
2.511	2.407	2.321	2.248	2.185	2.130	2.082	2.039	2.000

Table 5a: Studentized Range (q) Values, $\alpha = .05$

Degrees of freedom	k = 2	3	4	5	6	7	8	9	10	11	12
1	17.97	26.98	32.82	37.08	40.41	43.12	45.40	47.36	49.07	50.59	51.96
2	6.09	8.33	9.80	10.88	11.73	12.43	13.03	13.54	13.99	14.40	14.76
3	4.50	5.91	6.83	7.50	8.04	8.48	8.85	9.18	9.46	9.72	9.95
4	3.93	5.04	5.76	6.29	6.71	7.05	7.35	7.60	7.83	8.03	8.21
5	3.64	4.60	5.22	5.67	6.03	6.33	6.58	6.80	7.00	7.17	7.32
6	3.46	4.34	4.90	5.31	5.63	5.90	6.12	6.32	6.49	6.65	6.79
7	3.34	4.17	4.68	5.06	5.36	5.61	5.82	6.00	6.16	6.30	6.43
8	3.26	4.04	4.53	4.89	5.17	5.40	5.60	5.77	5.92	6.05	6.18
9	3.20	3.95	4.42	4.76	5.02	5.24	5.43	5.60	5.74	5.87	5.98
10	3.15	3.88	4.33	4.65	4.91	5.12	5.30	5.46	5.60	5.72	5.83
11	3.11	3.82	4.26	4.57	4.82	5.03	5.20	5.35	5.49	5.61	5.71
12	3.08	3.77	4.20	4.51	4.75	4.95	5.12	5.27	5.40	5.51	5.62
13	3.06	3.73	4.15	4.45	4.69	4.88	5.05	5.19	5.32	5.43	5.53
14	3.03	3.70	4.11	4.41	4.64	4.83	4.99	5.13	5.25	5.36	5.46
15	3.01	3.67	4.08	4.37	4.60	4.78	4.94	5.08	5.20	5.31	5.40
16	3.00	3.65	4.05	4.33	4.56	4.74	4.90	5.03	5.15	5.26	5.35
17	2.98	3.63	4.02	4.30	4.52	4.71	4.86	4.99	5.11	5.21	5.31
18	2.97	3.61	4.00	4.28	4.49	4.67	4.82	4.96	5.07	5.17	5.27
19	2.96	3.59	3.98	4.25	4.47	4.65	4.79	4.92	5.04	5.14	5.23
20	2.95	3.58	3.96	4.23	4.45	4.62	4.77	4.90	5.01	5.11	5.20
24	2.92	3.53	3.90	4.17	4.37	4.54	4.68	4.81	4.92	5.01	5.10
30	2.89	3.49	3.85	4.10	4.30	4.46	4.60	4.72	4.82	4.92	5.00
40	2.86	3.44	3.79	4.04	4.23	4.39	4.52	4.63	4.74	4.82	4.90
60	2.83	3.40	3.74	3.98	4.16	4.31	4.44	4.55	4.65	4.73	4.81
120	2.80	3.36	3.69	3.92	4.10	4.24	4.36	4.47	4.56	4.64	4.71
∞	2.77	3.31	3.63	3.86	4.03	4.17	4.29	4.39	4.47	4.55	4.62

						k						
13	14	15	16	17	18	19	20	30	40	60	80	100
53.20	54.33	55.36	56.32	57.22	58.04	58.83	59.56	65.15	68.92	73.97	77.40	79.98
15.09	15.39	15.65	15.92	16.14	16.38	16.57	16.78	18.27	19.28	20.66	21.59	22.29
10.15	10.35	10.52	10.69	10.84	10.98	11.11	11.24	12.21	12.86	13.76	14.36	14.82
8.37	8.52	8.66	8.79	8.91	9.03	9.13	9.23	10.00	10.53	11.24	11.73	12.10
7.47	7.60	7.72	7.83	7.93	8.03	8.12	8.21	8.88	9.33	9.95	10.37	10.69
6.92	7.03	7.14	7.24	7.34	7.43	7.51	7.59	8.19	8.60	9.16	9.55	9.84
6.55	6.66	6.76	6.85	6.94	7.02	7.10	7.17	7.73	8.11	8.63	8.99	9.26
6.29	6.39	6.48	6.57	6.65	6.73	6.80	6.87	7.40	7.76	8.25	8.59	8.84
6.09	6.19	6.28	6.36	6.44	6.51	6.58	6.64	7.14	7.49	7.96	8.28	8.53
5.94	6.03	6.11	6.19	6.27	6.34	6.41	6.47	6.95	7.28	7.73	8.04	8.28
5.81	5.90	5.98	6.06	6.13	6.20	6.27	6.33	6.79	7.11	7.55	7.85	8.08
5.71	5.80	5.88	5.95	6.02	6.09	6.15	6.21	6.66	6.97	7.39	7.69	7.91
5.63	5.71	5.79	5.86	5.93	6.00	6.06	6.11	6.55	6.85	7.27	7.55	7.77
5.55	5.64	5.71	5.79	5.85	5.92	5.97	6.03	6.46	6.75	7.16	7.44	7.65
5.49	5.57	5.65	5.72	5.79	5.85	5.90	5.96	6.38	6.67	7.07	7.34	7.55
5.44	5.52	5.59	5.66	5.73	5.79	5.84	5.90	6.31	6.59	6.98	7.25	7.46
5.39	5.47	5.54	5.61	5.68	5.73	5.79	5.84	6.25	6.53	6.91	7.18	7.38
5.35	5.43	5.50	5.57	5.63	5.69	5.74	5.79	6.20	6.47	6.85	7.11	7.31
5.31	5.39	5.46	5.53	5.59	5.65	5.70	5.75	6.15	6.42	6.79	7.05	7.24
5.28	5.36	5.43	5.49	5.55	5.61	5.66	5.71	6.10	6.37	6.74	6.99	7.19
5.18	5.25	5.32	5.38	5.44	5.49	5.55	5.59	5.97	6.23	6.58	6.82	7.01
5.08	5.15	5.21	5.27	5.33	5.38	5.43	5.48	5.83	6.08	6.42	6.65	6.83
4.98	5.04	5.11	5.16	5.22	5.27	5.31	5.36	5.70	5.93	6.26	6.48	6.65
4.88	4.94	5.00	5.06	5.11	5.15	5.20	5.24	5.57	5.79	6.09	6.30	6.46
4.78	4.84	4.90	4.95	5.00	5.04	5.09	5.13	5.43	5.64	5.93	6.13	6.28
4.69	4.74	4.80	4.85	4.89	4.93	4.97	5.01	5.30	5.50	5.76	5.95	6.09

Kind permission has been granted to reproduce this table from: Gleason, J. R. 1998. A Table of the Upper Quantile Points of the Studentized Range Distribution. *Stata Technical Bulletin*, 46: 6–10.

Table 5b: Studentized Range (q) Values, $\alpha = .01$

Degrees of freedom	k = 2	3	4	5	6	7	8	9	10	11	12
1	90.02	135.00	164.30	185.60	202.20	215.80	227.20	237.00	245.60	253.20	260.00
2	14.04	19.02	22.29	24.72	26.63	28.20	29.53	30.68	31.69	32.59	33.40
3	8.26	10.62	12.17	13.32	14.24	15.00	15.65	16.21	16.69	17.13	17.53
4	6.51	8.12	9.17	9.96	10.58	11.10	11.54	11.92	12.26	12.57	12.84
5	5.70	6.98	7.80	8.42	8.91	9.32	9.67	9.97	10.24	10.48	10.70
6	5.24	6.33	7.03	7.56	7.97	8.32	8.61	8.87	9.10	9.30	9.49
7	4.95	5.92	6.54	7.01	7.37	7.68	7.94	8.17	8.37	8.55	8.71
8	4.75	5.64	6.20	6.63	6.96	7.24	7.47	7.68	7.86	8.03	8.18
9	4.60	5.43	5.96	6.35	6.66	6.92	7.13	7.33	7.49	7.65	7.78
10	4.48	5.27	5.77	6.14	6.43	6.67	6.88	7.05	7.21	7.36	7.49
11	4.39	5.15	5.62	5.97	6.25	6.48	6.67	6.84	6.99	7.13	7.25
12	4.32	5.05	5.50	5.84	6.10	6.32	6.51	6.67	6.81	6.94	7.06
13	4.26	4.96	5.40	5.73	5.98	6.19	6.37	6.53	6.67	6.79	6.90
14	4.21	4.90	5.32	5.63	5.88	6.09	6.26	6.41	6.54	6.66	6.77
15	4.17	4.84	5.25	5.56	5.80	5.99	6.16	6.31	6.44	6.56	6.66
16	4.13	4.79	5.19	5.49	5.72	5.92	6.08	6.22	6.35	6.46	6.56
17	4.10	4.74	5.14	5.43	5.66	5.85	6.01	6.15	6.27	6.38	6.48
18	4.07	4.70	5.09	5.38	5.60	5.79	5.94	6.08	6.20	6.31	6.41
19	4.05	4.67	5.05	5.33	5.55	5.74	5.89	6.02	6.14	6.25	6.34
20	4.02	4.64	5.02	5.29	5.51	5.69	5.84	5.97	6.09	6.19	6.29
24	3.96	4.55	4.91	5.17	5.37	5.54	5.69	5.81	5.92	6.02	6.11
30	3.89	4.46	4.80	5.05	5.24	5.40	5.54	5.65	5.76	5.85	5.93
40	3.83	4.37	4.70	4.93	5.11	5.27	5.39	5.50	5.60	5.69	5.76
60	3.76	4.28	4.59	4.82	4.99	5.13	5.25	5.36	5.45	5.53	5.60
120	3.70	4.20	4.50	4.71	4.87	5.01	5.12	5.21	5.30	5.38	5.44
∞	3.64	4.12	4.40	4.60	4.76	4.88	4.99	5.08	5.16	5.23	5.29

k												
13	14	15	16	17	18	19	20	30	40	60	80	100
90.02	135.00	164.30	185.60	202.20	215.80	227.20	237.00	245.60	253.20	260.00	77.40	79.98
14.04	19.02	22.29	24.72	26.63	28.20	29.53	30.68	31.69	32.59	33.40	21.59	22.29
8.26	10.62	12.17	13.32	14.24	15.00	15.65	16.21	16.69	17.13	17.53	14.36	14.82
6.51	8.12	9.17	9.96	10.58	11.10	11.54	11.92	12.26	12.57	12.84	11.73	12.10
5.70	6.98	7.80	8.42	8.91	9.32	9.67	9.97	10.24	10.48	10.70	10.37	10.69
5.24	6.33	7.03	7.56	7.97	8.32	8.61	8.87	9.10	9.30	9.49	9.55	9.84
4.95	5.92	6.54	7.01	7.37	7.68	7.94	8.17	8.37	8.55	8.71	8.99	9.26
4.75	5.64	6.20	6.63	6.96	7.24	7.47	7.68	7.86	8.03	8.18	8.59	8.84
4.60	5.43	5.96	6.35	6.66	6.92	7.13	7.33	7.49	7.65	7.78	8.28	8.53
4.48	5.27	5.77	6.14	6.43	6.67	6.88	7.05	7.21	7.36	7.49	8.04	8.28
4.39	5.15	5.62	5.97	6.25	6.48	6.67	6.84	6.99	7.13	7.25	7.85	8.08
4.32	5.05	5.50	5.84	6.10	6.32	6.51	6.67	6.81	6.94	7.06	7.69	7.91
4.26	4.96	5.40	5.73	5.98	6.19	6.37	6.53	6.67	6.79	6.90	7.55	7.77
4.21	4.90	5.32	5.63	5.88	6.09	6.26	6.41	6.54	6.66	6.77	7.44	7.65
4.17	4.84	5.25	5.56	5.80	5.99	6.16	6.31	6.44	6.56	6.66	7.34	7.55
4.13	4.79	5.19	5.49	5.72	5.92	6.08	6.22	6.35	6.46	6.56	7.25	7.46
4.10	4.74	5.14	5.43	5.66	5.85	6.01	6.15	6.27	6.38	6.48	7.18	7.38
4.07	4.70	5.09	5.38	5.60	5.79	5.94	6.08	6.20	6.31	6.41	7.11	7.31
4.05	4.67	5.05	5.33	5.55	5.74	5.89	6.02	6.14	6.25	6.34	7.05	7.24
4.02	4.64	5.02	5.29	5.51	5.69	5.84	5.97	6.09	6.19	6.29	6.99	7.19
3.96	4.55	4.91	5.17	5.37	5.54	5.69	5.81	5.92	6.02	6.11	6.82	7.01
3.89	4.46	4.80	5.05	5.24	5.40	5.54	5.65	5.76	5.85	5.93	6.65	6.83
3.83	4.37	4.70	4.93	5.11	5.27	5.39	5.50	5.60	5.69	5.76	6.48	6.65
3.76	4.28	4.59	4.82	4.99	5.13	5.25	5.36	5.45	5.53	5.60	6.30	6.46
3.70	4.20	4.50	4.71	4.87	5.01	5.12	5.21	5.30	5.38	5.44	6.13	6.28
3.64	4.12	4.40	4.60	4.76	4.88	4.99	5.08	5.16	5.23	5.29	5.95	6.09

Table 6: Critical Values of r

df	α = .05, one-tailed -or- α = .10, two-tailed	α = .025, one-tailed -or- α = .05, two-tailed	α = .01, one-tailed -or- α = .02, two-tailed	α = .005, one-tailed -or- α = .01, two-tailed
1	.988	**.997**	1.000	1.000
2	.900	**.950**	.980	.990
3	.805	**.878**	.934	.959
4	.729	**.811**	.882	.917
5	.669	**.754**	.833	.875
6	.621	**.707**	.789	.834
7	.582	**.666**	.750	.798
8	.549	**.632**	.715	.765
9	.521	**.602**	.685	.735
10	.497	**.576**	.658	.708
11	.476	**.553**	.634	.684
12	.458	**.532**	.612	.661
13	.441	**.514**	.592	.641
14	.426	**.497**	.574	.623
15	.412	**.482**	.558	.606
16	.400	**.468**	.543	.590
17	.389	**.456**	.529	.575
18	.378	**.444**	.516	.561
19	.369	**.433**	.503	.549
20	.360	**.423**	.492	.537
21	.352	**.413**	.482	.526
22	.344	**.404**	.472	.515
23	.337	**.396**	.462	.505
24	.330	**.388**	.453	.496
25	.323	**.381**	.445	.487
26	.317	**.374**	.437	.479
27	.311	**.367**	.430	.471
28	.306	**.361**	.423	.463
29	.301	**.355**	.416	.456
30	.296	**.349**	.409	.449
31	.291	**.344**	.403	.442
32	.287	**.339**	.397	.436
33	.283	**.334**	.392	.430
34	.279	**.329**	.386	.424
35	.275	**.325**	.381	.418
36	.271	**.320**	.376	.413
37	.267	**.316**	.371	.408
38	.264	**.312**	.367	.403

df	α = .05, one-tailed -or- α = .10, two-tailed	α = .025, one-tailed -or- α = .05, two-tailed	α = .01, one-tailed -or- α = .02, two-tailed	α = .005, one-tailed -or- α = .01, two-tailed
39	.260	**.308**	.362	.398
40	.257	**.304**	.358	.393
41	.254	**.301**	.354	.389
42	.251	**.297**	.350	.384
43	.248	**.294**	.346	.380
44	.246	**.291**	.342	.376
45	.243	**.288**	.338	.372
46	.240	**.285**	.335	.368
47	.238	**.282**	.331	.365
48	.235	**.279**	.328	.361
49	.233	**.276**	.325	.358
50	.231	**.273**	.322	.354
55	.220	**.261**	.307	.339
60	.211	**.250**	.295	.325
65	.203	**.240**	.284	.313
70	.195	**.232**	.274	.302
75	.189	**.224**	.265	.292
80	.183	**.217**	.257	.283
85	.178	**.211**	.249	.275
90	.173	**.205**	.242	.267
95	.168	**.200**	.236	.260
100	.164	**.195**	.230	.254
120	.150	**.178**	.210	.232
140	.139	**.165**	.195	.216
160	.130	**.154**	.183	.202
180	.122	**.146**	.172	.190
200	.116	**.138**	.164	.181
250	.104	**.124**	.146	.162
300	.095	**.113**	.134	.148
350	.088	**.105**	.124	.137
400	.082	**.098**	.116	.128
450	.077	**.092**	.109	.121
500	.073	**.088**	.104	.115
600	.067	**.080**	.095	.105
700	.062	**.074**	.088	.097
800	.058	**.069**	.082	.091
900	.055	**.065**	.077	.086
1000	.052	**.062**	.073	.081

Table 7: Fisher's r to z Transformation

First digit of r	Second digit of r: .00	.01	.02	.03	.04	.05	.06	.07	.08	.09
.0	0.00	0.01	0.02	0.03	0.04	0.05	0.06	0.07	0.08	0.09
.1	0.10	0.11	0.12	0.13	0.14	0.15	0.16	0.17	0.18	0.19
.2	0.20	0.21	0.22	0.23	0.24	0.26	0.27	0.28	0.29	0.30
.3	0.31	0.32	0.33	0.34	0.35	0.37	0.38	0.39	0.40	0.41
.4	0.42	0.44	0.45	0.46	0.47	0.48	0.50	0.51	0.52	0.54
.5	0.55	0.56	0.58	0.59	0.60	0.62	0.63	0.65	0.66	0.68
.6	0.69	0.71	0.73	0.74	0.76	0.78	0.79	0.81	0.83	0.85
.7	0.87	0.89	0.91	0.93	0.95	0.97	1.00	1.02	1.05	1.07
.8	1.10	1.13	1.16	1.19	1.22	1.26	1.29	1.33	1.38	1.42
.9	1.47	1.53	1.59	1.66	1.74	1.83	1.95	2.09	2.30	2.65

Table 8: Fisher's z to r Transformation

First 2 digits of z value	Final digit of z value .00	.01	.02	.03	.04	.05	.06	.07	.08	.09
0.0	.00	.01	.02	.03	.04	.05	.06	.07	.08	.09
0.1	.10	.11	.12	.13	.14	.15	.16	.17	.18	.19
0.2	.20	.21	.22	.23	.24	.24	.25	.26	.27	.28
0.3	.29	.30	.31	.32	.33	.34	.35	.35	.36	.37
0.4	.38	.39	.40	.41	.41	.42	.43	.44	.45	.45
0.5	.46	.47	.48	.49	.49	.50	.51	.52	.52	.53
0.6	.54	.54	.55	.56	.56	.57	.58	.58	.59	.60
0.7	.60	.61	.62	.62	.63	.64	.64	.65	.65	.66
0.8	.66	.67	.68	.68	.69	.69	.70	.70	.71	.71
0.9	.72	.72	.73	.73	.74	.74	.74	.75	.75	.76
1.0	.76	.77	.77	.77	.78	.78	.79	.79	.79	.80
1.1	.80	.80	.81	.81	.81	.82	.82	.82	.83	.83
1.2	.83	.84	.84	.84	.85	.85	.85	.85	.86	.86
1.3	.86	.86	.87	.87	.87	.87	.88	.88	.88	.88
1.4	.89	.89	.89	.89	.89	.90	.90	.90	.90	.90
1.5	.91	.91	.91	.91	.91	.91	.92	.92	.92	.92
1.6	.92	.92	.92	.93	.93	.93	.93	.93	.93	.93
1.7	.94	.94	.94	.94	.94	.94	.94	.94	.94	.95
1.8	.95	.95	.95	.95	.95	.95	.95	.95	.95	.96
1.9	.96	.96	.96	.96	.96	.96	.96	.96	.96	.96
2.0	.96	.96	.97	.97	.97	.97	.97	.97	.97	.97
2.1	.97	.97	.97	.97	.97	.97	.97	.97	.97	.98
2.2	.98	.98	.98	.98	.98	.98	.98	.98	.98	.98
2.3	.98	.98	.98	.98	.98	.98	.98	.98	.98	.98
2.4	.98	.98	.98	.98	.98	.99	.99	.99	.99	.99
2.5	.99	.99	.99	.99	.99	.99	.99	.99	.99	.99
2.6	.99	.99	.99	.99	.99	.99	.99	.99	.99	.99
2.7	.99	.99	.99	.99	.99	.99	.99	.99	.99	.99
2.8	.99	.99	.99	.99	.99	.99	.99	.99	.99	.99
2.9	.99	.99	.99	.99	.99	.99	.99	.99	.99	.99
3.0	1.00	1.00	1.00	1.00	1.00	1.00	1.00	1.00	1.00	1.00

Table 9: Power for a Given *N* and a Given Observed or Hypothesized Correlation Value, $\alpha = .05$, Two-Tailed

N	.05	.10	.15	.20	.25	.30	.35	.40	.45	.50
5	.02	.03	.04	.04	.05	.06	.07	.08	.10	.11
6	.03	.03	.04	.05	.06	.07	.09	.11	.13	.15
7	.03	.03	.04	.06	.07	.08	.10	.13	.16	.19
8	.03	.04	.05	.06	.08	.10	.12	.15	.19	.23
9	.04	.04	.05	.07	.09	.11	.14	.17	.21	.26
10	.03	.04	.05	.07	.09	.12	.16	.20	.24	.30
11	.03	.04	.06	.08	.10	.13	.17	.22	.27	.34
12	.03	.04	.06	.08	.11	.15	.19	.24	.30	.37
13	.03	.05	.06	.09	.12	.16	.21	.26	.33	.41
14	.03	.05	.07	.09	.13	.17	.22	.28	.36	.44
15	.03	.05	.07	.10	.14	.18	.24	.31	.38	.47
16	.03	.05	.07	.10	.14	.19	.26	.33	.41	.50
17	.03	.05	.08	.11	.15	.21	.27	.35	.44	.53
18	.03	.05	.08	.12	.16	.22	.29	.37	.46	.56
19	.03	.05	.08	.12	.17	.23	.30	.39	.49	.59
20	.03	.06	.09	.13	.18	.24	.32	.41	.51	.61
21	.04	.06	.09	.13	.19	.25	.34	.43	.53	.64
22	.04	.06	.09	.14	.19	.27	.35	.45	.56	.66
23	.04	.06	.09	.14	.20	.28	.37	.47	.58	.69
24	.04	.06	.10	.15	.21	.29	.38	.49	.60	.71
25	.04	.06	.10	.15	.22	.30	.40	.51	.62	.73
26	.04	.06	.10	.16	.23	.31	.41	.52	.64	.74
27	.04	.07	.11	.16	.23	.32	.43	.54	.66	.76
28	.04	.07	.11	.17	.24	.34	.44	.56	.67	.78
29	.04	.07	.11	.17	.25	.35	.46	.57	.69	.79
30	.04	.07	.12	.18	.26	.36	.47	.59	.71	.81
32	.04	.07	.12	.19	.27	.38	.50	.62	.74	.84
34	.04	.08	.13	.20	.29	.40	.52	.65	.76	.86
36	.04	.08	.13	.21	.31	.42	.55	.68	.79	.88
38	.04	.08	.14	.22	.32	.44	.58	.70	.81	.90
40	.04	.08	.14	.23	.34	.46	.60	.73	.83	.91
42	.04	.09	.15	.24	.35	.48	.62	.75	.85	.92
44	.05	.09	.16	.25	.37	.50	.64	.77	.87	.94
46	.05	.09	.16	.26	.38	.52	.66	.79	.88	.94
48	.05	.09	.17	.27	.40	.54	.68	.81	.90	.95
50	.05	.10	.17	.28	.41	.56	.70	.82	.91	.96
55	.05	.10	.19	.30	.45	.60	.75	.86	.93	.97
60	.05	.11	.20	.33	.48	.64	.78	.89	.95	.98

NOTE: If cell is blank, power is greater than .99.

.55	.60	.65	.70	.75	.80	.85	.90	.95
.13	.16	.19	.23	.27	.34	.42	.54	.73
.18	.22	.26	.32	.39	.47	.58	.72	.88
.23	.28	.34	.41	.49	.59	.70	.83	.95
.28	.34	.41	.49	.58	.69	.80	.90	.98
.32	.39	.47	.56	.66	.76	.86	.95	
.37	.44	.53	.63	.73	.82	.91	.97	
.41	.50	.59	.68	.78	.87	.94	.98	
.45	.54	.64	.73	.83	.90	.96		
.49	.59	.68	.78	.86	.93	.97		
.53	.63	.72	.82	.89	.95	.98		
.57	.67	.76	.85	.92	.96			
.60	.70	.79	.87	.93	.97			
.63	.73	.82	.90	.95	.98			
.66	.76	.85	.91	.96	.98			
.69	.79	.87	.93	.97				
.72	.81	.89	.94	.97				
.74	.83	.90	.95	.98				
.76	.85	.92	.96	.98				
.78	.87	.93	.97					
.80	.88	.94	.97					
.82	.90	.95	.98					
.84	.91	.96	.98					
.85	.92	.96	.98					
.87	.93	.97						
.88	.94	.97						
.89	.94	.98						
.91	.96	.98						
.93	.97							
.94	.97							
.95	.98							
.96	.98							
.97								
.97								
.98								
.98								
.98								

Table 9: Power for a Given N and a Given Observed or Hypothesized Correlation Value, $\alpha = .05$, Two-Tailed (continued)

N	.05	.10	.15	.20	.25	.30	.35	.40	.45	.50
65	.05	.12	.22	.35	.52	.68	.82	.91	.96	
70	.06	.12	.23	.38	.55	.71	.84	.93	.97	
75	.06	.13	.24	.40	.58	.74	.87	.94	.98	
80	.06	.14	.26	.42	.61	.77	.89	.96	.98	
85	.06	.14	.27	.45	.63	.80	.91	.96		
90	.06	.15	.29	.47	.66	.82	.92	.97		
95	.06	.15	.30	.49	.68	.84	.93	.98		
100	.07	.16	.31	.51	.71	.86	.94	.98		
110	.07	.17	.34	.55	.75	.89	.96			
120	.07	.19	.37	.59	.78	.91	.97			
130	.08	.20	.39	.62	.82	.93	.98			
140	.08	.21	.42	.66	.84	.95	.98			
150	.08	.22	.44	.69	.87	.96				
160	.09	.24	.47	.71	.89	.97				
170	.09	.25	.49	.74	.90	.97				
180	.09	.26	.52	.76	.92	.98				
190	.10	.27	.54	.79	.93	.98				
200	.10	.29	.56	.81	.94					
220	.11	.31	.60	.84	.96					
240	.11	.33	.64	.87	.97					
260	.12	.36	.67	.90	.98					
280	.12	.38	.71	.92	.98					
300	.13	.40	.74	.93						
400	.16	.51	.85	.98						
500	.19	.60	.92							
600	.23	.68	.95							
700	.26	.75	.97							
800	.29	.80	.98							
900	.32	.85								
1000	.35	.88								
1500	.49	.97								
2000	.60									
3000	.78									
4000	.88									
5000	.94									
6000	.97									
7000	.98									
8000										

NOTE: If cell is blank, power is greater than .99.
Beta (β) = probability of a Type II error = 1 − power

	.55	.60	.65	.70	.75	.80	.85	.90	.95

Calculated via nQuery Advisor, 5.0.

Table 10: Critical Values of Chi-Square

	α level				α level		
df	.10	.05	.001	*df*	.10	.05	.001
1	2.706	**3.841**	10.828	24	33.196	**36.415**	51.179
2	4.605	**5.991**	13.816	25	34.382	**37.652**	52.620
3	6.251	**7.815**	16.266	26	35.563	**38.885**	54.052
4	7.779	**9.488**	18.467	27	36.741	**40.113**	55.476
5	9.236	**11.070**	20.515	28	37.916	**41.337**	56.892
6	10.645	**12.592**	22.458	29	39.087	**42.557**	58.301
7	12.017	**14.067**	24.322	30	40.256	**43.773**	59.703
8	13.362	**15.507**	26.124	31	41.422	**44.985**	61.098
9	14.684	**16.919**	27.877	32	42.585	**46.194**	62.487
10	15.987	**18.307**	29.588	33	43.745	**47.400**	63.870
11	17.275	**19.675**	31.264	34	44.903	**48.602**	65.247
12	18.549	**21.026**	32.909	35	46.059	**49.802**	66.619
13	19.812	**22.362**	34.528	36	47.212	**50.998**	67.985
14	21.064	**23.685**	36.123	37	48.363	**52.192**	69.346
15	22.307	**24.996**	37.697	38	49.513	**53.384**	70.703
16	23.542	**26.296**	39.252	39	50.660	**54.572**	72.055
17	24.769	**27.587**	40.790	40	51.805	**55.758**	73.402
18	25.989	**28.869**	42.312	41	52.949	**56.942**	74.745
19	27.204	**30.144**	43.820	42	54.090	**58.124**	76.084
20	28.412	**31.410**	45.315	43	55.230	**59.304**	77.419
21	29.615	**32.671**	46.797	44	56.369	**60.481**	78.750
22	30.813	**33.924**	48.268	45	57.505	**61.656**	80.077
23	32.007	**35.172**	49.728				

APPENDIX B

Solutions to Odd-Numbered End-of-Chapter Exercises

Chapter 1

1.01 summarize

1.03 cases

1.05 (a) *X* causes *Y*; (b) *Y* causes *X*; (c) some other variable (*Z* or a confounding variable) causes both *X* and *Y*.

1.07 random assignment

1.09 dependent variable (or outcome variable)

1.11 groups

1.13 independent = cause; effect = dependent

1.15 grouping

1.17 NOIR

1.19 ordinal

1.21 arbitrary; absolute

1.23 population

1.25 statistic; parameter

1.27 descriptive; inferential

1.29 two

1.31 (a) Do metal and plastic handcuffs differ in how much abrasion they cause?
(b) Type of handcuff is the explanatory variable; amount of abrasion is the outcome variable.
(c) Experimental
(d) N/A

1.33 (a) Are "conscientious" students smarter than "nonconscientious" students?
(b) Type of student (conscientious vs. nonconscientious) is the explanatory variable; IQ is the outcome variable.
(c) Quasi-experimental
(d) Socioeconomic status may be a confounding variable. Richer students may be able to afford to be conscientious (to buy all the books, not to have to work so they have more time for studying) and richer students may have gone to better schools, received a better education, and ended up with a higher IQ.

1.35 (a) Are a country's wealth and its greenhouse gas production related?
(b) GDP is the explanatory variable; tons of CO_2 produced is the outcome variable.
(c) Correlational
(d) Population could be a confounding variable. Over the last 50 years, as there were more people in the country, there would be more people to produce and consume goods, increasing both GDP and greenhouse gases.

1.37 (a) Does type of information received affect voting behavior?
(b) Type of information (none vs. positive vs. negative) is the explanatory variable; voting for or against is the outcome variable.
(c) Experimental
(d) N/A

1.39 Nominal

1.41 Ordinal

1.43 Interval

1.45 Ordinal

1.47 Nominal

1.49 (a) Sample
(b) Parameters
(c) Inferential

1.51 Parameter

1.53 $N = 6$

1.55 $\Sigma X^2 = 8^2 + 9^2 + 5^2 + 4^2 + 7^2 + 8^2 = 299.00$

1.57 $\Sigma X = 13 + 18 + 11 = 42.00$

1.59 $\frac{\Sigma X}{N} = \frac{(13 + 18 + 11)}{3} = \frac{42.0000}{3} = 14.00$

1.61 12.68

1.63 121.01

1.65 22.47

1.67 2.53

1.69 c

1.71 (a) Do higher levels of physical distress in cities lead to higher levels of social distress?
(b) Amount of graffiti (high, moderate, or low) is the explanatory variable, and teenage pregnancy rate is the outcome variable.
(c) quasi-experimental
(d) The proportion of the city population that has a high school degree could be a confounding variable. Less-educated teenage girls may be more likely to get pregnant. City residents who have not completed high school may be less likely to be employed and more likely to deface property.

Chapter 2

2.01 count

2.03 values; frequencies

2.05 f_c

2.07 nominal

2.09 grouped

2.11 details

2.13 midpoint

2.15 how many

2.17 nominal; ordinal

2.19 in-between (fractional)

2.21 interval; ratio

2.23 range

2.25 the interval width

2.27 bar graph

2.29 wider; tall

2.31 Y

2.33 touch

2.35 nominal

2.37 modality; skewness; kurtosis

2.39 bimodal

2.41 negative

2.43 ungrouped

2.45 Don't forget a title!

Frequency Distribution for Number of Psychiatric Diagnoses in 45 Residents of a Psychiatric Hospital

Number of Diagnoses	Frequency	Cumulative Frequency	Percentage	Cumulative Percentage
5	1	45	2.22	100.00
4	2	44	4.44	97.78
3	9	42	20.00	93.33
2	19	33	42.22	73.33
1	12	14	26.67	31.11
0	2	2	4.44	4.44

2.47

Grouped Frequency Distribution for Number of Chairs in 41 College Classrooms (Interval Width = 20)

Number of Chairs	Interval Midpoint	Frequency	Cumulative Frequency	Percentage	Cumulative Percentage
90–109	99.50	1	41	2.44	100.00
70–89	79.50	1	40	2.44	97.56
50–69	59.50	11	39	26.83	95.12
30–49	39.50	17	28	41.46	68.29
10–29	19.50	11	11	26.83	26.83

2.49 (a) Discrete

(b) Continuous

(c) Discrete

(d) Discrete

(e) Continuous

2.51

	Real Lower Limit	Real Upper Limit	Interval Width	Midpoint
a.	30.50	34.50	4.00	32.50
b.	25.00	45.00	20.00	35.00
c.	2.495	2.995	.500	2.745
d.	10.50	11.50	1.00	11.00

2.53

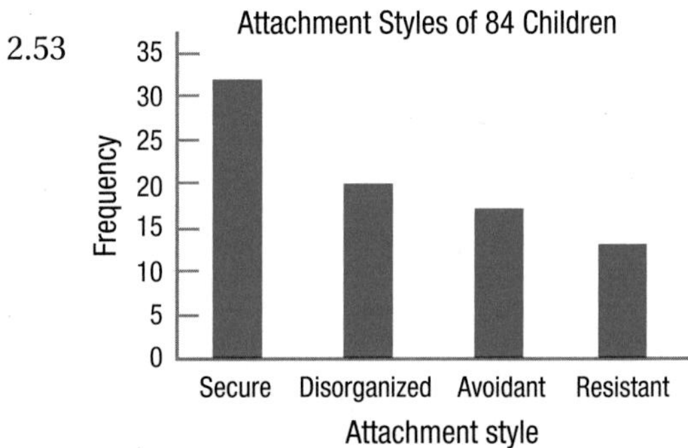

Since the data are nominal, the order is arbitrary and discussing the shape of the distribution does not make sense.

2.55 Either a histogram or frequency polygon can be made for these data.

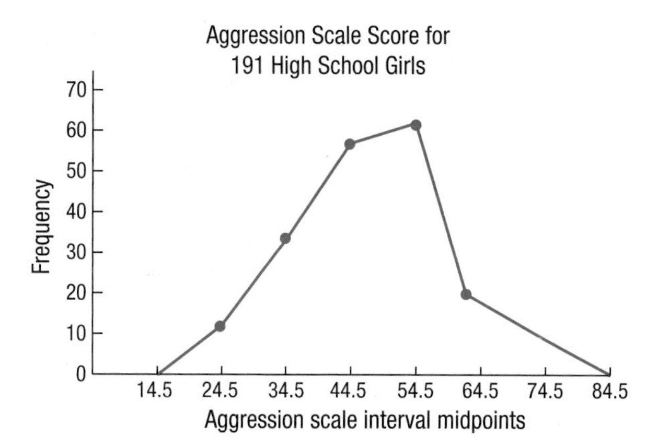

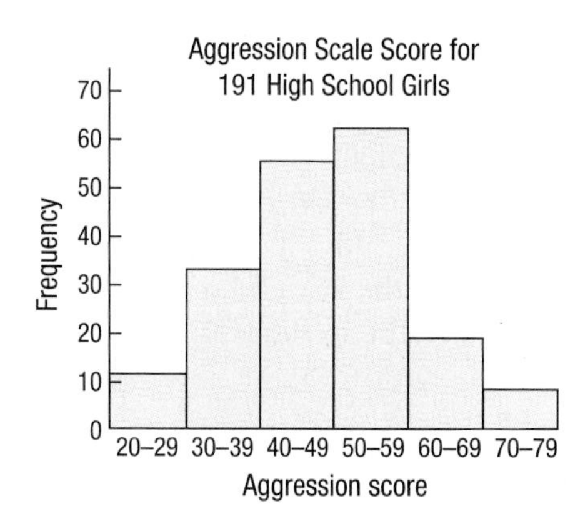

The distribution is unimodal, doesn't seem too peaked or too flat, and is not skewed.

2.57

1	12
2	13779
3	003459
4	357
5	33
6	7
7	5
8	4
9	6

The distribution is unimodal, doesn't seem too peaked or too flat, and is positively skewed.

2.59 b

2.61 e

2.63 Shortest time: 4 minutes: Leave at 10:16:59, return at 10:21:00. 241 seconds, or 4 minutes and one second.
Longest time: 6 minutes: Leave at 10:16:00, return at 10:21:59. 359 seconds, or 5 minutes and 59 seconds.

Chapter 3

3.01 central tendency; variability

3.03 variability

3.05 0

3.07 less

3.09 mean

3.11 mode

3.13 interval; ratio

3.15 is

3.17 more

3.19 population; sample

3.21 variance

3.23 cases

3.25 tightly

3.27 square

3.29 σ

3.31 corrected; population

3.33 deviation score

3.35 The data are ratio-level, but there is an outlier, so the median is the most appropriate measure of central tendency. $Mdn = 4.00$.

3.37 They are ratio-level data, and there isn't anything unusual in terms of skewness or modality, so the mean is the most appropriate measure of central tendency.

$$M = \frac{\Sigma X}{N} = 23.17$$

3.39 $M = \dfrac{80 + 88 + 76 + 65 + 59 + 77}{6} = 74.17$

3.41 $N = 47$, so median is the score associated with case number $\dfrac{47+1}{2} = 24$. Case number 24 has a value of 102, so $Mdn = 102.00$. The value of 102 is also the most commonly occurring score, so mode = 102.00.

3.43 (a) Range = 201 – 37 = \$164 million. (b) The interquartile range is a better measure of variability to report because it trims off the extreme scores. And the value of \$201 million seems to be an outlier, more than \$50 million higher than the next value.

3.45 (a) $N = 5$
(b) $\Sigma(X - M) = (47 - 50.00) + (53 - 50.00) + (67 - 50.00) + (45 - 50.00) + (38 - 50.00) = 0.00$
(c) $\Sigma(X - M)^2 = (47 - 50.00)^2 + (53 - 50.00)^2 + (67 - 50.00)^2 + (45 - 50.00)^2 + (38 - 50.00)^2 = 476.00$

3.47 $s^2 = \frac{\Sigma(X-M)^2}{N-1}$

$= \frac{(57-63)^2 + (60-63)^2 + (67-63)^2 + (68-63)^2}{4-1}$

$= 28.67$

3.49 $M = \frac{\Sigma X}{N} = \frac{21+18+12+9}{4} = 15.00$

$s^2 = \frac{\Sigma(X-M)^2}{N-1}$

$= \frac{(21-15.00)^2 + (18-15.00)^2 + (12-15.00)^2 + (9-15.00)^2}{4-1}$

$= 30.00$

3.51 $s = \sqrt{s^2} = \sqrt{6.55} = 2.56$

3.53 $s = \sqrt{\frac{\Sigma(X-M)^2}{N-1}}$

$= \sqrt{\frac{(24.00-37.20)^2 + (36.00-37.20)^2 + (42.00-37.20)^2 + (50.00-37.20)^2 + (34.00-37.20)^2}{5-1}}$

$= 9.65$

3.55 d

3.57 b

3.59 (a) The larger sample, $N = 100$, will have a smaller correction for s^2 to approximate the population variance. (b) As sample size gets larger, the sample probably captures more of the variability that exists in the population, so the formula "corrects" the standard deviation less.

Chapter 4

4.01 raw

4.03 mean; standard deviation

4.05 zero

4.07 mean; standard deviation

4.09 mean; median

4.11 decreases

4.13 rare

4.15 random

4.17 0.01

4.19 the same

4.21 less than

4.23 at or below

4.25 probability

4.27 the total number of possible outcomes

4.29 fraction; proportion; percentage

4.31 1.00; .00 (or, 100%; 0%)

4.33 .05; likely

4.35 3

4.37 $z = \frac{X-M}{s} = \frac{9.5-7}{2} = 1.25$

4.39 $X = M + (z \times s) = 12.75 + (1.45 \times 3.33) = 17.58$

4.41 $z = \frac{X-M}{s} = \frac{73-100}{15} = -1.80$

4.43 $X = M + (z \times s) = 100 + (0 \times 15) = 100.00$ (One could also complete this problem without doing any math as a z score of zero is right at the mean, and the mean for IQ scores is 100.)

4.45 $z_{SAT} = \frac{X-M}{s} = \frac{620-500}{100} = 1.20$

$z_{Spelling} = \frac{X-M}{s} = \frac{72-60}{15} = 0.80$

Hillary did better on the math subtest of the SAT than she did on the spelling test.

4.47 9.01%

4.49 80.23%

4.51 47.50%

4.53 $z = \frac{X-M}{s} = \frac{90-77}{11} = 1.18$

11.90% of cases in a normal distribution have a z score ≥ 1.18. Thus, 11.90% of Americans should have high diastolic blood pressure (90 or higher).

4.55 $PR = 1.70$

4.57 $z = \frac{X-M}{s} = \frac{113-100}{15} = 0.87$

$PR = 80.78$

$PR = 42.07$

4.59 PR of 88.5 = z score of 1.20

SAT = $X = M + (z \times s) = 500 + (1.20 \times 100) = 620.00$

4.61 $p(\text{boy}) = \frac{18}{28} = .64$

4.63 This is 5% on each side of the midpoint. Use column B: ±0.13

4.65 First find z scores for middle 84% (42% on each side of the midpoint). Use column B: ±1.41. Then calculate IQ scores: $100 + (-1.41 \times 15) = 78.85$ and $100 + (1.41 \times 15) = 121.15$.

The middle 84% of IQ scores fall from an IQ score of 78.85 to 121.15. (That could also be written as 100 ± 21.15.)

4.67 That's 2% in each tail. Use column C: ± 2.05

4.69 That's 7.5% in each tail. Using column C, *z* scores $= \pm 1.44$. Then calculate SAT subtest scores: $SAT = 500 + (-1.44 \times 100) = 356.00$ and $SAT = 500 + (1.44 \times 100) = 644.00$. The SAT scores are 356.00 and 644.00.

4.71 0.135%, using column C

4.73 Using column B, 14.80% of scores fall from the mean to a *z* score of 0.38, so $2 \times 14.80 = 29.60\%$ of scores fall within 0.38 standard deviations of the mean.

4.75 Using column B, 19.15% of scores fall from the mean to a *z* score of 0.5, so $2 \times 48.21 = 38.30\%$ of scores fall within a half standard deviations of the mean. Thus, *p*(Not with a half standard deviation) $= 1 - .3830 = .6170$, or 61.7%.

4.77 That's 5% in each tail. Using column C, the *z* score should be 1.645.

4.79 That's 5% in the upper tail. Using column C, the *z* score should be 1.645.

4.81 1.00

4.83 c

4.85 $z = -5.00$. (*z* scores are deviation scores and the sum of a set of deviation scores is always zero. The first four cases sum to 5, so the one missing case must have a value of -5 in order to bring the sum of all five *z* scores to zero.)

4.87 There are more genius dogs than genius cats. Cats, with a smaller standard deviation, have scores that cluster closer to the mean. Thus, they have fewer scores that fall far away from the mean, which means fewer cats will score in the genius range. (This also means that there are more stupid dogs than stupid cats.)

4.89 Since the normal distribution is symmetric, the mean is $\frac{15 + 45}{2} = 30$. Assuming a very, very rare event has a *z* score of 3, the standard deviation is $\frac{45 - 30}{3} = 5$.

Chapter 5

5.01 population; subset

5.03 attributes; proportion

5.05 convenience

5.07 random number table ("computer" is also an acceptable answer)

5.09 population

5.11 differ

5.13 70

5.15 random

5.17 small; large

5.19 standard error of the mean

5.21 30

5.23 mean

5.25 s

5.27 range; sample; population

5.29 1.96

5.31 $M = 10.5$, 95%CI [7.32,13.68]

5.33 07, 18, 32, 20, 39, 02, 31, 45, 14, and 24

5.35 Consent rate is less than 70%, so the director should not pay attention to the results (or should at least be skeptical).

5.37 $\sigma_M = \frac{\sigma}{\sqrt{N}} = \frac{4}{\sqrt{88}} = 0.43$

5.39 $s_M = \frac{s}{\sqrt{N}} = \frac{7.50}{\sqrt{72}} = 0.88$

5.41
$$\begin{aligned} 95\%CI_\mu &= M \pm (1.96 \times \sigma_M) \\ &= 17.00 \pm (1.96 \times 4) \\ &= 17.00 \pm 7.84 \\ M &= 17.00,\ 95\%CI\ [9.16, 24.84] \end{aligned}$$

5.43
$$\begin{aligned} 95\%CI_\mu &= M \pm (1.96 \times s_M) \\ &= M \pm \left(1.96 \times \frac{s}{\sqrt{n}}\right) \\ &= -12.00 \pm \left(1.96 \times \frac{17}{\sqrt{72}}\right) \\ &= -12.00 \pm 3.93 \\ M &= -12.00,\ 95\%CI\ [-15.93, -8.07] \end{aligned}$$

5.45 c

5.47 As the population standard deviation increases, so does the standard error of the mean as long as the sample size is held constant.

5.49
$$90\%CI_\mu = M \pm (1.645 \times \sigma_M)$$
$$= M \pm \left(1.645 \times \frac{\sigma}{\sqrt{n}}\right)$$
$$= 100.00 \pm \left(1.645 \times \frac{12}{\sqrt{81}}\right)$$
$$= 100.00 \pm 2.19$$
$$M = 100.00,\ 90\%CI\ [97.81, 102.19]$$

5.51 Increase the sample size, which would reduce the standard error of the mean.

Chapter 6

6.01 explanation; facts
6.03 hypothesis
6.05 H_0; H_1
6.07 specific; negative
6.09 alternative
6.11 disprove
6.13 true
6.15 sampling error
6.17 (1) Pick a test, (2) check the assumptions, (3) state the hypotheses, (4) set the decision rule, (5) calculate the value of the test statistic, (6) interpret the results.
6.19 assumptions
6.21 nondirectional
6.23 one; two
6.25 critical value
6.27 interpretation
6.29 population
6.31 influence
6.33 population mean
6.35 zero
6.37 rare zone (or extreme sections)
6.39 rare
6.41 -1.96; $+1.96$
6.43 z; $+1.96$
6.45 α
6.47 sample mean; population mean
6.49 has rejected
6.51 statistically different
6.53 $p < .05$
6.55 does not
6.57 be rejected
6.59 can't
6.61 5%
6.63 larger; easier
6.65 low; reject
6.67 5% (or .05)
6.69 Type II; Type I
6.71 it should be rejected
6.73 Type I; Type II
6.75 a single-sample z test
6.77 The random sample assumption is not violated. The independence assumption is not violated. School readiness is a psychological variable and probably is normally distributed. Since no assumptions are violated, it is OK to proceed with the single-sample z test.
6.79 H_0: $\mu = 60$
H_1: $\mu \neq 60$
6.81 If $z \leq -1.96$ *or* if $z \geq 1.96$, reject H_0.
If $-1.96 < z < 1.96$, fail to reject H_0.
6.83 $\sigma_M = \frac{\sigma}{\sqrt{N}} = \frac{10}{\sqrt{58}} = 1.31$
6.85 $z = \frac{M - \mu}{\sigma_M} = \frac{100 - 120}{17.5} = -1.14$
6.87 (a) $\sigma_M = \frac{\sigma}{\sqrt{N}} = \frac{5}{\sqrt{28}} = 0.94$
(b) $z = \frac{M - \mu}{\sigma_M} = \frac{12 - 10}{0.94} = 2.13$
6.89 (a) $2.37 \geq 1.96$, so H_0 rejected;
(b) $z\ (N=23) = 2.37, p < .05$
6.91 (a) There was a statistically significant difference; (b) the sample mean is higher than the population mean.
6.93 A researcher compared the mean weight of American women who want to lose weight (178 pounds) to the average weight of American women (164 pounds). The difference was statistically significant, $z\ (N = 123) = 3.68$, $p < .05$. Women in America who do want to lose weight are heavier than American women in general. In future studies, it would be important to make sure that the sample is representative of women who want to lose weight.
6.95 (a) He made a Type I error. (b) He should have concluded that there is not enough evidence to say a difference exists.
6.97 power $= 1 - \beta = 1 - .75 = .25$
6.99 **Step 1** Pick a Test
Comparing a sample mean to a population mean where the population standard deviation is known calls for a single-sample z test.

Step 2 Check the Assumptions

The random samples assumption is not violated. There is no evidence of violation of the independence of observations assumption. The normality assumption is not violated, as it seems reasonable to assume sodium consumption is normally distributed. No assumptions are violated, so it is OK to proceed with the single-sample z test.

Step 3 List the Hypotheses

The hypotheses are nondirectional.

H_0: $\mu_{\text{Dieters}} = 3{,}400$

H_1: $\mu_{\text{Dieters}} \neq 3{,}400$

Step 4 Set the Decision Rule

It is a two-tailed test and the dietician is willing to make a Type I error 5% of the time, so $\alpha = .05$:

Reject H_0 if $z \leq -1.96$ *or* if $z \geq 1.96$.

Fail to reject H_0 if $-1.96 < z < 1.96$.

Step 5 Calculate the Test Statistic

$$\sigma_M = \frac{\sigma}{\sqrt{N}} = \frac{270}{\sqrt{172}} = 20.59$$

$$z = \frac{M - \mu}{\sigma_M} = \frac{2{,}900 - 3{,}400}{20.59} = -24.28$$

Step 6 Interpret the Results

A dietitian compared the sodium intake of a random sample of dieting Americans ($M = 2{,}900$ mg sodium) to the mean daily sodium intake of Americans ($\mu = 3{,}400$ mg sodium). The difference was a statistically significant one ($z\,(N = 172) = -24.28$, $p < .05$). Dieters' sodium intake is lower than for Americans in general. From this study, it is not clear whether this is because they were dieting or if their sodium intake were lower to begin with. Future research should investigate this.

6.101 b (If $\beta = .60$, then power $= .40$.)

6.103 b (The size of the effect, the difference between the two means, is larger.)

6.105 correct

6.107 incorrect; Type II error

6.109 There is a way to calculate beta, but we haven't covered it. I hope you'll take a more advanced statistics class some time and learn how to calculate it. So, why ask this question? Because I don't want you to assume that $\beta = .05$, like α. And, I don't want you to assume that $\beta = .20$.

6.111 (a) This is a one-tailed test because he has made a prediction about the direction of the results before collecting any data.

(b) H_0: $\mu_{\text{Athletes}} \geq 78$

H_1: $\mu_{\text{Athletes}} < 78$

6.113 The statistician is comparing the mean of a sample to the mean of a population, and the population standard deviation is known. This sounds appropriate for a single-sample z test. However, before using the test, the assumptions must be checked. The random samples assumption is not violated, and neither is the independence of observations assumption. However, the normality assumption is violated. With random numbers, all numbers are equally likely. So, a 50 is as likely to occur as is a 0 or a 99. This assumption is robust to violations, but this distribution is not even close to normally distributed (it's flat, not normal), so the single-sample z test should not be done.

Chapter 7

7.01 does not

7.03 sampling error

7.05 is

7.07 independent

7.09 robust

7.11 two

7.13 does

7.15 $\neq$

7.17 increases

7.19 tails; I; sample size (or degrees of freedom)

7.21 one

7.23 2.5

7.25 N

7.27 s_M

7.29 facts

7.31 effect size; confidence interval

7.33 alternative hypothesis

7.35 sample mean; population mean

7.37 $p < .05$

7.39 not enough evidence

7.41 independent variable; dependent variable

7.43 no

7.45 decreases

7.47 II

7.49 outcome variable; explanatory variable

7.51 95; population

7.53 effect size

7.55 small

7.57 Single-sample *t* test

7.59 (a) The random sample assumption is violated. The independence assumption is not violated. The normality assumption is not violated. It is OK to proceed with the hypothesis test since the only assumption violated is a robust one. (b) With a convenience sample, it is doubtful that the sample is representative of the population at the college. Results can't be generalized to all students at the college.

7.61 H_0: $\mu_{\text{Nuns}} = 25$
H_1: $\mu_{\text{Nuns}} \neq 25$

7.63 H_0: $\mu_{\text{NewTreatment}} \geq 6.30$
H_1: $\mu_{\text{NewTreatment}} < 6.30$

7.65 $t_{cv} = 2.120$

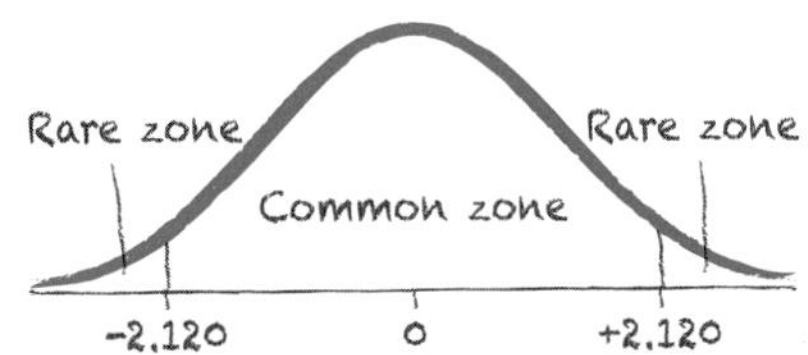

7.67 If $t \leq -2.000$ *or* if $t \geq 2.000$, reject H_0.
If $-2.000 < t < 2.000$, fail to reject H_0.

7.69 $$s_M = \frac{s}{\sqrt{N}} = \frac{12}{\sqrt{23}} = 2.50$$

7.71 $$t = \frac{M - \mu}{s_M} = \frac{10 - 12}{1.25} = -1.60$$

7.73 $$s_M = \frac{s}{\sqrt{N}} = \frac{1}{\sqrt{18}} = 0.24$$

$$t = \frac{M - \mu}{s_M} = \frac{12 - 10}{0.24} = 8.33$$

7.75 Rejected

7.77 Not rejected

7.79 $t(14) = 2.15, p < .05$

7.81 $t(68) = 1.99, p > .05$

7.83 $$d = \frac{M - \mu}{s} = \frac{90 - 100}{15} = -0.67$$

7.85 $$r^2 = \frac{t^2}{t^2 + df} \times 100 = \frac{3.45^2}{3.45^2 + 16} \times 100 = 42.66\%$$

7.87 $$95\%\ \text{CI}\mu_{\text{Diff}} = (M - \mu) \pm (t_{cv} \times s_M)$$
$$= (45 - 50) \pm (2.093 \times 2.24)$$
$$= [-9.69, -0.31]$$

7.89 A nurse practitioner found that people who are heavy salt users have a statistically significantly higher blood pressure ($M = 138$) than do people in the general population [$\mu = 120$; $t(23) = 5.50, p < .05$]. The 18-point increase in blood pressure is a large increase. This study suggests that heavy salt consumption puts one at risk for hypertension. In the larger population from which this sample came, the mean difference might be as small as 11 points or as large as 25 points. Even an 11-point increase in blood pressure has meaningful negative consequences for health. To increase confidence that salt consumption is associated with an increase in blood pressure, it would be a good idea to replicate this study. When doing so, a larger sample would allow the researcher to get a better estimate of the size of the effect in the population.

7.91 **Step 1** Pick a Test

Comparing a sample mean to a population mean when the population standard deviation is not known calls for a single-sample *t* test.

Step 2 Check the Assumptions

(1) The random sample assumption is violated, but the single-sample *t* test is robust to violations. (2) There is no evidence of violation of the independence of observations. (3) For the normality assumption, it seems plausible that hours spent in online social media is normally distributed.

Step 3 List the Hypotheses

The psychologist believes that valedictorians spend less time online in social media, so she has directional hypotheses.

H_0: $\mu_{\text{Valedictorians}} \geq 18.68$
H_1: $\mu_{\text{Valedictorians}} < 18.68$

Step 4 Set the Decision Rule

Doing a one-tailed test; using the standard alpha level, .05; $df = 31 - 1 = 30$. $t_{cv} = -1.697$

- Reject the null hypothesis if $t \leq -1.697$.
- Fail to reject the null hypothesis if $t > -1.697$.

Step 5 Calculate the Test Statistic

$$s_M = \frac{s}{\sqrt{N}} = \frac{6.80}{\sqrt{31}} = 1.22$$

$$t = \frac{M - \mu}{s_M} = \frac{16.24 - 18.68}{1.22} = -2.00$$

Step 6 Interpret the Results

(1) *Was the null hypothesis rejected?*
As $-2.00 \leq -1.697$, the null hypothesis is rejected, and the difference is statistically significant. Valedictorians spend significantly less time in online social media than the average high school student. $t(30) = -2.00, p < .05$ (one-tailed).

(2) *How big is the effect?* The d value of -0.36, calculated below, is a weak-to-medium effect:

$$d = \frac{M - \mu}{s} = \frac{16.24 - 18.68}{6.80} = -0.36$$

$$r^2 = \frac{t^2}{t^2 + df} \times 100 = \frac{2.00^2}{2.00^2 + 30} \times 100 = 12\%$$

(3) *How wide is the confidence interval?*

$$95\%\ CI\mu_{Diff} = (M - \mu) \pm (t_{cv} \times s_M)$$
$$= (16.24 - 18.68) \pm (2.042 \times 1.22)$$
$$= [-4.93, 0.05]$$

The confidence interval, calculated above, is roughly from -5 to 0. Thus, it is not clear how strong the effect is. If it is five hours a week less, that seems like a lot of time. Zero, obviously, is no difference.

Note that this confidence interval captures zero, even though the null hypothesis was rejected. How can that be? It is possible because the test was a one-tailed test, but confidence intervals are always two-tailed. When a researcher does a one-tailed test, he or she can change the degree of confidence to fit the test. With this one one-tailed test, all 5% of the rare zone was in one end. If the researcher put 5% in both ends, for a total of 10%, 90% would be left in the middle for a 90% confidence interval. To calculate the 90% confidence interval instead of the 95%, substitute the critical value of t found in Step 4, 1.697, into the confidence interval equation. This gives a confidence interval that doesn't capture 0, as it ranges from -4.51 to -0.37. Here are the calculations:

$$90\%\ CI\mu_{Diff} = (M - \mu) \pm (t_{cv} \times s_M)$$
$$= (16.24 - 18.68) \pm (1.697 \times 1.22)$$
$$= [-4.51, -0.37]$$

Putting it all together: "A psychologist investigating time management in high school students compared the average amount of time spent by valedictorians per week in online social media ($M = 16.24$) to the known average for the population of high school students in general ($\mu = 18.68$). The difference was statistically significant [$t(30) = -2.00, p < .05$ (one-tailed)], indicating that students who are doing very well in school, like valedictorians, spend less time with social media and, presumably, more time on school work.

"Unfortunately, the sample size, 31, in this study was relatively small, making the confidence interval relatively wide. To resolve this, I recommend replicating the study with a larger sample of valedictorians, chosen in such a way that they represent the United States."

7.93 The less variability there is, (a) the smaller the standard error of the mean, (b) the larger the t value, (c) the bigger the effect size, (d) and the narrower the confidence interval. Thus, when there is less variability it is easier to reject the null hypothesis, find a large effect, and have a more precise estimate of the population value.

7.95 The distance from the sample mean to the population mean has no impact on the standard error of the mean or the width of the confidence interval, but it does have an impact on the t value and the effect size. Having a greater distance between sample and population mean means there is a larger effect and that it is easier to reject the null hypothesis.

7.97 1.959 is the largest value of t that guarantees failure to reject the null hypothesis. (1.960 is the critical value of t for $\alpha = .05$, two-tailed, $df = \infty$.)

7.99 For the 90% confidence interval, use Equation 7.4, but instead of using t_{cv}, two-tailed, $\alpha = .05$, substitute the critical value of t, two-tailed, $\alpha = .10$. For the 99% confidence interval, substitute the critical value of t, two-tailed,

$\alpha = .05$. Here are the two equations for $N = 21$ (i.e., $df = 20$) and $s_M = 1$:
$90\%\ CI\mu_{Diff} = (M - \mu) \pm (t_{cv} \times s_M)$
$= (M - \mu) \pm (1.725 \times 1) = (M - \mu) \pm 1.725$
$99\%\ CI\mu_{Diff} = (M - \mu) \pm (t_{cv} \times s_M)$
$= (M - \mu) \pm (2.845 \times 1) = (M - \mu) \pm 2.845$

Chapter 8

8.01 mean
8.03 sample; population
8.05 independent; paired
8.07 paired
8.09 independent; mean
8.11 random samples
8.13 normally distributed
8.15 non
8.17 $\mu_1 \neq \mu_2$
8.19 rare; common; t
8.21 2
8.23 $\geq$
8.25 The pooled variance
8.27 means
8.29 rejected
8.31 5
8.33 $p < .05$; $p > .05$
8.35 0
8.37 outcome; explanatory
8.39 population means
8.41 small
8.43 t-statistic
8.45 a paired-samples t test (the student and parent are from the same family)
8.47 an independent-samples t test (no subject belongs to both groups)
8.49 The random samples assumption is violated. The independence assumption is not violated. The normality of distribution assumption is not violated. The homogeneity of variance assumption is violated, so much so that it is not OK to continue with the test.
8.51 (a) $H_0: \mu_1 = \mu_2$
$H_1: \mu_1 \neq \mu_2$
(b) The null hypothesis says that the two population means are the same, that is, there is no difference in effectiveness between the two treatments. Any difference between sample means is due to sampling error. The alternative hypothesis says that the two population means are different because the two populations are differentially effective. The difference between sample means is too large to be due to sampling error.
8.53 (a) $H_0: \mu_{Fluoride} \geq \mu_{Nonfluoride}$
$H_1: \mu_{Fluoride} < \mu_{Nonfluoride}$
(b) The null hypothesis says that the population mean for the number of cavities for the fluoride group will be greater than or the same as that for the nonfluoride group because fluoride has no effect, or a negative effect, on cavities. The alternative hypothesis states that the population mean for the number of cavities for the fluoride group will be less than that for the nonfluoride group because fluoride reduces cavities.
8.55 $df = n_1 + n_2 - 2 = 223 + 252 - 2 = 473$. Use $df = 450$. $t_{cv} = 1.965$.
8.57 $df = n_1 + n_2 - 2 = 46 + 46 - 2 = 90$. $t_{cv} = 2.632$.
8.59 (a) If $t \leq -2.086$ *or* if $t \geq 2.086$, then reject H_0.
(b) If $-2.086 < t < 2.086$, then fail to reject H_0.
8.61 $df = n_1 + n_2 - 2 = 12 + 13 - 2 = 23$

$$s^2_{pooled} = \frac{s_1^2(n_1 - 1) + s_2^2(n_2 - 1)}{df}$$

$$= \frac{7.4^2(12 - 1) + 8.2^2(13 - 1)}{23}$$

$$= 61.27$$

8.63 $N = n_1 + n_2 = 45 + 58 = 103$

$$s_{M_1-M_2} = \sqrt{s^2_{pooled}\left[\frac{N}{n_1 \times n_2}\right]}$$

$$= \sqrt{5.63\left[\frac{103}{45 \times 58}\right]} = 0.47$$

8.65 $N = n_1 + n_2 = 45 + 58 = 103$

$$df = N - 2 = 101$$

$$s_{M_1-M_2} = \sqrt{\left[\frac{s_1^2(n_1 - 1) + s_2^2(n_2 - 1)}{df}\right]\left[\frac{N}{n_1 \times n_2}\right]}$$

$$= \sqrt{\left[\frac{5.98^2(45 - 1) + 7.83^2(58 - 1)}{101}\right]\left[\frac{103}{45 \times 28}\right]}$$

$$= 1.41$$

8.67 $t = \frac{M_1 - M_2}{s_{M_1-M_2}} = \frac{68 - 57}{2.34} = 4.70$

8.69
$$N = n_1 + n_2 = 72 + 60 = 132$$
$$df = N - 2 = 130$$
$$s_{M_1-M_2} = \sqrt{\left[\frac{s_1^2(n_1-1) + s_2^2(n_2-1)}{df}\right]\left[\frac{N}{n_1 \times n_2}\right]}$$
$$= \sqrt{\left[\frac{4.6^2(72-1) + 3.3^2(60-1)}{130}\right]\left[\frac{132}{72 \times 60}\right]}$$
$$= 0.71$$

8.71 (a) H_0 rejected
(b) Statistically significant
(c) μ_2 is probably bigger than μ_1.

8.73 $t(21) = 2.07, p > .05$

8.75 $t(8) = 2.31, p < .05$

8.77 (a) $d = \frac{M_1 - M_2}{\sqrt{s^2_{\text{Pooled}}}} = \frac{17 - 12}{\sqrt{4.00}} = 2.50.$
(b) Very large effect

8.79 $df = N - 2 = 73 - 2 = 71$
$$r^2 = \frac{t^2}{t^2 + df} \times 100$$
$$= \frac{9.87^2}{9.87^2 + 71} \times 100 = 57.8\%$$

8.81 (a) $95\%\ \text{CI}\mu_{\text{Diff}} = (M_1 - M_2) \pm (t_{cv} \times s_{M_1-M_2})$
$$= (31 - 24) \pm (2.045 \times 2.88)$$
$$= [1.11, 12.89]$$
(b) Reject H_0.

8.83 (*Note:* One does not need to use everything that is calculated in an interpretation. Interpretation has a subjective element. What one person chooses to emphasize may differ from what another does.)

An education researcher investigated whether the color used to grade a test has an impact on first graders' self-esteem. The first graders took a third-grade math test and had 25% of their answers marked wrong, half of the students with a red pen and half of the students with pencil. They then took a self-esteem test. The "pencil" students had higher self-esteem ($M = 29.00$) than the "red ink" students ($M = 23.00$). This difference was statistically significant [$t(25) = 3.02, p < .05$]. The effect is a strong one and suggests that the color used to grade a test has an impact on self-esteem, with an effect of between 1.90 and 10.1 points with 95% confidence. Unfortunately, because there was no control group, it is unclear if grading in pencil raises self-esteem or using red ink lowers it. Future research should use larger sample sizes, include such a control group, and utilize random assignment of the students to the treatment conditions. Until that research is done, it seems advisable for teachers to use pencils, not red ink, to grade tests and papers.

8.85 **Step 1** Pick a Test.
Comparing the means of two independent samples calls for an independent-samples *t* test.

Step 2 Check the Assumptions.
- Random samples: violated, but robust to violations. Plus, sample is representative of Americans, which is the objective.
- Independence of observations: not violated, each participant is in the study only once.
- Normality: Cholesterol is a physical characteristic; reasonable to assume it is normally distributed. Plus, robust assumption, especially when $N > 50$.
- Homogeneity of variance: the 2 standard deviations are about equal—not violated.

Step 3 List the Hypotheses.
H_0: $\mu_{\text{American}} = \mu_{\text{Mediterranean}}$
H_1: $\mu_{\text{American}} \neq \mu_{\text{Mediterranean}}$

Step 4 Set the Decision Rule.
Set $\alpha = .05$, two-tailed, $df = 70$: $t_{cv} = 1.994$.
- If $t \leq -1.994$ *or* if $t \geq 1.994$, reject H_0.
- If $-1.994 < t < 1.994$, fail to reject H_0.

Step 5 Calculate the Test Statistic.
$$N = n_1 + n_2 = 36 + 36 = 72$$
$$df = N - 2 = 70$$
$$s_{M_1-M_2} = \sqrt{\left[\frac{s_1^2(n_1-1) + s_2^2(n_2-1)}{df}\right]\left[\frac{N}{n_1 \times n_2}\right]}$$
$$= \sqrt{\left[\frac{24^2(36-1) + 26^2(36-1)}{70}\right]\left[\frac{72}{36 \times 36}\right]}$$
$$= 5.90$$
$$t = \frac{M_1 - M_2}{s_{M_1-M_2}} = \frac{230 - 190}{5.90} = 6.78$$

Step 6 Interpret the Results.
Was the null hypothesis rejected?
6.78≥ 1.994, so H_0 rejected. Results are statistically significant. Cholesterol levels of those who follow a Mediterranean diet are lower than for those who follow an American diet.
How big is the effect?

$$d = \frac{M_1 - M_2}{\sqrt{s^2_{Pooled}}} = \frac{190 - 230}{\sqrt{626.0000}} = -1.60$$

How wide is the confidence interval?

$$95\%CI\mu_{Diff} = (M_1 - M_2) \pm (t_{cv} \times s_{M_1 - M_2})$$
$$= (190 - 230) \pm (1.994 \times 5.90)$$
$$= [-51.76, -28.24]$$

Write a four-point interpretation:

This study used a representative sample of Americans to compare the cholesterol levels of those who ate a Mediterranean diet to those who ate an American diet. The mean cholesterol level of those who followed the Mediterranean diet was 40 points lower with a mean of 190 compared to a mean cholesterol level of 230 for those following an American diet. This difference was statistically significant [$t(70) = 6.78, p < .05$] and a 40-point difference is a very meaningful difference. The Mediterranean diet is associated with lower cholesterol, a level sufficiently lower that it means the difference between high cholesterol and normal cholesterol.

This study, because it did not use random assignment, cannot give conclusions about cause and effect. The report suggests that the Mediterranean diet is associated with reduced cholesterol, but doesn't prove such. Based on this study, it would be reasonable to proceed to a study in which participants are randomly assigned to different diets.

8.87 d. There is probably a difference between the two population means.

8.89 (a) The first test is an independent-samples *t* test and the second test is a paired-samples *t* test. (b) The first test answers the question, "Does one store carry more expensive merchandise than the other?" The second test answers the question, "For the same items, is there a difference in price between the two stores?"

8.91 If the confidence interval is a 99% confidence interval but the alpha level for the test is set at .05, which is equivalent to a 95% confidence interval, the confidence interval might capture zero though the null hypothesis was rejected. Another way this could occur would be with a one-tailed test as confidence intervals are two-tailed. A one-tailed test with $\alpha = .05$ has a rare zone that encompasses the extreme 5%, while a 95% confidence interval reaches out into the tail an additional 2.5%.

8.93 (a) Yes
(b) There was evidence of an effect.
(c) Small.
(d) Because the confidence interval is so narrow, we are fairly certain that effect was small.
(e) Given how large the sample size was, it seems unlikely that a nonrepresentative sample was obtained. Not very worried about an erroneous conclusion being drawn.

Chapter 9

9.01 independent; paired-samples *t* test
9.03 confounding
9.05 individual differences
9.07 higher
9.09 sampling error
9.11 within; between
9.13 populations; samples
9.15 negative; small
9.17 one
9.19 s_{M_D}
9.21 does not
9.23 >
9.25 sample
9.27 individual differences; dependent variable
9.29 confidence interval for the difference between population means
9.31 small
9.33 paired-samples *t* test (each subject rated both types of coffee)
9.35 single-sample *t* test (the mean is being compared to the known USDA standard)
9.37 It is not OK to proceed with the paired-samples *t* test.

- The random samples assumption is violated, but this assumption is robust to violation.
- The nonrobust independence of observations assumption is violated, so it is not OK to proceed with the test. Why was it violated? All students are in the same classes, so presumably they were assigned the same homework, which would influence the online activity of all students.
- For the normality assumption, we can assume that the difference scores are normally distributed.

9.39 $H_0: \mu_{\text{School}} = \mu_{\text{Social}}$
$H_1: \mu_{\text{School}} \neq \mu_{\text{Social}}$

9.41 −5.00; 4.00; −2.00; 1.00; −1.00 or: 5.00; −4.00; 2.00; −1.00; 1.00

9.43 9

9.45 2.101

9.47 1.984

9.49 $s_{M_D} = \frac{s_D}{\sqrt{N}} = \frac{2.57}{\sqrt{17}} = 0.62$

9.51 $t = \frac{M_1 - M_2}{s_{M_D}} = \frac{18 - 14}{5.34} = 0.75$

9.53 $s_{M_D} = \frac{s_D}{\sqrt{N}} = \frac{7.00}{\sqrt{5}} = 3.13$

$$t = \frac{M_1 - M_2}{s_{M_D}} = \frac{28 - 25}{3.13} = 0.96$$

9.55 (a)

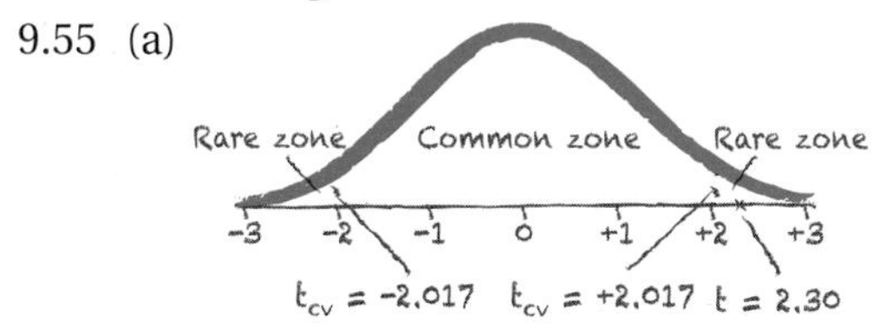

(b) The null hypothesis is rejected.

9.57 $t(4) = 3.21, p < .05$

9.59 $t(68) = 1.99, p > .05$

9.61 μ_2 is probably greater than μ_1.

9.63 There is not enough evidence to say a difference exists in the means of the two populations.

9.65 $95\%\text{CI}\mu_{\text{Diff}} = (M_1 - M_2) \pm (t_{cv} \times s_{M_D})$

$= (108 - 100) \pm (2.052 \times 5.00)$

$= [-2.26, 18.26]$

9.67 Rejected

9.69 A sleep therapist studied whether an herbal tea that was advertised as a sleep aid had any impact on sleep onset for people with insomnia. Using 23 pairs of insomniacs matched on a number of potential confounding variables, he found that those who used the tea went to sleep in 19.70 minutes and those who didn't took 21.20 minutes. This 1.5-minute difference was not statistically significant [$t(22) = 1.32, p > .05$], meaning that the study did not provide sufficient evidence that the tea had any impact on sleep onset.

Nonetheless, the study was suggestive that the tea might have some small effect, so it would be reasonable to replicate the study with a larger sample size. It would also be advisable to make the study double blind so that neither experimenter nor participant would know who was receiving the supposed sleep aid.

9.71 **Step 1** Pick a Test

There are two samples, feet receiving standard treatment and feet receiving experimental treatment, the samples are dependent, and the dependent variable is measured at the ratio level. This calls for a paired-samples *t* test.

Step 2 Check the Assumptions

- Random samples: Violated, but robust to violations.
- Independence of observations: Not violated.
- Normality: Willing to assume not violated.

Step 3 List the Hypotheses

- $H_0: \mu_E = \mu_C$
- $H_1: \mu_E \neq \mu_C$

Step 4 Set the Decision Rule

$df = N - 1 = 30 - 1 = 29$

$\alpha = .05$, two-tailed

$t_{cv} = \pm 2.045$

Decision rule:

- If $t \leq -2.045$ *or* if $t \geq 2.045$, reject H_0.
- If $-2.045 < t < 2.045$, fail to reject H_0.

Step 5 Calculate the Test Statistic

$$s_{M_D} = \frac{s_D}{\sqrt{N}} = \frac{8.65}{\sqrt{30}} = 1.58$$

$$t = \frac{M_1 - M_2}{s_{M_D}} = \frac{88 - 72}{1.58} = 10.13$$

Step 6 Interpret the Results

Was the null hypothesis rejected?

As $10.13 \geq 2.045$, the null hypothesis is rejected. The difference is statistically significant and is in favor of the new treatment working better. In APA format, the results are $t(29) = 10.13, p < .05$.

How wide is the confidence interval? How big is the effect?

$95\%\text{CI}\mu_{Diff} = (M_1 - M_2) \pm (t_{cv} \times s_{M_D})$
$= (88 - 72) \pm (2.045 \times 1.58) = [12.77, 19.23]$

Write a four-point interpretation:

A dermatologist evaluated a new treatment for athlete's foot by taking 30 people with athlete's foot on both feet and randomly assigning which foot received standard treatment and which received the experimental treatment. He measured outcome as the percentage of reduction of symptoms and found a statistically significant improvement in outcome with the new treatment [$t(29) = 10.13, p < .05$]. The new treatment led to an 88% reduction in symptoms vs. a 72% reduction in symptoms with the old, standard treatment. The new treatment seems to work substantially better than the standard treatment. But, to increase confidence in robustness of this finding, the study should be replicated.

9.73 e (the width does not depend on the differences in the mean)

9.75 d (all else being equal, a larger standard error of the mean gives a wider confidence interval)

9.77 $M_1 - M_2 = -2.00$ (that's the midpoint of the confidence interval)

$$t = \frac{M_1 - M_2}{s_{M_D}} = \frac{-2.00}{4.00} = -.50$$

Chapter 10

10.01 two
10.03 ANOVA
10.05 way; factor
10.07 Type I
10.09 post-hoc
10.11 variability
10.13 individual differences
10.15 Between-groups
10.17 individual differences
10.19 large
10.21 *F* ratio
10.23 one
10.25 *t*
10.27 an independent-samples *t* test
10.29 cannot
10.31 homogeneity of variance
10.33 non
10.35 rejected
10.37 between
10.39 $df_{Between}$; df_{Within}
10.41 sum of squares
10.43 SS_{Total}
10.45 treatment effect
10.47 ANOVA summary table
10.49 mean square
10.51 effect
10.53 the alternative hypothesis
10.55 >
10.57 between-groups
10.59 100; 0
10.61 statistically significant
10.63 Type II; Type I (order does matter)
10.65 statistically significant
10.67 smallest

10.69 (e) This is a between-subjects, one-way ANOVA because the means of three or more independent samples are being compared.

10.71 (f) This is none of the tests learned about so far. There are three levels of one independent variable being compared on an interval-level dependent variable, but the samples are dependent. This type of test, a repeated-measures ANOVA, is covered in the next chapter.

10.73 The random samples assumption is violated. The independence of observations assumption is not violated. There is no reason to think the homogeneity of variance assumption is violated, but it would be nice to know the standard deviation for each group. Because the only violated assumption is robust, we can proceed with the between-subjects, one-way ANOVA.

10.75 $H_0: \mu_1 = \mu_2 = \mu_3$.
H_1: At least one population mean differs from at least one other population mean.

10.77 (a) $df_{Total} = N - 1 = 48 - 1 = 47$
(b) $df_{Between} = k - 1 = 4 - 1 = 3$
(c) $df_{Within} = N - k = 48 - 4 = 44$

10.79 (a) $df_{Total} = N - 1 = 60 - 1 = 59$
(b) $df_{Between} = k - 1 = 4 - 1 = 3$
(c) $df_{Within} = N - k = 60 - 4 = 56$

10.81 $F_{cv} = 5.123$

10.83 $df_{Between} = 4 - 1 = 3$; $df_{Within} = 50 - 4 = 46$
$F_{cv} = 2.807$

10.85 $F_{cv} = 3.232$
If $F \geq 3.232$, reject H_0. If $F < 3.232$, fail to reject H_0.

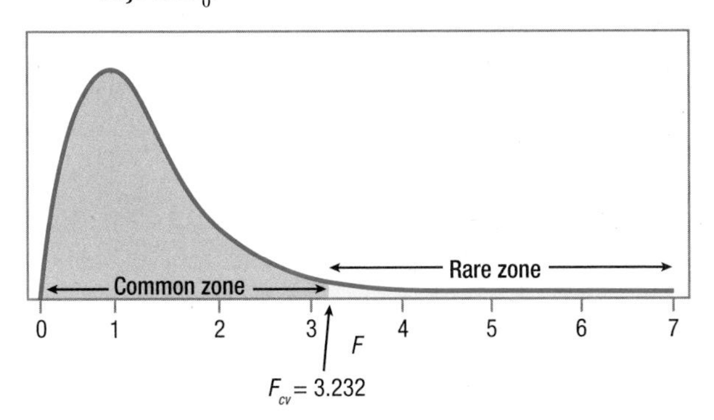

10.87

	Group 1		Group 2		Group 3			
	X	X²	X	X²	X	X²		
	108	11,664	100	10,000	99	9,801		
	102	10,404	105	11,025	95	9,025	Grand	
					91	8,281	X	X²
Sum	210	22,068	205	21,025	285	27,1067	700	70,200
n	2		2		3		7	

10.89 $$SS_{\text{Total}} = \Sigma X^2 - \frac{(\Sigma X)^2}{N}$$

$$= 75953 - \frac{821^2}{9} = 1059.556$$

$$SS_{\text{Between}} = \Sigma\left(\frac{(\Sigma X_{\text{Group}})^2}{n_{\text{Group}}}\right) - \frac{(\Sigma X)^2}{N}$$

$$= \left(\frac{216^2}{2} + \frac{276^2}{3} + \frac{329^2}{4}\right) - \frac{821^2}{9} = 886.81$$

$$SS_{\text{Within}} = \Sigma\left(\Sigma X^2_{\text{Group}} - \frac{(\Sigma X_{\text{Group}})^2}{n_{\text{Group}}}\right)$$

$$= \left(23360 - \frac{216^2}{2}\right) + \left(25448 - \frac{276^2}{3}\right)$$

$$= \left(27145 - \frac{329^2}{4}\right) = 172.75$$

10.91 $SS_{\text{Within}} = SS_{\text{Total}} - SS_{\text{Between}} = 98.75 - 40.33 = 58.42$

10.93 $$MS_{\text{Between}} = \frac{SS_{\text{Between}}}{df_{\text{Between}}} = \frac{2{,}378.99}{3} = 793.00$$

10.95 $$MS_{\text{Within}} = \frac{SS_{\text{Within}}}{df_{\text{Within}}} = \frac{78.95}{32} = 2.47$$

10.97 $$F = \frac{MS_{\text{Between}}}{MS_{\text{Within}}} = \frac{38.88}{17.44} = 2.23$$

10.99

Source of Variability	Sum of Squares	Degrees of Freedom	Mean Square	F ratio
Between groups	172.80	2	86.40	0.16
Within groups	6,410.80	12	534.23	
Total	6,583.60	14		

10.101 $F = 1.96$ falls in the common zone, so fail to reject the null hypothesis.

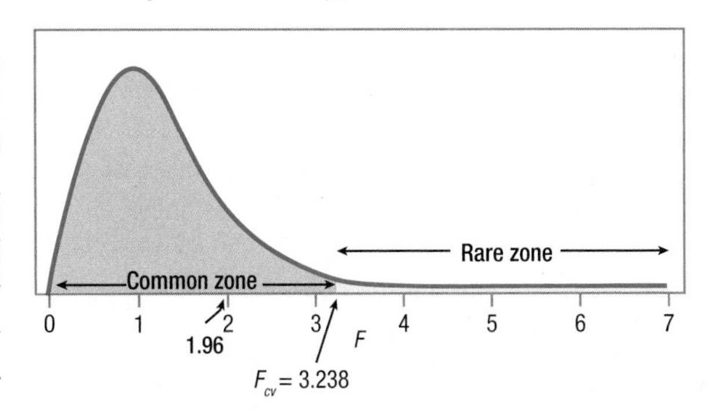

10.103 $7.64 \geq 3.467$, so reject H_0.

10.105 $F(3, 17) = 5.34, p < .05$

10.107 $F(3, 44) = 3.28, p < .05$

10.109 The results are statistically significant, so conclude that it is probable that at least one population has a mean different from at least one other population mean.

10.111 $$r^2 = \frac{SS_{\text{Between}}}{SS_{\text{Total}}} \times 100 = \frac{128.86}{413.67} \times 100 = 31.15\%$$

10.113 $$r^2 = \frac{SS_{\text{Between}}}{SS_{\text{Total}}} \times 100 = \frac{59.98}{328.58} \times 100 = 18.25\%$$

10.115 Yes, complete a post-hoc test because the results were statistically significant.

10.117 $q = 5.19$

10.119 $$HSD = q\sqrt{\frac{MS_{\text{Within}}}{n}} = 3.55\sqrt{\frac{10.44}{8}} = 4.06$$

10.121 $13.09 - 8.89 = 4.20$. $4.20 < 4.37$, so the difference is not statistically significant. There's not enough evidence to conclude that these two populations have a mean difference.

10.123
- M_1 vs. M_2: $67.86 - 53.56 = 14.30$, which is greater than 8.30, so it is a statistically significant difference. The mean of population 1 is probably greater than the mean of population 2.

- M_1 vs. M_3: 67.86 – 61.55 = 6.31, which is less than 8.30, so it is not a statistically significant difference. There is not enough evidence to conclude that these two population means are different.
- M_2 vs. M_3: 61.55 – 53.56 = 7.99, which is less than 8.30, so it is not a statistically significant difference. There is not enough evidence to conclude that the two population means are different.

10.125 "An addictions researcher obtained samples of alcoholics, smokers, and heroin addicts who were in treatment for the second time. From each participant she found out how long, in months, it took before relapse occurred after the first treatment. The alcoholics relapsed, on average, in 4.63 months, the smokers in 4.91 months, and the heroin addicts in 5.17 months. Using a between-subjects, one-way ANOVA, she found no statistically significant effect for the type of addiction on the mean time to relapse [$F(2, 22) = 0.07$, $p > .05$]. These data suggest, for those who fail in treatment, that there is no difference in the addictiveness of these three substances. Future research should use a larger number of participants and follow them prospectively after treatment ends."

10.127 **Step 1** Pick a Test

The test compares means of three independent samples, so use a between-subjects, one-way ANOVA.

Step 2 Check the Assumptions

- The random samples assumption is violated, but this assumption is robust. Need to be careful about population to which the results are generalized.
- Independence of observations is not violated.
- Normality is not violated. It's reasonable to assume that satisfaction scores are normally distributed.
- Homogeneity of variance is not violated. All standard deviations are equal.

Only the random samples assumption is violated, but it is robust, so it is OK to proceed with a between-subjects, one-way ANOVA.

Step 3 List the Hypotheses

H_0: $\mu_1 = \mu_2 = \mu_3$.
H_1: At least one population mean is different from at least one other.

Step 4 Set the Decision Rule

$df_{Between}$, numerator $df = k - 1 = 3 - 1 = 2$
df_{Within}, denominator $df = N - k = 12 - 3 = 9$
$df_{Total} = N - 1 = 12 - 1 = 11$
Set alpha at .05.
$F_{cv} = 4.256$
Decision rule: If $F \geq 4.256$, reject H_0; if $F < 4.256$, fail to reject H_0.

Step 5 Calculate the Test Statistic

$$SS_{Total} = \Sigma X^2 - \frac{(\Sigma X)^2}{N}$$

$$= 78350 - \frac{960^2}{12} = 1550$$

$$SS_{Between} = \Sigma\left(\frac{(\Sigma X_{Group})^2}{n_{Group}}\right) - \frac{(\Sigma X)^2}{N}$$

$$= \left(\frac{260^2}{4} + \frac{360^2}{4} + \frac{340^2}{4}\right) - \frac{960^2}{12} = 1400$$

$$SS_{Within} = \Sigma\left(\Sigma X^2_{Group} - \frac{(\Sigma X_{Group})^2}{n_{Group}}\right)$$

$$= \left(16950 - \frac{260^2}{4}\right) + \left(32450 - \frac{360^2}{4}\right)$$

$$= \left(28950 - \frac{340^2}{4}\right) = 150$$

Source of Variability	Sum of Squares	Degrees of Freedom	Mean Square	F ratio
Between groups	1,400.00	2	700.00	41.99
Within groups	150.00	9	16.67	
Total	1,550.00	11		

Step 6 Interpret the Results

- Was the null hypothesis rejected?
  - Yes, $41.99 \geq 4.256$
  - APA format: $F(2, 9) = 41.99$, $p > .05$

- Conclude that at least one population mean differs from at least one other mean.
- How big is the effect?
  - Big!

$$r^2 = \frac{SS_{Between}}{SS_{Total}} \times 100 = \frac{1{,}400.00}{1{,}550.00} \times 100 = 90.32\%$$

- Where is the effect?
  - $q = 3.95$

$$HSD = q\sqrt{\frac{MS_{Within}}{n}} = 3.95\sqrt{\frac{16.67}{4}} = 8.06$$

  - The premium brand in the premium bottle is statistically significantly more satisfying than the store brand in the store bottle.
  - The store brand in the premium bottle is statistically significantly more satisfying than the store brand in the store bottle.
  - There is not enough evidence to conclude that there is a difference in satisfaction between the premium brand and the store brand when both are in the premium bottle.
- Putting it all together: A consumer researcher investigated the effect of labeling on satisfaction with a product. Comparing a store brand shampoo in a store brand bottle, a premium brand shampoo in a premium brand bottle, and a store brand shampoo in a premium bottle, the mean satisfaction scores were, respectively, 65.00, 90.00, and 85.00. Using a between-subjects, one-way ANOVA, she found a strong and statistically significant effect of labeling [$F(2, 9) = 41.99$, $p < .05$]. There was no statistically significant difference between the store brand and the premium brand when both were in the premium bottle, but both were found to be statistically significantly more satisfying than the store shampoo in a store brand bottle. In rating shampoos, consumers seem to be more influenced by their perception of the shampoo from the bottle's label than they are by the shampoo in the bottle. It would be interesting, in future research, to use the same participants in all three conditions to see if the effect exists when they are directly comparing all three samples.

10.129 c. SS_{Total} must be largest; means squares are always no larger than corresponding sum of squares.

10.131 Could be true

10.133 False

10.135 False

10.137 Could be true

Chapter 11

11.01 independent-samples *t* test

11.03 Repeated-measures

11.05 rows; columns

11.07 powerful; greater

11.09 denominator

11.11 a repeated-measures ANOVA; a paired-samples *t* test

11.13 robust

11.15 normally distributed; population

11.17 is not

11.19 greater than or equal to

11.21 was

11.23 was

11.25 treatment; residual

11.27 n; k; N

11.29 df_{Total}

11.31 subjects

11.33 rare

11.35 variability

11.37 sum of squares; degrees of freedom

11.39 $MS_{Treatment}$; $MS_{Residual}$

11.41 decision rule

11.43 was

11.45 should

11.47 was not

11.49 eta squared

11.51 *r* squared

11.53 medium

11.55 rejected

11.57 not statistically

11.59 Between-subjects, one-way ANOVA (the means of three or more independent samples are being compared)

11.61 Independent-samples *t* test (the means of two independent samples are being compared)

11.63 (a) The random samples assumption is not violated. The independence of observations assumption is violated as participants met together in groups. The normality assumption is not violated.

(b) Should not proceed with the one-way, repeated-measures ANOVA because the independence of observations assumption is violated.

11.65 H_0: All population means are equal.
H_1: At least one population mean is different from at least one other.

11.67 $df_{\text{Subjects}} = (n-1) = 8-1 = 7$
$df_{\text{Treatment}} = (k-1) = 3-1 = 2$
$df_{\text{Residual}} = (n-1)(k-1) = (8-1)(3-1) = 14$
$df_{\text{Total}} = (N-1) = (24-1) = 23$

11.69 Remember, $N = n \times k$. So, $N = 15 \times 4 = 60$
$df_{\text{Subjects}} = (n-1) = 15-1 = 14$
$df_{\text{Treatment}} = (k-1) = 4-1 = 3$
$df_{\text{Residual}} = (n-1)(k-1) = (15-1)(4-1) = 42$
$df_{\text{Total}} = (N-1) = (60-1) = 59$

11.71 (a) $F_{cv} = 4.103$
(b)

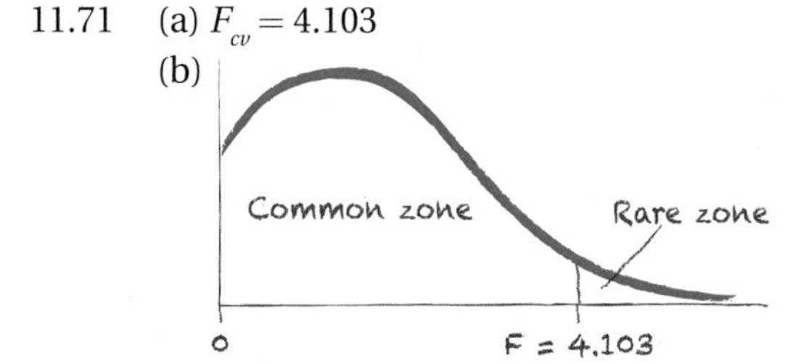

11.73 If $F \geq 3.287$, reject H_0.
If $F < 3.287$, fail to reject H_0.

11.75 $SS_{\text{Total}} = 301.48$

$$MS_{\text{Treatment}} = \frac{SS_{\text{Treatment}}}{df_{\text{Treatment}}} = \frac{48.48}{2} = 24.24$$

$$MS_{\text{Residual}} = \frac{SS_{\text{Residual}}}{df_{\text{Residual}}} = \frac{115.50}{22} = 5.25$$

$$F = \frac{MS_{\text{Treatment}}}{MS_{\text{Residual}}} = \frac{24.24}{5.25} = 4.62$$

Source of Variability	Sum of Squares	Degrees of Freedom	Mean Square	F ratio
Subjects	137.50	11		
Treatment	48.48	2	24.24	4.62
Residual	115.50	22	5.25	
Total	301.48	35		

11.77 H_0 was rejected.

11.79 Based on *df* in summary table, $F_{cv} = 3.232$. H_0 is rejected.

11.81 (a) $F(5, 45) = 7.84, p < .05$
(b) Statistically significant

11.83 Based on *df* in summary table, $F_{cv} = 3.555$.
(a) $F(2, 18) = 11.24, p < .05$
(b) Statistically significant

11.85 The results of the ANOVA are not statistically significant. There is not enough evidence to conclude that the population means are different.

11.87 (a) $$\eta^2 = \frac{SS_{\text{Treatment}}}{SS_{\text{Total}}} \times 100 = \frac{9.89}{86.98} \times 100 = 11.37\%$$

(b) Medium

11.89 (a) $$\eta^2 = \frac{SS_{\text{Treatment}}}{SS_{\text{Total}}} \times 100 = \frac{46.00}{336.00} \times 100 = 13.69\%$$

(b) Medium

11.91 (a) The effect size is small, so there is just the normal level of worry about a Type II error.
(b) Do not recommend replication with a larger sample size.

11.93 $q = 4.79$

11.95 $$HSD = q\sqrt{\frac{MS_{\text{Residual}}}{n}} = 2.90\sqrt{\frac{12.27}{8}} = 3.59$$

11.97 (a)
- M_1 vs. $M_2 = 14.98 - 12.83 = 2.15$. $2.15 < 3.78$: Not statistically significant.
- M_1 vs. $M_3 = 12.83 - 8.22 = 4.61$. $4.61 \geq 3.78$: Statistically significant.
- M_2 vs. $M_3 = 14.98 - 8.22 = 6.76$. $6.76 \geq 3.78$: Statistically significant.

(b)
- Not enough evidence to conclude there is a difference in means between population 1 and population 2.
- Population 1 probably has a bigger mean than population 3.
- Population 2 probably has a bigger mean than population 3.

11.99 "In this study, rats were reared in three different environments—a standard laboratory setting, an enriched laboratory setting, and a mimicked wild setting—to see if environment had an impact on the rats' intelligence. There was a statistically significant and strong effect of environment on mean intelligence [$F(2, 14) = 9.61, p < .05$]. Rats reared in the enriched environment earned statistically higher scores ($M = 108.00$) on a variety of behavioral tasks than either the rats reared in a standard lab setting ($M = 95.00$) or the rats reared in an environment meant to mimic the wild ($M = 94.00$). There was no difference in the mean intelligence of these latter two types of rats. An enriched environment has a positive impact on rat intelligence and a standard laboratory environment does not have a negative impact on rat intelligence, as compared to a mimicked wild environment. This research should be replicated to make sure the effect is stable. Future research should attempt to determine what aspects of the enriched environment lead to improved performance."

11.101 **Step 1** Pick a Test

Comparing the means of three dependent samples calls for a one-way, repeated-measures ANOVA.

Step 2 Check the Assumptions

- The random samples assumption is violated, but it is robust, so it is OK to proceed.
- The independence of observations assumption is not violated.
- The normality assumption is not violated. It is reasonable to assume a normal distribution for urges to drink alcohol.

Step 3 List the Hypotheses

H_0: $\mu_1 = \mu_2 = \mu_3$.

H_1: At least one population mean is different from at least one other.

Step 4 Set the Decision Rule

$df_{Subjects} = n - 1 = 10 - 1 = 9$

$df_{Treatment} = k - 1 = 3 - 1 = 2$

$df_{Residual} = (n - 1)(k - 1) = (10 - 1)(3 - 1) = 18$

$N = n \times k = 10 \times 3 = 30$

$df_{Total} = N - 1 = 30 - 1 = 29$

Set $\alpha = .05$.

$F_{cv} = 3.555$

If $F \geq 3.555$, reject H_0.

If $F < 3.555$, fail to reject H_0.

Step 5 Calculate the Test Statistic

$SS_{Total} = 646.67$

$$MS_{Treatment} = \frac{SS_{Treatment}}{df_{Treatment}} = \frac{26.67}{2} = 13.34$$

$$MS_{Residual} = \frac{SS_{Residual}}{df_{Residual}} = \frac{88.00}{18} = 4.89$$

$$F = \frac{MS_{Treatment}}{MS_{Residual}} = \frac{13.34}{4.89} = 2.73$$

Source of Variability	Sum of Squares	Degrees of Freedom	Mean Square	F ratio
Subjects	532.00	9		
Treatment	26.67	2	13.34	2.73
Residual	88.00	18	4.89	
Total	646.67	29		

Step 6 Interpret the Results

$2.73 < 3.555$, so fail to reject the null hypothesis.

In APA format: $F(2, 18) = 2.73, p > .05$.

$$\eta^2 = \frac{SS_{Treatment}}{SS_{Total}} \times 100 = \frac{26.67}{646.67} \times 100 = 4\%$$

This is more than a small effect, so some concern about Type II error does exist.

A study was conducted to see if different types of treatments for alcoholism were more or less effective in terms of an impact on urges to drink. Thirty alcoholics were matched on the severity of their addiction and then assigned to Alcoholics Anonymous, individual psychotherapy, or anti-urge medication. At the end of treatment the mean number of urges per day for the three treatments, respectively, was 8.00, 6.00, and 6.00. The results were not statistically significant, meaning that there was not enough evidence to show that one treatment was more effective than another in controlling urges [$F(2, 18) = 2.73, p > .05$.] However, the sample size in this study was small and might not have been sufficient to pick up the effect if one treatment was just slightly better or worse than the others. Thus, it is recommended that the study be replicated with a larger sample size.

11.103 If $df_{Treatment} = 3$, then $k = 4$. Given $df_{Residual} = 45$ and $df_{Treatment} = 3$, then $\frac{45}{3} = n - 1$. If $15 = n - 1$, then $n = 16$. $N = k \times n = 4 \times 16 = 64$.

11.105 There is no need to be concerned about Type II error. The null hypothesis was rejected, so there should be concern about the possibility of Type I error.

11.107 The difference between a *HSD* value calculated at the $\alpha = .01$ level and the $\alpha = .05$ level is the *q* value. *q* is bigger at .01 than .05, so *HSD* is bigger at .01 than .05. As a result of this, two means have to be farther apart to be considered a statistically significant difference at the .01 level. And, this makes sense—farther apart means are more likely to represent a real difference between populations, so there is less chance of a Type I error. And the probability of a Type I error is what alpha represents.

Chapter 12

12.01 Two
12.03 independent; within
12.05 two; one
12.07 main; interaction
12.09 not parallel
12.11 main
12.13 independent
12.15 independence of observations
12.17 three
12.19 at least; at least
12.21 interaction effect
12.23 the number of rows
12.25 degrees of freedom; degrees of freedom
12.27 between groups; within groups
12.29 squares
12.31 $F_{\text{Interaction}}$
12.33 statistically significant
12.35 .05 (or 5%)
12.37 eta squared (η^2)
12.39 9
12.41 statistically significant
12.43 statistically significant
12.45 between-subjects, one-way ANOVA
12.47 between-subjects, two-way ANOVA

12.49 (a)

$$\text{Row}_1 = \frac{\text{Add up all the cell means in a row}}{\text{The number of cells in a row}} = \frac{25 + 35 + 45}{3} = 45.00$$

$$\text{Row}_2 = \frac{\text{Add up all the cell means in a row}}{\text{The number of cells in a row}} = \frac{25 + 15 + 5}{3} = 15.00$$

(b)

$$\text{Column}_1 = \frac{\text{Add up all the cell means in a column}}{\text{The number of cells in a column}} = \frac{25 + 25}{2} = 25.00$$

$$\text{Column}_2 = \frac{\text{Add up all the cell means in a column}}{\text{The number of cells in a column}} = \frac{35 + 15}{2} = 25.00$$

$$\text{Column}_3 = \frac{\text{Add up all the cell means in a column}}{\text{The number of cells in a column}} = \frac{45 + 5}{2} = 25.00$$

12.51 (a) There is no evidence of an interaction.

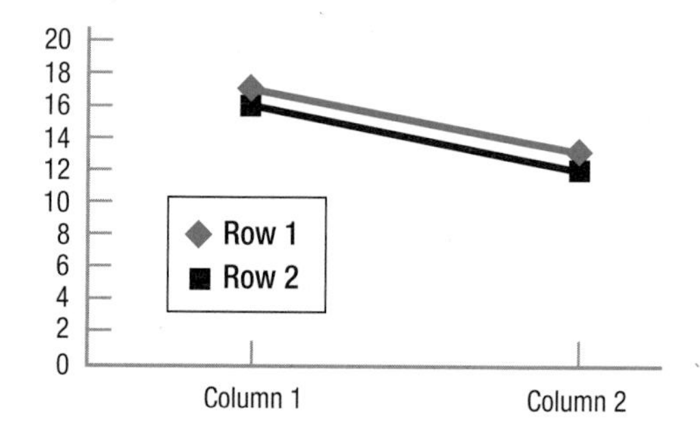

(b) There appears to be a statistically significant columns main effect.
(c) The main effect of columns should be interpreted.

12.53 (a) The random samples assumption was violated. The independence of observations assumption was not violated. It is assumed that the normality assumption was not violated. The homogeneity of variance assumption was violated.
(b) It is OK to proceed with the test. The random samples assumption for between-subjects, two-way ANOVA is robust to violations. And, the homogeneity of variance assumption is robust to violation as long as the total sample size is large.

12.55 (a) $H_{0\,\text{Rows}}$: All row population means are the same.
$H_{1\,\text{Rows}}$: At least one row population mean is different from at least one other row population mean.
(b) $H_{0\,\text{Columns}}$: All column population means are the same.
$H_{1\,\text{Columns}}$: At least one column population mean is different from at least one other column population mean.
(c) $H_{0\,\text{Interaction}}$: There is no interactive effect of the two explanatory variables on the dependent variable in the population.
$H_{1\,\text{Interaction}}$: The two explanatory variables, in the population, interact to affect the dependent variable in at least one cell.

12.57 (a) $df_{\text{Rows}} = R - 1 = 3 - 1 = 2$
(b) $df_{\text{Columns}} = C - 1 = 4 - 1 = 3$
(c) $df_{\text{Interaction}} = df_{\text{Rows}} \times df_{\text{Columns}} = 2 \times 3 = 6$
(d) $N = n \times R \times C = 8 \times 3 \times 4 = 96$
$df_{\text{Within}} = N - (R \times C) = 96 - (3 \times 4) = 84$
(e) $df_{\text{Between}} = df_{\text{Rows}} + df_{\text{Columns}} + df_{\text{Interaction}}$
$= 2 + 3 + 6 = 11$
(f) $df_{\text{Total}} = N - 1 = 96 - 1 = 95$

12.59 (a) $F_{cv\,\text{Rows}} = 3.259$
(b) $F_{cv\,\text{Columns}} = 3.259$
(c) $F_{cv\,\text{Interaction}} = 2.634$

12.61 If $F_{\text{Interaction}} \geq 2.668$, reject H_0.
If $F_{\text{Interaction}} < 2.668$, fail to reject H_0.

12.63 $MS_{\text{Within}} = \dfrac{SS_{\text{Within}}}{df_{\text{Within}}} = \dfrac{168.48}{40} = 4.21$

12.65 $F_{\text{Rows}} = \dfrac{SS_{\text{Rows}}}{df_{\text{Rows}}} = \dfrac{17.30}{3.33} = 5.20$

12.67 $df_{\text{Rows}} = R - 1 = 2 - 1 = 1$
$df_{\text{Columns}} = C - 1 = 2 - 1 = 1$
$df_{\text{Interaction}} = df_{\text{Rows}} \times df_{\text{Columns}} = 1 \times 1 = 1$
$N = n \times R \times C = 5 \times 2 \times 2 = 20$
$df_{\text{Within}} = N - (R \times C) = 20 - (2 \times 2) = 16$
$df_{\text{Between}} = df_{\text{Rows}} + df_{\text{Columns}} + df_{\text{Interaction}}$
$= 1 + 1 + 1 = 3$
$df_{\text{Total}} = N - 1 = 20 - 1 = 19$

$$MS_{\text{Rows}} = \frac{SS_{\text{Rows}}}{df_{\text{Rows}}} = \frac{15.00}{1} = 15.00$$

$$MS_{\text{Cols}} = \frac{SS_{\text{Cols}}}{df_{\text{Cols}}} = \frac{12.00}{1} = 12.00$$

$$MS_{\text{Interaction}} = \frac{SS_{\text{Interaction}}}{df_{\text{Interaction}}} = \frac{34.00}{1} = 34.00$$

$$MS_{\text{Within}} = \frac{SS_{\text{Within}}}{df_{\text{Within}}} = \frac{120.00}{16} = 7.50$$

$$F_{\text{Rows}} = \frac{MS_{\text{Rows}}}{MS_{\text{Within}}} = \frac{15.00}{7.50} = 2.00$$

$$F_{\text{Cols}} = \frac{MS_{\text{Cols}}}{MS_{\text{Within}}} = \frac{12.00}{7.50} = 1.60$$

$$F_{\text{Interaction}} = \frac{MS_{\text{Interaction}}}{MS_{\text{Within}}} = \frac{34.00}{7.50} = 4.53$$

Source of Variability	Sum of Squares	Degrees of Freedom	Mean Square	F ratio
Between groups	61.00	3		
Rows	15.00	1	15.00	2.00
Columns	12.00	1	12.00	1.60
Interaction	34.00	1	34.00	4.53
Within groups	120.00	16	7.50	
Total	181.00	19		

12.69 The null hypothesis was rejected.

12.71 $F(2, 66) = 3.70, p < .05$

12.73 $F(3, 168) = 2.45, p > .05$

12.75 (a) $F(2, 30) = 3.21, p > .05$
(b) $F(1, 30) = 1.08, p > .05$
(c) $F(2, 30) = 3.93, p < .05$

12.77 Interpret the interaction effect.

12.79 Interpret the interaction effect.

12.81 (a) $\eta^2_{\text{Rows}} = \dfrac{SS_{\text{Rows}}}{SS_{\text{Total}}} \times 100$

$$= \frac{3.78}{47.83} \times 100 = 7.90\%$$

(b) This is a medium effect.

12.83 (a) $\eta^2_{\text{Rows}} = \dfrac{SS_{\text{Rows}}}{SS_{\text{Total}}} \times 100$

$$= \frac{33.00}{344.00} \times 100 = 9.59\%$$

(b) $\eta^2_{\text{Columns}} = \dfrac{SS_{\text{Columns}}}{SS_{\text{Total}}} \times 100$

$$= \frac{54.00}{344.00} \times 100 = 15.70\%$$

(c) $\eta^2_{\text{Interaction}} = \frac{SS_{\text{Interaction}}}{SS_{\text{Total}}} \times 100$

$= \frac{45.00}{344.00} \times 100 = 13.08\%$

12.85 (a) $q_{\text{Rows}} = 3.44$
(b) $q_{\text{Columns}} = 2.85$
(c) $q_{\text{Interaction}} = 4.22$

12.87 $HSD_{\text{Rows}} = q_{\text{Rows}}\sqrt{\frac{MS_{\text{Within}}}{n_{\text{Rows}}}} = 3.49\sqrt{\frac{4.56}{12}} = 2.15$

12.89 *HSD* value should only be calculated for the interaction effect.

12.91 (a)
- Row 1 vs. Row 2 (117.66 – 113.63 = 4.03) is not a statistically significant difference.
- Row 1 vs. Row 3 (128.91 – 117.66 = 11.25) is a statistically significant difference.
- Row 2 vs. Row 3 (128.91 – 113.63 = 15.28) is a statistically significant difference.

(b) The mean of Row 3 is higher than that of both Row 1 and Row 2.

12.93 To aid in interpretation, make a graph.

Men and women who went shopping for groceries with and without children were surveyed regarding how much they enjoyed the shopping experience. How much shoppers enjoyed the experience depended on the interaction of one's sex with whether there were children on the shopping trip ($F(1, 16) = 41.15, p < .05$). Men shopping with children and women shopping by themselves rated their shopping trips, on average as somewhat enjoyable. These ratings were statistically significantly higher than men shopping without children or women shopping with children, who rated, on average, their shopping trips as somewhat less than enjoyable. It is possible that men and women feel differently about shopping with children, but another explanation is that novel situations are more enjoyable. For men, shopping with children tends to be novel and for women shopping without children is more likely to be novel. Future research should determine the novelty of the shopping experience as well as the presence of children.

12.95 **Step 1** Pick a Test

Comparing the means of four independent samples formed by two crossed independent variables calls for a between-subjects, two-way ANOVA.

Step 2 Check the Assumptions

OK to proceed with the test.
- The random samples assumption is violated unless the population is meant to be the local high school.
- The independence of observations assumption is not violated.
- It is reasonable to assume that this psychological variable is normally distributed, so the normality assumption is not violated.
- The homogeneity of variance assumption is not violated.

Step 3 List the Hypotheses

$H_{0\text{ Rows}}$: All row population means are the same.

$H_{1\text{ Rows}}$: At least one row population mean is different from at least one other row population mean.

$H_{0\text{ Columns}}$: All column population means are the same.

$H_{1\text{ Columns}}$: At least one column population mean is different from at least one other column population mean.

$H_{0\text{ Interaction}}$: There is no interactive effect of the two explanatory variables on the dependent variable in the population.

$H_{1\text{ Interaction}}$: The two explanatory variables, in the population, interact to affect the dependent variable in at least one cell.

Step 4 Set the Decision Rule

$df_{\text{Rows}} = 2 - 1 = 1$

$df_{\text{Columns}} = 2 - 1 = 1$

$df_{\text{Interaction}} = 1 \times 1 = 1$

$df_{\text{Within}} = 24 - (2 \times 2) = 20$

All three *F* ratios have $df_{\text{numerator}} = 1$ and $df_{\text{denominator}} = 20$. $F_{cv} = 4.351$.

The decision rule for each of the three effects is:

- If $F \geq 4.351$, reject H_0.
- If $F < 4.351$, fail to reject H_0.

Step 5 Calculate the Value of the Test Statistic

Source of Variability	Sum of Squares	Degrees of Freedom	Mean Square	F ratio
Between groups	447.12	3		
Rows	210.04	1	210.04	20.51
Columns	222.04	1	222.04	21.68
Interaction	15.04	1	15.04	1.47
Within groups	204.83	20	10.24	
Total	651.95	23		

Step 6 Interpret the Results

- Calculate the row means and the column means:
  - Row 1, positive reinforcement = $\frac{6.00 + 10.50}{2} = 8.25$
  - Row 2, punishment = $\frac{10.33 + 18.00}{2} = 14.17$
  - Column 1, restricted TV = $\frac{6.00 + 10.33}{2} = 8.17$
  - Column 2, unrestricted TV = $\frac{10.50 + 18.00}{2} = 14.25$
- Determine statistical significance:
  - The rows effect (use of reinforcement vs. punishment) is statistically significant:
    - $F(1, 20) = 20.51, p < .05$
    - The mean Approval of Violence Scale (AVS) score for teens who were raised with positive reinforcement is 8.25 and is statistically lower than the mean of 14.17 for teens who were raised with punishment.
    - Being raised with punishment leads to a statistically significant higher approval of the use of violence to solve disagreements.
  - The columns effect (Restrict TV or not) is statistically significant:
    - $F(1, 20) = 21.68, p < .05$
    - The mean AVS score for teens who were restricted in what they could watch on TV is 8.17 and is statistically lower than the mean of 14.25 for teens who were unrestricted in their TV watching.
    - Being unrestricted in what one watches on TV leads to a statistically significant higher approval of the use of violence to solve disagreements.
- The interaction effect is not statistically significant:
  - $F(1, 20) = 1.47, p > .05$
  - There is not enough evidence to conclude that type of discipline and restricting TV watching interact to affect the acceptance of violence as a way to solve disagreements.
  - Inspection of the graph to aid in interpretation
    - The lines are mostly parallel, so there is no interaction effect.
    - The "Punishment" line is higher than the "Positive reinforcement" line, suggesting a main effect for type of discipline.
    - The "Unrestricted" dots are higher than the "Restricted" dots, suggesting a main effect for restricted TV.

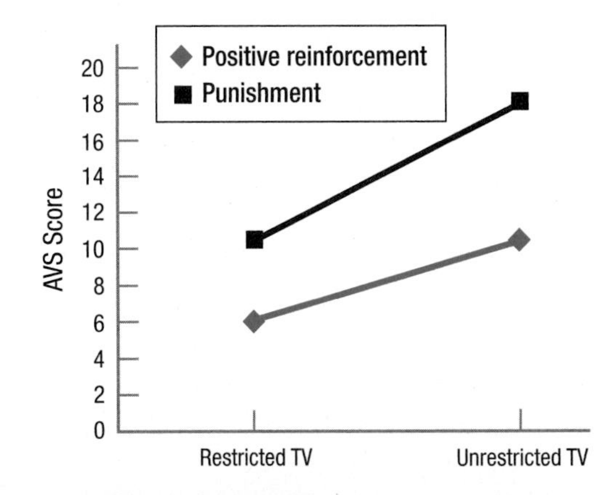

- Calculating effect sizes:
  - $$\eta^2_{Rows} = \frac{SS_{Rows}}{SS_{Total}} \times 100 = \frac{210.04}{651.95} \times 100 = 32.22\%$$
    - This a large effect.
  - $$\eta^2_{Columns} = \frac{SS_{Columns}}{SS_{Total}} \times 100 = \frac{222.04}{651.95} \times 100 = 34.06\%$$
    - This is a large effect.
  - $$\eta^2_{Interaction} = \frac{SS_{Interaction}}{SS_{Total}} \times 100 = \frac{15.04}{651.95} \times 100 = 2.31\%$$
    - This is a small effect.
    - There is little reason to worry about Type II error for the interaction effect.
- Determine whether post-hoc tests are needed:
  - There are only two rows, so there is no need to calculate a *HSD* value for the statistically significant row effect: the positive reinforcement row has a statistically lower mean than the punishment row.
  - There are only two columns, so there is no need to calculate a *HSD* value for the statistically significant column effect: the restricted column has a statistically lower mean than the unrestricted column.
  - The interaction effect was not statistically significant, so there is no need to calculate the *HSD* value of the cells for the interaction effect.

The interpretation:

A developmental psychologist investigated the impact of (a) type of parental discipline, positive reinforcement vs. punishment, and (b) restrictions on television viewing on teenagers' acceptance of violence as a way to solve disagreements. It was found that each variable by itself had a large and independent effect on acceptance of violence. Teens whose parents used punishment to discipline were statistically significantly more accepting of violence than were teens whose parents used positive reinforcement $F(1, 20) = 20.51$, $p < .05$. Teens whose television viewing was unrestricted were more accepting of the use of violence than were teens whose television viewing had been restricted $F(1, 20) = 21.68$, $p < .05$. This study suggests using positive reinforcement and restricting television viewing could lead to teenagers who are less likely to approve of the use of violence. But, nothing was manipulated in this study, so no conclusions regarding cause and effect can be drawn. Future research should attempt to manipulate either television viewing or type of discipline to see the impact on violence acceptance.

12.97 (a)
- A is given as 4.00.
- If $\frac{A+B}{2} = 8.00$ and $A = 4.00$, then $B = 12.00$.
- If $A = 4.00$ and $\frac{A+C}{2} = 6.00$, then $C = 8.00$.
- If $B = 12.00$ and $\frac{B+D}{2} = 9.00$, then $D = 6.00$.

(b) $$M_{Grand} = \frac{A+B+C+D}{4} = \frac{4.00 + 12.00 + 8.00 + 6.00}{4} = 7.50$$

12.99 (a)
- Given $A = 1.00$ and $\frac{A+D}{2} = 2.50$, then $D = 4.00$.
- The other cells can't be calculated.

(b) The grand mean, which is 4.00, can be calculated by finding the average of the row means or the column means.

12.101 e.

12.103 (a) $SS_{Between} = 17.50 + 13.75 + 22.25 = 53.50$

(b) $SS_{Total} = 53.50 + 44.75 = 98.25$

Chapter 13

13.01 two; a

13.03 association

13.05 systematically

13.07 cause and effect

13.09 Scatterplots

13.11 X; Y
13.13 circular shape
13.15 undetermined
13.17 .75
13.19 quantifies; two variables
13.21 weaker
13.23 positive (direct)
13.25 outlier
13.27 unrestricted
13.29 can
13.31 ρ (rho)
13.33 no
13.35 alternate
13.37 −1; 1; 0
13.39 I
13.41 reject
13.43 scatterplot
13.45 decision rule
13.47 10
13.49 negative; inverse
13.51 insufficient; relationship
13.53 r^2 (coefficient of determination)
13.55 100
13.57 ρ
13.59 was not
13.61 should
13.63 .54
13.65 partial correlation
13.67

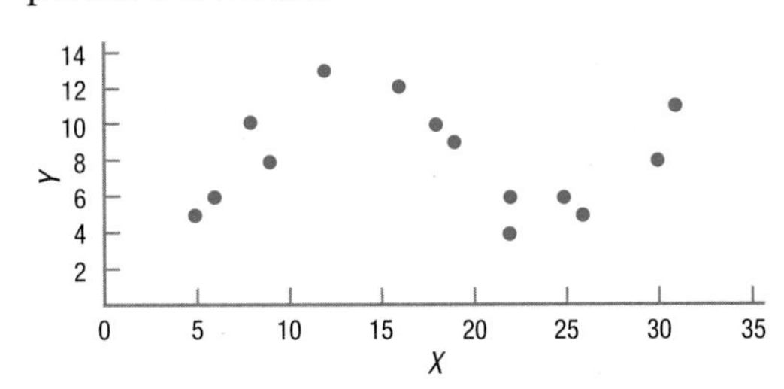

13.69 independent-samples t test
13.71 Pearson r

13.73 (a) Random samples is not violated; independence of observations is not violated; normality is probably violated for age in college students; linearity is not violated.
(b) The normality assumption is robust if the sample size is large, so it is OK to proceed with Pearson r.

13.75 $H_0: \rho = 0$; $H_1: \rho \neq 0$
13.77 $df = N - 2 = 63 - 2 = 61$
13.79 $r_{cv} = \pm.666$
13.81 $r_{cv} = .291$
13.83 If $r \leq -.438$ or $r \geq .438$, reject H_0;
If $-.438 < r < .438$, fail to reject H_0.
13.85

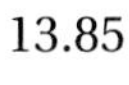

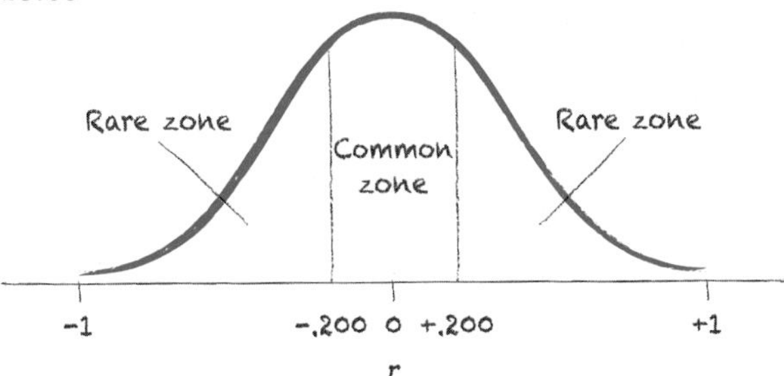

13.87

X	Y	X Deviation Scores	Y Deviation Scores
12	6	3.00	1.00
8	3	−1.00	−2.00
7	6	−2.00	1.00

13.89

X	Y	X Deviation Scores	Y Deviation Scores	Product of Deviation Scores
10	65	−2.33	11.67	−27.19
18	55	5.67	1.67	9.47
9	40	−3.33	−13.33	44.39
$M = 12.33$	$M = 53.33$			$\Sigma = 26.67$

13.91

X	Y	X Deviation Scores	Y Deviation Scores	Product of Deviation Scores	Squared X Deviation Scores	Squared Y Deviation Scores
21	110	−4.33	10.67	−46.20	18.75	113.85
38	90	12.67	−9.33	−118.21	160.53	87.05
17	98	−8.33	−1.33	11.08	69.39	1.77
$M = 25.33$	$M = 99.33$			$\Sigma = -153.33$	$\Sigma = 248.67$	$\Sigma = 202.67$

13.93

X	Y	X Deviation Scores	Y Deviation Scores	Product of Deviation Scores	Squared X Deviation Scores	Squared Y Deviation Scores
88	23	1.67	−5.00	−8.35	2.79	25.00
93	28	6.67	0.00	0.00	44.49	0.00
78	33	−8.33	5.00	−41.65	69.39	25.00
$M = 86.33$	$M = 28.00$			$\Sigma = -50.00$	$\Sigma = 116.67$	$\Sigma = 50.00$
					$SS_X SS_Y = 5{,}833.50$	
					$\sqrt{SS_X SS_Y} = 76.38$	

13.95 $r = \frac{\Sigma[(X - M_X)(Y - M_Y)]}{\sqrt{SS_X SS_Y}} = \frac{255.70}{337.29} = .76$

13.97 $\sqrt{SS_X SS_Y} = \sqrt{12.44 \times 28.81}$

$= \sqrt{358.3964} = 18.93$

$r = \frac{\Sigma[(X - M_X)(Y - M_Y)]}{\sqrt{SS_X SS_Y}} = \frac{-12.03}{18.93} = -.64$

13.99 (a) Failed to reject null hypothesis; (b) Results not statistically significant.

13.101 $r(20) = -.67, p < .05$

13.103 (a) Null hypothesis rejected; (b) Results statistically significant.

13.105 (a) There is not enough evidence to conclude a relationship exists between *X* and *Y* in the population; (b) There is no need to comment on the direction of the relationship because not enough evidence exists to conclude there is a relationship.

13.107 (a) $r^2 = r^2 \times 100 = .32^2 \times 100 = 10.24\%$; (b) Medium effect.

13.109 $z_r = -0.50$

13.111 $s_r = \frac{1}{\sqrt{N-3}} = \frac{1}{\sqrt{81-3}} = .11$

13.113 $95\%CI_{z_r} = z_r \pm 1.96s_r = -0.34 \pm 1.96(.24)$
$= [-0.81, 0.13]$

13.115 $z_r = 0.60$

$s_r = \frac{1}{\sqrt{N-3}} = \frac{1}{\sqrt{45-3}} = .15$

$95\%CI_{z_r} = z_r \pm 1.96s_r = 0.60 \pm 1.96(.15)$
$= [0.31, 0.89]$

13.117 [−.69, −.39]

13.119 (a) Null hypothesis was not rejected; (b) Results were not statistically significant; (c) Relationship in the population could be weak; (d) Unsure of population value of correlation.

13.121 Power $> .99$; $\beta < .01$

13.123 120 cases

13.125 "In this study, the mental health level of a large, random sample of American men was measured, and the men were surveyed regarding how much aerobic exercise they engaged in each week. A moderately strong, direct, and statistically significant relationship was found between the two variables [$r(1{,}000) = .36, p < .05$], indicating that men who had higher levels of mental health also had higher levels of exercise. Because this is a correlational study, it is not clear if being mentally healthy leads to exercise, exercising leads to mental health, or a third variable, like physical health, leads to both. Future research should try to tease apart the direction of this relationship."

13.127 $r_{XY-Z} = \frac{r_{XY} - (r_{XZ} \times r_{YZ})}{\sqrt{(1 - r_{XZ}^2)(1 - r_{YZ}^2)}}$

$= \frac{.28 - (.20 \times .22)}{\sqrt{(1 - .20^2)(1 - .22^2)}} = .25$

13.129 After adjusting for socioeconomic status, there is a statistically significant, but small, relationship between the IQ of the student and the number of absences.

13.131 **Step 1** Pick a Test

Examining the relationship between two interval and/or ratio variables: Pearson *r*.

Step 2 Check the Assumptions

- Random samples: Violated.
- Independence of observations: Not violated.

- Normality: Willing to assume normality for both variables.
- Linearity: Not violated—see the scatterplot below. It is OK to proceed with Pearson *r*.

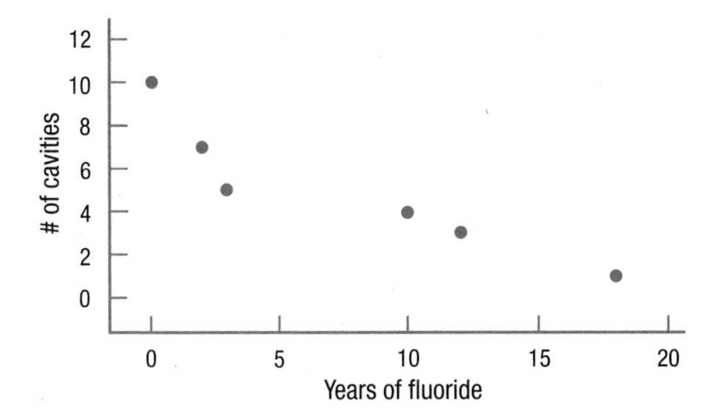

Step 3 List the Hypotheses

- H_0: $\rho = 0$.
- H_1: $\rho \neq 0$.

Step 4 Set the Decision Rule

A nondirectional (two-tailed) test, willing to have a 5% chance of a Type I error, so $\alpha = .05$, $df = N - 2 = 6 - 2 = 4$, so $r_{cv} = \pm.811$.

- If $r \leq -.811$ or $r \geq .811$, reject H_0.
- If $-.811 < r < .811$, fail to reject H_0.

Step 5 Calculate the Test Statistic

First, calculate means for *X* and *Y* so that deviation scores can be calculated.

$$r = \frac{\Sigma[(X - M_X)(Y - M_Y)]}{\sqrt{SS_X SS_Y}} = \frac{-102.00}{110.34}$$

$$= -.9244 = -.92$$

Step 6 Interpret the Results

Was the null hypothesis rejected?

- $-.92 \leq -.811$, so reject H_0.
- APA format: $r(4) = -.92$, $p < .05$.

How big is the effect?

- $r^2 = r^2 \times 100 = -.92^2 \times 100 = .8464 \times 100 = 84.64\%$.
- This is a very large effect.

How wide is the confidence interval?

- $z_r = -1.59$.
- $s_r = \frac{1}{\sqrt{N-3}} = \frac{1}{\sqrt{6-3}} = \frac{1}{\sqrt{3.00}} = \frac{1}{1.7321}$
  $= .5773 = .58$
- $95\%CI_{z_r} = z_r \pm 1.96 s_r = -1.59 \pm 1.96(.58)$
  $= -1.59 \pm 1.1368 =$ from -2.7268 to $-0.4532 =$ from -2.73 to -0.45.
- In *r* units: from $-.99$ to $-.42$.

A four-point interpretation:

This study examined, in a small sample of adults from one dental practice, the relationship between the number of years of fluoride use in childhood and the number of cavities in permanent teeth. There was a very strong, statistically significant, inverse relationship

X	Y	X Deviation Scores	Y Deviation Scores	Product of Deviation Scores	Squared X Deviation Scores	Squared Y Deviation Scores
0	10	−7.50	5.00	−37.50	56.25	25.00
18	1	10.50	−4.00	−42.00	110.25	16.00
2	7	−5.50	2.00	−11.00	30.25	4.00
12	3	4.50	−2.00	−9.00	20.25	4.00
3	5	−4.50	0.00	0.00	20.25	0.00
10	4	2.50	−1.00	−2.50	6.25	1.00
$M = 7.50$	$M = 5.00$			$\Sigma = -102.00$	$\Sigma = 243.50$	$\Sigma = 50.00$
					$SS_X SS_Y = 243.50 \times 50.00$	
					$= 12{,}175.00$	
					$\sqrt{SS_X SS_Y} = 110.34$	

[$r(4) = -.92$, $p < .05$]. One plausible interpretation is that the more years of fluoride one has as a child, the fewer cavities one develops. However, as this is a correlational study, other explanations are possible. Perhaps parents who make sure their children receive fluoride also ensure that their children brush their teeth. Though the effect found in this sample was large, the sample size was small and the data derived from only one dental practice. Future research should replicate this study with a much larger and more representative sample.

13.133 There is insufficient evidence to conclude that X and Y are related in the population; thus, she should not draw any conclusion regarding the direction of the relationship.

13.135 g. r ranges from -1 to 1. -1.26 is an impossible value.

13.137 A small effect requires at least 800 cases. A medium effect requires at least 85 cases. A large effect requires at least 30 cases.

Chapter 14

14.01 association; predicting; variable; variable

14.03 statistically significant

14.05 mean

14.07 minimizes; squared errors

14.09 error

14.11 predicted score; predictor

14.13 positive

14.15 Y-axis

14.17 point

14.19 the same

14.21 error

14.23 width

14.25 multiple regression; simple regression

14.27 ≈ 45

14.29 $b = r\left(\frac{s_Y}{s_X}\right) = -.47\left(\frac{3.65}{9.66}\right) = -.47 \times 0.3778$

$= -0.1776 = -0.18$

14.31 Every penny increase in the price of gas was associated with families driving, on average, 35 fewer miles on their summer vacation.

14.33 $a = M_Y - bM_X = 17.50 - (4.33 \times 5.00)$
$= 17.50 - 21.6500 = -4.1500 = -4.15$

14.35 $Y' = bX + a = 12.98\,X + (-5.00)$
$= 12.98X - 5.00$

14.37 First, calculate b and a:

$$b = r\left(\frac{s_Y}{s_X}\right) = -.24\left(\frac{3.98}{11.00}\right)$$

$$= -.24 \times 0.3618 = -0.0868 = -0.09$$

$a = M_Y - bM_X = 25.00 - (-0.09 \times 55.00)$
$= 25.00 - (-4.9500) = 25.00 + 4.9500$
$= 29.9500 = 29.95$

Then complete the regression equation:
$Y' = bX + a = -0.09X + 29.95$

14.39 $Y' = bX + a = 0.37X + 15 = (0.37 \times 25) + 15$
$= 9.2500 + 15 = 24.2500 = 24.25$

14.41 (a)

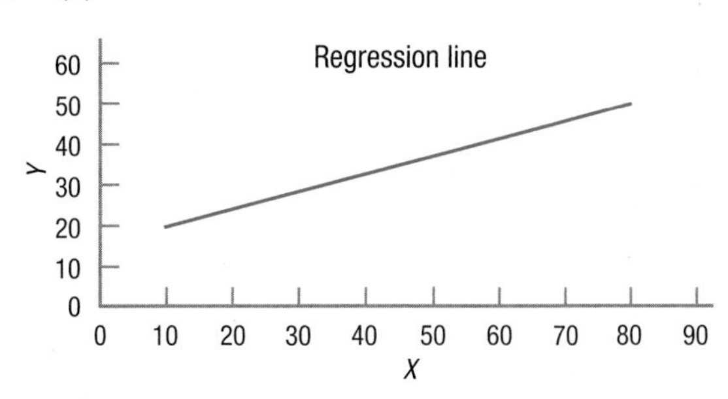

(b) Y' values can be calculated for X values from 10 to 80.

14.43 First calculate the Y' values for the two endpoints:
$Y' = 2.50X - 12.50 = (2.50 \times 30) - 12.50$
$= 75.0000 - 12.50 = 62.5000 = 62.50$
$Y' = 2.50X - 12.50 = (2.50 \times 70) - 12.50$
$= 175.0000 - 12.50 = 162.5000 = 162.50$
The two endpoints are (30, 62.50) and (70, 162.50)
Then draw the regression line:

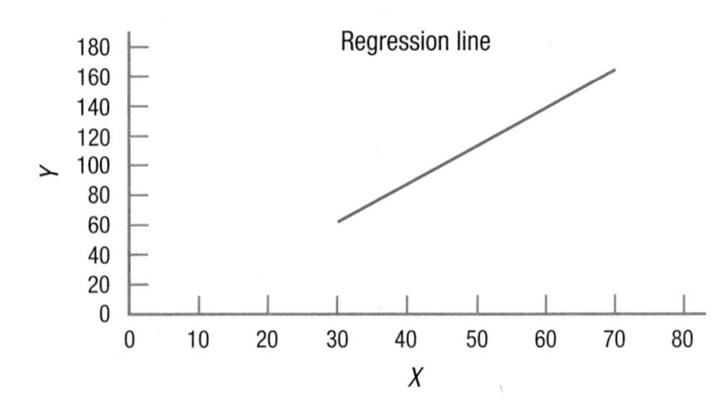

14.45 Residual $= Y - Y' = 12.96 - 13.43 = -0.4700$
$= -0.47$

14.47 $s_{Y-Y'} = s_Y\sqrt{1-r^2} = 10.55\sqrt{1-.28^2}$

$= 10.55\sqrt{1-.0784} = 10.55\sqrt{.9216}$

$= 10.55 \times .9600 = 10.1280 = 10.13$

14.49 The range of scores, from 400 to 1600, covers 1,200 points. If the average error is 40 points, that's off by 3% of the range. This seems to be a relatively small amount of error.

14.51 $Y' = (3.42 \times \text{variable 1}) + (-0.76 \times \text{variable 2}) + 5.74 = (3.42 \times 56.66) + (-0.76 \times 88.99) + 5.74 = 193.7772 + (-67.6324) + 5.74 = 193.7772 - 67.6324 + 5.74 = 131.8848 = 131.88$

14.53 b

14.55 c

14.57 The college applauds Shemekia's knowledge of statistics. But, the error could just as easily go the other way and Shemekia's first year GPA would be 2.52, well below their admission cutoff score.

14.59 Yes. Start with the value on the *y*-axis. Then move to the right horizontally until you intersect the regression line. Then move vertically downward until you hit the *x*-axis. This seems to be your predicted value of *X* for a given value of *Y*.

Chapter 15

15.01 interval; ratio

15.03 parametric

15.05 fall back

15.07 power; parametric

15.09 observed; expected

15.11 random sample

15.13 sample; population

15.15 chi-square value; rare zone

15.17 not rejected

15.19 *N*

15.21 degrees of freedom

15.23 statistically significant

15.25 relationship; difference

15.27 case; cell

15.29 dependent variable; independent variable; independent

15.31 rows; columns

15.33 100

15.35 ranks

15.37 samples; group; rank

15.39 Examining the relationship between two nominal-level variables: chi-square test of independence. (Or, comparing two independent samples on a nominal-level dependent variable: chi-square test of independence.)

15.41 Looking at the relationship between two ordinal-level variables: Spearman *r*

15.43 (a) Random samples assumption is not violated; (b) Independence of observations assumption is not violated; (c) All cells have expected frequencies of at least 5 is violated, so can't proceed with the planned test.

15.45 H_0: In the population of small towns, 2% experienced a murder in the past 10 years. H_1: In the population of small towns, the percentage experiencing a murder in the past 10 years is not equal to 2%.

15.47 $df = k - 1 = 4 - 1 = 3$

15.49 $df = (R-1) \times (C-1) = (3-1) \times (3-1) = 2 \times 2 = 4$

15.51 (a) $\chi^2_{cv} = 5.991$ (b) If $\chi^2 \geq 5.991$, reject H_0. If $\chi^2 < 5.991$, fail to reject H_0.

15.53
$$f_{\text{ExpectedA}} = \frac{\%_{\text{ExpectedA}} \times N}{100} = \frac{63 \times 97}{100} = 61.11$$

$$f_{\text{ExpectedB}} = \frac{\%_{\text{ExpectedB}} \times N}{100} = \frac{37 \times 97}{100} = 35.89$$

15.55
$$\chi^2 = \Sigma\frac{(f_{\text{Observed}} - f_{\text{Expected}})^2}{f_{\text{Expected}}}$$

$$= \frac{(9-6.97)^2}{6.97} + \frac{(14-15.58)^2}{15.58} + \frac{(18-18.45)^2}{18.45}$$

$$= 0.76$$

15.57
$$f_{\text{ExpectedA}} = \frac{\%_{\text{ExpectedA}} \times N}{100} = \frac{35 \times 90}{100} = 31.50$$

$$f_{\text{ExpectedB}} = \frac{\%_{\text{ExpectedB}} \times N}{100} = \frac{65 \times 90}{100} = 58.50$$

$$\chi^2 = \Sigma\frac{\left(f_{\text{Observed}} - f_{\text{Expected}}\right)^2}{f_{\text{Expected}}}$$

$$= \frac{(40 - 31.50)^2}{31.50} + \frac{(50 - 58.50)^2}{58.50}$$

$$= 3.53$$

15.59 $\chi^2(1, N = 43) = 4.81, p < .05$

15.61 $\chi^2(4, N = 55) = 9.49, p < .05$

15.63 There are more cases than expected in Category I and fewer than expected in Category II.

15.65 "This study investigated if fortunetellers had the extrasensory ability to tell whether a randomly drawn card was red or black. Though the fortunetellers judged the color accurately 58% of the time, there was insufficient evidence to show that they performed better than chance [$\chi^2(1, N=124) = 3.23, p > .05$]. The possibility exists that fortune tellers have some slight extrasensory abilities, too slight to be found in this sample. The study should be replicated with a larger sample size in order to have a better chance of finding the effect if it exists."

15.67 **Step 1** Pick a Test

Seeing if the selection of one's favorite cola by the sample matches what would occur if the selection were made randomly. This calls for a chi-square goodness-of-fit test.

Step 2 Check the Assumptions

1. Random samples is not violated.
2. Independence of observations is not violated.
3. All expected frequencies more than 5 is not violated.

Step 3 List the Hypotheses

- H_0: The distribution of cola choices in the population is random.
- H_1: The distribution of cola choices in the population differs from random.

Step 4 Set the Decision Rule

$df = k - 1 = 2 - 1 = 1$

$\chi^2_{cv} = 3.841$ with $\alpha = .05$

If $\chi^2 \geq 3.841$, then reject H_0. If $\chi^2 < 3.841$, then fail to reject H_0.

Step 5 Calculate the Test Statistic

$$f_{\text{ExpectedCorrect}} = \frac{\%_{\text{ExpectedCorrect}} \times N}{100}$$

$$= \frac{25 \times 880}{100} = 220.00$$

$$f_{\text{ExpectedIncorrect}} = \frac{\%_{\text{ExpectedIncorrect}} \times N}{100}$$

$$= \frac{75 \times 880}{100} = 660.00$$

$$\chi^2 = \Sigma\frac{\left(f_{\text{Observed}} - f_{\text{Expected}}\right)^2}{f_{\text{Expected}}}$$

$$= \frac{(279 - 220.00)^2}{220.00} + \frac{(601 - 660.00)^2}{660.00}$$

$$= 21.10$$

Step 6 Interpret the Results

- $21.10 \geq 3.841$, so reject H_0.
- $\chi^2(1, N = 880) = 21.10, p < .05$

Comparing observed frequency to expected frequency shows that more people than expected correctly identified their favorite cola:

The ability of consumers to differentiate their favorite cola from three competing brands was tested in this study. If consumers could not differentiate and were making decisions randomly, then they would have selected their preferred cola 25% of the time. They showed a statistically significant improvement over this [$\chi^2(1, N = 880) = 21.10, p < .05$], correctly choosing their cola 32% of the time. Though an effect exists, it doesn't seem very powerful as the majority of consumers could not correctly identify their preferred cola. It would be interesting to retest this sample and find out whether the people who identified the cola correctly can do so consistently.

15.69

	Support Gay Marriage: Yes	Support Gay Marriage: No
20s		
40s		
60s		

15.71 $N_{Row1} = 34 + 86 = 120$

$N_{Row2} = 67 + 113 = 180$

$N_{Column1} = 34 + 67 = 101$

$N_{Column2} = 86 + 113 = 199$

15.73 $f_{ExpectedA} = \frac{N_{Row1} \times N_{Column1}}{N} = \frac{34 \times 30}{72}$

$= 14.1667 = 14.17$

$f_{ExpectedB} = \frac{N_{Row1} \times N_{Column2}}{N} = \frac{34 \times 42}{72}$

$= 19.8333 = 19.83$

$f_{ExpectedC} = \frac{N_{Row2} \times N_{Column1}}{N} = \frac{38 \times 30}{72}$

$= 15.8333 = 15.83$

$f_{ExpectedD} = \frac{N_{Row2} \times N_{Column2}}{N} = \frac{38 \times 42}{72}$

$= 22.1667 = 21.17$

15.75 $\chi^2 = \Sigma\frac{(f_{Observed} - f_{Expected})^2}{f_{Expected}}$

$= \frac{(28 - 25.00)^2}{25.00} + \frac{(12 - 15.00)^2}{15.00}$

$+ \frac{(12 - 15.00)^2}{15.00} + \frac{(12 - 9.00)^2}{9.00}$

$= 2.56$

15.77 $f_{Expected} = \frac{N_{Row1} \times N_{Column1}}{N}$

$= \frac{35 \times 32}{71} = 15.7746 = 15.77$

$f_{Expected} = \frac{N_{Row1} \times N_{Column2}}{N}$

$= \frac{35 \times 39}{71} = 19.2253 = 19.22$

$f_{Expected} = \frac{N_{Row2} \times N_{Column1}}{N}$

$= \frac{36 \times 32}{71} = 16.2253 = 16.23$

$f_{Expected} = \frac{N_{Row2} \times N_{Column2}}{N}$

$= \frac{36 \times 39}{71} = 19.7746 = 19.77$

$\chi^2 = \Sigma\frac{(f_{Observed} - f_{Expected})^2}{f_{Expected}}$

$= \frac{(14 - 15.77)^2}{15.77} + \frac{(21 - 19.22)^2}{19.22}$

$+ \frac{(18 - 16.23)^2}{16.23} + \frac{(18 - 19.77)^2}{19.77}$

$= 0.72$

15.79 Type I

15.81 Outcome A occurs less frequently than expected for Group I and more frequently than expected for Group II, while Outcome B occurs more frequently than expected for Group I and less frequently than expected for Group II.

15.83 (a) $V = \sqrt{\frac{\chi^2}{N \times df_{RC}}} = \sqrt{\frac{8.92}{78 \times 1}} = .34$

(b) Medium effect

(c) No need to worry about Type II error. (As the null hypothesis was rejected, worry about Type I error.)

15.85 "A tennis journalist gathered a random sample of points played at major tournaments in order to investigate whether the server's advantage diminished as the length of the rally increased. She classified rallies as short (the ball went back and forth no more than 2 times) or long (the ball went back and forth more than 2 times). She found that when the rally was short, the server won the point about 62% of the time. In contrast, the server won only 48% of the time on long rallies. This difference is statistically significant [$\chi^2(1, N = 209) = 3.93, p < .05$], and it is a meaningful difference. Future research might examine if the same effect is found in nonprofessional players."

15.87 **Step 1** Pick a Test

Comparing a nominal-level dependent variable between two independent samples calls for a chi-square test of independence.

Step 2 Check the Assumptions

- Random samples is violated. Can proceed, but limitations on generalizability.
- Independence of observations is not violated.

- Expected frequencies ≥ 5 is not violated.

Step 3 List the Hypotheses

H_0: There is no difference in satisfaction with the scar left by sutures vs. staples.

H_1: There is some difference in satisfaction with the scar left by sutures vs. staples.

Step 4 Set the Decision Rule

$df = (R - 1) \times (C - 1) = (2 - 1) \times (2 - 1)$
$= 1 \times 1 = 1$

$\chi^2_{cv} = 3.841$ with $\alpha = .05$

If $\chi^2 \geq 3.841$, then reject H_0. If $\chi^2 < 3.841$, then fail to reject H_0.

Step 5 Calculate the Test Statistic

$$f_{Expected} = \frac{N_{Row1} \times N_{Column1}}{N} = \frac{17 \times 21}{42} = 8.50$$

$$f_{Expected} = \frac{N_{Row1} \times N_{Column2}}{N} = \frac{21 \times 25}{42} = 12.50$$

$$f_{Expected} = \frac{N_{Row2} \times N_{Column1}}{N} = \frac{21 \times 17}{42} = 8.50$$

$$f_{Expected} = \frac{N_{Row2} \times N_{Column2}}{N} = \frac{21 \times 25}{42} = 12.50$$

$$\chi^2 = \Sigma \frac{(f_{Observed} - f_{Expected})^2}{f_{Expected}}$$

$$= \frac{(8 - 8.50)^2}{8.50} + \frac{(13 - 12.50)^2}{12.50}$$

$$+ \frac{(9 - 8.50)^2}{8.50} + \frac{(12 - 12.50)^2}{12.50}$$

$$= 0.0988 = 0.10$$

Step 6 Interpret the Results

- $0.10 < 3.841$; fail to reject H_0.
- $\chi^2(1, N = 42) = 0.10, p > .05$

$$V = \sqrt{\frac{\chi^2}{N \times df_{RC}}} = \sqrt{\frac{0.10}{42 \times 1}} = .05$$

In this study, patient satisfaction with scars left by sutures vs. staples was measured. There was no statistical difference in level of satisfaction [$\chi^2(1, N = 42) = 0.10, p > .05$], suggesting that either could be used to close incisions, at the preference of the surgeon and/or the patient. A positive of this study is that patients were randomly assigned to receive sutures or staples, but only one surgeon provided both. Thus, it is possible that he is better at one technique over the other. Therefore, this study should be replicated using multiple surgeons and with the quality of the suturing and the stapling assessed.

15.89

12	8	20
14	16	30
26	24	

15.91 This table can't be completed. If the marginal frequencies are known, a table can be completed if there are as many cells filled in as degrees of freedom (in this case, 2).

15.93 If observed frequencies and expected frequencies are all the same, then $\chi^2 = 0$ and the null hypothesis is not rejected.

Chapter 16

16.01 describe
16.03 two variables
16.05 explanatory variable (or independent variable)
16.07 context
16.09 shape; level of measurement
16.11 is
16.13 Descriptive
16.15 Correlational
16.17 Experimental
16.19 Descriptive
16.21 (a) Ratio; (b) Continuous
16.23 (a) Nominal; (b) Discrete
16.25 (a) Interval; (b) Continuous
16.27 Standard deviation is the best answer. Also acceptable are interquartile range, variance, and range.
16.29 Bar graph
16.31 chi-square goodness-of-fit test
16.33 between-subjects, one-way ANOVA
16.35 independent-samples *t* test
16.37 single-sample *t* test
16.39 Pearson *r*
16.41 independent-samples *t* test
16.43 independent-samples *t* test
16.45 No such test appears in the book.
16.47 Pearson *r*
16.49 chi-square test of independence

APPENDIX C

Solutions to Practice Problems

Chapter 1

Practice Problems 1.1

1.01 Statistics are techniques used to summarize data in order to answer questions.

1.02 Examples of questions to be asked of the sample IQ data:
- How smart is the average sixth grader?
- How smart is the smartest sixth grader?
- What is the range of IQ scores for sixth graders?
- How many sixth graders have IQ scores that classify them as geniuses?
- Is there a difference in the IQs of boys and girls?

Practice Problems 1.2

1.03 Correlational, experimental, and quasi-experimental

1.04 Experimental

1.05 Experimental

1.06 Correlational and quasi-experimental

1.07 (a) Do different types of studying, spaced vs. massed, have an impact on how much one learns?
(b) Type of studying is the explanatory (independent) variable; number of nonsense syllables recalled is the outcome (dependent) variable.
(c) Experimental
(d) N/A

1.08 (a) Does amount of sleep affect school performance?
(b) Amount of sleep is the explanatory (grouping) variable; average grade is the outcome (dependent) variable.
(c) Quasi-experimental
(d) Breakfast may be a confounding variable. People who don't get enough sleep may not have time to eat breakfast. Not eating breakfast may impair school performance.

1.09 (a) Is fiber consumption related to GI health?
(b) Fiber consumption is the explanatory (predictor) variable; number of episodes of GI distress is the outcome (criterion) variable.
(c) Correlational
(d) Intelligence may be a confounding variable. More intelligent people keep up on the news and the news recently has been pushing consumption of fiber. More intelligent people may also take better care of themselves in general, so they have better health, thus fewer episodes of GI distress.

Practice Problems 1.3

1.10 Nominal, ordinal, interval, and ratio

1.11 The scale now provides information about how much distance separates two scores.

1.12 An absolute zero point represents an absence of the characteristic.

1.13 Ratio

1.14 Ordinal

1.15 Nominal

1.16 Interval

Practice Problems 1.4

1.17 Greek letters are abbreviations for population values (parameters); Latin letters are abbreviations for sample values (statistics).

1.18 Inferential statistic

1.19 (a) Parameter. (It is a statement about the whole population of softball players at that college.)
(b) Descriptive

1.20 (a) Sample
(b) Inferential

Practice Problems 1.5

1.21 Σ

1.22 Math within parentheses or brackets

1.23 Two

1.24 (a) $N = 4$
(b) $\Sigma X = 12 + 8 + 4 + 6 = 30.0000 = 30.00$
(c) $\Sigma X^2 = 12^2 + 8^2 + 4^2 + 6^2 = 144.0000 + 64.0000 + 16.0000 + 36.0000 = 260.0000 = 260.00$
(d) $(\Sigma X)^2 = (12 + 8 + 4 + 6)^2 = 30.0000^2 = 900.0000 = 900.00$
(e) $\Sigma X + 1 = \Sigma(12 + 8 + 4 + 6) + 1 = 30.0000 + 1 = 31.0000 = 31.00$
(f) $\Sigma(X + 1) = (12 + 1) + (8 + 1) + (4 + 1) + (6 + 1) = 13.0000 + 9.0000 + 5.0000 + 7.0000 = 34.0000 = 34.00$

1.25 (a) 17.79
(b) 9.74
(c) 12.98
(d) 8.35
(e) 7.12
(f) 1.67
(g) 2.00

Chapter 2

Practice Problems 2.1

2.01 If the variable takes on a limited number of values, make an ungrouped frequency distribution. If the variable has a large number of values and it is OK to collapse them into intervals, make a grouped frequency distribution. If the variable has a large number of values and there is need to retain information about the frequency of all the values, make an ungrouped frequency distribution.

2.02 Cumulative frequency tells how many cases in the data set have a given value or a lower value.

2.03 One needs to impose some logical order on a frequency distribution for a nominal variable, perhaps arrange the categories in alphabetical order or in ascending or descending frequency order.

2.04 One should aim to have 5–9 intervals.

2.05 One calculates the midpoint by finding the point halfway between the upper limit and the lower limit of an interval.

2.06 Don't forget a title!

Frequency Distribution for Time for Heart Rate to Return to Normal after Exercise for 30 College Students

Minutes for Heart Rate to Return to Normal	Frequency	Cumulative Frequency	Percentage	Cumulative Percentage
14	1	30	3.33	100.00
13	0	29	0.00	96.67
12	1	29	3.33	96.67
11	2	28	6.67	93.33
10	1	26	3.33	86.67
9	1	25	3.33	83.33
8	4	24	13.33	80.00
7	3	20	10.00	66.67
6	2	17	6.67	56.67
5	2	15	6.67	50.00
4	3	13	10.00	43.33
3	5	10	16.67	33.33
2	4	5	13.33	16.67
1	1	1	3.33	3.33

2.07

Grouped Frequency Distribution for Final Grades in Psychology Class of 58 Students (Interval Width = 10)

Grade Interval	Interval Midpoint	Frequency	Cumulative Frequency	Percentage	Cumulative Percentage
90–99	94.5	9	58	15.52	100.00
80–89	84.5	20	49	34.48	84.48
70–79	74.5	16	29	27.59	50.00
60–69	64.5	9	13	15.52	22.41
50–59	54.5	4	4	6.90	6.90

Practice Problems 2.2

2.08 Continuous numbers answer the question "How much?" and discrete numbers answer the question "How many?" Continuous numbers can be made more specific, for example, measuring weight to the nearest ounce, not pound, by using a more precise measuring instrument. Discrete numbers only take whole number values.

2.09 Assign them the value of the midpoint, 47.00.

2.10 (a) Continuous
(b) Discrete
(c) Continuous

2.11

	Real Lower Limit	Real Upper Limit	Interval Width	Midpoint
a.	19.50	24.50	5.00	22.00
b.	250.00	300.00	50.00	275.00
c.	1.25	1.45	0.20	1.35
d.	2,500.00	4,500.00	2,000.00	3,500.00

Practice Problems 2.3

2.12 With continuous data, use a histogram or a frequency polygon. With discrete data, use a bar graph.

2.13 Bars go up and come down at the real limits of the interval for a histogram.

2.14 The data, which are nominal and so discrete, need to be ordered in some way. Here, they are organized in descending order of frequency. Note that the sample size is included in the title and the axes are labeled.

Religious Faith of 53 University Students

Frequency

20
15
10
5
0

Christian Jewish Muslim Atheist Other Buddhist Hindu

Religion

2.15 One can make either a histogram or a frequency polygon for these continuous data. The requency polygon is a little awkward as the point at which it goes to zero, at both ends of the scale, occurs at values that don't exist.

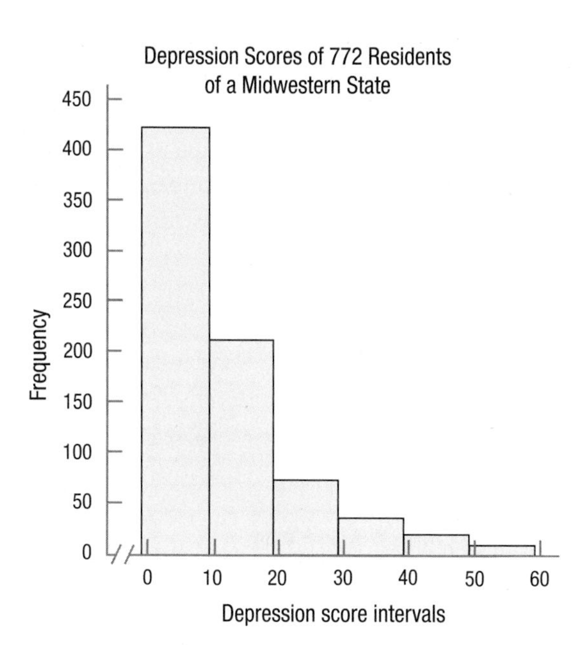

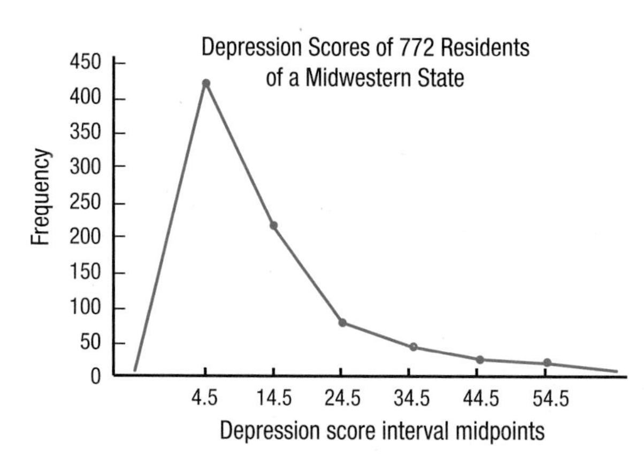

Practice Problems 2.4

2.16 One can describe the shape of a frequency distribution for ordinal-, interval-, or ratio-level variables.

2.17 The normal curve is symmetric, unimodal, and neither too peaked nor too flat.

2.18 Positive skewness means that the tail of the frequency distribution extends further on the right-hand side.

2.19 Religious faith is a nominal-level variable. Because the order in which a nominal variable is arranged is arbitrary, it is inappropriate to discuss shape.

2.20 Depression level is unimodal. The peak may be a little higher than normal. The distribution is positively skewed.

2.21

18	3
19	2
20	33
21	19
22	07889
23	0344
24	89
25	4
26	6

The stem-and-leaf plot is unimodal. The peak may be a little flatter than normal. The distribution is approximately symmetric.

Chapter 3

Practice Problems 3.1

3.01 Mean, median, and mode

3.02 (a) Mode; (b) mode or median; (c) mode or median or mean

3.03 A skewed data set makes the mean an inappropriate choice as a measure of central tendency. It drags the mean in the direction of the skew. For example, if there is positive skew, the mean will be bigger than the median.

3.04 IQ is an interval-level measure, so the "go to" measure of central tendency is the mean. A frequency distribution shows nothing unusual in terms of skewness or kurtosis, so it is OK to proceed with the mean.

$$M = \frac{\Sigma X}{N} = \frac{94 + 100 + 110 + 112 + 100 + 98 + 100}{7}$$

$$= \frac{714.00}{7} = 102.00$$

3.05 The number of times a neuron fires is a ratio-level variable, so the "go to" measure of central tendency is the mean. However, a frequency distribution shows that the data are bimodal, making the mean inappropriate. Report the two modes, 2 and 10, instead.

3.06 The data are nominal, so the only option for a measure of central tendency is the mode. The average taste bud in the sample responded to sweet (5).

Practice Problems 3.2

3.07 The set of scores with more variability is less tightly clustered together; it stretches out over a wider range.

3.08 A disadvantage of the range is that it is influenced by outliers. Another disadvantage is that it only uses information from two scores.

3.09 An advantage of the interquartile range is that it is not influenced by extreme scores. Another advantage is that it also functions as a measure of central tendency.

3.10 The variance and standard deviation, unlike the range and interquartile range, use information from all the scores in the data set to determine variability.

3.11 The lowest MMPI score is 60 and the highest is 84. Range = 84 – 60 = 24.00.

3.12 The mean is 19.00.

Size in Acres	Deviation Score (X – M)	Squared Deviation Score $(X – M)^2$
8	−11.00	121.00
10	−9.00	81.00
22	3.00	9.00
23	4.00	16.00
32	13.00	169.00
		$\Sigma = 396.00$

(a)

$$s^2 = \frac{\Sigma(X - M)^2}{N - 1} = \frac{396.00}{5 - 1} = \frac{396.00}{4} = 99.00$$

(b) $s = \sqrt{s^2} = \sqrt{99.00} = 9.95$

Chapter 4

Practice Problems 4.1

4.01 z scores express raw scores in standard deviation units.

4.02 $\Sigma z = 0.00$

4.03 A positive z score indicates that the raw score falls above the mean.

4.04 $z = \frac{X - M}{s} = \frac{3.20 - 2.75}{0.40} = \frac{0.4500}{0.40} = 1.1250$

$= 1.13$

4.05 GPA $= X = M + (z \times s) = 2.75 + (-2.30 \times 0.40)$
$= 2.75 + (-0.9200) = 2.75 - 0.9200 = 1.8300$
$= 1.83$

Practice Problems 4.2

4.06 The normal distribution is symmetrical, with the highest spot at the midpoint, and decreasing frequencies as one moves away from the midpoint. Though many distributions have this shape, the normal distribution is a specific version defined by the percentage of cases that fall in specified regions.

4.07 The midpoint of a normal distribution is the mean, median, and mode.

4.08 34, 14, and 2 represent the approximate percentage of cases that fall in each of the first three standard deviations as one moves away from the mean in a normal distribution.

4.09 Using column A, 90.82%

4.10 Using column C, 0.38%

4.11 Using column B, 30.23%

Practice Problems 4.3

4.12 A percentile rank is a score's case expressed as the percentage of cases with a score at the same level or lower.

4.13 $PR = 50$

4.14 Using column A, $z = 0.84$.

4.15 Using column C, $PR = 32.64$.

Practice Problems 4.4

4.16 Probability is the likelihood of the occurrence of an outcome.

4.17 The smallest probability for an outcome is zero; the greatest is 1.

4.18 ± 0.84

4.19 $p(\text{Monday}) = \frac{1}{7} = .1429 = .14$

4.20 $p(\text{not Monday}) = 1.00 - p(\text{Monday}) = 1 - \frac{1}{7}$

$= 1 - .1429 = .8571 = .86$

4.21 $\frac{34\%}{2} = 17\%$. Using column B, $z = \pm 0.44$.

4.22 $z = \frac{123 - 100}{15} = 1.53$, 6.3% (from column C) of cases in a normal distribution have a z score ≥ 1.53. Thus, the probability of someone having an IQ of 123 or higher is .0630.

Chapter 5

Practice Problems 5.1

5.01 A representative sample is an accurate reflection of the population. It contains all the attributes of the population in a similar proportion as they exist in the population.

5.02 Consent rate is the percentage of cases targeted to be in the sample that end up in it.

5.03 Sampling error is caused by random factors.

5.04 There are many ways a random sample could be drawn. All numbers could be put in a hat, the hat shaken, and then numbers drawn. A random number table could be used.

5.05 Drawing a large and random sample can minimize sampling error.

Practice Problems 5.2

5.06 A sampling distribution is a frequency distribution of some statistic, like a mean, that has been calculated for repeated random samples of a given size from a population.

5.07 The central limit theorem predicts three things about a sampling distribution of the mean as long as the size of the samples is large enough: (a) the sampling distribution will be normally distributed; (b) the mean of the sampling distribution is the mean of the population; and (c) the standard deviation of the sampling distribution, called the standard error of the mean, is calculated from N, the sample size, and σ, the population standard deviation.

5.08 There are 21 unique samples as shown below:

AA	AB	AC	AD	AE	AF
BB	BC	BD	BE	BF	
CC	CD	CE	CF		
DD	DE	DF			
EE	EF				
FF					

5.09 Thanks to the central limit theorem, which holds true if sample size is large, Researcher Y's sampling distribution should have an approximate normal shape. Because Researcher X's sample size is not large, we can't be sure that his or her sampling distribution will be normally distributed.

5.10 $\sigma_M = \frac{12}{\sqrt{78}} = \frac{12}{8.8318} = 1.3587 = 1.36$

Practice Problems 5.3

5.11 A point estimate is a discrete, single value estimate of a population value. An interval estimate is a range of values that estimate the population value.

5.12 95% of the time

5.13 100% – 95% = 5% of the time

5.14
$$\begin{aligned} 95\%CI_\mu &= M \pm (1.96 \times \sigma_M) \\ &= M \pm \left(1.96 \times \frac{\sigma}{\sqrt{n}}\right) \\ &= 17.00 \pm \left(1.96 \times \frac{8}{\sqrt{55}}\right) \\ &= 17.00 \pm 2.11 \\ M &= 17.00,\ 95\%CI\ [14.89, 19.11] \end{aligned}$$

5.15
$$\begin{aligned} 95\%CI_\mu &= M \pm (1.96 \times s_M) \\ &= M \pm \left(1.96 \times \frac{s}{\sqrt{n}}\right) \\ &= 250.00 \pm \left(1.96 \times \frac{60}{\sqrt{180}}\right) \\ &= 250.00 \pm 8.77 \\ M &= 250.00,\ 95\%CI\ [241.23, 258.77] \end{aligned}$$

Chapter 6

Practice Problems 6.1

6.01 The null hypothesis is a specific statement (e.g., this population has a mean of 5.00) and it is a negative statement (e.g., the independent variable has no impact). The alternative hypothesis is nonspecific (e.g., the population mean is something other than 5.00) and is a positive statement (e.g., the independent variable has some impact).

6.02 A researcher predicts what he or she thinks should happen if the null hypothesis is true. This is the expected outcome. One then sees what actually happens, the observed outcome.

If what is observed is what was expected, then there is no reason to question the null hypothesis. However, if what is observed is not what was expected, then there is reason to question, or reject, the null hypothesis.

Practice Problems 6.2

6.03 The six steps of hypothesis testing:

1. Pick a test.
2. Check the assumptions.
3. List the hypotheses.
4. Set the decision rule.
5. Calculate the test statistic.
6. Interpret the results.

6.04 $\sigma_M = \frac{\sigma}{\sqrt{N}} = \frac{12}{\sqrt{55}} = 1.62$

6.05 $z = \frac{M - \mu}{\sigma_M} = \frac{19.40 - 22.80}{4.60} = -0.74$

6.06 (a) $\sigma_M = \frac{\sigma}{\sqrt{N}} = \frac{4}{\sqrt{64}} = 0.50$

$z = \frac{M - \mu}{\sigma_M} = \frac{20.75 - 20}{0.50} = 1.50$

(b) $z\,(N = 64) = 1.50,\ p > 0.05$

Practice Problems 6.3

6.07 Type II errors occur when a researcher, by mistake, fails to reject the null hypothesis. For example, he or she might conclude that no evidence of a difference between M and μ exists, but there really is a difference between the two.

6.08 Power is the probability of being able to reject the null hypothesis when it should be rejected.

6.09 There are two correct conclusions in hypothesis testing:

- One concludes that the null hypothesis is wrong and it really is wrong.
- One concludes that there is insufficient reason to reject the null hypothesis and the null hypothesis really is true.

6.10 There are two incorrect conclusions in hypothesis testing:

- One concludes that the null hypothesis is wrong and it really is right.
- One fails to reject the null hypothesis and it really is wrong.

Chapter 7

Practice Problems 7.1

7.01 Sampling error could explain why the sample mean is different from the population mean.

7.02 (1) Pick a test; (2) check the assumptions; (3) list the hypotheses; (4) set the decision rule; (5) calculate the test statistic; (6) interpret the results.

7.03 When comparing a sample mean to a population mean and the population standard deviation is unknown

7.04 Random sample, independence of observations, and normal distribution of the dependent variable in the population

7.05 One-tailed test

7.06 $H_0: \mu_{\text{NobelPrizeWinners}} \leq 100$

$H_1: \mu_{\text{NobelPrizeWinners}} > 100$

7.07 If $t \leq -2.012$ *or* if $t \geq 2.012$, reject the null hypothesis.

If $-2.012 < t < 2.012$, fail to reject the null hypothesis.

7.08 $s_M = \frac{s}{\sqrt{N}} = \frac{6}{\sqrt{17}} = 1.46$

7.09 $t = \frac{M - \mu}{s_M} = \frac{24 - 30}{8} = -0.75$

Practice Problems 7.2

7.10 $t(18) = 2.23,\ p < .05$

7.11 $t(6) = 2.31,\ p > .05$

7.12 $t(35) = 2.03,\ p < .05$

7.13 $t(339) = 3.68,\ p < .05$

Practice Problems 7.3

7.14 $d = \frac{M - \mu}{s} = \frac{66 - 50}{10} = 1.60$

7.15 $d = \frac{M - \mu}{s} = \frac{93 - 100}{30} = -0.23$

7.16 $r^2 = \frac{t^2}{t^2 + df} \times 100 = \frac{2.37^2}{2.37^2 + 17} \times 100 = 25\%$

7.17 $r^2 = (.32)^2 = .1024 = 10\%$; therefore, 10% of the variability in the outcome variable, Y, is accounted for by the explanatory variable, X.

Practice Problems 7.4

7.18 $95\%\ \text{CI}\mu_{\text{Diff}} = (M - \mu) \pm (t_{cv} \times s_M)$

$= (70 - 60) \pm (2.086 \times 4.36)$

$= 10.0000 \pm 9.0950$

$= [0.91, 19.10]$

7.19 $95\%\ CI\mu_{Diff} = (M - \mu) \pm (t_{cv} \times s_M)$
$= (55 - 50) \pm (2.052 \times 1.89)$
$= 5.0000 \pm 3.8783$
$= [1.12, 8.88]$

7.20 The sample does not seem representative of the population in terms of academic performance. The sample's mean GPA, 3.16, is statistically significantly higher than the population mean of 3.02 [$t(80) = 3.50$, $p < .05$]. The students in the sample are doing better in school than the general population. Their feelings about academic policies may also be different than those of the general student population.

Chapter 8

Practice Problems 8.1

8.01 a paired-samples *t* test (each subject was measured before and after the program)

8.02 an independent-samples *t* test (the subjects from the east and west coast are different subjects)

8.03 an independent-samples *t* test (the restaurants with and without tablecloths are different restaurants)

Practice Problems 8.2

8.04 H_0: $\mu_{Veterans} \leq \mu_{Nonveterans}$
H_1: $\mu_{Veterans} > \mu_{Nonveterans}$

8.05 $N = n_1 + n_2 = 12 + 16 = 28$

$$df = N - 2 = 26$$

$$s_{Pooled} = \frac{s_1^2(n_1 - 1) + s_2^2(n_2 - 1)}{df}$$

$$= \frac{4^2(12 - 1) + 3^2(16 - 1)}{26}$$

$$= 11.96$$

$$s_{M1-M2} = \sqrt{s^2_{Pooled}\left[\frac{N}{n_1 \times n_2}\right]}$$

$$= \sqrt{11.96\left[\frac{28}{12 \times 16}\right]}$$

$$= 1.32$$

8.06 $t = \frac{M_1 - M_2}{s_{M_D}} = \frac{99 - 86}{8.64}$

$$= 1.50$$

Practice Problems 8.3

8.07 (a) As $10.77 \geq 1.997$, reject H_0.
(b) The difference is statistically significant.
(c) The no sunscreen (control group) mean is higher than the sunscreen (experimental group) mean.
(d) $t(65) = 10.77$, $p < .05$
(e) $d = \frac{M_1 - M_2}{\sqrt{s^2_{Pooled}}} = \frac{17 - 10}{\sqrt{7.11}}$

$$= 2.63$$

$$r^2 = \frac{t^2}{t^2 + df} \times 100$$

$$= \frac{10.77^2}{10.77^2 + 65} \times 100 = 63.9\%$$

(f) $95\%CI\mu_{Diff} = (M_1 - M_2) \pm (t_{cv} \times s_{M_1-M_2})$
$= (10.00 - 17.00) \pm (1.997 \times 0.65)$
$= [-8.30, -5.70]$

8.08 This study compared the resting heart rate of people who do their own chores ($M = 72.00$) to those who pay others to do their chores ($M = 76.00$). The results were not statistically significant [$t(33) = 0.79$, $p > .05$], meaning that there wasn't enough evidence to conclude a difference in the population means existed. There appears to be no exercise benefit to doing one's own chores. Future research should measure whether people also exercise and the body mass of the individuals involved to see if these two groups differ on these potential confounding variables.

Chapter 9

Practice Problems 9.1

9.01 If the second score is subtracted from the first score: −3.00; 24.00; 3.00; 1.00; −6.00; −11.00; 2.00; −14.00. If the first score is subtracted from the second score: 3.00; −24.00; −3.00; −1.00; 6.00, 11.00; −2.00; 14.00.

9.02 $s_{M_D} = \frac{s_D}{\sqrt{N}} = \frac{8.43}{\sqrt{64}} = 1.05$

9.03 $t = \frac{M_1 - M_2}{s_{M_D}} = \frac{19.98 - 18.65}{2.45}$

$$= 0.54$$

Practice Problems 9.2

9.04 (a) $t(45) = 3.67$, $p < .05$. (b) Sample 2 has a mean that is significantly higher than the mean of Sample 1.

9.05 (a) $t(19) = 2.01$, $p > .05$. (b) There's not enough evidence to suggest that a difference exists between the two population means.

9.06 $95\%CI\mu_{Diff} = (M_1 - M_2) \pm (t_{cv} \times s_{M_D})$
$= (55 - 48) \pm (2.023 \times 2.86) = [1.21, 12.79]$

9.07 Data were analyzed from a study in which a tennis instructor compared having students practice by hitting against a wall vs. practicing by playing against another player. Students were matched in terms of ability and were assigned to use one of the methods of practice for two weeks before they were tested in terms of the percentage of time they could hit targets on the court. The students who hit against the wall hit the targets a significantly higher percentage of the time than the students who played against other players (57% vs. 48%; $t(11) = 5.20$, $p < .05$.). Hitting against the wall seems to have a moderate to large effect in improving the skills of beginning players. In the larger population from which this sample came, it probably leads to a 5 to 13 percentage points improvement in beginning players placing the ball accurately on the court. To obtain a better estimate of the size of the effect, it is recommended that this study be replicated with a larger sample size.

Chapter 10

Practice Problems 10.1

10.1 Individual differences

10.2 Individual differences and treatment effect

10.3 Between-group variability divided by within-group variability

10.4 When the results of the ANOVA are statistically significant (also OK, when the F ratio is statistically significant)

Practice Problems 10.2

10.5 Comparing the means of two independent samples: independent-samples t test because each subject is assigned to only one exercise program and there are only two groups.

10.6 Comparing the means of three independent samples: between-subjects, one-way ANOVA because each subject is assigned to open only one type of cereal box and there are more than two groups.

10.7 If $F \geq 2.866$, reject H_0; if $F < 2.866$, fail to reject H_0.

10.8 $df_{Between} = k - 1 = 4 - 1 = 3$
$df_{Within} = N - k = 32 - 4 = 28$
$df_{Total} = N - 1 = 32 - 1 = 31$

10.9

	Group 1		Group 2		Group 3			
	X	X²	X	X²	X	X²		
	16	256	12	144	13	169		
	17	289	14	196	15	225	Grand	
	20	400			18	324	X	X²
Σ	53	945	26	340	46	718	125	2,003
N	3		2		3		8	

10.10 $$SS_{Total} = \Sigma X^2 - \frac{(\Sigma X)^2}{N}$$

$$= 5435 - \frac{251^2}{12} = 184.91$$

$$SS_{Between} = \Sigma\left(\frac{(\Sigma X_{Group})^2}{n_{Group}}\right) - \frac{(\Sigma X)^2}{N}$$

$$= \left(\frac{66^2}{4} + \frac{85^2}{4} + \frac{100^2}{4}\right) - \frac{251^2}{12} = 145.17$$

$$SS_{Within} = \Sigma\left(\Sigma X^2_{Group} - \frac{(\Sigma X_{Group})^2}{n_{Group}}\right)$$

$$= \left(\left(1100 - \frac{66^2}{4}\right) + \left(1821 - \frac{85^2}{4}\right) - \left(2514 - \frac{100^2}{4}\right)\right)$$

$$= 39.75$$

10.11

ANOVA Summary Table				
Source of Variability	Sum of Squares	Degrees of Freedom	Mean Square	F ratio
Between groups	716.00	3	238.67	12.56
Within groups	228.00	12	19.00	
Total	944.00	15		

Practice Problems 10.3

10.12 $F(4, 30) = 7.37, p < .05$

10.13 (a) $r^2 = \frac{SS_{\text{Between}}}{SS_{\text{Total}}} \times 100 = \frac{1{,}827.50}{4631.50} \times 100$

$= 39\%$

(b) The effect is a large one.

10.14 $q = 3.85, n = 10$

$$HSD = q\sqrt{\frac{MS_{\text{Within}}}{n}} = 3.85\sqrt{\frac{77.89}{10}}$$

$= 10.75$

(a) $M_1 - M_2 = 29.00 - 18.00 = 11.00$. $11.00 > 10.75$, so the difference between these two sample means is a statistically significant one. The mean of population 1 is probably higher than the mean of population 2.

(b) $M_1 - M_3 = 29.00 - 22.00 = 7.00$. $7.00 \leq 10.75$, so the difference between these two sample means is not statistically significant. There is not enough evidence to conclude that the two population means differ.

10.15 First, determine if the result is statistically significant. F_{cv} with 2 and 24 degrees of freedom is 3.403. As 5.25 (F) $\geq$ 3.403, the results are statistically significant and at least one mean differs from at least one other. r^2 has already been calculated as 30.43%, a large effect, and the *HSD* value is 2.35. Because the *F* was statistically significant, it is OK to complete post-hoc tests.

- The difference between the mean raise for employees of managers who use rewards vs. those who use punishments is 3.00 percentage points. This is greater than 2.35 percentage points, so the difference is statistically significant. One can conclude that using rewards leads to statistically significantly better performance than using punishments, as judged by the size of the raise.
- The differences between (a) rewards vs. a mixture of rewards and punishments, 1.00 percentage point, and (b) punishments vs. a mixture of rewards and punishments, 2.00 percentage points, are not statistically significant. There is not enough evidence to conclude that (a) rewards alone work better than rewards and punishment, or (b) that rewards and punishments work better than punishments alone.

Here is an interpretation:

An industrial/organizational psychologist compared management styles based on using rewards, using punishments, or using a combination of rewards and punishments on employee performance. Employee performance was measured by the size of raise the employee received. The effect of management style was statistically significant, $F(2, 24) = 5.25, p < .05$, and was a strong one. Type of management does have an impact on employee mean performance.

The effect is due to the difference in raises between employees of managers who use rewards ($M = 8\%$ raise) and those who use punishments ($M = 5\%$ raise). Rewarding good performance leads to statistically better performance than does punishing poor performance. This study offered insufficient evidence to show that using a mixture of rewards and punishments ($M = 7\%$ raise) led to worse performance than rewards alone or better performance than punishment alone.

This study took place at only one company, so the results should not be generalized beyond this one site. Future research should extend this study to other companies. If managers want to improve the performance of employees at this company, there is evidence that using only rewards is a better strategy than using only punishments.

Chapter 11

Practice Problems 11.1

11.01 Repeated-measures ANOVA is used to compare the means of two or more dependent samples.

11.02 A one-way, repeated-measures ANOVA divides variability into two parts: (1) that due to the independent variable, the treatment effect, and (2) that due to individual differences.

11.03 When comparing the means of two dependent samples, a paired *t* test should be used. This test allows you to account for the relationship between the two dependent measurements. (A repeated-measures, one-way ANOVA could be used in this case, but it isn't necessary with only two dependent samples.)

Practice Problems 11.2

11.04 (a) The means of three, matched (dependent) samples are being compared: one-way, repeated-measures ANOVA.
(b) Means of two independent samples are being compared: independent-samples *t* test.

11.05 (a) Random samples; (b) independence of observations; (c) normality

11.06 H_0: All population means are equal
H_1: At least one population mean is different from at least one other.

11.07 (a) Numerator degrees of freedom is
$df_{\text{Treatment}} = k - 1 = 4 - 1 = 3.$
(b) Denominator degrees of freedom is
$df_{\text{Residual}} = (n-1)(k-1) = (12-1)(4-1) = 33.$

11.08 If $F \geq 3.072$, reject H_0.
If $F < 3.072$, fail to reject H_0.

11.09 $SS_{\text{Total}} = 749.40$

$$MS_{\text{Treatment}} = \frac{SS_{\text{Treatment}}}{df_{\text{Treatment}}} = \frac{216.00}{3} = 72.00$$

$$MS_{\text{Residual}} = \frac{SS_{\text{Residual}}}{df_{\text{Residual}}} = \frac{410.40}{57} = 7.20$$

$$F = \frac{MS_{\text{Treatment}}}{MS_{\text{Residual}}} = \frac{72.00}{7.20} = 10.00$$

Source of Variability	Sum of Squares	Degrees of Freedom	Mean Square	F ratio
Subjects	123.00	19		
Treatment	216.00	3	72.00	10.00
Residual	410.40	57	7.20	
Total	749.40	79		

Practice Problems 11.3

11.10 (a) $F(3, 36) = 4.36, p < .05$
(b) The results are statistically significant as $F_{cv} = 2.866$.

11.11 (a) $$\eta^2 = \frac{SS_{\text{Treatment}}}{SS_{\text{Total}}} \times 100 = \frac{35.76}{124.64} \times 100$$

$$= 29\%$$

(b) Large effect

11.12 (a) $k = 5$ and $df_{\text{Residual}} = (n-1)(k-1) = (16-1)(5-1) = 60.$
$q = 3.98$

(b)

$$HSD = q\sqrt{\frac{MS_{\text{Residual}}}{n}} = 3.98\sqrt{\frac{12.98}{16}} = 3.58$$

11.13 A software designer examined how comfort with technology varied across generations by timing how long it took, when installing new software, to click on the "Agree to Terms of Installation" button for grandparents, parents, and teenagers in the same family. The teenagers agreed in a mean of 1.50 seconds, the parents took 3 times as long ($M = 4.50$ seconds), and the grandparents ($M = 20.00$ seconds) took more than 4 times as long as the parents. This was a very strong and statistically significant effect for age/generation [$F(2, 18) = 219.08, p < .05$], with the difference between each generation being a statistically significant one. The most likely explanation for the results is that younger generations have more experience with computers and give less thought to following—and agreeing to—computer directions. It is also possible that as people age their reaction time slows, their eyesight becomes poorer, and their movements slow and that these changes could account for the effect observed in this study. Future research should measure the physical abilities of the participants.

Chapter 12

Practice Problems 12.1

12.01 (a) 2 × 2 × 2 (Gender by handedness by razor type)
(b) 2 × 2 (Gender by type of career)

12.02 (a) Row mean for Group 1 =
$$\frac{12.00 + 18.00}{2} = 15.00$$
Row mean for Group 2 = $\frac{3.00 + 9.00}{2} = 6.00$
Column mean for Condition 1 =
$$\frac{12.00 + 3.00}{2} = 7.50$$

Column mean for Condition 2 = $\frac{18.00 + 9.00}{2} = 13.50$

(b) There is a main effect for group, with Group 1 ($M = 15.00$) having a higher mean than Group 2 ($M = 6.00$). There is also a main effect for condition, with Condition 1 ($M = 7.50$) having a lower mean than Condition 2 ($M = 13.50$).

12.03 (a)

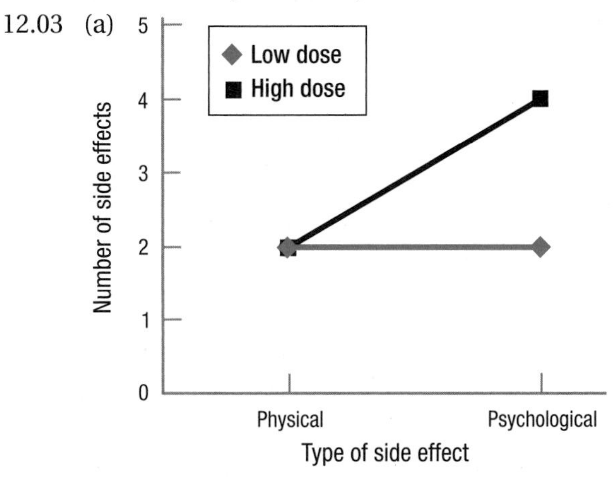

(b) Yes, there is an interaction.

(c) How many side effects one has depends on the dose taken and the type of side effects one is talking about. With a low dose, few physical or psychological side effects occur, while with a high dose, there are just as few physical side effects but more psychological side effects.

Practice Problems 12.2

12.04 (a) Between-subjects, two-way ANOVA

(b) Between-subjects, one-way ANOVA

12.05 $H_{0\text{ Rows}}$: All row population means are the same (or $\mu_{\text{Row 1}} = \mu_{\text{Row 2}} = \mu_{\text{Row 3}} = \mu_{\text{Row 4}}$).

$H_{1\text{ Rows}}$: At least one row population mean is different from at least one other row population mean.

$H_{0\text{ Columns}}$: All column population means are the same (or $\mu_{\text{Column 1}} = \mu_{\text{Column 2}} = \mu_{\text{Column 3}}$).

$H_{1\text{ Columns}}$: At least column population mean is different from at least one other column population mean.

$H_{0\text{ Interaction}}$: There is no interactive effect of the two explanatory variables on the dependent variable in the population.

$H_{1\text{ Interaction}}$: The two explanatory variables, in the population, interact to affect the dependent variable in at least one cell.

12.06 $F_{cv\text{ Rows}} = 3.053$; $F_{cv\text{ Columns}} = 2.428$; $F_{cv\text{ Interaction}} = 1.997$

12.07 $df_{\text{Rows}} = R - 1 = 2 - 1 = 1$

$df_{\text{Columns}} = C - 1 = 2 - 1 = 1$

$df_{\text{Interaction}} = df_{\text{Rows}} \times df_{\text{Columns}} = 1 \times 1 = 1$

$N = n \times R \times C = 7 \times 2 \times 2 = 28$

$df_{\text{Within}} = N - (R \times C) = 28 - (2 \times 2) = 24$

$df_{\text{Between}} = df_{\text{Rows}} + df_{\text{Columns}} + df_{\text{Interaction}} = 1 + 1 + 1 = 3$

$df_{\text{Total}} = N - 1 = 28 - 1 = 27$

$SS_{\text{Between}} = SS_{\text{Rows}} + SS_{\text{Columns}} + SS_{\text{Interaction}} = 250.00 + 300.00 + 100.00 = 650.00$

$SS_{\text{Total}} = SS_{\text{Between}} + SS_{\text{Within}} = 650.00 + 800.00 = 1{,}450.00$

$$MS_{\text{Rows}} = \frac{SS_{\text{Rows}}}{df_{\text{Rows}}} = \frac{250.00}{1} = 250.00$$

$$MS_{\text{Cols}} = \frac{SS_{\text{Cols}}}{df_{\text{Cols}}} = \frac{300.00}{1} = 300.00$$

$$MS_{\text{Interaction}} = \frac{SS_{\text{Interaction}}}{df_{\text{Interaction}}} = \frac{100.00}{1} = 100.00$$

$$MS_{\text{Within}} = \frac{SS_{\text{Within}}}{df_{\text{Within}}} = \frac{800.00}{24} = 33.33$$

$$F_{\text{Rows}} = \frac{MS_{\text{Rows}}}{MS_{\text{Within}}} = \frac{250.00}{33.33} = 7.50$$

$$F_{\text{Cols}} = \frac{MS_{\text{Cols}}}{MS_{\text{Within}}} = \frac{300.00}{33.33} = 9.00$$

$$F_{\text{Interaction}} = \frac{MS_{\text{Interaction}}}{MS_{\text{Within}}} = \frac{100.00}{33.33} = 3.00$$

Source of Variability	Sum of Squares	Degrees of Freedom	Mean Square	F ratio
Between groups	650.00	3		
Rows	250.00	1	250.00	7.50
Columns	300.00	1	300.00	9.00
Interaction	100.00	1	100.00	3.00
Within groups	800.00	24	33.33	
Total	1,450.00	27		

Practice Problems 12.3

12.08 (a) Rows: $F(3, 60) = 3.25$, $p < .05$ (using $F_{cv} = 2.758$)

Columns: $F(2, 60) = 1.22$, $p > .05$ (using $F_{cv} = 3.150$)
Interaction: $F(6, 60) = 0.83$, $p > .05$ (using $F_{cv} = 2.254$)

(b) Rows effect is statistically significant. Columns effect is not statistically significant. Interaction effect is statistically significant.

12.09 (a)

$$\eta^2_{\text{Rows}} = \frac{SS_{\text{Rows}}}{SS_{\text{Total}}} \times 100 = \frac{3{,}725.00}{12{,}749.00} \times 100 = 29.22\%$$

$$\eta^2_{\text{Columns}} = \frac{SS_{\text{Columns}}}{SS_{\text{Total}}} \times 100 = \frac{1{,}312.00}{12{,}749.00} \times 100 = 10.29\%$$

$$\eta^2_{\text{Interaction}} = \frac{SS_{\text{Interaction}}}{SS_{\text{Total}}} \times 100 = \frac{320.00}{12{,}749.00} \times 100 = 2.51\%$$

(b) η^2_{Rows} is a large effect; η^2_{Columns} is a medium effect; $\eta^2_{\text{Interaction}}$ is a small effect.

12.10 The interaction effect is not statistically significant, so there is no need to do post-hoc tests on that. Both the row effect and column effect are statistically significant. But, there are only two rows, so it is obvious which row mean (Row 1 mean = 30.00) is statistically different than (and higher than) the other row (Row 2 mean = 22.00.)
There are three columns, so it's necessary to calculate a *HSD* value to compare column means:

$$q_{\text{Columns}} = 3.49$$

$$HSD_{\text{Columns}} = q\sqrt{\frac{MS_{\text{Within}}}{n_{\text{Columns}}}} = 3.49\sqrt{\frac{18.67}{14}} = 4.03$$

- Column 2 – Column 1 = 26.00 – 22.00 = 4.00. 4.00 < 4.03, so the difference is not statistically significant. One can't conclude that the population from which Column 2 is drawn has a mean greater than the Column 1 population.
- Column 3 – Column 1 = 30.00 – 22.00 = 8.00. 8.00 ≥ 4.00, so the difference is statistically significant. The population from which Column 3 is drawn probably has a mean greater than the Column 1 population.
- Column 3 – Column 2 = 30.00 – 26.00 = 4.00. 4.00 < 4.03, so the difference is statistically significant. The population from which Column 3 is drawn probably has a mean greater than the mean for the Column 2 population.

12.11 "In this study, the effects singly and together of increased temperature and increased humidity on human performance were investigated. The outcome variable was how far participants walked on a treadmill, at their normal pace, in 60 minutes. There was a statistically significant interaction effect of temperature and humidity on distance walked. The effect was moderately strong $F(1, 24) = 9.00$, $p < .05$. It took *both* a change in humidity from normal to high *and* a change in temperature from normal to high to lead to a statistically significant decline in performance. When temperature and humidity were normal, the average distance walked in 60 minutes was 3.00 miles; it dropped only to 2.80 miles when either temperature or humidity became higher. But, when both temperature and humidity were high, the distance walked fell to 2.00 miles. This is a large and meaningful decrease. Temperature and humidity interact to affect physical performance. Future research should vary the changes in temperature and humidity in order to determine more exactly how these two variables affect physical performance. Studies looking at their impact on cognitive performance should also be undertaken."

Chapter 13

Practice Problems 13.1

13.01

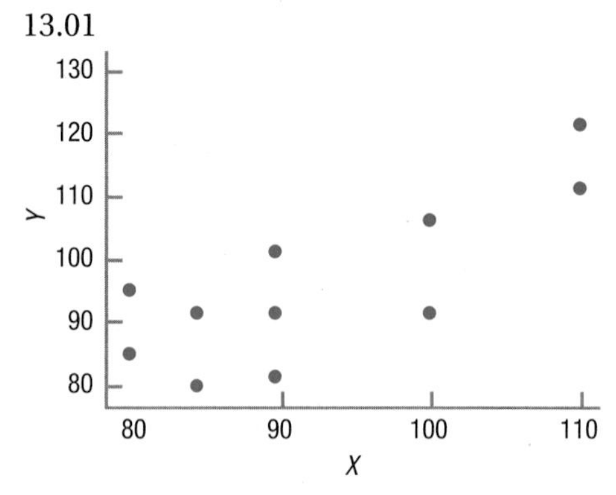

13.02 b, c, d, g, h, i
13.03 a, f
13.04 d, i
13.05 c, g, h
13.06 b
13.07 b, g, h

Practice Problems 13.2

13.08 (a) Yes (relationship between two interval/ratio level variables).
(b) No (one variable is ordinal).
(c) No (one variable is nominal).

13.09 (a) Random samples is violated; independence of observations is not violated; willing to assume both variables are normally distributed; unable to check linearity.
(b) No nonrobust assumptions violated, so it is OK to proceed.

13.10 (a) $H_0: \rho \leq 0$; $H_1: \rho > 0$
(b) One-tailed test, $\alpha = .05$, df=35: $r_{cv} = .275$

- If $r \geq .275$, reject H_0.
- If $r < .275$, fail to reject H_0.

13.11

	Col. 1	Col. 2	Col. 3	Col. 4	Col. 5	Col. 6	Col. 7
	X	Y	$X - M_X$	$Y - M_Y$	$(X - M_X)(Y - M_Y)$	$(X - M_X)^2$	$(Y - M_Y)^2$
	10	20	−1.00	−7.00	7.00	1.00	49.00
	14	24	3.00	−3.00	−9.00	9.00	9.00
	9	37	−2.00	10.00	−20.00	4.00	100.00
$M=$	11.00	27.00			$\Sigma = -22.00$	$SS_X = 14.00$	$SS_Y = 158.00$

13.12

$$r = \frac{\Sigma[(X - M_X)(Y - M_Y)]}{\sqrt{SS_X SS_Y}} = \frac{80.00}{\sqrt{200.00 \times 90.00}} = .60$$

Practice Problems 13.3

13.13 (a) Null hypothesis was rejected; (b) results are statistically significant; (c) $r(20) = .63$, $p < .05$.

13.14 (a) $r^2 = r^2 \times 100 = (-.34)^2 \times 100 = 11.56\%$
(b) Medium effect.

13.15 Convert r to z_r: If $r = .40$, $z_r = 0.42$.
Calculate the confidence interval in z_r units:

$$s_r = \frac{1}{\sqrt{N - 3}} = \frac{1}{\sqrt{12 - 3}} = .33$$

$95\%\text{CI}_{zr} = z_r \pm 1.96\, s_r = 0.42 \pm 1.96\,(.33)$
$= [-0.23, 1.07]$
Convert confidence interval back into r units: from −.23 to .79.

13.16 $\beta = .28$

13.17 power = .50

13.18 power = .34, so there is a 66% chance of making a Type II error

13.19 85 cases; 140 cases

13.20 "In this study, the relationship between prenatal parental marital harmony and children's mental health was studied. The level of marital harmony was measured in a large, random sample of married couples who were about to have a child. Eighteen years later, the mental health of all the children was assessed. There was a medium-strength, statistically significant relationship found between the two variables [$r(400) = .38$, $p < .05$]. Parents whose marriages were more harmonious tended to have children who, as young adults, were mentally healthier.

It is tempting, as marital harmony was measured 18 years before mental health was measured, to conclude that more harmonious marriages result in mentally healthier children. But, this is a correlational study, so cause and effect cannot be determined. Perhaps, children who are mentally healthier are the result of easier pregnancies and easier pregnancies promote marital harmony. Perhaps living in nicer climates leads both to marital harmony and well-adjusted teens. Future research should try to address some of these other explanations."

Practice Problems 13.4

13.21
$$r_{XY-Z} = \frac{r_{XY} - (r_{XZ} \times r_{YZ})}{\sqrt{(1 - r_{XZ}^2)(1 - r_{YZ}^2)}} = \frac{.53 - (.46 \times .25)}{\sqrt{(1 - .46^2)(1 - .25^2)}} = .48$$

13.22
$$r_{XY-Z} = \frac{r_{XY} - (r_{XZ} \times r_{YZ})}{\sqrt{(1 - r_{XZ}^2)(1 - r_{YZ}^2)}} = \frac{.35 - (.18 \times .17)}{\sqrt{(1 - .18^2)(1 - .17^2)}} = .33$$

Chapter 14

Practice Problems 14.1

14.01 $b = r\left(\frac{s_Y}{s_X}\right) = -.37\left(\frac{9.33}{12.88}\right) = -0.27$

14.02 $a = M_Y - bM_X$
$= 45.64 - (4.54 \times 10.65)$
$= -2.71$

14.03 $Y' = bX + a$
$= 1.80X + (-12.42)$
$= 1.80X - 12.42$

14.04 (a) $Y' = 1.50X + 4.50$
$= (1.50 \times 2) + 4.50$
$= 7.50$

$Y' = 1.50X + 4.50$
$= (1.50 \times 12) + 4.50$
$= 22.50$

(b)

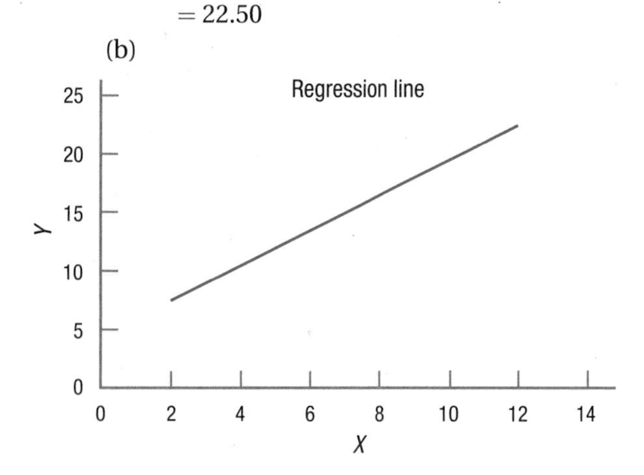

14.05 ≈35

Practice Problems 14.2

14.06 $s_{Y-Y'} = s_Y\sqrt{1-r^2} = 5.64\sqrt{1-.42^2} = 5.12$

14.07 There are two ways to think of this. By both criteria, this is a large standard error of the estimate. (1) The range of IQ scores, from 55 to 145, is 90 points wide. Being off, on average, by 14 points is being off by almost 16% of the possible range. That seems like a lot. (2) If someone's IQ is predicted, it is off, on average, by 14 points. Misjudging IQ by 14 points seems like a lot.

Practice Problems 14.3

14.08 Because multiple regression combines the predictive power of multiple variables, each of which probably explains some unique bit of the predicted variable, it will almost certainly explain more variability in the predicted variable than a sole variable.

14.09 $Y' = (13.17 \times \text{variable 1}) + (4.55 \times \text{variable 2}) + 55.12 = (13.17 \times 12) + (4.55 \times 33) + 55.12 = 363.31$

Chapter 15

Practice Problems 15.1

15.01

Black	Blonde	Brown	Red	Auburn	Chestnut	Grey	White

$df = k - 1 = 8 - 1 = 7$

15.02 $\chi^2_{cv} = 9.210$

15.03 $f_{\text{ExpectedBoy}} = \frac{\%_{\text{ExpectedBoy}} \times N}{100} = \frac{51.69 \times 5{,}873}{100} = 3{,}035.75$

$f_{\text{ExpectedGirl}} = \frac{\%_{\text{ExpectedGirl}} \times N}{100} = \frac{48.31 \times 5{,}873}{100} = 2{,}837.25$

15.04 $\chi^2 = \Sigma\frac{(f_{\text{Observed}} - f_{\text{Expected}})^2}{f_{\text{Expected}}} = \frac{(75-62.80)^2}{62.80} + \frac{(82-94.20)^2}{94.20} = 3.95$

15.05 $\chi^2(1, N = 73) = 4.72, p < .05$

Practice Problems 15.2

15.06 (a) $df = (R-1) \times (C-1) = (4-1) \times (3-1) = 3 \times 2 = 6$

(b) $\chi^2_{cv} = 12.592$

15.07
$N_{\text{Row 1}} = 5 + 10 + 12 = 27;\ N_{\text{Row 2}} = 5 + 8 + 10 = 23;$
$N_{\text{Col 1}} = 5 + 5 = 10;\ N_{\text{Col 2}} = 10 + 8 = 18;$
$N_{\text{Col 3}} = 12 + 10 = 22$

15.08 $N = 39 + 49 = 88$

$f_{\text{ExpectedA}} = \frac{N_{\text{Row1}} \times N_{\text{Col1}}}{N} = \frac{39 \times 28}{88} = 12.41$

$f_{\text{ExpectedB}} = \frac{N_{\text{Row1}} \times N_{\text{Col2}}}{N} = \frac{39 \times 60}{88} = 26.59$

$f_{\text{ExpectedC}} = \frac{N_{\text{Row2}} \times N_{\text{Col1}}}{N} = \frac{49 \times 28}{88} = 15.59$

$f_{\text{ExpectedD}} = \frac{N_{\text{Row2}} \times N_{\text{Col2}}}{N} = \frac{49 \times 60}{88} = 33.41$

15.09
$$\chi^2 = \sum \frac{\left(f_{\text{Observed}} - f_{\text{Expected}}\right)^2}{f_{\text{Expected}}}$$

$$= \frac{(50 - 47.22)^2}{47.22} + \frac{(36 - 38.78)^2}{38.78}$$

$$+ \frac{(45 - 47.77)^2}{47.77} + \frac{(42 - 39.23)^2}{39.23}$$

$$= 0.72$$

Practice Problems 15.3

15.10 $\chi^2(1, N = 36) = 4.76, p < .05$

15.11 $V = \sqrt{\frac{\chi^2}{N \times df_{\text{RC}}}} = \sqrt{\frac{6.78}{45 \times 2}} = .27$

15.12 "An experiment was conducted in a driving simulator to see whether a heads-up display had any effect on the proportion of drivers who crashed while driving on a twisty mountain road. Using a heads-up display, compared to driving with a normal dashboard display, led to a statistically and practically significant reduction in the proportion of crashes [$\chi^2(1, N = 180) = 5.27, p < .05$]. Drivers assigned to the dashboard display condition crashed about half the time, while drivers in the heads-up display condition only crashed one third of the time. These results certainly suggest that heads-up displays are safer to use. But, using a heads-up display is novel, so drivers in that condition may have given more attention to what they were doing. Future research should make sure that the novelty effect has worn off before comparing heads-up displays to normal driving conditions."

Practice Problems 15.4

15.13 Examining the relationship between two interval and/or ratio variables calls for a Pearson *r*. But, the normality assumption is badly violated, so fall back to the Spearman *r*.

15.14 Comparing two independent samples on an interval-level dependent variable calls for an independent-samples *t* test.

15.15 Comparing two independent samples on an ordinal-level dependent variable calls for a Mann–Whitney *U* test.

Chapter 16

Practice Problems 16.1

16.01 (a) quasi-experimental; (b) descriptive; (c) correlational

16.02 (a) ordinal; (b) nominal; (c) ordinal

16.03 (a) *z* or PR; (b) frequency distribution, histogram, or frequency polygon; (c) mode; (d) 95% confidence interval for the population mean

Practice Problems 16.2

16.04 independent-samples *t* test

16.05 between-subjects, one-way ANOVA

16.06 dependent-samples *t* test

Practice Problems 16.3

16.07 chi-square test of independence

16.08 Pearson *r*

16.09 Spearman *r*

APPENDIX D

Solutions to Part Tests

Part I: Chapters 1–5

1. a. Use Column B of Appendix A, Table 1: 49.01%.

b. Use Column C: 17.62%.

c. Use Column A: 91.47%.

d. Use Column B for −0.23 and 1.84: 9.10% + 46.71% = 55.81%.

e. $z = \frac{150 - 100}{15} = 3.33$; use Column C: 0.043%.

f. $z = -0.44$; $x = M + (s \times z) = 100 + (15 \times -0.44) = 93.40$

g. SAT: $z = \frac{X - M}{s} = \frac{670 - 500}{100} = 1.70$;
ACT: $X = M + (s \times z) = 21.50 + (5 \times 1.70) = 30.00$

h. Use Column B for −2.23 and −1.23: 48.71% − 39.07% = 9.64%.

2. Answer each section.

a.

6	12
5	2458
4	44578
3	1233366
2	2468
1	19

b. $i = 4$

c. 25 – 28

d. 12.50

e. 12.50

f. 22.50

g. $N = 150$

h. 1.00

3. a. $\Sigma X = 7 + 9 + 4 + 11 + 14 = 45.0000 = 45.00$

b. $M = \frac{\Sigma X}{N} = \frac{7 + 9 + 4 + 11 + 14}{5} = \frac{45.00}{5} = 9.00$

c. $\Sigma(X - M) = (7 - 9.00) + (9 - 9.00) + (4 - 9.00) + (11 - 9.00) + (14 - 9.00) = 0.00$

d. $\Sigma(X - M)^2 = (7 - 9.00)^2 + (9 - 9.00)^2 + (4 - 9.00)^2 + (11 - 9.00)^2 + (14 - 9.00)^2 = 58.00$

e. $s = \frac{\Sigma(X - M)^2}{N - 1} = \frac{58.00}{4} = 14.50$

f. Median = score number $\frac{N + 1}{2} = \frac{6}{2} = 3$; scores in order: 4 7 9 11 14; Mdn = 9.00

4. a. s, σ, s^2, σ^2

b. $s = 5.00$

5. a. $s_M = \frac{s}{\sqrt{N}} = \frac{32}{\sqrt{75}} = 3.6950$

95%CI = $M \pm (1.96 \times s_M) = 157 \pm (1.96 \times 3.6950) = 157 \pm 7.2423 = [149.76, 164.24]$

b. We are 95% confident that the mean number of times teenagers in the United States check their phones for texts daily is between 149.76 and 164.24.

6. a. QE; GV

b. E; IV

c. QE; GV

d. E; DV

e. QE; DV

7.

Grouped Frequency for an Interval Variable

Interval	Midpoint	Frequency	Cumulative Frequency	Percentage	Cumulative Percentage
70–79	74.5	3	60	5.00%	100.00%
60–69	64.5	8	57	13.33%	95.00%
50–59	54.5	22	49	36.67%	81.67%
40–49	44.5	18	27	30.00%	45.00%
30–39	34.5	8	9	13.33%	15.00%
20–29	24.5	1	1	1.67%	1.67%

8. a. Nominal

b. Ratio

c. Ratio (could argue for Interval)

d. Interval

e. Ordinal

9. a. 12.4550 = 12.46

b. 13.4105 = 13.41

c. 14.7812345 = 14.78

d. 15.9937 = 15.99

e. 16.99 = 16.99

f. $\frac{2}{3} = 0.\overline{6666} = 0.67$

g. $\frac{3}{2} = 1.50$

h. $\frac{10}{2} = 5.00$

i. $\frac{4.00}{2.00} = 2.0000 = 2.00$

10.

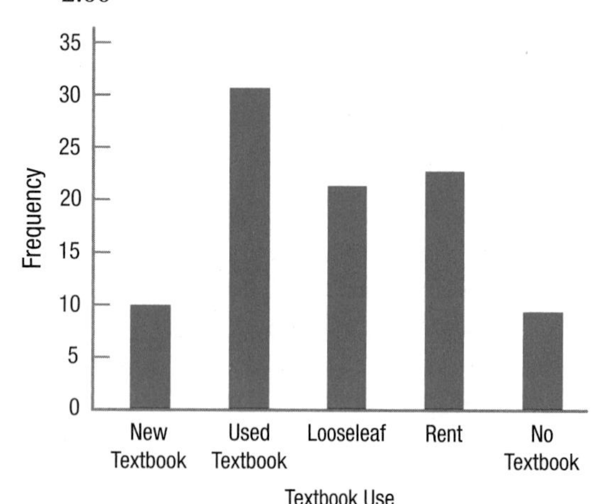

11. a. The distribution is unimodal, doesn't seem too peaked or too flat, and is not skewed.
b. The distribution is unimodal, doesn't seem too peaked or too flat, and is positively skewed.
c. The distribution is unimodal, is quite peaked, and is not skewed.
d. The distribution is bimodal, doesn't seem too peaked or too flat, and is not skewed.
e. The distribution is unimodal, is quite flat, and is not skewed.
f. The distribution is unimodal, doesn't seem too peaked or too flat, and is negatively skewed.

Part II: Chapters 6–9

1. $$s_{M_1-M_2} = \sqrt{s^2_{\text{Pooled}} \frac{N}{n_1 \times n_2}}$$

where $s^2_{\text{Pooled}} = \dfrac{s_1^2(n_1 - 1) + s_2^2(n_2 - 1)}{df}$.

We have $df = N - 2 = (34 + 40) - 2 = 72$:

$$s^2_{\text{Pooled}} = \frac{3.55^2(34 - 1) + 5.42^2(40 - 1)}{72} = 21.69$$

$$s_{M_1-M_2} = \sqrt{21.69 \frac{74}{34 \times 40}} = 1.09$$

$$s_{M_1-M_2} = 1.09$$

2. $$\begin{aligned} 95\%\ CI_{\mu\text{Diff}} &= (M_1 - M_2) \pm (t_{cv} \times s_{M_D}) \\ &= (52 - 48) \pm (1.966 \times 0.73) \\ &= [2.56, 5.44] \end{aligned}$$

(*Note:* The subtraction of means has been arranged to arrive at a positive number. A CI from –5.44 to –2.56 is also correct.)

3. First, calculate the standard error of the mean:

$$s_M = \frac{s}{\sqrt{N}} = \frac{5.82}{\sqrt{18}} = 1.37.$$

Then, calculate *t*:

$$t = \frac{M - \mu}{s_M} = \frac{23.42 - 25}{1.37} = -1.15$$

4. $$s^2_{\text{Pooled}} = \frac{s_1^2(n_1 - 1) + s_2^2(n_2 - 1)}{df}$$

where $df = N - 2 = (12 + 10) - 2 = 20$

$$= \frac{4.50^2(12 - 1) + 6.40^2(10 - 1)}{20} = 29.57$$

5. First, recognize that this is a dependent-samples *t* test, since the information states that we have *pairs* of cases:

$$N = 25,\ df = N - 1 = 24$$

$$t_{cv} = 2.064$$

$$\begin{aligned} 95\%\ CI_{\mu\text{Diff}} &= (M_1 - M_2) \pm (t_{cv} \times s_{M_D}) \\ &= (3.70 - 3.50) \pm (2.064 \times 0.05) \\ &= [0.10, 0.30] \end{aligned}$$

(*Note:* Subtraction of means has been arranged to arrive at a positive number. A CI from –0.30 to –0.10 is also correct.)

6. $$\sigma_M = \frac{\sigma}{\sqrt{N}} = \frac{12}{\sqrt{36}} = 2.00$$

$$z = \frac{M - \mu}{\sigma_M} = \frac{55 - 57}{2} = -1.00$$

7. $z(N = 55) = 1.70,\ p > .05$

8. Adele is using the wrong test. She should be conducting an independent-samples *t* test with $df = 25$.

9. The confidence interval is very wide, so the size of the effect is uncertain. Meghan should suggest that future research increase the sample size, *N*, and repeat the study.

10.

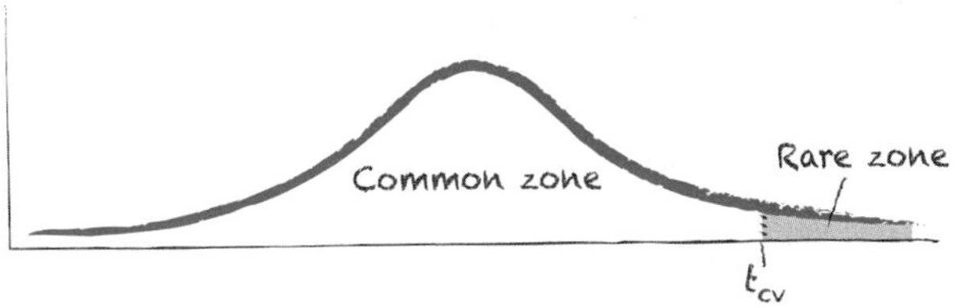

Common zone

Rare zone

t_{cv}

Note that the numerator is (men – women) and should be a positive number. If it is a high positive number, reject the null hypothesis.

11. Failed to reject the null hypothesis.
12. The researcher should use a paired-samples *t* test.
13. The researcher should use an independent-samples *t* test.
14. Realize that this is an independent-samples *t* test, then figure out *df*(75), find t_{cv}, calculate $t = 4/2.5 = 1.60$, and realize that we fail to reject. Report

$$t(75) = 1.60,\ p > 0.05$$

15. No. There is a homogeneity of variance problem: One *s* is more than twice the other. Also, a vast discrepancy exists between the two sample sizes. Homogeneity of variance is robust only if the value of the *n*'s are roughly equal.
16. No. We will need to know σ if we are going to perform a one-sample *z* test.
17. In the populations, Druids are probably more spiritual than Wiccans.
18. False. The value of the test statistic as compared to the critical value determines whether or not the null hypothesis should be rejected. However, if the null is not rejected, there is the possibility of a Type II error occurring. Power refers to the probability of rejecting the null hypothesis when the null hypothesis should be rejected, and we know $\beta + power = 1.00$. β, the probability of making a Type II error, is commonly set at 0.20, or 20%, meaning that we generally want power to be 80% or higher.
19. The researcher should use a paired-samples *t* test.
20. The researcher should use a single-sample *z* test.
21. The storeowner should use an independent-samples *t* test.
22. A study was conducted to see if altruism or self-interest motivates people more. Members of a high school band doing a fundraiser were randomly assigned to one of two groups: an altruism group that was told it was raising money to help others, and a self-interest group that was told it could keep 10% of the money raised. The self-interest group raised an average of $78.66 per member, while the altruism group raised, on average, almost $27 more per member, with a mean of $106.00. The difference in amount raised is statistically significant [$t(70) = 6.72$, $p < .05$]. The effect size is large and $r^2 = 40\%$, indicating that the type of motivation accounts for a large portion of the variability in how much money is raised. The size of the difference in the larger population probably ranges from a $19.22 to a $35.46 increase in fundraising with altruism as a motivation. This study suggests that altruism is a more powerful motivator than self-interest. Replication of the study with different age groups and different types of tasks is recommended to determine the limits of the effect.
23. A study was conducted to examine the impact of organic foods on intelligence. Using a matched-pairs design, 25 children who were fed organic food as babies were matched against 25 children who were not fed organic food as babies. Families were matched on SES and other potential confounding variables. In the organic food group, mean IQ in first grade was 109, whereas it was 102 in the nonorganic food group. The 7-point difference in means is statistically significant [$t(24) = 5.00, p < .05$]. The effect is meaningful; being fed organic food as babies seems to have a positive effect on intelligence, enough potentially to make a difference in the child's life, possibly allowing the child to do better in school, attend a better college, and so on. Since the study participants were matched for SES and other potentially confounding variables, the effect is more likely to be due to the organic food than to, for instance, SES, child rearing style, or parental intelligence. It is recommended that the study be replicated with a larger sample size to determine if the same effect is observed. Additionally, participants should be tested at a future point, such as after high school, to see if the effect is sustained.

Part III: Chapters 10–12

1.

Source of Variability	Sum of Squares	Degrees of Freedom	Mean Square	*F* Ratio
Between groups	728.00	2	364.00	68.29
Within groups	48.00	9	5.33	
Total	776.00	11		

2. Overall *F* is statistically significant, so it is appropriate to do a post-hoc test. To use the *HSD* formula, first find $q = 3.61$: $HSD = 3.61\sqrt{\frac{34.98}{6}} = 8.72$. Therefore, any pair of means that differ by ≥ 8.72 points statistically differ.

- Groups 1 and 2 differ by 3.57 points, not a statistically significant difference.
- Groups 1 and 3 differ by 13.63 points, a statistically significant difference.
- Groups 2 and 3 differ by 10.06 points, a statistically significant difference.

3. a.

	+ Expect	– Expect	
Env A	100	100	100.00
Env B	130	70	100.00
	115.00	85.00	

b.

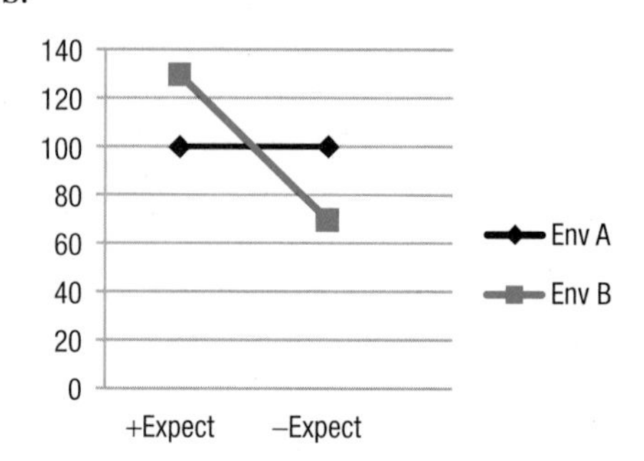

1. The effect of the environment on outcome depends on the type of expectations.
2. The effect of the expectations on outcome depends on the environment.

4. a. This is a two-way design with exercise preference and preference for fruit or chips being the "ways." Each way has two levels. Both ways have independent samples.

b. This is a one-way design with type of instruments being the "way." This way has four levels. The samples are independent.

c. This is a one-way design with the year of college being the "way." This way has four levels. The samples are dependent.

d. This is a three-way design with type of symptom, type of medication, and when measured being the "ways." The ways have two, two, and three levels, respectively. Two of the ways have independent samples; the third way, when outcome is measured, involves dependent samples.

e. This is a one-way design. Type of major has three levels. The samples are dependent because each student belongs to one and only one treatment and is measured only once.

5. Answers will vary, but should display the characteristics indicated below:

a. A between-subjects, one-way ANOVA:
- One grouping variable with at least two levels.
- Independent samples.
- Dependent variable should be either interval- or ratio-level.
- Example: Compare body mass for lacrosse players, soccer players, and cross country runners.

b. A one-way, repeated-measures ANOVA:
- One grouping variable with at least two levels.
- Dependent samples.
- Dependent variable should be either interval- or ratio-level.
- Example: Compare body mass at the beginning, middle, and end of the first semester. Each student is measured 3 times.

c. A between-subjects, two-way ANOVA:
   - Two crossed independent variables, each with at least two levels.
   - Independent samples.
   - Dependent variable should be either interval- or ratio-level.
   - Example: Compare boys vs. girls and right-handers vs. left-handers in terms of age (in months) and when they first walked.

6. a. Paired-samples *t* test because each student is measured twice and the outcome variable is interval or ratio. Could also use repeated-measures ANOVA.

b. Paired-samples *t* test because each student is measured twice. (Or, repeated-measures ANOVA.) Between-subjects, one-way ANOVA because each person is assigned to only one of the drugs and is measured only once.

c. Single-sample *z* test because the target mean is specified. The known value of sigma is not used because the hospitals in the rural communities may have a different standard deviation.

7. The critical value for the *F* statistic is 3.35:

$$HSD = 3.53\sqrt{\frac{4.32}{10}} = 2.32$$

Groups 1 and 2 differ by 1.5 years, and this is not a statistically significant difference. Groups 1 and 3 differ by 3.8 years, and this is a statistically significant difference. Groups 2 and 3 differ by 2.3 years, and this is a statistically significant difference. $r^2 = 39\%$.

I compared the age at which women had their first child depending on whether they had previously been frequent or regular babysitters, or not babysitters at all. The effect was strong and statistically significant—being a regular babysitter was associated with waiting longer to bear a child. This study suggests that regular babysitting functions as a sort of contraceptive. But, please, to draw a cause-and-effect conclusion, young women would have to be assigned to do different amounts of babysitting.

Part IV: Chapters 13–16

1. Because *df* is 38, the sample size, $N_{,} = 40$. The *z*-transformed value is found using Table 7 to be −.472.

The standard error on the *z* scale is

$$s_{z_r} = \frac{1}{\sqrt{N-3}} = \frac{1}{\sqrt{40-3}} = 0.164$$

$$95\%\ CI_{z_r} = z_r \pm 1.96 s_{z_r}$$

$$= -.472 \pm 1.96(.164) = [-.79, -.15]$$

Transforming back to correlation units using Table 8 gives (−.66, −.15).

2. First, compute the slope:

$$b = r\left(\frac{s_Y}{s_X}\right) = .68\left(\frac{17}{16}\right) = 0.7225$$

Then, compute the intercept:

$$a = M_Y - bM_x = 55 - 0.7225(60) = 11.65$$

Finally,

$$Y' = bX + a$$

$$= 0.7225(50) + 11.65 = 47.78$$

3. a. Yes, because (1) the assumption of a random sample is satisfied; (2) the assumption of independence of scores is presumably satisfied if there is no collusion among students; (3) the assumption of normality seems reasonable; and (4) the assumption of linearity of relationship seems reasonable.

b. A direct relationship is shown.

c. The predictor variable is the score on Test 1; the criterion variable is the score on the final exam.

d. You should not predict outside the range of the data, or from about 30 to 80 on Test 1.

4. a. Pearson correlation coefficient

b. Paired *t* test or repeated-measures ANOVA

c. Between-subjects, one-way ANOVA

d. Between-subjects, one-way ANOVA

e. Chi-square test of independence

5.

X	Y	X − M_X	Y − M_Y	(X − M_X)(Y − M_Y)	$(X - M_X)^2$	$(Y - M_Y)^2$
93	8	40	−4	−160	1,600	16
48	11	−5	−1	5	25	1
18	17	−35	5	−175	1,225	25
$M = 53$	$M = 12$			$\Sigma = -330$	$\Sigma = SS_X = 2{,}850$	$\Sigma = SS_Y = 42$
					$SS_X SS_Y = 119{,}700$	
					$\sqrt{SS_X SS_Y} = 345.98$	

$$r = \frac{\Sigma[(X - M_X)(Y - M_Y)]}{\sqrt{SS_X SS_Y}} = \frac{-330.00}{345.98} = -.95$$

6. a. When comparing a specific value to the population value and the population standard deviation is not known

b. When looking at the relationship between an ordinal variable and ordinal-, interval-, or ratio-level variable

c. When comparing an ordinal-level variable between two independent samples

d. When comparing the means of two or more dependent samples

7. A dietician looked at the relationship between the use of a Mediterranean diet and degree of memory impairment in a representative sample of Americans. He found a significant inverse relationship: The more closely people followed the diet, the less impairment they displayed. The relationship was not a strong one, but it was statistically significant. We think the effect is small yet meaningful. The study included a representative sample of Americans, so we can generalize these results to Americans at large.

However, because correlation is not causation, we cannot say that following the diet results in better memory. Perhaps a certain type of person, say, someone of Mediterranean ancestry, is more likely to follow the diet. And these people, because of their genetic heritage, are also less likely to experience memory impairment in their old age. Additionally, because the sample was age 55 at the start of the study, we do not know about the diet's effect on younger people. Finally, we don't know how well each individual followed the diet during intervening years. We only assessed whether the diet was followed at the start of the study.

There are limitations to this study, but it does suggest that the Mediterranean diet may be of some benefit in preventing/reducing short-term memory impairment. It is certainly worthy of further study.

8. We will use a chi-square goodness-of-fit test. The assumptions appear to be satisfied, namely: (1) it was not a random sample of children, but this assumption is robust; (2) independence of the observations (assuming no multiple births); and (3) all expected frequencies at least 5 (see below). We will test for a 50:50 sex ratio.

The expected frequencies are both

$$f_{\text{Expected, Male}} = f_{\text{Expected, Female}} = \frac{125}{2} = 67.50$$

The observed frequencies are

$$f_{\text{Observed, Male}} = 81$$

$$f_{\text{Observed, Female}} = 125 - 81 = 44$$

The test statistic is computed as

$$\chi^2 = \Sigma\frac{\left(f_{\text{Observed}} - f_{\text{Expected}}\right)^2}{f_{\text{Expected}}}$$

$$= \frac{(81 - 67.5)^2}{67.5} + \frac{(44 - 67.5)^2}{67.5}$$

$$= 10.88$$

The df is $1 = (2 - 1)$. The critical value is 3.841. The observed test statistic exceeds the critical value. There is evidence that the observed number of boys in the classroom is unusual.

9. We will use a chi-square test of independence. The assumptions appear to be satisfied, namely: (1) it was not a random sample of parents, but this assumption is robust; (2) independence of the observations because each parent performed the test only once and presumably did not collude over the results; and (3) all expected frequencies are at least 5 (see below). The observed frequencies are given. The expected frequencies are as follows:

	Say "Fever"	Say "No fever"	
Actual fever	$\frac{1{,}033 \times 1{,}043}{2{,}100} = 513.06$	$\frac{1{,}033 \times 1{,}057}{2{,}100} = 519.94$	1,033
Actual no fever	$\frac{1{,}067 \times 1{,}043}{2{,}100} = 529.94$	$\frac{1{,}067 \times 1{,}057}{2{,}100} = 537.06$	1,067
	1,043	1,057	

The test statistic is

$$\chi^2 = \Sigma\frac{\left(f_{\text{Observed}} - f_{\text{Expected}}\right)^2}{f_{\text{Expected}}}$$

$$= \frac{(610 - 513.06)^2}{513.06} + \frac{(423 - 519.94)^2}{519.94}$$

$$+ \frac{(433 - 529.94)^2}{529.94} + \frac{(634 - 537.06)^2}{537.06}$$

$$= 71.63$$

The df is $1 = (2 - 1)(2 - 1)$. The critical value is 3.841. The observed test statistic exceeds the critical value. There is evidence the hand on the forehead technique does predict whether there is a fever. But the size of the effect is small and the parents made a lot of errors. When the doll's temperature was elevated, the parents correctly detected fever only 59% of the time; when the doll's temperature was normal, they incorrectly perceived a fever 41% of the time.

10. First, compute the slope:

$$b = r\left(\frac{s_Y}{s_X}\right)$$

$$= .47\left(\frac{6}{12}\right)$$

$$= 0.23$$

Then, compute the intercept:

$$a = M_Y - bM_x$$

$$= 42 - 0.235(128)$$

$$= 11.92$$

Finally,

$$Y' = bX + a$$

$$= 0.235(73) + 11.92$$

$$= 29.08$$

GLOSSARY

alpha (or alpha level) – the probability of making a Type I error; the probability that a result will fall in the rare zone and the null hypothesis will be rejected when the null hypothesis is true; often called significance level; abbreviated α; usually set at .05 or 5%.

alternative hypothesis – abbreviated H_1; a statement that the explanatory variable has an effect on the outcome variable in the population; usually, a statement of what the researcher believes to be true.

analysis of variance (ANOVA) – a family of statistical tests for comparing the means of two or more groups.

apparent limits – what seem to be the upper and lower bounds of an interval in a grouped frequency distribution.

bar graph – a graph of a frequency distribution for discrete data that uses the heights of bars to indicate frequency; the bars do not touch.

beta – the probability of making a Type II error; abbreviated β.

between-group variability – variability in scores that is primarily due to the different treatments that different groups receive.

between-subjects – ANOVA terminology for independent samples.

between-subjects, one-way ANOVA – a statistical test used to compare the means of two or more independent samples when there is just one explanatory variable.

cases – the participants in or subjects of a study.

central limit theorem – a statement about the shape that a sampling distribution of the mean takes if the size of the samples is large and every possible sample were obtained.

central tendency – a value used to summarize a set of scores; also known as the average.

chi-square goodness-of-fit test – a nonparametric, single-sample test used to compare the distribution of a categorical (nominal- or ordinal-level) outcome variable in a sample to a known population value.

chi-square test of independence – a nonparametric test used to determine whether two or more populations of cases differ on a categorical (nominal- or ordinal-level) outcome variable.

clinical significance (or practical significance) – whether the size of the effect is large enough to say the explanatory variable has a meaningful impact on clinical outcome.

coefficient of determination – formal name for the effect size r^2.

Cohen's *d* – a standardized measure of effect used to measure the difference between means.

common zone – the section of the sampling distribution of a test statistic in which the observed outcome should fall if the null hypothesis is true; typically set to be the middle 95%.

confidence interval – a range within which it is estimated, based on a sample value, that a population value falls.

confounding variable – a third variable in correlational and quasi-experimental designs that is not controlled for and that has an impact on *both* of the other variables.

consent rate – the percentage of targeted subjects who agree to participate in a study.

contingency table – a table showing the degree to which a case's value on the outcome variable depends on its category on the explanatory variable.

continuous number – number that answers the question "how much" and can have "in-between" values; the specificity of the number, the number of decimal places reported, depends on the precision of the measuring instrument.

convenience sample – a sampling strategy in which cases are selected for study based on the ease with which they can be obtained.

correlation coefficient – a statistic that summarizes, in a single number, the strength of a relationship between two variables.

correlational design – a scientific study in which the relationship between two variables is examined without any attempt to manipulate or control them.

criterion variable – the outcome variable in a correlational design.

critical value – the value of the test statistic that forms the boundary between the rare zone and the common zone of the sampling distribution of the test statistic.

critical value of *t* – value of *t* used to determine whether a null hypothesis is rejected or not; abbreviated t_{cv}.

crossed – a factorial ANOVA in which each level of each explanatory variable occurs with each level of the other explanatory variable.

cumulative frequency – a count of how often a given value, or a lower value, occurs in a set of data.

cumulative percentage – cumulative frequency expressed as a percentage of the number of cases in the data set.

degrees of freedom (*df*) – the number of values in a sample that are free to vary.

dependent samples – samples in which the selection of cases for one group is related to, influences, or is determined by case selection for another group.

dependent variable – the variable where the effect is measured in an experimental or quasi-experimental study; an outcome variable.

descriptive statistic – a summary statement about a set of cases.

descriptive statistics – statistics used to describe a set of observations.

deviation score – a measure of how far away a score falls from the mean.

difference tests – statistical tests that look for differences among groups of cases.

direct relationship – a relationship in which high scores on *X* are associated with high scores on *Y*. Also called a positive relationship.

discrete number – numbers that answer the question "how many," take whole number values, and have no "in-between" values.

effect size – a measure of the degree of impact of the explanatory variable on the outcome variable.

eta squared (η^2) – an effect size that calculates the percentage of variability in the outcome variable accounted for by the explanatory variable.

experimental design – a scientific study in which an explanatory variable is manipulated or controlled by the experimenter and the effect that is measured in a dependent variable allows for a cause and effect conclusion.

explanatory variable – the variable that causes, predicts, or explains the outcome variable.

extreme percentage – percentage of the normal distribution that is found in the two tails and is evenly divided between them.

factor – term for an explanatory variable in ANOVA.

factorial ANOVA – an analysis of variance in which there is more than one explanatory variable.

frequency distribution – a tally of how often different values of a variable occur in a set of data.

frequency polygon – a frequency distribution for continuous data, displayed in graphical format, using a line connecting dots above interval midpoints to indicate frequency.

grouped frequency distribution – a count of how often the values of a variable, grouped into intervals, occur in a set of data.

grouping variable – the variable that is the explanatory variable in a quasi-experimental design.

histogram – a frequency distribution for continuous data, displayed in graph form, using the heights of bars to indicate frequency; the bars touch each other.

hypothesis – a proposed explanation for observed facts; a statement or prediction about a population value.

hypothesis testing – a statistical procedure in which data from a sample are used to evaluate a hypothesis about a population.

independence – in probability, when the occurrence of one outcome does not have any impact on the occurrence of a second outcome.

independent samples – when the selection of cases for one sample has no impact on the selection of cases for another sample.

independent-samples *t* test – an inferential statistical test used to compare two independent samples on an interval- or ratio-level outcome variable.

independent variable – the variable that is controlled by the experimenter in an experimental design.

individual differences – attributes that vary from case to case.

inferential statistic – using observations from a sample to draw a conclusion about a population.

interaction effect – situation, in factorial ANOVA, in which the impact of one explanatory variable on the outcome variable depends on the level of another explanatory variable.

interquartile range – a measure of variability for interval- or ratio-level data; the distance covered by the middle 50% of scores; abbreviated *IQR*.

interval estimate – an estimate of a population value that says the population value falls somewhere within a range of values.

interval-level numbers – numbers that provide information about how much of an attribute is possessed, as well as information about same/different and more/less; interval-level numbers have equality of units and an arbitrary zero point.

inverse relationship – a relationship in which high scores on *X* are associated with low scores on *Y*. Also called a negative relationship.

kurtosis – how peaked or flat a frequency distribution is.

least squares criterion – prediction errors are squared and the best-fitting regression line is the one that has the smallest sum of squared errors.

level – ANOVA terminology for a category of an explanatory variable.

linear regression – a predictor variable is used to predict a case's score on another variable and the prediction equation takes the form of a straight line.

longitudinal research (or repeated-measures design) – a study in which the same participants are measured at two or more points in time.

main effect – the impact of an explanatory variable, by itself, on the outcome variable.

Mann–Whitney *U* test – a nonparametric test used to compare two independent samples on an ordinal-level outcome variable.

matched pairs – participants are grouped into sets of two based on their being similar on potential confounding variables.

mean – an average calculated for interval- or ratio-level data by summing all the values in a data set and dividing by the number of cases; abbreviated *M*.

median – an average calculated by finding the score associated with the middle case, the case that separates the top half of scores from the bottom half; abbreviated *Mdn;* can be calculated for ordinal-, interval-, or ratio-level data.

middle percentage – percentage of the normal distribution found around the midpoint, evenly divided into two parts, one just above the mean and one just below it.

midpoint – the middle of an interval in a grouped frequency distribution.

modality – the number of peaks that exist in a frequency distribution.

mode – the score that occurs with the greatest frequency.

multiple linear regression – prediction in which multiple predictor variables are combined to predict an outcome variable.

negative relationship – a relationship in which high scores on *X* are associated with low scores on *Y;* also called an inverse relationship.

negative skew – an asymmetrical frequency distribution in which the tail extends to the left, in the direction of lower scores.

nominal-level numbers – numbers used to place cases in categories; numbers are arbitrary and only provide information about same/different.

nonparametric test – a statistical test for use with nominal- or ordinal-level outcome variables, and for which assumptions about the shape of the population don't have to be met.

nonrobust assumption – an assumption for a statistical test that must be met in order to proceed with the test.

normal distribution – also called the normal curve; a specific bell-shaped curve defined by the percentage of cases that fall in specific areas under the curve.

null hypothesis – abbreviated H_0; a statement that in the population the explanatory variable has no impact on the outcome variable.

one-tailed hypothesis test – hypothesis that predicts the explanatory variable has an impact on the outcome variable in a specific direction.

ordinal-level numbers – numbers used to indicate if more or less of an attribute is possessed; numbers provide information about same/different and more/less.

outcome variable – the variable that is caused, predicted, or influenced by the explanatory variable; the variable in a relationship test, *Y*, that is predicted from the other variable, *X*. Sometimes called the dependent variable.

outlier – an extreme (unusual) score that falls far away from the rest of the scores in a set of data.

***p* value** – the probability of Type I error; the same as alpha level or significance level.

paired samples – case selection for one sample is influenced by, depends on, the cases selected for another sample.

paired-samples *t* test – hypothesis test used to compare the means of two dependent samples; also known as dependent-samples *t* test, correlated-samples *t* test, related-samples *t* test, matched-pairs *t* test, within-subjects *t* test, or repeated-measures *t* test.

parameter – a value that summarizes a population.

parametric test – a statistical test for use with interval- or ratio-level outcome variables, and for which assumptions about the shape of the population must be met.

partial correlation – a correlation between two variables from which the influence of a third variable has been mathematically removed.

Pearson correlation coefficient – a statistical test that measures the degree of linear relationship between two interval/ratio-level variables.

percentile rank – percentage of cases with scores at or below a given level in a frequency distribution.

perfect relationship – a relationship between two variables in which the value of one can be exactly predicted from the other.

point estimate – an estimate of a population value that is a single value.

pooled variance – the average variance for two samples.

population – the larger group of cases a researcher is interested in studying.

positive relationship – a relationship in which high scores on X are associated with high scores on Y; also called direct relationship.

positive skew – an asymmetrical frequency distribution in which the tail extends to the right, in the direction of higher scores.

post-hoc test – a follow-up test to a statistically significant ANOVA, engineered to find out which pairs of means differ while keeping the overall alpha level at the chosen level.

power – the probability of rejecting the null hypothesis when the null hypothesis should be rejected.

practical significance (or clinical significance) – the size of the effect is large enough to say the explanatory variable has a meaningful impact on the outcome variable (or the clinical outcome).

prediction interval – a range around Y' within which there is some certainty that a case's real value of Y falls.

predictor variable – the variable in a relationship test, X, that is used to predict the other variable, Y; the explanatory variable in a correlation design.

pre-post design – participants are measured on the dependent variable before and after an intervention or manipulation.

probability – how likely an outcome is; the number of ways a specific outcome can occur, divided by the total number of possible outcomes.

quasi-experimental design – a scientific study in which cases are classified into naturally occurring groups and then compared on a dependent variable.

r^2 – an effect size that reveals the percentage of variability in one variable that is accounted for by the other variable; formally called coefficient of determination.

random assignment – every case has an equal chance of being assigned to any group in an experiment; random assignment is the hallmark of an experiment.

random sample – a sampling strategy in which each case in the population has an equal chance of being selected.

range – a measure of variability for interval- or ratio-level data; the distance from the lowest score to the highest score.

rare zone – the section of the sampling distribution of a test statistic in which it is unlikely an observed outcome will fall if the null hypothesis is true; typically, 5% of the sampling distribution.

ratio-level numbers – numbers that have all the attributes of interval-level numbers, plus a real zero point; numbers that provide information about same/different, more/less, how much of an attribute is possessed, and that can be used to calculate a proportion.

real limits – what are really the upper and lower bounds of a single continuous number or of an interval in a grouped frequency distribution.

regression line – the best-fitting straight line for predicting Y from X.

relationship tests – statistical tests that determine if two variables in a group of cases covary.

repeated-measures ANOVA – a statistical test used to compare two or more dependent samples on an interval- or ratio-level–dependent variable; also called within-subjects ANOVA, dependent-samples ANOVA, or related-samples ANOVA.

repeated-measures design (or longitudinal research) – a study in which the same participants are measured at two or more points in time.

replicate – to repeat a study, usually introducing some change in procedure to make it better.

representative – the attributes of the population are present in the sample in approximately the same proportion as in the population.

residual – the difference between an actual score and a predicted score; the size of the error in prediction.

robust assumption – an assumption for a statistical test that can be violated to some degree and it is still OK to proceed with the test.

sample – a group of cases selected from a population.

sampling distribution – a frequency distribution generated by taking repeated, random samples from a population and generating some value, like a mean, for each sample.

sampling error – discrepancies, due to random factors, between a sample statistic and a population parameter.

self-selection bias – a nonrepresentative sample that may occur when the subjects who agree to participate in a research study differ from those who choose not to participate.

significance level – the probability of Type I error; the same as alpha level or *p* value.

simple linear regression – prediction in which Y' is predicted from a single predictor variable.

single-sample test – a statistical test used to compare the results in a sample to a known population value or a specified value.

single-sample *t* test – a statistical test that compares a sample mean to a population mean when the population standard deviation is not known.

skewness – the degree to which a set of scores is not symmetric but tails off in one direction or the other.

slope – the tilt of the line; rise over run; how much up or down change in *Y* is predicted for each 1-unit change in *X*.

Spearman rank-order correlation coefficient – a nonparametric test that examines the relationship between two ordinal-level variables or one ordinal and an interval/ratio variable.

standard deviation – a measure of variability for interval- or ratio-level data; the square root of the variance; a measure of the average distance that scores fall from the mean.

standard error of the estimate – the standard deviation of the residual scores, a measure of error in regression.

standard error of the mean – the standard deviation of a sampling distribution of the mean.

standard error of the mean difference for difference scores – the standard deviation of the sampling distribution of difference scores, abbreviated s_{M_D}; used as the denominator in the paired-samples *t* test equation.

standard score – raw score expressed in terms of how many standard deviations it falls away from the mean; also known as a *z* score.

statistic – a value that summarizes data from a sample.

statistical significance – the observed difference between sample means is large enough to conclude that it represents a difference between population means.

statistically significant – when a researcher concludes that the observed sample results are different from the null-hypothesized population value.

statistics – techniques used to summarize data in order to answer questions.

stem-and-leaf display – a data summary technique that combines features of a table and a graph.

sum of squares – squaring a set of scores and then adding together the squared scores; abbreviated *SS*.

sum of squares between (SS_{Between}) – a sum of the squared deviation scores representing the variability between groups.

sum of squares total (SS_{Total}) – a sum of the squared deviation scores representing all the variability in the scores.

sum of squares within (SS_{Within}) – a sum of the squared deviation scores representing the variability within groups.

treatment effect – the impact of the explanatory variable on the dependent variable.

two-samples *t* test – an inferential statistical test used to compare the mean of one sample to the mean of another sample.

two-tailed hypothesis test – hypothesis that predicts the explanatory variable has an impact on the outcome variable, but doesn't predict the direction of the impact.

Type I error – the error that occurs when the null hypothesis is true but is rejected; $p(\text{Type I error}) = \alpha$.

Type II error – the error that occurs when one fails to reject the null hypothesis but should have rejected it; p(Type II error) $= \beta$.

underpowered – term for a study with a sample size too small for the study to have a reasonable chance to reject the null hypothesis given the size of the effect.

ungrouped frequency distribution – a count of how often each individual value of a variable occurs in a set of data.

variability – how much variety (spread or dispersion) there is in a set of scores.

variables – characteristics measured by researchers.

variance – a measure of variability for interval- or ratio-level data; the mean of the squared deviation scores.

way – term for an explanatory variable in ANOVA.

within-group variability – variability within a sample of cases, all of which have received the same treatment.

within-subjects – ANOVA terminology for dependent samples.

within-subjects design – the same participants are measured in two or more different situations or under two or more different conditions.

***Y*-intercept** – the spot where the regression line would pass through the Y-axis.

***Y* prime** – the value of Y predicted from X by a regression equation; Y'.

***z* score** – raw score expressed in terms of how many standard deviations it falls away from the mean; also known as a standard score.

REFERENCES

Abelson, Reed. (2006, August 18). Heart procedure is off the charts in an Ohio city. *The New York Times.* Retrieved from http://www.nytimes.com

American Psychological Association. (2010). *Publication manual of the American Psychological Association* (6th ed.). Washington, DC: American Psychological Association.

Aspelmeier, J. E., & Pierce, T. W. (2015). *SPSS: A user-friendly approach* (3rd ed.). New York, NY: Worth.

Babbie, E. R. (1973). *Survey research methods.* Belmont, CA: Wadsworth.

Bakalar, N. (2008, January 22). Heeding familiar advice may add years to your life. *The New York Times.* Retrieved from http://www.nytimes.com

Baker, P., & Sussman, D. (2012, May 14). Obama's switch on same-sex marriage stirs skepticism. *The New York Times.* Retrieved from http://www.nytimes.com

Bandura, A., Ross, D., & Ross, S. A. (1961). Transmission of aggression through imitation of aggressive models. *Journal of Abnormal and Social Psychology, 63,* 575–582.

Bargh, J. A., Chen, M., & Burrows, L. (1996). Automaticity of social behavior: Direct effects of trait construct and stereotype activation on action. *Journal of Personality and Social Psychology, 71,* 230–244.

Bartz, A. E. (1999). *Basic statistical concepts* (4th ed.). Upper Saddle River, NJ: Prentice Hall.

Bureau of Labor Statistics. (2011, May). *Occupational employment statistics. State occupational employment and wage estimates.* Retrieved from http://www.bls.gov/oes/current/oessrcst.htm

Cohen, J. (1988). *Statistical power analysis for the behavioral sciences* (2nd ed.). Hillsdale, NJ: Lawrence Erlbaum Associates.

Cohen, J. (1994). The earth is round ($p < .05$). *American Psychologist, 49,* 997–1003.

College Board. (2012, June 1). *Admission validity report for Sample One University.* Retrieved from http://professionals.collegeboard.com/higher-ed/validity/aces/study

Craik, F. I. M., & Tulving, E. (1975). Depth of processing and the retention of words in episodic memory. *Journal of Experimental Psychology: General, 104,* 268–294.

Cumming, G., & Finch, S. (2005). Inference by eye: Confidence intervals and how to read pictures of data. *American Psychologist, 60,* 170–180.

Diekhoff, G. M. (1996). *Basic statistics for the social and behavioral sciences.* Upper Saddle River, NJ: Prentice Hall.

Dodge, L. G. (2003). *Dr. Laurie's introduction to statistical methods.* Los Angeles, CA: Pyrczak Publishing.

Eron, L. P., Huesmann, L. R., Lefkowitz, M. M., & Walder, L. O. (1972). Does television violence cause aggression? *American Psychologist, 27,* 253–263.

Festinger, L., & Carlsmith, J. M. (1959). Cognitive consequences of forced compliance. *Journal of Abnormal and Social Psychology, 58,* 203–210.

Hite, S. (1987). *Women and love.* New York, NY: Knopf.

Huntley, D. A., Cho, D. W., Christman, J., & Csernansky, J. G. (1998). Predicting length of stay in an acute psychiatric hospital. *Psychiatric Services, 49,* 1049–1053.

Kiecolt-Glaser, J. K., Marucha, P. T., Malarkey, W. B., Mercado, A. M., & Glaser, R. (1995). Slowing of wound healing by psychological stress. *The Lancet, 346,* 1194–1196.

Laumann, E. O., Gagnon, J. H., Michael, R. T., & Michaels, S. (1994). *The social organization of sexuality.* Chicago: University of Chicago Press.

Loftus, E. F., & Palmer, J. C. (1974). A reconstruction of automobile destruction: An example of the interaction between language and memory. *Journal of Verbal Learning and Verbal Behavior, 13,* 585–589.

Michael, R. T., Gagnon, J. H., Laumann, E. O., & Kolata, G. (1994). *Sex in America.* Boston, MA: Little, Brown and Company.

Milgram, S. (1963). A behavioral study of obedience. *Journal of Abnormal and Social Psychology, 67,* 371–378.

Morgan, T. M., Tang, D., Stratton, K. L., Barocas, D. A., Anderson, C. B., Gregg, J. R., . . . Clark, P. E. (2011). Preoperative nutritional status is an important predictor of survival in patients undergoing surgery for renal cell carcinoma. *European Urology, 59,* 923–928.

National Center for Chronic Disease Prevention and Health Promotion. (2013). *Behavioral risk factor surveillance system.* Retrieved from Centers for Disease Control and Prevention website: http://www.cdc.gov/brfss/

Ott, J. C. (2011). Government and happiness in 130 nations: Good government fosters higher level and more equality of happiness. *Social Indicators Research, 102,* 3–22.

Rohwedder, S., & Willis, R. J. (2010). Mental retirement. *Journal of Economic Perspectives, 24,* 119–138.

Scheiber, N., & Sussman, D. (2015, June 3). Inequality troubles Americans across party lines, Times/CBS poll finds. *The New York Times.* Retrieved from http://www.nytimes.com

Substance Abuse and Mental Health Services Administration. (2012–2013). *National survey of drug use and health.* Retrieved from http://www.samhsa.gov/data/sites/default/files/NSDUHresultsPDFWHTML2013/Web/NSDUHresults2013.pdf

Wilkinson, L., & the Task Force on Statistical Inference. (1999). Statistical methods in psychology journals: Guidelines and explanations. *American Psychologist, 54,* 594–604.

INDEX

Note: Page numbers followed by f indicate figures; those followed by t indicate tables.

Stat Sheet: Chapter 1 Introduction to Statistics

Explanatory and Outcome Variables

Explanatory variable: the variable that causes, predicts, or explains the outcome variable.
Outcome variable: the variable that is caused, predicted, or influenced by the explanatory variable.

Deciding What Type of Study Is Being Done

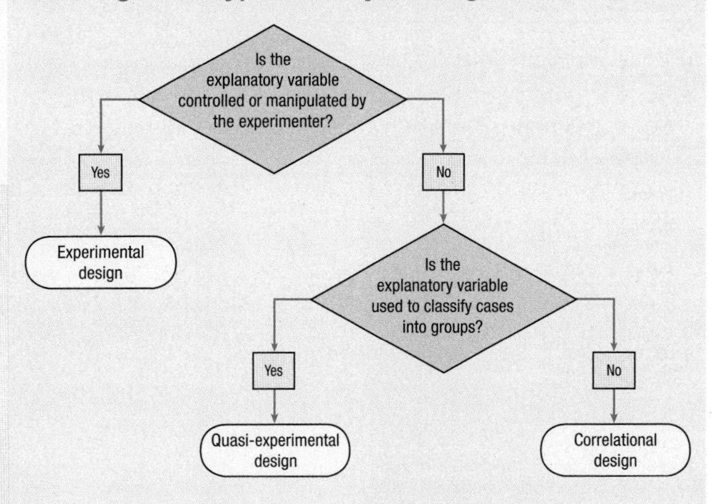

Types of Studies

There are three types of experimental designs: correlational, experimental, and quasi-experimental.

	Correlational	Experimental	Quasi-Experimental
Explanatory variable is called	Predictor variable	Independent variable	Grouping variable
Outcome variable is called	Criterion variable	Dependent variable	Dependent variable
Cases are . . .	Measured for naturally occurring variability on both variables.	Assigned to groups by experimenter using random assignment.	Classified in groups on basis of naturally occurring status.
Is the explanatory variable manipulated/ controlled by the experimenter?	No	Yes	No
Can one draw a firm conclusion about cause and effect?	No	Yes	No
Is there a need to worry about confounding variables?	Yes	No	Yes
What is the question being asked by the study?	Is there a relationship between the two variables?	Do the different groups possess different amounts of the dependent variable?	Do the different groups possess different amounts of the dependent variable?
What is an advantage of this type of study?	Researchers can study conditions that can't be manipulated.	Researchers can draw a conclusion about cause and effect.	Researchers can study conditions that can't be manipulated.

Levels of Measurement

There are four levels of measurement: nominal, ordinal, interval, and ratio.

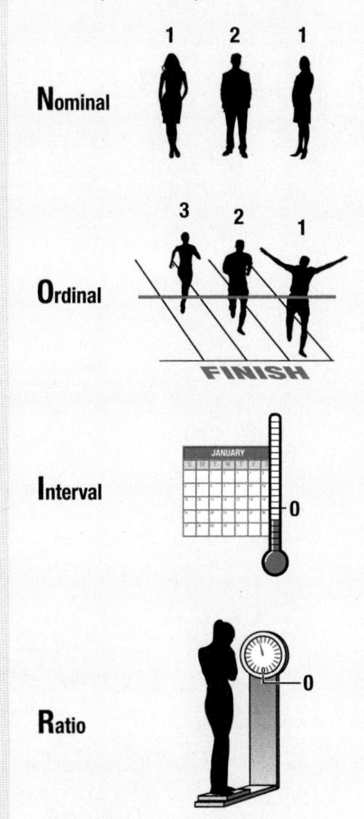

As numbers move from nominal to ratio, they contain more information.

Stat Sheet: Chapter 1 Introduction to Statistics (cont.)

Information Contained in Numbers

	Same/ Different	Direction of Difference (more/less)	Amount of Distance (equality of units)	Proportion (absolute zero point)
Nominal	✓			
Ordinal	✓	✓		
Interval	✓	✓	✓	
Ratio	✓	✓	✓	✓

Determining Level of Measurement

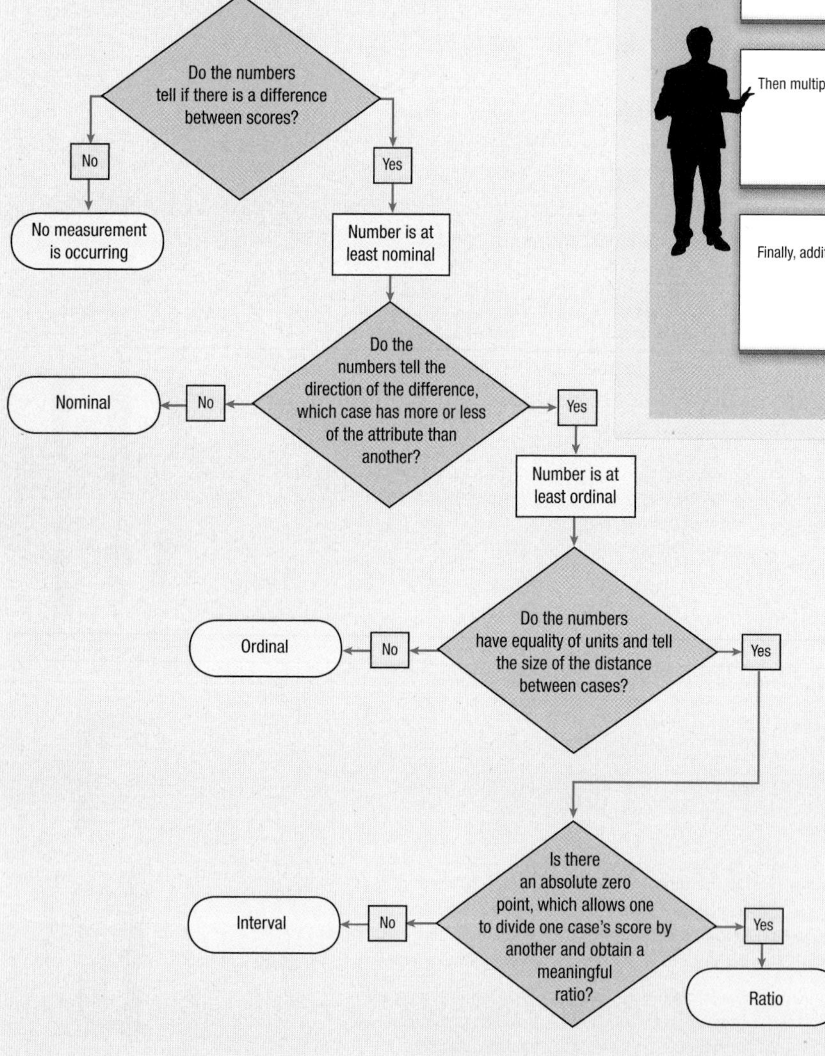

Order of Operations and Rounding

The order in which mathematical operations should be done must be followed in order to get the right answers.

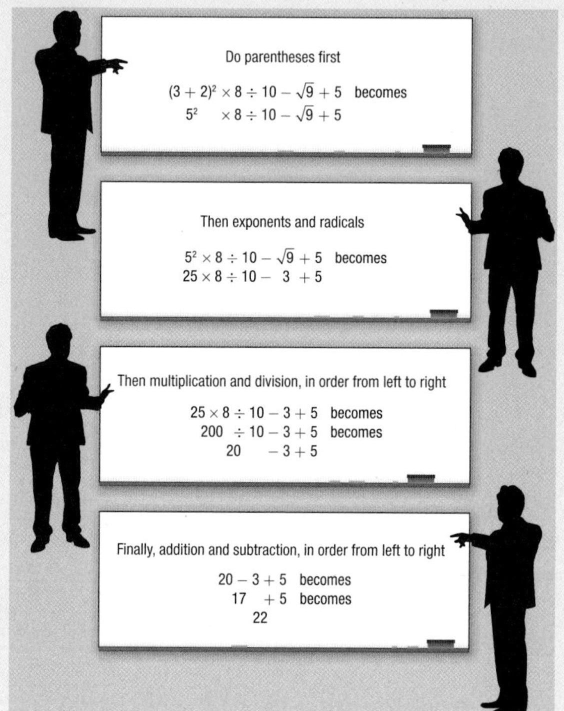

Rules of Rounding

1. Round final answers to two decimal places.
2. Don't round until the very end. (If one does round as one goes, carry four decimal places.)
3. If the unrounded number is centered exactly between the two rounding options, round up.

Stat Sheet: Chapter 2 Frequency Distribution Tables

Determining Whether to Make a Grouped or Ungrouped Frequency Distribution

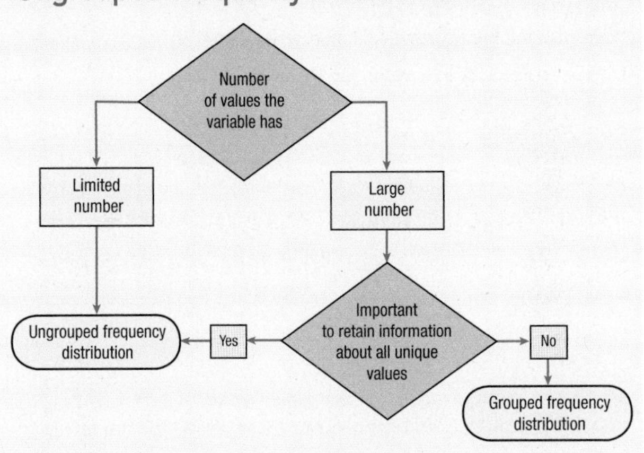

Equation 2.1 Formula for Calculating Frequency Percentage (%) for a Frequency Distribution

$$\% = \frac{f}{N} \times 100$$

where % = frequency percentage
f = frequency
N = number of cases

Equation 2.2 Formula for Calculating Cumulative Percentage ($\%_c$) for a Frequency Distribution

$$\%_c = \frac{f_c}{N} \times 100$$

where $\%_c$ = cumulative percentage
f_c = cumulative frequency
N = total number of cases

Determining if a Measure Is Continuous or Discrete

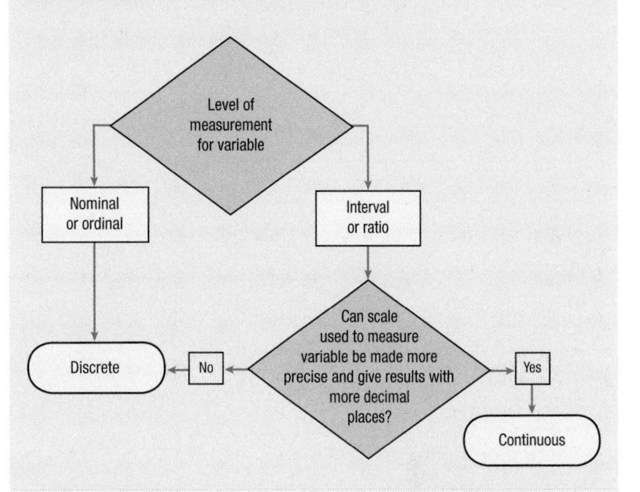

Determining Type of Graph to Make

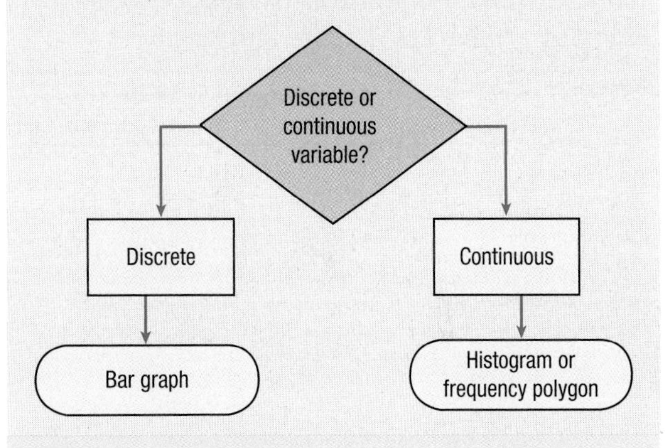

Differences Among Graphs

	Bar Graphs	Histograms	Frequency Polygons
Type of data	Discrete	Continuous	Continuous
Physical characteristics of the graph	Bars don't touch each other.	Bars touch each other. Bars go up and down at the real limits of the interval.	Frequencies are marked with dots at the midpoints of the intervals. Dots are connected by lines. Frequencies "go to zero" at the far left and far right of the graph.

Describing the Shape of Frequency Distributions

Modality	Skewness	Kurtosis
How many high points there are in a data set	Whether the data set is symmetric	Whether a data set is peaked or flat

Stat Sheet: Chapter 3 Measures of Central Tendency and Variability

Measures of Central Tendency

Equation 3.1 Formula for Sample Mean (*M*)

$$M = \frac{\Sigma X}{N}$$

Equation 3.2 Formula for the Median (*Mdn*)

Step 1 Put the scores in order from low to high and number them (1, 2, 3, etc.).

Step 2 The median is the *X* value associated with score number $\frac{N + 1}{2}$ where N = number of cases.

Which Measure of Central Tendency Can Be Used for Which Level of Measurement?

	Measure of Central Tendency		
Level of Measurement	**Mode**	**Median**	**Mean**
Nominal	✓		
Ordinal	✓	✓	
Interval or ratio	✓	✓	✓

Not all measures of central tendency can be used with all levels of measurement. When more than one measure of central tendency may be used, choose the one that uses more of the information available in the numbers. If planning to calculate a mean, be sure to check the shape of the data set for skewness and modality.

Measures of Variability

Equation 3.3 Formula for Range

$$\text{Range} = X_{High} - X_{Low}$$

Equation 3.6 Formula for Sample Variance (s^2)

$$s^2 = \frac{\Sigma(X - M)^2}{N - 1}$$

where s^2 = sample variance
X = raw score
M = the sample mean
N = the number of cases in the sample

Step 1 Subtract the mean from each score to calculate deviation scores.

Step 2 Take each deviation score and square it.

Step 3 Add up all the squared deviation scores.

Step 4 Take the sum of the squared deviation scores and divide it by the number of cases minus 1 to find the sample variance.

Equation 3.7 Formula for Sample Deviation (s)

$$s = \sqrt{s^2}$$

where s = sample standard deviation
s^2 = sample variance

Stat Sheet: Chapter 4 Standard Scores, the Normal Distribution, and Probability

Equation 4.1 Formula for Calculating Standard Scores (z Scores)

$$z = \frac{X - M}{s}$$

where z = the standard score
X = the raw score
M = the mean score
s = the standard deviation for the scores

Equation 4.2 Formula for Calculating a Raw Score (X) from a Standard Score (z)

$$X = M + (z \times s)$$

where X = the raw score
M = the mean
z = the standard score for which the raw score is being calculated
s = the standard deviation

Flowchart for Finding Areas Under the Normal Curve

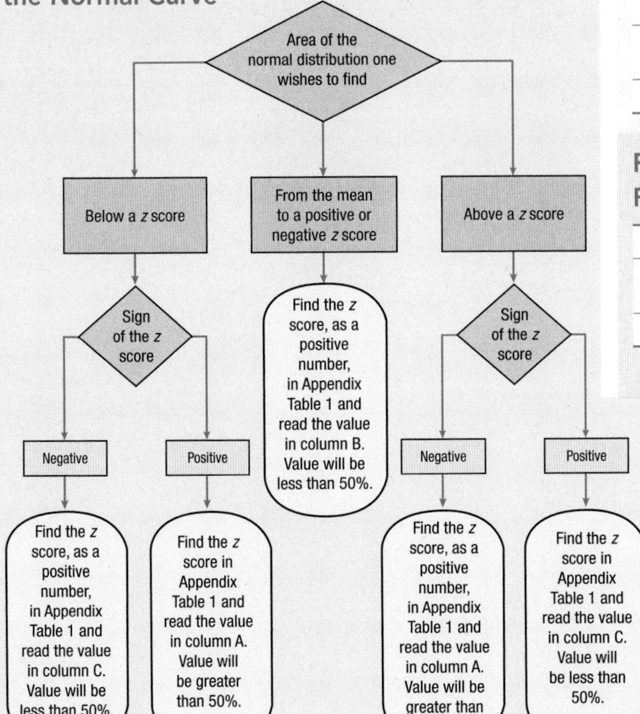

Hint

When finding an area under the normal curve or converting a percentile rank to a z score, make a sketch of the normal distribution, shade in the area being looked for, and make an estimate.

Flowchart for Calculating z Scores from Percentile Ranks

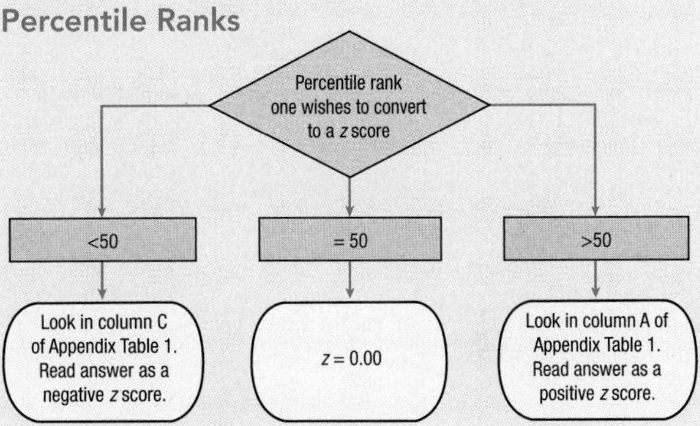

Finding z Scores Associated with a Middle Percentage of the Normal Distribution

Step 1	Take the middle percentage and cut it in half.
Step 2	Use column B of Appendix Table 1 to find the percentage closest to the ½ value calculated in Step 1.
Step 3	Report the z score for the percentage in $\pm$ format.

Finding z Scores Associated with an Extreme Percentage of the Normal Distribution

Step 1	Take the extreme percentage and cut it in half.
Step 2	Use column C of Appendix Table 1 to find the percentage closest to the ½ value calculated in Step 1.
Step 3	Report the z score for the percentage in $\pm$ format.

Stat Sheet: Chapter 4 Standard Scores, the Normal Distribution, and Probability (cont.)

Simplified Version of the Normal Distribution

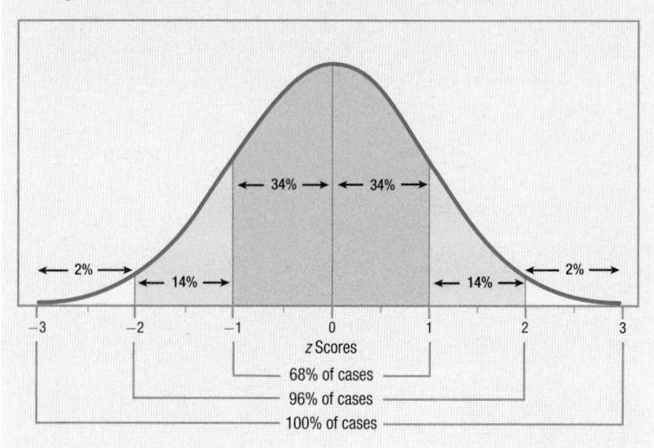

This version of the normal distribution limits it to scores ranging from −3 to +3.

Probability

Equation 4.3 Formula for Calculating Probability

$$p(A) = \frac{\text{Number of ways outcome } A \text{ can occur}}{\text{Total number of possible outcomes}}$$

where $p(A)$ = the probability of outcome A occurring

Stat Sheet: Chapter 5 Sampling and Confidence Intervals

Equation 5.1 Calculating the Standard Error of the Mean

$$\sigma_M = \frac{\sigma}{\sqrt{N}}$$

where σ_M = the standard error of the mean
σ = the standard deviation of the population
N = the number of cases in the sample

Equation 5.2 Estimated Standard Error of the Mean

$$s_M = \frac{s}{\sqrt{N}}$$

where s_M = the estimated standard error of the mean
s = the sample standard deviation
N = the number of cases in the sample

Equation 5.3 Formula for 95% Confidence Interval for the Population Mean

$$95\%CI_\mu = M \pm (1.96 \times \sigma_M)$$

where $95\%\ CI_\mu$ = the 95% confidence interval for the population mean being calculated
M = sample mean
σ_M = standard error of the mean (If σ_M is unknown, substitute s_M.)

Stat Sheet: Chapter 6 Introduction to Hypothesis Testing

Step 1 Pick a Test

- A single-sample z test is used to compare the mean of a sample (M) to a specific value like the mean of a population (μ) when the population standard deviation (σ) is known.

Step 2 Check the Assumptions

- *Random samples (robust):* The cases in the sample are a random sample from the sample's population.
- *Independence of observations (not robust):* Cases in the sample don't influence each other in terms of the dependent variable.
- *Normality (robust):* The dependent variable is normally distributed in the population.

Step 3 List the Hypotheses

Nondirectional (two-tailed):

H_0: μ = some specific value

H_1: $\mu \neq$ that specific value

Step 4 Set the Decision Rule

- Critical value of z for a two-tailed test with the alpha set at .05 is ±1.96.
- Set decision rule:
  - Reject H_0 if $z \leq -1.96$ or if $z \geq 1.96$.
  - Fail to reject H_0 if $-1.96 < z < 1.96$.
- See the table below for other critical values of z.

Step 5 Calculate the Test Statistic

First, calculate standard error of the mean (σ_M) using Equation 5.1:

$$\sigma_M = \frac{\sigma}{\sqrt{N}}$$

where σ = the standard deviation of the population
N = the number of cases in the sample

Then, calculate z using Equation 6.1:

$$z = \frac{M - \mu}{\sigma_M}$$

where M = the sample mean
μ = the population mean
σ_M = the standard error of the mean

Step 6 Interpret the Results

The Goal of Interpretation

Write an interpretative paragraph, (1) telling what the study did, (2) presenting some facts, (3) explaining the results, and (4) making suggestions for future research.

Was the Null Hypothesis Rejected?

Apply the decision rule to determine this.

- If rejected (i.e., results fall in the rare zone), then:
  - The difference between the sample mean and population mean is statistically significant.
  - Conclude there is probably a difference between the population mean and the mean of the population from which the sample was drawn.
  - Compare the direction of difference between the sample mean and population mean to draw a conclusion about how the sample's population differs.
  - Report the results in APA format [e.g., $z (N = 72) = 2.54$, $p < .05$].
- If failed to reject the null hypothesis (i.e., results fall in the common zone), then:
  - The difference between the sample mean and population mean is not statistically significant.
  - There is not enough evidence to conclude a difference exists (and so no need to comment on the direction of a difference).
  - Report the results in APA format [e.g., $z (N = 81) = 1.50$, $p > .05$].

Critical Values of z for Commonly Used Alpha Levels

	Critical values of z			
df	$\alpha = .10$, two-tailed *or* $\alpha = .05$, one-tailed	$\alpha = .05$, two-tailed *or* $\alpha = .025$, one-tailed	$\alpha = .02$, two-tailed *or* $\alpha = .01$, one-tailed	$\alpha = .01$, two-tailed *or* $\alpha = .005$, one-tailed
	$z_{cv} = 1.65$	$z_{cv} = 1.96$	$z_{cv} = 2.33$	$z_{cv} = 2.58$

Stat Sheet: Chapter 7 The Single-Sample t Test

Step 1 Pick a Test

A single-sample t test is used to compare a sample mean to a population mean or some specified value when the population standard deviation is unknown.

Step 2 Check the Assumptions

1. *Random samples (robust):* The cases in the sample are a random sample from the sample's population.
2. *Independence of observations (not robust):* Cases in the sample don't influence each other in terms of the dependent variable.
3. *Normality (robust):* The dependent variable is normally distributed in the population.

Step 3 List the Hypotheses

For a two-tailed test:

H_0: μ = a specific value

H_1: $\mu \neq$ the specific value

For a one-tailed test:

H_0: $\mu >$ a specific value

or

H_0: $\mu <$ a specific value

H_1: $\mu \leq$ the specific value

or

H_1: $\mu \geq$ the specific value

Step 4 Set the Decision Rule

- Is the test one-tailed or two-tailed?
- Set the alpha level.
- Calculate degrees of freedom: $df = N - 1$.
- Use Appendix Table 3 to find t_{cv}.
- For a two-tailed test, here's the decision rule:
  - If $t \leq -t_{cv}$ or if $t \geq t_{cv}$, reject H_0.
  - If $-t_{cv} < t < t_{cv}$, fail to reject H_0.

Step 5 Calculate the Test Statistic

First, use Equation 5.2 to calculate the estimated standard error of the mean:

$$s_M = \frac{s}{\sqrt{N}}$$

where s_M = the estimated standard error of the mean
s = the sample standard deviation
N = the number of cases in the sample

Then, use Equation 7.2 to calculate the single-sample t value:

$$t = \frac{M - \mu}{s_M}$$

where t = t value
M = sample mean
μ = population mean (or a specified value)

Step 6 Interpret the Results

Write a four-point interpretation:

1. What was done?
2. What was found?
3. What does it mean?
4. What suggestions are there for future research?

Was the Null Hypothesis Rejected?

- Compare the calculated value of t to the critical value of t, t_{cv}, using the decision rule:
  - For a two-tailed test,
    - If $t \leq -t_{cv}$ or if $t \geq t_{cv}$, then reject H_0.
    - If $-t_{cv} < t < t_{cv}$, then fail to reject H_0.
- If H_0 is rejected, call the results statistically significant, and compare M to μ to determine the direction of the difference.
- If H_0 is not rejected, call the results not statistically significant, and conclude that there is not enough evidence to say a difference exists.
- Report the results in APA format [e.g., $t(140) = 8.81$, $p < .05$].
- If a one-tailed test, add "(one-tailed)" at the end.

How Big Is the Effect?

Use Equation 7.3 to calculate Cohen's d:

$$d = \frac{M - \mu}{s}$$

where d = the effect size
M = sample mean
μ = hypothesized population mean
s = sample standard deviation

- No effect: $d = 0.00$
- Small effect: $d \approx 0.20$
- Medium effect: $d \approx 0.50$
- Large effect: $d \geq 0.80$

Use Equation 7.4 to calculate r^2:

$$r^2 = \frac{t^2}{t^2 + df} \times 100$$

where r^2 = the percentage of variability in the outcome variable that is accounted for by the explanatory variable
t^2 = the squared value of t
df = the degrees of freedom for the t value

No effect: $r^2 = 0$
Small effect: $r^2 \approx 1\%$
Medium effect: $r^2 \approx 9\%$
Large effect: $r^2 \geq 25\%$

How Wide Is the Confidence Interval?

Use Equation 7.5 to calculate the 95% confidence interval for the difference between population means:

$$95\%\ CI\mu_{Diff} = (M - \mu) \pm (t_{cv} \times s_M)$$

where $95\%\ CI\mu_{Diff}$ = the 95% confidence interval for the difference between two population means

M = sample mean from one population

μ = mean for other population

t_{cv} = critical value of t, two-tailed, $\alpha = .05$, $df = N - 1$

s_M = estimated standard error of the mean

Use Equation 7.6 to calculate the width of the confidence interval:

$$CI_W = CI_{UL} - CI_{LL}$$

where CI_W = the width of the confidence interval

CI_{UL} = the upper limit of the confidence interval

CI_{LL} = the lower limit of the confidence interval

Interpreting a Confidence Interval for the Difference Between Population Means

	Confidence Interval Captures Zero	**Confidence Interval Is Near Zero**	**Confidence Interval Is Far from Zero**
Confidence Interval Is Narrow	There is not enough evidence to conclude an effect exists. A researcher can't say the two population means are different.	The effect is likely weak. It is plausible that the two population means are different. Cohen's d represents the size of the effect.	The effect is likely strong. It is plausible that the two population means are different. Cohen's d represents the size of the effect.
Confidence Interval Is Wide	There is not enough evidence to conclude an effect exists. There is little information about whether the two population means are different. Replicate the study with a larger sample.	The effect is likely weak to moderate. It is plausible that the two population means are different. Calculate Cohen's d for both ends of the CI. Replicate the study with a larger sample.	The effect is likely moderate to strong. It is plausible that the two population means are different. Calculate Cohen's d for both ends of the CI. Replicate the study with a larger sample.

Stat Sheet: Chapter 8 Independent-Samples *t* Tests

Differentiating Independent Samples from Paired Samples

Independent samples: Each sample is, or could be, a random sample from its population.

Dependent samples: Samples are same cases measured at two points of time or two conditions; samples are matched, yoked, or paired together in some way.

Conducting an Independent-Samples *t* Test

Step 1 Pick a Test

Independent-samples *t* tests are used to compare the means of two independent samples.

Step 2 Check the Assumptions

- Random samples (robust)
- Independence of observations (not robust)
- Normality (robust)
- Homogeneity of variance (robust, especially if *N* is large)

Step 3 List the Hypotheses

Nondirectional (two-tailed):

$$H_0: \mu_1 = \mu_2$$

$$H_1: \mu_1 \neq \mu_2$$

Step 4 Set the Decision Rule

- Find t_{cv} in Appendix Table 3, based on (1) number of tails, (2) alpha level, and (3) degrees of freedom ($df = N - 2$).
- Two-tailed test decision rule:
  - If $t \leq -t_{cv}$ or if $t \geq t_{cv}$, reject H_0.
  - If $-t_{cv} < t < t_{cv}$, fail to reject H_0.

Step 5 Calculate the Test Statistic

First, use Equation 8.2 to calculate the pooled variance:

$$s^2_{Pooled} = \frac{s^2_1(n_1 - 1) + s^2_2(n_2 - 1)}{df}$$

where s^2_{Pooled} = the pooled variance
n_1 = the sample size for Group (sample) 1
s^2_1 = the variance for Group 1
n_2 = the sample size for Group (sample) 2
s^2_2 = the variance for Group 2
df = the degrees of freedom ($N - 2$)

Then, use Equation 8.3 to calculate the standard error of the mean $s_{M_1-M_2}$:

$$s_{M_1-M_2} = \sqrt{s^2_{Pooled}\left(\frac{N}{n_1 \times n_2}\right)}$$

Finally, use Equation 8.4 to calculate the independent-samples *t* value:

$$t = \frac{M_1 - M_2}{s_{M_1-M_2}}$$

Step 6 Interpret the Results

Write a four-point interpretation:

1. What was done?
2. What was found?
3. What does it mean?
4. What suggestions exist for future research?

Was the Null Hypothesis Rejected?

- Apply the decision rule from Step 4:
  - If H_0 is rejected (i.e., results fall in the rare zone): (1) The difference between sample means is statistically significant, (2) conclude that there is probably a difference between population means, (3) use sample means to draw a conclusion about the direction of the difference for the population means, (4) report the results in APA format [e.g., $t(36) = 6.58$, $p < .05$].
  - If H_0 is not rejected (i.e., results fall in the common zone): (1) The difference between sample means is not statistically significant, (2) there is not enough evidence to conclude a difference exists between populations means, (3) report the results in APA format [e.g., $t(59) = 1.50$, $p > .05$].

How Big Is the Effect?

- Use Equation 8.5 to calculate Cohen's *d* and Equation 7.4 to calculate r^2:

$$d = \frac{M_1 - M_2}{\sqrt{s^2_{Pooled}}}$$

where d = Cohen's *d* value
M_1 = the mean for Group (sample) 1
M_2 = the mean for Group (sample) 2
s^2_{Pooled} = the pooled variance

- Interpreting *d*
  - No effect: $d = 0.00$
  - Small effect: $d \approx 0.20$
  - Medium effect: $d \approx 0.50$
  - Large effect: $d \geq 0.80$

$$r^2 = \frac{t^2}{t^2 + df} \times 100$$

where r^2 = the percentage of variability in the outcome variable that is accounted for by the explanatory variable
t^2 = the squared value of *t*
df = the degrees of freedom for the *t* value

No effect: $r^2 = 0$
Small effect: $r^2 \approx 1\%$
Medium effect: $r^2 \approx 9\%$
Large effect: $r^2 \geq 25\%$

How Wide Is the Confidence Interval?

- Use Equation 8.6 to calculate the 95% confidence interval for the difference between population means:

$$95\%\ CI\mu_{Diff} = (M_1 - M_2) \pm (t_{cv} \times s_{M_1-M_2})$$

  - Interpret the confidence interva[illegible] based on (1) whether it captur[illegible] zero, (2) how close it comes t[illegible] zero, and (3) how wide it is.
  - Can calculate *d* for the up[illegible] and lower ends of a confi[illegible] interval.

Stat Sheet: Chapter 9 Paired-Samples *t* Tests

Dependent Samples

With dependent samples, each case consists of a pair of data points. The samples are also called paired samples or matched samples. These research designs are also called within subjects, longitudinal, or pre-post.

Step 1 Pick a Test

Use a paired-samples *t* test to compare the means of two dependent samples.

Step 2 Check the Assumptions

1. Random samples (robust)
2. Independence of observations (not robust)
3. Normality (robust)

Step 3 List the Hypotheses

For a two-tailed test:

$$H_0: \mu_1 = \mu_2$$

$$H_1: \mu_1 \neq \mu_2$$

Step 4 Set the Decision Rule

- Find t_{cv} in Appendix Table 3, based on (1) the number of tails, (2) alpha level, and (3) degrees of freedom ($df = N - 1$, where N = how many pairs of cases).
- Two-tailed test decision rule:

  If $t \leq -t_{cv}$ or if $t \geq t_{cv}$, reject H_0.

  If $t_{cv} < t < t_{cv}$, fail to reject H_0.

Step 5 Calculate the Test Statistic

First, calculate the standard error of the mean difference using Equation 9.3:

$$s_{M_D} = \frac{s_D}{\sqrt{N}}$$

where: s_{M_D} = the standard error of the mean difference for difference scores

s_D = the standard deviation (*s*) of the difference scores

N = the number of pairs of cases

Then, use Equation 9.4 to calculate the paired-samples *t* test:

$$t = \frac{M_1 - M_2}{s_{M_D}}$$

where t = the value of the test statistic for a paired-samples *t* test

M_1 = the mean of one sample

M_2 = the mean of the other sample

s_{M_D} = standard error of the mean difference for difference scores

Step 6 Interpret the Results

Write a four-point interpretation:

1. What was done?
2. What was found?
3. What does it mean?
4. What suggestions exist for future research?

Was the Null Hypothesis Rejected?

Apply the decision rule from Step 4:

- If H_0 is rejected (i.e., results fall in the rare zone): (1) The difference between sample means is statistically significant, (2) conclude that there is probably a difference between populations means, (3) use sample means to draw a conclusion about the direction of the difference for the population means, (4) report the results in APA format [e.g., $t(36) = 6.58$, $p < .05$].
- If H_0 is not rejected (i.e., results fall in the common zone): (1) The difference between sample means is not statistically significant, (2) there is not enough evidence to conclude a difference exists between populations means (so no need to comment on the direction of the difference), (3) report the results in APA format [e.g., $t(59) = 1.50$, $p > .05$].

How Big Is the Effect?

Use Equation 9.5 to calculate the 95% confidence interval for the difference between population means (95% $CI\mu_{Diff}$) to use as an effect size.

$$95\%\ CI\mu_{Diff} = (M_1 - M_2) \pm (t_{cv} \times s_{M_D})$$

where M_1 = the mean of the experimental condition sample

M_2 = the mean of the control condition sample

t_{cv} = the critical value of *t*, two-tailed, $\alpha = .05$, $df = N - 1$

s_{M_D} = the standard error of the mean difference for difference scores

Interpret the confidence interval:

- If the confidence interval captures zero, then it is possible that the null hypothesis is true.
- The distance that the confidence interval falls from zero provides information about the size of effect.
- The narrower the confidence interval, the more precisely the effect can be specified in the population.

Stat Sheet: Chapter 10 Between-Subject, One-Way Analysis of Variance

Step 1 Pick a Test

A between-subjects, one-way ANOVA is used to compare the means of two or more independent samples when there is only one explanatory variable.

Step 2 Check the Assumptions

1. Random samples (robust)
2. Independence of observations (not robust)
3. Normality (robust)
4. Homogeneity of variance (robust, especially if $N \geq 50$ and *n*s are about equal)

Step 3 List the Hypotheses

H_0: $\mu_1 = \mu_2 \cdots \mu_k$

H_1: At least one population mean is different from at least one other.

Step 4 Set the Decision Rule

Decision rule:

- If $F \geq F_{cv}$, reject H_0.
- If $F < F_{cv}$, fail to reject H_0.

Find F_{cv} in Appendix Table 4 based on degrees of freedom (see the ANOVA summary table template) and selected α level.

Step 5 Calculate the Test Statistic

Source of Variability	Sum of Squares	Degrees of Freedom	Mean Square	F Ratio
Between groups	Equation 10.3 $SS_{Between} = \Sigma\left(\frac{(\Sigma X_{Group})^2}{n_{Group}}\right) - \frac{(\Sigma X)^2}{N}$	$k - 1$	$\frac{SS_{Between}}{df_{Between}}$	$\frac{MS_{Between}}{MS_{Within}}$
Within groups	Equation 10.4 $SS_{Within} = \Sigma\left(\Sigma X^2_{Group} - \frac{(\Sigma X_{Group})^2}{n_{Group}}\right)$	$N - k$	$\frac{SS_{Within}}{df_{Within}}$	
Total	Equation 10.2 $SS_{Total} = \Sigma X^2 - \frac{(\Sigma X)^2}{N}$	$N - 1$		

Note:
k = the number of cells
N = the total number of observations

Step 6 Interpret the Results

Write a four-point interpretation:

1. What was done?
2. What was found?
3. What does it mean?
4. What suggestions exist for future research?

Was the Null Hypothesis Rejected?

Apply the decision rule from Step 4:

- If $F \geq F_{cv}$, reject H_0: Conclude that at least one population mean is different from at least one other population mean.
- If $F < F_{cv}$, fail to reject H_0: Not enough evidence exists to conclude any difference among population means.

Report the results in APA format [e.g., $F(2, 7) = 31.50, p < .05$].

How Big Is the Effect?

- Calculate r^2 using Equation 10.8 to find the percentage of variability in a dependent variable that is accounted for by the independent variable:

$$r^2 = \frac{SS_{Between}}{SS_{Total}} \times 100$$

- Cohen's effect sizes:
  Small effect $\approx$ 1%
  Medium effect $\approx$ 9%
  Large effect $\geq$ 25%

Where Is the Effect?

- Only conduct post-hoc tests if the null hypothesis is rejected.
- Find q in Appendix Table 5 for k groups and $df = df_{Within}$.
- Calculate the Tukey *HSD* value using Equation 10.9:

$$HSD = q\sqrt{\frac{MS_{Within}}{n}}$$

- If two means differ by $\geq$ *HSD* value, then the difference is statistically significant.

Stat Sheet: Chapter 11 One-Way, Repeated-Measures ANOVA

Step 1 Pick a Test

A repeated-measures ANOVA compares the means of two or more dependent samples.

Step 2 Check the Assumptions

1. Random samples (robust)
2. Independence of observations (not robust)
3. Normality (robust)

Step 3 List the Hypotheses

H_0: $\mu_1 = \mu_2 \cdots \mu_k$
H_1: At least one population mean is different from at least one other.

Step 4 Set the Decision Rule

Find F_{cv} in Appendix Table 4:

- Use $df_{\text{Treatment}}$ as numerator *df*.
- Use df_{Residual} as denominator *df*.

Decision rule:

- If $F \geq F_{cv}$, reject H_0.
- If $F < F_{cv}$, fail to reject H_0.

Step 5 Calculate the Test Statistic

Source of Variability	Sum of Squares	Degrees of Freedom	Mean Square	*F* Ratio
Subjects	See Chapter Appendix	$n - 1$		
Treatment	See Chapter Appendix	$k - 1$	$\frac{SS_{\text{Treatment}}}{df_{\text{Treatment}}}$	$\frac{MS_{\text{Treatment}}}{MS_{\text{Residual}}}$
Residual	See Chapter Appendix	$(n - 1)(k - 1)$	$\frac{SS_{\text{Residual}}}{df_{\text{Residual}}}$	
Total	See Chapter Appendix	$N - 1$		

Note: The sums of squares for subjects, treatment, and residual will be provided.
n = the number of cases per cells
k = the number of cells
N = the total number of observations

Step 6 Interpret the Results

Write a four-point interpretation:

- What was done?
- What was found?
- What does it mean?
- What suggestions exist for future research?

Was the Null Hypothesis rejected?

Apply the decision rule from Step 4:

- If $F \geq F_{cv}$, reject H_0: Conclude that at least one population mean is different from at least one other population mean.
- If $F < F_{cv}$, fail to reject H_0: Not enough evidence exists to conclude any difference among population means.

Report the results in APA format [e.g., $F(3, 24) = 15.11$, $p < .05$].

How Big Is the Effect?

Use Equation 11.2 to calculate η^2, the percentage of variability in the dependent variable that is explained by the independent (treatment) variable:

$$\eta^2 = \frac{SS_{\text{Treatment}}}{SS_{\text{Total}}} \times 100$$

where $SS_{\text{Treatment}}$ = sum of squares treatment
SS_{Total} = sum of squares total

- Cohen's guidelines for η^2:
  Small effect $\approx$ 1%
  Medium effect $\approx$ 9%
  Large effect $\geq$ 25%

If the effect is more than small and the null hypothesis was not rejected, then raise the possibility of Type II error and suggest replication with a larger sample size.

Where Are the Effects and What Is Their Direction?

- Only conduct post-hoc tests if the ANOVA was statistically significant.
- Use Equation 11.3 to calculate the Tukey *HSD* value:

$$HSD = q\sqrt{\frac{MS_{\text{Residual}}}{n}}$$

where q = value from Appendix Table 5, where k = the number of groups and $df = df_{\text{Residual}}$
MS_{Residual} = mean square residual
n = number of cases per cell

- Use sample means to comment on the direction of the difference of population means for statistically significant post-hoc tests.

Stat Sheet: Chapter 12 Between-Subjects, Two-Way ANOVA

Step 1 Pick a Test

A between-subjects, two-way ANOVA is used when there are two independent variables, independent samples, and an interval- or ratio-level dependent variable.

It tests for two main effects, the row independent variable and the column independent variable, and for their interaction.

Step 2 Check the Assumptions

1. Random samples (robust)
2. Independence of observations (not robust)
3. Normality (robust)
4. Homogeneity of variance (robust)

Step 3 List the Hypotheses

$H_{0\text{ Rows}}$ and $H_{0\text{ Columns}}$: All row/column population means are the same.

$H_{0\text{ Rows}}$ and $H_{0\text{ Columns}}$: At least one row/column population mean is different from at least one other row/column population mean.

$H_{0\text{ Rows}}$ and $H_{0\text{ Columns}}$: The two independent variables have no interactive effect on the dependent variable in the population.

$H_{0\text{ Rows}}$ and $H_{0\text{ Columns}}$: The two independent variables in the population interact to affect the dependent variable in at least one cell.

Step 4 Set the Decision Rule

Find F_{cv}s for rows, columns, and interaction in Appendix Table 4, based on α and *df*:

Numerator *df* = *df* for the effect being tested

Denominator $df = df_{\text{Within}}$

Decision rule:

- If observed *F* for the effect $\geq F_{cv}$ for the effect, reject H_0.
- If not, fail to reject H_0.

Step 5 Calculate the Test Statistic

Source of Variability	Sum of Squares	Degrees of Freedom	Mean Square	*F* Ratio
Between groups	See Chapter Appendix	$df_{\text{Rows}} + df_{\text{Columns}} + df_{\text{Interaction}}$		
Rows	See Chapter Appendix	$R - 1$	$\frac{SS_{\text{Rows}}}{df_{\text{Rows}}}$	$\frac{MS_{\text{Rows}}}{MS_{\text{Within}}}$
Columns	See Chapter Appendix	$C - 1$	$\frac{SS_{\text{Columns}}}{df_{\text{Columns}}}$	$\frac{MS_{\text{Columns}}}{MS_{\text{Within}}}$
Interaction	See Chapter Appendix	$df_{\text{Rows}} \times df_{\text{Columns}}$	$\frac{SS_{\text{Interaction}}}{df_{\text{Interaction}}}$	$\frac{MS_{\text{Interaction}}}{MS_{\text{Within}}}$
Within groups	See Chapter Appendix	$N - (R \times C)$	$\frac{SS_{\text{Within}}}{df_{\text{Within}}}$	
Total	See Chapter Appendix	$N - 1$		

Note: R = the number of rows
C = the number of columns
N = the total number of cases

Step 6 Interpret the Results

Write a four-point interpretation:

1. What was done?
2. What was found?
3. What does it mean?
4. What suggestions exist for future research?

Was the Null Hypothesis Rejected?

Apply the decision rule from Step 4:

- If *F* for the effect $\geq F_{cv}$ for the effect, reject H_0.
- If *F* for the effect $< F_{cv}$ for the effect, fail to reject H_0.

Report the results in APA format [e.g., $F(1, 24) = 116.46, p < .05$].

How Big Is the Effect?

If the interaction effect is statistically significant, then interpret that and do not interpret main effects.

Use Equation 12.3 to calculate η^2, the percentage of variability in the dependent variable that is explained by an effect:

$$\eta^2_{\text{Rows}} = \frac{SS_{\text{Rows}}}{SS_{\text{Total}}} \times 100$$

$$\eta^2_{\text{Columns}} = \frac{SS_{\text{Columns}}}{SS_{\text{Total}}} \times 100$$

$$\eta^2_{\text{Interaction}} = \frac{SS_{\text{Interaction}}}{SS_{\text{Total}}} \times 100$$

- Cohen's guidelines for η^2:
  Small effect $\approx 1\%$
  Medium effect $\approx 9\%$
  Large effect $\geq 25\%$

Where Are the Effects and What Is Their Direction?

Use Equation 12.4 to calculate the Tukey *HSD* values for statistically significant effects. Two means that differ by at least the *HSD* value have a statistically significant difference:

$$HSD_{\text{Rows}} = q_{\text{Rows}}\sqrt{\frac{MS_{\text{Within}}}{n_{\text{Rows}}}}$$

$$HSD_{\text{Columns}} = q_{\text{Columns}}\sqrt{\frac{MS_{\text{Within}}}{n_{\text{Columns}}}}$$

$$HSD_{\text{Cells}} = q_{\text{Cells}}\sqrt{\frac{MS_{\text{Within}}}{n_{\text{Cells}}}}$$

Stat Sheet: Chapter 13 The Pearson Correlation Coefficient

Conducting the Pearson r

Step 1 Pick a Test

Pearson r is used to examine the degree of linear relationship between two interval- or ratio-level variables.

Step 2 Check the Assumptions

1. Random samples (robust)
2. Independence of observations (not robust)
3. Normality (robust)
4. Linearity (not robust)

Step 3 List the Hypotheses

- Nondirectional:

$$H_0: \rho = 0$$
$$H_1: \rho \neq 0$$

- Directional: For example,

$$H_1: \rho > 0$$
$$H_0: \rho \leq 0$$

Step 4 Set the Decision Rule

- Find r_{cv} in Appendix Table 6 based on α, number of tails, and df.
  - $df = N - 2$
- Decision rule for a two-tailed test:
  - If $r \leq -r_{cv}$ or if $r \geq r_{cv}$, reject H_0.
  - If $-r_{cv} < r < r_{cv}$, fail to reject H_0.

Step 5 Calculate the Test Statistic

Calculate r using Equation 13.2:

$$r = \frac{\Sigma[(X - M_X)(Y - M_Y)]}{\sqrt{SS_X SS_Y}}$$

Use this table to organize the calculations for r:

	Column 1	Column 2	Column 3	Column 4	Column 5	Column 6	Column 7
Case	X	Y	$X - M_X$	$Y - M_Y$	$(X - M_X)(Y - M_Y)$	$(X - M_X)^2$	$(Y - M_Y)^2$
1							
.							
N							
	$M_X =$	$M_Y =$			Σ	$\Sigma = SS_X$	$\Sigma = SS_Y$

Column 5 Σ: (Numerator for Equation 13.2)

Columns 6 and 7: (Denominator for Equation 13.2)

Step 6 Interpret the Results

Write a four-point interpretation:

1. What was done?
2. What was found?
3. What does it mean?
4. What suggestions exist for future research?

Was the Null Hypothesis Rejected?

Apply the decision rule from Step 4:

- If H_0 is rejected, then report the results as statistically significant; conclude there is a relationship in the population between X and Y; and use the sign of r to determine the direction of the relationship.
- If H_0 is not rejected, then the results are called not statistically significant, and not enough evidence is available to conclude a relationship exists between X and Y in the population.

Report the results in APA format [e.g., $r(8) = .73, p < .05$].

How Big Is the Effect?

To determine how big the effect is, calculate effect size r^2 using Equation 13.3:

$$r^2 = (r)^2 \times 100$$

- r^2 calculates the percentage of variability in Y accounted for by X (and vice versa).
- Cohen's guidelines for r^2:
  Small effect $\approx$ 1%
  Medium effect $\approx$ 9%
  Large effect $\geq$ 25%
- If the results are not statistically significant and the effect size is meaningful, consider the possibility of a Type II error. Recommend replication with a larger sample size.

How Wide Is the Confidence Interval?

Determine the width of the confidence interval, the range within which ρ probably falls:

- First, use Appendix Table 7 to convert r to z_r.
- Then, calculate the standard error of r, s_r, using Equation 13.5:

$$s_r = \frac{1}{\sqrt{N - 3}}$$

- Use s_r to calculate the confidence interval in z_r units using Equation 13.4:

$$95\%\ CI_{z_r} = z_r \pm 1.96 s_r$$

- Use Appendix Table 8 to convert the confidence interval back to r units.

Interpreting the confidence interval:

- If the interval captures zero, then it is possible that there is no relationship between the two variables.
- If the interval comes close to zero, then it is possible that the relationship is weak.
- If the interval is wide, then it is uncertain what ρ is. Recommend replication with a larger sample size.

Use Appendix Table 10 to calculate power and sample size.

Use Equation 13.6 to calculate a partial correlation:

$$r_{XY-Z} = \frac{r_{XY} - (r_{XZ} \times r_{YZ})}{\sqrt{(1 - r^2_{XZ})(1 - r^2_{YZ})}}$$

where

r_{XY-Z} = the partial correlation of X and Y controlling for the influence of Z

r_{XY} = the correlation between X and Y

r_{XZ} = the correlation between X and Z

r_{YZ} = the correlation between Y and Z

Stat Sheet: Chapter 14 Simple Linear Regression

Simple Linear Regression

If the correlation between X and Y is a statistically significant one, then given a score on X, a predicted Y score can be calculated. This predicted score is called Y'.

To Calculate Y'

- First, calculate slope, b, using Equation 14.2. Slope is "rise over run," how much change occurs in Y for each unit change in X.

$$b = r\left(\frac{s_Y}{s_X}\right)$$

where
b = slope of the regression line
r = observed correlation between X and Y
s_Y = standard deviation of the Y scores
s_X = standard deviation of the X scores

- Then, using Equation 14.3, calculate the Y-intercept, the spot where the regression line would pass through the Y-axis:

$$a = M_Y - bM_X$$

where
a = Y-intercept for the regression line
M_Y = mean of the Y scores
b = slope of the regression line
M_X = mean of the X scores

- Use the slope and Y-intercept to calculate Y' using Equation 14.1:

$$Y' = bX + a$$

where
Y' = predicted value of Y
b = slope of the regression line
X = value of X for which one wants to find Y'
a = Y-intercept of the regression line

Estimating Error in Prediction

- The standard error of the estimate, Equation 14.4, reveals the amount of error in prediction:

$$s_{Y-Y'} = s_Y\sqrt{1 - r^2}$$

where
$s_{Y-Y'}$ = standard error of the estimate
s_Y = standard deviation of the Y scores
r = the Pearson r value

Stat Sheet: Chapter 15 Nonparametric Statistical Tests: Chi-Square

Chi-Square Goodness-of-Fit Test

Step 1 Pick a Test

The chi-square goodness-of-fit test is used to compare the distribution of a categorical (nominal or ordinal) variable in a sample to the expected distribution.

Step 2 Check the Assumptions

- Random samples (robust)
- Independence of observations (not robust)
- All cells have expected frequencies of at least 5 (not robust)

Step 3 List the Hypotheses

H_0: The distribution of the dependent variable in the sample is exactly the same as in the population.

H_1: The distribution of the dependent variable in the sample differs from the distribution in the population.

Step 4 Set the Decision Rule

- Set α based on the willingness to make a Type I error.
  - Calculate degrees of freedom: $df = k - 1$.
- Find χ^2_{cv} in Appendix Table 9.
- Decision rule:
  - If $\chi^2 \geq \chi^2_{cv}$, reject H_0.
  - If $\chi^2 < \chi^2_{cv}$, fail to reject H_0.

Step 5 Calculate the Test Statistic

- Calculate the expected frequencies using Equation 15.2:

$$f_{\text{Expected}} = \frac{\%_{\text{Expected}} \times N}{100}$$

- Calculate the chi-square value using Equation 15.3:

$$\chi^2 = \Sigma \frac{\left(f_{\text{Observed}} - f_{\text{Expected}}\right)^2}{f_{\text{Expected}}}$$

Step 6 Interpret the Results

Write a four-point interpretation:

1. What was done?
2. What was found?
3. What does it mean?
4. What suggestions exist for future research?

Was the Null Hypothesis Rejected?

Apply the decision rule from Step 4. If H_0 is rejected, what is the direction of the difference?

Report the results in APA format [e.g., $\chi^2(1, N = 72) = 5.18, p < .05$].

Chi-Square Test of Independence

Step 1 Pick a Test

The chi-square test of independence is used to determine whether two or more samples differ on a categorical (nominal or ordinal) variable. To start, arrange the data in a contingency table showing the observed frequencies.

Step 2 Check the Assumptions

- Random samples (robust)
- Independence of observations (not robust)
- All cells have expected frequencies of at least 5 (not robust)

Step 3 List the Hypotheses

H_0: $\rho = 0$

H_1: $\rho \neq 0$

Step 4 Set the Decision Rule

- Find χ^2_{cv} based on α and df in Appendix Table 9.
  - Set α depending on the willingness to make a Type I error.
  - Find degrees of freedom: $df = (R - 1) \times (C - 1)$.
- Decision rule:
  - If $\chi^2 \geq \chi^2_{cv}$, reject H_0.
  - If $\chi^2 < \chi^2_{cv}$, fail to reject H_0.

Step 5 Calculate the Test Statistic

- Calculate the expected frequencies using Equation 15.5:

$$f_{\text{Expected}} = \frac{N_{\text{Row}} \times N_{\text{Column}}}{N}$$

- Calculate the chi-square value using Equation 15.3:

$$\chi^2 = \Sigma \frac{(f_{\text{Observed}} - f_{\text{Expected}})^2}{f_{\text{Expected}}}$$

Step 6 Interpret the Results

Write a four-point interpretation:

1. What was done?
2. What was found?
3. What does it mean?
4. What suggestions exist for future research?

Was the Null Hypothesis Rejected?

- Apply the decision rule from Step 4. If H_0 is rejected, what is the direction of the difference?
- Report the results in APA format [e.g., $\chi^2(1, N = 72) = 5.18, p < .05$].
- Calculate Cramer's V, Equation 15.6, to determine how big the effect is:

$$V = \sqrt{\frac{\chi^2}{N \times df_{RC}}}$$

Guidelines for Interpreting Cramer's V:

	Small Effect Size	Medium Effect Size	Large Effect Size
$df_{RC} = 1$	.10	.30	.50
$df_{RC} = 2$	.07	.21	.35
$df_{RC} = 3$	.06	.17	.29
$df_{RC} = 4$	.05	.15	.25

Stat Sheet: Chapter 16 Selecting the Right Test

Determining Type of Study: Descriptive, Experimental, Quasi-Experimental, or Correlational

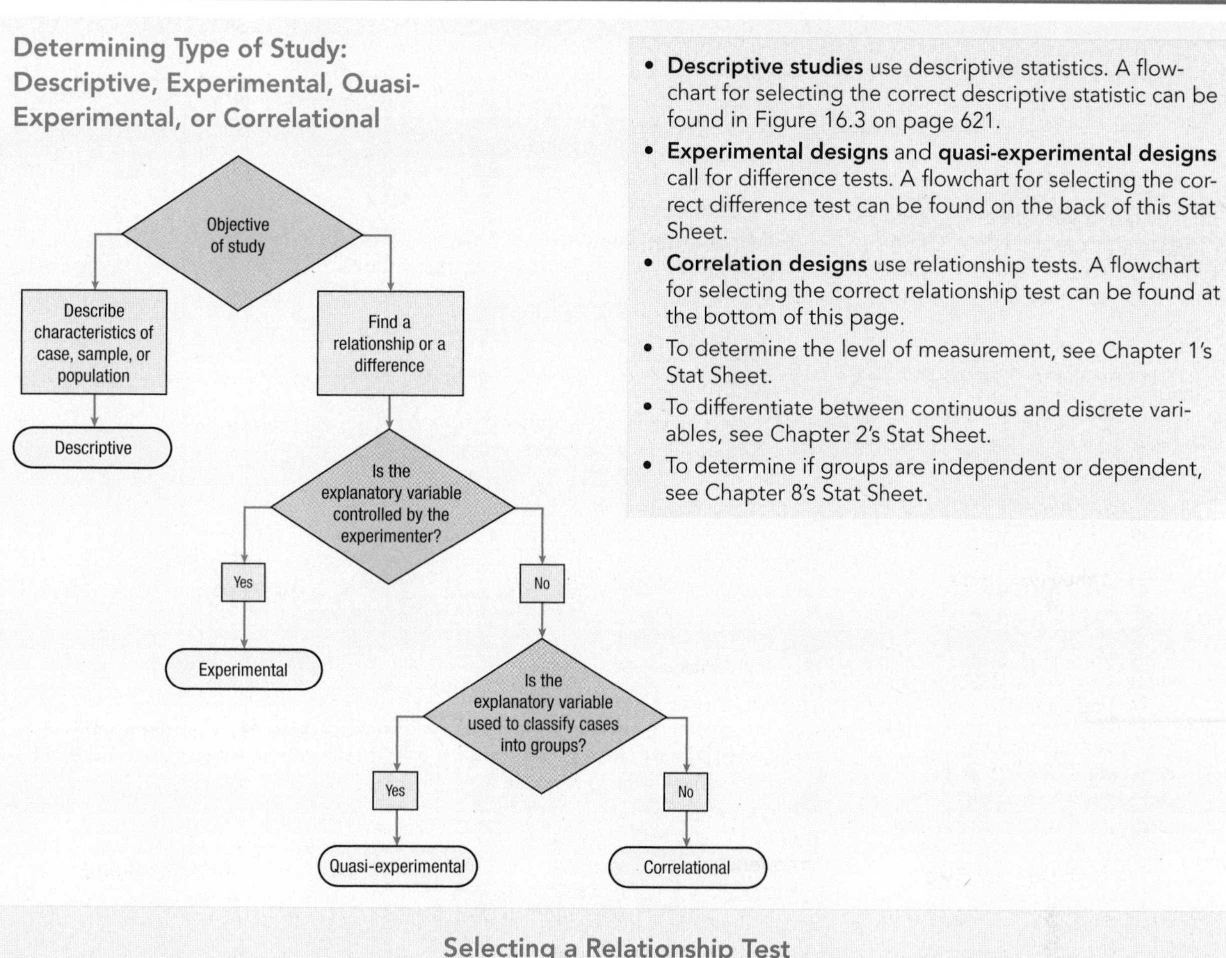

- **Descriptive studies** use descriptive statistics. A flowchart for selecting the correct descriptive statistic can be found in Figure 16.3 on page 621.
- **Experimental designs** and **quasi-experimental designs** call for difference tests. A flowchart for selecting the correct difference test can be found on the back of this Stat Sheet.
- **Correlation designs** use relationship tests. A flowchart for selecting the correct relationship test can be found at the bottom of this page.
- To determine the level of measurement, see Chapter 1's Stat Sheet.
- To differentiate between continuous and discrete variables, see Chapter 2's Stat Sheet.
- To determine if groups are independent or dependent, see Chapter 8's Stat Sheet.

Selecting a Relationship Test

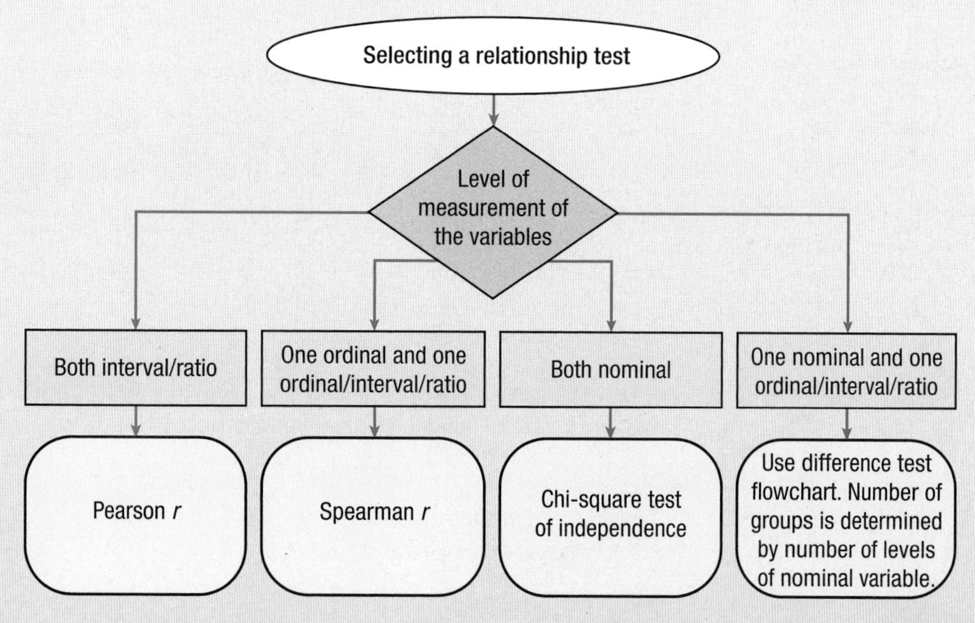

Selecting the Correct Difference Test

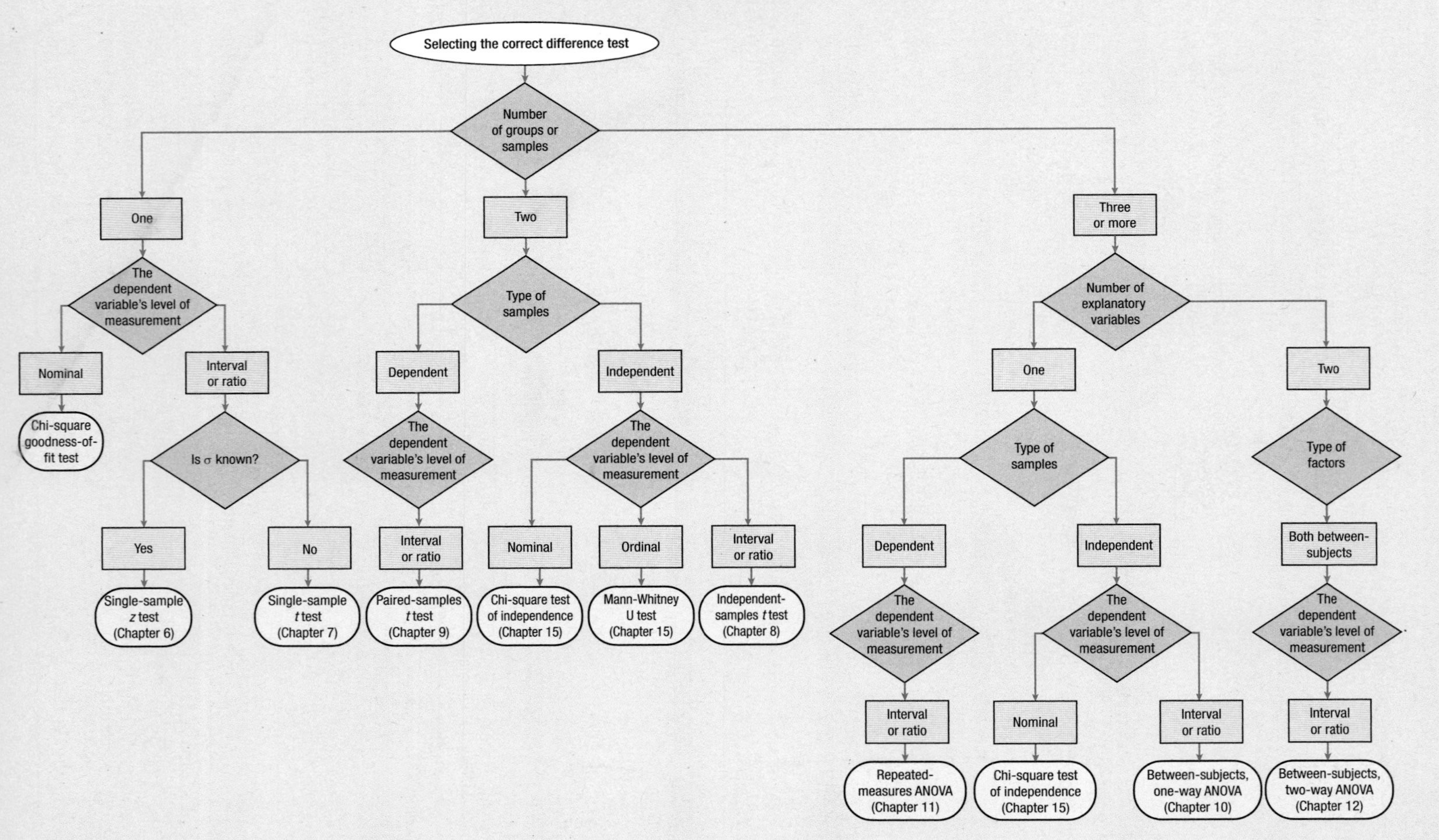